The CRC Handbook of

SOLID STATE

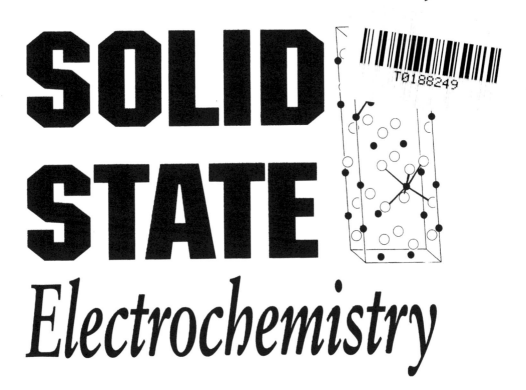

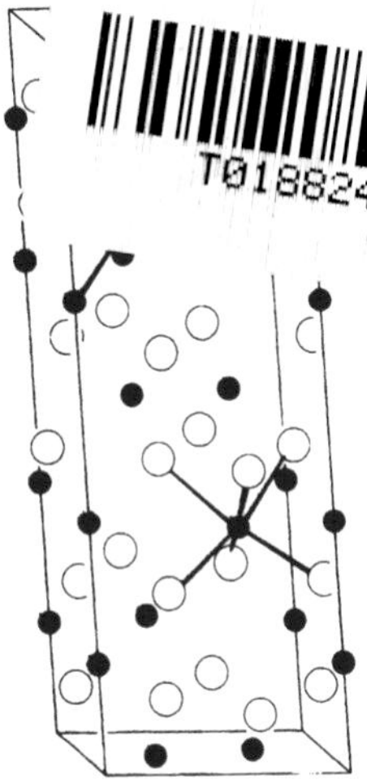

Electrochemistry

The CRC Handbook of

SOLID STATE

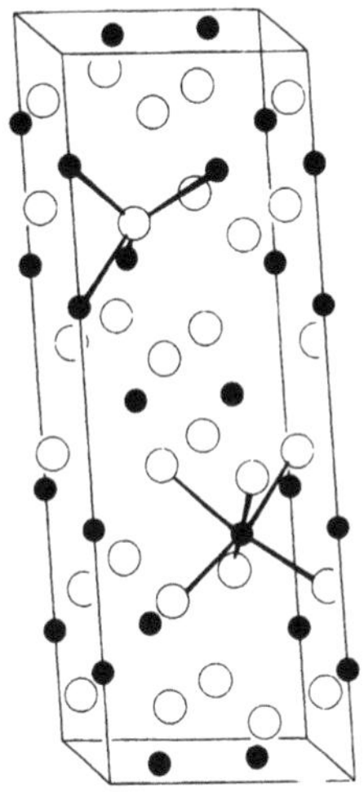

Electrochemistry

Edited by

P.J. Gellings • H.J.M. Bouwmeester

Laboratory for Inorganic Materials Science

University of Twente

Enschede, The Netherlands

CRC Press

Taylor & Francis Group

Boca Raton London New York

CRC Press is an imprint of the
Taylor & Francis Group, an **informa** business

CRC Press
Taylor & Francis Group
6000 Broken Sound Parkway NW, Suite 300
Boca Raton, FL 33487-2742

First issued in paperback 2019

ISBN-13: 978-0-367-45586-6 (pbk)
ISBN-13: 978-0-8493-8956-6 (hbk)

Visit the Taylor & Francis Web site at
http://www.taylorandfrancis.com

and the CRC Press Web site at
http://www.crcpress.com

Cover design: Denise Craig

Library of Congress Card Number 96-31466

Library of Congress Cataloging-in-Publication Data

The CRC handbook of solid state electrochemistry / edited by P.J. Gellings
and H.J.M. Bouwmeester.
p. cm.
Includes bibliographical references and index.
ISBN 0-8493-8956-9
1. Solid state chemistry—Handbooks, manuals, etc. I. Gellings, P.J. II. Bouwmeester, H.J.M.
QD478.C74 1996
541.3′7—dc20

96-31466
CIP

ABOUT THE EDITORS

Prof. Dr. P.J. Gellings. After studying chemistry at the University of Leiden (the Netherlands), Prof. Gellings received his degree in physical chemistry in 1952. Subsequently he worked as research scientist in the Laboratory of Materials Research of Werkspoor N.V. (Amsterdam, the Netherlands). He obtained his Ph.D. degree from the University of Amsterdam in 1963 on the basis of a dissertation titled: "Theoretical considerations on the kinetics of electrode reactions."

In 1964 Prof. Gellings was appointed professor of Inorganic Chemistry and Materials Science at the University of Twente. His main research interests were coordination chemistry and spectroscopy of transition metal compounds, corrosion and corrosion prevention, and catalysis. In 1991 he received the Cavallaro Medal of the European Federation Corrosion for his contributions to corrosion research. In 1992 he retired from his post at the University, but has remained active as supervisor of graduate students in the field of high temperature corrosion.

Dr. H.J.M. Bouwmeester. After studying chemistry at the University of Groningen (the Netherlands), Dr. Bouwmeester received his degree in inorganic chemistry in 1982. He received his Ph.D. degree at the same university on the basis of a dissertation titled: "Studies in Intercalation Chemistry of Some Transition Metal Dichalcogenides." For three years he was involved with industrial research in the development of the ion sensitive field effect transistor (ISFET) for medical application at Sentron V.O.F. in the Netherlands.

In 1988 Dr. Bouwmeester was appointed assistant professor at the University of Twente, where he heads the research team on Dense Membranes and Defect Chemistry in the Laboratory of Inorganic Materials Science. His research interests include defect chemistry, order-disorder phenomena, solid state thermodynamics and electrochemistry, ceramic surfaces and interfaces, membranes, and catalysis. He is involved in several international projects in these fields.

CONTRIBUTORS

Isaac Abrahams
Department of Chemistry
Queen Mary and Westfield College
University of London
London, United Kingdom

Symeon I. Bebelis
Department of Chemical Engineering
University of Patras
Patras, Greece

Henny J.M. Bouwmeester
Laboratory for Inorganic Materials Science
Faculty of Chemical Technology
University of Twente
Enschede, The Netherlands

Peter G. Bruce
School of Chemistry
University of St. Andrews
St. Andrews, Fife, United Kingdom

Anthonie J. Burggraaf
Laboratory for Inorganic Materials Science
Faculty of Chemical Technology
University of Twente
Enschede, The Netherlands

Hans de Wit
Materials Institute Delft
Delft University of Technology
Faculty of Chemical Technology and Materials Science
Delft, The Netherlands

Pierre Fabry
Université Joseph Fourier
Laboratoire d'Electrochimie et de Physicochimie des Matériaux et Interfaces (LEPMI)
Domaine Universitaire
Saint Martin d'Hères, France

Thijs Fransen
Laboratory for Inorganic Materials Science
University of Twente
Enschede, The Netherlands

Paul J. Gellings
Laboratory for Inorganic Materials Science
Faculty of Chemical Technology
University of Twente
Enschede, The Netherlands

Heinz Gerischer‡
Scientific Member Emeritus of the Fritz Haber Institute
Department of Physical Chemistry
Fritz-Haber-Institut der Max-Planck-Gesellschaft
Berlin, Germany

Claes G. Granqvist
Department of Technology
Uppsala University
Uppsala, Sweden

Jacques Guindet
Université Joseph Fourier
Laboratoire d'Electrochimie et de Physicochimie des Matériaux et Interfaces (LEPMI)
Domaine Universitaire
Saint Martin d'Hères,
France

Abdelkader Hammou
Université Joseph Fourier
Laboratoire d'Electrochimie et de Physicochimie des Matériaux et Interfaces (LEPMI)
Domaine Universitaire
Saint Martin d'Hères,
France

Christian Julien
Laboratoire de Physique des Solides
Université Pierre et Marie Curie
Paris, France

Tetsuichi Kudo
Institute of Industrial Science
University of Tokyo
Tokyo, Japan

Janusz Nowotny
Australian Nuclear Science & Technology Organisation
Advanced Materials Program
Lucas Heights Research Laboratories
Menai, Australia

Ilan Riess
Physics Department
Technion — Israel Institute of Technology
Haifa, Israel

‡ Deceased

Joop Schoonman
Laboratory for Applied Inorganic Chemistry
Delft University of Technology
Faculty of Chemical Technology and Materials Science
Delft, The Netherlands

Elisabeth Siebert
Université Joseph Fourier
Laboratoire d'Electrochimie et de Physicochimie des Matériaux et Interfaces (LEPMI)
Domaine Universitaire
Saint Martin d'Hères, (France)

Constantinos G. Vayenas
Department of Chemical Engineering
University of Patras
Patras, Greece

Werner Weppner
Chair for Sensors and Solid State Ionics
Technical Faculty, Christian-Albrechts University
Kiel, Germany

IN MEMORIAM

Heinz Gerischer
1919–1994

On September 14, 1994, Professor Heinz Gerischer died from heart failure. With his death, the international community of electrochemistry lost the man who most probably was its most eminent representative. Professor Gerischer was one of the founders of modern electrochemistry, having contributed to nearly all modern extensions and improvements of this science.

He was born in 1919 and studied chemistry at the University of Leipzig from 1937 to 1944, presenting his Ph.D. thesis, under the supervision of Professor Bonhoeffer, in 1946. He worked throughout Germany, was professor of physical chemistry at the Technical University–Munich, and director of the Fritz-Haber-Institut der Max-Planck-Gesellschaft in Berlin. He made great contributions to the kinetics of electrode reactions and to the electrochemistry at semiconductor surfaces. He also initiated the application of a wide range of modern experimental methods to the study of electrochemical reactions, including nonelectrochemical techniques such as optical and electron spin resonance spectroscopy, and advocated the use of synchroton radiation in surface research. His scientific work was published in more than 300 publications and was notable for its great originality, clarity of exposition, and high quality.

We are grateful that we can publish as Chapter 2 of this handbook, what may be Professor Gerischer's last publication, in which he again shows his ability to give a very clear exposition of the basic principles of modern electrochemistry.

PREFACE

The idea for this book arose out of the realization that, although excellent surveys and handbooks of electrochemistry and of solid state chemistry are available, there is no single source covering the field of solid state electrochemistry. Moreover, as this field gets only limited attention in most general books on electrochemistry and solid state chemistry, there is a clear need for a handbook in which attention is specifically directed toward this rapidly growing field and its many applications.

This handbook is meant to provide guidance through the multidisciplinary field of solid state electrochemistry for scientists and engineers from universities, research organizations, and industries. In order to make it useful for a wide audience, both fundamentals and applications are discussed, together with a state-of-the-art review of selected applications.

As is true for nearly all fields of modern science and technology, it is impossible to treat all subjects related to solid state electrochemistry in a single textbook, and choices therefore had to be made. In the present case, the solids considered are mainly confined to inorganic compounds, giving only limited attention to fields like polymer electrolytes and organic sensors.

The editors thank all those who cooperated in bringing this project to a successful close. In the first place, of course, we thank the authors of the various chapters, but also those who advised us in finding these authors. We are also grateful to the staff of CRC Press — in particular associate editor Felicia Shapiro and project editor Gail Renard, who were of great assistance to us with their help and experience in solving all kinds of technical problems.

It is a great loss for the whole electrochemical community that Professor Heinz Gerischer died suddenly in September 1994 and we remember with gratitude his great services to electrochemistry. We consider ourselves fortunate to be able to present as Chapter 2 of this handbook one of his last important contributions to this field.

P.J. Gellings
H.J.M. Bouwmeester

TABLE OF CONTENTS

Chapter 1

INTRODUCTION

Henny J. M. Bouwmeester and Paul J. Gellings

I. INTRODUCTION

As in aqueous electrochemistry, research interest in the field of solid state electrochemistry can be split into two main subjects:

Ionics: in which the properties of electrolytes have the central attention
Electrodics: in which the reactions at electrodes are considered.

Both fields are treated in this handbook. This first chapter gives a brief survey of the scope and contents of the handbook. Some elementary ideas about these topics, which are often unfamiliar to those entering this field, are introduced, but only briefly. In general, textbooks and general chemical education give only minor attention to elementary issues such as defect chemistry and kinetics of electrode reactions. Ionics in solid state electrochemistry is inherently connected with the chemistry of defects in solids, and some elementary considerations about this are given in Section III. Electrodics is inherently concerned with the kinetics of electrode reactions, and therefore some elementary considerations about this subject are presented in Section IV. In an attempt to lead into more professional discussions as provided in subsequent chapters, some of these considerations are presented in this first chapter.

II. GENERAL SCOPE

The distinction made between *ionics* and *electrodics* is translated into detailed discussions in various chapters on the following topics:

- electrochemical properties of solids such as oxides, halides, cation conductors, etc., including ionic, electronic, and mixed conductors
- electrochemical kinetics and mechanisms of reactions occurring on solid electrolytes, including gas-phase electrocatalysis.

0-8493-8956-9/97/$0.00+$.50

An important point to note in solid state electrochemistry is that electrolyte and electrode behavior may coincide in compounds showing both ionic and electronic conduction, the so-called *mixed ionic–electronic conductors* (often abbreviated to *mixed conductors*).

A review of the necessary theoretical background in electrochemistry and solid state chemistry is given in Chapters 2 through 5. The fundamentals of these topical areas, which include structural and defect chemistry, diffusion and transport in solids, conductivity and electrochemical reactions, adsorption and reactions on solid surfaces, get due attention, starting with a discussion of fundamental concepts from aqueous electrochemistry in Chapter 2. Also discussed are fast ionic conduction in solids, the structural features associated with transport, such as order–disorder phenomena, and interfacial processes. Because of the great variety in materials and relevant properties, a survey of the most important types of solid electrolytes is presented separately in Chapter 6. In addition, a detailed account is provided, in Chapter 7, of the electrochemistry of mixed conductors, which are becoming of increasing interest in quite a number of applications. Finally, attention is given to electrode processes and electrodics in Chapter 8, while the principles of the main experimental methods used in this field are presented in Chapter 9.

In view of the many possible applications in various fields of common interest, a discussion of a number of characteristic and important applications emerging from solid state electrochemistry follows the elementary and theoretical chapters. In Chapter 10, electrochemical sensors for the detection and determination of the constituents of gaseous (and for some liquid) systems are discussed. Promising applications in the fields of generation, storage, and conversion of energy in fuel cells and in solid state batteries are treated in Chapters 11 and 12, respectively. The application of solid state electrochemistry in chemical processes and (electro)catalysis is considered in detail in Chapter 13, followed by a discussion of (dense) ceramic mixed conducting membranes for the separation of oxygen in Chapter 14. The fundamentals of high-temperature corrosion processes and tools to either study or prevent these are deeply connected with solid state electrochemistry and are considered in Chapter 15. The application and properties of optical, in particular electrochromic, devices are discussed in Chapter 16.

We have not attempted to rigorously avoid all overlap between the different chapters, nor to alter carefully balanced appraisals of fundamental or conceptual issues given in a number of chapters by different authors. In particular, most chapters devoted to applications also treat some of the background and underlying theory.

III. ELEMENTARY DEFECT CHEMISTRY

Some elementary considerations on defect chemistry are presented here, but within the limits of this introductory chapter, only briefly. For a more extensive treatment, see, in particular, Chapters 3 and 4 of this handbook.

A. TYPES OF DEFECTS

Ion conductivity or diffusion in oxides can only take place because of the presence of imperfections or defects in the lattice. A finite concentration of defects is present at all temperatures above 0°K arising from the entropy contribution to the Gibbs free energy as a consequence of the disorder introduced by the presence of the defects.

If x is the mole fraction of a certain type of defect, the entropy increase due to the formation of these defects is

$$\Delta s = -R\big(x\ln(x) + (1-x)\ln(1-x)\big) \tag{1.1}$$

which is the mixing entropy of an (ideal) mixture of defects and occupied lattice positions. If the energy needed to form the defects is E Joule per mole, the corresponding increase of the enthalpy is equal to:

$$\Delta h = x \cdot E \tag{1.2}$$

The change in free enthalpy (or Gibbs free energy) then becomes:

$$\Delta g = \Delta h - T \cdot \Delta s = x \cdot E + RT\big(x\ln(x) + (1-x)\ln(1-x)\big) \tag{1.3}$$

Because in equilibrium g (and of course also Δg) must be minimal, we find, by partial differentiation:

$$\frac{\partial(\Delta g)}{\partial x} = 0 = E + RT\big(\ln(x) - \ln(1-x)\big) \tag{1.4}$$

so that, in equilibrium:

$$\frac{x}{1-x} = \exp\left(-\frac{E}{RT}\right) \tag{1.5}$$

or, if $x \ll 1$:

$$x = \exp\left(-\frac{E}{RT}\right) \tag{1.6}$$

From this we find that, for example, if E = 50 kJ/mol then at 300°K one would have a defect mole fraction $x \approx 2 \times 10^{-9}$, which increases to $x \approx 2 \times 10^{-3}$ at 1000°K. At each temperature a finite, albeit often small, concentration of defects is found in any crystal.

Because the energies needed for creating different defects usually differ greatly, it is often a good approximation to consider only one type of defect to be present: the *majority defect.* For example, a difference of 40 kJ/mol in the formation energy of two defects leads to a difference of a factor of about 10^7 between their concentrations at 300°K and of about 10^2 at 1000°K.

The defects under consideration here may be

- vacant lattice sites, usually called vacancies
- ions placed at normally unoccupied sites, called interstitials
- foreign ions present as impurity or dopant
- ions with charges different from those expected from the overall stoichiometry

In the absence of macroscopic electric fields and of gradients in the chemical potential, charge neutrality must be maintained throughout an ionic lattice. This requires that a charged defect be compensated by one (or more) other defect(s) having the same charge, but of opposite sign. Thus, these charged defects are always present in the lattice as a combination of two (or more) types of defect(s), which in many cases are not necessarily close together.

Two common types of disorder in ionic solids are Schottky and Frenkel defects. At the stoichiometric composition, the presence of Schottky defects (see Figure 1.1a) involves equivalent numbers of cation and anion vacancies. In the Frenkel defect structure (see

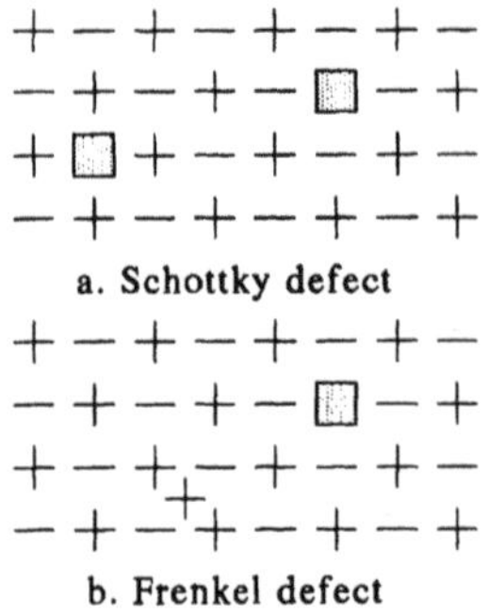

FIGURE 1.1. Schottky and Frenkel defects.

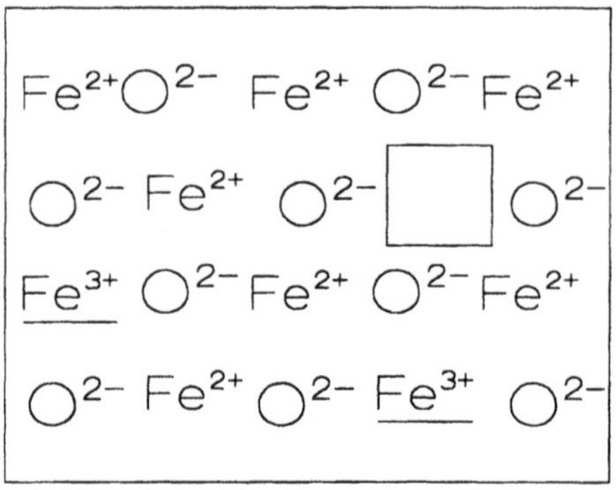

FIGURE 1.2. $Fe_{1-x}O$ as example of a compound with a metal deficit.

Figure 1.1b) defects are limited to either the cations or the anions, of which both a vacancy and an interstitial ion are present. Ionic defects which are present due to the thermodynamic equilibrium of the lattice are called intrinsic defects.

Nonstoichiometry occurs when there is an excess of one type of defect relative to that at the stoichiometric composition. Since the ratio of cation to anion lattice sites is the same whether a compound is stoichiometric or nonstoichiometric, this means that complementary electronic defects must be present to preserve electroneutrality. A typical example is provided by FeO, which always has a composition $Fe_{1-x}O$, with $x > 0.03$. As shown schematically in Figure 1.2, we see that this is accomplished by the formation of two Fe^{3+} ions for each Fe^{2+} ion removed from the lattice.

Electronic defects may arise as a consequence of the transition of electrons from normally filled energy levels, usually the valence band, to normally empty levels, the conduction band. In those cases where an electron is missing from a nominally filled band, this is usually called a hole (or electron hole).

The number of electrons and holes in a nondegenerate semiconductor is determined by the value of the electronic band gap, E_g. The intrinsic ionization across the band gap can be expressed by:

$$e' + h^{\bullet} \rightleftarrows 0$$
$$K_{el} = [e'] \times [h^{\bullet}] \tag{1.7}$$

When electrons or electron holes are localized on ions in the lattice, as in $Fe_{1-x}O$, the semiconductivity arises from electrons or electron holes moving from one ion to another, which is called *hopping-type semiconductivity.*

TABLE 1.1
Kröger–Vink Notation for Point Defects in Crystals

Type of defect	Symbol	Remarks
Vacant M site	V_M''	Divalent ions are chosen as example with MX as compound formula
Vacant X site	$V_X^{\cdot\cdot}$	M^{2+}, X^{2-}: cation and anion
Ion on lattice site	M_M^x, X_X^x	^{x}: uncharged
L on M site	L_M'	L^+ dopant ion
N on M site	$N_M^{\cdot}$	N^{3+} dopant ion
Free electron	e'	
Free (electron) hole	$h^{\cdot}$	
Interstitial M ion	$M_i^{\cdot\cdot}$	$^{\cdot\cdot}$: effective positive charge
Interstitial X ion	X_i''	$'$: effective negative charge

B. DEFECT NOTATION

The charges of defects and of the regular lattice particles are only important with respect to the neutral, unperturbed (ideal) lattice. In the following discussion the charges of all point defects are defined relative to the neutral lattice. Thus only the effective charge is considered, being indicated by a dot ($^{\cdot}$) for a positive excess charge and by a prime ($'$) for a negative excess charge. The notation for defects most often used has been introduced by Kröger and Vink[1] and is given in Table 1.1. Only fully ionized defects are indicated in this table. For example, considering anion vacancies we could, besides doubly ionized anion vacancies, $V_X^{\cdot\cdot}$, also have singly ionized or uncharged anion vacancies, $V_X^{\cdot}$ or V_X^x, respectively.

C. DEFECT EQUILIBRIA

The extent of nonstoichiometry and the defect concentrations in solids are functions of the temperature and the partial pressure of their chemical components, which are treated more fully in Chapters 3 and 4 of this handbook.

Foreign ions in a lattice (substitutional ions or foreign ions present on interstitial sites) are one type of extrinsic defect. When aliovalent ions (impurities or dopes) are present, the concentrations of defects of lattice ions will also be changed, and they may become so large that they can be considered a kind of extrinsic defect too, in particular when they form minority defects in the absence of foreign ions. For example, dissolution of CaO in the fluorite phase of zirconia (ZrO_2) leads to Ca^{2+} ions occupying Zr^{4+} sites, and an effectively positively charged oxygen vacancy is created for each Ca^{2+} ion present to preserve electroneutrality. The defect reaction can then be written as:

$$CaO \rightarrow Ca_{Zr}'' + O_O^{\times} + V_O^{\bullet\bullet} \tag{1.8}$$

with the electroneutrality condition or charge balance:

$$\left[Ca_{Zr}''\right] = \left[V_O^{\bullet\bullet}\right] \tag{1.9}$$

where the symbol of a defect enclosed in brackets denotes its mole fraction. In this case the concentrations of electrons and holes are considered to be negligible with respect to those of the substituted ions and vacancies. Consequently, in this situation the mole fraction of ionic defects is fixed by the amount of dopant ions present in the oxide.

As another example, we consider an oxide MO_2 with Frenkel defects in the anion sublattice. As the partial pressure of the metal component is negligible compared with that of oxygen under most experimental conditions, nonstoichiometry thus is a result of the interaction of the oxide with the oxygen in the surrounding gas atmosphere. The Frenkel defect equilibrium for the oxygen ions can be written as:

$$O_O^\times \rightleftarrows O_i'' + V_O^{\bullet\bullet}$$
$$K_F \rightleftarrows [O_i''] \times [V_O^{\bullet\bullet}] \quad (1.10)$$

for fully ionized defects, as is usually observed in oxides at elevated temperature. The thermal equilibrium between electrons in the conduction band and electron holes in the valence band is represented by Equation (1.7). Taking into account the presence of electrons and electron holes, the electroneutrality condition reads:

$$[h^\bullet] + 2[V_O^{\bullet\bullet}] \rightleftarrows [e'] + 2[O_i''] \quad (1.11)$$

If ionic defects predominate, the concentrations of oxygen interstitials O_i'' and oxygen vacancies $V_O^{\bullet\bullet}$ ($[V_O^{\bullet\bullet}] \gg [h^\bullet]$ and $[O_i''] \gg [e']$) are equal and independent of oxygen pressure.

As the oxygen pressure is increased, oxygen is increasingly incorporated into the lattice. The corresponding defect equilibrium is

$$\frac{1}{2} O_2 \rightleftarrows O_i'' + 2\ h^\bullet$$
$$K_{ox} \times p_{O_2}^{1/2} = [O_i''] \times [h^\bullet]^2 \quad (1.12)$$

This type of equilibrium, which involves p-type semiconductivity, is only possible if cations are present which have the capability of increasing their valence. As the oxygen pressure is decreased, oxygen is being removed from the lattice. The corresponding defect equilibrium is

$$O_O^\times = \frac{1}{2} O_2(g) + V_O^{\bullet\bullet} + 2\ e'$$
$$[V_O^{\bullet\bullet}] \times [e']^2 = K_{red} \times p_{O_2}^{-1/2} \quad (1.13)$$

noting that $[O_O^\times] \approx 1$.

When only lower oxidation states are available, as in ZrO_2, an n-type semiconductor is obtained. Reduction increases the conductivity, and this type of compound is called a reduction-type semiconductor. Oxidation would involve the creation of electron holes, e.g., in the form of Zr^{5+}, which is energetically very unfavorable because the corresponding ionization energy is very high, although this could occur in principle at very high oxygen partial pressures.

IV. ELEMENTARY CONSIDERATIONS OF THE KINETICS OF ELECTRODE REACTIONS

In this section a simplified account of some basic concepts of the kinetics of electrode processes is given. We consider a simple electrode reaction:

$$Ox + ne^- \rightleftarrows Red \quad (1.14)$$

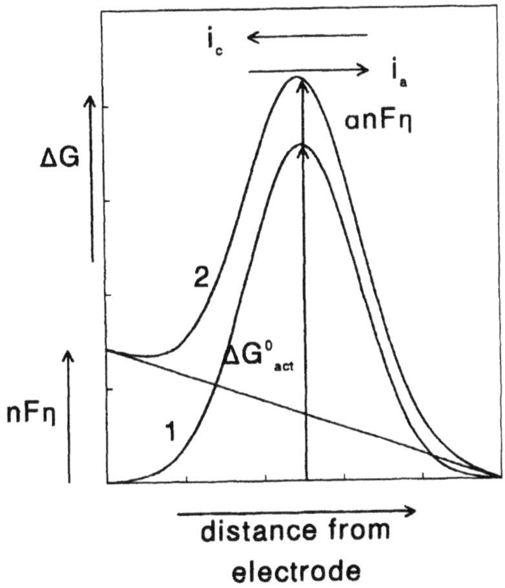

FIGURE 1.3. Schematic of free enthalpy–distance curves at equilibrium and with an externally applied potential $+\eta$ V with respect to solution.

where n is the number of electrons transferred in the reaction, Ox is the oxidized form of a redox couple, e.g., Fe^{2+}, O_2, and H^+, while Red is the corresponding reduced form, thus respectively: Fe(metal), OH^- in aqueous solution or O^{2-} in a solid oxide, and H_2.

The rate of reaction in the two opposite directions is proportional to the anodic current density i_a (>0) and the cathodic current density i_c (<0). In equilibrium these are numerically equal and the balanced term is called the exchange current density i_0. The total current density $i_{total} = i_a + i_c$ is equal to zero.

The reaction situation is represented schematically by curve 1 in Figure 1.3. At equilibrium, we obtain from the well-known Arrhenius equation for reaction rates:

$$i_a = -i_c = A \cdot \exp\left(-\Delta G^0_{act}/RT\right) \tag{1.15}$$

where A is the pre-exponential factor which contains the factor nF and in general also a factor depending on the concentration [Ox] for the cathodic and [Red] for the anodic reaction.

When an electric potential difference or *overvoltage* of magnitude η V is applied between the electrode and the solution relative to the equilibrium condition, the corresponding situation is then represented by curve 2 in Figure 1.3. Because of the charge nF transferred in the reaction, this means that the energy of the reacting species at the electrode is increased by $nF\eta$. Now the activation enthalpies for the anodic and cathodic reaction are no longer equal and have become, respectively: $(\Delta G^0_{act} - \alpha nF\eta)$ and $(\Delta G^0_{act} + (1-\alpha)nF\eta)$. Thus the current densities are given by:

$$i_a = A \cdot \exp\left(-\left(\Delta G^0_{act} - \alpha nF\eta\right)/RT\right) \tag{1.16}$$

$$i_c = -A \cdot \exp\left(-\left(\Delta G^0_{act} + (1-\alpha)nF\eta\right)/RT\right)$$

where α is usually called the *symmetry factor*. For the total current this gives:

$$i_{total} = i_a + i_c = i_0\left[f([Ox])\exp(\alpha nF\eta/RT) - f([Red])\exp(-(1-\alpha)nF\eta/RT)\right] \tag{1.17}$$

The factors f([Ox]) and f([Red]) are the concentration-dependent factors mentioned above. An equation of this form was originally derived by Butler and Volmer for the kinetics of the hydrogen evolution reaction. Therefore it is called by many authors the "Butler–Volmer equation," even when other reactions are considered. Other authors prefer to use more general names like "current–voltage equation" or "i–V-equation." For a more complete treatment, see Chapter 2, Section VIII of this handbook.

For an extensive treatment of electrode processes and electrodics for the case of solid state electrochemistry, see Chapter 8 of this handbook.

REFERENCES

1. Kröger, F.A. and Vink, H.J., *Solid State Phys.* 3, 307, 1956.

Chapter 2

PRINCIPLES OF ELECTROCHEMISTRY

Heinz Gerischer

CONTENTS

0-8493-8956-9/97/$0.00+$.50

I. THE SUBJECT OF ELECTROCHEMISTRY

Electrochemistry covers all phenomena in which a chemical change is the result of electric forces and, vice versa, where an electric force is generated by chemical processes. It includes the properties and behavior of electrolytic conductors in liquid or solid form. A great many of these phenomena occur at interfaces between electronic and electrolytic conductors where the passage of electric charge is connected with a chemical reaction, a so-called *redox reaction*. The rate of such a reaction can be followed with great sensitivity as an electric current. Such contacts constitute the electrodes of galvanic cells which can be used for the conversion of chemical into electrical energy in the form of batteries or for the generation of chemical products by electric power (electrolysis).

The properties of the electrodes are at the center of scientific interest. They qualitatively and quantitatively control the electrochemical reactions in galvanic cells. The properties of electrolytes are controlled by the concentrations of ions, their mobilities, and the interactions between ionic particles of opposite charge as well as their interactions with other constituents of the respective phases (e.g., solvent, membrane matrix, solid matrix). This latter area will, however, not be dealt with in any detail in this chapter. Emphasis will be given to interfacial processes. Since the basic laws of electrochemistry were developed in systems with liquid electrolytes, these relations will be derived and presented in this introduction for such systems.

II. FARADAY'S LAW AND ELECTROLYTIC CONDUCTIVITY

Electrochemical experiments are performed in electrolysis cells which consist of two electrodes in contact with an electrolyte. The electrodes are electronic conductors which can be connected to a voltage source in order to drive an electric current through the electrolyte. In the years 1833–1834, Faraday discovered that the chemical change resulting from the current is proportional to the amount of electricity having passed through the cell and that the mass of chemicals produced at the electrodes is in relation to their chemical equivalent:

$$\frac{dm}{dt} = \frac{1}{\mathrm{F}} \cdot I \tag{2.1}$$

where m is the number of chemical equivalents, t the time, F the Faraday constant, and I the current ($C \cdot s^{-1}$). The value of the Faraday constant is F = 96,485 C/mole equivalent, corresponding with the charge of 1 mole of electrons.

The electrolyte contains at least two types of ions with opposite charge. In liquids, all ions are mobile and contribute to the conductivity. Their mobilities are, however, different, and their individual contributions to the conductivity can therefore vary over wide ranges. In solid electrolytes, often only one of the ions is mobile. The transference number, t_i, characterizes the contribution of each ion to the current in an electrolyte. The knowledge of t_i for all components i is therefore important for an understanding of electrolytic processes.

Cations move to the cathode, anions to the anode. If they are not consumed at the respective electrode at the same rate as they arrive there by ionic migration, they accumulate or deplete in front of this electrode, and the composition of the electrolyte changes in the region close to both electrodes in opposite directions. Such changes in composition can be used for the determination of transference numbers.

The specific conductivity of an electrolyte, κ (Ω^{-1} cm^{-1}), is connected with the mobilities of the ions, u_i (cm^2 s^{-1} V^{-1}) and their concentrations, N_i (mole cm^{-3}), by the relation

$$\kappa = \mathrm{F} \cdot \sum_{i} z_i u_i \; N_i \; \left(\Omega^{-1} cm^{-1}\right) \tag{2.2}$$

The transference number of an ion j depends on the mobilities of the other ions

$$t_j = \frac{|z_j| u_j}{\sum_i |z_i| u_i} \tag{2.3}$$

One can determine the individual mobilities, u_j, if the transference numbers, t_i, and the concentrations of all mobile ions, N_i, are known.

The mobility of an ion is connected with its diffusion coefficient, D_i, by the Nernst–Einstein relation

$$D_i = \frac{RT}{|z_i| \mathrm{F}} u_i \ \left(\mathrm{cm^2\,s^{-1}}\right). \tag{2.4}$$

This diffusion coefficient describes only the self-diffusion of an ion in a homogeneous phase (obtainable by tracer experiments). If a concentration gradient exists, as is often the case in electrolyte solutions, the diffusion of all ions is balanced by their electrostatic interaction. Potential gradients are generated between faster- and slower-moving ions (diffusion potentials) which retard the first and accelerate the latter ions and lead to equal rates of diffusion. The common diffusion coefficient of a binary salt at low concentration may be used here as an example. This coefficient is connected to the individual mobilities by

$$D_{\pm} = \frac{z^+ + z^-}{z^+ \cdot z^-} \cdot \frac{u_+ \cdot u_-}{u_+ + u_-} \cdot \frac{RT}{\mathrm{F}^2} \tag{2.5}$$

where z^+ and z^- are the charges of the ions and u_+, u_- their mobilities. For salt solutions of higher concentration the relation is more complicated.[1]

Ionic mobilities and diffusion coefficients vary with the composition of the respective phase. In solutions they depend on the concentration and the solvent, in mixtures of salts on the mole fraction. The interpretation of the properties of electrolytes is a special area of electrochemistry,[2,3] which shall, however, not be discussed here in detail.

The properties of solid electrolytes are discussed more fully in Chapters 6 and 7 of this handbook. They of course also get attention in many of the other chapters, in particular Chapters 3 and 5, while interfacial phenomena are treated more extensively in Chapter 4.

III. THE GALVANIC CELL AT THERMODYNAMIC EQUILIBRIUM

A galvanic cell can be used to measure the free energy difference of a chemical reaction if this reaction can be performed in separate steps at two different electrodes. Examples are the reactions

$$H_2 + Cl_2 \rightarrow 2\,HCl \tag{2.6A}$$

$$2\,Ag + Br_2 \rightarrow 2\,AgBr \tag{2.6B}$$

$$2\,H_2 + O_2 \rightarrow 2\,H_2O \tag{2.6C}$$

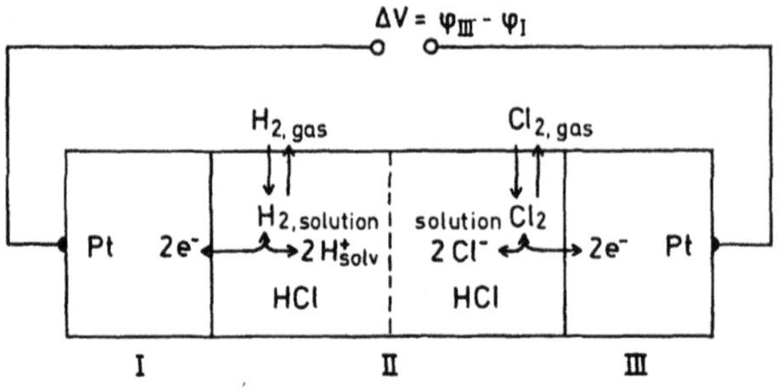

FIGURE 2.1. Phase scheme of a galvanic cell for the reaction $H_2 + Cl_2 \rightleftarrows 2\,HCl$.

Reaction (2.6A) can occur in the following steps at the electrodes I and III and in the electrolyte II, as shown in Figure 2.1.

$$\text{Electrode I:} \quad H_{2,II} \rightleftarrows 2H^+_{II} + 2e^-_{I} \tag{2.6A$_1$}$$

$$\text{Electrode III:} \quad Cl_{2,II} + 2e^-_{III} \rightleftarrows 2Cl^-_{II} \tag{2.6A$_2$}$$

$$\text{Electrolyte II:} \quad 2H^+_{II} + 2Cl^-_{II} \rightleftarrows 2HCl_{II} \tag{2.6A$_3$}$$

$$\text{Net reaction:} \quad H_{2,II} + Cl_{2,II} + 2e^-_{III} \rightleftarrows 2HCl_{II} + 2e^-_{I} \tag{2.6A$'$}$$

If these processes occur reversibly, as indicated in the reaction equations above, the result is a voltage difference between the electrodes III and I which corresponds with the driving force of the net reaction (2.6A′). The difference in the formulation of Equation (2.6A′) from Equation (2.6A) is indicated by the appearance of the electrons on the two different electrodes in the net reaction (2.6A′). The formation of two HCl molecules is connected with a transfer of two electrons from phase III to phase I. We know from thermodynamics that the change in free enthalpy (or Gibbs free energy) ΔG for Process (2.3A) is negative. In order to bring Reaction (2.6A′) to equilibrium, this free enthalpy difference must be compensated by the free enthalpy for the transfer of the corresponding amount of electric charge from phase III to phase I. This is the energy $-2\,F\,\Delta V$ per mole electrons. Consequently,

$$-2F \cdot \Delta V + 2\Delta G_{HCl} = 0$$

or:

$$\Delta V = -\frac{\Delta G_{HCl}}{F}. \tag{2.7}$$

This derivation demonstrates the general relation for the voltage of a galvanic cell, ΔV, in which a chemical reaction with the driving force ΔG_{chem} has reached equilibrium by the appropriate charge separation. This voltage is described by the Nernst equation for the electromotive force (emf) of a galvanic cell at equilibrium:

$$\Delta V = -\frac{\Delta G_{chem}}{zF} \tag{2.8}$$

where z is the number of charge equivalents needed for the formation of 1 mol of product.

ΔG_{chem} depends on the concentrations (more exactly the activities a_i) of the reactants and products. The equilibrium emf for the reactions (2.6A), (2.6B) and (2.6C) is therefore

$$\Delta V_A = -\frac{\Delta G^0_{HCl}}{F} + \frac{RT}{2F}\ln\left[\frac{a_{H_2}}{a^0_{H_2}}\frac{a_{Cl_2}}{a^0_{Cl_2}}\left(\frac{a_{HCl}}{a^0_{HCl}}\right)^2\right] \quad (2.9)$$

$$\Delta V_B = -\frac{\Delta G^0_{AgBr}}{F} + \frac{RT}{2F}\ln\left[\frac{a_{Br_2}}{a^0_{Br_2}}\left(\frac{a_{AgBr}}{a^0_{AgBr}}\right)^2\right] \quad (2.10)$$

$$\Delta V_C = -\frac{\Delta G^0_{H_2O}}{2F} + \frac{RT}{4F}\ln\left[\frac{a_{O_2}}{a^0_{O_2}}\left(\frac{a_{H_2}}{a^0_{H_2}}\right)\left(\frac{a_{H_2O}}{a^0_{H_2O}}\right)^2\right] \quad (2.11)$$

In these equations the ΔG^0 values are the free energy change of the respective reactions when all participants are in their standard state with the activity a^0.*

For the cell (2.6B) the metal is in its standard state if it is pure. If the electrolyte is saturated with AgBr, its activity is constant and the emf depends only on the activity of Br_2.

The temperature dependence of the cell voltage is related to the entropy of the reactions. This relation can be derived directly from Equation (2.8)

$$\left(\frac{\partial \Delta V}{\partial T}\right)_{p,n_j} = -\frac{1}{zF}\left(\frac{\partial \Delta G_{chem}}{\partial T}\right)_{p,n_j} = \frac{1}{zF}\Delta S_{chem} \quad (2.12)$$

The reaction enthalpy is determined by

$$\Delta H_{chem} = \Delta G_{chem} + T\Delta S_{chem} = -zF\left[\Delta V - T\left(\frac{\partial \Delta V}{\partial T}\right)_{p,n_j}\right] \quad (2.13)$$

and the pressure dependence is expressed according to the thermodynamic relations

$$\left(\frac{\partial \Delta V}{\partial p}\right)_{T,n_j} = -\frac{1}{zF}\Delta v_{chem} \quad (2.14)$$

with Δv the volume change in the reaction. From cell voltage measurements one can often very precisely determine thermodynamic data of chemical reactions.

IV. ELECTROSTATIC POTENTIALS: GALVANI POTENTIAL, VOLTA POTENTIAL, SURFACE POTENTIAL

While electric potential differences inside a homogeneous phase can be directly measured by the work needed for moving an electrically charged probe from one point to another, this

* a^0 is often set to one and is omitted in the thermodynamic equations. The formulation above should remind the reader that the standard state can be defined arbitrarily and has to be stated in the thermodynamic data.

is not possible if the phase is inhomogeneous in composition and, in particular, if a phase boundary has to be passed by this probe. The reason is that any probe is exposed not only to the influence of an electric field, but also to a chemical interaction with the surrounding matter. If this interaction varies, its variation contributes to the work for the movement of the probe between two locations. This contribution can be small in single phases with variable composition; it is, however, of extreme importance for the transfer of a charged particle through a phase boundary. What can be measured is only the combination of chemical and electric forces to which a charged particle is exposed. This combination can be represented by the *electrochemical potential*, $\tilde{\mu}_i$,

$$\tilde{\mu}_i = \mu_i + z_i F\varphi \tag{2.15}$$

where μ_i is the chemical potential in its usual definition:

$$\mu_i = (\partial G/\partial n_i)_{T,p,n_j \neq n_i} \tag{2.16}$$

for a single phase. The definition of $\tilde{\mu}_i$, is, accordingly,

$$\tilde{\mu}_i = \left(\frac{\partial G}{\partial n_i}\right)_{T,p,\varphi,n_j \neq n_i} + F\varphi \tag{2.17}$$

where φ is the Galvani potential. Any measurement of work for the transfer of a particle with the charge z_i to another phase is a determination of

$$\Delta\tilde{\mu}_i = {}_{II}\tilde{\mu}_i - {}_{I}\tilde{\mu}_i = {}_{II}\mu_i - {}_{I}\mu_i + z_i F(\varphi_{II} - \varphi_I) \tag{2.18}$$

For a neutral salt the electrical term is canceled in the sum of anions and cations, and one obtains the chemical potential of the salt. Distribution of a neutral salt between two different solvents only requires that at equilibrium the chemical potentials become equal, although electric potential differences may exist at the interface.

Equilibrium at an interface requires, for all charged species which can pass the phase boundary, that $\Delta\tilde{\mu}_i$ is zero or

$$\varphi_{II} - \varphi_I = \frac{{}_{II}\mu_i - {}_{I}\mu_i}{z_i F} \tag{2.19}$$

where z_i has the sign of the charge of the species. At the contact between two different metals, a contact potential, $\Delta\varphi$, arising from the different chemical potentials of the electrons in these metals is generated. Balance of $\Delta\tilde{\mu}_i$ between the two metals requires an excess of electrons on the phase with the lower chemical potential and a corresponding loss of electrons on the other side of the contact. This is illustrated in Figure 2.2.

The real measurement of a voltage difference between two electrodes, as in Figure 2.1, occurs between electronic conductors of equal composition. If the electrodes I and III of Figure 2.1 are different metals, a contact between metals I and III has to be made on the other side, and the potential difference would be determined between the open ends of the same metal.

The reason for the appearance of a Galvani potential difference at the contact between two phases is not only the accumulation of electric charge of opposite sign at both sides of

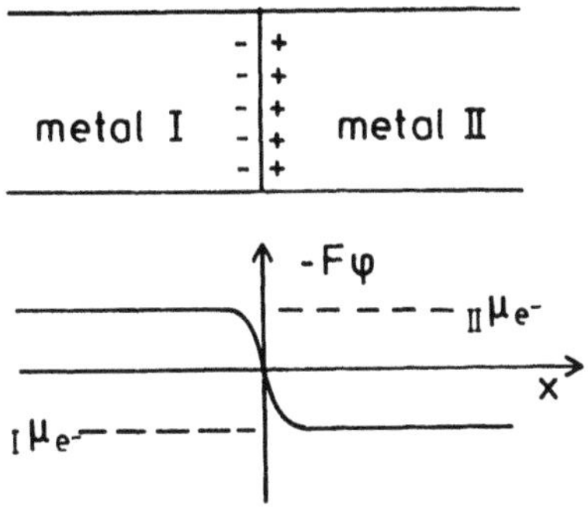

FIGURE 2.2. Formation of a contact potential between two metals.

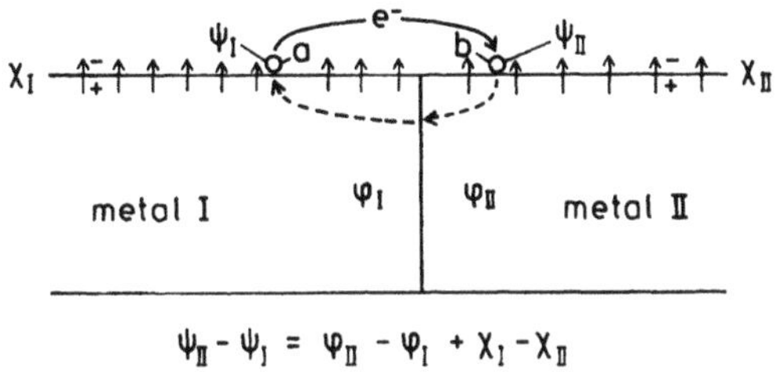

FIGURE 2.3. Origin of Volta potential differences between two metals.

an interface; it is also due to the formation of dipole layers in one or both phases directly at the contact. Polar molecules of the solvent of an electrolyte can be oriented, e.g., by interaction with a metal surface, and form a dipole layer at the interface. Chemisorption of one type of ion at the surface of the contacting phase can also result in a dipole layer if the counter ions remain at a larger distance from the contact. Dipole layers exist on the surface of metals in contact with vacuum, because the kinetic energy of the electrons in the conduction band allows the electrons to extend somewhat beyond the last row of positive nuclei of a crystal. The size of this effect depends on the surface structure and its orientation with respect to the crystal. One sees this in work function measurements and the so-called *Volta potentials* above different surfaces in vacuum or inert gases.[4]

Volta potential differences are determined by the work required for moving an electrically charged probe in vacuum or in an inert gas from one point close to the surface of a condensed phase I to a point close to the surface of another phase II. The distance of the probe from the surface should be large enough that all chemical interaction and the effect of electric polarization (image force) can be neglected. This situation is illustrated for a contact between two metals in vacuum in Figure 2.3.

The probe may be an electron which shall be moved from point *a* to point *b* in Figure 2.3. The same energy will be required if the probe is moved from point *a* through the surface of phase I, through the contact between phase I and phase II, and from there through the surface of phase II to point *b*. At the contact we have equilibrium and need no work for the transfer of the electron between I and II. With the introduction of the electron from point *a* into phase I, we gain the chemical binding energy, $_{I}\mu_{e-}$, and the electrical work for the passage of the surface dipole layer, $-F\chi_I$. The passage through the surface of phase II to point *b* requires the work to overcome the binding energy $_{II}\mu_{e-}$ and the electrostatic work for passing the dipole layer $F\chi_{II}$. The result yields the definition of the Volta potential difference, $\Delta\psi$,

$$\Delta G_{a\to b} = {}_{II}\mu_{e^-} - {}_{I}\mu_{e^-} + F(\chi_{II} - \chi_I) = -F(\psi_{II}(b) - \psi_I(a)) \tag{2.20}$$

Here, χ is the so-called surface potential corresponding with the electrostatic potential step resulting from the inhomogeneous charge distribution on the surface. This surface dipole can be calculated for metals on the basis of more or less refined theories and depends very much on the band structure. In order to get some idea of the possible size of such surface potentials, one should realize that a layer of water molecules with each molecule oriented perpendicular to an atom of a surface with 10^{15} atoms per square centimeter would correspond with a χ potential of about 3.8 V.

Since ${}_{II}\mu_{e^-} - {}_{I}\mu_{e^-} = F(\varphi_{II} - \varphi_{I})$, Equation (2.20) yields the connection between the Volta potential difference between two points in front of different surfaces in vacuum and the terms $\Delta\varphi$ and $\Delta\chi$

$$\Delta\psi = \Delta\varphi + \Delta\chi \tag{2.21}$$

The surface dipole layer is also essential for the measurement of the work function, which is the energy required to remove an electron from a state in the conduction band where the probability of occupation by an electron is one half, the Fermi energy, E_F, into vacuum. Since the electron has to pass the surface dipole layer in this process, the work function ϕ is defined by

$$\phi = -\frac{\mu^0_{e^-}}{N_A} - e_0\chi \tag{2.22}$$

where N_A is Avogadro's number and e_0 the electron charge, since ϕ is usually determined in electron volts.

On metal surfaces χ is negative and, therefore, $N_A \times \phi$ is usually larger than $-\mu^0_{e^-}$. The work function can be obtained from photoelectron emission experiments[5] or in relative terms with the Kelvin method.[6] All such measurements determine the energy to take an electron from the Fermi energy in the bulk to a position in front of the surface where the chemical and image force interaction has disappeared, but sufficiently close that the influence of the surface dipole layer can be considered as the effect of a layer of infinite extension. If one could measure the energy for the transfer of the electron from the bulk of a closed body to a point in vacuum at infinite distance where the dipole layer around the body would no longer affect the energy, one could directly measure $\mu^0_{e^-}$. This is indicated in Figure 2.4 for a spherical body, but, unfortunately, such an experiment cannot be performed.

V. ELECTROCHEMICAL EQUILIBRIUM AT INTERFACES

The simplest electrode reactions are those in which either an ion passes the interface between an electrolyte and a metal and is incorporated into the metal together with the uptake of electrons, or in which only the electron passes the interface from an electron donor in the electrolyte to the metal or from the metal to an electron acceptor. The first case occurs at the silver electrode of Process (2.6B), the other case occurs in some steps of the electrode reactions of the Processes (2.6A), (2.6B), and (2.6C). The latter reactions are more complex and will therefore be discussed later. Instead, the simple redox reaction $Ox^+ + e^- = Red$ shall be used here as an example for the second case.

Equilibrium for the electrode reaction with ion transfer through the interface

$$M_I \rightleftarrows M^{z+}_{II} + ze^-_I \tag{2.23}$$

where I is the metal phase and II the electrolyte, requires

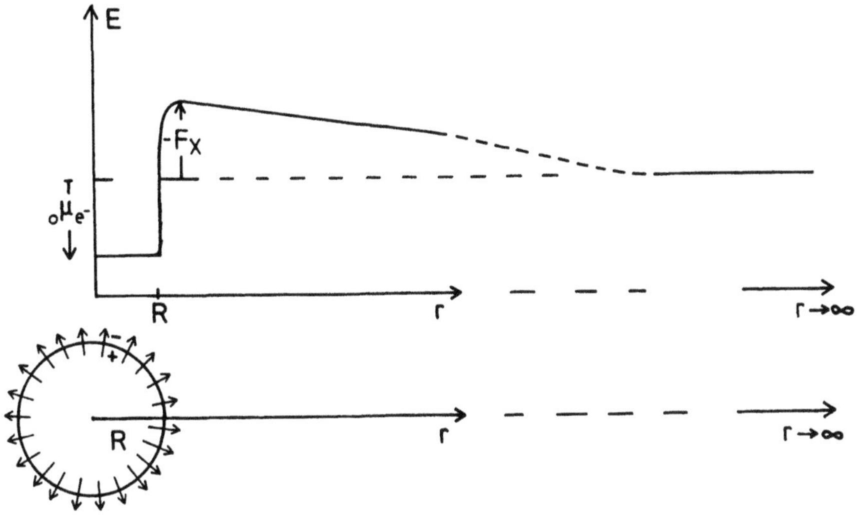

FIGURE 2.4. *Gedanken* experiment for the determination of the chemical potential of electrons in a single phase by transfer to infinite distance in vacuum.

$$_{I}\mu_{M} = _{II}\tilde{\mu}_{M^{z+}} + z\,_{I}\tilde{\mu}_{e^-} \tag{2.24}$$

or

$$_{I}\mu^{0}_{M} = _{II}\mu^{0}_{M^{z+}} + RT\ln\left(\frac{a_{M^{z+}}}{a^{0}_{M^{z+}}}\right) + z\,_{I}\mu^{0}_{M_{e^-}} + zF(\varphi_{II} - \varphi_{I}) \tag{2.25}$$

The difference of the Galvani potential is determined by the chemical potential of the reactants in their standard states and depends on the activity of the ions in the electrolyte

$$\varphi_{II} - \varphi_{I} = \Delta\varphi^{0} + \frac{RT}{zF}\ln\left(\frac{a_{M^{z+}}}{a^{0}_{M^{z+}}}\right) \tag{2.26A}$$

with

$$\Delta\varphi^{0} = \frac{_{II}\mu^{0}_{M^{z+}} + z\,_{I}\mu^{0}_{M_{e^-}} - _{I}\mu^{0}_{M}}{zF} \tag{2.26B}$$

For a redox reaction of the type,

$$Ox^{n+}_{solv} + e^{-}_{M} \rightleftarrows Red^{(n-1)+}_{solv} \tag{2.27}$$

where phase I may be an inert metal and phase II the electrolyte, the equilibrium condition is

$$_{II}\tilde{\mu}_{Ox^+} + _{I}\tilde{\mu}_{e^-} = _{II}\tilde{\mu}_{Red} \tag{2.28}$$

or

$$ {}_{II}\mu^0_{Ox} + RT\ln\left(\frac{a_{Ox}}{a^0_{Ox}}\right) + nF\varphi_{II} + {}_{I}\mu^0_{e^-} - F\varphi_I = {}_{II}\mu^0_{Red} + RT\ln\left(\frac{a_{Red}}{a^0_{Red}}\right) + (n-1)F\varphi_{II}. \quad (2.29) $$

The result for $\Delta\varphi$ is

$$ \varphi_I - \varphi_{II} = \Delta\varphi^0 + \frac{RT}{F}\ln\left(\frac{a_{Ox}}{a_{Red}} \cdot \frac{a^0_{Red}}{a^0_{Ox}}\right) \quad (2.30) $$

with

$$ \Delta\varphi^0_{redox} = \frac{{}_{II}\mu^0_{Ox} - {}_{I}\mu^0_{Red} + {}_{I}\mu^0_{e^-}}{F} \quad (2.30A) $$

As a third example, one of the electrode reactions with a gaseous reaction partner of the cells in Section III may be considered. We choose the hydrogen electrode with the process

$$ H_{2,gas} \rightleftarrows 2\ H^+_{II} + 2\ e^-_I $$

where II represents the electrolyte and I the metal. If the H_2 gas has some solubility in the electrolyte, as in liquid solutions, there exists an equilibrium distribution between the gaseous phase and the liquid phase II.

$$ {}_{II}\mu_{H_2} = {}_{gas}\mu_{H_2} = {}_{gas}\mu^0_{H_2} + RT\ln\left(\frac{a_{H_2}}{a^0_{H_2}}\right)_{gas} \quad (2.31) $$

Consequently, one obtains for the electrode equilibrium

$$ {}_{gas}\mu_{H_2} = 2\,{}_{II}\tilde{\mu}_{H^+} + 2\,{}_{I}\tilde{\mu}_{e^-} \quad (2.32) $$

and as a result for $\Delta\varphi$

$$ \varphi_I - \varphi_{II} = \Delta\varphi^0 + \frac{RT}{2F}\ln\left[\left(\frac{{}_{II}a_{H+}}{{}_{II}a^0_{H^+}}\right)^2\left(\frac{{}_{gas}a^0_{H_2}}{{}_{gas}a_{H_2}}\right)\right] \quad (2.33) $$

with

$$ \Delta\varphi^0_{redox} = \frac{{}_{gas}\mu^0_{H_2} - 2\,{}_{II}\mu^0_{H^+} - 2\,{}_{I}\mu^0_{e^-}}{2\,F} \quad (2.33A) $$

If the gas is not soluble in the electrolyte, but can come in contact with the metal surface, e.g., at a three-phase boundary between metal, gas, and electrolyte, equilibrium requires that the gas is adsorbed on the metal. In this case, the adsorption equilibrium has to be introduced into the thermodynamic relations instead of the equilibrium with the electrolyte, and the adsorbed molecules may be considered as a part of the electrode phase I.

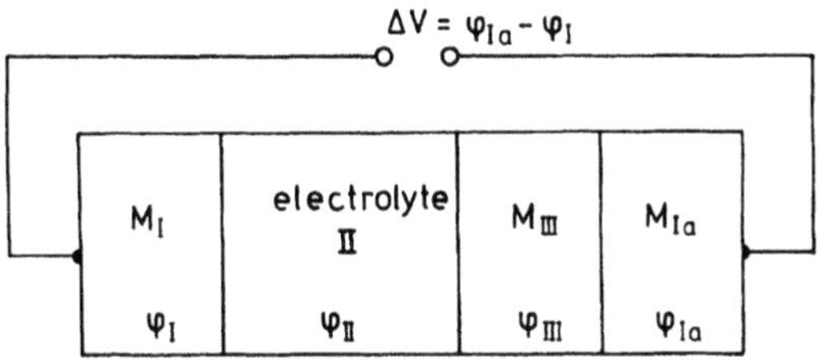

FIGURE 2.5. Phase scheme of a galvanic cell with two different metal electrodes and the necessary third contact for the measurement of the cell voltage.

$$H_{2,gas} \rightleftarrows H_{2,adsorbed} \tag{2.34}$$

$$_{gas}\mu_{H_2} = {}_I\mu_{H_{2,ad}}$$

One can see that equilibrium at the electrode leads to the same Relation (2.32), if Relation (2.34) is fulfilled.

The Relations (2.26), (2.30), and (2.33) demonstrate that the difference of the Galvani potential between two conductive phases, where a charge carrier can pass the interface, is at equilibrium controlled by the chemical composition of the phases in contact with each other. Since the chemical potential of the electron appears in Equations (2.26A), (2.30A), and (2.33A), one realizes that the absolute value of $\Delta\varphi$ depends on the nature of the specific metal which forms the electrode for the particular reaction. It is, however, impossible to absolutely determine $\Delta\varphi$, as pointed out in Section IV. Measured is the voltage of a galvanic cell consisting of two electrodes in contact with an electrolyte. This is schematically shown in Figure 2.5. The voltage, ΔV, is a measure of the difference of the electrochemical potential of the electrons between the two electrodes and is only in this case equal to the Galvani potential difference if the electrodes are of the same metal. In order to achieve this, in Figure 2.5, a contact between the metals M_{III} and M_{Ia} is added, but the contact potential difference $\Delta\varphi_{MIa/MIII}$ now appears in the measured voltage

$$\Delta V = \varphi_{III} - \varphi_I + \Delta\varphi_{M_{Ia}/M_{III}} \tag{2.35}$$

$$\varphi_{Ia} - \varphi_{III} = \Delta\varphi_{M_{Ia}/M_{III}} = \frac{\mu^0_{e^-,M_{Ia}} - \mu^0_{e^-,M_{III}}}{F} \tag{2.36}$$

VI. STANDARD POTENTIALS AND ELECTROMOTIVE SERIES

A. REFERENCE ELECTRODES

The simple scheme of a galvanic cell depicted in Figures 2.1 and 2.5 can be verified only in rare cases. When both metal electrodes are in contact with the same electrolyte, irreversible reactions will often occur. For instance, if in the cell of Figure 2.1 the metal of phase III is silver, AgCl would be formed spontaneously. Or if the electrolyte HCl contained $CuCl_2$, copper would be deposited on phase I. Even the simple cell of Figure 2.1 requires that H_2 and Cl_2 must be kept apart from the counter electrodes, i.e., H_2 should only be dissolved in the electrolyte next to the electrode I, Cl_2 only in the electrolyte next to electrode II. Otherwise, Cl_2 would be reduced at electrode I, and H_2 could be oxidized at electrode III. This means that the electrolytes in contact with the two electrodes cannot be exactly equal in composition.

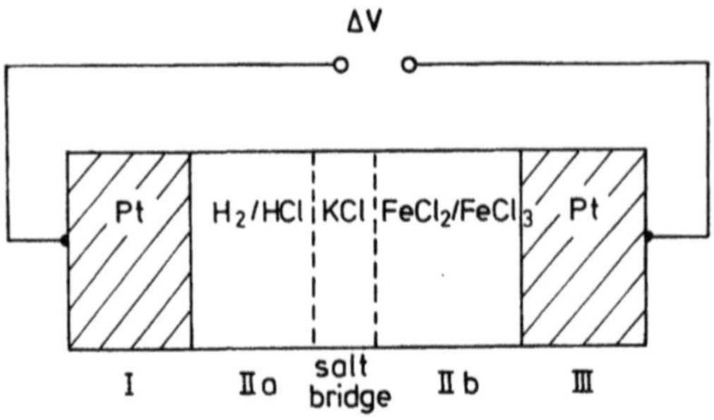

$$\Delta V = \varphi_{III} - \varphi_{I} = \Delta\varphi_{III/IIb} + \Delta\varphi_{IIb/IIa} + \Delta\varphi_{IIa/I}$$

FIGURE 2.6. Phase scheme of a galvanic cell with a salt bridge.

If the electric properties are not changed by a different composition of the electrolyte and if a separation by a diaphragm or other means can prevent the interfering reactant from reaching the counter electrode, the cell voltage is not affected. This is the case for gases of low solubility as in the system of Figure 2.1. In other cases, one can try to calculate the potential difference which arises at the contact between two electrolytes in the zone where they mix by diffusion. These are diffusion potentials due to different mobilities of the anions and cations. Because such calculations are in most cases inaccurate, one usually tries to minimize such potential differences. This can be achieved between electrolytic solutions of the same solvent by salt bridges connecting the two different electrolytes. The salt in the electrolyte of the bridge has to consist of ions with nearly equal mobility so that anions and cations diffuse at the same rate into the neighboring electrolytes. Aside from this requirement, the salt concentration in the bridge should be high in comparison to the concentration of salts in both electrolytes, the ions of which may have very different mobilities. The remaining diffusion potentials between the electrolyte of the salt bridge and the two electrolytes in contact with the electrodes are in this way kept small and compensate each other to a large extent. Such a cell is represented in Figure 2.6. Suitable salts for such bridges between aqueous electrolytes are KCl and KNO_3.

As a reference electrode for such equilibrium potentials of electrode reactions in aqueous solution, the hydrogen electrode was initially introduced. This was a very unfortunate choice because the standard state of H^+ ions in solution of concentrated acids is difficult to realize, and the standard potential with activity 1 of H^+ ions has to be derived from measurements at low concentration by extrapolation. In addition, the mobilities of cations and anions in concentrated acids are very different, and therefore large diffusion potentials arise at the contact to any other electrolyte — a situation which can hardly be minimized by salt bridges. Furthermore, only a few noble metals catalyze the hydrogen electrode reaction to such an extent that it becomes reversible. Therefore, in practical measurements, other reference electrodes are used which are much more stable, can easily be reproduced, and are much better suited to depressing diffusion potentials. The most commonly applied reference electrodes are the so-called *calomel electrode* and the *silver/silver chloride electrode*. Since the potential difference between any other reference electrode and the standard hydrogen electrode can be determined experimentally, the hypothetical cell voltage vs. the hydrogen electrode can easily be calculated by adding this difference. A cell with an Ag/AgCl reference electrode is depicted in Figure 2.7, where the cell voltage is a relative measure of the equilibrium potential of the Cu electrode in contact with a Cu salt solution.

The Ag/AgCl reference electrode contains as an electrolyte a KCl solution which is saturated with AgCl. AgCl is present in a solid form as a porous layer on the Ag electrode. The chemical potential of AgCl in the solution is therefore constant and with it the sum of

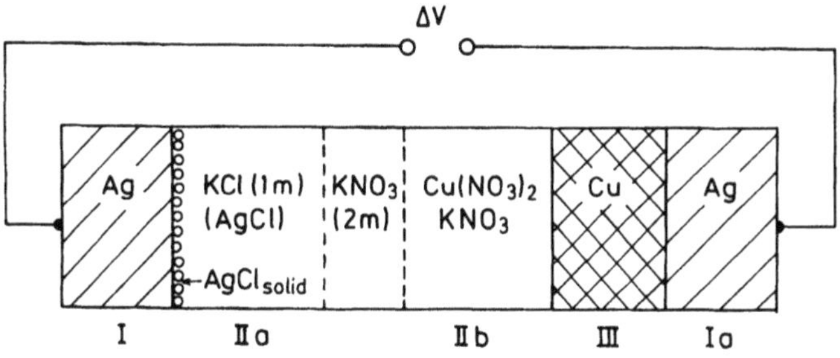

$$\Delta V = \varphi_{Ia} - \varphi_{I} = \Delta\varphi_{IIa/I} + \Delta\varphi_{IIb/IIa} + \Delta\varphi_{III/Ib} + \Delta\varphi_{Ia/III}$$

FIGURE 2.7. Phase scheme of a galvanic cell with an Ag/AgCl reference electrode.

the individual chemical potentials of the ions Ag^+ and Cl^-, which both depend on their concentration.

$$\mu_{AgCl} = \mu_{Ag^+} + \mu_{Cl^-} = \mu^0_{Ag^+} + \mu^0_{Cl^-} + RT\ln\left(\frac{a_{Ag^+} \cdot a_{Cl^-}}{a^0_{Ag^+} \cdot a^0_{Cl^-}}\right) \tag{2.37}$$

The potential difference $\Delta\varphi$ between the Ag metal and the electrolyte is given by the reaction $Ag_{I} \leftarrow Ag^+_{II} + e^-_{I}$. Equation (2.26), after taking into account Relation (2.37), yields

$$\Delta\varphi_{Ag/AgCl} = \Delta\varphi^0_{Ag/AgCl} - \frac{RT}{F}\ln\left(\frac{a_{Cl^-}}{a^0_{Cl^-}}\right) \tag{2.38}$$

where

$$\Delta\varphi^0_{Ag/AgCl} = \frac{{}_I\mu^0_{e^-} - {}_I\mu^0_{Ag} - {}_{II}\mu^0_{Cl^-} + \mu_{AgCl}}{F} \tag{2.38A}$$

One sees that the potential difference at the electrode depends on the Cl^- activity which can easily be kept constant by a constant concentration of KCl.

B. ELECTROMOTIVE SERIES

With the help of such cell constructions, as described by Figures 2.6 and 2.7, assisted by calculations on the basis of thermodynamic data where measurements were not possible, the electromotive series of electrode reactions could be determined for ion transfer and electron transfer reactions relative to a particular reference electrode. In order to get some idea of the factors which control the position of an electrode reaction in such a scheme, an ion transfer reaction of type (2.23) and a redox reaction of type (2.27) shall be analyzed.

The transfer of an ion from the metal to the electrolyte can, in a *gedanken* experiment, be performed by a cycle with the following steps where all species involved are in their standard state. This cycle is shown in Figure 2.8 for the free energy changes ΔG_{chem} without the electrostatic contributions. M represents a metal atom, I the metal, and II the electrolyte phase. In the following list of steps, the electric energy terms for the charged species resulting from the Galvani potentials of phases I and II are, however, included.

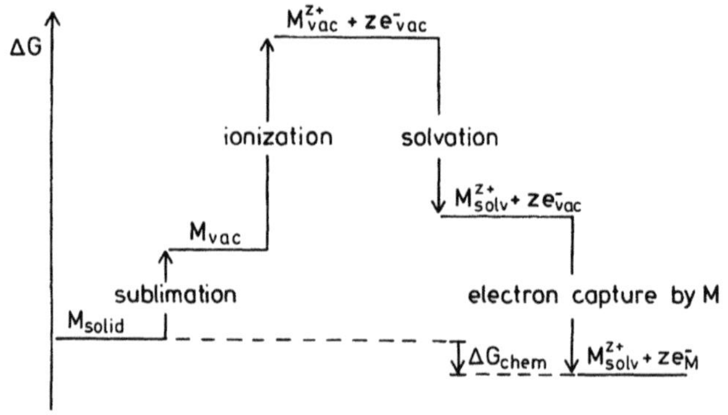

FIGURE 2.8. Thermodynamic cycle for a metal electrode in equilibrium with its ions in solution.

$$1.\quad M_I \rightarrow M_{vacuum} \qquad ;\ \Delta G_1 = \mu^0_{sublimation}$$

$$2.\quad M_{vac} \rightarrow M^{z+}_{vac} + ze^-_{vac} \qquad ;\ \Delta G_2 = \mu^0_{ionization}$$

$$3.\quad M^{z+}_{vac} \rightarrow M^{z+}_{II} \qquad ;\ \Delta G_3 = {}_{II}\mu^0_{M^{z+}} + zF\varphi^0_{II}$$

$$4.\quad ze^-_{vac} \rightarrow ze^-_{I} \qquad ;\ \Delta G_4 = {}_{I}\mu^0_{e^-} - zF\varphi^0_{I}$$

$$(1\ldots4)\, M_I \rightleftarrows M^{z+}_{II} + ze^-_I \qquad \sum_i \Delta G_i = 0 = \Delta G_{chem} - zF\left(\varphi^0_I - \varphi^0_{II}\right) \tag{2.39}$$

with $\Delta G_{chem} = \sum_i \mu^0_i$. In these equations ${}_{II}\mu^0_{M^{z+}}$ is the solvation free enthalpy of the ion, ${}_{I}\mu^0_{e^-}$ is the binding free enthalpy of the electron in the metal.

The sum of ΔG_i corresponds with the difference in the electrochemical potentials for the transfer of an M^{z+} ion from the metal into the electrolyte which must be zero, if equilibrium is established. The difference of the chemical potential is compensated at equilibrium by the difference of the electrostatic energy of the charged species in both phases. This is the energy $zF(\varphi^0_{II} - \varphi^0_{I})$. The result for the Galvani potential difference is therefore

$$\Delta\varphi^0_{M/M^{z+}} = \varphi^0_I - \varphi^0_{II} = \frac{{}_{I}\mu^0_{subl} + {}_{vac}I^0_M + {}_{II}\mu^0_{M^{z+}} + z\,{}_{I}\mu^0_{e^-}}{zF} \tag{2.40}$$

It is seen that aside from the evaporation energy and the ionization energy of the metal atom — both of which are positive — and the chemical potential of the electrons in the metal, a negative term, the voltage difference depends on the chemical potential of the ion in the electrolyte, i.e., its solvation energy. This is also a negative term which can, however, differ to a large extent for different electrolytes (e.g., if complexes are formed with ions or molecules in the solution or if the solvent is different). This is the reason why the ordering of metal electrodes within the electromotive series depends on the type of salts and the solvents used in the measurements. For instance, the process $Ag_I \rightleftarrows Ag^+_{II} + e^-_I$ has a much more negative position in the electromotive scale with acetonitrile as the solvent than in water due to the strong interaction of Ag^+ ions with CH_3CN molecules. Similar effects occur at electrodes in contact with molten salts or with solids as electrolytes.

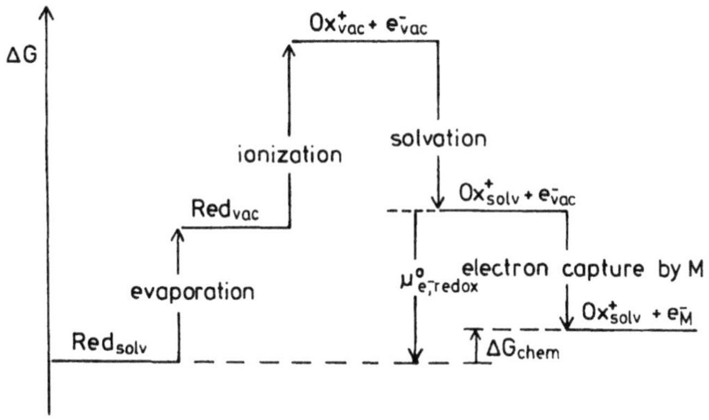

FIGURE 2.9. Thermodynamic cycle for a redox system in equilibrium with an inert electrode and the chemical potential of electrons in a redox couple.

An analogous cycle can be formulated for electrodes in contact with redox systems of the type (2.27). This is outlined in Figure 2.9. In this case, the potential difference is controlled by the ionization energy between the two oxidation states performed in the gas phase, the standard chemical potential of the two redox species in the electrolyte, and the chemical potential of the electrons in the metal used as the electrode.

$$\Delta\varphi^0_{red/ox^+} = \varphi^0_I - \varphi^0_{II} = \frac{_{vac}I^0_{red} - {}_I\mu^0_{red} + {}_{II}\mu^0_{ox^+} + {}_M\mu^0_{e^-}}{F} \tag{2.41}$$

Equation (2.41) shows again that the Galvani potential difference of an electrode in contact with a redox electrolyte depends on the chemical potential of the electrons in the metal used as the electrode. This difference does not, however, appear in the measured cell voltage vs. a reference electrode, because it is compensated at the external connection to the metal of the reference electrode by the contact potential between the two metals as explained in Figure 2.5 and Equations (2.35) and (2.36). The consequence for the two cycles of ion transfer (type 2.23) and redox reactions (type 2.27) is that the chemical potentials of the electrons in the individual electrodes have to be replaced by the chemical potential of the electrons in the metal of the reference electrode. The relative position of an electrode reaction in the electromotive series is therefore exclusively controlled by the sum of all the other chemical potentials in these cycles.

Electrode reactions in different electrolytes, particularly in nonaqueous electrolytes, require different reference electrodes. Often, a reaction which is possible in one electrolyte cannot be performed in another or is not reversible therein. It is therefore attractive to seek a uniformly applicable reference state which appears in all electrode reactions. This is the quest for an *absolute scale* of redox potentials. The cycles of Figures 2.8 and 2.9 suggest the use of an electron in vacuum at infinite distance as the common reference state. Equilibrium for the electrode reaction will be obtained if the free energy of the electron compensates the difference between the states $M^{z+}_{solv} + ze^-_{vac}$ and M in Reaction (2.23) or between $Ox^+_{solv} + e^-_{vac}$ and Red_{solv} in Reaction (2.27) (cf. Figures 2.8 and 2.9). This free energy difference can be considered as the chemical potential of the electron in the individual electrode reaction, a concept which is particularly useful for redox reactions. In Figure 2.9, this energy difference is shown and designated $\mu^0_{e^-,redox}$. Electrons from this free energy level relative to the vacuum can be added to the Ox^+ species in the electrolyte or can be taken away from the Red species to this energy level without changing the free energy of the redox system. The chemical potential of the electron in a redox system is in this concept defined by

$$\mu^0_{e^-,\mathrm{redox}} = \mu^0_{\mathrm{red,solv}} - \mu^0_{\mathrm{ox^+,solv}} - {}_{vac}I_{\mathrm{red}} \tag{2.42}$$

$\mu^0_{e^-,\mathrm{redox}}$ is negative and represents the averaged binding energy of the electron in a redox system.

VII. THE ELECTRIC DOUBLE LAYER AT INTERFACES

Up to this point, the origin of the difference of the Galvani potential at a contact between an electronic and an electrolytic conductor was attributed to excess charge of opposite sign on either side of the interface and to dipole layers between the two phases without considering the spatial charge distribution or the atomic or molecular structure of the interface. In this section, the main models describing such interfaces and the experimental basis of these models shall be presented.

A. METAL/ELECTROLYTE INTERFACES

Electrolytes have a definite voltage range of stability in a galvanic cell before they are decomposed by electrolysis in which one component is oxidized, the other reduced, provided the electrodes themselves remain inert in this voltage range. These voltage limits can be controlled by the solvent itself, which may be oxidized at the anode instead of the anion and/or be reduced at the cathode instead of the cation. The electrodes behave beyond these limits like a capacitor with a definite capacity for the storage of electric charge. This capacitive behavior can be studied by polarizing an inert electrode in contact with a suitable electrolyte in a galvanic cell, the potential being measured vs. a reference electrode. The main information pertaining to capacities comes from such experiments, which are summarized in the following.

The simplest model is that of a plate capacitor developed very early by Helmholtz.[7] The idea is that the ions of the electrolyte, which form the excess charge there, can approach the metal surface only up to the distance of the radius which includes the inner solvation sphere in liquid solutions. Measurements of the differential capacity of smooth electrodes yielded values for the Helmholtz double-layer capacity, C_H, on the order of 20 to 30 μF cm^{-2}. The model of a plate capacitor gives for the differential capacity

$$\frac{\partial Q}{\partial \Delta\varphi} = C_H = \frac{\varepsilon_H \cdot \varepsilon_0}{d} \tag{2.43}$$

where Q is the charge in C cm^{-2}, ε_H the dimensionless dielectric constant relative to vacuum, ε_0 the permittivity in vacuum ($\varepsilon_0 = 8.85 \times 10^{-14}$ C V^{-1} cm^{-1}), and d the distance between the plates in centimeters. For $d \approx 2 \times 10^{-8}$ cm, the capacity is $C_H \approx \varepsilon_H \times 4.4$ μF cm^{-2}. The effective dielectric constant therefore has to be on the order of 5 to 7. That ε_H is much smaller than in the bulk of the electrolyte solution is reasonable, since the molecules of the solvation shell are oriented and no solvent layer exists between the ion and the metal surface, except when the solvent molecules strongly adsorb thereon. In such a case, however, these molecules will also be oriented and much less polarizable than in the bulk of the solution. Assuming a variability of ε_H and a variability of d, depending on the field strength in the double layer, which changes the orientation of the polar molecules and can distort the solvation shell or even break up part of it, this crude model can to some extent explain the dependence of C_H on the voltage applied. The real situation is, however, much more complex in the molecular picture, and still open to debate.[8]

It has been observed that the capacity is usually larger on the branch where the metal has a positive excess charge than on the negative branch (cf. Figure 2.10).[9] There are two reasons for this behavior. On the positive branch, the counter ions are anions which have a larger radius and a more weakly bound solvation shell. They will come closer to the interface at

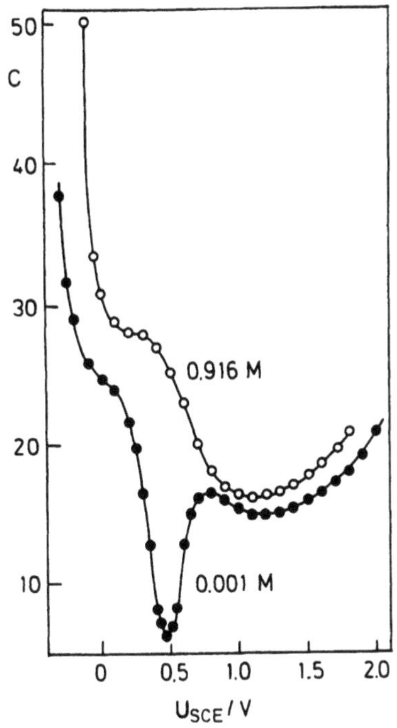

FIGURE 2.10. Differential capacity of a mercury electrode in aqueous solutions of NaF as a function of the potential vs. a calomel electrode. (Adapted from Grahame, D.C., *J. Am. Chem. Soc.*, 1954, 76, 4819.)

higher positive field strength by losing part of their solvation shell. Cations have less distortable solvation shells, and their distance should remain less affected by the field strength. Another effect with consequences in the same direction is connected with the excess charge distribution on the metal. The negative charge of electrons extends somewhat over the last layer of the positive atomic nuclei, the more the higher the negative excess charge. The repulsive forces between these electrons and the solvent or ion molecules increase the average distance of the counter charge from the surface. On the positive branch, the electrons are drawn back into the bulk while the nuclei remain fixed.[10] The counter ions can come closer to the surface, and the capacity increases. Aside from such effects, the dipole formed between the surface electrons and the ionic atom cores varies also with the potential, and this can contribute considerably to the differential capacity in some potential ranges.[11]

The adsorption of specific ions from the electrolyte has an important influence upon the charge distribution in the double layer. Such ions, which are stabilized in solution by a solvation shell, can lose part of their solvation shell and come into direct contact with the metal surface. This happens when the adsorption forces overcompensate the loss of interaction energy with the solvent molecules. This is shown in Figure 2.11a. The counter ions will remain at a larger distance from the surface and can form part of the opposite charge if the charge of the adsorbed ions is not fully compensated by the excess charge on the metal. Such a charge distribution is represented in Figure 2.11b for the case that there is no excess charge on the metal.

The amount of adsorbed ions varies with the voltage applied. The adsorption of anions increases with positive excess charge on the metal and is completely suppressed at some critical negative excess charge on the metal. Cations behave in the opposite way. The differential capacity increases if specific ion adsorption contributes to the charge distribution across the interface. The quantitative relations are very complex because the local position of the adsorbed ions and the structure of their remaining solvation shell varies with the field strength in the inner Helmholtz double layer. In addition to this, a partial charge transfer between the

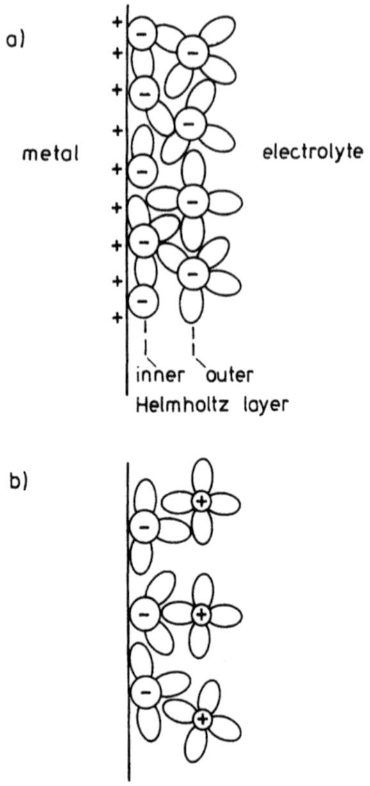

FIGURE 2.11. Double layer at a metal electrode with specific adsorption of anions: (a) large counter charge on the metal; (b) no excess charge on the metal.

ions and the metal is often connected with the adsorption which is equivalent to the formation of a polar bond between adsorbed ions and the metal surface. The amount of charge transfer (or the dipole moment of the adsorption bond) also varies with the field strength. The quantitative interpretation of capacity data is therefore in such cases extremely model dependent, and is not discussed here in more detail.

Another situation which can be analyzed more clearly is important for the understanding of the electrode interface. This is a contact to an electrolyte with a small concentration of mobile ions as in dilute electrolyte solutions. In this case, the counter charge in the electrolyte cannot be represented by a layer of ions in the outer Helmholtz plane, but extends over some distance into the space of the electrolyte. The reason is that the accumulation of the excess charge in the Helmholtz double layer leads to a relatively large local increase in the concentration which, in turn, exerts a chemical driving force for back diffusion into the bulk of the electrolyte. Therefore, the excess ions and with it the electric field extend into the bulk until the electric and chemical forces are balanced. This situation was analyzed by Gouy[12] and by Chapman.[13] Stern[14] combined their ideas with the Helmholtz model. His model is represented by two capacitors in series. The charge and potential distribution for this model is shown in Figure 2.12.

In this model, it is assumed that the space of the extension of the Helmholtz double layer, d, with the averaged dielectric constant ε_H is free of charge while the counter charge Q balancing the surface charge Q_M on the metal is in the diffuse layer $x \geq d$.

$$Q = \int_{\infty}^{d} \rho(x)dx = -Q_M \quad (2.44)$$

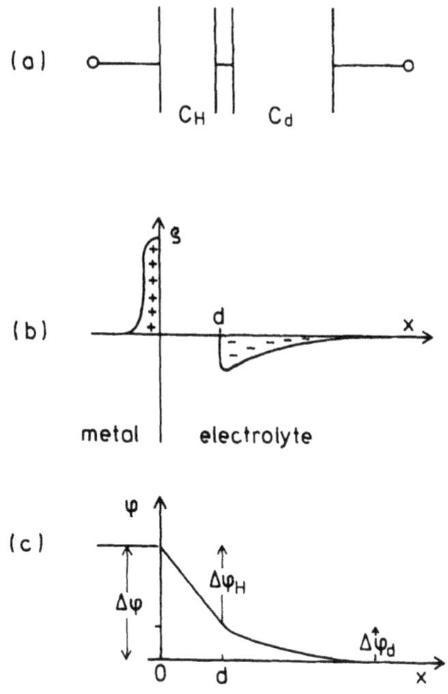

FIGURE 2.12. Model of the double layer in dilute electrolytes: (a) scheme for the differential capacity; (b) charge distribution; (c) potential distribution. (Adapted from Stern, O., *Z. Electrochem.*, 1924, *30*, 508.)

The charge distribution in this layer is controlled by a combination of the Poisson equation and the Boltzmann distribution function

$$\frac{d^2\varphi}{dx^2} = -\frac{1}{\varepsilon\varepsilon_0}\rho(x) \tag{2.45}$$

$$\rho(\varphi) = e_0\left[z^+ N_c^0 \exp\left(-z^+ e_0\,\varphi/kT\right) - z^- N_a^0 \exp\left(-z^-\,e_0\varphi/kT\right)\right] \tag{2.46}$$

where z^+ and z^- are the charges of the cations and anions, N_c^0 and N_a^0 are their concentrations per cubic centimeter in atomic units, and e_0 the elementary charge. Electroneutrality requires

$$z^+ N_C^0 = z^- N_A^0 \tag{2.47}$$

The charge distribution in the diffuse double layer can be calculated from the Poisson–Boltzmann model explicitly for the case $z^+ = z^- = z$. For asymmetric electrolytes the solution of these differential equations requires numerical methods. For the simpler case above where $N_c^0 = N_a^0 = N^0$, the relation between charge density and potential, Equation (2.46), becomes

$$\rho(\varphi) = -2ze_0\,N^0 \sinh\left(ze_0\varphi/kT\right) \tag{2.48}$$

Integration of Equation (2.45) with the charge density of Equation (2.48) yields a connection between the excess charge Q in the space charge layer and the field strength at x = d (cf. Figure 2.12, which also shows the boundary conditions for the integration)

$$-\varepsilon\varepsilon_0 \int_{\infty}^{d} \frac{d^2\varphi}{dx^2} dx = -\varepsilon\varepsilon_0 \frac{d\varphi}{dx} dx \bigg|_{x=d} = \int_{\infty}^{d} \rho(x) \mathrm{d}x = \mathrm{Q} \tag{2.49}$$

After some transformations one obtains

$$Q = \left(\varepsilon_s \varepsilon_0 kTN^0\right)^{\frac{1}{2}} \ 4 \sinh\left(\frac{ze_0}{kT}\varphi_d\right) = -Q_M \tag{2.50}$$

with $\varphi_d = \varphi$ (x = d).

It is possible to calculate the local distribution of φ in the space charge layer by a second integration of $d\varphi/dx$ up to φ_d, but this cannot be checked by direct measurements. However, the relation between space charge and differential capacity can be derived from this model and gives a tool to compare measurements with the theory.

The potential drop between the metal and the bulk of the electrolyte is $\Delta\varphi = \Delta\varphi_H + \Delta\varphi_d$ according to the model of Figure 2.12. $\Delta\varphi_H$ is determined by

$$\Delta\varphi_H = \frac{Q_M}{\varepsilon_H \varepsilon_0} d \tag{2.51}$$

and $\Delta\varphi_d = \varphi_d$ is related to Q by Equation (2.50).

The differential capacity is

$$\mathrm{C} = \frac{\mathrm{dQ_M}}{\mathrm{d}\Delta\varphi} \tag{2.52}$$

which can be related to Equations (2.50) and (2.51) by

$$\frac{1}{C} = \frac{d\Delta\varphi_H}{dQ_M} + \frac{d\Delta\varphi_d}{dQ_M} = \frac{1}{C_H} + \frac{1}{C_d} \tag{2.53}$$

One sees that the capacity can be represented by a series of two capacities, the Helmholtz capacity $C_H = \varepsilon_H \times \varepsilon_0/d$ and a capacity of the diffuse layer $C_d = d\,Q_M/d\,\Delta\varphi_d$

$$C_d = 2\left(\frac{\varepsilon_s \varepsilon_0 N^0}{kT}\right)^{\frac{1}{2}} ze_0 \cosh\left(\frac{ze_0}{kT}\Delta\varphi_d\right) \tag{2.54}$$

C_d has a minimum for $\Delta\varphi_d = 0$ or $Q_M = 0$.

$$_{\min}C_d = 2ze_0\left(\frac{\varepsilon_s \varepsilon_0 N^0}{kT}\right)^{\frac{1}{2}} = 1.5 \times 10^{-9}\, z\sqrt{\varepsilon N^0} \ \ \mu\mathrm{F.cm^{-2}} \tag{2.55}$$

Since the capacity of the diffuse double layer is in series with the Helmholtz double layer, the best chance to see it in the measurements requires $C_d \leq C_H$. Equations (2.50) and (2.54) indicate that C_d increases rapidly with Q_M. In accordance with this conclusion, a minimum of the differential capacity was found in measurements on metals in very dilute electrolyte solutions ($\leq 10^{-3}$ M), which is an indication for the absence of excess charge on the electrode, provided that there is also no specific adsorption of ions. In Figure 2.10, examples of capacity measurements in dependence on the voltage applied to a mercury electrode in two different concentrations of NaF are shown. One sees the pronounced minimum of the capacity at the low concentration which disappears at high electrolyte concentrations.

The description of the double layer properties by the Stern–Gouy model is a very crude one. A very weak point is the assumption that the dielectric contact suddenly changes from that of the solution to that of the Helmholtz double layer. The main information comes, therefore, from the minimum which indicates the potential of zero excess charge on the metal. This is, however, only correct in the absence of specific adsorption of ions. If ions are adsorbed, the counter charge for the diffuse double layer is the sum of the surface charge in the metal and of the adsorbed ions. Since the concentration of adsorbed ions also varies with the applied potential, this effect increases the apparent capacity of the Helmholtz double layer.

The potential of zero charge (pzc) can in such a situation be determined by the maximum of the interfacial tension between a metal and the electrolyte, because its potential dependence is dominated by the interaction between the excess charge on the metal and the ions of the electrolyte.[15] The adsorption of ions reduces the interfacial energy. This causes a shift of the potential where the interfacial tension reaches its maximum. In this situation the adsorbed ions and the counter charge in the electrolyte form a dipole layer. Therefore, anion adsorption results in a shift of the pzc into negative direction; cation adsorption has the opposite effect. Figure 2.13 gives an example for the interfacial tension dependence on the voltage applied in the case of specific adsorption of anions.

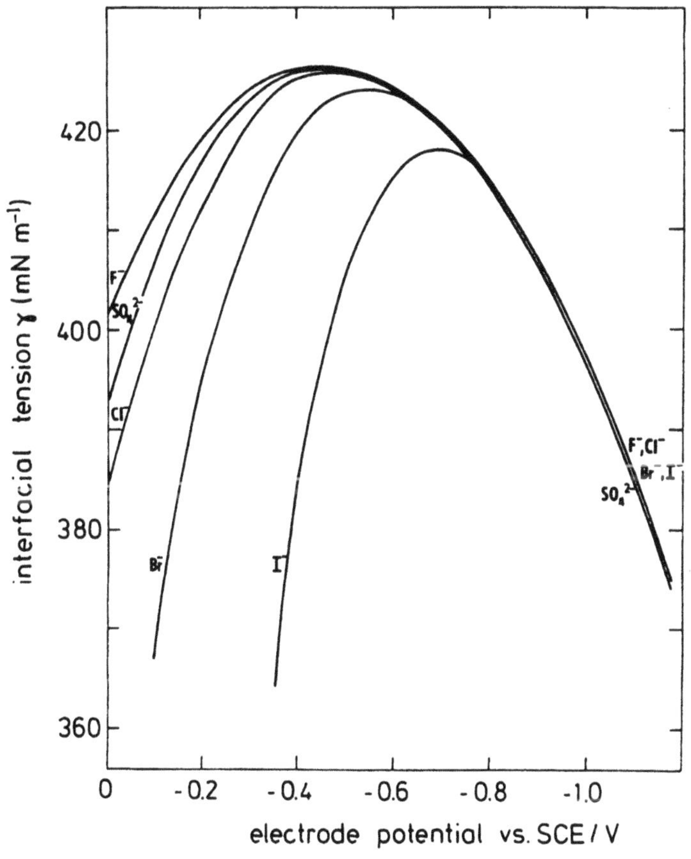

FIGURE 2.13. Interfacial tension of a mercury electrode in KF, K_2SO_4, KCl, KBr, and KI (0.1 *M*) solutions as a function of the electrode potential (Adapted from Novotny, L., in *Principles of Electrochemistry,* John Wiley & Sons, Chichester, 1993, 208.)

The pzc of solid electrodes depends on the crystallographic orientation of the surface. This is a consequence of the surface dipole dependence on the crystal orientation as in the measurements of work functions. One sees that measurements of the pzc and of work functions are closely related to each other. The main difference, apart from the different reference state, is the additional presence of a dipole layer in the electrolyte which modifies the overall dipole moment and may also change the dipole on the metal surface.

B. SEMICONDUCTOR/ELECTROLYTE INTERFACES

Semiconductors have a much lower concentration of mobile electronic charge carriers than metals, and their accumulation at the contact to an electrolyte can, therefore, not be treated like a surface charge. Instead, a space charge is formed as in the case of the Gouy layer in dilute electrolytes. The mobile electronic charge carriers are electrons in the conduction band and holes in the valence band. Their charge has opposite sign and they are distributed accordingly in an electric field. In the bulk of n-type materials, the charge of the mobile electrons is compensated by positively charged immobile donors; in p-type materials one has mobile holes and negatively charged immobile acceptors. Within some limits, the electronic properties can be adjusted by doping, depending on the materials and the method of preparation.

Intrinsic semiconductors, where electrons and holes are present at equal concentration in the bulk, behave like dilute 1-1-electrolytes with respect to the formation of a space charge if a voltage is applied to a blocking contact with an electrolyte. The charge distribution and the capacity behavior are the same as derived for the Gouy layer of such an electrolyte (see Equation [2.46]).[16]

Practically important is the space charge distribution in a doped semiconductor of n- or p-type character. Such materials have a much higher conductivity and are therefore much more useful as electrodes in electrolytic cells than intrinsic semiconductors. In naturally or intentionally doped semiconductors there is only one mobile charge carrier. The immobility of the compensating charge in the bulk, however, makes the space charge distribution very different from that in a liquid electrolyte if a voltage is applied.

As an example, the charge distribution for an n-type semiconductor and its dependence of the voltage applied shall be discussed here. We choose n-type semiconductors because they are much more common than p-type materials. The situation with p-type semiconductors is fully analogous; one only has to invert all signs of charge and voltage.

In the bulk of an n-type semiconductor the concentration of electrons n(x) is everywhere equal to the concentration of ionized donors, N_D. If a voltage is applied to a contact with an electrolyte, only electrons can be moved, and the net excess charge Q is (cf. Equation [2.44])

$$Q = \int_0^\infty \rho(x)dx \tag{2.56}$$

where:

$$\rho(x) = e_0\left(N_D - n(x)\right) \tag{2.57}$$

The potential distribution is again the result of the combination of the Poisson equation (Equation [2.45]) and the Boltzmann distribution function for the mobile charge carriers (cf. Equation [2.46]). The result for $\rho(x)$ is here

$$\rho(x) = e_0 N_D\left[1 - \exp\left(\frac{e_0\varphi}{kT}\right)\right] \tag{2.58}$$

This relation is only valid for the range where the electronic system is not degenerate which requires $n(x) \ll N_{eff}$. N_{eff} is the effective density of states which is for most semiconductors on the order of 10^{19} to 10^{20} cm^{-3}. This means, in the case of accumulation where the concentration of electronic charge carriers reaches the order of N_{eff}, Boltzmann statistics have to be replaced by Fermi statistics, and the behavior of the semiconductor approaches more and more that of a metal. This case is much more complicated and outside the scope of this introduction. We shall concentrate on the case of electron depletion in the space charge layer.

The exact integration of the Poisson equation with Equation (2.58) can only be achieved numerically. However, a simplified approach can give direct insight into the behavior of the semiconductor if the mobile charge carriers are removed from the interface by the voltage applied. Since the term $\exp(e_0\varphi/kT)$ decreases rapidly when φ relative to the bulk becomes negative, one can assume a constant charge density of $e_0 N_D$ for a region where $\Delta\varphi(x) = \varphi(x)$ to $\varphi(\infty) < -kT/e_0$ (≈25 mV). Such a model is shown in Figure 2.14. In this model the potential in the space charge layer of an extension w follows a relation

$$\varphi(x) \approx \Delta\varphi_{sc}\left(1 - x/w\right)^2 \tag{2.59}$$

with

$$w \approx \left[\frac{2\varepsilon\varepsilon_0}{e_0 N_D}\Delta\varphi_{sc}\right]^{\frac{1}{2}} \tag{2.60}$$

when the total space charge is

$$Q = e_0 N_D \cdot w = \left(2\varepsilon\varepsilon_0 e_0 N_D\right)^{\frac{1}{2}} \cdot \left(\Delta\varphi_{sc}\right)^{-\frac{1}{2}} \tag{2.61}$$

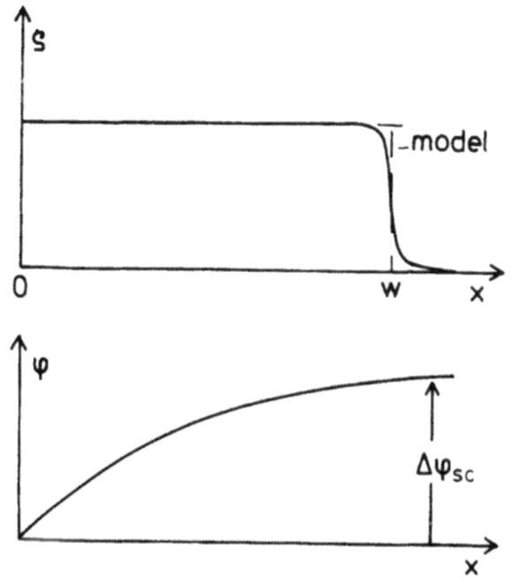

FIGURE 2.14. Charge density and potential in depletion layer of an n-type semiconductor.

Assuming that the counter charge of opposite sign would be just in front of the semiconductor surface, the differential capacity C_{sc} of this space charge layer is found to be

$$C_{sc} = \frac{dQ}{d\Delta\varphi_{sc}} = \left(\frac{\varepsilon\varepsilon_0 e_0 N_D}{2}\right)^{\frac{1}{2}} \cdot \left(\Delta\varphi_{sc}\right)^{-\frac{1}{2}} \tag{2.62}$$

The exact solution of this problem for the differential capacity is only slightly different.[17] In Equation (2.62) one has to replace $(\Delta\varphi_{sc})^{-1/2}$ by $(\Delta\varphi_{sc} + kT/e_0)^{-1/2}$.

In reality, the counter charge arises from the ions in the electrolyte which can approach the surface only up to a distance which corresponds with the thickness of the Helmholtz double layer d_H. The result is an increase of the potential drop between the electrolyte and the bulk of the semiconductor by a voltage

$$\Delta\varphi_H = \frac{Q_{sc}}{C_H} \tag{2.63}$$

The situation is analogous to the Stern model for the combination of the Helmholtz layer with a diffuse double layer in the electrolyte described in Figure 2.12. The only difference here is that the diffuse part of the charge appears in the semiconductor. The total voltage drop between the bulk of the semiconductor and the electrolyte due to the excess electric charge in both phases is

$$\varphi(x = \infty) - \varphi(x = -d_H) = \Delta\varphi = \Delta\varphi_{sc} + \Delta\varphi_H \tag{2.64}$$

It is difficult to measure the amount of excess charge directly. Therefore, as in the case of a metal electrode, the differential capacity can be used to yield information upon the charge distribution within the framework of a suitable model. Since there are two contributions to $\Delta\varphi$, the appropriate model is represented by two capacitors in series:

$$\frac{d\Delta\varphi}{dQ_{sc}} = \frac{d\Delta\varphi_H}{dQ_{sc}} + \frac{d\Delta\varphi_{sc}}{dQ_{sc}} = \frac{1}{C_H} + \frac{1}{C_{sc}} = \frac{1}{C} \tag{2.65}$$

While C_H is on the order of 20 $\mu F/cm^2$, C_{sc} is usually much smaller and decreases with $\Delta\varphi_{sc}$ (cf. Equation [2.62]). With $\Delta\varphi_{sc} = 2\,\frac{kT}{e_0} = 0.05$ V, C_{sc} would have a value of

$$C_{sc} = \left(\frac{\varepsilon_0 e_0}{5\times 10^{-2}}\right)^{\frac{1}{2}} \cdot (\varepsilon N_D)^{\frac{1}{2}} = 2.7\times 10^{-10} \cdot (\varepsilon N_D)^{\frac{1}{2}}\ \mu F\cdot cm^{-2} \tag{2.66}$$

With N_D on the order of 10^{16} to 10^{18} and ε of the semiconductor between 3 and 10, C_{sc} will be smaller than 1 $\mu F/cm^2$ and decrease at higher voltages. Therefore, any change of the potential drop in the semiconductor due to an external voltage is in the depletion range practically equal to the change of $\Delta\varphi_{sc}$, and the differential capacity corresponds with C_{sc}. This offers a way to determine the potential of zero charge for a semiconductor electrode by measuring the capacity as a function of an external voltage referred to a suitable reference electrode.

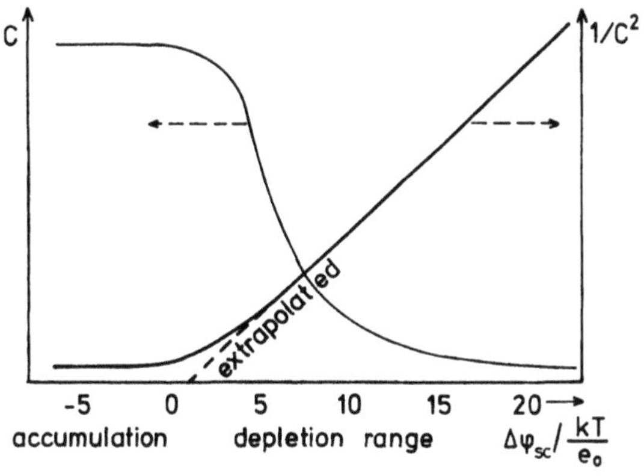

FIGURE 2.15. Differential capacity of the space charge layer of an n-type semiconductor and the Mott–Schottky plot.

One obtains from Equations (2.62) and (2.65) with a constant $\Delta\varphi_H$ the relation

$$\frac{1}{C^2} = \frac{2}{\varepsilon\varepsilon_0 e_0 N_D}\Delta\varphi_{sc} \tag{2.67}$$

This relation and the capacity of the space charge layer are shown in Figure 2.15. Since only the externally applied voltage U can be varied, one has to measure the differential capacity as a function of U. If $\Delta\varphi_H$ remains constant in the depletion range, $\Delta\varphi_{sc}$ varies in parallel with U:

$$\Delta\varphi_{sc} = U - U_{fb} \quad \text{for } \Delta\varphi_{sc} \geq 100\,\text{mV} \tag{2.68}$$

The potential U_{fb} is called the *flatband potential* because the energy of the electrons at the band edges is constant from the bulk up to the surface. This corresponds with the potential of zero excess charge in the space charge layer of the semiconductor. It can be determined by an extrapolation of the so-called *Mott–Schottky plot* for the capacity measurements, namely a plot of $1/C^2$ vs. the externally applied voltage.[18,19] Figure 2.16 shows results obtained for the n-type semiconductor CdS in two different electrolytes.

The extrapolation to $1/C^2 \rightarrow 0$ yields the flatband potential, and from Equation (2.68) one obtains $\Delta\varphi_{sc}$ for the range where the Mott–Schottky relation is linear. The slope of the line yields the donor concentration, N_D, if ε is known. One sees in Figure 2.16 that the flatband potential depends on the electrolyte. The reason is that ion adsorption can generate dipole layers which change $\Delta\varphi_H$ as in the case of the pzc at metals. Such effects are very pronounced on semiconductor electrodes. Figure 2.16 suggests a strong adsorption of S^{2-} or SH^- ions on CdS with a corresponding shift of U_{fb} in the negative direction. Semiconducting oxides and all semiconductors having an oxide layer on their surface in contact with aqueous electrolytes show this phenomenon, a shift of U_{fb} in the negative direction with increasing pH by specific adsorption of OH^- or H^+ ions.[20,21]

The flatband potential of a semiconductor electrode also depends on the orientation of the crystal surface, because the surface dipole of the crystal contributes to $\Delta\varphi_H$ as in the metal case. Additionally, other complications cause deviations from the ideal picture which was the basis of the previous calculations. The most serious and very common deviation results from electronic surface states with energies in the band gap.[18,22] They can accumulate additional charge on the surface which affects the potential drop in the Helmholtz layer. Because

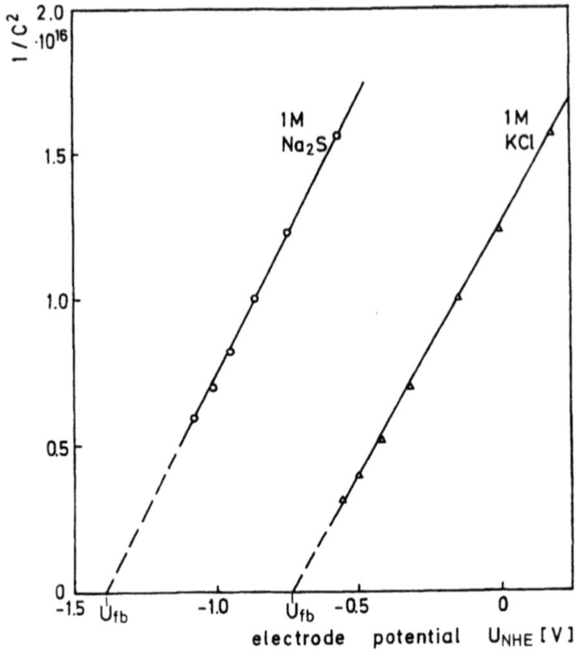

FIGURE 2.16. Mott–Schottky plots for a CdS electrode in two different electrolytes and determination of the flatband potential U_{fb}.

this charge varies with the applied voltage, deviations from the linearity of the Mott–Schottky plot result, particularly when one approaches the flatband potential. Similar consequences arise from donor or acceptor states with energies deeper in the band gap which are not ionized at the flatband potential, but become charged at a larger potential drop in the depletion layer. This is also the situation with noncrystalline semiconductors where no sharp band edge exists. However, such complications are outside the scope of this introduction.

It should be noted that the potential distribution between a metal and a solid electrolyte, in which only one of the ions is mobile, corresponds in the electrolyte exactly to the space charge distribution in n- or p-type semiconductors, depending on whether the anions or cations are mobile in this electrolyte. For a consideration of these effects in solid electrolytes see in particular also Chapter 4 of this handbook.

C. MEMBRANE/ELECTROLYTE INTERFACES

Membranes are used to separate two electrolytes. If they are porous and all ions and the solvent can permeate the membrane, only a diffusion potential is generated in the pores as long as the concentrations are different on both sides of the membrane. The situation is different if the membrane is impermeable for one or several ions. Equilibrium can then be established only for ions which can pass the membrane. For those ions the equilibrium condition is the equality of the electrochemical potentials at both sides of the membrane in phases I and II

$$ {}_{I}\mu_i + z_i F\varphi_I = {}_{II}\mu_i + z_i F\varphi_{II} \tag{2.69} $$

With ${}_{k}\mu_i = {}_{k}\mu_i^0 + RT \ln {}_{k}a_i$ and the solvent being the same on both sides such that ${}_{I}\mu_i^0 = {}_{II}\mu_i^0$, one obtains the condition

$$ \varphi_{II} - \varphi_I = \frac{RT}{F} \ln \left(\frac{{}_{I}a_i}{{}_{II}a_i} \right)^{1/z_i} \tag{2.70} $$

The factor

$$\left(\frac{{}_{\mathrm{I}}a_i}{{}_{\mathrm{II}}a_i}\right)^{1/z_i}$$

is the so-called *Donnan distribution coefficient* which controls the equilibrium distribution of the ions between the two sides.[23] Since nonpermeable ions are present on both sides at a concentration set by the initial conditions and electroneutrality is required for the electrolytes in the bulk, Equation (2.70) can be fulfilled for all ions only at a definite concentration distribution for all salts present on both sides. The consequences are different salt concentrations and a different osmotic pressure on both sides of the membrane which has to be balanced by a hydrostatic pressure in order to reach real equilibrium. If the solvent is different on both sides, the different standard chemical potentials influence the distribution of the permeable ions and the potential difference at equilibrium. It is, however, possible to define the activities in such a way that the μ_i^0 values remain equal and Equation (2.70) remains valid.

The potential difference between the two sides of a membrane is the result of the formation of two electric double layers on both sides of the membrane. If the membrane itself does not contain fixed charged species, only those ions exist in its pores which can pass the membrane (the impermeability may be caused in such cases by the size of the pores which prevents large ions from passage). The derivation of the individual potential differences at each interface is, however, too complicated for a general treatment in this context. A common case is that the membrane itself contains immobile ionic components of one sign and acts as an ion exchanger for ions with the opposite sign. In this case we have the formation of two Donnan potentials at both interfaces of the membrane and a diffusion potential inside the membrane as long as the equilibrium distribution between both electrolytes has not yet been reached.[23] If all ions can enter the ion exchanger, the final state is an equal concentration distribution on both sides, and the two Donnan potentials cancel each other. If some ions cannot pass the membrane, the same situation is obtained as discussed before for a permselective membrane with pores.

For analytical purposes, membranes with specific selectivity for one ion are especially useful. This is the basis for the glass electrode where the proton activity is kept constant inside the glass by the proton exchange equilibrium between silicate ions and water. In this case, the interfacial equilibrium is reached by a small exchange of H^+ ions, resulting in the formation of a potential difference corresponding with the Nernst equation. High concentrations of alkali ions in the electrolyte, which can also be exchanged with the glass, cause deviations from the Nernst formula for the pH difference because the proton activity in the glass is no longer constant and the Donnan equilibrium condition for both ions (cf. Equation [2.70]) must be fulfilled in the final ion distribution.

VIII. KINETICS OF ELECTRON TRANSFER REACTIONS AT INTERFACES

Charge transfer reactions at interfaces are strongly affected by the electric forces which are present in the electric double layer. Since these forces can be varied by external voltages, the kinetics are phenomenologically characterized by the relations between current and voltage applied to the electrolytic cell where such interfacial reactions occur at both electrodes. In order to measure the influence of the voltage for a single electrode, one has to refer the voltage to a reference electrode which is in contact with the same electrolyte, but not passed by the current. It remains therefore at its equilibrium potential.

As for all interfacial reactions, transport of the reactants to the interface from the bulk and, vice versa, removal of the products from the interface, are inherent steps in the overall

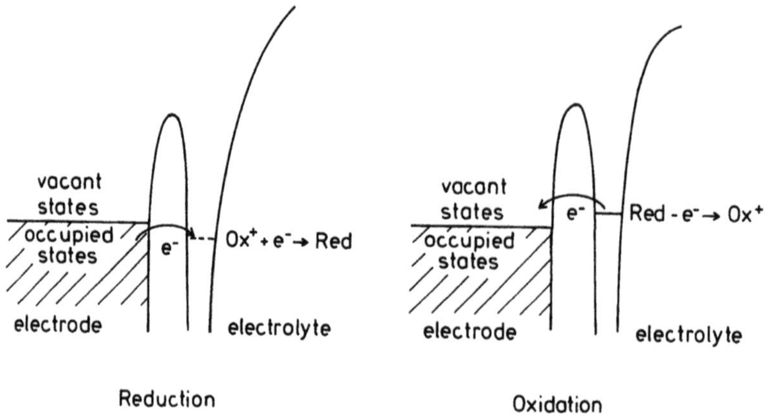

FIGURE 2.17. Electron transfer between a metal electrode and a redox couple in solution via tunneling through the energy barrier at the interface.

kinetics of electrode reactions. If the charge transfer step at the interface is fast, the transport processes become rate controlling very shortly after the start of the reaction. Special techniques had to be developed for the separation of the charge transfer step from all other reaction steps involved. Where this was successful, an exponential correlation between reaction rate (current) and overvoltage was observed. Overvoltage means the potential difference between the potential of the electrode under current and the potential at equilibrium for the appropriate charge transfer reaction, the potential being in both cases referred to the same reference electrode. However, for electron transfer and ion transfer reactions, this exponential relation results from different physical principles. While electrons can pass the energy barrier in the electric double layer via tunneling, ions have to go over this barrier. It is therefore necessary to discuss the basic concepts for these two processes separately.

A. GENERAL CONCEPTS OF ELECTRON TRANSFER

Electron transfer occurs between two quantum states, one being occupied, the other vacant. Since the electron is a much lighter particle than any atom, the transition will be fast in comparison to atomic movements. Therefore, the Franck–Condon principle, well known for optical transitions, has to be applied in a theoretical description. Along with this, the principle of energy conservation has to be taken into account.

Electron transfer from an electrode being an electronic conductor, metal or semiconductor, to a redox couple in an electrolyte, i.e., the cathodic process, occurs from an occupied quantum state in the electrode to a vacant state in the electrolyte. This state is offered by the oxidized species of the redox couple, denoted here as Ox^+. In the anodic process the electron comes from an occupied state in the electrolyte, from the reduced species, denoted as Red, and is transferred to a vacant state in the electrode. The conservation of energy is fulfilled if the binding energy, E, of the electron in both quantum states is the same. Figure 2.17 illustrates the process of tunneling through the energy barrier in front of an electrode for the cathodic and anodic process.

Electron transfer in both directions can be described by the following two equations

$$i^- = \iint N_{ox}(x) \cdot \kappa(E,x) \cdot D_{occ}(E) \cdot W_{ox}(E) \cdot dE \cdot dx \tag{2.71}$$

$$i^+ = \iint N_{red}(x) \cdot \kappa(E,x) \cdot D_{vac}(E) \cdot W_{red}(E) \cdot dE \cdot dx \tag{2.72}$$

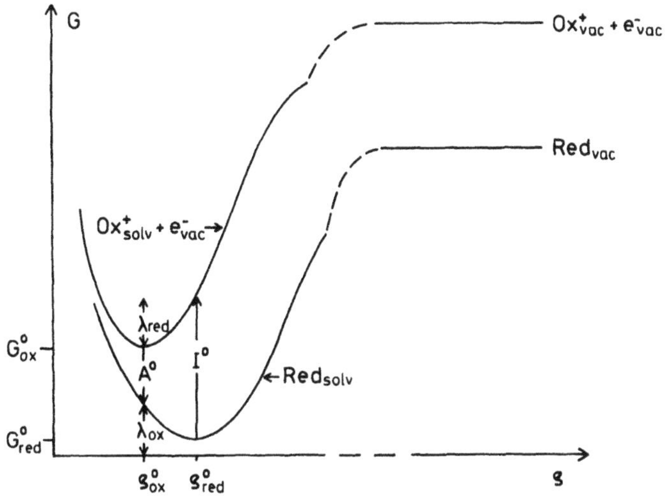

FIGURE 2.18. Free energy of the two components of a redox couple as a function of a reaction coordinate ρ, characterizing the interaction of these components with a surrounding solvent.

In these equations, $N_{ox}(x)$, $N_{red}(x)$ are the concentrations of the respective redox species at a distance x from the interface in terms of ions (or molecules) per cubic centimeter. $D_{occ}(E)$ and $D_{vac}(E)$ are the concentrations of occupied or vacant states in the electrode at an energy level E, usually expressed as states per cubic centimeter and electron volts. The energy has to be related to the same reference state, e.g., the electron in vacuum. $W_{ox}(E)$, $W_{red}(E)$ are the probabilities that the electronic energy level in the respective redox species reaches, by thermal fluctuations, an energy level E, and $\kappa(E,x)$ is the tunneling probability for the electron transfer as a function of the energy and the distance.

Energy states of electrons in solid (or liquid) electronic conductors are well known from the band theory. Electronic energy levels of ions in solution have first been discussed by Gurney.[24] They depend on the quantum state in the particular redox species and the interaction with its surrounding matter (liquid or solid). The concept from which the position of energy levels in redox systems can be developed[25] is explained in Figure 2.18. In this figure, the ionization energy of Red in vacuum corresponds with the binding energy of the electron in this species. The coordinate ρ describes the interaction with the surrounding matter as a function of the averaged distance and orientation of its individual components (molecules of the solvent in liquids, ligands in complexes, etc.). This coordinate represents, in a crude simplification, the multicomponent interaction with the surroundings. The distance between the two energy curves at a particular ρ represents the energy for the removal of an electron to vacuum from the occupied state (Red) or the energy gained in adding an electron from vacuum to the vacant state (Ox^+) at constant ρ (Franck–Condon principle). This energy difference varies with ρ and corresponds with the binding energy of the electron in the redox system relative to vacuum.

This energy depends on the interaction with the surrounding matter which results from electrical polarization, dipole orientation, van der Waals forces, and chemical bond formation. Ligands strongly bound to a central ion or atom which remain connected with the redox species independent of its state of charge can be treated as a unit already existing in vacuum. For an illustration of the concepts, a polar liquid will be considered as the interacting medium.

The most probable state of both species of the redox couple is given by the minima of G in Figure 2.18 at ρ^0_{ox} and ρ^0_{red}, respectively. The energy levels of the electron in these states are, following the Franck–Condon principle,

$$E^0_{red} = -I^0 \text{ and } E^0_{ox} = A^0 \tag{2.73}$$

where I^0 is the ionization energy of Red and A^0 the electron affinity of Ox^+ in the minima of G_{red} and G_{ox}. Let us assume that an electron has been removed to vacuum from the reduced form at $\rho = \rho^0_{red}$ and the oxidized state has been created without a change of its surroundings. As a result, Ox^+ is now in a state of higher energy which will be dissipated by a reorganization of its surroundings. The parameter ρ will approach ρ^0_{ox}. The dissipated energy, λ_{ox}, is the so-called *reorganization energy* (see Figure 2.18). In full analogy, if we add an electron from vacuum to the oxidized state at ρ^0_{ox}, the resulting reduced form will dissipate its excess energy and reach ρ^0_{red} by releasing the reorganization energy λ_{red}. These reorganization energies play a decisive role in the kinetics of electron transfer reactions. They were introduced by Marcus[26,27] when he derived a statistical theory of homogeneous electron transfer reactions which he later extended to redox reactions at electrodes. Another theory, which was based on quantum mechanical formulations and resulted in very similar conclusions, was developed by Levich.[28]

The energy difference between the minima of G in Figure 2.18 represents the process

$$Ox^+\left(\rho^0_{ox}\right) + e^-_{vac} \rightleftarrows Red\left(\rho^0_{red}\right)$$

when equilibrium between the redox species and their surroundings is established. This free energy difference therefore corresponds with the redox potential of the electron in the redox couple with the reference state for the electron in vacuum. This is equal to the chemical potential of the electron in the redox system in its standard state (equal concentration of Ox^+ and Red) and can be designated as the Fermi energy of the electron in a redox system in the same way as it was defined for metals (cf. Figure 2.4),

$$\Delta G^0_{redox} = G^0_{ox} - G^0_{red} = -E^0_{F,redox} = -\mu^0_{e^-,redox} \tag{2.74}$$

The chemical potential of the electron in a redox system has already been introduced from a thermodynamic point of view in Section VI via Equation (2.42) with the vacuum level as reference state. The Fermi energy and chemical potential of electrons are identical in a single phase of uniform composition.

For the practical application of the kinetic Equations (2.71) and (2.72), one needs a knowledge of the distribution functions $W_{ox}(E)$ and $W_{red}(E)$. A simple version of this type of function can be derived on the basis of Figure 2.18. In order to simplify the mathematics, it is assumed that the reorganization energies λ_{ox} and λ_{red} are equal to λ. This is correct for the interaction due to electric polarization which is important in polar solvents. It is not such a good approximation for vibrational interactions where the force constants can change considerably for different states of oxidation. Nevertheless, neglecting such complications is helpful for the basic understanding.

We have seen that the distance between G_{ox} and G_{red} at constant ρ is a measure of the electronic energy level in the redox system

$$G_{ox}(\rho) - G_{red}(\rho) = -E(\rho) \tag{2.75}$$

Assuming a parabolic relation of the same shape for $G_{ox}(\rho)$ and $G_{red}(\rho)$ around their minima, one can describe the parabola by

$$\Delta G_{ox}(\rho) = G_{ox}(\rho) - G_{ox}^{0} = \lambda\left(\frac{\rho - \rho_{ox}^{0}}{\Delta\rho^{0}}\right)^{2} \tag{2.76}$$

$$\Delta G_{red}(\rho) = G_{red}(\rho) - G_{red}^{0} = \lambda\left(\frac{\rho - \rho_{red}^{0}}{\Delta\rho^{0}}\right)^{2} \tag{2.77}$$

with

$$\Delta\rho^{0} = \rho_{red}^{0} - \rho_{ox}^{0}$$

This yields for Equation (2.75)

$$-E(\rho) = G_{ox}^{0} - G_{red}^{0} + \lambda\left[\left(\frac{\rho - \rho_{ox}^{0}}{\Delta\rho^{0}}\right)^{2} - \left(\frac{\rho - \rho_{red}^{0}}{\Delta\rho^{0}}\right)^{2}\right] \tag{2.78}$$

$$E(\rho) = E_{red}^{0} - 2\lambda\frac{\rho - \rho_{red}^{0}}{\Delta\rho^{0}} = E_{ox}^{0} - 2\lambda\frac{\rho - \rho_{ox}^{0}}{\Delta\rho^{0}} \tag{2.79}$$

$$E_{red}^{0} = E\left(\rho_{red}^{0}\right) = E_{F,redox}^{0} + \lambda \tag{2.80}$$

$$E_{ox}^{0} = E\left(\rho_{ox}^{0}\right) = E_{F,redox}^{0} - \lambda$$

Equation (2.79) exhibits a linear relation between E(ρ) and ρ in the range of the two parabola.

The probability to reach a state with an excess energy $\Delta G = G - G^0$ by a thermal fluctuation is

$$W(\Delta G) = W_0 \cdot \exp\left(-\frac{\Delta G}{kT}\right) \tag{2.81}$$

where W_0 is a normalization factor. For each redox species, ΔG can be related to the deviation of its electron energy from the most probable state $\Delta E_{ox} = E_{ox} - E_{ox}^{0}$ and $\Delta E_{red} = E_{red} - E_{red}^{0}$.

One obtains from Equations (2.76) to (2.81)

$$\Delta G_{ox} = \frac{\Delta E_{ox}^{2}}{4\lambda} \quad ; \quad \Delta G_{red} = \frac{\Delta E_{red}^{2}}{4\lambda} \tag{2.82}$$

This gives the distribution functions for the fluctuation energy levels

$$W_{ox}\left(\Delta E_{ox}\right) = W_0 \cdot \exp\left(-\frac{\Delta E_{ox}^{2}}{4\lambda kT}\right) \tag{2.83}$$

$$W_{red}\left(\Delta E_{red}\right) = W_0 \cdot \exp\left(-\frac{\Delta E_{red}^{2}}{4\lambda kT}\right) \tag{2.84}$$

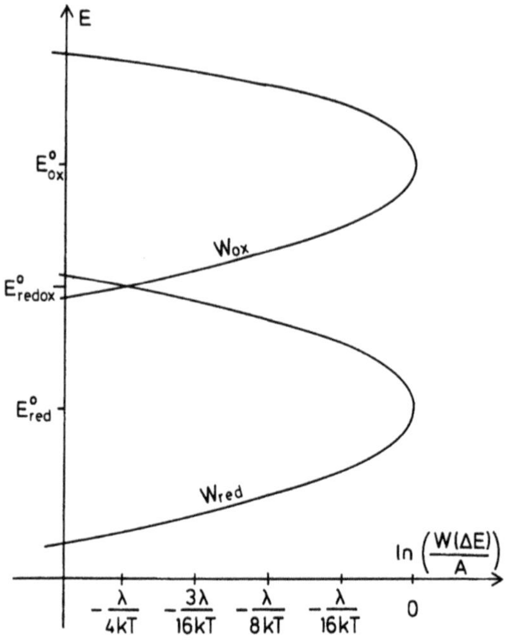

FIGURE 2.19. Distribution function for the thermal fluctuation of the energy levels E in redox couples in a logarithmic scale in relation to their deviation from the most probable energies $\Delta E = E - E^0_{red}$ or $E - E^0_{ox}$, respectively.

The normalization factor W_0 is equal to

$$W_0 = (4\pi\lambda kT)^{-\frac{1}{2}} \tag{2.85}$$

Figure 2.19 gives an impression of the dependence of these distribution functions on the energy levels in the redox species. The logarithmic energy scale is normalized to λ. The probability of reaching the energy level of $E^0_{F,redox}$ requiring $\Delta E = \lambda$ is therefore for both species

$$W(\Delta E = \lambda) = W_0 \cdot \exp\left(-\frac{\lambda}{4kT}\right) \tag{2.86}$$

All parameters appearing in Equations (2.71) and (2.72) are herewith defined. In applying these equations to electron transfer at electrodes, the difference in the density of states distribution between metals and semiconductors suggests a separate treatment of these cases.

B. ELECTRON TRANSFER AT METAL ELECTRODES

The density of electronic states in metals has a large concentration in its broad conduction band which is partially filled by electrons. There is a sharp transition between occupied and vacant states at the Fermi energy E_F. At equilibrium between a metal and a redox couple in an electrolyte, the Fermi energy of the metal and the free energy of electrons in the redox system coincide. The Galvani potential difference, $\Delta\varphi_0$, between the metal and the electrolyte at equilibrium causes, due to the accompanying charging of the electric double layer, a shift of the electronic states between both sides of the interface relative to each other such that this condition is fulfilled. At equilibrium we have the relation (cf. Equation [2.74]):

$$E_{F,metal} - e_0\Delta\varphi^0 = E^0_{F,redox} \tag{2.87}$$

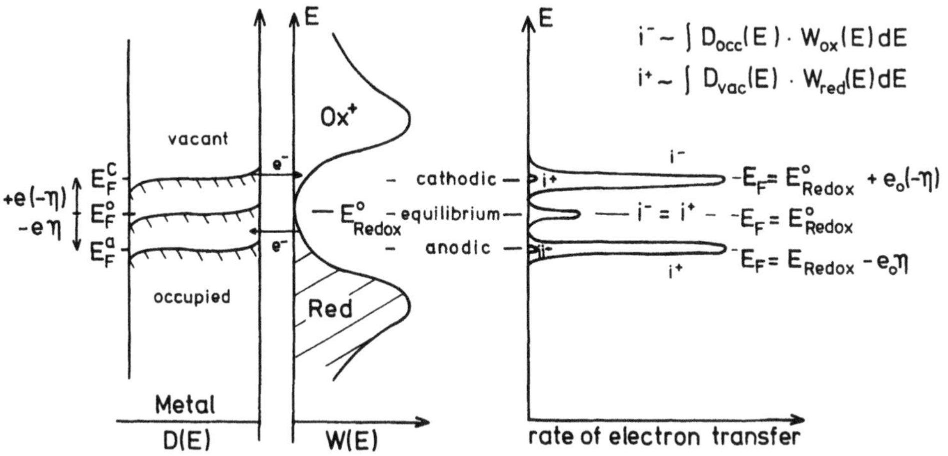

FIGURE 2.20. Electron transfer at a metal electrode. Left: position of vacant and occupied states in the metal and the electrolyte at equilibrium and anodic or cathodic overvoltage η. Right: energy distribution of electron transfer rates for these cases.

and the rate of electron transfer must be equal in both directions. This is the so-called *exchange current at equilibrium.*

A graphic illustration of this situation is seen in Figure 2.20, which displays the corresponding distribution of occupied and vacant states in both phases at equilibrium and at anodic or cathodic polarization. Electron transfer occurs where occupied states are present in one phase and vacant states are present at the same energy in the other phase. The sharp separation between occupied and vacant states at E_F in the metal has the consequence that electron transfer can only occur in a narrow energy range around the Fermi energy in the metal.[29] This is indicated in the right-hand figure by the energy distribution of the exchange current.

It has turned out in the kinetic investigations that the exponential term in Equations (2.83) and (2.84) for the probability to reach an energy level around E_F of the metal electrode is the decisive factor for the exchange current density. Obviously, the coupling between the multielectronic system of the metal surface with the electronic states of the redox species in the energy range of the Fermi level is strong enough to guarantee a much faster rate for electron exchange than the rate of energy level fluctuation in the redox species. This is described in the theory of chemical bond as *electronic resonance.* The process therefore occurs adiabatically, provided the redox species approaches the interface close enough for the electronic coupling to be sufficiently appreciable. As a result, the individual parameters which appear in Equations (2.71) and (2.72) cannot be separated, and the process can be treated within the transition state theory description,[27] where the transition state is the activation of the redox species to a configuration in which these energy levels reach the Fermi energy of the metal. There remains a preexponential frequency factor which averages all the other contributions in the fundamental Equations (2.71) and (2.72).

The exchange current density for equal concentrations of Ox^+ and Red can therefore be approximated by

$$i_0 = k_0 \cdot e_0 \cdot N_0 \cdot \exp\left(-\frac{\lambda}{4kT}\right) \quad \text{A cm}^{-2} \qquad (2.88)$$

where k_0 is a rate constant with the dimension [cm s^{-1}], N_0 the concentration of each redox species [ions cm^{-3}], and λ the reorganization energy of the redox system (cf. Figure 2.18). The density of states of the metal at the Fermi energy does not appear in this relation due to the *resonance* between the interacting electronic states.

If an external voltage is applied, the potential of the metal electrode relative to the electrolyte is changed by a potential difference, the so-called *overvoltage*, η.

$$\eta = \Delta\varphi^0 - \Delta\varphi \tag{2.89}$$

This has the effect that the electronic levels of the electrode are shifted upward or downward relative to the energy states of the redox system by an amount of $-e_0\,\eta$. This is shown in Figure 2.20 for anodic ($\eta > 0$) and cathodic ($\eta < 0$) polarization. The position of $E_{F,metal}$ relative to the reference state in the electrolyte is given by

$$E_{\text{F,metal}}(\eta) = E^0_{\text{redox}} - e_0 \cdot \eta \tag{2.90}$$

This means that the energy differences between the Fermi level in the metal and the most probable states E^0 of the redox components are also changed by $-e_0\,\eta$.

In the adiabatic model for the electron transfer, the redox species have to reach the energy level $E_{F,metal}$ by thermal activation in order to allow electron transfer to or from the metal electrode.[27,28] According to Equations (2.83) and (2.84), the probability factors for this event are in the description given here

$$W_{\text{ox}}\left(E = E_{\text{F,metal}}\right) = W_0 \cdot \exp\left(-\frac{\left(\lambda + e_0 \cdot \eta\right)^2}{4\lambda kT}\right) \tag{2.91}$$

and

$$W_{\text{red}}\left(E = E_{\text{F,metal}}\right) = W_0 \cdot \exp\left(-\frac{\left(\lambda - e_0 \cdot \eta\right)^2}{4\lambda kT}\right) \tag{2.92}$$

These probability factors can be approximated for the case $|e_0\,\eta| \ll \lambda$ by

$$W_{\text{ox}}\left(E = E_{\text{F,metal}}\right) = W_0 \cdot \exp\left(-\frac{\lambda}{4kT} - \frac{e_0 \cdot \eta}{2\lambda kT}\right) \tag{2.93}$$

$$W_{\text{red}}\left(E = E_{\text{F,metal}}\right) = W_0 \cdot \exp\left(-\frac{\lambda}{4kT} + \frac{e_0 \cdot \eta}{2\lambda kT}\right) \tag{2.94}$$

In analogy to Equation (2.88), one obtains as a result for the anodic and cathodic current

$$i^+ = i_0 \cdot \exp\left(+\frac{1}{2}\frac{e_0\,\eta}{kT}\right) \tag{2.95}$$

$$i^- = i_0 \cdot \exp\left(-\frac{1}{2}\frac{e_0\,\eta}{kT}\right) \tag{2.96}$$

These partial currents are represented in Figure 2.20 by the energy range in which electron transfer occurs around the Fermi level in the metal.

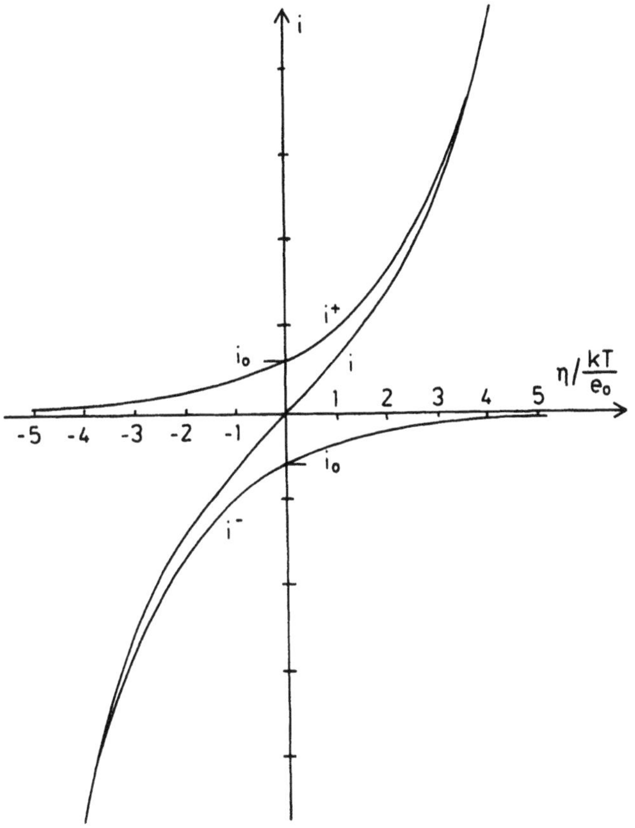

FIGURE 2.21. Partial and net currents of a one-electron transfer redox reaction as a function of the overvoltage.

The total current is

$$i = i_0 \cdot \left[\frac{N_{\text{red}}}{N_{\text{red},0}} \exp\left(\frac{\alpha e_0 \eta}{kT} \right) - \frac{N_{\text{ox}}}{N_{\text{ox},0}} \exp\left(-\frac{\beta e_0 \eta}{kT} \right) \right] \tag{2.97}$$

and is shown in Figure 2.21. The factor 1/2 in the exponents of Equations (2.95) and (2.96) is replaced in Equation (2.97) by the more general factors α and β, which are the so-called (apparent) *charge transfer coefficients*. These coefficients are 0.5 only for the symmetrical energy level distribution used in the simplified model of Figures 2.18 and 2.19. They can be different for the anodic and cathodic process. However, for a one-electron transfer reaction, the charge transfer coefficients for the anodic process α and the cathodic process β must follow the relation $\alpha + \beta = 1$. All models conclude that α and β should vary with the overvoltage (cf. the approximations of Equations [2.91] and [2.92] by Equations [2.93] and [2.94]). α should decrease and β increase with high positive overvoltage while the opposite behavior is expected at large negative overvoltage, but the relation $\alpha + \beta = 1$ remains valid. Deviations from this rule indicate changes in the mechanism of the electron transfer reaction. Such complications occur frequently in real systems by preceding chemical reactions such as loss of ligands or association of counter ions prior to the charge transfer. In such cases, the apparent charge transfer coefficients may deviate widely from each other.

Because an equation of the form of Equation (2.97) was first derived by Butler and Volmer for the kinetics of the hydrogen evolution reaction, equations of this form are often called *Butler–Volmer equations*.

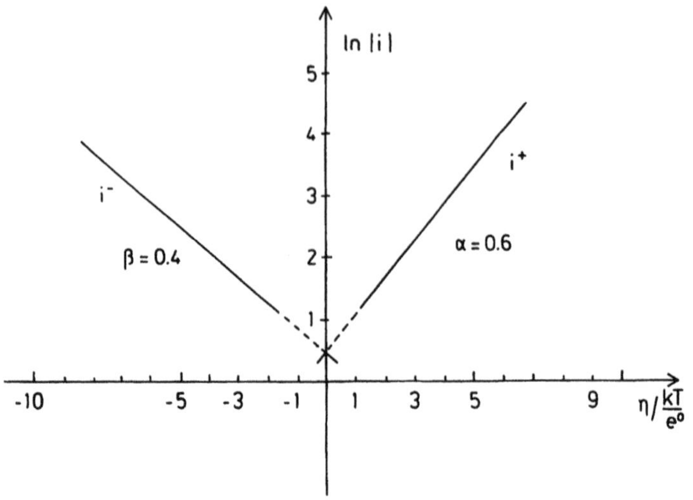

FIGURE 2.22. Logarithmic representation of the partial currents of a redox reaction as a function of the overvoltage.

In Equation (2.97), the factors N/N_0 have been introduced in the partial currents in order to take into account that the concentration of the redox components can change at the interface when a current passes through the electrode. These factors are important for the analysis of the current–voltage relations in real experiments which are discussed in Section X.

The range of overvoltages where the charge transfer coefficients begin to change with the overvoltage is rarely accessible for simple one-electron transfer redox reactions, because they are too fast and the net rate is at higher current densities controlled by transport processes. If an electrode reaction is slow enough that high overvoltages can be reached without transport control, a logarithmic plot of the anodic and cathodic currents is useful. The extrapolation of the current to the equilibrium potential ($\eta = 0$) gives a measure of the apparent exchange current density. Such a plot is shown in Figure 2.22. It is a clear indication for a change in mechanism between the anodic and the cathodic range when the extrapolated values from the anodic and the cathodic current do not coincide.

Finally, it should be mentioned that the derivations given here are also valid for redox reactions where the concentrations of both components are not equal. If $N_{ox,0} \neq N_{red,0}$, the equilibrium potential is changed according to the Nernst equation (cf. Equation [2.30]). The Fermi energy of the metal has then at equilibrium another position relative to the energy levels of the redox system, and the activation energy necessary to reach this level is altered accordingly. This affects the exchange current density, which now becomes

$$i_0 = k_0 \cdot e_0 \cdot \left(N_{ox,0} \cdot N_{red,0}\right)^{\frac{1}{2}} \exp\left(-\frac{\lambda}{4kT}\right) \tag{2.98}$$

With this modification of the exchange current density, the form of Equation (2.97) remains unaffected.

Redox reactions in which more than one electron is exchanged in the net process can usually be described by several consecutive one-electron transfer steps. The kinetics are complicated by the very small concentrations of the intermediates, which are the reduced component for one and the oxidized component for another redox reaction. Their concentrations vary with the flow of current, and the analysis of current–voltage curves becomes much more complicated. Such processes shall not be discussed here.

C. ELECTRON TRANSFER AT SEMICONDUCTOR ELECTRODES

The general Equations (2.71) and (2.72) are also valid for semiconductor electrodes. The difference compared with metal electrodes is caused by the splitting of the electronic energy bands in semiconductors into the valence band and the conduction band. These bands are separated by an energy difference, E_{gap}. The consequence is that Equations (2.71) and (2.72) have to be split into two integrals, one over the energy range of the valence band, the other over the energy range of the conduction band. One has to distinguish between charge transfer reactions via the conduction band and via the valence band.[30,31]

A theoretical description can be given in close analogy to the theory for metal electrodes. The condition is again that occupied and vacant states are available at the same energy on both sides of the interface. This is possible on the semiconductor either at and above the band edge, E_C, of the conduction band or at and below the band edge, E_V, of the valence band. In contrast to metals, the Fermi energy E_F, as long as it is located in the band gap, is excluded from electron transfer. If E_F is above E_C or below E_V — a situation which is called *degeneration of the semiconductor* — the properties of the semiconductor come close to those of a metal, and the theory for metals can be applied with some modification. Here, however, only the situation where the Fermi level remains in the band gap will be discussed.

In the discussion of the kinetics we begin as in the metal case with the exchange current at equilibrium. Equilibrium means that the Fermi energies coincide for the semiconductor and the redox system in the electrolyte. We have seen that this is achieved by a charge separation at the interface. The result at a metal is the generation of a potential drop in the Helmholtz double layer, which causes a shift of all electronic energy levels of the metal relative to the electrolyte. A charge separation at the semiconductor/electrolyte interface does not, however, primarily result in a voltage drop in the Helmholtz double layer, as we have seen in Section VII.B and in Figure 2.15. The potential drop extends over a space charge layer far into the bulk. As long as the variation of $\Delta\varphi_H$ with a changing voltage difference between the bulk of the semiconductor and the electrolyte is negligible in comparison with the variation of $\Delta\varphi_{sc}$, the band edge positions remain practically constant. Only the energy levels in the bulk of the semiconductor relative to the electrolyte are moved with the local variation of $\Delta\varphi(x)$ by an amount of $-e_0\,\Delta\varphi(x)$. The consequence of the charge separation is, in this case, a change of the position of the Fermi energy in the semiconductor relative to the band edge positions at the surface. This affects the occupation of the electronic states by electrons at the surface.

Since electrodes must have a high enough conductivity in order to avoid large ohmic energy losses under electrolytic operation, only n-type or p-type materials are usually employed. For this discussion we shall consider n-type materials. The behavior of p-type electrodes can be derived in full analogy.

In Figure 2.23 the course of the band edges in an n-type semiconductor for different amounts of excess charge is represented. Figure 2.23a shows the semiconductor at the zero point of excess charge, the flatband situation. The distance between the position of E_F and the band edges reflects the local concentration of the electronic charge carriers. As long as Boltzmann statistics can be applied to the occupation of electronic states in semiconductors, the electron concentration n(x) is given by

$$n(x) = N_C \cdot \exp\left(-\frac{E_C(x) - E_F}{kT}\right) \tag{2.99}$$

where N_C is the effective density of states at the band edge. If a positive bias is applied to the n-type semiconductor, $\Delta\varphi_{sc}$ becomes positive and the position of the band edges is shifted downwards in the bulk relative to the surface, as shown in Figure 2.23b. The result is a

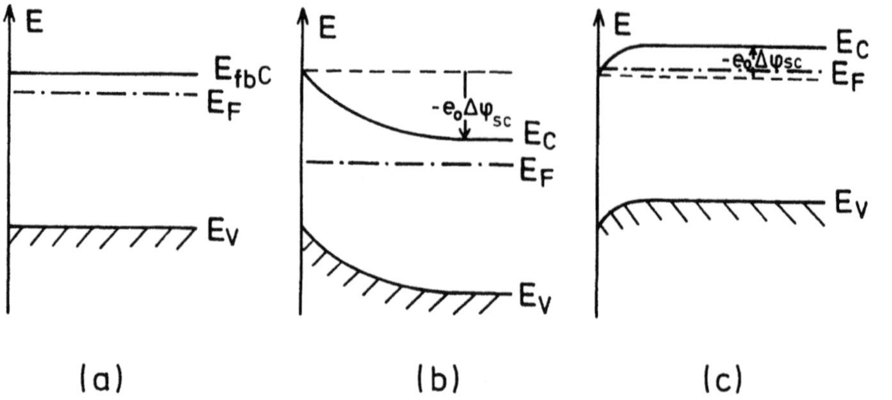

FIGURE 2.23. Course of the band edges in an n-type semiconductor electrode for different space charge situations: (a) no space charge = flatband potential; (b) positive space charge = depletion layer; (c) negative space charge = accumulation layer.

depletion layer and a decrease of n at the surface. If the electron concentration in the bulk is n_b, the surface concentration n_s becomes

$$n_s = n_b \cdot \exp\left(-\frac{e_0 \Delta\varphi_{sc}}{kT}\right) \tag{2.100}$$

In an accumulation layer, depicted in Figure 2.23c with a negative $\Delta\varphi_{sc}$, the electrons accumulate at the surface. The range of the validity of Equation (2.100) is here, however, limited by the degeneration of the electronic system at the surface when n_s approaches N_C.

For p-type electrodes with a hole concentration of p_b in the bulk the analogous relation is

$$p_s = p_b \cdot \exp\left(\frac{e_0 \Delta\varphi_{sc}}{kT}\right) \tag{2.101}$$

In Section VII.B it was shown that the flatband potential of a semiconductor U_{fb} can be derived from capacity measurements. If U_{fb} is known, one can predict the surface concentration of electrons or holes for all different states of polarization, provided that $\Delta\varphi_H$ remains constant or its variation can be determined.

Equilibrium requires $E_{F,sem} = E_{F,redox}$. Figure 2.24 shows two possible situations for the contact between an n-type semiconductor and two different redox systems. In Figure 2.24a the redox system has a redox potential near the conduction band edge of the semiconductor. The concentration of electrons at the surface is very low, but they easily find vacant states of the redox system. Electrons of the reduced species have to be activated up to the conduction band edge where they find plenty of vacant states in the conduction band. Electron exchange occurs at the energy of the conduction band edge at an equal rate in both directions, as indicated in the figure by the energy distribution for $i^-(E)$.

The redox system in Figure 2.24b has a much more positive redox potential, and the balance of the Fermi energies causes a deep depletion layer at the semiconductor. Since $E^0_{F,redox}$ is closer to E_V, electron exchange with the valence band becomes possible, which corresponds with hole exchange. The holes accumulate at the surface because the depletion barrier prevents their movement into the bulk and their recombination with electrons. The result is an inversion layer where electrons in the bulk are separated from holes at the surface.

For redox systems with a Fermi energy in the middle of the band gap, the exchange current becomes negligibly small.

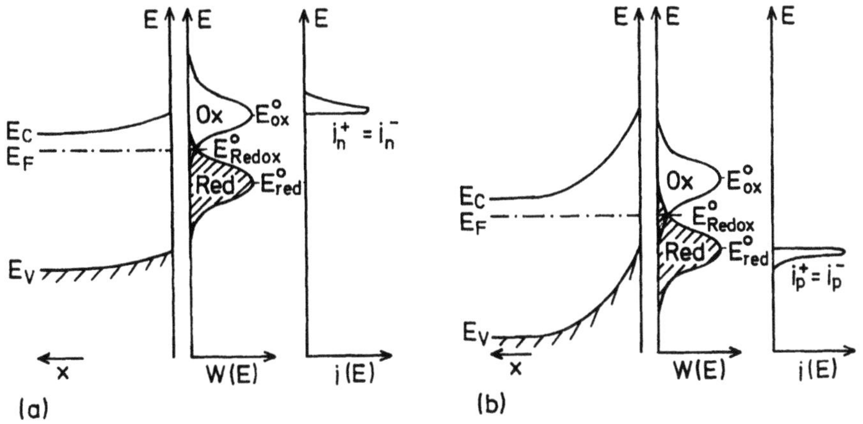

FIGURE 2.24. Electron exchange at equilibrium between an n-type semiconductor and two redox systems: (a) electron exchange via the conduction band; (b) electron exchange via the valence band.

For a derivation of the rate of electron transfer, we shall make use of the same approximation as applied to metal electrodes, namely, that the redox species require thermal excitation of their energy level to the energy of the band edges where energy states on the semiconductor are accessible.[32] For electron transfer via the conduction band, this approximation yields for the rate of charge transfer at equilibrium in cathodic direction

$$i_{c,0}^{-} = k_c^{-} n_s^0 N_{ox} \exp\left[-\frac{\left(E_C - E_{ox}^0\right)^2}{4\lambda kT}\right] = i_{c,0} \tag{2.102}$$

and in anodic direction

$$i_{c,0}^{+} = k_c^{+}\left(N_C - n_s^0\right) N_{red} \exp\left[-\frac{\left(E_C - E_{red}^0\right)^2}{4\lambda kT}\right] = i_{c,0} \tag{2.103}$$

with $n_s^0 \ll N_C$ (n_s^0 is determined by $\Delta\varphi_{sc}^0$, the potential drop at equilibrium cf. Equation [2.100]).

For electron transfer via the valence band, the relations at equilibrium are

$$i_{v,0}^{-} = k_v^{-}\left(N_V - p_s^0\right) N_{ox} \exp\left[-\frac{\left(E_{ox}^0 - E_V\right)^2}{4\lambda kT}\right] = i_{v,0} \tag{2.104}$$

and

$$i_{v,0}^{+} = k_v^{+} p_s^0 N_{red} \exp\left[-\frac{\left(E_{red}^0 - E_V\right)^2}{4\lambda kT}\right] = i_{v,0} \tag{2.105}$$

with $p_s^0 \ll N_V$ (p_s^0 is determined by Equation (2.101) at equilibrium). The rate constants k_c and k_v contain all the other factors of Equations (2.71) and (2.72) which are not specified in this approximation.

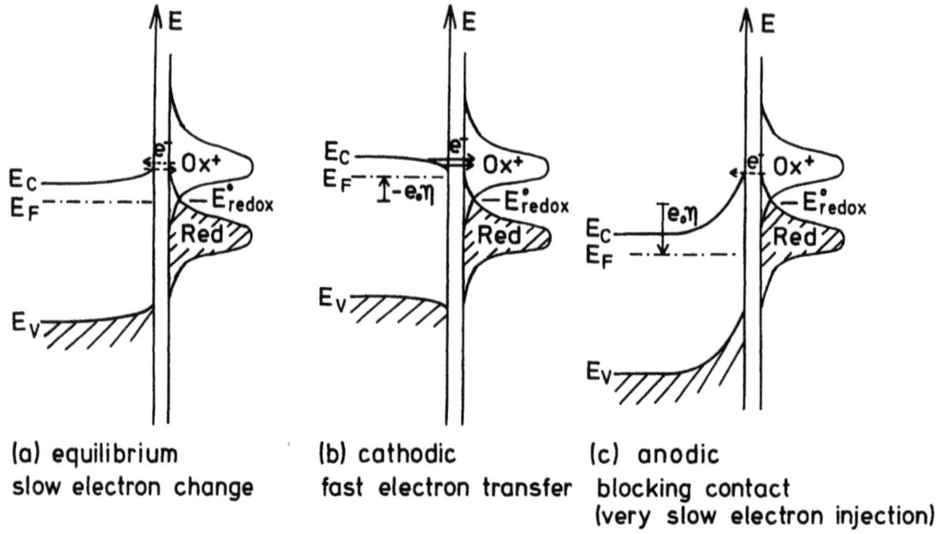

FIGURE 2.25. Correlation between electronic energy levels of an n-type semiconductor and a redox couple with electron exchange via the conduction band at different states of polarization: (a) at equilibrium (slow electron exchange); (b) cathodic bias (fast electron transfer); (c) anodic bias (slow electron injection).

One can immediately conclude that the exchange current densities are much smaller than on metals. For instance, in the situation of Figure 2.24a, the activation energy needed to reach the conduction band edge is very large for the reduced species according to Equation (2.103), much larger than to reach $E^0_{F,redox}$ which is required on metals. The smaller activation energy for the oxidized species to reach E_C (cf. Equation [2.104]) is compensated by the low surface concentration of electrons. An analogous argument leads to the same conclusion for the hole exchange with the valence band. Only if the Fermi energies of the redox system are located in one of the energy bands can exchange current densities comparable to metals be expected for semiconductor electrodes.

If a bias is applied, the main effect is a change of $\Delta\varphi_{sc}$. In the case that only the space charge is affected by the bias while $\Delta\varphi_H$ remains constant, the result is a change of the band bending, as shown in Figure 2.25, while the position of the band edges at the surface remains unaffected. Therefore, n_s varies according to Equation (2.100). Since the cathodic current via the conduction band is proportional to n_s while the anodic current depends on the concentration of vacant states which is hardly affected by the bias, one obtains a very simple relation for the dependence of the current on the voltage, defining the overvoltage by

$$\eta = \Delta\varphi - \Delta\varphi_0 \approx \Delta\varphi_{sc} - \Delta\varphi^0_{sc} \tag{2.106}$$

For an n-type semiconductor with a redox system corresponding to the situation of Figure 2.24a, one obtains

$$i_c(\eta) = i_c^+ - i_c^- = i_{c,0}\left[1 - \exp\left(-\frac{e_0\eta}{kT}\right)\right] \tag{2.107}$$

Such a current–voltage curve requires that only the interfacial charge transfer is rate controlling and the transport of the components of the redox couple and of the electrons to the interface is so fast that concentration polarization can be neglected.

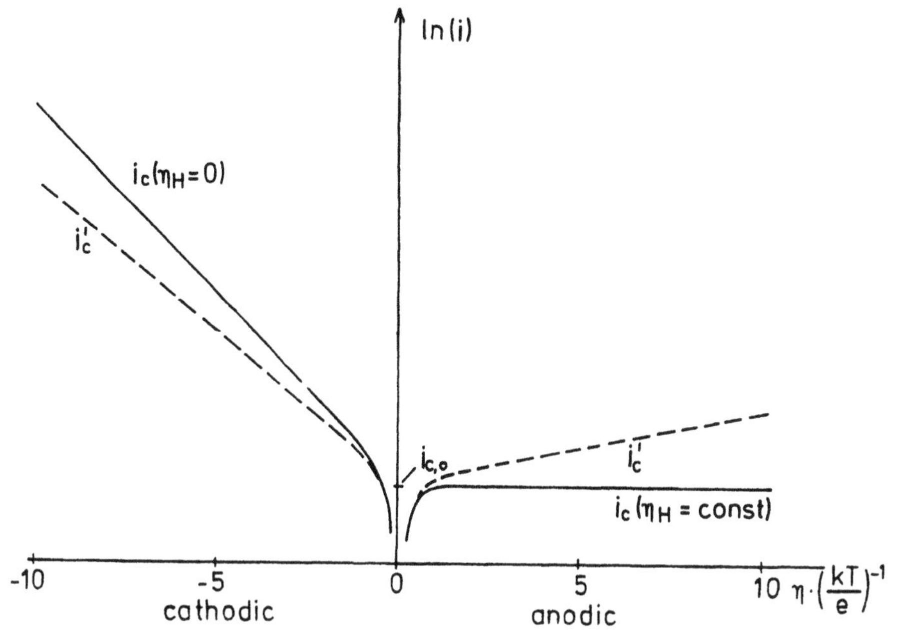

FIGURE 2.26. Current–voltage curves for an n-type semiconductor in contact with a redox couple where electron exchange occurs via the conduction band. Full line for $\Delta\varphi_H$ = constant; broken line for slightly varying $\Delta\varphi_H$.

Equation (2.107) fails to describe the current–voltage curve in the case of Figure 2.24b where the exchange current occurs in the valence band. The main barrier for the current is, in this case, the passage of electrons and holes through the inversion layer. The effective exchange current is in this case the exchange current through the inversion layer, which is much smaller than $i_{v,0}$ at the interface. Only at high cathodic bias are electrons driven from the bulk through the inversion layer and recombine therein with holes. If this recombination occurs radiatively, one sees electroluminescence. This can be used as a proof for the injection of minority carriers from redox systems into a semiconductor.[33] Equation (2.107) can therefore only be applied to n-type semiconductors and redox systems with predominant electron transfer via the conduction band. Figure 2.26 shows such a current–voltage curve for an n-type electrode as well as the effect of a variation of $\Delta\varphi_H$ with the bias which, however, will not necessarily result in a straight line for ln i vs. η.

The reason for such deviations from the ideal behavior is the existence of electronic surface states on semiconductors which can store excess charge. Part of the overvoltage then appears in $\Delta\varphi_H$. This contribution causes a shift of the band edge positions relative to the electrolyte as in the case of metals. The result is a mixed appearance of the current–voltage curves which is difficult to analyze in detail. As a general feature, the cathodic current increases much more steeply with the overvoltage than the anodic branch when a redox process occurs on n-type materials.

The situation of p-type materials is completely analogous, and the ideal current–voltage curve for redox systems with hole exchange at equilibrium can be described by

$$i_v(\eta) = i_v^+ - i_v^- = i_{v,0}\left[\exp\left(-\frac{e_0\eta}{kT}\right) - 1\right] \tag{2.108}$$

Deviations from this ideal behavior due to a variation of $\Delta\varphi_H$ with η result in a slow increase of the cathodic current with negative η and a less steep increase of the anodic current with positive η.

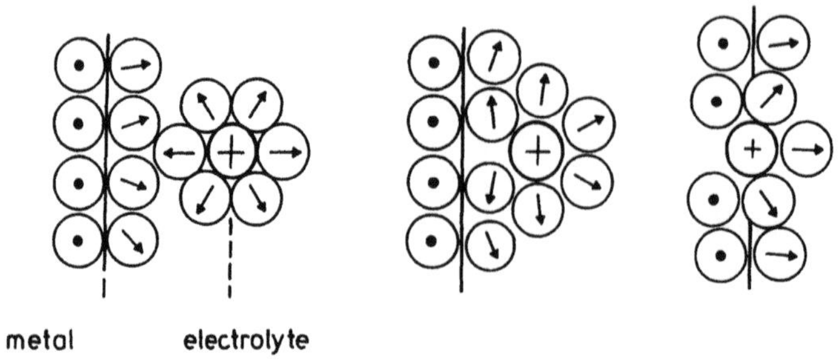

FIGURE 2.27. Three situations in the deposition of a metal ion on a liquid metal.

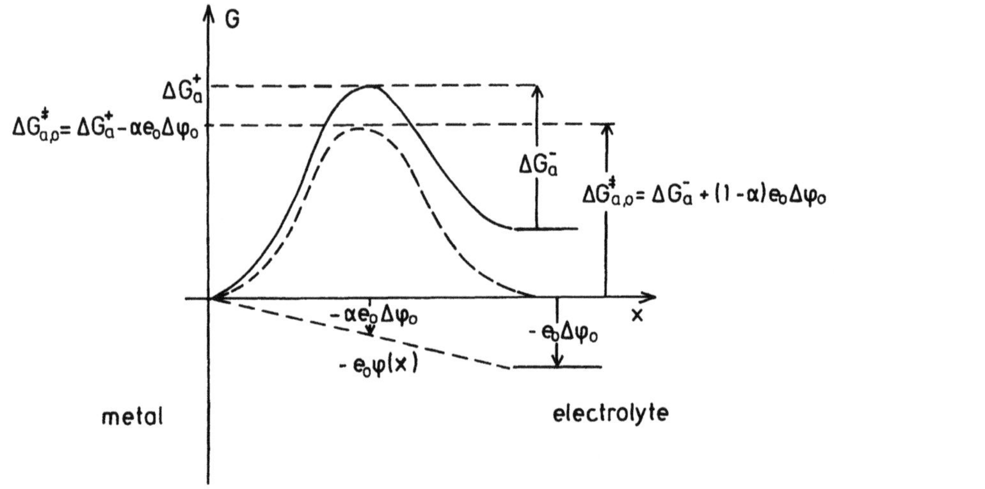

FIGURE 2.28. Free energy profiles for the discharge of a metal ion at a liquid metal. Solid line: $\Delta\varphi = 0$; broken line: $\Delta\Gamma = \Delta\varphi_0$.

IX. KINETICS OF ION TRANSFER REACTIONS AT INTERFACES

A. LIQUID METALS

The principles of ion transfer reactions can best be discussed for liquid metals as electrodes in contact with a liquid electrolyte containing the respective metal ions. On a liquid interface there are no special sites where atoms have different properties as on solid surfaces. In the kinetics of ion discharge or formation at a metal, the ion has to pass the electric double layer. The metal ion to be deposited has to lose its solvation shell — or, in general, its interaction with the components of the electrolyte — and has to be incorporated into the metal by the interaction with the metal electrons. There are intermediate stages where the interaction with the electrolyte has already been weakened while the interaction with the metal electrons is still incomplete. The ion has to overcome an energy barrier. This energy barrier is affected by the electric field in the double layer. Figure 2.27 shows three stages of the ion in the double layer. The energy profile resulting from this model is depicted in Figure 2.28. The full line represents a hypothetical free energy profile in the absence of a Galvani potential difference $\Delta\varphi = \varphi_M - \varphi_{El} = 0$. The modification of this energy profile by the electric field in the double layer at equilibrium, where $\Delta\varphi = \Delta\varphi_0$, is indicated by the broken line. The rate of ion transfer must, at equilibrium, be equal in both directions.

The kinetics of ion transfer for the anodic and cathodic process can be formulated in terms of the transition state theory by the following equations. For the situation of $\Delta\varphi = 0$:

$$i^+ = {}^0k^+ \cdot e_0 \cdot N_M \cdot \delta_M \exp\left(-\frac{\Delta G_a^+}{kT}\right) \tag{2.109}$$

$$i^- = {}^0k^- \cdot e_0 \cdot N_{M^+} \cdot \delta_{M^+} \exp\left(-\frac{\Delta G_a^-}{kT}\right) \tag{2.110}$$

where ${}^0k^+$ and ${}^0k^-$ are rate constants with the dimension [s^{-1}], N_M and N_{M^+} are concentrations of atoms or ions per cubic centimeter; δ_M and δ_{M+} are the depth of the layer at the interface from where the reactants enter the double layer (about the diameter of the ions with their solvation shell). The factors ${}^0k^+ \times \delta_M$ and ${}^0k^- \times \delta_{M^+}$ are often combined to a rate constant k^+, k^-, with the dimension [cm s^{-1}]. ΔG_a^+ and ΔG_a^- are the free energies of activation for a single atom or ion, as designated in Figure 2.28.

The rates of the anodic and the cathodic processes will normally be very different at $\Delta\varphi = 0$. The consequence is the charging of the double layer until both rates become equal and $\Delta\varphi$ has reached its equilibrium value $\Delta\varphi_0$. The additional electrostatic energy of an ion with the positive charge $z = +1$ throughout the Helmholtz double layer is indicated in Figure 2.28 with the assumption of a linear relation between φ and the distance x. This changes the height of the activation barriers for both processes by a fraction of the energy difference $e_0 \Delta\varphi_0$, but in opposite direction. The free energies of activation become for the anodic process: $\Delta G_{a,0}^+ = \Delta G_a^+ - \alpha \cdot e_0 \cdot \Delta\varphi_0$, for the cathodic process: $\Delta G_{a,0}^+ = \Delta G_a^- + (1 - \alpha) e_0 \cdot \Delta\varphi_0$. In this case α is the anodic *charge transfer coefficient* with values between 0 and 1, formally equal to the same coefficient for redox reactions, but with a very different physical meaning. Here, α describes the fraction of the variation of the electrostatic energy difference of an ion between the metal and the electrolyte which changes the free energy of activation for the anodic process. Due to the principle of microscopic reversibility, the fraction of this energy affecting the cathodic process is $(1 - \alpha)$. This correlation is independent of the real course of the Galvani potential in the double layer.

The exchange current density is obtained from Equations (2.109) and (2.110) by replacing the activation energies for the situation at $\Delta\varphi = \Delta\varphi_0$.

$$\begin{aligned} i_0 &= {}^0k^+ \cdot e_0 \cdot N_{M,0} \cdot \delta_M \cdot \exp\left(-\frac{\Delta G_a^+ - \alpha e_0 \Delta\varphi_0}{kT}\right) \\ &= {}^0k^- \cdot e_0 \cdot N_{M^+,0} \cdot \delta_{M^+} \cdot \exp\left(-\frac{\Delta G_a^- + (1-\alpha) e_0 \Delta\varphi_0}{kT}\right) \end{aligned} \tag{2.111}$$

Since absolute $\Delta\varphi$ values are not accessible, the exchange current density characterizes the rate of an electrode reaction.[34] The current–voltage curves can be related to i_0 and the overvoltage $\eta = \Delta\varphi - \Delta\varphi_0$, as it was done for redox reactions. The variation of the activation energies with η is illustrated in Figure 2.29. One obtains for the partial currents

$$i^+(\eta) = i_0 \frac{N_{M,\delta}}{N_{M,0}} \exp\left(-\frac{\alpha e_0 \eta}{kT}\right) \tag{2.112}$$

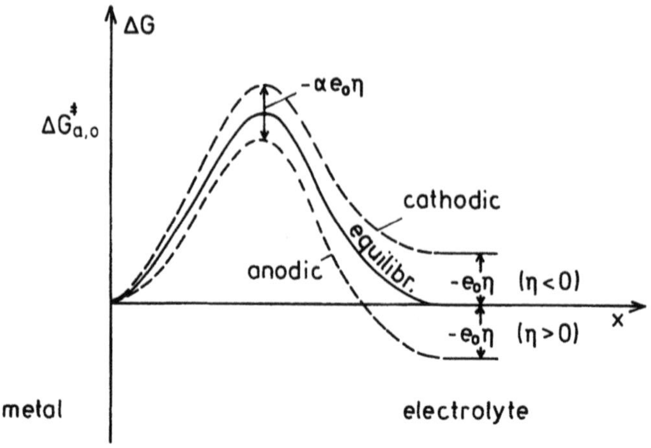

FIGURE 2.29. Influence of an overvoltage on the activation barriers for ion transfer.

$$i^-(\eta) = i_0 \frac{N_{M^+,\delta}}{N_{M^+,0}} \exp\left(-\frac{(1-\alpha)e_0\eta}{kT}\right) \tag{2.113}$$

and for the net rate of ion transfer

$$i(\eta) = i^+(\eta) - i^-(\eta) \tag{2.114}$$

The terms $\frac{N_\delta}{N_0}$ have again been introduced in Equations (2.112) and (2.113) in order to take into account the possibility that the concentration of the reactants at the interface can deviate from their equilibrium values. The influence of the transport processes upon the current–voltage curves consists in a change of N_δ. The term $\frac{N_\delta}{N_0}$ represents the so-called *concentration polarization* which is often rate controlling if the ion transfer step is fast.

Figure 2.30 shows current–voltage curves for the ion transfer reaction as a function of α with constant N_M and N_{M^+} at the interface. The exchange current density determines the slope of the current–voltage curves at $\eta = 0$ and constant N_δ

$$\left.\frac{di}{d\eta}\right|_{\eta=0} = \frac{e_0}{kT} i_0 \tag{2.115}$$

The real slope of the current–voltage curve at $\eta = 0$ is lower if the change in the terms $N_{M,\delta}/N_{M,0}$ and $N_{M^+,\delta}/N_{M^+,0}$ allows the current to decrease. Analysis of the transport rates of the reactants to the interface plays an important role in the analysis of electrode reactions.[35] Systems for which this can be done with great accuracy and experimental reproducibility, such as the dropping mercury electrode, have therefore been instrumental for the development of a knowledge of electrode reaction kinetics. Fortunately, mercury forms liquid alloys with many metals (in low concentration) which are less noble in their electrochemical character. Their reduction and oxidation can therefore be studied in potential ranges where the oxidation of mercury itself does not interfere in the kinetics. The role of transport steps will be discussed in Section X.

The reaction paths become more complicated if the ions in solution are present in the form of complexes with strong chemical bonds between the ions and their ligands. It could

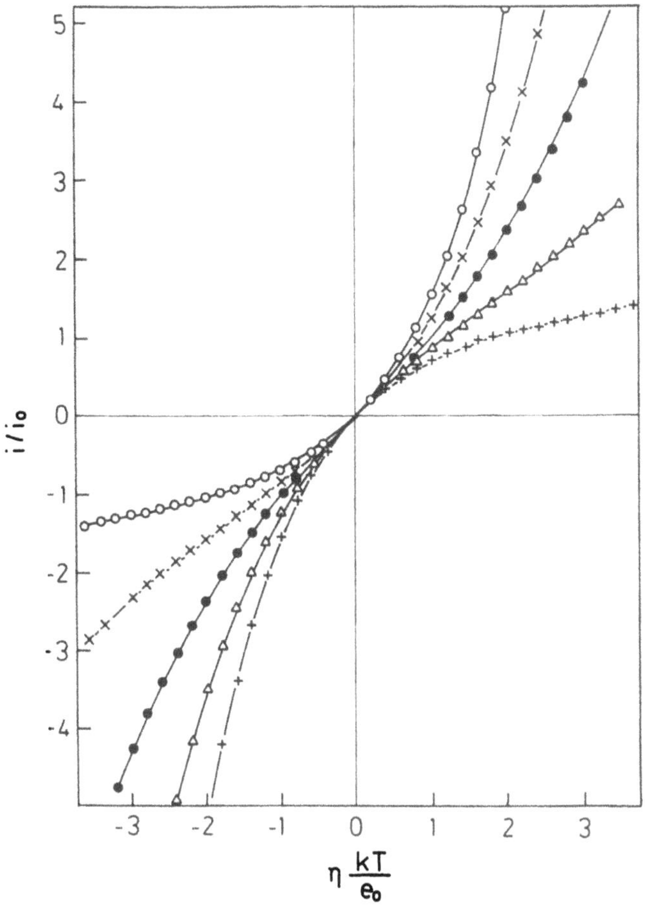

FIGURE 2.30. Theoretical current–voltage curves for ion transfer reactions with different charge transfer coefficients α. $\alpha = 0.9$ (○); 0.7 (×); 0.5 (●); 0.3 (Δ); 0.1 (+).

be shown that intermediates with a lower number of ligands than those present as the majority species in the solution are often the real reactants in the charge transfer step.[36] Such intermediates are formed by preceding or consecutive chemical reactions.

B. SOLID METALS

On solid metals the situation for ion deposition or dissolution in electrode reactions is much more complicated. The models for crystal growth from the vapor phase or atomic evaporation have to be applied, being modified by ion discharge or ion formation in passing the electrical double layer at the interface. Figure 2.31 represents the main positions of atoms on the surface of a low index face of a metal with one monoatomic step. It is assumed that the edge of the step is not smooth and contains several kink sites.

The kink site is the decisive position for the building of a crystal and its dissolution. At this site, an atom has half the number of the next neighbors which it has in the bulk. This corresponds with the average binding energy of the atoms in the crystal if the interaction with the molecules of the surrounding phase is not counted. The rate of crystal growth and dissolution depend on the number of kink sites on the crystal surface, which causes an enormous variability of the net rate. Therefore, only the kinetics of charge transfer for selected configurations on the surface can be discussed here.

Let us assume that the crystal is in contact with a liquid electrolyte containing ions of the same metal. The incorporation of an ion into the crystal at a kink site can occur either

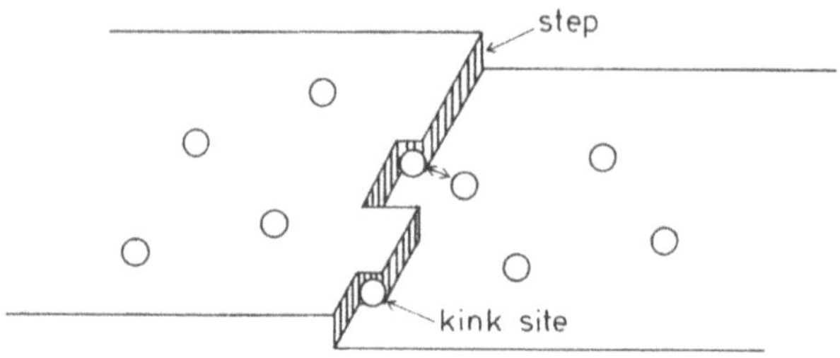

FIGURE 2.31. Monoatomic step with kink sites on a low-index face of a metal with adsorbed ions on the plane

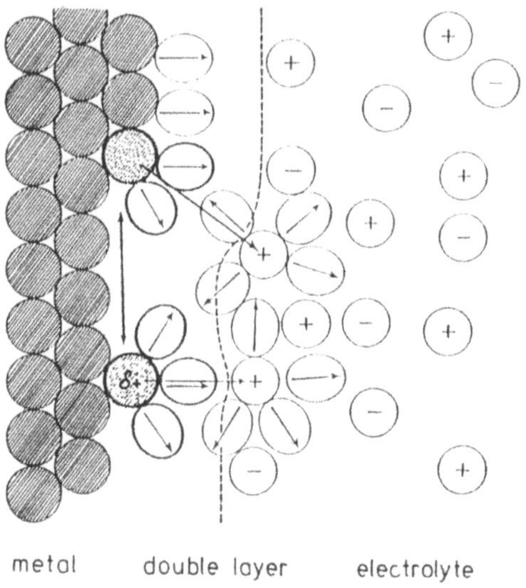

FIGURE 2.32. Model for metal ion deposition on a solid surface.

by a direct discharge at this position or by diffusion of adsorbed atoms on the surface to a kink site.[37] These processes are illustrated in Figure 2.32. Between ions in the electrolyte and adsorbed atoms on the surface, the so-called *adatoms*, there will be an equilibrium which depends on the potential difference in the electric double layer in front of the surface. An adatom is still much more exposed to the components of the electrolyte and will therefore keep some of its ligands which stabilize the ions in solution. A total discharge in the position of an adatom is therefore unlikely. Its formerly vacant electronic quantum states will be partially shared by metal electrons, and an average partial charge of (δ^+) will remain on the adatom. Figure 2.32 indicates this for the equilibrium between ions in the solution and adatoms on a low-index face.

At the equilibrium potential with $\Delta\varphi_0$, the concentration of adatoms may on a particular face be ${}_0N_{ad}$ cm^{-2}, which depends on the interaction of the adatoms with the surface atoms and the molecules of the solution. The free energy necessary to remove species of the solution (e.g., solvent molecules) from adsorption sites on the surface is another factor controlling the adsorption equilibrium. The concentration of adatoms will vary with $\Delta\varphi$ and reach a new equilibrium far from steps with kink sites as long as the concentration does not increase at cathodic bias to such an extent that two-dimensional nucleation can occur. However, close to a step with kink sites, the equilibrium will be distorted by a flux of atoms to or from these steps via surface diffusion. From the model of Figure 2.32, one can postulate the dependence of the adatom equilibrium concentration upon the overvoltage by the formula

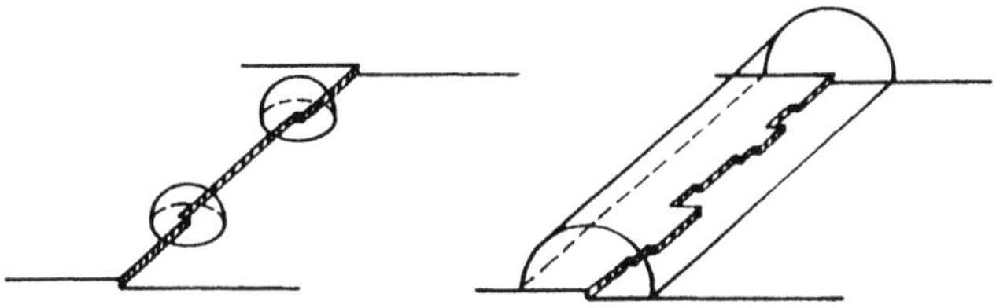

FIGURE 2.33. Models for metal deposition at steps on low index planes and the role of diffusion. (a) Steps with isolated kink sites, diffusion occurs from a hemisphere; (b) steps with high concentration of kink sites, diffusion occurs from a semicylinder.

$$N_{ad}(\eta) = {}_0N_{ad} \cdot \exp\left[-\frac{(z-\delta^+)e_0\beta\eta}{kT}\right] \tag{2.116}$$

where it is assumed that ${}_0N_{ad}$ is small in comparison with the number of adsorption sites on the surface. δ^+ is the remaining charge on the adatom (cf. Figure 2.32), and β is the fraction of the change of $\Delta\varphi$ to which the adatom is exposed.

For the direct discharge of ions at kink sites, the kinetic Equations (2.112) to (2.114) can be applied with some modification. $N_M \times \delta_M$ has to be replaced by N_k, the concentration of kink sites on the surface, and the term $N_{M^+} \times \delta_{M^+}$ has to be multiplied with a factor N_k/N_{max} where N_{max} is the number of kink sites which should be present in order to give all ions in the layer of $N_{M^+} \times \delta_{M^+}$ the chance to be directly discharged at kink sites. Since the area of discharge is very restricted if kink sites are rare, transport of ions in the electrolyte to these areas can limit the rate.

Models for diffusion-limited reaction have been worked out and can be combined with a discharge rate at steps.[38] Figure 2.33a gives an example for a model with discharge at kink sites at a large distance from each other where diffusion is hemispherical. Figure 2.33b shows a model for steps with a high concentration of kink sites where diffusion can be treated as hemicylindrical. For hemispherical diffusion the maximum rate is

$$i^-_{diff} = ze_0 \cdot N_k \cdot 2\pi r_0 \cdot DN_{M^{z+}} \tag{2.117}$$

where r_0 is the distance from which ions can be discharged ($r_0 \approx \delta_{M^+}$) and D is the diffusion coefficient. For hemicylindrical diffusion to a step, a real steady state can only be reached if some convection keeps the concentration constant at a distance L from the step. In this case, the diffusion-controlled current is approximately

$$i^-_{diff} = ze_0 \cdot N_{st} \cdot \ell \cdot \pi DN_{M^{z+}} \left[\ln\left(\frac{L}{r_0}\right)\right]^{-1} \tag{2.118}$$

N_{st} is here the number of steps with length l per square centimeter, the distance between steps being larger than 2 L, and r_0 will have the same order of magnitude as for hemispherical diffusion.

The contribution of surface diffusion to electrolytic crystal growth can only be considerable if the concentration in the solution is low and the concentration of adatoms is high. Diffusion coefficients in liquid solutions will be larger than on surfaces, a reason for the preference of direct discharge from the solution. The contribution of surface diffusion can formally be taken into account by a slightly larger value for r_0 in Equations (2.117) and (2.118).

The simple model of crystal growth and dissolution via kink sites, their number being dependent on the prehistory of surface preparation, is a picture which can rarely be verified in reality and even then only at very low rates of deposition or dissolution.[37] Under practical conditions of electrolysis, new kink sites are generated by two- and three-dimensional nucleation of new atom layers or crystallites in the case of deposition, by the nucleation of vacancies and cavities in atomic layers in the case of dissolution. This is the reason why the microstructure of metal surfaces is so variable after electrolytic deposition or dissolution. Several models for a combination of nucleation and growth of the nuclei, until they conglomerate, were developed and current–voltage curves for such models were calculated.[39] However, the variability of these models is too large to be discussed here.

Only the conditions for nucleation on a smooth, low-index face shall be presented. In principle, the growth of an atom layer on a surface with steps should come to a halt when all steps have moved over the whole area and the face is completely flat. Further growth requires the nucleation of a new layer by the aggregation of adatoms. This is unnecessary only if screw dislocations are present on the surface, where the growth can continue in the form of spirals of steps.[40] Nucleation requires an excess energy which can be calculated from crystallographic considerations. This excess energy can be expressed by a critical overvoltage at which nuclei of a critical size can be formed statistically and begin to grow instead of dissociating again.

The theory of nucleation was worked out by Erdey-Gruz and Volmer.[41] The result is based on the following assumptions. Nucleation requires an oversaturation. The equivalent of the increase of the free energy by the excess vapor pressure Δp for nucleation from the gas phase $\Delta G = kT \ln \Delta p/p_0$ with p_0 the equilibrium pressure, is the increase of the adatom concentration with a negative overvoltage (cf. Equation [2.116])

$$\Delta G_{\text{adatom}} = kT \cdot \ln \frac{\Delta N_{\text{ad}}}{{}_0N_{\text{ad}}} = -\left(z - \delta^+\right) e_0 \beta \eta \tag{2.119}$$

In the formation of the nucleus the adatoms will be fully discharged. This means that $\delta^+ \rightarrow 0$ and $\beta \rightarrow 1$ if the nucleus is large enough that the atoms have the same number of neighbors as in a complete atomic layer on the surface. Only the atoms on the edge have fewer neighbors and, therefore, an excess free energy is needed to form an edge of the length of the periphery of the nucleus. The result is that the rate to generate nuclei of a critical size — for which further growth at the same overvoltage decreases the free energy of the aggregates — is controlled by an exponential term which has the overvoltage in the denominator,

$$v_{\text{2D-nucl}} \propto \exp\left(-\frac{\text{const}'}{\text{kT}\eta}\right) \tag{2.120}$$

where the constant in the exponent contains the excess energy for the formation of the critical size of the nucleus.

A similar derivation for three-dimensional nucleation could be confirmed by Kaishev and Mutaftschiev[42] with

$$v_{\text{3D-nucl}} \propto \exp\left(-\frac{\text{const}'}{\text{kT}\eta^2}\right) \tag{2.121}$$

Two-dimensional nucleation requires ideally smooth crystal surfaces which exist in reality only under exceptional circumstances. In reality, imperfections of the crystal surface play the predominating role for nucleation in electrolytic crystal growth and dissolution. The presence of dislocations on the surface enhances the formation of nuclei for growth and dissolution drastically. The real process consists, therefore, of an alternating combination of layer growth and nucleation. The relation between these two processes depends very much on components of the solution and can be widely modified by the presence of adsorbates. The same situation is found in electrolytic dissolution of crystals.

In deposition of a metal on another substrate, nucleation of the new crystal is the first step. If the interaction of the new phase with the substrate is weak, as is usual on semiconductors or metals covered by oxide layers, nucleation requires relatively large overvoltages. There are, however, cases in which the interaction of the deposited metal atoms with the substrate surface is stronger than with a surface of their own material. In such cases, deposition in the form of submonolayers and full monolayers occurs already at potentials more positive than the equilibrium potential for the ion exchange on their own solid phase. This is the so-called *underpotential deposition*.[43]

This occurs when the metallic substrate has a higher work function than the deposited metal.[44] This is an indication that the interaction of the electrons of the metal atoms in the adsorption layer with the electronic states of the substrate makes a decisive contribution to the binding energy of adatoms on foreign metals. This phenomenon is related to the observation that metal deposition from the gas phase on foreign metals in the form of submonolayers often results in ordered two-dimensional phases in registration with the geometrical order of the substrate atoms.[45,46] Underpotential deposition is seen in cyclic voltammograms where the current reflects the rate and amount of deposition and dissolution. The integral, $\int i dt$, for a potential step from an anodic potential where no atoms are deposited to a potential in the range of underpotential deposition gives the amount of deposited metal. As an example, a cyclic voltammogram for the deposition of copper on a (111) face of gold is shown in Figure 2.34 together with the integral of the deposited amount as a function of the potential and the structure of the overlayer in the respective potential ranges.[47] This structure was obtained by LEED (low-energy electron diffraction) studies of the surface after transfer from the solution to a vacuum chamber.

Many ordered surface phases were found on electrodes by electrolytic metal deposition at underpotentials, often having different structures than those observed by deposition from the gas phase. The reason is that the interactions of the substrate and the deposits with components of the electrolyte influence the forces of interaction between deposited atoms. This is demonstrated particularly well by the influence of anion adsorption on the potential and structure of underpotentially deposited metals.[43] The structure of the full monolayer is in some cases different from the bulk phase of the deposited metal. The monolayer structure can be controlled by the structure of the substrate, and such differences can extend even to the second layer. Three-dimensional nucleation, however, seems to play a minor role in the growth of the bulk phase on foreign substrates when underpotential deposition takes place.[48]

C. SEMICONDUCTORS

Electrochemical deposition of semiconductors has been achieved in only a very few cases. Cadmium chalcogenides are examples which can be formed by the reduction of Cd^{2+} ions in parallel with chalcogenides present as oxi-anions or from nonaqueous electrolytes containing the chalcogenes dissolved in elementary form. The more important is the anodic dissolution of semiconductors used practically for etching and structuring of semiconductor devices or occurring unintentionally as corrosion in contact with liquids or a moist atmosphere. The electrolytic oxidation of semiconductors such as Ge, Si or InP, GaAs, GaP, and CdS, etc., has been studied intensively. The kinetics are controlled by the concentration of holes at the

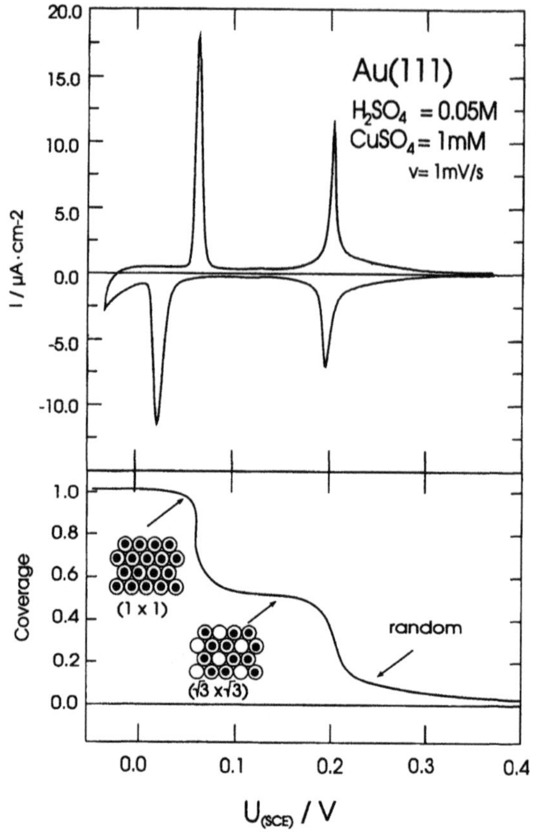

FIGURE 2.34. Cyclic voltammogram and amount of coverage of copper on a Au(111) surface in the range of underpotential deposition. The structure of the deposit is indicated in the lower figure. (Adapted from Kolb, D.M., *Z. Phys. Chem. (Frankfurt)*, 1987, 154, 179.)

interface.[49] One consequence is that p-type semiconductors can easily be oxidized at anodic bias, while n-type materials require the generation of holes by illumination or by internal electron emission from the valence band to the conduction band by high electric fields in the depletion layer at high bias.

Semiconductors have localized bonds between next neighbors, represented in the chemical description by a pair of electrons with opposite spin. An electron hole means that one electron of this bond is missing. If this happens in bonds between atoms on the surface, this bond is weakened and there is a local excess of positive charge. Such a site is attractive for electronegative components of the electrolyte, particularly for anions, which can form a new bond with one of the surface atoms. The electron remaining in the bond of this atom, which binds it to the crystal, will be at a higher energy. If the band gap is small, this energy increase may be sufficient to inject this electron into the conduction band by thermal excitation. On semiconductors with a wide band gap, this state will act as a trap for another hole. In both cases, this bond is broken and will be replaced by a bond to the anionic ligand from the electrolyte.[50] The four-step oxidation of a kink site atom on a semiconductor with four bonds per atom in the bulk is represented in Figure 2.35.

The removal of the second electron by electron injection was found in the oxidation of germanium and silicon, not on GaAs and CdS or other semiconductors with band gaps >1.2 eV. The injection of electrons could be seen in the quantum yield of anodic photo currents at n-type specimens which exceeded one and reached nearly two in several cases. The energetics of the two steps in the photooxidation of n-type substrates are shown in Figure 2.36

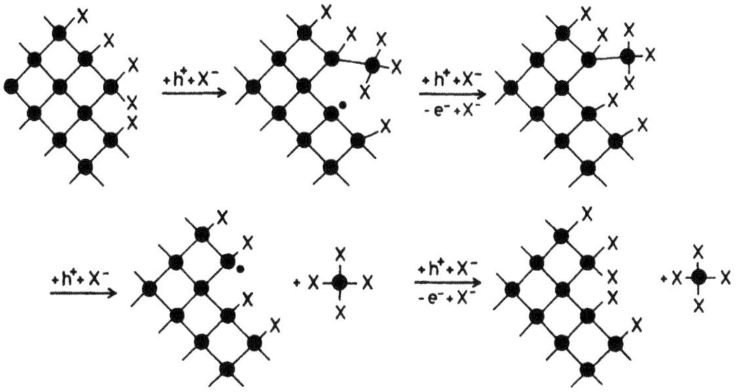

FIGURE 2.35. Mechanism of the oxidation of a semiconductor with tetrahedral band structure by the action of holes in four stages.

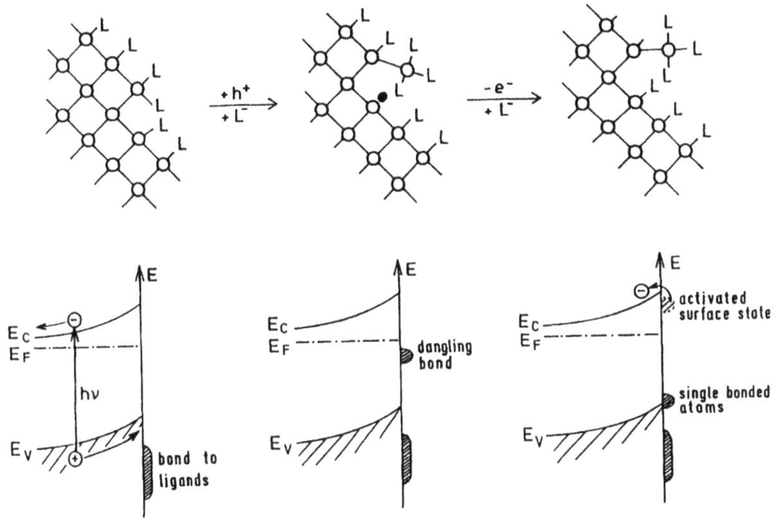

FIGURE 2.36. Mechanism of bond breaking at semiconductors in two steps. Energetics for the possibility of electron injection in the second step.

for the energy positions of the involved electronic states in relation to the position of the band edges.

In the total process of oxidation of surface atoms on semiconductors, several bonds have to be broken, as Figure 2.36 suggests. If the first bond is dissolved, the remaining bonds will, however, be weakened to such an extent that they react much faster with holes, and their consecutive breaking will follow quickly. The assumption is therefore reasonable that the first step will be rate determining. In this case, the kinetics can be simply described as being proportional to p_s, the concentration of holes at the surface

$$i^+ = k^+ \cdot p_s \tag{2.122}$$

The dissolution will occur predominantly at kink sites on the surface, a factor which again makes the rate very surface structure dependent. The factor p_s is controlled on p-type materials by the potential difference in the space charge layer $\Delta\varphi_{sc}$ (cf. Equation [2.101]). If $\Delta\varphi_H$ remains practically constant, the anodic current will be given by

$$i^{+} = z \cdot e_0 \cdot k^{+} \cdot \exp\left(\frac{e_0 \Delta\varphi_{sc}}{kT}\right) \tag{2.123}$$

where z is the number of equivalents for the total oxidation of the semiconductor unit. The rate constant k^+ contains all the unknown parameters like the number of kink sites, the concentration of holes on the surface at $\Delta\varphi_{sc} = 0$, and the frequency factor of the rate-determining step. If the potential drop in the Helmholtz double layer is not constant, a factor $\alpha < 1$ has to be added to the exponent of Equation (2.123). $\Delta\varphi_H$ will change with the bias if electronic surface states exist on the semiconductor or are generated thereon by the anodic process itself, which produces intermediates on the surface. These intermediates can also change the surface dipole moment which affects $\Delta\varphi_H$.

As an illustration, the reaction mechanisms for some semiconductors in contact with aqueous electrolytes are given below.[49] Germanium is oxidized in alkaline solution to $Ge(OH)_4$ or $GeO(OH)_2$ in four steps, two induced by holes, alternating with two steps in which electrons are injected into the conduction band. GaAs is oxidized in acidic solution to Ga^{3+} ions and AsO_3^- by six oxidation steps which all consume holes. CdS is oxidized to Cd^{2+} ions and elemental S with the consumption of two holes. Silicon is a very complicated case. In an aqueous solution containing HF or NH_4F, the dangling bonds of the surface atoms are saturated by hydrogen atoms. The electrochemical oxidation requires only one hole and is followed by the injection of one electron. The next step of breaking the back bonds to the crystal occurs by hydrolysis, in this way restoring the hydrogen-covered surface. The species which enters the solution contains the Si atom in the formal oxidation state of two. A chemical reaction with H_2O or HF completes the oxidation to the Si(IV) state, and the final product is SiF_4 (or SiF_6^{2-}, respectively).[51]

X. TECHNIQUES FOR THE INVESTIGATION OF ELECTRODE REACTION KINETICS

A general problem in the study of electrode reactions is the separation of the charge transfer reaction step at the interface from preceding and consecutive transport processes and chemical reactions. Transport processes are inherently involved. They influence the true charge transfer step by a time-dependent change of the reactant concentrations at the interface, after an electrolysis has been initiated (concentration polarization). This can be corrected by an extrapolation to the time zero when the electrolysis had begun, if a mathematical calculation of the concentration changes during the electrolysis can be made under the particular conditions of the transport kinetics — i.e., the cell geometry and the externally controlled parameters (voltage, current). Another possibility is to analyze the frequency response between the current and a periodically alternating voltage applied to the electrode, i.e., to measure the impedance of the electrode. These two methods shall be treated. For electrodes in contact with liquid electrolytes, a great variety of other methods are available in which a steady state of the transport processes is reached by convection in the electrolyte.[52] Since this is not applicable to solid electrolytes, such techniques will not be discussed here. For a more detailed discussion of experimental methods, see Chapter 9 of this handbook.

A. CURRENT AND POTENTIAL STEP

Applying to the electrolysis cell a constant current step is technically the easiest method of operation. The electrode to be investigated may be inert and at equilibrium with a redox reaction $Ox^+ + e^- \rightleftarrows Red$. An external anodic current will oxidize the Red components immediately after its onset and their concentrations will decrease at the interface, while that of the Ox^+ components will increase in parallel. For pure diffusion control of the transport, the concentrations at the interface will, at constant current i, follow an equation first derived by Sand,[53]

$$N_{red} = N_{red,0}\left(1 - \frac{1}{|i|}\sqrt{\frac{t}{\tau_{red}}}\right) \tag{2.124}$$

where $N_{red,0}$ is the initial concentration at the interface and τ_{red} is the so-called *transit time* when N_{red} would become zero at an anodic current,

$$\tau_{red} = \frac{\sqrt{\pi D_{red}} e_0}{2|i|} N_{red,0} \tag{2.125}$$

The analogous equations for N_{ox} are

$$N_{ox} = N_{ox,0}\left(1 - \frac{1}{|i|}\sqrt{\frac{t}{\tau_{ox}}}\right) \tag{2.126}$$

with

$$\tau_{ox} = \frac{\sqrt{\pi D_{ox}} e_0}{2|i|} N_{ox,0} \tag{2.127}$$

where τ_{ox} is the transit time after which N_{ox} would reach zero at a cathodic current i.

Inserting these relations into Equation (2.97), one obtains a relation for the time response of the overvoltage at constant current when the rate is controlled by charge transfer and diffusion without convection. This relation can be written as

$$i = i_{ct}(\eta) - \frac{i}{|i|} i_0 \left[\frac{1}{\sqrt{\tau_{red}}} \exp\left(\frac{\alpha e_0 \eta}{kT}\right) + \frac{1}{\sqrt{\tau_{ox}}} \exp\left(\frac{(1-\alpha) e_0 \eta}{kT}\right)\right] \sqrt{t} \tag{2.128}$$

with

$$i_{ct}(\eta) = i_0 \left[\exp\left(\frac{\alpha e_0 \eta}{kT}\right) - \exp\left(-\frac{(1-\alpha) e_0 \eta}{kT}\right)\right]$$

In this equation, $i_{ct}(\eta)$ represents the charge transfer rate dependence on η for a constant concentration of the redox components at the interface: $N_{red} = N_{red,0}$ and $N_{ox} = N_{ox,0}$. Equation (2.128) represents the so-called *galvanostatic transient* of the overvoltage. One sees that an extrapolation of the transient overvoltage η (t) to $t \to 0$ performed for different constant currents should yield direct information on the relation between charge transfer current and overvoltage.

In some cases, Equation (2.128), which contains η only in an implicit form, can be simplified. For instance, when $i > i_0$ and η is larger than $kT/\alpha e_0$, one of the terms with η in the brackets can be neglected, and the respective exponential factor can be separated. Or, for great differences in the concentrations of $N_{red,0}$ and $N_{ox,0}$, one of the transition times τ becomes

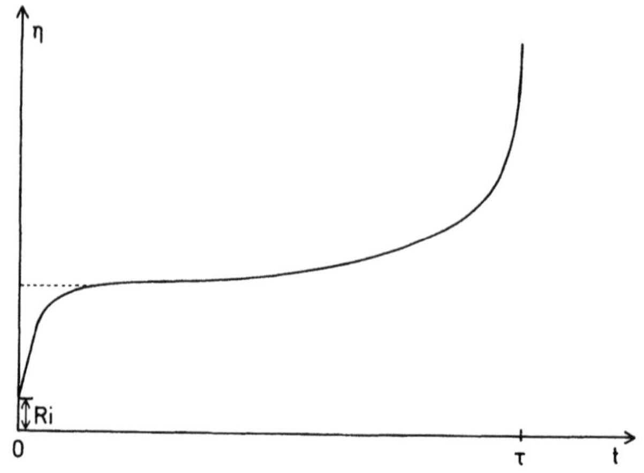

FIGURE 2.37. Galvanostatic transient of the overvoltage.

very large in relation to the other, and one term in the second bracket of Equation (2.128) can be neglected.

There is, however, a principal difficulty in the analysis of overvoltage transients at a constant current. This is indicated in Figure 2.37, where the real transient exhibits an initial delay in the increase of η due to the current needed for the charging of the capacity of the electrode. This part of the current, $I_C = \frac{1}{C}\frac{d\eta}{dt}$, is lost for the charge transfer current and causes a deviation from the theoretical transient. A correction for this deviation can be made by a more sophisticated mathematical analysis.[54] In order to reduce this deviation, the double layer charging can be accelerated by superposition of an initial current pulse, the so-called *double pulse method*.[55] This gives an improvement, but some uncertainty remains because the exact adjustment of the additional current pulse is not possible.

The mathematical analysis of current transients at constant overvoltage, the so-called *potentiostatic transient*, is somewhat easier. Again, the concentration changes with time under this condition have to be calculated for the analysis of the current transient. For the interplay of charge transfer and transport by diffusion as rate controlling in a redox reaction at an inert electrode, as discussed for the galvanostatic transient, the following relation can be derived:[56]

$$i(t) = i_{ct}(\eta)\exp\left(g^2 t\right)\operatorname{erfc}\left(g\sqrt{t}\right) \tag{2.129}$$

with

$$g = \frac{i_0}{e_0}\left[\frac{\exp\left(\frac{\alpha e_0 \eta}{kT}\right)}{N_{\mathrm{red},0}\sqrt{D_{\mathrm{red}}}} + \frac{\exp\left(\frac{(1-\alpha)e_0 \eta}{kT}\right)}{N_{\mathrm{ox},0}\sqrt{D_{\mathrm{ox}}}}\right] \tag{2.130}$$

where erfc is the complementary error function (1 – erf). For the initial part of the transient, this relation can be approximated by

$$g\sqrt{t} \ll 1 \;:\; i(t) \approx i_{ct}(\eta)\left(1 - \frac{2g}{\sqrt{\pi}}\sqrt{t}\right) \tag{2.131}$$

for the last part of the transient by

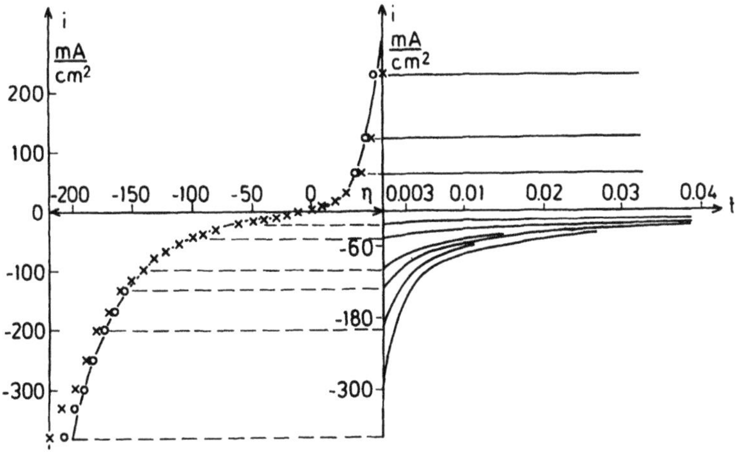

FIGURE 2.38. Right: potentiostatic transients for a Zn-amalgam electrode in 0.02 *M* $Zn(ClO_4)_2$ + 1 *M* $NaClO_4$ solution for different potentials. Left: the current–voltage curve constructed from the initial currents (○ points corrected for the ohmic voltage drop).

$$g\sqrt{t} \gg 1 \; : \; i(t) \approx i_{ct}(\eta)\left(\sqrt{\pi}\, g\sqrt{t}\right)^{-1} \tag{2.132}$$

Again, the limitation for the realization of the theoretical transient in its initial part is caused by the double layer charging. However, modern potentiostats can reach the constant overvoltage in a very short time ($\sim 10^{-6}$ s). Thus, this problem becomes critical only for extremely fast charge transfer reactions. But another problem becomes more serious in potentiostatic transient studies. This problem is the ohmic potential drop between the electrode and the location of the contact of the reference electrode in the electrolyte. This ohmic potential drop varies with the current and restricts the time for charging the double layer up to the intended overvoltage. In the galvanostatic transient the correction for the ohmic potential drop can easily be made, as long as the electrolyte resistance remains constant. For the potentiostatic transients, potentiostats have been constructed which compensate an ohmic potential drop by an additional feedback voltage for a constant electrolyte resistance.[57] The exact adjustment to the real ohmic potential drop remains, nonetheless, a problem which restricts the accuracy of the transient analysis. Nevertheless, if the charge transfer rate is not extremely high, this method can by applied with great success. Figure 2.38 gives an example for such transients and the extrapolated current–voltage curve for the rate of charge transfer without concentration polarization.[58]

B. IMPEDANCE SPECTROSCOPY

A very frequently used technique for the study of electrode reactions is measuring the impedance of an electrode at variable frequency.[59,60] This technique can be applied to electrodes at equilibrium where the external ac current causes concentration changes of both components of the redox reaction in opposite directions. The ac current can also be superimposed upon a constant current, provided a steady state can be reached for this dc current. This requires the presence of convection in the transport process. Since in solid electrolytes convection is impossible, such cases will not be discussed here.

Figure 2.39 represents the equivalent circuit of an electrode impedance, Z. It consists of an ohmic series resistance R_{el} in the electrolyte between the reference electrode and the interface of the electrode and of two impedances in parallel, one for charging the Helmholtz double layer capacity, Z_H, and the other representing the impedance for the charge transfer

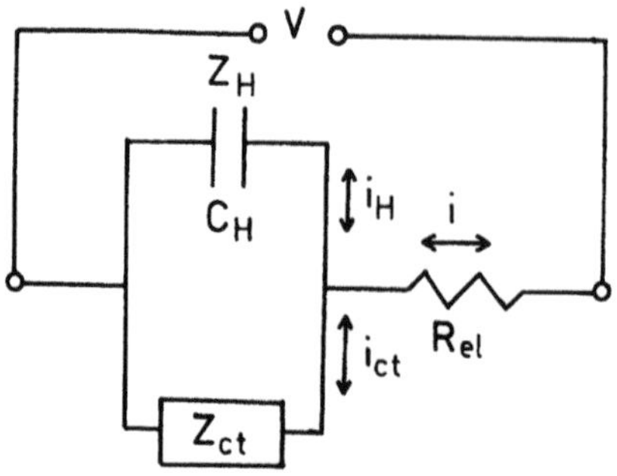

FIGURE 2.39. Equivalent circuit of the electrode impedance.

process, Z_{ct}. These two impedances cause a phase shift with respect to the externally applied voltage V, the ac character of which may be described by a complex function of the time t: $V = V_0\, e^{j\omega t}$, where j is the imaginary unit $\sqrt{-1}$ and $\omega = 2\,\pi\,f$; f = frequency. The impedance of the double layer capacity has only an imaginary term in this description:

$$Z_H = -\frac{j}{\omega C_H},$$

while the charge transfer impedance Z_{ct} has a real and an imaginary component. The task in the analysis of impedances is to develop models for the behavior of Z_{ct} as a function of the circular frequency ω and the reactant transport to and from the electrode which can simulate the experimental data. Only one example demonstrating the procedure and the limitations will be discussed here.

As our example, we will again use a simple redox reaction at equilibrium and, as in the previous section, assume that the transport occurs exclusively by diffusion. The current may not have a dc component, so that in the time average the concentrations of the redox components remain constant in front of the electrode. The oscillating current causes sinusoidal concentration waves moving from the interface into the bulk and decaying therein. The amplitude is largest at the interface. Figure 2.40 shows such a damped concentration wave which follows the periodic voltage with a phase shift.

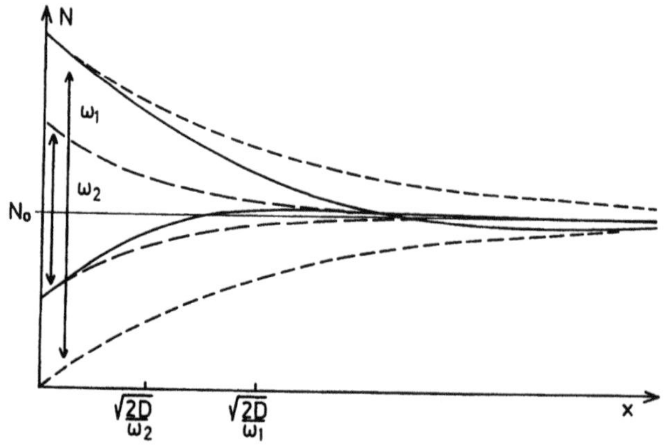

FIGURE 2.40. Concentration profiles in front of an electrode for two different values of the frequency of the ac current. Solid lines: profiles when the difference $N - N_0$ has reached a maximum at the interface. Broken lines: amplitude of the periodic concentration variations. The distance for which the amplitude has decayed to 1/e of its maximum is $\sqrt{(2D)/\omega}$.

For a small amplitude of the current, the amplitude of the overvoltage and of the concentration changes will be small. Assuming that $\eta \ll kT/e_0$ we obtain

$$\Delta N_{ox} = N_{ox} - N_{ox,0} \ll N_{ox,0} \tag{2.133}$$

$$\Delta N_{red} = N_{red} - N_{red,0} \ll N_{red,0}$$

and remembering that $\beta = 1 - \alpha$, one can approximate Equation (2.97) by a linear relation between the charge transfer current i_{ct} and η

$$i_{ct} \approx i_0 \left[\frac{e_0 \eta}{kT} + \frac{\Delta N_{red}}{N_{red,0}} - \frac{\Delta N_{ox}}{N_{ox,0}} \right] \tag{2.134}$$

The variations of ΔN are consequences of the charge transfer current and will be proportional to i_{ct} with a phase shift.

$$\Delta N_{red} = b_{red} \cdot i_{ct}; \; \Delta N_{ox} = b_{ox} \cdot i_{ct} \tag{2.135}$$

where b_{red}, b_{ox} are complex quantities when i_{ct} is described by

$$i_{ct} = i_{max} \exp(j\omega t) \tag{2.136}$$

The result for the factors b in the case of transport by diffusion alone is[61]

$$b_{red} = \frac{j-1}{e_0 \sqrt{2 D_{red} \omega}}; \; b_{ox} = \frac{1-j}{e_0 \sqrt{2 D_{ox} \omega}} \tag{2.137}$$

The total electrode impedance has three parts,

$$Z_{ct} = \frac{d\eta}{di_{ct}} = \frac{kT}{e_0 i_0} + \frac{kT}{e_0} \left(\frac{1}{e_0 N_{ox,0} \sqrt{2 D_{ox} \omega}} + \frac{1}{e_0 N_{red,0} \sqrt{2 D_{red} \omega}} \right) (1-j) \tag{2.138}$$

$$Z_{ct} = R_{ct} + Z_{diff} \tag{2.139}$$

The first term in Equation (2.138) represents the contribution by the charge transfer reaction alone and has ohmic character. The two other contributions to Equation (2.138) added in Equation (2.139) to Z_{diff} represent the contribution of the diffusion processes. Z_{diff} has a real (ohmic) and a negative imaginary (capacitive) component of equal size. That means that a phase shift of 45° in the negative direction exists between current and overvoltage. Z_{diff} decreases with $\omega^{-1/2}$. If an extrapolation to $\omega \to \infty$ is possible for Z_{ct}, one can determine the exchange current density from

$$R_{ct} = \frac{kT}{e_0 i_0} \tag{2.140}$$

If the capacitive current could be neglected, the impedance would be determined by the charge transfer impedance plus the ohmic resistance in the electrolyte, and a plot of the real

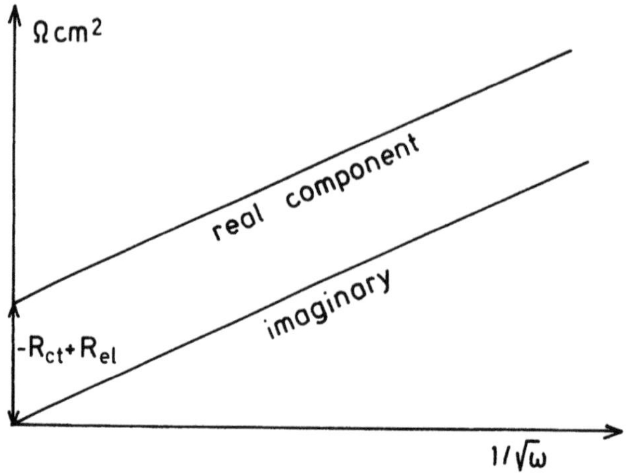

FIGURE 2.41. Electrode impedance for charge transfer plus diffusion of the redox components without the contribution of the Helmholtz capacity as a function of the angular frequency.

and imaginary component vs. $\omega^{-1/2}$ should give a linear run for both parts as shown in Figure 2.41. The extrapolation of the real component to $\omega^{-1/2} \rightarrow 0$ gives the value of $R_{ct} + R_{el}$.

The difficulty is, however, again the double layer capacity. Figure 2.39 shows that the current has a parallel path to Z_{ct} via Z_H. The impedance of the electrode Z_E results therefore from

$$Z_E = \frac{Z_{ct} \cdot Z_H}{Z_{ct} + Z_H} \tag{2.141}$$

The double layer capacity appears in the real and imaginary components of this impedance. The result for R_{ct} thus depends on the value assumed for the double layer capacity which limits the accuracy of kinetic parameters obtained for fast electron transfer reactions with this technique.

With regard to the capacitive currents, the galvanostatic and potentiostatic techniques are superior because the double layer charging can be accelerated in these cases. The attraction of impedance measurements lies in the high accuracy of the data thus obtained. They are particularly useful for systems in which adsorption of the reactants or of intermediates plays an important role. The discussion of such cases is, however, outside the scope of this introduction.

XI. MECHANISMS OF ELECTRODE REACTIONS AND ELECTROCATALYSIS

Elementary redox reactions with the transfer of one electron are usually fast if the components are present in normal concentration. They can appear to be slow if the component which controls the electron transfer is present only in a very low concentration because it is in chemical equilibrium with a more stable compound in the electrolyte. Catalysis of the charge transfer can occur by adsorbed species which act as a bridge for the electron transfer. But such effects are in most cases modest. More often, adsorbed species act as inhibitors for the electron transfer.

In metal deposition processes, the ion discharge can be very slow if the surface is covered by oxide layers or by strongly chemisorbed ions or molecules. Ligands of metal ions in solution, if they favor adsorption of these species at the electrode, can increase the rate of ion transfer. However, the field for which so-called electrocatalysis is decisive for the reaction rate is the large group of electrode reactions in which the net process occurs in several steps and one or several intermediate species remain adsorbed on the electrode. The classical example is the hydrogen electrode for which adsorbed H atoms are the intermediates in the reaction $2H^+_{solv} + 2e^- \rightleftarrows H_2$.

A more complicated case is the electrolytic generation or reduction of oxygen which occurs in four steps. These two reactions shall be discussed as examples for processes where catalysis plays a decisive role in the kinetics.

A. HYDROGEN ELECTRODE

Starting with the reduction of an electrolyte which contains protons in some kind of chemical bond, the first step is the formation of a H atom by capturing an electron from the electrode. This may be a metal M. Since all metals interact in some way with H atoms, these atoms remain in an adsorbed state on the electrode surface, according to the Volmer reaction

$$H^+_{solv} + e^-_M \rightleftarrows H_{ad} \quad v_1 \tag{2.142}$$

In the second step the recombination reaction

$$H_{ad} + H_{ad} \rightleftarrows H_2 \quad v_2 \tag{2.143}$$

is competing with the Heyrovsky reaction (or electrochemical desorption)

$$H_{ad} + H^+_{solv} + e^-_M \rightarrow H_2 \quad v_3 \tag{2.144}$$

It was found that Reaction (2.144) cannot be reversed. Therefore, for the oxidation of H_2, only the Path (2.143) followed by (2.142) in the reverse direction appears to be realistic.

If the intermediate product, the H atoms, are not adsorbed, the reaction

$$H^0_{solv} + e^- \rightleftarrows H_{solv} \tag{2.145}$$

would require an excess of free energy, ΔG_H, corresponding with one half of the free energy of dissociation of the H_2 molecules (2.3 eV) minus the free energy of solvation of H atoms (about 0.2 eV in aqueous solution). This excess energy will be released in the consecutive reaction forming the H_2 molecule, which would follow extremely fast. The adsorption energy reduces the overvoltage of the first step and decreases the driving force of the two consecutive reactions. If the adsorption becomes too strong, the Reactions (2.143) and/or (2.144) will become rate determining. This can be seen in the relation between the exchange current density, i_0, at the equilibrium potential and the averaged adsorption energy of different metals for H atoms,[62] which is represented in Figure 2.42. One sees a maximum of i_0 for an adsorption energy of about 230 kJ/mol (=2.3 eV/mol) which corresponds with about half the dissociation energy of the H_2 molecule. For this adsorption energy, the activation energies of the Reactions (2.142) and (2.143) or (2.144) will be very similar and the rate-determining step can be either of them (the solvation energy of H atoms seems to be compensated by the energy required to remove H_2O molecules from the adsorption sites for H atoms).

The relations in Figure 2.42 correspond with the so-called *volcano curves* in heterogeneous catalysis.[63] The exchange current densities of Figure 2.43 are a measure of the activation energy of the rate-determining step. They were obtained by an extrapolation of the cathodic current–voltage curves from large overpotentials to the equilibrium potential, with the exception of the highly catalytic metals where equilibrium can be established and the exchange

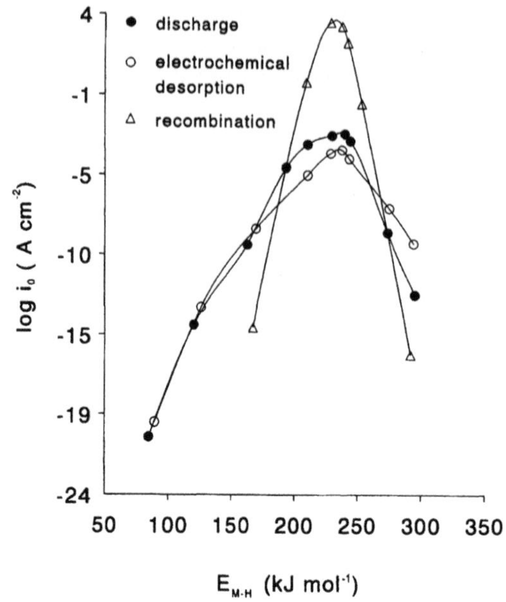

FIGURE 2.42. Exchange current densities at the equilibrium potential for hydrogen evolution in relation to the adsorption energy of H atoms on metals. (Data from Krishtalik., L.I., in *Advances in Electrochemistry and Electrochemical Engineering*, Interscience, New York, 1970, 283-339.)

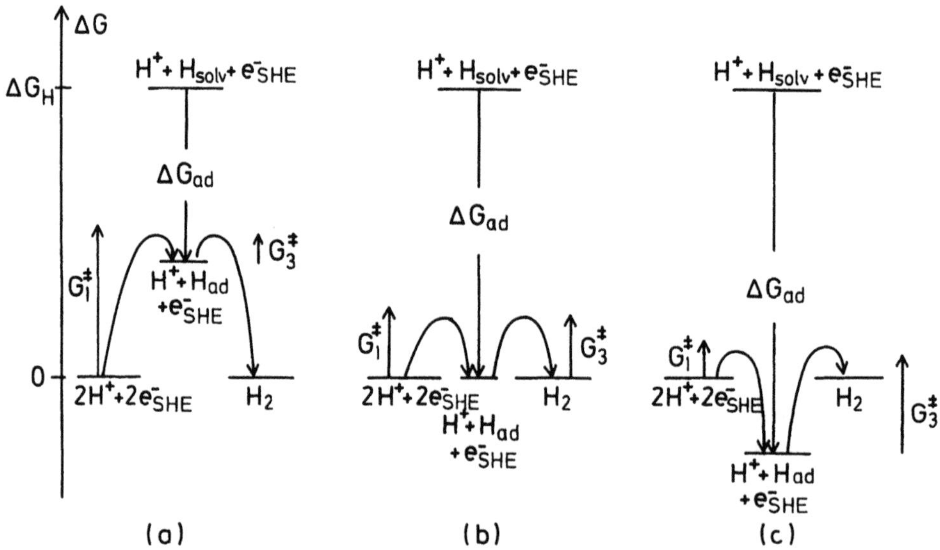

FIGURE 2.43. Qualitative representation of the influence of the adsorption energy, ΔG_{ad}, of H atoms on the activation energies of the two electrochemical steps in the electrolytic formation of hydrogen.

current density can be obtained more directly. For the net cathodic current, the steady-state condition for the reaction rate is

$$v_1 = 1/2\, v_2 + v_3 \tag{2.146}$$

For a qualitative understanding of the role of the adsorption energy, one can use a kinetic approach for the three reactions which is oversimplified, but shows the principal correlations

$$v_1 = \vec{k}_1(1-\Theta)\exp\left(-\frac{\beta_1 e_0 \eta}{kT}\right) - \overleftarrow{k}_1 \Theta \exp\left(\frac{\alpha_1 e_0 \eta}{kT}\right) \tag{2.147}$$

$$v_2 = \vec{k}_2 \Theta^2 - \overleftarrow{k}_2(1-\Theta)^2 \tag{2.148}$$

$$v_3 = \vec{k}_3 \Theta \exp\left(-\frac{\beta_3 e_0 \eta}{kT}\right) \tag{2.149}$$

where Θ is the degree of coverage of the surface by H atoms. Reaction (2.144) is considered to be irreversible. The above formulation of the rate of Reaction (2.143) would require a high mobility of the adsorbed H atoms. Any change of the concentration of H_2 in front of the electrode is also neglected, which is irrelevant at a large cathodic overvoltage η (η is negative) where the second term in Equation (2.147) can be disregarded, and the rate of the recombination reaction will be negligible because it does not depend on the overvoltage. The coverage by H atoms is then determined by the competition of Reactions (2.142) and (2.144).

The rate constants $\vec{k}_1$ and $\vec{k}_3$ vary with the adsorption energy in the opposite direction. This is indicated in the diagram of Figure 2.43 where the activation energies for the formation of the intermediate and its consecutive electrochemical formation of H_2 are qualitatively shown for the three cases: (a) $-\Delta G_{ad} < \Delta G_H$; (b) $-\Delta G_{ad} = \Delta G_H$; and (c) $\Delta G_{ad} > \Delta G_H$. ΔG_H is the free energy for the formation of one H atom from H_2 without interaction with a surface. With $\vec{k} = \exp(-\Delta G^{\neq}/kT)$ and the charge transfer coefficients $\beta_1 \approx \beta_3$, one can easily derive that $\Theta \to 1$ in case (a) and $\Theta \to 0$ in case (c). In case (b) Θ will have an intermediate value. In case (a) the Volmer reaction is rate determining, in case (c) the Heyrovsky reaction. For case (b) the recombination can become rate determining near the equilibrium potential if the Heyrovsky reaction is slow. At higher cathodic overvoltage, however, the Heyrovsky reaction is accelerated and will in any case control the reaction path, since for parallel reactions the faster process predominates.

The cathodic current–voltage curves have an exponential dependence on the overvoltage, but the exponent depends on the mechanism. When the recombination (Equation [2.148]) can be neglected and the Volmer or the Heyrovsky reaction is rate determining, Θ barely varies with η, and the cathodic current in the logarithmic scale will follow the relation

$$\ln|i| \propto -\beta \frac{e_0 \eta}{kT} \tag{2.150}$$

where β is β_1 for rate control by the Volmer reaction or β_3 for the rate control by the Heyrovsky reaction.

If the recombination controls the rate (and the Heyrovsky reaction is slow), the influence of η upon the Volmer reaction leads to an increase of Θ in order to reach the steady-state condition of Equation (2.146). The relation between cathodic current and overvoltage can in this case become

$$\ln|i| \propto -2\frac{e_0 \eta}{kT} \tag{2.151}$$

which is, however, only possible if Θ remains small. This can only occur at low overvoltages.

The real kinetics are much more complicated due to the great variability of the dependence of the adsorption energy on crystal orientation and defect structure, on the interactions between the adsorbed species themselves and on the influence of the electrical potential.

Hence, the individual properties of the electrode modify the real behavior of electrodes to a very large extent and the above discussion can only give the general trend.

B. OXYGEN ELECTRODE

The anodic oxidation of water to oxygen is an important process in electrolysis as well as the cathodic reduction of oxygen for the corrosion of metals in aqueous solutions or a moist atmosphere. The net reaction

$$2H_2O \rightleftarrows O_2 + 4H^+_{aq} + 4e^- \quad \text{with}\, U^0 = 1.23\, V_{NHE} \tag{2.152a}$$

or in alkaline solution

$$4OH^- \rightleftarrows O_2 + 2H_2O + 4e^- \quad \text{with } U^0 = 0.4\, V_{NHE} \tag{2.152b}$$

includes four elementary steps and several possible intermediates.[64] The formation of those intermediates requires excess energies which can be diminished by chemisorption as in the case of the H atoms in the hydrogen reaction kinetics. For example, if the first step of the oxidation of H_2O would be the formation of a free OH radical,

$$H_2O \rightarrow OH + H^+_{aq} + e^- \tag{2.153}$$

the equilibrium potential for this reaction at pH = 0 would be +2.8 V_{NHE}, about 1.6 V more than the equilibrium potential of the net reaction.

The role of electrocatalysis is again to reduce this excess energy by adsorption. However, the optimization requires in this case the optimal adsorption energy of several intermediates. For example, the second step in the reaction path could be

$$OH_{ad} + OH_{ad} \rightarrow H_2O_2 \tag{2.154a}$$

or

$$OH_{ad} + H_2O \rightarrow H_2O_2 + H^+_{aq} + e^- \tag{2.154b}$$

Reaction (2.154a) is independent of the electrode potential and should be close to optimal if the adsorption free energy of the OH radicals with $\Delta G_{ad} \approx -1.6$ eV would compensate the difference between the equilibrium potential of reaction and the net reaction Equation (2.152a). However, the oxidation of H_2O_2 has to pass another intermediate, the HO_2 radical

$$H_2O_2 \rightarrow HO_2 + H^+_{aq} + e^- \tag{2.155}$$

with another equilibrium potential, $U_0 = 1.44\, V_{NHE}$, again different from the equilibrium potential of the net reaction. This could be compensated by the adsorption of the HO_2 radicals with an adsorption free energy of about –0.21 eV if the final step is the disproportionation of HO_2

$$2HO_{2,ad} \rightarrow O_2 + H_2O_2 \tag{2.156}$$

The net reaction could be optimized with these two adsorption energies for OH and HO_2. This mechanism, however, requires building up a large H_2O_2 concentration as an intermediate in the solution and is not realistic.

If the second step is Reaction (2.154b), which has an equilibrium potential of $U^0 \approx 0.72 - \Delta G_{OH_{ad}}$, this path would be very unfavorable with the optimal adsorption free energy of the OH radicals for reaction, namely, $U^0 \approx +2.3\ V_{NHE}$. An adsorption of H_2O_2 could to some extent compensate this unfavorable situation, but a total compensation would require a ΔG_{ad} for H_2O_2 of −1.1 eV, which is impossible.

The last step of the oxidation path via H_2O_2 after Reaction (2.155) will more probably occur by

$$HO_2 \rightarrow O_2 + H^+_{aq} + e^- \quad \text{with } U_0 = 0.13\ V_{NHE} \tag{2.157}$$

which has a very negative redox potential if the HO_2 radicals are not adsorbed. Adsorption would shift the redox potential by $-\Delta G_{HO_{2,ad}}$ in the positive direction. However, in order to reach the equilibrium potential of the net reaction (Equation [2.152a]), an adsorption energy of −1.26 eV would be required, a very different amount from the optimum for the disproportionation reaction (Equation [2.156]).

The conclusion of this discussion is that the optimum of catalysis of the oxygen electrode reaction depends on a critical compromise of adsorption energies of the intermediates. It is therefore not surprising that no electrode has been found where this reaction occurs reversibly at room temperature. In both directions rather large overvoltages are required to oxidize H_2O to O_2 or to reduce O_2 to H_2O. Only in very alkaline solutions does the partial reaction

$$2\ OH^-_{aq} \rightleftarrows H_2O_2 + 2\ e^- \tag{2.158}$$

come close to equilibrium.[65]

Several other mechanisms in addition to those mentioned here have been postulated.[66] Many of them include adsorbed O atoms produced by the reaction

$$OH_{ad} \rightarrow O_{ad} + H^+ + e^- \tag{2.159}$$

On oxide electrodes or in oxide formation on metal electrodes O atoms or O^- radicals certainly play an important role, particularly at high temperatures where the overvoltages become much lower. The kinetics, however, depend in all cases on the individual system, and no general conclusion can be given for the oxygen electrode reaction mechanism except for the role of the chemisorption energy of the intermediates, as outlined in the preceding discussion.

C. GENERAL REMARKS

Nearly all electrode reactions of technical interest proceed in several steps via intermediates. An example is the large group of electrochemical processes with organic molecules in which the interaction of the intermediates with the electrode surface often controls not only the reaction rate, but also the reaction path and the final product. The electrode reactions in primary cells and in batteries include complex transformations between different solids, and the reaction rates can be influenced by small amounts of additives. In all such cases, purely electrochemical methods are insufficient for a definite elucidation of the reaction mechanisms. It requires spectroscopic techniques over the whole range of electromagnetic radiation to detect the intermediate molecules or structures which determine the reaction path on the atomic and molecular levels. The present developments in electrochemistry go in this direction, allied with parallel developments in surface science and heterogeneous catalysis.[67]

ACKNOWLEDGMENT

H. G. is most grateful to K. W. Kolasinski, Ph.D., Alexander von Humboldt Foundation and Max Planck Society Fellow at the Fritz Haber Institute, 1992 to 1994, for improving the English text and to Ms. I. Reinhardt for critical typesetting.

REFERENCES

1. Horvath, A.L. *Handbook of Aqueous Electrolyte Solutions*, Ellis Horwood Limited, Chichester, 1985.
2. Harned, M.S. and Owen, B.B., *The Physical Chemistry of Electrolyte Solutions*, Reinhold, New York, 1958.
3. Robinson, R.A. and Stokes, R.H., *Electrolyte Solutions*, Butterworths, London, 1959.
4. *Handbook of Chemistry and Physics*, 73rd ed., CRC Press, Boca Raton, FL, 1992/93, 12–108/109.
5. Cardona, M. and Ley, L., in *Photoemission in Solids*, Part I, Topics in Applied Physics Vol. 26, Cardona, M. and Ley, L., Eds., Springer-Verlag, Berlin, 1978, 1-48.
6. Craig, P.P. and Radeka, V., *Rev. Sci. Instr.* 1970, *41*, 258.
7. Helmholtz, H. von, *Wied. Ann.*, 1879, *7*, 337.
8. Schmickler, W. Models for the interface between metal and electrolyte solutions, in *Structure of Electrified Interfaces*, Lipkowski, J. and Ross, P.N., Eds., VCH Publishers, New York, 1993, 201–275.
9. Grahame, D.C., *J. Am. Chem. Soc.*, 1954, *76*, 4819.
10. Price, D. and Halley, J.W., *J. Electroanal. Chem.*, 1983, *150*, 347.
11. Amokrane, S. and Badiali, J.P., *J. Electroanal. Chem.*, 1989, *266*, 21.
12. Gouy, G., *J. Phys. (Paris)*, 1910, *9*, 457.
13. Chapman, D.C., *Philos. Mag.*, 1913, *25*, 475.
14. Stern, O., *Z. Elektrochem.*, 1924, *30*, 508.
15. Frumkin, A.N., Petrii, O.A., and Damaskin, B.B., Potentials of zero charge, in *Comprehensive Treatise of Electrochemistry*, Vol. 1, Bockris, J.O'M., Conway, B.E., and Yeager, E.B., Eds., Plenum Press, New York, 1980, 221–289.
16. Sparnaay, M.J., *The Electrical Double Layer*, Pergamon Press, Oxford, 1972, 57–62.
17. Hammett, A. Semiconductor electrochemistry, in *Comprehensive Chemical Kinetics*, Vol. 27, Compton, R.G., Ed., Elsevier, Amsterdam, 1987, 61–246.
18. Mott, N.F., *Proc. R. Soc. (London)*, 1939, *A 171*, 27.
19. Schottky, W., *Z. Phys.*, 1939, *113*, 367.
20. Hofmann-Perez, M. and Gerischer, H., *Ber. Bunsenges. Phys. Chem.*, 1961, *65*, 771.
21. Lohmann, F., *Ber. Bunsenges. Phys. Chem.*, 1966, *70*, 428.
22. Pleskov, Yu.V., Electric double layer on semiconductor electrodes, in *Comprehensive Treatise of Electrochemistry*, Vol. 1, Bockris, J.O'M., Conway, B.E., and Yeager, E.B., Eds., Plenum Press: New York, 1980, 291–328.
23. Koryta, J., Dvorák, J., and Kavan, L., Ion selective electrodes, in *Principles of Electrochemistry*, John Wiley & Sons, Chichester, 1993, 425–433.
24. Gurney, R.W., *Proc. R. Soc. (London)*, 1931, *A 134*, 137.
25. Gerischer, H., *Z. Phys. Chem. (Frankfurt)*, 1960, *26*, 223.
26. Marcus, R.A., *J. Chem. Phys.*, 1956, *24*, 955.
27. Marcus, R.A., *Can. J. Chem.*, 1959, *37*, 155; *J. Chem. Phys.*, 1965, *43*, 679.
28. Levich, V.G., Present state of the theory of oxidation-reduction in solution in *Advances in Electrochemistry and Electrochemical Engineering*, Vol. 4, Delahay, P. and Tobias, C.W., Eds., Interscience, New York, 1965, 249–371.
29. Gerischer, H., *Z. Phys. Chem. (Frankfurt)*, 1960, *26*, 325.
30. Gerischer, H., Semiconductor electrode reactions, in *Advances in Electrochemistry and Electrochemical Engineering*, Vol. 4, Delahay, P. and Tobias, C.W., Eds., Interscience, New York, 1965, 139–232.
31. Memming, R., Processes at semiconductor electrodes, in *Comprehensive Treatise of Electrochemistry*, Vol. 7, Conway, B.E., Bockris, J.O'M., and Yeager, E.B., Eds., Plenum Press, New York, 1983, 529–591.
32. Gerischer, H., *Z. Phys. Chem. (Frankfurt)*, 1961, *27*, 47.
33. Decker, F., Pettinger, B., and Gerischer, H., *J. Electrochem. Soc.* 1983, *130*, 1335.
34. Randles, J.E.B. and Somerton, K.W., *Trans. Faraday Soc.*, 1952, *48*, 951.
35. Vetter, K.J., *Elektrochemische Kinetik;* Springer-Verlag, Berlin, 1961; *Electrochemical Kinetics;* Academic Press, New York, 1967.
36. Gerischer, H., *Z. Elektrochem.*, 1953, *57*, 604.

37. Budevski, E.B., Electrocrystallisation, in *Comprehensive Treatise of Electrochemistry*, Vol. 7, Conway, B.E., Bockris, J.O'M., and Yeager, E.B., Eds., Plenum Press, New York, 1983, 399–450.
38. Gerischer, H., in *Proc. Protection against Corrosion by Metal Finishing. Surface 66*, Forster Verlag, Zürich, 1967, 11–23.
39. Fleischmann, M. and Thirsk, H.R., Metal deposition and electrocrystallisation, in *Advances in Electrochemistry and Electrochemical Engineering*, Vol. 3, Delahay, P. and Tobias, C.W., Eds., Interscience, New York, 1963, 123–210.
40. Burton, K.W., Cabrera, N., and Frank, F.C., *Phil. Transact. Royal Soc. (London)*, 1951, *A 243*, 299.
41. Erdey-Gruz, T. and Volmer, M., *Z. Phys. Chem.*, 1931, *A 157*, 165.
42. Kaischev, R. and Mutaftschiev, B., *Z. Phys. Chem. (Frankfurt)*, 1955, *204*, 334.
43. Kolb, D.M., Physical and electrochemical properties of metal monolayers on metallic substrates, in *Advances in Electrochemistry and Electrochemical Engineering*, Vol. 11, Gerischer, H. and Tobias, C.W., Eds., Wiley & Sons, New York, 1978, 127–271.
44. Kolb, D.M., Przasnyski, M., and Gerischer, H., *J. Electroanal. Chem.*, 1974, *54*, 25.
45. Kolb, D.M., Surface reconstruction at metal-electrolyte interfaces, in *Structure of Electrified Interfaces*, Lipkowski, J. and Ross, P.N., Eds., VCH Publishers, New York, 1993, 65–102.
46. Ohtani, H., Kao, C.T., Van Hove, M.A., and Somorjai, G.A., *Progr. Surf. Sci.*, 1987, *23*, 155.
47. Kolb, D.M., *Z. Phys. Chem. (Frankfurt)*, 1987, *154*, 179.
48. Toney, M.F. and Melroy, O.R., Surface X-ray scattering, in *Electrochemical Interfaces: Modern Techniques for In-Situ Interface Characterisation*, Abruña, H.D., Ed., VCH Publishers, New York, 1991, 55–129.
49. Gerischer, H., Semiconductor electrochemistry, in *Physical Chemistry*, Vol. IXA, Eyring, H., Henderson, D., and Jost, W., Eds., Academic Press, New York, 1970, 463–542.
50. Gerischer, H. and Mindt, W., *Electrochim. Acta*, 1968, *13*, 1329.
51. Gerischer, H., Allongue, P., and Costa-Kieling, V., *Ber. Bunsenges. Phys. Chem.*, 1993, *97*, 753.
52. Yeager, E.B. and Kuta, J., Techniques for the study of electrode processes, in *Physical Chemistry*, Vol. IXA, Eyring, H., Henderson, D., and Jost, W., Eds., Academic Press, New York, 1970, 3453–461.
53. Sand, H.J.S., *Z. Phys. Chem.*, 1900, *35*, 641.
54. Berzins, T. and Delahay, P., *J. Am. Chem. Soc.*, 1955, *77*, 6448.
55. Gerischer, H. and Krause, M., *Z. Phys. Chem. (Frankfurt)*, 1957, *10*, 264.
56. Gerischer, H. and Vielstich, W., *Z. Phys. Chem. (Frankfurt)*, 1955, *3*, 16.
57. Britz, G., *J. Electroanal. Chem.*, 1978, *88*, 309.
58. Vielstich, W. and Gerischer, H., *Z. Phys. Chem. (Frankfurt)*, 1955, *4*, 10.
59. Dolin, P. and Ershler, B.V., *Acta Physicochim. U.S.S.R.*, 1940, *13*, 747.
60. Randles, J.E.B., *Trans. Faraday Soc.*, 1948, *44*, 327.
61. Gerischer, H., *Z. Phys. Chem.*, 1951, *198*, 286.
62. Krishtalik, L.I., Hydrogen overvoltage and adsorption phenomena in *Advances in Electrochemistry and Electrochemical Engineering*, Vol. 7, Delahay, P. and Tobias, C.W., Eds., Interscience, New York, 1970, 283–339.
63. Balandin, A.A., Modern state of the multiplet theory in heterogeneous catalysis, in *Advances in Catalysis*, Vol. 19, Eley, D.D., Pines, M., and Weisz, P.B., Eds., Academic Press, New York, 1969, 103–210.
64. Tarasevich, M.R., Sadkowski, A., and Yeager, E.B., Oxygen electrochemistry, in *Comprehensive Treatise of Electrochemistry*, Vol. 7; Conway, B.E., Bockris, J.O'M., and Yeager, E.B., Eds., Plenum Press, New York, 1983, 301–398.
65. Berl, W.G., *Trans. Electrochem. Soc.*, 1939, *76*, 359; 1943, *83*, 253.
66. Gnanamuthu, D.S. and Petrocelli, J.V., *J. Electrochem. Soc.*, 1967, *114*, 1036.
67. Gerischer, H., *Ber. Bunsenges. Phys. Chem.*, 1980, *92*, 1325; *Spectroelectrochemistry*, Gale, R.J., Ed., Plenum Press, New York, 1988; *Electrochemical Interfaces: Modern Techniques for In-Situ Interface Characterisation*, Abruña, H.D., Ed., VCH Publishers, New York, 1991; Kolb, D.M., *Ber. Bunsenges. Phys. Chem.*, 1994, *98*, 1421.

Chapter 3

SOLID STATE BACKGROUND

Isaac Abrahams and Peter G. Bruce

CONTENTS

0-8493-8956-9/97/$0.00+$.50

I. INTRODUCTION

The purpose of this chapter is to provide a background to solid state chemistry, particularly in the field of solid state ionics. Much of the material used has been presented at undergraduate and postgraduate lectures, and hence the chapter is designed to take a reader with basic chemical knowledge and leave them with an understanding of the concepts and terminology used in this field. The reader is first introduced to the solid state, crystalline, and amorphous solids and the differences between molecular and nonmolecular solids. Basic concepts such as close packing, bonding, and crystallography are introduced and then used to interpret structures of mainly inorganic solids. The field of solid state ionics is, however, often concerned with solids which are defective or are solid solutions, and so the final sections of this chapter are devoted to discussing these topics. While we recognize it is impossible to give detailed coverage to all aspects of solid state chemistry within the pages of this chapter we have focused on those areas which are likely to be of interest to researchers and students in the field of solid state ionics. For a more comprehensive treatment of solid state chemistry we recommend other texts to readers such as West[1] or Rao and Gopalakrisnan.[2]

II. THE SOLID STATE

There are a number of ways of classifying solids, but perhaps the most useful division is into *crystalline* and *amorphous* solids. Crystalline solids have a regular repeating array of atoms characterized by a repeat unit known as the *unit cell.* Solids which do not show this regular repeating structure are classified as amorphous. It is important to consider here that many solids are incorrectly described as being amorphous, but are in fact microcrystalline or nanocrystalline with small crystallite sizes which fail to give crystalline X-ray diffraction patterns. However, often these materials can be confirmed as crystalline using electron diffraction, with lattice images routinely obtained from particles in the range of 5 nm. The fundamental differences between crystalline and amorphous solids are summarized in Table 3.1.

TABLE 3.1
Differences Between Crystalline and Amorphous Solids

Crystalline	Amorphous
Regular repeating structure characterized by unit cell	Irregular repeating structure with no defined unit cell
High degree of symmetry on atomic scale	Low symmetry on atomic scale
Long-range order	Short-range order only
Sharp melting point	Large melting point range
Sharp diffraction pattern	Diffuse diffraction pattern

A. AMORPHOUS SOLIDS

Unlike crystalline solids, noncrystalline or amorphous solids have no regular repeating structure, and X-ray or electron diffraction results in a broad diffuse pattern with no sharp peaks. Under certain cooling conditions from the melt, some compounds or mixtures of compounds form a supercooled liquid or glass. The atomic arrangement in glasses is truly amorphous with no regular repeating array of atoms, and the structure effectively represents a frozen liquid. Glasses can be synthesized with wide-ranging properties, including semiconductivity as in $Te_{0.8}Ge_{0.2}$ and superconductivity as in $Pb_{0.9}Cu_{0.1}$. Oxides such as SiO_2 and P_2O_5 are known as *network formers,* which give rise to a covalent framework to which *network modifiers* such as Li_2O and Ag_2O can be added to introduce ionic bonds and hence ionic conduction. Ionic conductivity in glasses is well known; for example, Li^+ ion conductivity in $LiAlSiO_4$ glasses, and Ag^+ ion conductivity in the AgI-Ag_2SeO_4 system.[3] Silver ion conducting glasses are of particular interest because of their very high ionic conductivities at low temperature, for example, 6×10^{-2} Scm^{-1} in 75% AgI– 25% Ag_2SeO_4.[4] Because there is no regular repeating structure, many of the properties of amorphous solids, for example, ionic conductivity, are not dependent on the orientation of the solid material and are hence said to be isotropic. The lack of long-range order means that pathways for ionic conduction are less well defined than in crystalline solids.

1. Glasses

Glasses can be defined as amorphous solids which show a glass transition.[5] On cooling a glass-forming compound from the melt it is possible under certain conditions to bypass the normal crystallization point and form a supercooled liquid or glass which is metastable to the crystalline phase. These transitions can be followed by a number of techniques, including dilatometry. Figure 3.1 shows a schematic representation of a hypothetical glass formation followed by dilatometry. The melting point of the solid T_m is indicated on the diagram. On slow cooling to this temperature, the volume rapidly decreases as the crystalline phase is formed. If, however, the liquid is rapidly cooled past T_m, it becomes gradually more viscous until it eventually solidifies at T_g, the glass transition temperature. Unlike crystallization, in glass formation there is no major discontinuity in volume, merely a change in slope. Further heating at temperatures below T_g can result in a structural relaxation of the glass which is accompanied by a contraction in volume. The exact position of T_g depends on the rate of cooling, and consequently may vary not only with compositional changes in the glass, but also with the preparative method. A convenient and relatively accurate way of monitoring glass formation is by thermal analysis; for example, differential thermal analysis (DTA). A schematic DTA trace for a glass-forming system on heating is shown in Figure 3.2. T_m is marked by a sharp endothermic peak, while T_g is a broader endotherm.

The exact reason why some systems form glasses and others do not is still the subject of debate. Zachariasen[6] in 1932 examined glass-forming oxides known at that time, i.e., SiO_2, GeO_2, B_2O_3, As_2O_3, and P_2O_5.[6] He argued that the coordination polyhedra in glasses would be similar to those found in crystalline structures. He suggested four rules for glass formation in oxides of the type A_xO.

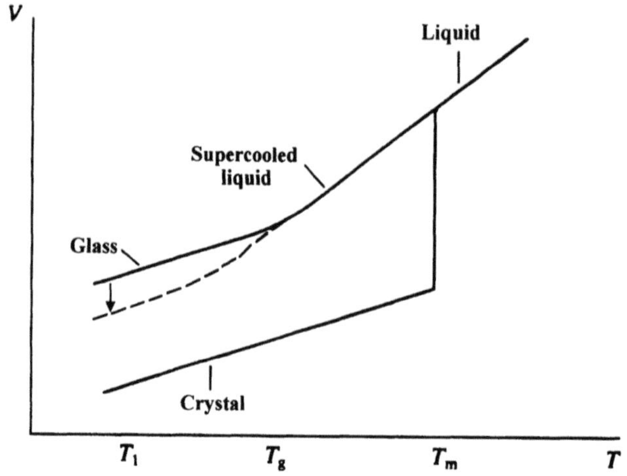

FIGURE 3.1. Schematic representation of the change in volume on glass formation of a supercooled liquid, where T_g and T_m are the glass and melting transition temperatures, respectively. The vertical arrow shows the change in volume due to structural relaxation when the temperature is held at T_1. (From Elliot, S.R., *Physics of Amorphous Materials*, 2nd ed., Longman, Harlow, England, 1990. With permission.)

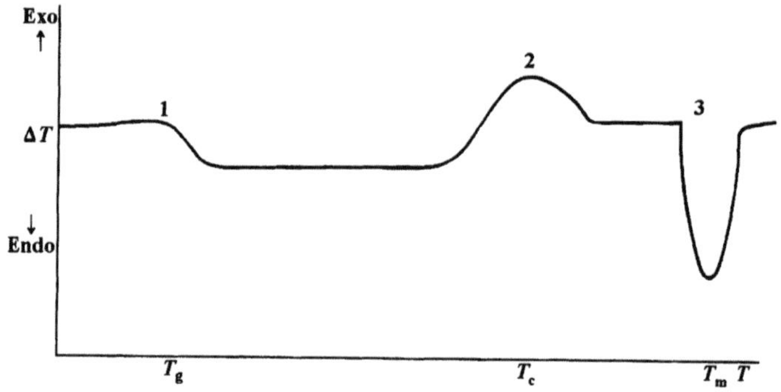

FIGURE 3.2. Schematic representation of a typical DTA trace on heating for a glass-forming system where points 1, 2, and 3 represent glass transition (T_g), crystallization (T_c), and melting point (T_m) temperatures, respectively. (From Elliot, S.R., *Physics of Amorphous Materials*, 2nd ed., Longman, Harlow, England, 1990. With permission.)

1. An oxygen atom may be linked to not more than two A atoms.
2. The coordination number (CN) of A must be small.
3. The oxygen polyhedra share only corners.
4. At least three corners in each oxygen polyhedron must be shared to result in a three-dimensional framework.

These rules therefore explain why AO and A_2O are nonglass forming. For example, MgO has the rock salt structure (Section V) with octahedral coordination for both Mg^{2+} and O^{2-}, thus violating rules 1 and 2; the MgO_6 polyhedra share edges, violating rule 3. If we now examine vitreous silica as an example of a system which obeys the rules, oxygen is linked to only two Si atoms (rule 1); the Si CN is four (rule 2); the oxygen polyhedra share only corners (rule 3), each polyhedron usually shares four corners to give the three-dimensional network (rule 4). Rule 3 has subsequently been shown to be redundant and in some cases false. Zachariasen[6] himself later modified his rules to take into account the addition of nonglass-forming oxides to glass systems of the type A_xB_yO. Although since Zachariasen's time glasses have been prepared which violate his rules (particularly rule 3), they are still useful in the understanding of simple glass systems.

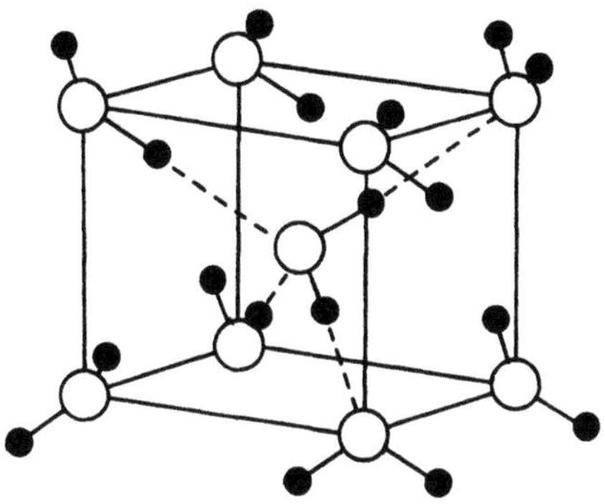

FIGURE 3.3. The structure of Ice–VIII, where H and O atoms are black and open circles, respectively. Intermolecular contacts are indicated by dashed lines. (Adapted from Wells, A.F., *Structural Inorganic Chemistry,* 5th ed., Clarendon Press, Oxford, 1984.)

B. CRYSTALLINE SOLIDS

Solids which possess a regular repeating crystal structure form the majority of solid materials in common use. These solids may be *nonmolecular* where the three-dimensional structure is formed from atoms or ions coming together with primary bonding (ionic, covalent or metallic) extending in one, two, or three dimensions. In contrast, *molecular solids* are characterized by the molecule remaining as a discrete entity within the solid, where intramolecular interactions are through primary bonding (usually covalent) and intermolecular contacts are weaker secondary interactions such as hydrogen bonding.

1. Molecular Solids

Molecular solids are characterized by having lower binding energies than nonmolecular solids, typically 0.04 to 0.4 eV mol^{-1}. This can lead to high rates of transport through molecular solids. Nonmolecular solids have binding energies typically an order of magnitude higher. Several types of electrostatic forces hold molecules in place in these solids and are collectively known as van der Waals forces. The most important types of interactions are permanent dipole and induced dipole; these are best illustrated by considering some simple examples.

Ice may adopt several crystalline forms dependent on temperature and pressure. The structure of ice-VIII is shown in Figure 3.3 and indicates that the water molecules remain discrete, while being held in the solid state by permanent dipole interactions between oxygen and hydrogen on neighboring molecules. The permanent dipole of the $^{\delta+}H–O^{\delta-}$ bond allows for electrostatic attraction between a hydrogen atom on one molecule and an oxygen on a neighboring molecule. The potential energy of attraction V_A between two dipoles of moments μ_A and μ_B at a distance r apart and at temperature T is proportional to:

$$V_A \propto \mu_A \mu_B r^{-6} T^{-1} \qquad (3.1)$$

The molecular halogens X_2 (where X = Cl, Br, and I) have no permanent dipole. However, in the solid state the close approach of neighboring molecules results in an induced dipole on the X-X bond vector. The layered structures of solid Cl_2, Br_2, and I_2 (Figure 3.4) show that the X_2 molecules remain discrete with shorter X-X intramolecular contacts (1.98, 2.27, 2.72 Å for Cl_2, Br_2, and I_2, respectively),[7] compared to the induced dipole interactions which show significantly longer contacts (3.32, 3.31, 3.50 intralayer and 3.74, 3.99, 4.27 Å interlayer for Cl_2, Br_2, and I_2, respectively). Induced dipole interactions depend very much on the polarizability of the atoms, and hence are dependent on atomic number.

Most organic and metal-organic compounds crystallize as molecular solids where the three-dimensional structure is constructed by the packing of molecules in the solid state,

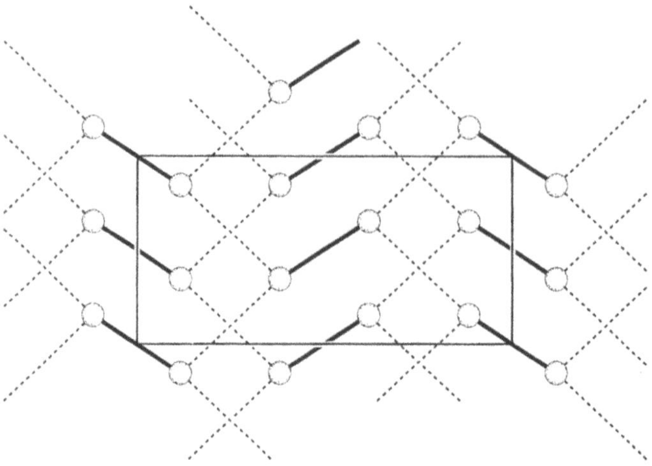

FIGURE 3.4. Structure of solid I_2: (a) two-dimensional projection showing intermolecular contacts (dashed lines) and intramolecular bonds (solid lines).

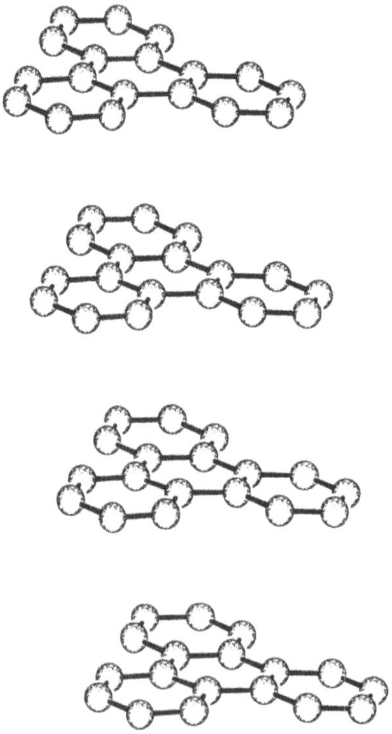

FIGURE 3.5. Structure of triphenylene showing stacking of aromatic rings.

which itself is influenced by the molecular stereochemistry and the nature of intermolecular interactions. For example, orthorhombic benzene at 138°K has a structure with benzene molecules almost perpendicular to each other.[9] This is in contrast to the structure of triphenylene $C_{18}H_{12}$, which packs in the solid state with unsaturated rings parallel and facing each other, which facilitates electron overlap of the π electron clouds (Figure 3.5).

When molecular ions have spherical or near spherical stereochemistry they often pack in a similar way to classical ionic solids. For example, in the fulleride K_6C_{60} the spherical C_{60}^{6-} ions pack in a standard body-centered cubic structure with potassium clusters located in the interstitial sites.[10]

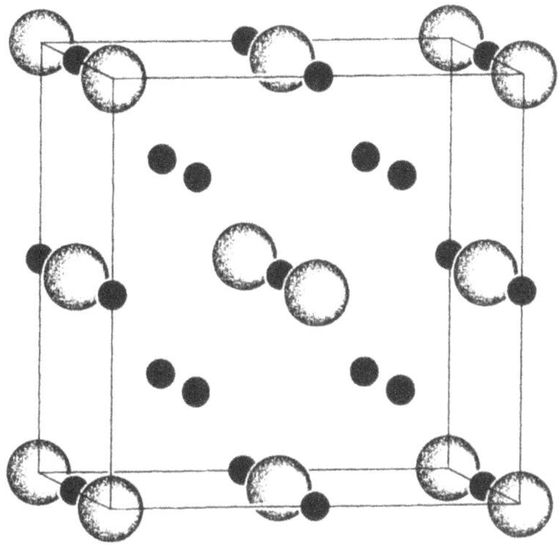

FIGURE 3.6. Unit cell contents of Li_3Bi. Large shaded circles are Bi atoms; smaller filled circles are Li atoms.

2. Nonmolecular Solids

Many crystalline inorganic solids are nonmolecular; these can be classified as metallic, covalent, or ionic. Let us deal briefly with each of these.

a. Metallic Solids

Metallic solids are generally metallic elements or alloys. These solids are usually formed between electropositive elements. The structures adopted by metals are usually close packed (Section V), for example, Ag, Au, Fe, and Pb show cubic close packing; Be, Co, and Mg show hexagonal close packing, while Ba, Cr, and K adopt the less dense body-centered cubic packing. In the case of alloys, where there is a significant difference in size between metal atoms, the smaller metal atoms may locate themselves in the interstitial sites, e.g., Li_3Bi may be described as cubic close-packed Bi with Li in all the interstitial sites (Figure 3.6).

b. Covalent Solids

Covalent bonding in solids results in highly directional bonds, with preferences for certain coordination geometries by particular elements. Covalent bonds are formed between elements where the difference in electronegativity is not as great as in ionic solids. In many real systems the true bonding character is somewhere in between the ionic and covalent extremes. Structures formed by covalent solids depend on valency and the desire to achieve maximum overlap of bonding orbitals. The relative atom/ion size plays a less important part than in ionic solids. For example, carbon has an outer electronic configuration of $2s^2\ 2p^2$. Carbon may form four sp^3 hybrid orbitals tetrahedrally displaced around the atom and achieves maximum overlap by adopting a tetrahedral arrangement of other carbon atoms around it. This results in the well-known diamond structure (Figure 3.7). Other forms of hybridization may also occur in carbon; for example, in graphite, carbon is considered to adopt sp^2 hybridization, where the trigonal planar arrangement of sp^2 hybrid orbitals overlap with neighboring carbon orbitals to give the characteristic hexagonal two-dimensional layer structure. The unhybridized p_z orbital is perpendicular to the hexagonal planes and can overlap with orbitals on neighboring carbon atoms to give a delocalized band structure, resulting in high electronic conductivity.

c. Ionic Solids

Ionic solids are formed between elements which have a large difference in electronegativity. The particular structures adopted by ionic solids are driven by a number of factors, including the desire to achieve maximum CN while maintaining local electroneutrality and

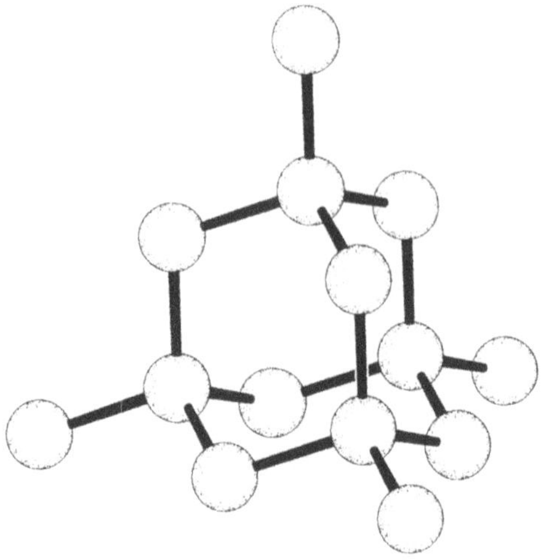

FIGURE 3.7. Diamond structure.

TABLE 3.2
Radius Ratio Limits for Polyhedral Coordinations

Maximum r^+/r^-	Predicted coordination
0.155	Linear
0.225	Trigonal planar
0.414	Tetrahedral
0.732	Octahedral/square planar
1.0	Cubic

minimizing like-charge repulsion. These factors are summarized by Pauling's rules for complex ionic crystals.[7]

Pauling's first rule states that the particular coordination adopted by ions in an ionic solid depends on the ratio of cation radius to anion radius r^+/r^-; this is often termed the *radius ratio rule*. Limits for regular polyhedral coordinations may be derived and are summarized in Table 3.2.

From Table 3.2 it can be seen that for values of r^+/r^- below 0.155 a linear coordination is predicted. As the cation size increases with respect to anion size, more anions are able to fit around the cation and hence higher coordinations are possible. The maximum limit for tetrahedral coordination is $r^+/r^- = 0.414$; above this, octahedral or square planar coordination is preferred up to a limit of $r^+/r^- = 0.732$. For solids where $r^+/r^- = 1$, a close-packed structure is predicted with a CN of 12. For solids where the anion is smaller than the cation it is more useful to use the anion to cation ratio r^-/r^+; for example, CsF has $r^+/r^- = 1.82/1.31 = 1.38$, while $r^-/r^+ = 0.72$;[11] CsF in fact adopts the rock salt structure (Section V).

Consider, for example, LiCl, where the radius ratio may be calculated as $r^+/r^- = 0.41$. The minimum radius ratio for octahedral coordination is 0.414, while that for cubic coordination is 0.732 and tetrahedral coordination 0.225. Indeed, LiCl adopts the six coordinate rock salt structure rather than that of eight coordinate CsCl or the four coordination of ZnS (see Section V). If, however, we consider RbCl, a radius ratio of 0.82 may be calculated and the CsCl structure predicted. However, RbCl is seen to retain the rock salt structure at ambient conditions. It is interesting to note that a phase transition to the CsCl structure can be achieved at relatively low pressures, reflecting the small energy difference between these two structures. Structures with radius ratios close to borderlines often show polymorphism, e.g., GeO_2 has

TABLE 3.3
Electrostatic Bond Strengths for Cations M^{n+}

	Coordination number					
Cation	**3**	**4**	**6**	**8**	**9**	**12**
M^{+}	1/3	1/4	1/6	1/8	1/9	1/12
M^{2+}	2/3	1/2	1/3	1/4	2/9	1/6
M^{3+}	1	3/4	1/2	3/8	1/3	1/4
M^{4+}	4/3	1	2/3	1/2	4/9	1/3

two structural forms, one in which Ge has a CN = 4 and the other with CN = 6. Although radius ratio rules are not strictly adhered to, particularly at values close to the r^+/r^- limits or where there is significant covalency present, they do serve as a useful guide in rationalizing ionic structures.

Pauling's second rule states that the sum of individual electrostatic bond strengths (e.b.s.) around a particular ion are equal in magnitude to the ion charge, i.e., that local electroneutrality is observed. The e.b.s. for a cation M^{m+} surrounded by nX^{x-} is given by:

$$\text{e.b.s.} = \frac{m}{n} \tag{3.2}$$

Consider, for example, rutile TiO_2 (see Section V). Each Ti^{4+} is surrounded by six O^{2-} ions; therefore, the e.b.s. = 4/6 or 2/3. Each oxygen is bonded to three titaniums, the sum of the individual e.b.s. values is 6/3 = 2, i.e., equal in magnitude to the formal charge on O. If the structure contained tetrahedral Ti, i.e., only 4 Ti–O bonds, the e.b.s. would be 4/4 = 1. If the oxygen was still bonded to three titaniums, then the sum of the e.b.s. would be three, i.e., greater in magnitude than the formal oxidation state of oxygen, and there would be a local imbalance of charges. If, however, each oxygen was only bonded to two Ti, then the sum of the e.b.s. would be two and local electroneutrality would be maintained. Hence the e.b.s. allows for some rationalization of polyhedral linkages. Some e.b.s. values are summarized in Table 3.3.

Pauling's third rule concerns the minimization of repulsions between like charges, particularly between atom-sharing polyhedra. Thus vertex sharing is energetically more favorable than edge sharing, which is more favorable than face sharing. This is particularly important in structures where ions are highly charged. Of course, edge and face sharing regularly occur in inorganic solids, but often some distortion of regular site symmetry is observed. In more complex structures, those ions with high charge and small CN tend not to be linked to one another, i.e., maximize their separation (Pauling's fourth rule).

III. LATTICE ENERGY

Lattice energies represent a key factor in the formation of ionic solids. The lattice energy U may be defined as the energy required at 0 K to convert 1 mol of a crystal into its constituent ions at infinite separation in the gas phase.

$$MX(s) \rightarrow M^{+}(g) + X^{-}(g) \qquad \Delta H = U$$

The lattice energy of an ionic crystal is derived from a balance of attractive and repulsive forces within the crystal. The strength of attraction and repulsion between ions are represented by potential functions. The simplest of these is based on Coulomb's law which essentially states that like charges repel and unlike charges attract. So for two ions A and B at an internuclear separation r, the coulombic force of attraction for unlike ions and repulsion for like ions is given by:

$$F = \frac{Z_A e Z_B e}{r^2} \tag{3.3}$$

where Z_Ae and Z_Be are the charges for atoms A and B, respectively. If we now consider only the attractive force between an anion of charge Z^-e and a cation of charge Z^+e, the potential energy of attraction V is given by:

$$V = \int_\infty^r F\ dr = -\frac{Z^+Z^-e^2}{r} \tag{3.4}$$

In the solid state it is necessary to take into account the interactions with next nearest neighbor ions, and this will depend on crystal structure. Several methods have evolved for calculating these interactions. The simplest method, known as Evjen's method,[12] relies on the summation of the Coulomb potentials between a designated central ion and all of its neighbors. For example, in NaCl, consider a Na^+ ion located at the origin. Its nearest neighbors are six Cl^- ions located at distance r (half the unit cell edge). From Equation 3.4 the attractive potential energy between these ions is given by:

$$V = -6\frac{Z^+Z^-e^2}{r} \tag{3.5}$$

The next nearest neighbors are 12 Na^+ ions at a distance $\sqrt{2}$ r, which exert a repulsive force with potential energy given by:

$$V = 12\frac{Z^+Z^-e^2}{\sqrt{2}\ r} \tag{3.6}$$

The third nearest neighbors are eight Cl^- ions at a distance $\sqrt{3}$ r with a potential energy of:

$$V = -8\frac{Z^+Z^-e^2}{\sqrt{3}\ r} \tag{3.7}$$

The net potential is therefore an infinite series:

$$V = -\frac{Z^+Z^-\ e^2}{r}\left(6 - \frac{12}{\sqrt{2}} + \frac{8}{\sqrt{3}} - \frac{6}{\sqrt{4}} + \dots\dots\dots\right) \tag{3.8}$$

The term in brackets is known as the Madelung constant, A. This series, however, shows poor convergence. Better convergence is achieved by multiplying individual terms in the series by the volume fraction of each ion that lies within the coordination shell being considered. The terms corresponding to lower coordination shells contribute wholly, while the terms corresponding to the ions at the shell edge are multiplied by the volume fraction that lies within that shell. Thus in NaCl, if we consider all the ions up to the second coordination shell, the Madelung constant becomes:

TABLE 3.4
Madelung Constants for Selected Structure Types

Structure type	A	Structure type	A	Structure type	A
CsCl	1·76267	CaF_2	5·03879	TiO_2 (rutile, M^{2+} X_2^-)	4·816
NaCl	1·74756	$CaCl_2$	4·730	TiO_2 (anatase, M^{2+} X_2^-)	4·800
ZnS (Wurtzite)	1·64132	$CdCl_2$	4·489	SiO_2 (β-quartz, M^{2+} X_2^-)	4·4394
ZnS (sphalerite)	1.63805	CdI_2	4·383	Al_2O_3 (corundum)	25·0312
PdO (*c/a* = 2·0)	1·60494	Cu_2O	4·44249		

Data from Greenwood, N.N., *Ionic Crystals Lattice Defects and Nonstoichiometry*, Butterworths, London, 1968. With permission.

$$A = 6 - \frac{12}{\sqrt{2}} + \frac{8}{\sqrt{3}} - \frac{6}{2\sqrt{4}} + \frac{24}{2\sqrt{5}} - \frac{24}{2\sqrt{6}} + \frac{12}{4\sqrt{8}} - \frac{24}{4\sqrt{9}} + \frac{8}{8\sqrt{12}}$$
$$= 1.750$$

By increasing the dimensions of the shell, a rapid convergence is achieved to a value of 1.748 in NaCl.

An alternative to the Evjen method is described by Ewald[13] and formulated by Tosi.[14] This method involves a summation over the reciprocal lattice and, unlike Evjen's method, is valid over any point in the crystal, not just the lattice sites, and so is important for calculations involving lattice defects or interstitial ions. Van Gool and Piken[15] have developed a program for automatically calculating lattice energies and Madelung constants based on this method and have used this on compounds such as β-alumina.[16] The Madelung term remains constant for a particular structure type. Some values are listed in Table 3.4.[17]

Equation 3.4 can now be modified to take into account the geometric factors described by the Madelung constant. Multiplication by Avogadro's number N gives the net attraction per mole of crystal.

$$V = -\frac{NAZ^+Z^-e^2}{r} \tag{3.9}$$

If we now include a term for the short-range repulsive forces we get the Born–Mayer equation for lattice energy:

$$U = \frac{NAZ^+Z^-e^2}{r_o}\left(1 - \rho/r_o\right) \tag{3.10}$$

where r_o is the equilibrium interatomic distance and ρ is a constant with a typical value of 0.35. Further refinements on the Born–Mayer equation are possible to account for crystal zero point energy and van der Waals forces; readers are here referred to an early review on lattice energy calculation by Waddington.[18]

The simple two-body approach to interionic potentials (Equation 3.4) is only really applicable to simple ideal ionic solids. In most solids, other factors such as lattice dynamics and polarization interactions need to be taken into account. The calculation of accurate lattice energies in these systems therefore depends very much on the use of improved models for interionic potentials. Several methods have been proposed for the calculation of these potentials and have been reviewed elsewhere.[19] One model which has been extensively used is based on the method proposed by Dick and Overhauser,[20] and has become known as the shell model, in which each ion is composed of a charged core and an electronic shell which are coupled harmonically and isotropically. In this model the repulsive forces are considered to be a function not only of internuclear separation r, but also the dipole moments and polarizabilities of the individual ions. Data for analysis are available in the form of elastic and dielectric constants which yield information on the attractive forces between nearest neighbor ions and repulsive forces between next nearest neighbor ions, and ion polarizabilities.

Lattice energies cannot generally be measured directly. An alternative to the direct calculations outlined above is to indirectly calculate lattice energy through measurement of other physical properties. A thermochemical cycle may be constructed and is known as the Born–Haber cycle. Consider the lattice energy of a crystal M^+X^-. Several individual steps may be defined:

Sublimation of solid M,	$M(s) \rightarrow M(g)$	S
Ionization of gas M,	$M(g) \rightarrow M(g)^+$	IP
Dissociation of gas X_2,	$1/2\ X_2(g) \rightarrow X(g)$	1/2D
– Electron affinity,	$X(g) \rightarrow X(g)^-$	–EA
Heat of formation,	$M(s) + 1/2\ X_2(g) \rightarrow MX(s)$	ΔH_f

Arranging these in a thermochemical cycle:

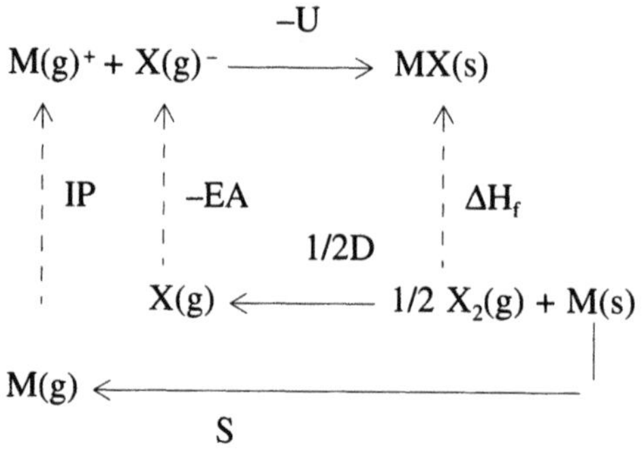

Therefore from Hess's law:

$$U = S + 1/2\,D + IP - EA - \Delta H_f \tag{3.11}$$

Using NaCl as an example, U may be calculated from the following data:

$-\Delta H_f$	=	410.9 kJ mol^{-1}
S	=	109 kJ mol^{-1}
1/2D	=	121 kJ mol^{-1}
IP	=	493.7 kJ mol^{-1}
–EA	=	–356 kJ mol^{-1}

Therefore, U for NaCl may be calculated as –778.6 kJ mol^{-1}

Calculation of lattice energies in this way not only provides a good check on values derived from direct calculations, but also allows for calculation of lattice energies of hypothetical compounds.

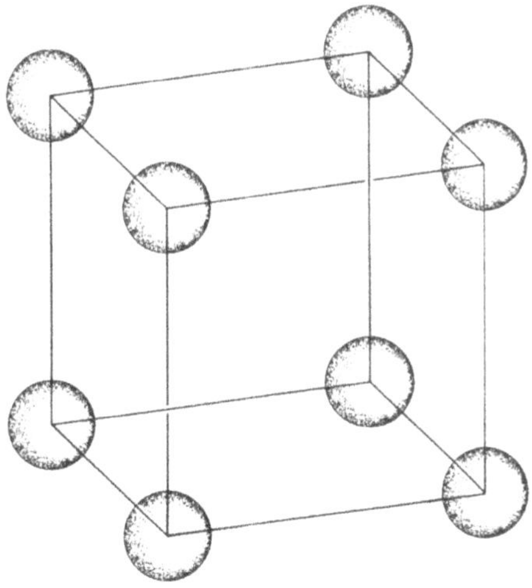

FIGURE 3.8. Primitive unit cell.

IV. THE CRYSTAL LATTICE AND UNIT CELLS

The regular repeating structure of crystalline solids may be represented by imposing an imaginary three-dimensional grid onto the crystal structure. This grid, known as the *crystal lattice,* has points of intersection or *lattice points* which act as points of origin for symmetry within the crystal. In the simplest or *primitive* lattice, the basic repeating unit, defined by eight lattice points, is known as the unit cell (Figure 3.8). Unit cells are the smallest repeating units that possess all the symmetry of the crystal structure and, when repeated in all directions, cover all the space in the crystal. The unit cell may be described by six parameters; the axial lengths of the parallel sides a, b, and c and the interaxial angles α, β, and γ, where α is the angle between b and c, β is the angle between a and c, and γ is the angle between a and b. Location of the origin of the unit cell is arbitrary, but is normally chosen to coincide with the center of symmetry (in centrosymmetric structures).

The *crystal system* relates the shape and symmetry of the unit cell. Table 3.5 summarizes the seven crystal systems. The simplest and lowest symmetry is triclinic, while the highest is cubic with four threefold axes. It is important to note that the symbol ≠ used in this context means not necessarily unequal, as in some cases coincidental pseudosymmetry may arise with cell dimensions resembling those of a higher symmetry crystal system, but without the required symmetry elements within the unit cell.

TABLE 3.5
The Seven Crystal Systems

Crystal system	Unit cell shape	Essential symmetry
Triclinic	$a \neq b \neq c$; $\alpha \neq \beta \neq \gamma \neq 90°$	None
Monoclinic (standard setting)	$a \neq b \neq c$; $\alpha = \gamma = 90°$, $\beta \neq 90°$	One twofold axis or mirror plane
Orthorhombic	$a \neq b \neq c$; $\alpha = \beta = \gamma = 90°$	Three twofold axes or mirror planes
Tetragonal	$a = b \neq c$; $\alpha = \beta = \gamma = 90°$	One fourfold axis
Trigonal		One threefold axis
(rhombohedral setting)	$a = b = c$; $\alpha = \beta = \gamma \neq 90°$	
(hexagonal setting)	$a = b \neq c$; $\alpha \neq \beta \neq 90°, \gamma = 120°$	
Hexagonal	$a = b \neq c$; $\alpha \neq \beta \neq 90°, \gamma = 120°$	One sixfold axis
Cubic	$a = b = c$; $\alpha = \beta = \gamma = 90°$	Four threefold axes

TABLE 3.6
Types of Centering Shown by Crystalline Solids

Symbol	Centering	Extra lattice points	Lattice points per cell
P	Primitive	None	1
A	A face centered	At center of b/c face	2
B	B face centered	At center of a/c face	2
C	C face centered	At center of a/b face	2
F	Face centered	At center of all faces	4
I	Body centered	At center of unit cell	2
R	Rhombohedral (hexagonal setting)	At (1/3,2/3/,2/3) and (2/3,1/3,1/3)	3

In some structures the lattice has additional lattice points which occur either at the center of unit cell faces or at the body center of the unit cell. This condition, known as *centering,* occurs in several forms (Table 3.6).

Up to 14 unique combinations of crystal systems with centering conditions are possible, and these combinations are known as the 14 *Bravais lattices* (Figure 3.9). The Bravais lattices therefore give general information on overall symmetry in the crystal, but contain no detail on point or translational symmetry. This further symmetry information when combined with the 14 Bravais lattices results in 230 unique combinations known as *space groups*. The 230 space groups and their symmetries are tabulated in *International Tables for Crystallography.*[21] Every crystal structure belongs to one and only one space group.

The positions of atoms in a crystal can be described with respect to the unit cell using *fractional coordinates*. Consider a point P located within a unit cell (Figure 3.10). To reach P one has to travel a distance X along a, Y along b, and Z along c. Thus, position P may be described by three fractional coordinates x, y, z, where:

$$x = X/a \ (X = \text{distance along a-axis})$$
$$y = Y/b \ (Y = \text{distance along b-axis})$$
$$z = Z/c \ (Z = \text{distance along c-axis})$$

In this scheme the origin is defined as 0,0,0 and the body center as 0.5,0.5,0.5. For a more detailed treatment of basic crystallography, see for example References 7, 22, and 23.

V. CLOSE PACKING

The crystal structures of nonmolecular solids can often be described in simple terms with reference to the close packing of spheres. In particular, many metals adopt the structures of simple close packing of identical spheres. In the case of metals, close-packed structures enable close contact between metal atoms and allow for the conduction of the outer electrons, which gives rise to basic metallic properties such as electronic and thermal conduction.

Consider a row of identical spheres (Figure 3.11a). Each sphere is in close contact with two others, i.e., the CN for each sphere is 2. If several of these one-dimensional close-packed arrays are brought together, a two-dimensional close-packed array (Figure 3.11b) results, with each sphere now immediately surrounded by six equidistant neighbors, CN = 6. This contrasts with the nonclose-packed situation in a square-packed array where CN = 4 (Figure 3.11c). If two of these two-dimensional, close-packed arrays are stacked one on another, then the spheres from the upper layer fit into the recesses of the lower layer and vice versa. If a third layer is introduced the orientation of the third layer can be such that its spheres are aligned with those in the first layer to give an ...ABABABA... stacking sequence known as hexagonal close packing, hcp. Alternatively, the third layer may enter a new orientation to give an

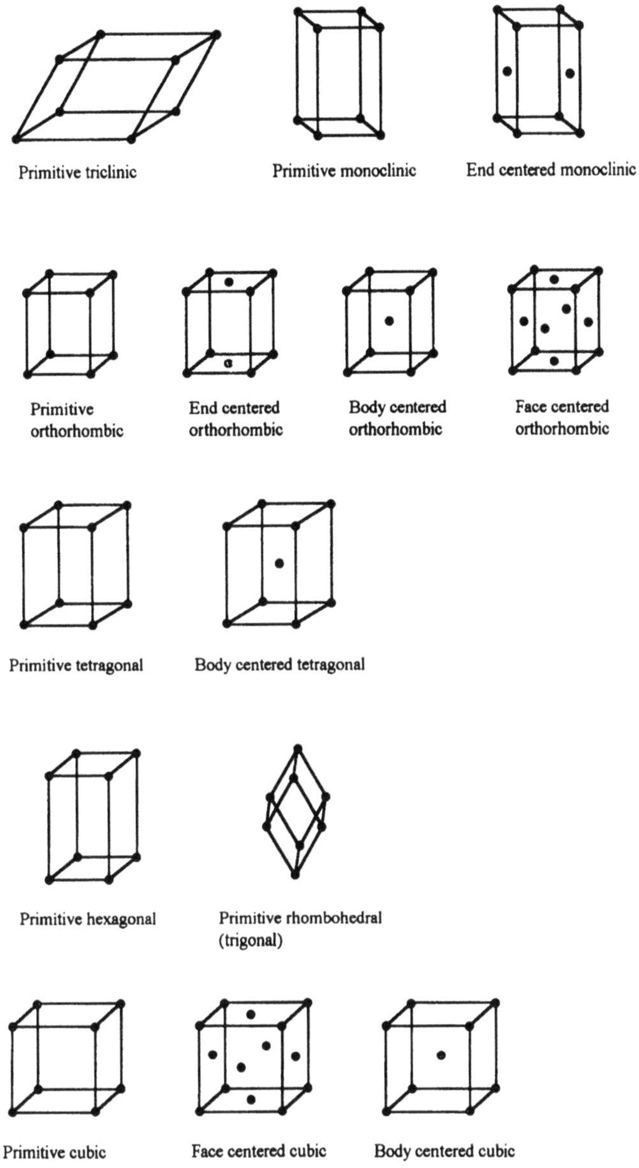

FIGURE 3.9. The 14 Bravais lattices.

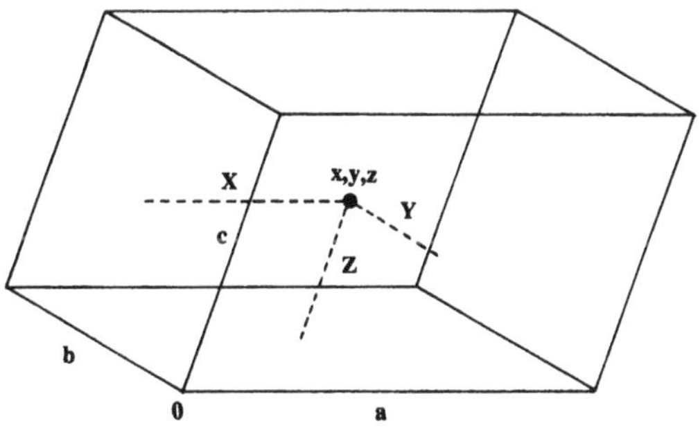

FIGURE 3.10. Fractional coordinate system with respect to unit cell dimensions a, b, c. Fractional coordinates x,y,z are defined by $x = X/a$, $y = Y/b$, $z = Z/c$.

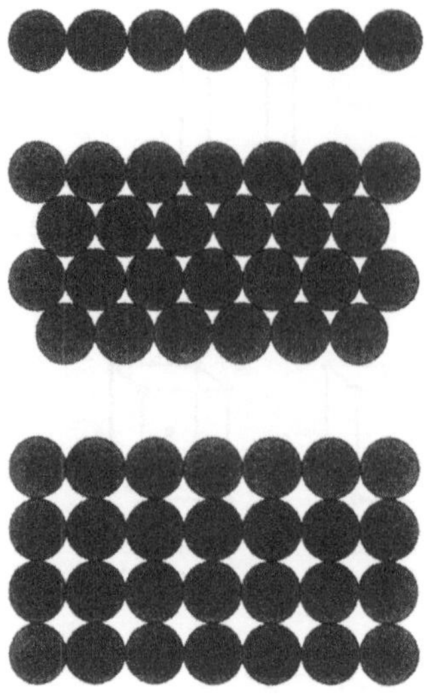

FIGURE 3.11. Close packing in spheres. (a) One-dimensional close packing CN = 2; (b) two-dimensional close packing CN = 6; (c) nonclose-packed arrangement of spheres (square packing) CN = 4.

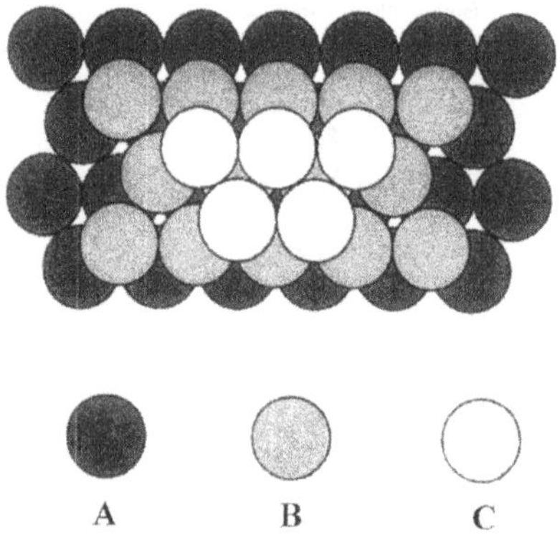

FIGURE 3.12. Close packing in three dimensions. Layer A (dark shading) represents lowest layer; layer B (medium shading) is second layer; layer C (light shading) is the arrangement adopted for the third layer in cubic close packing. Alternatively, the third layer may adopt a position directly in line with layer A as in hexagonal close packing.

...ABCABCABC... stacking sequence known as cubic close packing, ccp; in both cases CN = 12. The stacking sequences are summarized in Figure 3.12.

Like all crystalline structures, those based on close packing can be described entirely by a basic repeating unit or unit cell. Considering first hcp, the hexagonal unit cell is easily visualized with the c-axis running perpendicular to the close-packed layers. Thus the unit cell has a close-packed atom at each vertex and one in the unit cell body with fractional coordinates 1/3,2/3,1/2, i.e., two close-packed atoms per hexagonal unit cell (Figure 3.13). The ccp unit cell is slightly more difficult to visualize with respect to the close-packed layers which run perpendicular to the body diagonals of the cubic unit cell. The unit cell is therefore face centered, i.e., an atom at each vertex and at the center of each face giving a total unit cell content of four close-packed atoms (Figure 3.14). Both hcp and ccp have the same amount of space occupied by spheres with a packing density of 74.02%, which is therefore the highest that can be achieved for close packing of identical spheres. One important difference between the two forms of close packing is that in hcp there is only one close-packing direction,

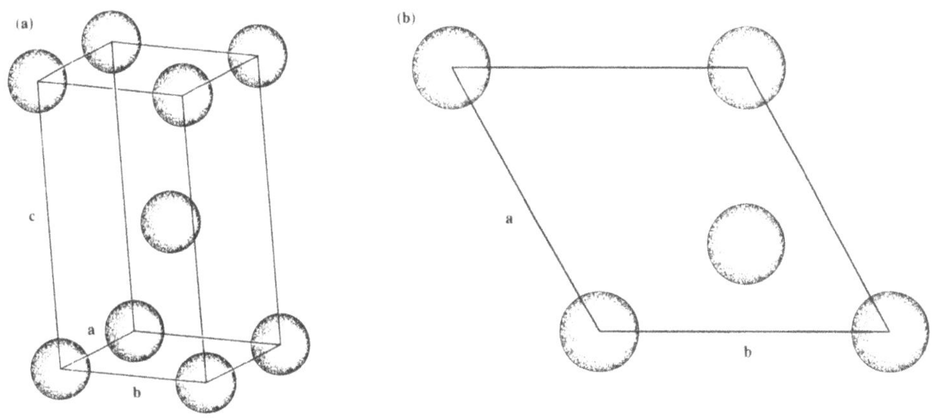

FIGURE 3.13. Unit cell contents for hcp atoms. (a) Three-dimensional view; (b) projection down c-axis.

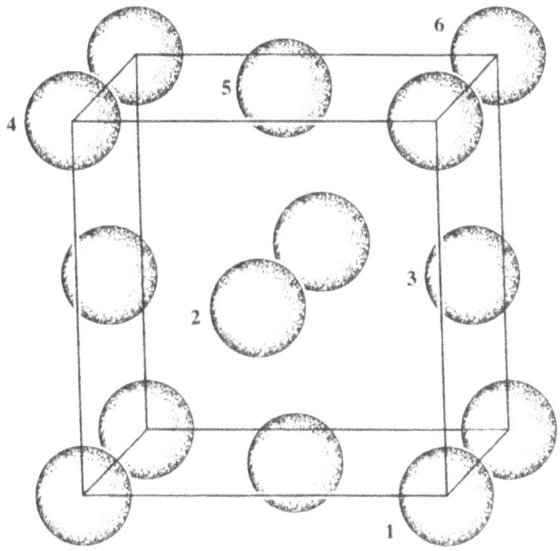

FIGURE 3.14. Unit cell contents for a cubic close-packed system. A single close-packed layer is indicated by atoms numbered 1 to 6.

whereas in ccp there are four equivalent close-packed directions; therefore, these configurations are termed anisotropic and isotropic packing, respectively.

Another form of packing adopted by inorganic structures is body-centered cubic (bcc) packing (Figure 3.15). It is not a close-packed arrangement, with only eight coordination being achieved and a 68.02% packing density. The cubic unit cell has an atom at each vertex and one at the body center, giving a total of two atoms per cell. Some metals adopt bcc-packed structures, for example, α-Fe, Na, and W. In addition, the structures of some ionic compounds can be derived from bcc packing; for example, the structure of CsCl can be derived from the bcc unit cell, with Cl atoms at the cell vertices and Cs at the body-centered position. Thus each Cs is surrounded by eight equidistant Cl and each Cl is coordinated to eight equidistant Cs, hence the structure is often termed 8:8. Other compounds which adopt this structure include CsBr, TlCl, TlBr, NH_4Cl, $CsNH_2$, and TlCN.

A. INTERSTITIAL SITES

As described above, even in true close-packed structures only 74.02% of the available space is used by the close-packed atoms. This leaves 25.98% of interstitial space which is

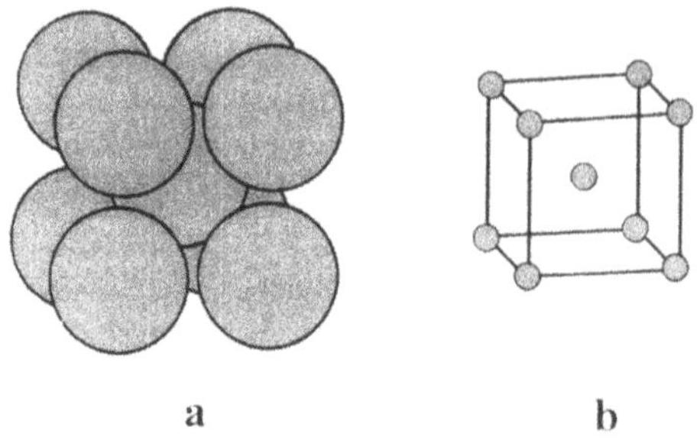

FIGURE 3.15. Body-centered packing: (a) space filling model; (b) unit cell contents.

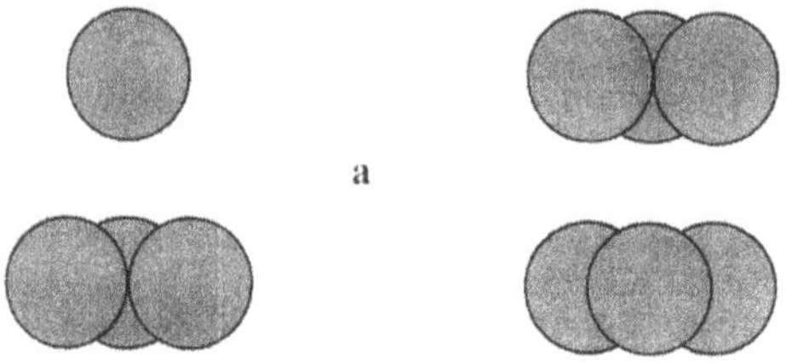

FIGURE 3.16. Interstitial sites between close-packed layers. (a) Tetrahedral site, (b) octahedral site.

available for occupation by suitably sized atoms or ions to give inorganic compounds. In regular close-packed arrays, two types of interstitial sites are found (Figure 3.16), viz, tetrahedral and octahedral interstitial sites. A tetrahedral site occurs where a sphere from one layer fits into a recess of the next layer; where two recesses come together in facing layers, an interstitial octahedral site results. It is helpful to relate these sites to the close-packed unit cells. In the hexagonal close-packed cell, two tetrahedral sites occur along the c-axis with fractional coordinates 0,0,3/8 and 0,0,5/8 and a further two sites occur within the unit cell body at 1/3,2/3,1/8 and 1/3,2/3,7/8 (Figure 3.17a). The octahedral sites are located above and below the close-packed atom at 2/3,1/3,1/4 and 2/3,1/3,3/4 (Figure 3.17b). Thus there are four tetrahedral and two octahedral interstitial sites per hexagonal close-packed unit cell. In the ccp cell the tetrahedral sites are located in the center of each octant (Figure 3.18a), while the octahedral sites are located halfway along each cell edge and at the body center (Figure 3.18b). Therefore, there are eight tetrahedral and four octahedral interstitial sites per ccp cell. Hence, the number and type of interstitial sites per close-packed atom is independent of the close-packed arrangement adopted, i.e., two tetrahedral and one octahedral interstitial site per close-packed atom for both ccp and hcp (Table 3.7).

In close-packed lattices, the relative sizes of octahedral and tetrahedral sites are important when considering the energetics of compound formation. The radius of an interstitial site, r_I, is dependent on that of the close-packed atom r_{cp} such that $r_I/r_{cp} = 0.225$ and 0.414 for tetrahedral and octahedral sites, respectively.

B. POLYHEDRAL REPRESENTATIONS OF CLOSE PACKING

The models described so far have emphasized the close-packed atoms. The solid state electrochemist, however, is very often more concerned with the passage of mobile ions from site to site. The use of polyhedral representations to emphasize the interstitial sites rather than the close-packed atoms allows for easier interpretation of intersite connectivities and conduction pathways. In this type of model the center of a polyhedron represents the center

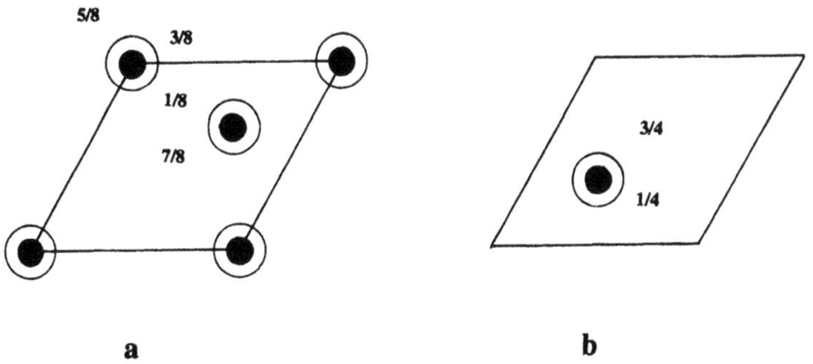

FIGURE 3.17. Interstitial sites in hcp unit cell, projections down c-axis. (a) Tetrahedral sites, (b) octahedral sites. c-Axis heights are given.

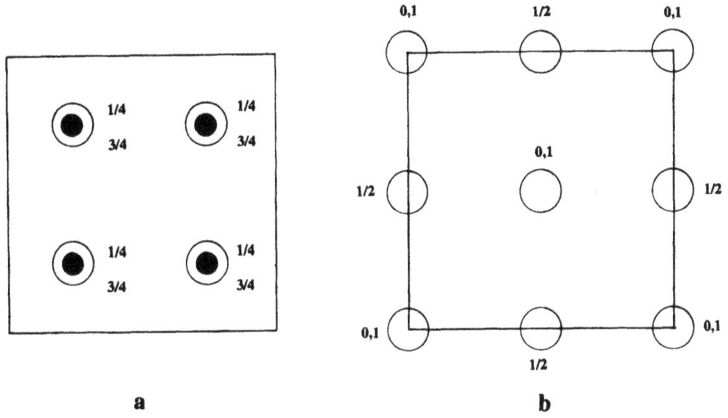

FIGURE 3.18. Interstitial sites in ccp unit cell, two-dimensional projection. (a) Tetrahedral sites, (b) octahedral sites. Vertical axis heights are given.

TABLE 3.7
Numbers of Interstitial Sites in Close-Packed Geometries

Packing	Close-packed atoms per cell	Tetrahedral sites per cell	Octahedral sites per cell	Tetrahedral sites per close-packed atom	Octahedral sites per close-packed atom
ccp	4	8	4	2	1
hcp	2	4	2	2	1

of a site, while the close-packed atoms are reduced to become the vertices of the polyhedron. Therefore, two close-packed layers are represented by a single layer of polyhedra. Each polyhedral layer contains tetrahedra and octahedra in a 2:1 ratio. Within a single layer, tetrahedra share edges with other tetrahedra and faces with octahedra. Similarly, octahedra share only edges with other octahedra. Addition of a second layer of polyhedra, i.e., a third close-packed layer, results in face sharing of like polyhedra in hcp, but only edge sharing in ccp. Thus in hcp chains of face sharing octahedral sites are formed with chains running parallel to the close-packing direction. In a similar way the tetrahedral sites alternate between

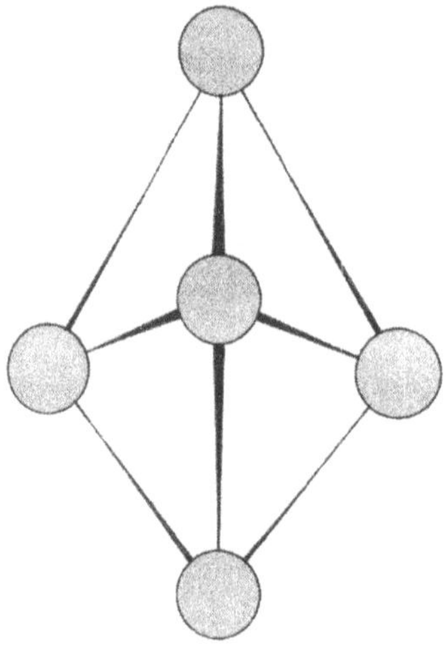

FIGURE 3.19. Interstitial five coordinate site formed by face sharing of two tetrahedra in hcp.

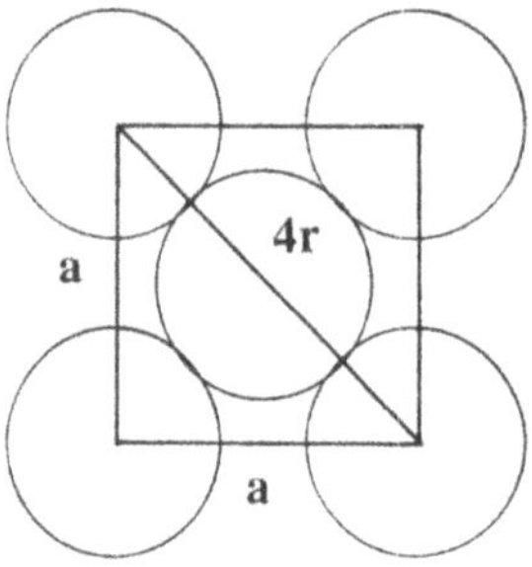

FIGURE 3.20. Two-dimensional projection of fcc unit cell. Face diagonal = $4r_{cp\ ion}$.

face and vertex, sharing in the close-packing direction. This is important in some structures because it results in a unique five-coordinate site formed by a face-sharing pair of tetrahedra (Figure 3.19).

C. STRUCTURES BASED ON CLOSE-PACKING

In real solids, true or near close packing is rare and is generally limited to metallic or intermetallic compounds. Ionic solids based on close-packing geometry do not show true close packing, because this would necessitate close contact between like charges. In a true ccp unit cell, the length of the face diagonal would be equivalent to four times the radius of the close-packed atom r_{cp} (Figure 3.20). In NaCl the cubic unit cell dimension a = 5.6402 Å. Using Pythagoras the face diagonal = $\sqrt{2a^2}$ = 7.9764 Å, which makes rCl^- = 1.994 Å. This compares with a literature value for rCl^- of 1.81 Å.[12] It can therefore be concluded that the structure of NaCl does not show true close packing of Cl^- ions. Despite this, many structures can be described in terms of close packing geometry of one sublattice, with the other sublattice occupying fractions of the interstitial sites. Some important structures are described below.

1. Structures Based on hcp

a. NiAs

Nickel arsenide (Figure 3.21) can be described in terms of hexagonal close-packed arsenic atoms with nickel in all the octahedral sites. This results in columns of face-sharing $NiAs_6$ octahedra, with columns sharing edges with each other. Each Ni atom has six As atoms as

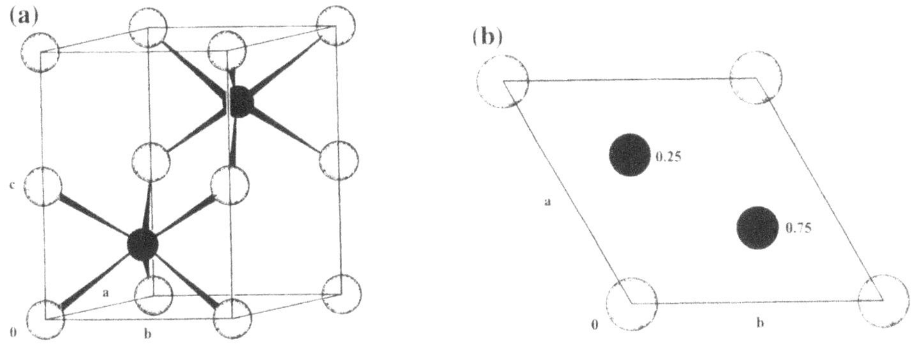

FIGURE 3.21. NiAs structure showing Ni (black) and As (shaded) atoms. (a) Unit cell contents, (b) view down c-axis with vertical heights given.

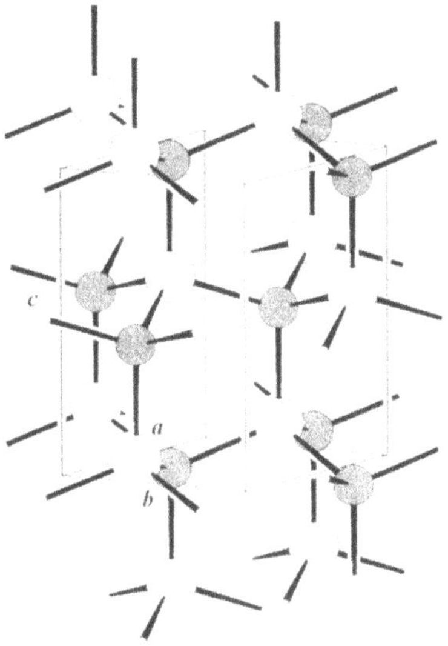

FIGURE 3.22. Structure of wurtzite showing Zn (shaded circles) and S (open circles) coordination.

nearest neighbors at the corners of an octahedron and two Ni atoms immediately above and below. The Ni–Ni distance of 2.50 Å (c/2) is only marginally longer than the Ni–As distance of 2.43 Å. This facilitates Ni–Ni bonding and probably accounts for the low c/a cell parameter ratio of 1.39 compared with a value of 1.63 for an ideal close-packed system. The arsenic atoms are also six coordinate, but are in trigonal prismatic sites. The As atoms are far from true close-packed — the As–As distance of 3.6 Å is much greater than the value predicted from the sum of the atomic radii of 2.50 Å. Other structures which adopt the NiAs structure include VS, FeS, and NiS.

b. ZnS (Wurtzite)

The hexagonal close-packed form of ZnS known as wurtzite (Figure 3.22) consists of hcp S^{2-} ions with Zn^{2+} in half the tetrahedral sites. The sulfide coordination is also tetrahedral. The corner-sharing Zn tetrahedra all point in the same direction. The structure is commonly adopted by II–VI and III–V solids such as CdS and InN. Other compounds with the wurtzite structure include ZnO (zincite), BeO, CdSe, AgI, MgTe, and MnS.

2. Structures Based on ccp

a. NaCl (Rock Salt)

The NaCl or rock salt structure (Figure 3.23) may be described as ccp Cl^- ions with Na^+ in all the octahedral sites (or equally, ccp Na^+ with Cl^- in the octahedral sites). The rock salt structure may be regarded as the ccp equivalent to NiAs. Rock salt is therefore a 6:6 structure, i.e., with both Na^+ and Cl^- six coordinate. The resulting structure contains $NaCl_6$ octahedra which share all 12 edges with other octahedra. Each octahedral face is parallel to a close-packed layer and marks the shared face with a vacant tetrahedral site. The important difference between NiAs and NaCl lies in the distance between the centers of the octahedral sites. In NiAs where the octahedral sites share faces rather than edges, the short contact between the centers of the octahedral sites is stabilized by Ni–Ni interactions. In NaCl no such interaction is present. Many binary structures adopt the NaCl framework and are summarized in Table 3.8.

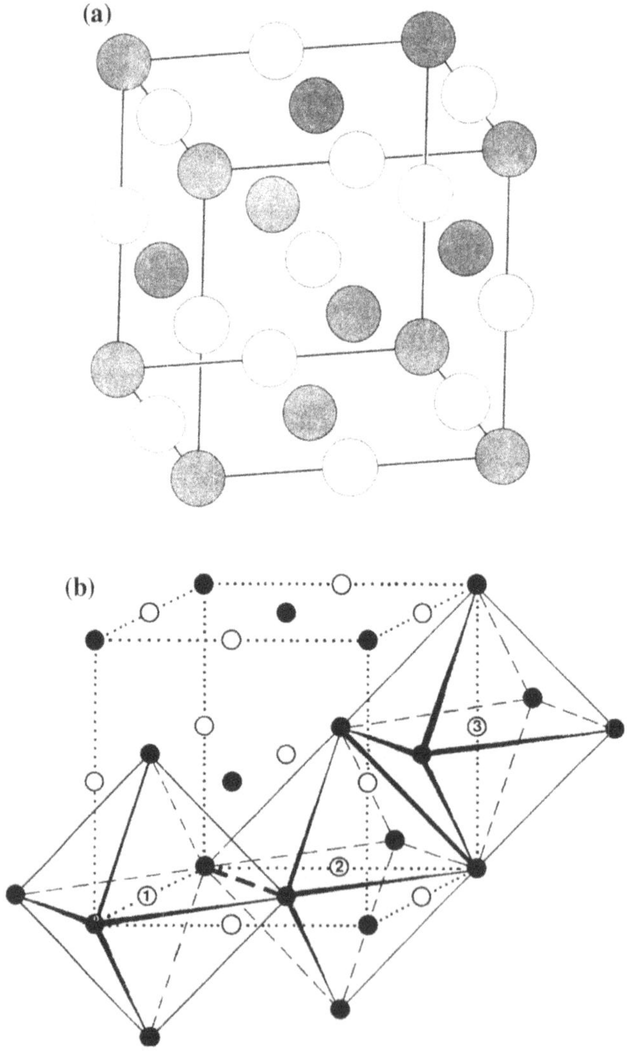

FIGURE 3.23. NaCl structure. (a) Unit cell contents of NaCl, shaded atoms represent Na, and unshaded Cl ions. (b) Edge-sharing octahedra in NaCl. (From West, A.R., *Solid State Chemistry and Its Applications*, John Wiley & Sons, New York, 1984. With permission.)

TABLE 3.8
Compounds with the NaCl Structure

	a(Å)		a(Å)		a(Å)		a(Å)
MgO	4.213	MgS	5.200	LiF	4.0270	KF	5.347
CaO	4.8105	CaS	5.6948	LiCl	5.1396	KCl	6.2931
SrO	5.160	SrS	6.020	LiBr	5.5013	KBr	6.5966
BaO	5.539	BaS	6.386	LiI	6.00	KI	7.0655
TiO	4.177	αMnS	5.224	LiH	4.083	RbF	5.6516
MnO	4.445	MgSe	5.462	NaF	4.64	RbCl	6.5810
FeO	4.307	CaSe	5.924	NaCl	5.6402	RbBr	6.889
CoO	4.260	SrSe	6.246	NaBr	5.9772	RbI	7.342
NiO	4.1769	BaSe	6.600	NaI	6.473	AgF	4.92
CdO	4.6953	CaTe	6.356	NaH	4.890	AgCl	5.549
SnAs	5.7248	SrTe	6.660	ScN	4.44	AgBr	5.7745
TiC	4.3285	BaTe	7.00	TiN	4.240	CsF	6.014
UC	4.955	LaN	5.30	UN	4.890		

From West, A.R., *Solid State Chemistry and Its Applications*, John Wiley & Sons, New York, 1984. With permission.

b. CaF_2 (Fluorite)

The structure of fluorite (Figure 3.24a) is derived from ccp Ca^{2+} with fluoride occupying all the tetrahedral sites. Thus each fluoride has tetrahedral coordination, but each Ca^{2+} is surrounded by eight fluorides in a cubic coordination. The tetrahedra share all their edges with other tetrahedra (Figure 3.24b). An alternative description is to consider the structure as being constructed from CaF_8 cubes which edge share with adjacent cubes as shown in Figure 3.24c. Alternate cubes are vacant, but may become occupied in solid solution formation (see Section VII). Other compounds which adopt the fluorite structure include SrF_2, BaF_2, ThO_2, and UO_2. Na_2O adopts an inverse fluorite structure with ccp anions and cations in the tetrahedral sites.

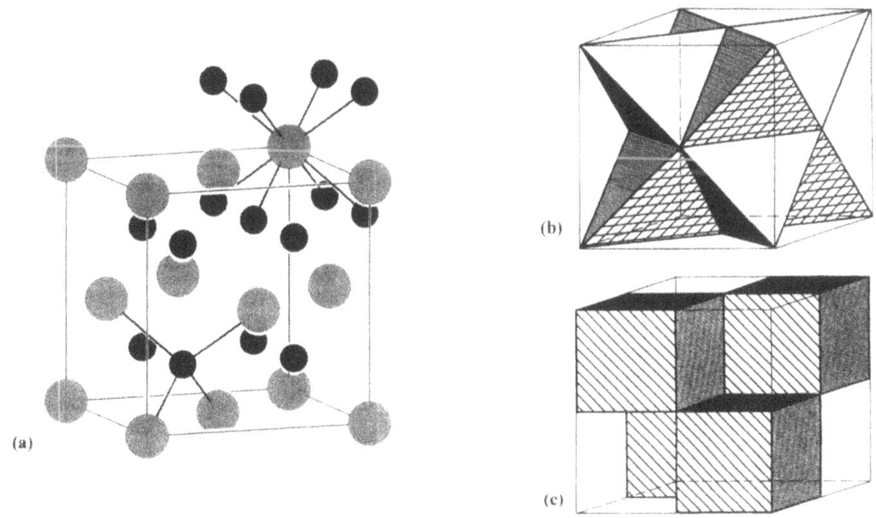

FIGURE 3.24. Structure of CaF_2. (a) Unit cell projection, (b) edge-sharing tetrahedra in CaF_2, (c) description of CaF_2 with alternate cubes filled. (From West, A.R., *Solid State Chemistry and Its Applications*, John Wiley & Sons, New York, 1984. With permission.)

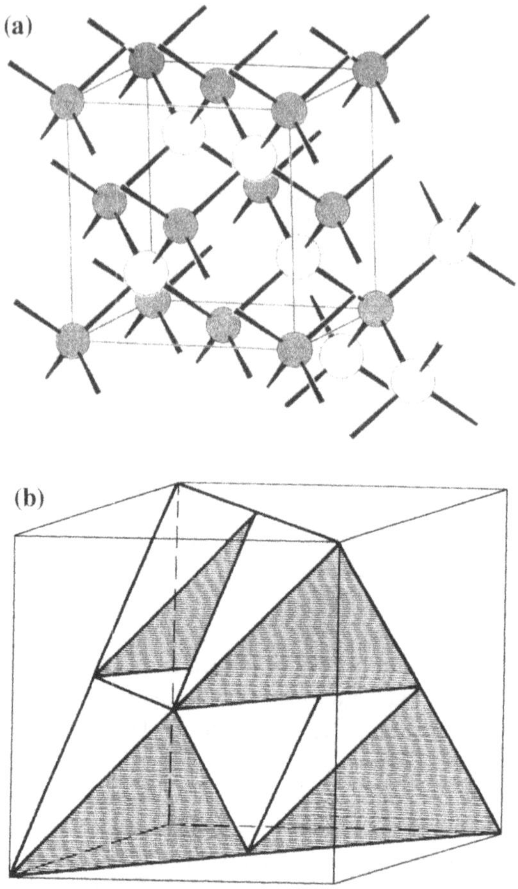

FIGURE 3.25. Structure of zinc blende. (a) Unit cell projection showing Zn (shaded circles) and S (open circles) coordination, (b) corner-sharing tetrahedral arrangement in zinc blende. (Adapted from West, A.R., *Solid State Chemistry and Its Applications*, John Wiley & Sons, New York, 1984.)

c. *ZnS (Zinc Blende or Sphalerite)*

The ccp polymorph of ZnS, zinc blende (Figure 3.25a), consists of ccp S^{2-} with Zn^{2+} in half the tetrahedral sites. The tetrahedra share corners with three other tetrahedra, all pointing in the same direction (Figure 3.25b). The faces of the tetrahedral sites are parallel to the close-packed layers. The S^{2-} ions are also tetrahedrally coordinated to zinc. Replacement of the Zn and S by C results in the diamond structure (see Figure 3.7). Other compounds which adopt the zinc blende structure include BeS, CuF, and α-CdS. Like the wurtzite structure, zinc blende is a common structure for II–VI and III–V compounds such as GaAs and InSb.

3. Layered Structures Based on Close Packing

a. *$CdCl_2$ and CdI_2*

The structures of $CdCl_2$ and CdI_2 are based on ccp and hcp halide lattices with alternate layers of octahedra filled (Figure 3.26). The structure of $CdCl_2$ consists of ccp Cl^- ions with Cd^{2+} filling all the octahedral sites in alternate layers. This results in a layered compound with layers held together by van der Waals forces. The $CdCl_6$ octahedra edge share within the filled layer. $CdBr_2$, ZnI_2, and $CoCl_2$ are known to adopt this structure. Cs_2O packs in an anti-$CdCl_2$ structure with ccp Cs^+ and O^{2-} in all the octahedral sites in alternate layers.

CdI_2 (Figure 3.27) has a similar structure to $CdCl_2$, but in this case the structure is based on hcp I^- ions, with Cd^{2+} in all the octahedral sites in alternate layers. The octahedra share

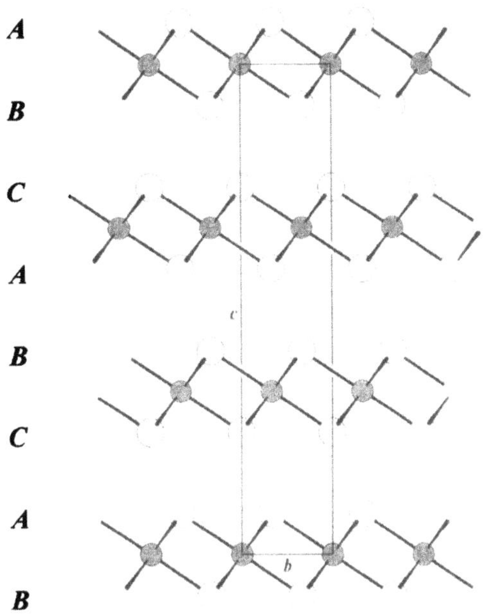

FIGURE 3.26. View down a-axis of $CdCl_2$ structure. Cd and Cl atoms are represented by shaded and open circles, respectively. Cubic close packed layers are indicated.

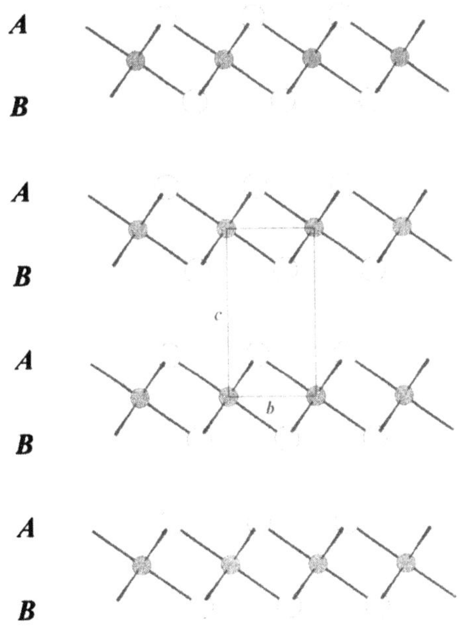

FIGURE 3.27. View down a-axis of CdI_2 structure. Cd and CI atoms are represented by shaded and open circles, respectively. Hexagonal close packed layers are indicated.

six of their edges with other octahedra in the same layer. Each I^- ion is coordinated to three cadmiums. CaI_2, VBr_2, $TiCl_2$, $Ni(OH)_2$, and $Ca(OH)_2$ all adopt the CdI_2 structure. Many of these compounds including CdI_2 itself show polymorphism.

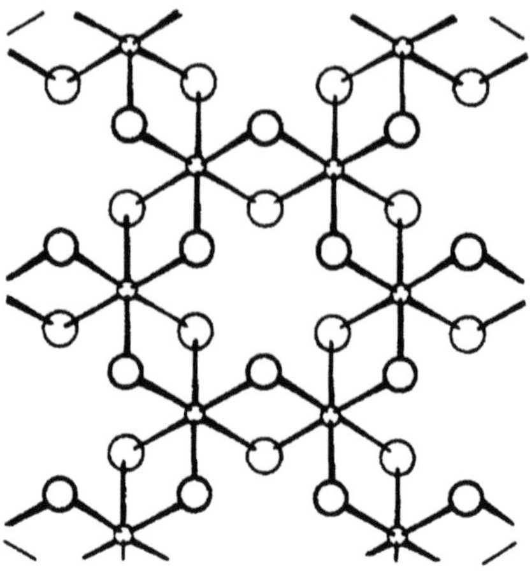

FIGURE 3.28. Two-dimensional projection of $CrCl_3$ structure. (From Wells, A.F., *Structural Inorganic Chemistry*, 2nd ed., Clarendon Press, Oxford, 1950. With permission.)

b. $CrCl_3$ and BiI_3

The structures of $CrCl_3$ and BiI_3 are based on ccp Cl^- and hcp I^-, respectively. In both structures one third of the available octahedral sites are occupied with two thirds of the sites in alternate layers filled by cations, resulting in layered structures. Each octahedron shares three edges with other octahedra (Figure 3.28).

4. Other Important Structures

a. TiO_2 (Rutile)

The structure of rutile (Figure 3.29) is based on a distorted hcp array of oxide ions with titanium in half the octahedral sites. The filled octahedra are arranged so that every alternate octahedron is filled. This results in chains of edge-sharing TiO_6 octahedra. The chains share corners to give the resulting tunnel structure with tunnels running parallel to the c-axis. There is a buckling of the close-packed layers to give a tetragonal unit cell with a = 4.5937, c = 2.9587 Å. The titanium coordination is slightly distorted from regular octahedral geometry, with two Ti–O bond lengths of 1.95 and 1.98 Å. Each oxygen is coordinated to three titaniums in a planar arrangement. The structure is adopted by several MO_2 compounds, including M = Se, Sn, Pb, Ti, Cr, Mn, Ta, Tc, Re, Ru, Os, Ir, Te, as well as some MF_2 compounds (M = Mg, Mn, Fe, Co, Ni, Zn, Pd). Orthorhombic distortions of the rutile structure are seen in $CaCl_2$ and $CaBr_2$. Anti-rutile structures are adopted by δ-Co_2N and Ti_2N.

b. α-Al_2O_3 (Corundum)

The corundum structure (Figure 3.30) is formed from hcp oxide ions with Al^{3+} in two thirds of the octahedral sites. The Al positions are displaced, resulting in distorted tetrahedral coordination for O^{2-}. Each O^{2-} has four Al neighbors. The Al–Al distance is reduced by distortion of regular octahedral geometry with Al–O bond lengths of 1.93 and 1.89 Å. Corundum is noted for its hardness and high melting point. Doping with Cr or Ti results in the gemstones ruby and sapphire, respectively. Ti_2O_3, V_2O_3, Cr_2O_3, Fe_2O_3, Rh_2O_3, and Ga_2O_3 are all isostructural with Al_2O_3.

c. ReO_3

ReO_3 (Figure 3.31) consists of a ccp array of oxide ions with one fourth of the oxide ions missing. Re is located in one fourth of the octahedral sites. This results in each octahedron sharing all six vertices with other octahedra and linear Re-O-Re linkages. ScF_3, NbF_3, TaF_3, and MoF_3 all show the same structure.

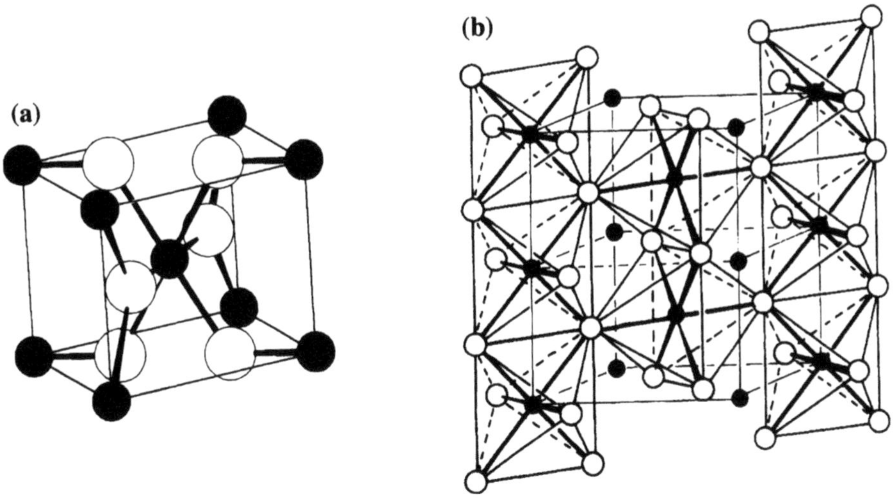

FIGURE 3.29. Rutile structure (a) unit cell projection, (b) edge-sharing octahedral arrangement in rutile. Ti and O atoms are represented by black and open circles, respectively. (From Adams, D.M., *Inorganic Solids,* John Wiley & Sons, New York, 1974. With permission.)

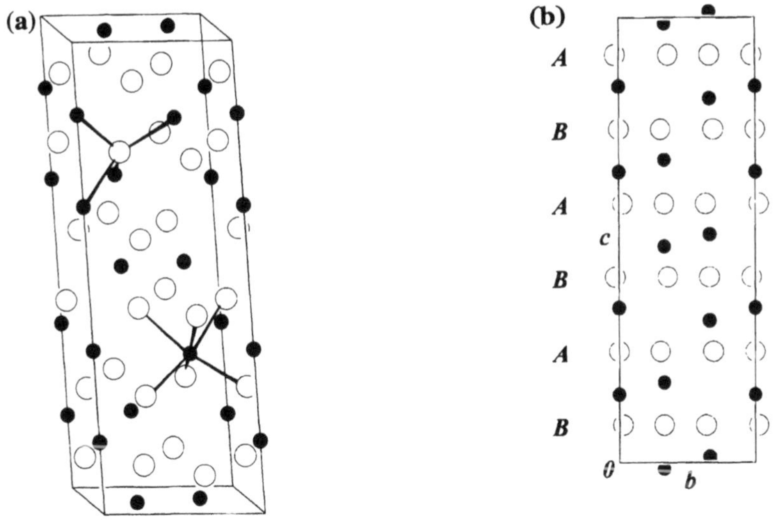

FIGURE 3.30. Structure of corundum. (a) Unit cell contents showing Al (black) and O (open circles) coordinations; (b) view down a-axis showing octahedral site filling sequence between hcp layers.

d. $CaTiO_3$ (Perovskite)

The ideal perovskite structure (Figure 3.32) is related to that of ReO_3 and may be thought of as a ccp array of oxide ions with one fourth of the oxygens missing. Ti occupies one fourth of the octahedral sites with Ca located in the oxide ion vacancy. The TiO_6 octahedra share corners to give the characteristic three-dimensional framework. This gives a CN of 12 for Ca; however, distortions of the lattice reduce this to 8.

e. $MgAl_2O_4$ (Spinel)

The spinel structure (Figure 3.33) can be described as ccp O^{2-} ions with Al^{3+} in half the octahedral sites and Mg^{2+} in one eighth of the tetrahedral sites. The structure is therefore built up of edge-sharing ribbons of octahedra which are joined by parallel ribbons in adjacent layers by further edge sharing. The tetrahedra share vertices with the octahedra. Fe_2MgO_4

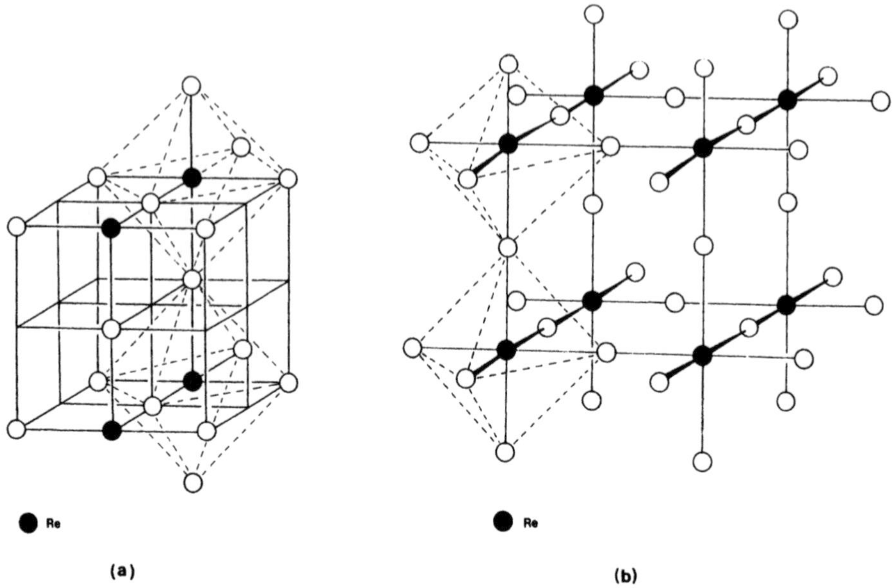

FIGURE 3.31. Structure of ReO_3. Re and O atoms are black and open circles, respectively; (a) unit cell contents and Re coordination (b) 3-D structure. (From Adams, D.M., *Inorganic Solids,* John Wiley & Sons, New York, 1974. With permission.)

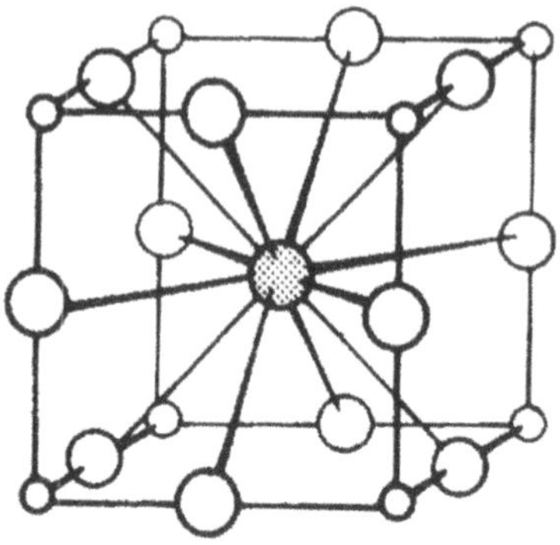

FIGURE 3.32. Structure of perovskite showing Ba (shaded), O (large open circles), and Ti (small open circles). (From Wells, A.F., *Structural Inorganic Chemistry,* 5th ed., Clarendon Press, Oxford, 1984. With permission.)

adopts an inverse spinel structure with half the Fe in tetrahedral sites and the other half sharing the octahedral sites with Mg. Similarly, Fe_3O_4 (hematite) has Fe(III) in one eighth of the tetrahedral sites and Fe(III) and Fe(II) randomly distributed in half the octahedral sites.

VI. CRYSTAL DEFECTS

As described in Section II, the crystalline state is derived from a regular repeating array of atoms in three dimensions. The net result of this construction is a perfect crystalline solid with all atoms present and located in their ideal positions. However, the perfect crystalline state has a zero entropy and can therefore only exist at absolute zero temperature. Above this temperature there is a finite probability that defects from ideality will exist. These defects from the perfect crystal, although typically small in number, affect greatly the properties of the crystal, such as electronic conductivity and mechanical strength. Of most importance in the context of this chapter is the crucial role defects play in controlling ionic transport.

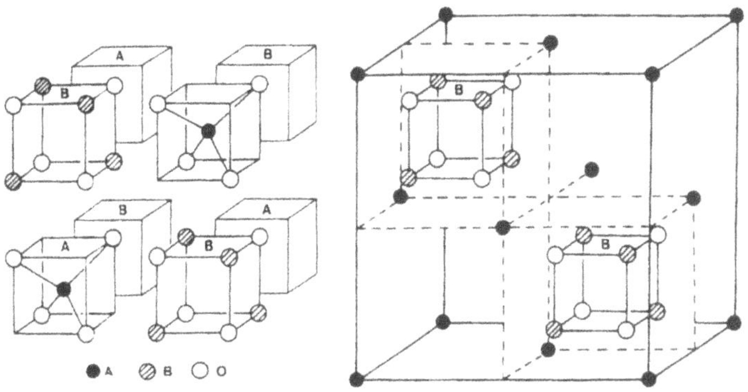

(a) (b)

FIGURE 3.33. Structure of spinel. The structure is composed of alternating octants of AO_4 tetrahedra and B_4O_4 cubes (a) to build the fcc unit cell (b). (From Greenwood, N.N., *Ionic Crystals Lattice Defects and Nonstoichiometry*, Butterworths, London, 1968. With permission.)

A. ENERGETICS OF DEFECT FORMATION

As the number of defects in a crystalline solid increases, the degree of disorder and hence the entropy, S, increases. The change in entropy on introducing defects into a perfect crystal is given by:

$$\Delta S = k \ \ln \ W \tag{3.12}$$

where k is the Boltzmann constant and W is the number of different possible arrangements for a particular point defect. It can be shown that for n defects distributed over N sites:

$$W = \frac{N!}{(N-n)!n!} \tag{3.13}$$

What prevents the system proceeding to a completely random distribution of atoms is the energy associated with defect formation ΔH_f. Using the Gibbs equation:

$$\Delta G = \Delta H - T\Delta S \tag{3.14}$$

the change in Gibbs free energy (ΔG) of a crystal containing n defects at a particular temperature and pressure is given by:

$$\Delta G = n\Delta H_f - T(\Delta S + n\Delta S') \tag{3.15}$$

where S′ is the entropy associated with the change in atomic vibration around a defect.

At a particular temperature, starting from a zero defect concentration, introduction of a single defect increases the entropy significantly such that there is a drop in free energy. This continues as more defects are introduced, and the enthalpy and entropy terms increase. At a certain point introduction of further defects causes little change in the overall disorder and hence entropy, while enthalpy continues to increase. This would result in an increase in free energy, and so further defect formation is no longer energetically favored. Thus an equilibrium defect concentration has been achieved (Figure 3.34). Change in temperature causes a shift in this equilibrium position.

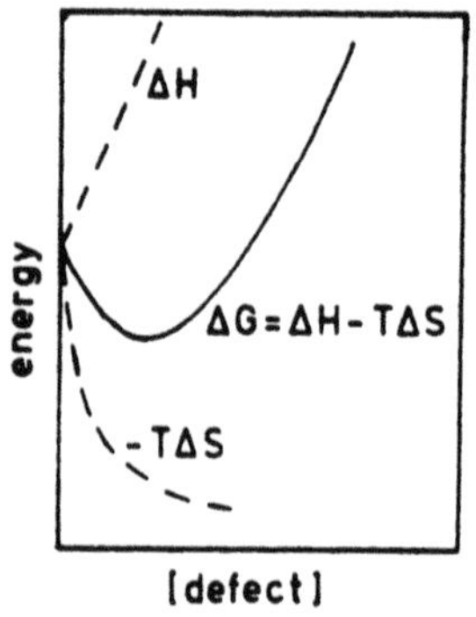

FIGURE 3.34. Variation of thermodynamic parameters with defect concentration. (From West, A.R., *Solid State Chemistry and Its Applications,* John Wiley & Sons, New York, 1984. With permission.)

Creation of a defect in a perfect crystalline solid results in a strain energy in the surrounding lattice. Relaxation of atoms or ions surrounding a defect helps to compensate for this strain energy. A number of methods have evolved to calculate defect formation enthalpies and entropies which take into account lattice relaxation and other factors such as polarization. These have been reviewed extensively[24] and are outlined only briefly here. Simulation methods are based on calculations of interatomic potentials, for example, using the shell model (see Section III).[25] Relaxation around a defect can be modeled via static lattice simulations at constant volume to yield both defect enthalpies and entropies.[26,27] A Mott–Littleton approach[28] is generally used in which two regions are established, an inner region containing the defect and the immediately surrounding lattice in which defect forces are strong and an outer region where the long-range defect forces are weaker. The inner region is energy minimized through modeling of the lattice relaxation through interatomic potentials until equilibrium is achieved. In the outer region a more approximate approach is satisfactory. A number of computer programs have been developed for modeling defects through static lattice calculations, including HADES[29,30] and CASCADE.[31] These calculations have been applied to a number of systems, including relatively simple systems such as the alkali halides and alkaline earth fluorides,[32,33] and more complex systems such as the pyrochlore $Gd_2Zr_2O_7$.[34]

An alternative to the static lattice approach is to use quantum mechanical methods in the calculation of defect energies. These methods essentially rely on solution of the Schrödinger equation using the Hartree–Fock approximation. The two main techniques involve calculation on an embedded defect cluster, i.e., the defect and surrounding lattice. Alternatively, calculations may be performed on a defect supercell in which the defect is periodically repeated on a superlattice. *Ab initio* Hartree–Fock calculations are computationally intensive, and a number of approximations have been made in order to simplify these calculations to obtain reasonable computer processing times. For a more detailed description of *ab initio* calculations readers are referred to the recent review by Pyper.[35]

For further information on defect formation, readers are referred to Chapter 1.

B. CLASSIFICATION OF CRYSTAL DEFECTS

Crystal defects can take several forms and be classified in a number of ways. The solid state electrochemist, however, is mainly concerned with five types of crystal defect as follows:

1. Lattice vacancies/interstitials
2. Defect clusters
3. Dislocations
4. Stacking faults
5. Grain boundaries

Let us examine these classes of crystalline defects in more detail.

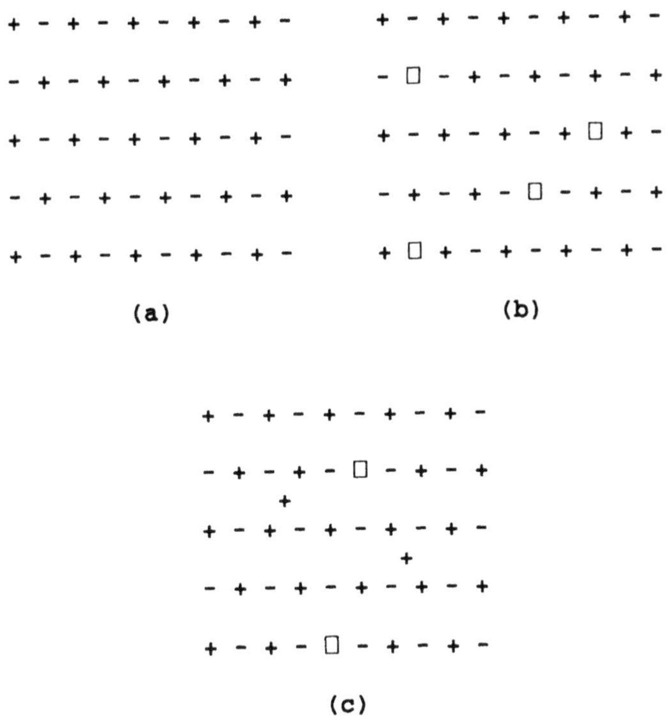

FIGURE 3.35. Two-dimensional representation of point defects in ionic lattices. (a) Perfect lattice, (b) Schottky defect, (c) Frenkel defect.

1. Lattice Vacancies/Interstitials

a. Intrinsic Defects

Intrinsic defects such as lattice vacancies or interstitials are present in the pure crystal at thermodynamic equilibrium. The simplest of these crystalline defects involve single or pairs of atoms or ions and are therefore known as point defects. Two main types of point defect have been identified: Schottky defects,[36] in which an atom or ion pair are missing from the lattice (Figure 3.35a), and Frenkel defects,[37] in which an atom or ion is displaced from its ideal lattice position into an interstitial site (Figure 3.35b).

To illustrate these, let us consider two isostructural solids, NaCl and AgCl. Both these solids adopt the fcc rock salt structure (Section V), with ccp Cl^- and Na^+ or Ag^+ in the octahedral sites. In NaCl, Schottky defects are observed, with pairs of Na^+ and Cl^- ions missing from their ideal lattice sites. As equal numbers of vacancies occur in the anion and cation sublattices, overall electroneutrality and stoichiometry are preserved. In AgCl a Frenkel defect is preferred with some of the silver ions displaced from their normal octahedral sites into interstitial tetrahedral sites. This leaves the anion sublattice intact, as for every cation vacancy introduced a cation interstitial is formed. The defects in AgCl and NaCl are illustrated schematically in Figure 3.36.

Why should two isostructural solids display two different kinds of point defect? The explanation lies in the nature of the bonding in these two solids. NaCl has a high degree of ionic character in its bonding, the cations being very electropositive in nature obey well Pauling's rules and are not easily accommodated in the small tetrahedral site. Furthermore, they would exhibit significant cation–cation repulsion if sodium ions were to occupy the interstitial tetrahedral sites, which share faces with occupied octahedral sites. The Ag–Cl bond has a much higher degree of covalency, with Ag^+ considered to be far less electropositive

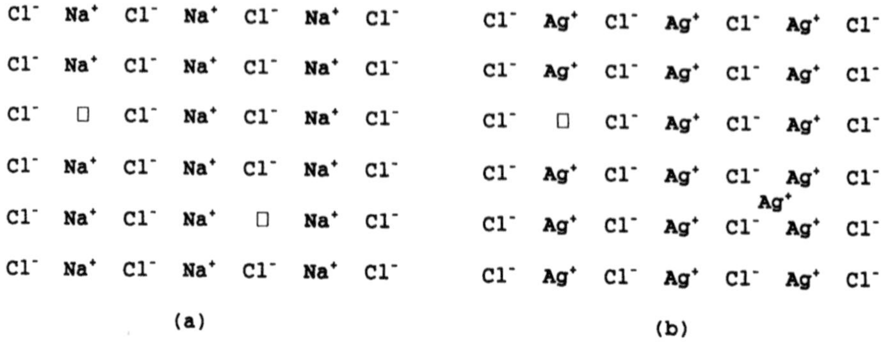

FIGURE 3.36. Schematic representations of (a) Schottky defect in NaCl, and (b) Frenkel defect in AgCl.

than Na^+. This means that the close cation–cation contact required for occupation of the interstitial tetrahedral site is more favorable in AgCl than in NaCl, and the lower CN of 4 is more easily accommodated.

The concentration of Schottky defects in a binary salt MX can be readily calculated. From Equation (3.12) it can be seen that the configurational entropy change for introduction of a Schottky pair of anion and cation vacancies is

$$\Delta S = k \ \ln W^+W^- \tag{3.16}$$

where W^+ is probability associated with cations and W^- is that for anions. From Equation (3.13):

$$\Delta S = 2 \ k \ \ln \frac{N!}{(N - n_s)!n_s!} \tag{3.17}$$

where n_S is the number of Schottky defects. This can be simplified using the super-Stirling Approximation:

$$\ln x! \approx x \ \ln x \tag{3.18}$$

thus:

$$\Delta S = 2k\left[N \ \ln N - (N - n_s) \ \ln (N - n_s) - n_s \ \ln n_s\right] \tag{3.19}$$

From Equation (3.15):

$$\Delta G = n_s\Delta H_f - 2kT \ \left[N \ \ln N - (N - n_s) \ \ln (N - n_s) - n_s \ \ln n_s\right] \tag{3.20}$$

Now for equilibrium at constant T

$$\frac{\delta \Delta G}{\delta n_s} = 0 \tag{3.21}$$

Hence:

$$\Delta H_f = 2kT \frac{\delta}{\delta n_s}\left[N \ln N - (N - n_s) \ln (N - n_s) - n_s \ln n_s\right]$$
$$= 2 kT \ln \frac{N - n_s}{n_s} \qquad (3.22)$$

Taking exponentials and rearranging gives:

$$n_s = (N - n_s)\exp[-\Delta H_f/2kT] \qquad (3.23)$$

In general, $n_S \ll N$ and so:

$$n_s \approx N \exp[-\Delta H_f/2kT] \qquad (3.24)$$

Thus 1 mol of Schottky defects is given by

$$n_s \approx N \exp[-\Delta H_m/2RT] \qquad (3.25)$$

where ΔH_m is the energy of formation of 1 mol of Schottky defects.

In a similar way the number of Frenkel defects at equilibrium n_F, in a binary ionic crystal MX, can be found from:

$$\Delta S = 2 k \left\{\ln \frac{N!}{(N - n_F)!n_F!} + \ln \frac{N_i!}{(N_i - n_F)!n_F!}\right\} \qquad (3.26)$$

where N_i is the number of interstitial sites over which n_F ions may be distributed and N is the number of framework lattice sites.

Therefore, in a similar way to the calculation of Schottky defects:

$$n_F \approx \sqrt{(N N_i)} \exp[-\Delta H_f/2kT] \qquad (3.27)$$

Table 3.9 shows calculated defect concentrations in NaCl at various temperatures.

b. Extrinsic Defects

Intrinsic defect concentration is typically very small at temperatures well below the melting point. For example, the room temperature concentration of Schottky defects in NaCl is in the order of 10^{-17} mol^{-1}. Defect concentration, however, may be increased by the inclusion of impurity or dopant atoms. Extrinsic defects occur when an impurity atom or ion is incorporated into the lattice either by substitution onto the normal lattice site or by insertion into interstitial positions. Where the impurity is aliovalent with the host sublattice, a compensating charge must be found within the lattice to preserve electroneutrality. For example, inclusion of Mg^{2+} in the NaCl crystal lattice results in an equal number of cation vacancies. These defects therefore alter the composition of the solid. In many systems the concentration

TABLE 3.9
Number of Schottky Defects at Thermodynamic Equilibrium in Sodium Chloride

t°C	T°K	n_s/N	n_s per cm³ [a]
–273	0	$10^{-\infty}$	0
25	298	3×10^{-17}	5×10^{5}
200	473	4×10^{-11}	6×10^{8}
400	673	5×10^{-8}	8×10^{14}
600	873	3×10^{-6}	4×10^{16}
800[b]	1073[b]	3×10^{-5}	4×10^{17}

[a] Calculated using a value of 1.6×10^{22} ion pairs per cm³ for N as obtained from the density (1.544 g cm³), the molecular weight (58.45 g/mole) and Avogadro's number (S = 6.023×10^{23} ion pairs per mole).
[b] i.e., 1° below the m.p.

From Greenwood, N.N., *Ionic Crystals Lattice Defects and Non-stoichiometry,* Butterworths, London, 1968. With permission.

of the dopant ion can vary enormously and can be used to tailor specific properties. These systems are termed solid solutions and are discussed in more detail in Section VII.

2. Defect Clusters

The simple point defects discussed above make the assumption that structures are unperturbed by the presence of vacancies or interstitials. This, however, is an oversimplification and has been found to be untrue in a number of cases, where atoms or ions immediately surrounding a defect are found to be shifted away from their ideal sites. The defect now involves two or more atoms and may be considered to be a defect cluster or aggregate. In ionic crystals the ions surrounding an interstitial are distorted away from the interstitial ion due to charge interactions. Vacancies in ionic crystals are effectively charged, anion vacancies possessing an overall positive charge and cation vacancies an overall negative charge. Thus cation vacancies and interstitial cations attract each other to form simple clusters. Although these clusters show overall electroneutrality, they do possess a dipole and hence may attract other defects to form even larger clusters. Thus individual point defects in favorable situations may coalesce to form defect clusters. These defect clusters allow for an overall lowering of free energy with respect to the formation of the individual defects. Solid electrolytes are generally massively defective systems with high ionic conductivities (see Section VII), and it is believed that defect clustering is very significant in these systems. It has recently been proposed that in the solid electrolyte LISICON and its analogs, where defect clustering has been characterized by neutron diffraction,[38] the defect clusters are mobile and that ionic conduction involves the effective movement of these clusters through an interstitialcy mechanism rather than a simple ion-hopping model.[39,40]

A now classic example of defect clustering is the proposed defect structure of wüstite $Fe_{1-x}O$. Fully stoichiometric FeO would be predicted to adopt the rock salt structure with iron fully occupying the octahedral sites. Nonstoichiometric wüstite contains a significant amount of Fe^{3+}, some of which has been shown by neutron and X-ray diffraction studies to occupy tetrahedral sites. Since occupation of the tetrahedral site would generate short intercation contacts between Fe^{3+} in the tetrahedral sites and iron in the octahedral sites, it is unlikely that they would be simultaneously occupied in any one part of the structure. The so-called Koch cluster[42] results when alternate tetrahedral sites are occupied in the fcc cell and all the surrounding octahedral sites are vacant (Figure 3.37), i.e., 4 Fe^{3+} interstitials and 13 vacancies giving a net formal charge of –14. Overall electroneutrality is maintained by

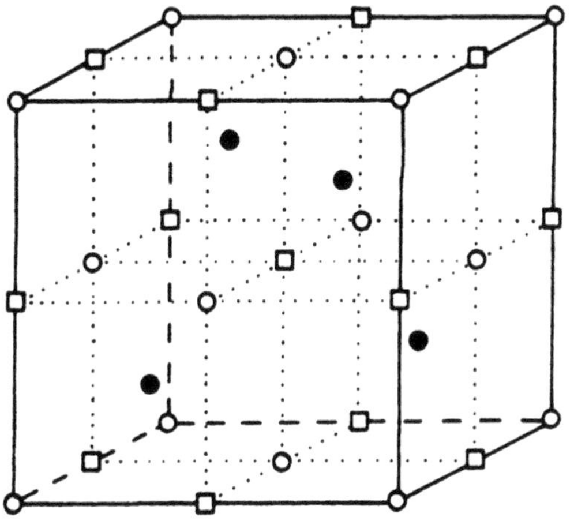

FIGURE 3.37. Koch cluster in wüstite. (From West, A.R., *Solid State Chemistry and Its Applications*, John Wiley & Sons, New York, 1984. With permission.)

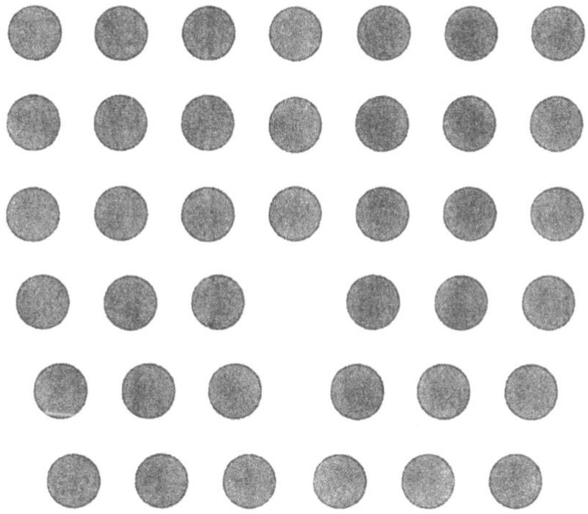

FIGURE 3.38. Schematic two-dimensional representation of atom positions around an edge dislocation.

Fe^{3+} ions in neighboring octahedral sites. As x increases in $Fe_{1-x}O$, the concentration of Fe^{3+} increases and hence the separation between clusters decreases, eventually leading to ordering, and ultimately a superlattice structure is generated. Theoretical calculations on defect clusters in wüstite are described by Catlow and Mackrodt.[30]

3. Dislocations

The description of simple point defects leaves us with the impression that point defects or small defect clusters occur as isolated features in an otherwise perfect crystal. This is not strictly true. In many crystals individual point defects come together to create an extended defect or dislocation. The simplest of these is an edge dislocation, in which an extra half plane of atoms occurs within the lattice (Figure 3.38). The atoms in the layers above and below the half plane distort beyond its edge and are no longer planar. An analogy is when a half page is torn out of a notebook. The pages above and below the removed half page relax

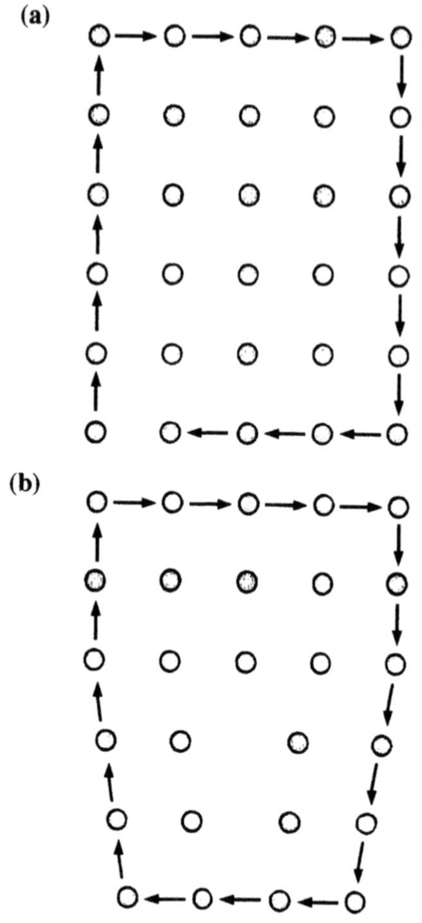

FIGURE 3.39. Burger's vectors; (a) perfect lattice, (b) lattice containing an edge dislocation.

to occupy the space left by the torn-out page. The direction of the edge of the half plane into the crystal is known as the line of dislocation.

Dislocations are characterized by a vector known as the Burger's vector. If a circular path is taken from lattice point to lattice point in a region of perfect crystal, the end point will be the same lattice point as the starting point. If, however, the region encompasses an edge dislocation, the starting and finishing points will not coincide and the distance and direction between these points correspond to the magnitude and direction of the Burger's vector (Figure 3.39). For an edge dislocation the Burger's vector is perpendicular to the line of dislocation and is also parallel to the motion of the dislocation under an applied stress.

Another form of dislocation, known as a screw dislocation, occurs when an extra step is formed at the surface of a crystal, causing a mismatch which extends spirally through the crystal. If a circular path is taken through lattice points around a screw dislocation, a helix is formed. The resulting Burger's vector is now parallel to the line of dislocation (Figure 3.40).

4. Stacking Faults

Stacking faults, as the name implies, are misaligned layers. In close-packed systems a stacking fault may occur on stacking of individual layers, for example, in a ccp system the omission of a C layer, i.e., ...ABCABABC... (Figure 3.41a). This results in a region of hcp within the ccp system. Conversely, a stacking fault may arise in hcp where a C layer is included, and thus a region of ccp is formed within the hcp lattice (Figure 3.41b). Various stacking faults may occur within close-packed systems, but the formation of AA, BB, or CC

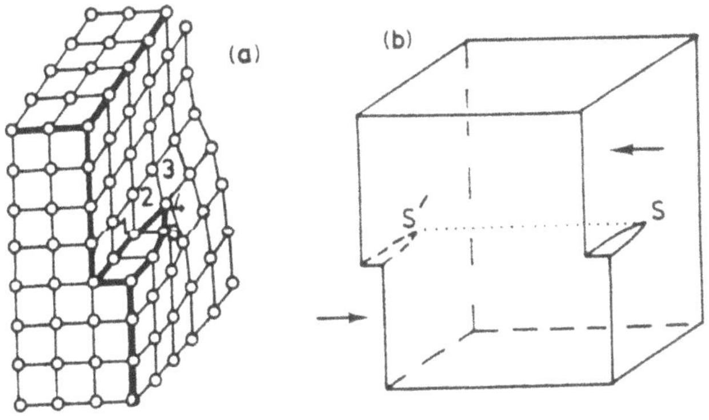

FIGURE 3.40. (a) Schematic representation of a screw dislocation. Consider the circuit 12345 which passes around the dislocation. The direction of the Burger's vector is defined by 1 to 5 and is parallel to the line of dislocation S–S′, shown in (b). (From West, A.R., *Solid State Chemistry and Its Applications,* John Wiley & Sons, New York, 1984. With permission.)

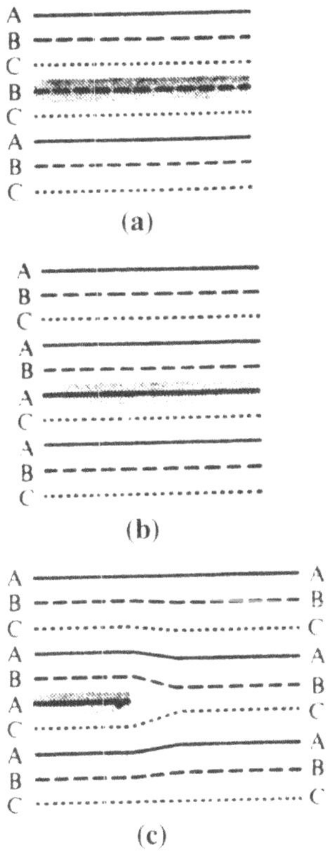

FIGURE 3.41. Stacking faults. (a) Missing A layer in ccp system, (b) insertion of an extra A layer, (c) partial stacking fault. (From Rosenberg, H.M., *The Solid State,* Oxford University Press, Oxford, 1988. With permission.)

faults in most systems is very energetically unfavorable and rarely observed. Where the extra plane does not extend through the whole crystal a partial dislocation is formed (Figure 3.41c).

5. Grain Boundaries

So far our descriptions of crystalline solids have been restricted to discussions based on single crystals. In practice, most materials, although made up of a single chemical phase, are polycrystalline; for example, a metal wire or a ceramic component. The individual crystallites or grains are single crystals in there own right and may be randomly oriented with respect to each other. Within a particular grain the crystal lattice therefore shows one orientation

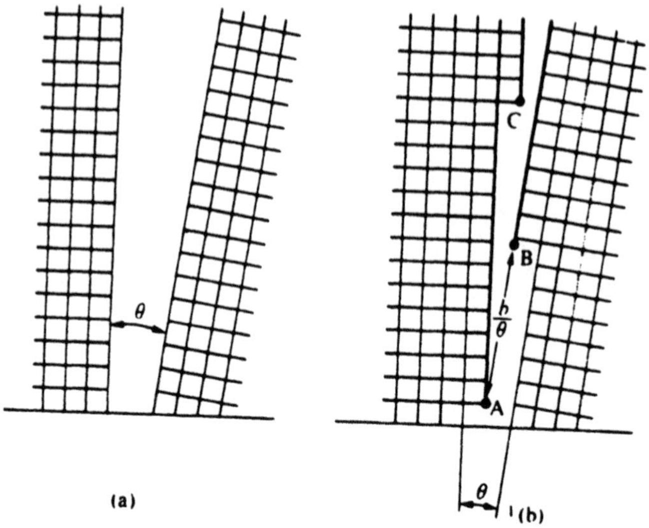

FIGURE 3.42. (a) Schematic two-dimensional representation of a low-angle grain boundary. (b) The space between the two crystallites is now filled, forming an array of edge dislocations. (From Rosenberg, H.M., *The Solid State*, Oxford University Press, Oxford, 1988. With permission.)

which may be different from neighboring grains. Between grains there are regions which are not aligned with either neighbor. These transition regions are termed grain boundaries, and their size and complexity greatly influence many properties of the material.

Consider two crystallites where the crystal lattices are at a small angle with respect to each other. The space between the two crystallites will also be filled by atoms. However, because of the angle, whole planes of atoms will not fit and the space is filled by part planes, i.e., edge dislocations (Figure 3.42). This low-angle grain boundary can therefore be considered to be constructed from an array of edge dislocations. This model is really only satisfactory at low-angle grain boundaries, and more complex structural models are required for angles greater than 10°. Low-angle grain boundaries are often found in so-called single crystals where small misalignments of the crystal lattice occur due to the inclusion of defects. This has been used to explain observations such as the broadening of lines on X-ray powder diffraction photographs. Examination of X-ray rocking curve widths yields information on low-angle grain boundaries; for example, in the IV–VI semiconductor $Pb_{1-x}Ge_xTe$ grown by vapor deposition techniques, low-angle grain boundaries of 1–3° were measured by this method.[43] An alternative and powerful technique for examining low-angle grain boundaries and dislocations is X-ray topography which can give information on dislocation densities and Burger's vectors as well as more qualitative information about these features. For more detailed information on defect visualization through X-ray topography, readers are referred to Reference 44.

C. MOVEMENT OF DEFECTS

Solid state ionic conductivity is observed when ions and hence defects are free to move through the solid. In order to do this, ions must have sites available for occupation, i.e., vacancies and these must be connected by suitable conduction pathways. Site vacancies may be present due to intrinsic defects such as Schottky or Frenkel defects, or may be extrinsic through solid solution formation (see Section VII). The conduction pathways may be interstitial, i.e., solely involve ions in interstitial sites, or may involve an interstitialcy mechanism where framework ions are also involved. In both cases ions of one sublattice will approach ions of the other sublattice in high-energy bottlenecks. The most common type of bottleneck

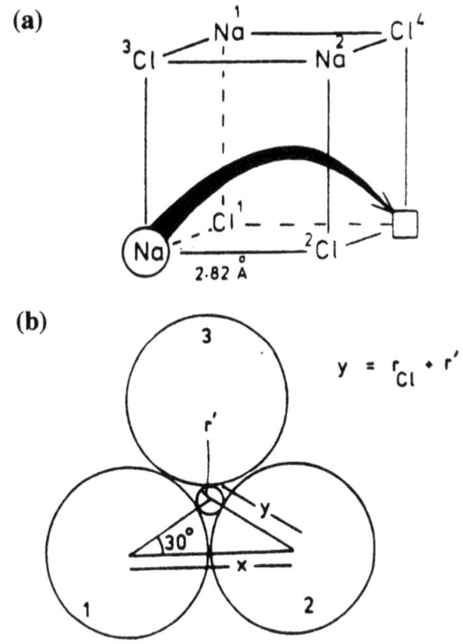

FIGURE 3.43. (a) Pathway for ionic motion in NaCl. (b) Bottleneck formed by Cl^- ions 1, 2, 3. (From West, A.R., *Solid State Chemistry and Its Applications,* John Wiley & Sons, New York, 1984. With permission.)

is the triangle formed by face sharing between octahedral and or tetrahedral sites. In order to enter the site an ion must effectively squeeze through the face from a neighboring site, i.e., the pathways for motion may be considered as a continuous set of face-sharing octahedral and tetrahedral sites.

Consider the Schottky defect in NaCl. In order for ionic conduction to be observed a Na^+ ion must vacate its normal octahedral lattice site and move into a neighboring octahedral vacancy. In order to achieve this, the Na^+ ion must first pass through the two bottlenecks formed by the chloride ions at the faces of the interstitial tetrahedral site (Figure 3.43). The bottlenecks and indeed the interstitial site itself are considerably smaller than the size of the Na^+ ion. This means that there is a considerable activation energy barrier to overcome in order to observe ionic motion in NaCl.

The modeling of defect migration has been the subject of many studies. As indicated in Section III, static lattice calculations can yield important information on defect energies. However, these calculations specifically omit thermal motion. However, ion migration can be modeled using these methods by repeating the calculation in steps along the migration pathway, which allows for the determination of migration energies. This method has been used, for example, in the determination of migration pathways in pyrochlores such as $Gd_2Zr_2O_7$.[34] For more detailed information on defect migration and thermal behavior, other computational methods are employed. These methods have been reviewed extensively[45] and are only mentioned briefly here.

The molecular dynamics approach is a simulation method which essentially solves the equations of motion for a collection of particles within a periodic boundary. This allows for specific transport properties such as the diffusion coefficient to be calculated; for example, the effect of nonstoichiometry on ionic conductivity in Na $\beta''Al_2O_3$ has been investigated using a molecular dynamics approach.[46] An alternative simulation involves a Monte Carlo approach and relies on statistical sampling by random number generation, yielding information on defect migration. This has been used successfully in the modeling of migration pathways in UO_{2+x}.[47,48]

VII. SOLID SOLUTIONS

As we have seen in the previous section, defects in ionic solids can be introduced without being thermally created. This can occur either by replacement of an ion of one sublattice with an aliovalent ion, with a compensating charge made up either by interstitials or vacancies; or by change in oxidation state on the cation sublattice such as occurs in transition metal solids. In these structures the compound remains a single discrete phase, but the composition may vary and is termed a solid solution. Thus a solid solution may be defined as a single crystalline phase with variable composition. In general, these solids maintain a basic structural framework throughout the solid solution range. The extent of solid solution formation is very much dependent on the chemical system and can vary from fractions of a percent to 100%. Many naturally occurring minerals are in fact solid solutions; for example, common olivines, $Mg_{2-x}Fe_xSiO_4$, vary in composition from Mg_2SiO_4 (fosterite) to Fe_2SiO_4 (fayalite), with the composition $Mg_{1.8}Fe_{0.2}SiO_4$ commonly found. The effects of solid solution formation can be to introduce or enhance physical properties, such as mechanical strength, conductivity, ferromagnetism, etc. Of particular interest to the solid state electrochemist is the enhancement of ionic and ionic/electronic conductivity in solids.

In ionic systems, solid solutions may be formed in a number of ways. Let us deal with each of these in turn.

A. SUBSTITUTIONAL SOLID SOLUTIONS

Substitutional solid solutions involve direct substitution of ions, or atoms by different, but isovalent ions or neutral atoms, respectively. For example, the rock salt structures of AgBr and AgCl allow a solid solution to be formed between these two solids of the general formula $AgCl_{1-x}Br_x$. Similarly, complete solid solution ranges can be formed by substitution of Na^+ in NaCl by K^+ as in $Na_{1-x}K_xCl$, or by Ag^+ in $Na_{1-x}Ag_xCl$, in both cases $0 \leq x \leq 1.0$ and the rock salt structure is maintained throughout. Thus this type of solid solution formation does not normally alter the number of interstitials or vacancies, but merely involves gradual substitution of one sublattice. Substitutional solid solution mechanisms are employed widely in alloy formation; for example, in brass, zinc atoms replace copper atoms in the fcc copper lattice. The requirements for formation of a substitutional solid solution are firstly that the ion being substituted is isovalent with the substituting ion; and secondly that both ions should be roughly similar in size, typically less than 15% size difference. This second requirement, however, is not always strictly adhered to. For example, potassium and sodium are very different in size; K^+, 1.33; Na^+, 0.97 Å,[11] and we have already seen that a full solid solution range exists in the $Na_{1+x}K_xCl$ system. It is often easier to substitute smaller ions for larger ions, e.g., the solid solution between Na_2SiO_3 and Li_2SO_3 has two different ranges, viz., $Na_{2-x}Li_xSiO_3$ where $0 \leq x \leq 0.5$ and $Li_{2-x}Na_xSiO_3$ where $x \leq 0.1$. As already mentioned, complete solid solution ranges require end members to be isostructural. However, even in systems where there is a difference between the structures of end members, incomplete solid solution ranges may be obtained. For example, Mg_2SiO_4 and Zn_2SiO_4 have quite different structures. Mg_2SiO_4 adopts the olivine structure, with octahedral Mg and tetrahedral Si, while Zn_2SiO_4 has tetrahedral coordination for both Zn and Si. One might expect good solid solution ranges as Mg^{2+} and Zn^{2+} are isovalent and similar in size. However, because of the different structures adopted by the end members, only limited solid solution ranges are obtainable, viz., $Mg_{2-x}Zn_xSiO_4$, $0 \leq x \leq 0.2$; $Zn_{2-x}Mg_xSiO_4$, $0 \leq x \leq 0.3$.

B. INTERSTITIAL/VACANCY SOLID SOLUTIONS

In ionic solids where substitution is by an aliovalent ion, electroneutrality is maintained either by formation of vacancies or by introduction of interstitials. Four types of interstitial/vacancy solid solution mechanisms may be defined.

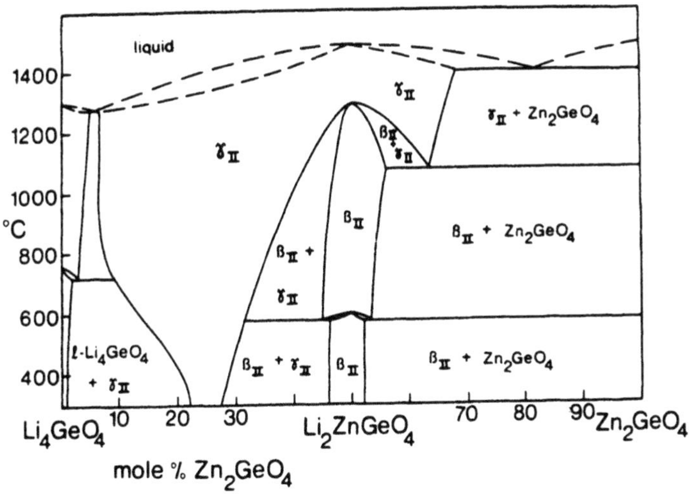

FIGURE 3.44. Phase diagram for the LISICON system $Li_{2+2x}Zn_{1-x}GeO_4$. Notice large γ_{II} solid solution range. (From Bruce, P.G. and West, A.R., *Mater. Res. Bull.*, 1980, 15, 379. With permission.)

1. Cation Vacancies

Cation vacancies may be introduced when a cation of higher charge partially substitutes on the cation lattice. Alternatively, the substitution of an anion by one of lower charge may also achieve this in certain systems. For example, low levels of doping of silver halides by ions such as Ca^{2+} introduces vacancies on the cation sublattice. $Ag_{1-2x}Ca_x\square_xX$ (where $\square$ denotes a vacancy). Similarly, for NaCl doped with Ca^{2+}, $Na_{1-2x}Ca_x\square_xCl$, $x \lessapprox 0.15$. wüstite $Fe_{1-x}O$ (Section VI) is essentially a cation vacancy solid solution where Fe^{2+} has been replaced by Fe^{3+} to give $Fe^{2+}{}_{1-3x}Fe^{3+}{}_{2x}\square_xO$.

2. Cation Interstitials

If a framework cation is exchanged for one of lower valency, then electrical neutrality may be satisfied by introduction of an interstitial cation. In these solid solutions the relative site sizes are important, as is the availability of suitable interstitial sites. Substitution of the framework ion is more easily facilitated when the substituting ion is of similar or smaller size to the ion being replaced. The size of the interstitial cation will depend on the size of the available interstitial sites. The LISICON system is a solid solution between Li_2ZnGeO_4 and Li_4GeO_4. The end members are not isostructural, but there is nevertheless an extensive solid solution range (Figure 3.44).[49] Solid solution members have a general formula $Li_{2+2x}Zn_{1-x}GeO_4$. The end member, Li_2ZnGeO_4, has a high-temperature structure of distorted hcp oxide ions with half the tetrahedral sites filled. Solid solution formation involves the replacement of Zn^{2+} cations by two Li^+ ions, one of which substitutes directly onto the Zn tetrahedron, while the other enters interstitial octahedral sites resulting in high Li^+ ion conductivity for this material. Many other important cation-conducting solid electrolytes are cation interstitial solid solutions, for example the sodium ion conductors, NASICON, and sodium β-alumina.

3. Anion Vacancies

Where cations are replaced with ones of lower charge, vacancies are often introduced into the anion sublattice in order to preserve electroneutrality. An important system where this occurs is in the stabilized zirconias, such as calcium- or yttrium-stabilized zirconia. In these structures the high-temperature cubic zirconia phase, which has the fluorite structure (Section V), is stabilized by substitution of Zr^{4+} by either Ca^{2+} or Y^{3+}, thus introducing anion

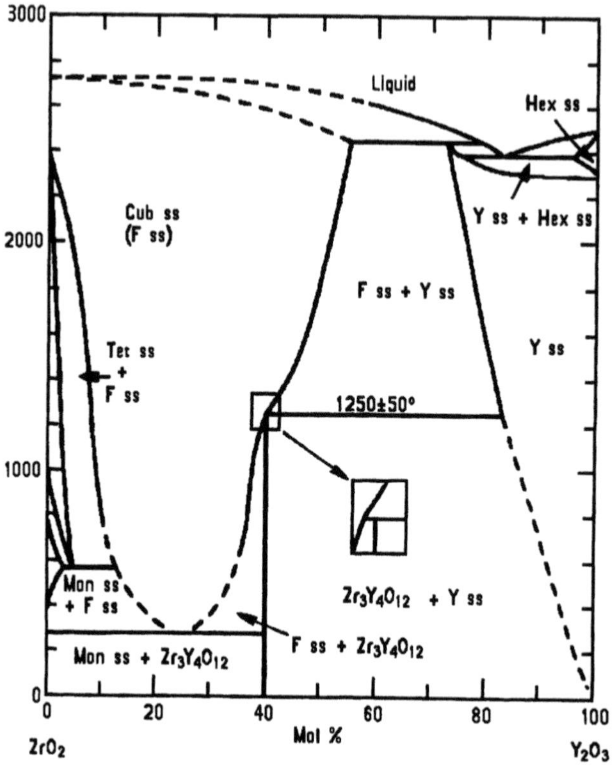

FIGURE 3.45. Phase diagram for the ZrO_2-Y_2O_3 system. (From *Phase Diagrams for Ceramists*, Vol. VI, Clevinger, M.A. and Ondik, H.M., Eds., The American Ceramic Society, Westerville, Ohio, 1987. With permission.)

vacancies, i.e., $Zr_{1-x}Ca_xO_{2-x}$ or $Zr_{1-x}Y_xO_{2-x/2}$. Figure 3.45 shows the phase diagram for the ZrO_2-Y_2O_3 system and illustrates the large solid solution range of the stabilized cubic zirconia phase. These materials show high oxide ion conductivity at temperatures around 1000°C and have been used in a number of applications including fuel cells and O_2 sensors. Another important recent example is the BIMEVOX system based on divalent substitution into $Bi_4V_2O_{11}$, the basic structure of which contains layers of bismuth oxide $[Bi_2O_2]_n^{n2-}$ separated by defect-distorted perovskite layers of general formula $[VO_{3.5}\,\square_{0.5}]_n^{n2+}$ (where □ denotes a vacancy). The number of oxide vacancies can be increased by substitution of V^{5+} by ions such as Cu^{2+}, and in doing so the highly conducting high-temperature γ-phase is stabilized at room temperature.[51]

4. Anion Interstitials

A number of fluorite-based systems show anion interstitials. For example, fluorite itself, CaF_2, exhibits a small solid solution range with trivalent fluorides such as YF_3, to give a solid solution $Ca_{1-x}Y_xF_{2+x}$ which contains interstitial F^- ions. The interstitial ions are accommodated in the vacant cubic sites (see Figure 3.24c). Similarly, the fluorite-based polymorphs of PbF_2 and BaF_2 may also be doped with suitable trivalent cations to give interstitial solid solutions, e.g., $Pb_{1-x}Bi_xF_{2+x}$ and $Ba_{1-x}La_xF_{2+x}$. Nonstoichiometric UO_{2+x} also has the fluorite structure with interstitial O^{2-} ions.

C. MONITORING OF SOLID SOLUTION FORMATION

Solid solution formation may be followed by a number of techniques such as density change, thermal analysis, and X-ray diffraction. Alternatively, specific properties such as ferroelectric behavior and ionic conductivity may be used. X-ray powder diffraction represents

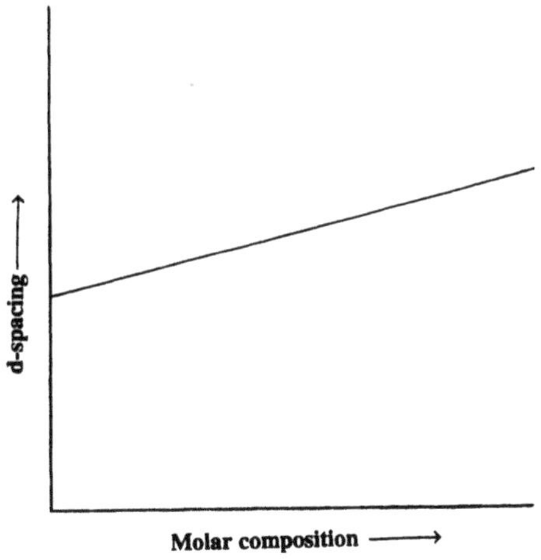

FIGURE 3.46. Schematic representation of a system showing ideal Vegard behavior.

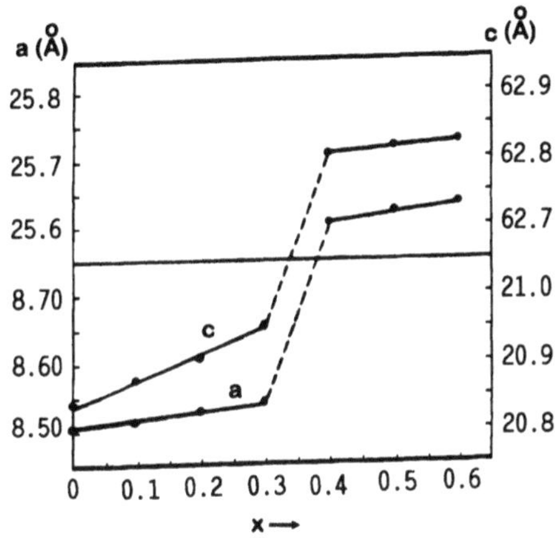

FIGURE 3.47. Unit cell variation in $Li_{1+x}Ti_{2-x}GaP_3O_{12}$. Dashed lines indicate phase transition. (Adapted from Zu-xiang, L., Hui-Jun, Y., Shui-chun, L., and Shun-pao, T., *Solid State Ionics,* 1986, 18/19, 549.)

one of the most generally applicable methods and is commonly used to monitor the progress of solid state reactions through phase identification. The unique powder diffraction fingerprints of starting materials and products enable a qualitative monitoring of the solid state reaction. Analysis of d-spacings of individual peaks allows for a more detailed interpretation, giving unit cell dimensions.

In an ideal solid solution, A_xB_yX, if ion A is replaced by a larger ion B, then a linear expansion in the unit cell dimensions is predicted, and, conversely, if ion B is smaller than ion A, a linear contraction is predicted. This is known as Vegard's law (Figure 3.46). Vegard's law states that in a solid solution unit cell parameters should change linearly with composition. For example, the cation interstitial system $Li_{1+x}Ti_{2-x}Ga(PO_4)_3$ shows a linear expansion in the rhombohedral cell dimensions a and c (hexagonal axes) up to $x = 0.3$ (Figure 3.47).[52] Above $x = 0.3$ a phase transition to a triple cell superstructure occurs, and is indicated by a discontinuity in the cell expansion.

Vegard's law is, however, a generalization, and many systems show significant variation from linearity. For example, in the solid solution A_xB_yX, if the substitution of A by B is not totally random and clusters or regions of A- and B-substituted material are formed, then a positive deviation from Vegard's law results. Similarly, if the A–B interactions are stronger

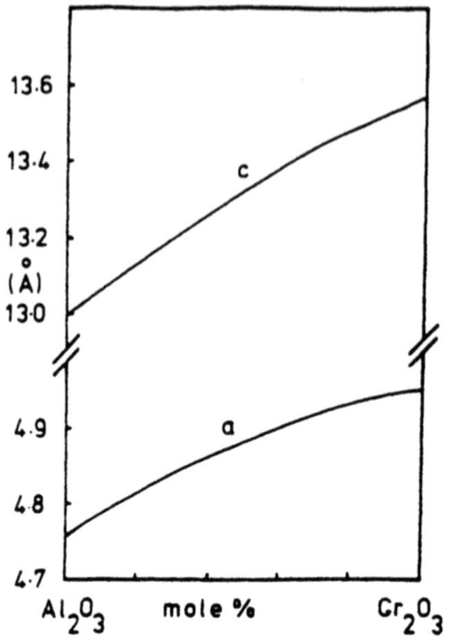

FIGURE 3.48. Positive deviation from Vegard's law in Al_2O_3-Cr_2O_3 system. (From West, A.R., *Solid State Chemistry and Its Applications*, John Wiley & Sons, New York, 1984. With permission.)

than the AA or BB interactions, then a negative deviation results. It is possible to prepare a complete solid solution range in the Al_2O_3-Cr_2O_3 system, with the hexagonal corundum structure (Section V) maintained throughout. However, a positive deviation from Vegard's law is seen in the unit cell parameters (Figure 3.48).[1] This has been explained in terms of the solid solution-forming regions of alumina-rich and chrome-rich microdomains. As seen in Figure 3.47, more severe deviations from Vegard's law occur when solid solutions undergo phase transitions over the composition range studied.

REFERENCES

1. West, A.R., *Solid State Chemistry and Its Applications,* John Wiley & Sons, New York, 1984.
2. Rao, C.N.R. and Gopalakrisnan, J., *New Directions in Solid State Chemistry,* Cambridge University Press, Cambridge, 1986.
3. Ravaine, D. and Souquet, J.L., in *Solid Electrolytes: General Principles, Characterization, Materials and Applications,* Hagenmuller, P. and van Gool, W., Eds., Academic Press, New York, 1978, 277.
4. Kunze, D., in *Fast Ion Transport in Solids,* van Gool, W., Ed., North-Holland, Amsterdam, 1972.
5. Elliot, S.R., *Physics of Amorphous Materials,* 2nd ed., Longman, Harlow, England, 1990.
6. Zachariasen, W.H., *J. Am. Ceram. Soc.,* 1932, *54,* 3841.
7. Wells, A.F., *Structural Inorganic Chemistry,* 5th ed., Clarendon Press, Oxford, 1984.
8. Adams, D.M., *Inorganic Solids,* John Wiley & Sons, New York, 1974.
9. Wyckoff, R.W.G., *Crystal Structures,* 2nd ed., Vol. 6, Part 1, John Wiley & Sons, New York, 1969.
10. Rosseinsky, M.J., Murphy, D.W., Fleming, R.M., Tycko, R., Ramirez, A.P., Siegrist, T., Dabbagh, G., and Barrett, S.E., *Nature,* 1992, *356,* 416.
11. Shannon, R.D. and Prewitt, C.T., *Acta Crystallogr.,* 1969, *B25,* 925; 1970, *B26,* 1046.
12. Evjen, H.M., *Phys. Rev.,* 1932, *39,* 675.
13. Ewald, P.P., *Ann. Physik,* 1921, *64,* 253.
14. Tosi, M.P., *Solid State Physics,* 1964, *16,* 1.
15. van Gool, W. and Piken, A.G., *J. Mater. Sci.,* 1969, *4,* 95.

16. van Gool, W. and Piken, A.G., *J. Mater. Sci.*, 1969, *4*, 105.
17. Greenwood, N.N., *Ionic Crystals Lattice Defects and Nonstoichiometry*, Butterworths, London, 1968.
18. Waddington, T.C., *Adv. Inorg. Radiochem.*, 1959, *1*, 157.
19. Catlow, C.R.A., Diller, K.M., and Norgett, M.J., *J. Phys. C*, 1977, *10*, 1395.
20. Dick, B.G. and Overhauser, A.W., *Phys. Rev.*, 1958, *112*, 90.
21. *International Tables for Crystallography*, Vol. A, Hahn, T., Ed., Kluwer, Dordrecht, 1992.
22. Dent-Glasser, L.S., *Crystallography and Its Applications*, Van Nostrand Reinhold, Wokingham, England, 1977.
23. Giacovazzo, C., *IUCr Texts on Crystallography 2, Fundamentals of Crystallography*, IUCr, Oxford University Press, New York, 1992.
24. Catlow, C.R.A., in *Defects in Solids*, NATO-ASI Series B, Vol. 147, Chadwick, A.V. and Terenzi, M., Eds., Plenum Press, New York, 1986, 269–303.
25. Catlow, C.R.A. and Norgett, M.J., *J. Phys. C*, 1973, *6*, 1325.
26. Harding, J.H., *Physica*, 1985, *131B*, 13.
27. Gillan, M.J., Harding, J.H., and Leslie, M., *J. Phys. C*, 1988, *21*, 5465.
28. Mott, N.F. and Littleton, M.J., *Trans. Faraday Soc.*, 1938, *34*, 485.
29. Norgett, M.J., *UKAEA Report*, AERE-R.7650, 1974.
30. Catlow, C.R.A. and Mackrodt, W.C., *Computer Simulation of Solids. Lecture Notes in Physics*, Springer-Verlag, Berlin, 1982.
31. Leslie, M., *SERC Daresbury Laboratory Report*, DL-SCI-TM3 1T, 1982.
32. Catlow, C.R.A., Corish, J., Diller, K.M., Jacobs, P.W.M., and Norgett, M.J., *J. Phys. C*, 1979, *12*, 451.
33. Catlow, C.R.A., Norgett, M.J., and Ross, T.A., *J. Phys. C*, 1977, *10*, 1627.
34. vanDijk, M.P., Burggraaf, A.J., Cormack, A.N., and Catlow, C.R.A., *Solid State Ionics*, 1985, *17*, 159.
35. Pyper, N.C., in *Advances in Solid State Chemistry*, Vol. 2, Catlow, C.R.A., Ed., JAI Press, London, 1991, 223.
36. Schottky, W. and Wagner, C., *Z. Phys. Chem.*, 1930, *11B*, 163; and Schottky, W., *Z. Phys. Chem.*, 1935, *29B*, 353.
37. Frenkel, I., *Z. Phys.*, 1926, *35*, 652.
38. Abrahams, I. and Bruce, P.G., *Acta Crystallogr.*, 1991, *B47*, 696.
39. Abrahams, I. and Bruce, P.G., *Philos. Mag. A.*, 1991, *64*, 1113.
40. Bruce, P.G. and Abrahams, I., *J. Solid Stat. Chem.*, 1991, *94*, 74.
41. Rosenberg, H.M., *The Solid State*, Oxford University Press, Oxford, 1988.
42. Koch, F. and Cohen, J.B., *Acta Crystallogr.*, 1969, *B25*, 275.
43. Leszczynski, M., in *X-Ray and Neutron Structure Analysis in Materials Science*, Hasek, J., Ed., Plenum Press, New York, 1989, 253.
44. Lang, A.R., in *Characterisation of Crystal Growth Defects by X-ray Methods*, Tanner, B.K. and Bowen, D.K., Eds., Plenum Press, New York, 1980, 161.
45. Catlow, C.R.A., *Ann. Rev. Mater. Sci.*, 1986, *16*, 517.
46. Wolf, M.L., Walker, J.R., and Catlow, C.R.A., *Solid State Ionics*, 1984, *13*, 33.
47. de Bruin, H.J. and Murch, G.E., *Philos. Mag.*, 1973, *27*, 1475.
48. Murch, G.E., *Philos. Mag.*, 1975, *32*, 1129.
49. Bruce, P.G. and West, A.R., *Mater. Res. Bull.*, 1980, *15*, 379.
50. *Phase Diagrams for Ceramicists*, Vol. VI, Clevinger, M.A. and Ondik, H.M., Eds., The American Ceramic Society, Westerville, Ohio, 1987.
51. Boivin, J.C., Vannier, R.N., Mairesse, G., Abraham, F., and Nowogrocki, G., *ISSI Lett.*, 1992, *3*, 14.
52. Zu-xiang, L., Hui-jun, Y., Shui-chun, L., and Shun-pao, T., *Solid State Ionics*, 1986, *18/19*, 549.

Chapter 4

INTERFACE ELECTRICAL PHENOMENA IN IONIC SOLIDS

Janusz Nowotny

CONTENTS

0-8493-8956-9/97/$0.00+$.50

LIST OF SYMBOLS AND ABBREVIATIONS

AES	Auger electron spectroscopy
AFM	atom force microscope
LEIS	low-energy ion scattering
CPD	contact potential difference
LEED	low-energy electron diffraction
PTCR	positive temperature coefficient of resistance
SIMS	secondary ion mass spectrometry
WF	work function
XPS	X-ray photoelectron spectroscopy
YSZ	yttria-stabilized zirconia
A_n	kinetic term related to electrons
A_h	kinetic term related to electron holes
D	diffusion coefficient
e	elementary charge
$[e']$	concentration of electrons
E_C	energy level of the conduction band
E_i	energy level of species i
E_F	Fermi energy level
E_g	band gap
E_v	energy level of the valence band
f_n	distribution function related to electrons
f_p	distribution function related to electron holes
$[h^{\bullet}]$	concentration of electron holes
k	Boltzmann constant
k^*	rate constant

n	parameter related to ionization degree of defects
N	density of states
N_n	electron density
N_p	electron hole density
$[V_O^{\bullet\bullet}]$	concentration of doubly ionized oxygen vacancies
O_O	oxygen in its lattice site
S	thermopower
S_n	thermopower component related to electrons
S_h	thermopower component related to electron holes
t	time
w	mass
WF	work function
XPS	X-ray photoelectron spectroscopy
$[V_M]$	concentration of neutral metal vacancy
$[V_M']$	concentration of singly ionized metal vacancy
$[V_M'']$	concentration of doubly ionized metal vacancy
$[Y_{Zr}']$	yttrium on Zr site (singly ionized)
δ	thickness
ε_V	distance between Fermi energy and the conduction band
μ	mobility
μ_n	mobility of electrons
μ_p	mobility of electron holes
π	3.1415926
Σ	parameter characterizing the degree of equilibration
σ	electrical conductivity
σ_{ion}	ionic component of σ
σ_n	the component of σ related to electrons
σ_h	the component of σ related to electron holes
$\Delta\sigma_t$	change of σ after time t
$\Delta\sigma_\infty$	change of σ after infinite time
ϕ	work function
ϕ_s	the work function component related to the surface charge
ϕ_{dip}	the work function component related to surface dipoles
$\Delta\phi$	work function changes
χ	external work function
ψ	electrical potential

ABSTRACT

This chapter considers the electrical properties of the gas/solid interface. The paper is focused on the electrical phenomena accompanying the processes which take place during the equilibration at elevated temperatures in systems composed of oxygen and a metal oxide crystal.

The experimental approaches in studies of interface properties are analyzed. Several electrical methods based on measurements of work function, thermopower, and electrical conductivity and their applications in studies of defect-related properties are described. Several applied aspects of interface electrical phenomena are also briefly considered.

I. INTRODUCTION

Awareness is growing that properties of ceramic materials are determined by interfaces, their chemical composition, and structure. Therefore, a search for new materials with improved properties, which may meet the demand of new technologies, requires better understanding of interface properties of nonstoichiometric compounds, such as structure, chemistry, and the local transport kinetics. So far, little is known in this matter.

Literature reports on interfaces are mainly limited to metallic solids while little is known on ceramic materials, which are mainly ionic solids of nonstoichiometric compounds. The reason for the scarcity of literature reports on ceramic interfaces results from the substantial experimental difficulties in studies of these compounds. Even the most advanced surface-sensitive techniques have experimental limitations in the surface studies of materials. Most of these techniques are based on ion and electron spectroscopy, such as XPS, SIMS, LEED, AFM, and LEIS, and are still not adequate to characterize the complex nature of compounds. Namely, these surface techniques require an ultra-high vacuum and therefore may not be applied to determine surface properties during the processing of materials which takes place at elevated temperatures and under controlled gas phase composition. Consequently, the resultant experimental data allow one to derive only an approximate picture of the interface layer of compounds.

There has been an increasing interest in theoretical methods in studies of interface properties and related phenomena in compounds, such as adsorption and segregation. Again, these methods have several limitations in solving problems relevant to the interface layer. Most of these methods are still not adequate to determine the properties of nonstoichiometric compounds.

Electrical methods are very sensitive to the properties of solids, especially ionic solids.[1] Even changes of composition on the ppm level may result in substantial changes in semiconducting properties, such as n–p-type transitions, which may be determined by both electrical conductivity and thermopower. The WF is sensitive to surface properties on an atomic level. This is the reason the electrical methods are finding increasing applications in studies of defect-related properties of materials, based on nonstoichiometric compounds, such as transport properties. Electrical methods have been widely used with high accuracy in studies of nonstoichiometry and related concentrations of lattice defects at elevated temperatures and under gas phase of controlled composition.[1]

The electrical conductivity is probably the most frequently used electrical property in studies of ionic solids despite substantial complications related to its interpretation. The major complication is due to a complex physical constitution of this property, which involves both mobility and concentration terms. Moreover, the electrical conductivity is sensitive to intergranular contacts which, in the case of polycrystalline materials, result in difficulties in the determination of these properties, which are related to point defects. It has been realized that the construction of defect models cannot be based on a single property, such as electrical conductivity, especially in the case of ceramic materials.

The high sensitivity of electrical methods to materials properties and the possibility of their application at elevated temperatures have resulted in the development of several techniques that are sensitive to the interface layer, such as impedance spectroscopy[1-3] and the WF.[4,5]

Impedance spectroscopy is a unique electrical method that allows *in situ* studies of grain boundaries. This method, however, does not allow the monitoring of processes and related changes of properties as a function of time. This method was developed in more detail in the book by J. Ross MacDonald.[2]

So far, determination of the WF remains the most powerful method that enables the *in situ* monitoring of electrical effects accompanying chemical processes at the gas/solid interface, including adsorption, segregation, and related phenomena, such as lattice transport in the boundary layer. This method also allows the monitoring of the charge transfer at the gas/solid interface.

This chapter will be limited to the phenomena that take place at the gas/solid interface and will focus on WF measurements in studies of interface properties of ionic solids at elevated temperatures and under controlled gas phase composition. The illustrating experimental details will be limited to the oxygen/metal oxide interface.

II. DEFINITION OF TERMS

In this section several specific terms, to be used later in the text, are defined. This involves the terms that have been widely used in the literature in describing interface properties and interface phenomena with a large degree of freedom in the interpretation of their physical meaning.

A. INTERFACE

Both gas/solid and solid/solid interfaces and related electrical phenomena are considered in this chapter in more detail. Figure 4.1a schematically illustrates the gas/solid interfaces involving the adsorption layer, the surface charge layer, the space charge layer, and the bulk phase. Figure 4.1b illustrates the solid/solid interface. Also, several other terms have been applied in the consideration of local properties of interfaces.

1. Interphase and Interface

Interphase denotes a planar defect that demarcates two different phases, such as gas/solid, solid/solid, and liquid/solid interphases. The term interface involves interphases as well as all kinds of planar defects within individual solid phases such as grain boundaries.

2. Surface

This term has been used in the literature in relation to both the geometrical surface and the near-surface region of arbitrarily assumed thickness. Here, this term will be used in relation to the gas/solid interphase.

3. Adsorption Layer

This layer is localized at the surface and involves predominantly the species coming from the gas phase. The characteristic lateral interactions between the adsorbed species are essentially different from those in the crystal (Figure 4.1a).

4. Interface Layer

This term essentially corresponds to the layer in the vicinity of the interface, which involves a gradient of some property, such as chemical composition.

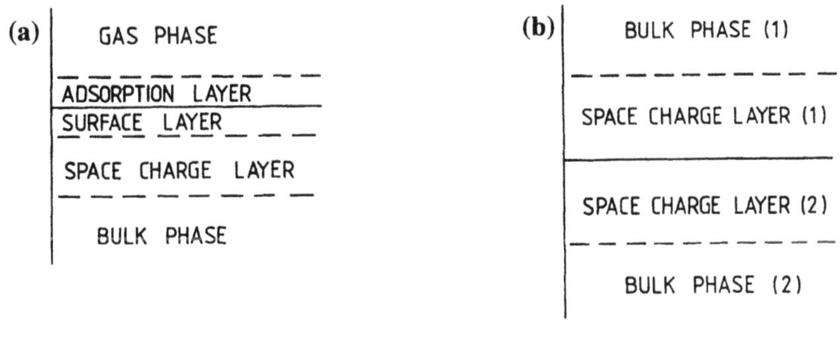

FIGURE 4.1. Schematic illustration of the (a) gas/solid and (b) solid/solid interfaces for an ionic solid involving the adsorbed layer, the surface layer, and the space charge layer.

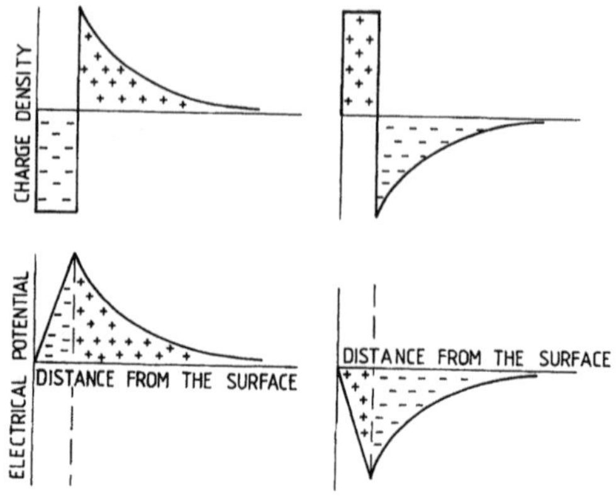

FIGURE 4.2. Schematic illustration of the surface and space charge layers of a nonstoichiometric compound corresponding to charge density (above) and electrical potential (below) for a negative and positive surface charge (left and right, respectively).

5. Near-Surface Layer

This layer corresponds to the vicinity of the surface within the solid phase. The thickness of this layer, involving the segregation-induced concentration gradients, is limited to several atomic layers.

6. Surface Layer

This is the outermost lattice layer, which is essentially limited to the presence of the surface charge formed as a result of ionization of defects within this layer. In the first approximation, the thickness of this layer is limited to dimensions of the outer crystal layer (Figure 4.1a).

7. Space Charge Layer

This is a diffusion layer involving the electric charge that compensates the surface charge localized in the surface layer (Figure 4.2). In the case of metals, this layer is limited to one or two crystal layers.[6] In the case of semiconducting crystals, the thickness of this layer, L, depends on the concentration of charge carriers, $[e', h^{\bullet}]$, and temperature, T, according to:

$$L = \left\{ \left(\varepsilon\varepsilon_o kT \right) / 2e^2 \left[e', h^{\bullet} \right] \right) \right\}^{1/2} \tag{4.1}$$

where ε and ε_o denote the dielectric permittivity of the sample and of vacuum, respectively; e is the electron charge; and k is the Boltzmann constant. For well-conducting crystals, such as Li-doped NiO, the boundary layer is limited to one crystal layer. For insulating crystals, such as Cr-doped NiO[7], or undoped MgO and Al_2O_3, the boundary layer thickness may assume substantial values.

8. Boundary Layer

Its meaning is similar to that of the near-surface layer and may be applied to both gas/solid and solid/solid interfaces. This term essentially corresponds to the crystal thickness containing electrical potential gradients.

9. Grain Boundary

This term corresponds to a bidimensional planar defect between two grains in a polycrystalline material.

10. Grain Boundary Layer

This is the crystal layer in the vicinity of the grain boundary containing segregation-induced gradients in crystal properties, such as chemical composition and electrical potential. Essentially, its meaning is the same as that in (8) with respect to the grain boundary.

11. Structural Deformation Layer (Bidimensional Interface Structure)

Segregation-induced concentration gradients of defects may result in structural deformations of the crystal layers in the direct vicinity of interfaces. These structural deformations lead to the formation of bidimensional interface structures which are not displayed in the bulk phase.

B. SEGREGATION

Both segregation and adsorption result in an enrichment (or depletion) of the interface layer. The driving force of both processes is the tendency to decrease the excess interface energy. Segregation may also be called adsorption from the solid phase.

Segregation is a process involving transport of lattice defects from the bulk phase toward the interface and, in consequence, results in an enrichment of the interface layer in these defects. Since the rate of segregation is determined by lattice diffusion, the segregation equilibrium may be established at elevated temperatures where the lattice transport becomes effective and relatively fast.

Segregation results in the formation of gradients in several properties within the interface layer, such as chemical composition, structure, and related transport properties. Figure 4.3 schematically illustrates the concentration gradient, c, and related electrical potential gradient, ψ, in the boundary layer. Several driving forces for segregation may be relevant, such as strain energy and electrostatic energy. The principles of the thermodynamics of segregation are described by J. Cabane and F. Cabane.[6]

C. DEFECT STRUCTURE

There are several types of structural lattice defects, such as point defects, line defects, and plane defects. The theory of defect chemistry concerns the point defects which are thermodynamically reversible.

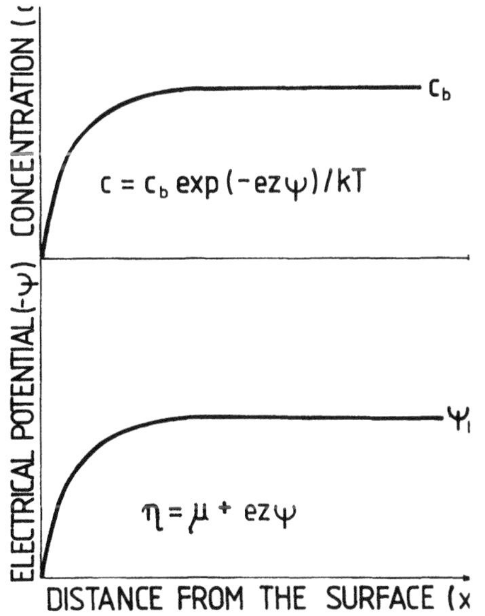

FIGURE 4.3. Schematic representation of the space charge layer, illustrating changes of concentration, c, and electrical potential, Ψ (η, electrochemical potential; μ, chemical potential; z, valency; e, elementary charge).

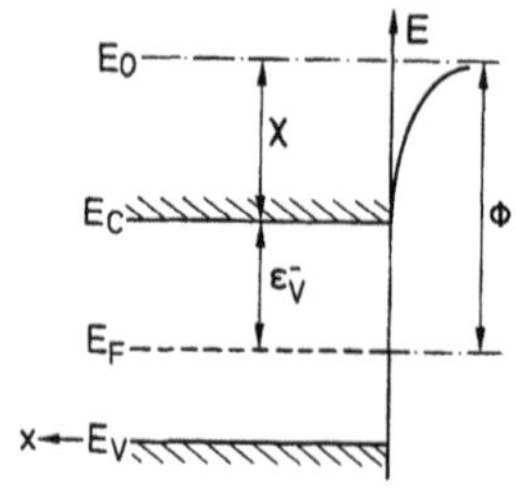

FIGURE 4.4. Surface band model of a p-type semiconductor.

In contrast to the term "structure," which describes a crystallographic order, the term "defect structure" corresponds to a description of both ionic and electronic point defects and their concentration as a function of temperature and partial pressure of the lattice constituents. The defect structure may be considered in terms of the principles of thermodynamics applied to defect equilibria.[1] Equilibrium constants for several defect equilibria for binary metal oxides have been reported by Kofstad.[1] These equilibria have been commonly considered in terms of bulk properties.

It will be shown below that the defect structure of the interface layer may be entirely different from that of the crystalline bulk. This difference has a substantial impact on material properties.

III. BASIC ELECTRICAL METHODS AND PROCEDURES IN STUDIES OF IONIC SOLIDS AT ELEVATED TEMPERATURES

This section describes several electrical properties, such as work function, thermopower, and electrical conductivity. These electrical properties have found wide application in studies of ionic solids at elevated temperatures.

A. WORK FUNCTION

The WF is defined as the energy required to remove an electron from a solid to a level outside its surface. The WF can also be defined as the difference between the Fermi energy level of the crystal and a reference level outside the crystal surface, where the electron is not subject to any interactions.

Figure 4.4 illustrates the band model of a p-type semiconductor surface without a surface charge. In this case, the WF (ϕ) includes the component related to the Fermi energy, ε_v (which is also called the internal WF), and the external WF, χ. For a charged surface (e.g., due to the presence of chemisorbed species) the WF involves an additional component, ϕ_S, corresponding with the surface charge (Figure 4.5). Finally, when the adsorbed molecules have a dipole moment, an additional component, ϕ_{dip}, is involved (Figure 4.6). Thus the WF assumes the following general form:

$$\phi = \varepsilon_v^- + \phi_S + \phi_{dip} + \chi \tag{4.2}$$

The most convenient method of measuring the WF of solids at elevated temperatures is based on the vibrating condenser method.[4,5,8,9] This method, initially developed by Kelvin,[10] was later substantially improved by Zisman,[11] Mignolet,[12] Chrusciel et al.,[13] and Besocke and Berger.[14] Most of the available vibrating condensers enable the determination of WF changes at or near room temperature. The construction of a vibrating condenser that enables WF measurements at elevated temperatures (High-Temperature Kelvin Probe) is described in References 5, 9, and 13 (Figure 4.7). This construction allows performance of measurements up to 1000°C under a controlled oxygen partial pressure.

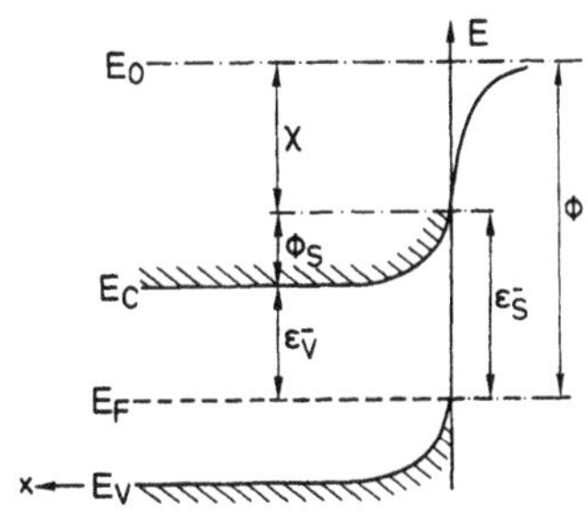

FIGURE 4.5. Surface band model of a p-type semiconductor involving a positive surface charge.

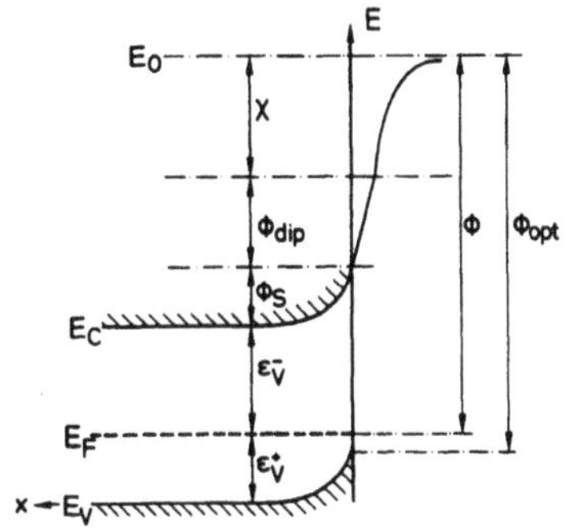

FIGURE 4.6. Surface band model of a p-type semiconductor illustrating the dipole moment component of the WF.

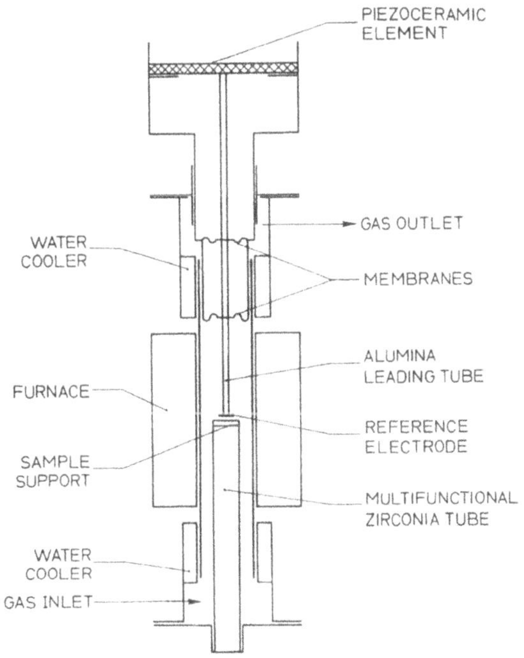

FIGURE 4.7. Schematic illustration of the high-temperature Kelvin probe.

The principle of the Kelvin method and its application in the study of surface properties of compounds have been reported in References 4, 5, and 9.

The WF is extremely sensitive to the surface state on an atomic level. Figure 4.8 shows, according to Haas et al.,[15] the WF changes of the barium–tungsten system, involving a Ba monolayer on W(100) at different stages of oxidation, along with the respective AES oxygen peak height, at several stages of oxidation. The AES peak height increases monotonously, while the WF changes allow the demarcation of regions corresponding to different

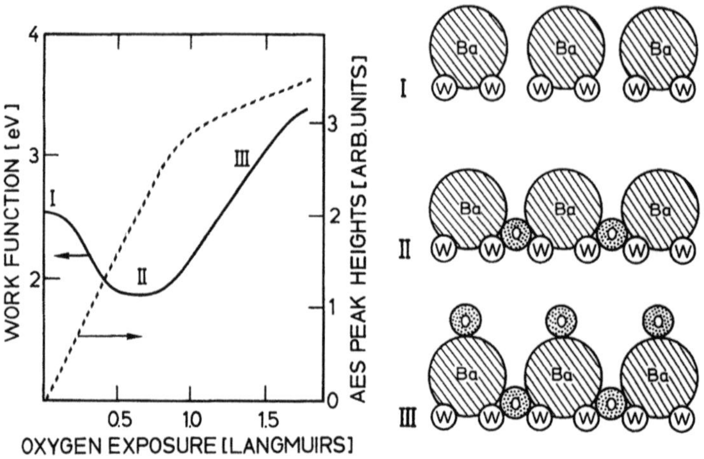

FIGURE 4.8. Changes in WF and corresponding AES peak heights vs. oxygen exposure for a Ba monolayer on W (100) (left), as well as the surface structures at stages I, II, and III of the oxidation (right), according to Haas et al.[15]

mechanisms of oxygen uptake. The right-hand side of the figure shows the proposed different configurations of the surface involving (I) the initial surface before oxygen adsorption, (II) the surface at an initial stage of oxidation involving the first oxygen layer and resulting in the formation of surface dipoles, and (III) the surface at the final oxidation stage involving oxygen build-up on top of Ba.

The absolute WF value of compounds, especially of ceramic materials, has a complex physical meaning and, moreover, it is difficult to determine this experimentally, especially at elevated temperature. On the other hand, measurement of WF changes enables the monitoring of several processes that take place at the surface, such as adsorption and segregation.

The evaluation of absolute WF values is possible using a semiempirical model derived by Gordy and Thomas,[16] who have correlated the WF with electronegativity, X, of metallic elements (Figure 4.9). Yamamoto et al.[17] have used the model of Gordy and Thomas[16] to evaluate the WF of compounds such as carbides, nitrides, silicides, borides, and oxides (Figure 4.10). As seen, the correlation is much better for metals than in the case of compounds. The main complication for compounds is that their WF values depend substantially on the deviation from stoichiometry. This is not taken into account in the model of Gordy and Thomas.[16] The effect of nonstoichiometry on the WF is apparently the reason for the large scatter shown in Figure 4.10.

It has been shown that well-defined and reproducible surface properties of nonstoichiometric compounds should correspond with the gas/solid equilibrium.[7,18,19] In the case of metal oxides, the gas/solid equilibrium can be achieved at elevated temperatures when the mobility of lattice defects is sufficiently high. The equilibrium is then defined by the temperature and oxygen activity of the gas phase. In this case, the WF data can be considered to be material data that are independent of the experimental procedure and are determined only by the equilibrium conditions. Figure 4.11 illustrates the CPD between Pt and FeO within the stability range of the FeO phase. In the first approximation, the change in the CPD is the same as that of the WF of the FeO specimen. As seen, the WF data are identical for both reduction and oxidation experiments. Also, the WF change exhibits a slope that corresponds with the change in oxide nonstoichiometry and related change in Fermi energy.

Figure 4.12 illustrates the changes in the thermionic current of YSZ, which corresponds with the change of WF, as a function of equilibrium partial pressure of O_2.[20] The slope, n, of this dependence is in good agreement with the defect model of zirconia, which requires that:

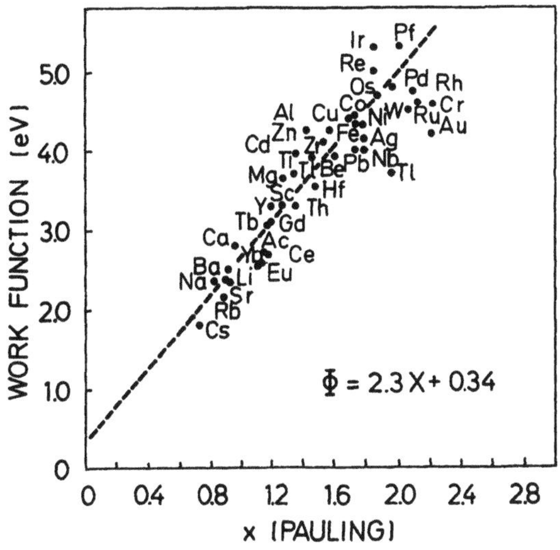

FIGURE 4.9. WF of metals as a function of their electronegativity. (Adapted from Gordy, W. and Thomas, W., *J. Chem. Phys.*, 1956, 24, 439–446.)

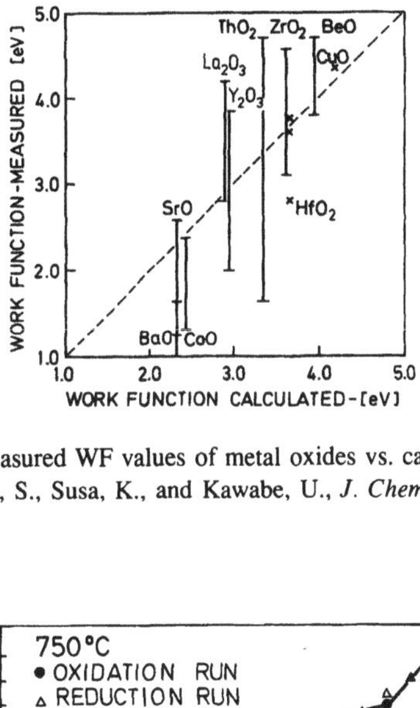

FIGURE 4.10. Experimentally measured WF values of metal oxides vs. calculated values using the Gordy and Thomas equation. (From Yamamoto, S., Susa, K., and Kawabe, U., *J. Chem. Phys.*, 1974, 60, 4076–4080. With permission.)

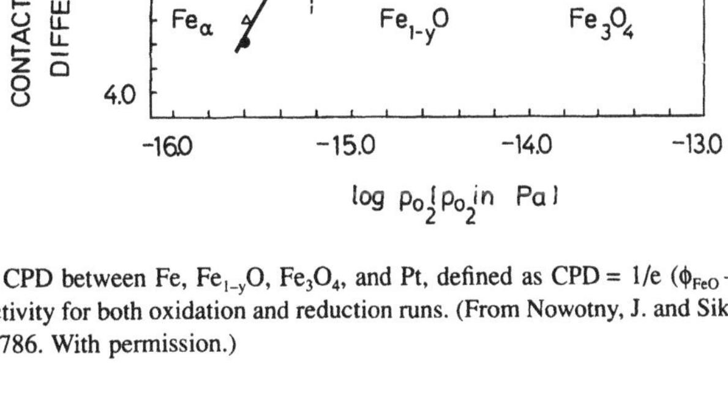

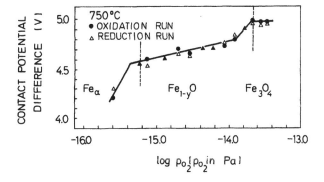

FIGURE 4.11. The CPD between Fe, $Fe_{1-y}O$, Fe_3O_4, and Pt, defined as CPD = 1/e ($\phi_{FeO} - \phi_{Pt}$), as a function of equilibrium oxygen activity for both oxidation and reduction runs. (From Nowotny, J. and Sikora, I., *J. Electrochem. Soc.*, 1978, 125, 781–786. With permission.)

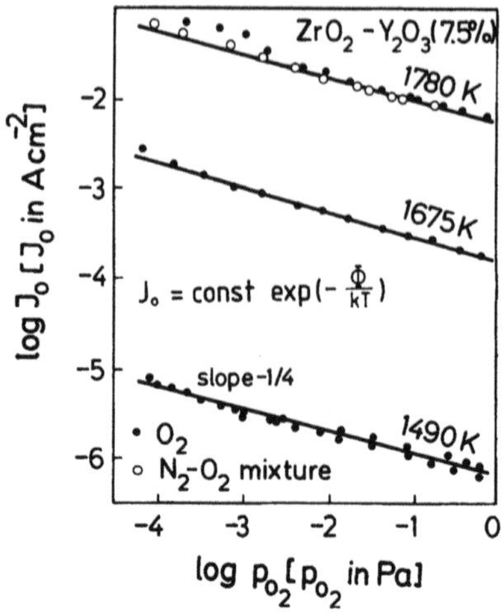

FIGURE 4.12. Logarithm of thermionic current of yttria-stabilized zirconia, corresponding with its WF, as a function of equilibrium oxygen activity. (From Odier, P. Riflet, J.C., and Loup, J.P., in *Reactivity of Solids,* Haber, J., Dyrek, K., and Nowotny, J., Eds., Elsevier, Amsterdam, 1982, 458–466. With permission.)

$$[e'] = \text{const } p(O_2)^{1/n} \tag{4.3}$$

where n may be determined from the relation:

$$1/n = (1/kT)\ \{\partial\phi/\partial \ln\ p(O_2)\} \tag{4.4}$$

and [e′] is the concentration of electrons. Assuming that the lattice charge neutrality condition of YSZ requires that:

$$[V_O^{\bullet\bullet}] = 1/2\ [Y'_{Zr}] \tag{4.5}$$

where $[V_O^{\bullet\bullet}]$ is the concentration of oxygen vacancies and $[Y'_{Zr}]$ is the concentration of yttria incorporated into Zr sites of ZrO_2, the $p(O_2)$ exponent is:

$$\frac{1}{n} = 1/4 \tag{4.6}$$

As a result of the above considerations, only the WF data taken at the gas/solid equilibrium may be considered to be characteristic material data. Then the crystal nonstoichiometry is well defined. Unfortunately, most of the WF data reported so far for compounds were determined below equilibrium and, therefore, are not well defined.[16,17]

B. THERMOPOWER

The measurement of thermopower is based on the Seebeck effect, which consists of the generation of a difference in the electrical potential across imposed temperature gradient.[21] For a nondegenerate semiconductor, in which one type of electronic charge carrier predominates, one may formulate interrelationships between thermopower (S) and the concentration of electronic carriers. For n-type and p-type regimes the following expressions may be written:[18]

TABLE 4.1
Physical Meaning of Symbols in Equations (4.7) and (4.8)

Symbol	Description	
k	Boltzmann constant, $k = 8.6167 \times 10^{-5}$ eV K^{-1}	
e	Elementary charge, $e = 1.60206 \times 10^{-19}$ C	
[e′]	Concentration of electron carriers	
[h˙]	Concentration of electron holes	
	Band model	**Hopping model**
N_n	Density of states in the conduction band	$N_n = \beta_e\{1 - [e']\} \approx \beta_e$
N_h	Density of states in the valence band	$N_h = \beta_h\{1 - [h^{\cdot}]\} \approx \beta_h$ β is the degeneracy factor including both spin and orbital degeneracy of electron carriers {β = 2 (spin degeneracy), β = 1 (Hikes Formula)}[24]
A_n A_h	Parameters depending on scattering mechanism of electrons (A_n) and electron holes (A_h) A = 2,4[21]	$A_n = S_v/k$ $A_h = S_v/k$ (S_v is vibrational entropy associated with ions surrounding the polaron; $S_v \approx 10$ μeV/K)[25] $A_n = A_h \approx 0$

$$S_n = -\frac{k}{e}\left[\ln\left(\frac{N_n}{[e']}\right) + A_n\right] \tag{4.7}$$

$$S_h = \frac{k}{e}\left[\ln\left(\frac{N_h}{[h^{\cdot}]}\right) + A_h\right] \tag{4.8}$$

where the meaning of the symbols in Equations (4.7) and (4.8) is given in Table 4.1.[21-25] Assuming Maxwell–Boltzmann statistics of electrons in a nondegenerate semiconductor, the following relationships between the concentration of electron carriers and E_F exist:[26-28]

$$[e'] = N_n \exp\left(\frac{E_F - E_C}{kT}\right) \tag{4.9}$$

$$[h^{\bullet}] = N_h \exp\left(\frac{E_V - E_F}{kT}\right) \tag{4.10}$$

where E_C and E_V denote, respectively, the energy of the bottom of the conduction band and the top of the valence band and E_F is the Fermi energy level, which is a parameter in the Fermi–Dirac statistics:

$$f_n(E) = \left\{1 + \exp\left(E_i - E_F\right)/kT\right\}^{-1} \tag{4.11}$$

where $f_n(E)$ is the probability of occupation of the level E_i. There is the following relation between $f_n(E)$ and that for electron holes:

$$f_h(E) = 1 - f_n(E) \tag{4.12}$$

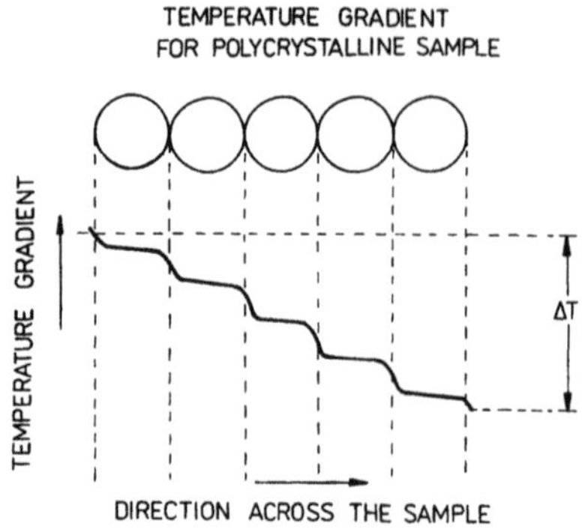

FIGURE 4.13. Schematic illustration of a temperature gradient distribution across a polycrystalline specimen.

When the Fermi energy level, E_F, is only slightly populated $\{(E_i - E_F) \gg kT\}$ then the number of electrons and electron holes, respectively in the conduction and the valence band, may be expressed by Equations (4.9) and (4.10). Then Equations (4.7) and (4.8) assume the following forms:

$$S_n = \frac{E_F - E_C}{eT} - \frac{k}{e} A_n \tag{4.13}$$

$$S_h = \frac{E_F - E_V}{eT} + \frac{k}{e} A_h \tag{4.14}$$

The thermopower is essentially a bulk-sensitive property. However, in the case of polycrystalline materials, the temperature gradient across the specimen is not uniform and thermopower exhibits higher values at grain boundaries (Figure 4.13). Consequently, the thermopower is sensitive to grain boundaries. The sensitivity depends on the grain size and the bulk-to-interface thermal conductivity coefficient ratio.

Figure 4.14 illustrates the thermopower of undoped $BaTiO_3$ as a function of its equilibrium oxygen activity within the n- to p-type transition range.[26] As seen, the thermopower indicates the transition state (S = O) with high precision.

The experimental determination of the thermopower performed by using two different approaches is illustrated in Figures 4.15 and 4.16. The first (Figure 4.15) involves the determination of the Seebeck coefficient across a sample for several temperature gradients imposed by microheaters located on both sides of the sample. The second approach is based on the temperature gradient in the experimental chamber. The gradient can be changed by changing the position of the specimen in the chamber. The Seebeck coefficient is determined from several Seebeck voltages imposed across the specimen by different temperature gradients. The measurement of thermopower of ionic solids at elevated temperatures is described in References 26, 27, and 29.

The thermopower becomes predominantly sensitive to grain boundaries when the grain size becomes very small.[19]

C. ELECTRICAL CONDUCTIVITY

The electrical conductivity (σ) of a semiconducting compound can be expressed as a sum of the electronic and ionic components:

$$\sigma = \sigma_n + \sigma_h + \sigma_{ion} \tag{4.15}$$

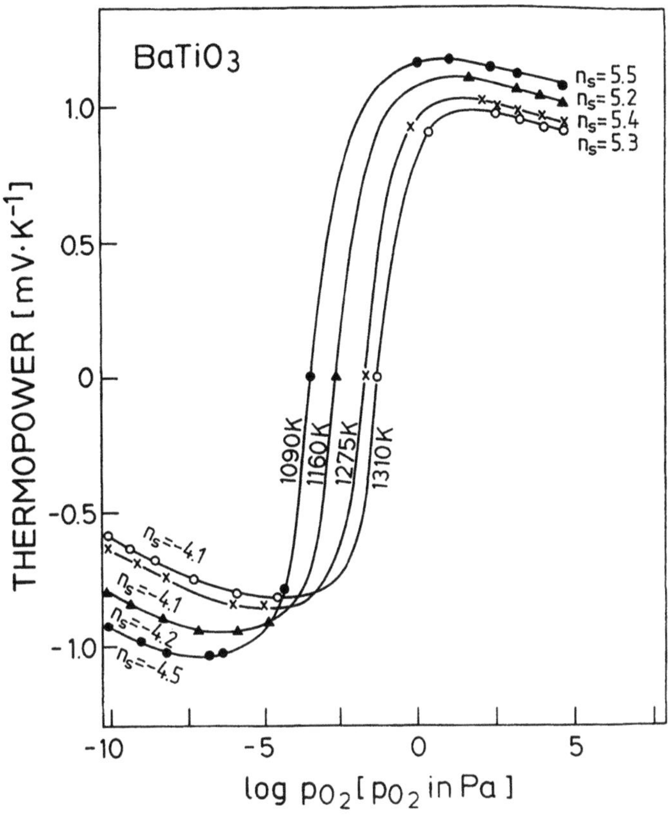

FIGURE 4.14. Thermopower of undoped $BaTiO_3$ as a function of log $p(O_2)$ in the temperature range 1090 to 1310 K. (From Nowotny, J. and Rekas, M., *Ceram. Int.*, 1994, 20, 217. With permission.)

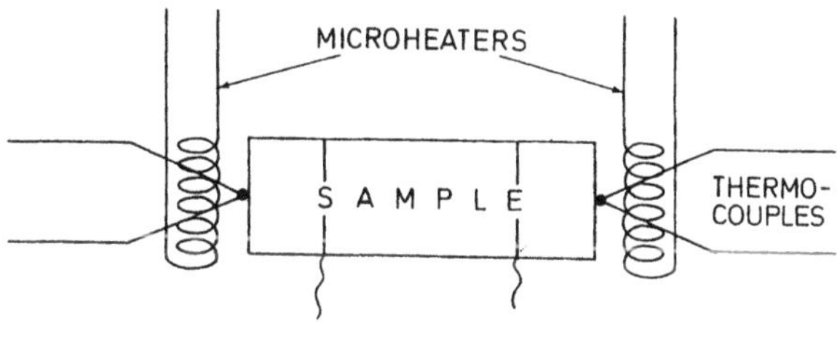

FIGURE 4.15. The principle of the determination of thermopower by imposition of a desired temperature gradient using microheaters. The two thermocouples serve also as external current probes for measurement of electrical conductivity. The internal probes serve for the determination of the voltage drop along the specimen.

where the subscripts n, h, and ion correspond with electrons, electron holes, and ions, respectively. The electronic components can be expressed as:

$$\sigma_n = e\mu_n[e'] \tag{4.16}$$

$$\sigma_h = e\mu_h[h^\bullet] \tag{4.17}$$

where $[e']$ and $[h^\bullet]$ are the concentrations of electrons and electron holes, respectively, and μ_n and μ_h are the respective mobility terms. As seen, the electrical conductivity is a complex

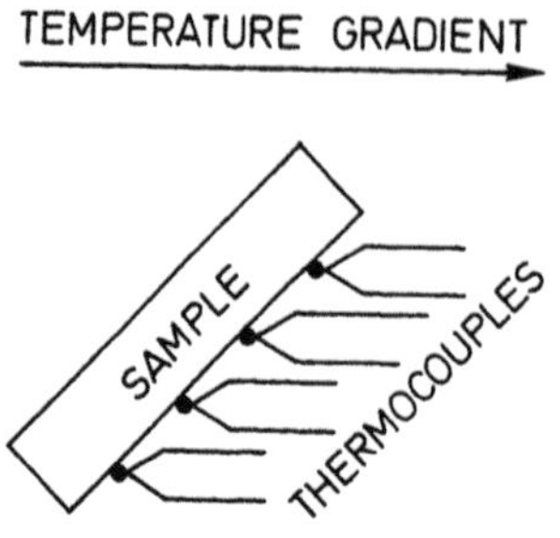

FIGURE 4.16. The principle of the determination of thermopower in the temperature gradient of a furnace.

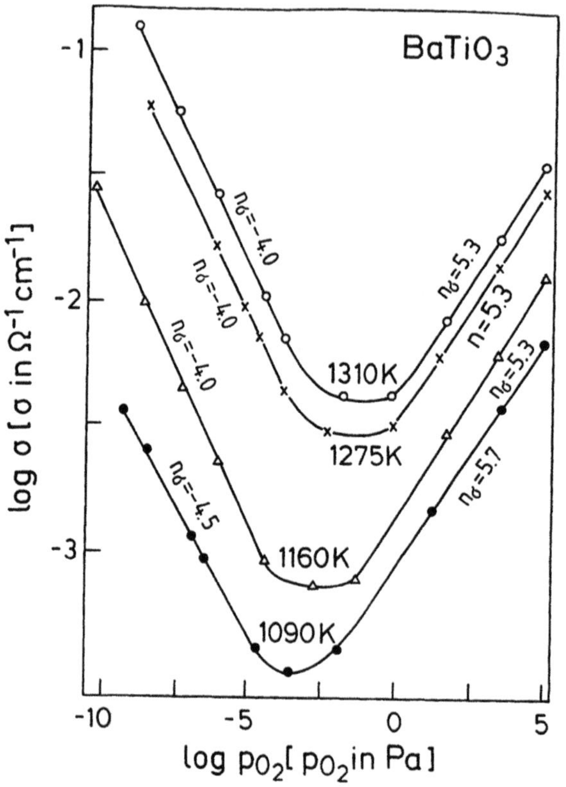

FIGURE 4.17. Electrical conductivity of undoped $BaTiO_3$ as a function of oxygen partial pressure in the temperature range 1090–1310 K. (From Nowotny, J. and Rekas, M., *Ceram. Int.*, 1994, 20, 217. With permission.)

property involving concentration and mobility terms of all charge carriers. Consequently, its interpretation is difficult. Parallel measurements of both thermopower and electrical conductivity enable the determination of the two terms (see Section III.D).

Figure 4.17 illustrates the electrical conductivity as a function of equilibrium oxygen activity for undoped $BaTiO_3$ in the n- to p-type transition range.[26] As seen, the conductivity exhibits a minimum at a $p(O_2)$ close to where $S = 0$ (Figure 4.14). Precise determination of the transition point from the minimum in the conductivity requires knowledge of the mobility terms (see Section III.D).

An extensive analysis of the electrical conductivity of binary metal oxides in terms of their defect structures has been given by Kofstad.[1]

The most common method for the determination of dc electrical conductivity involves the four-probe method. Figure 4.15 schematically illustrates the setup for simultaneous measurement of thermopower and electrical conductivity. In the determination of electrical conductivity by using the setup in Figure 4.16, one may ignore the small temperature gradient required to determine thermopower (1 to 3 K).

Impedance spectroscopy analysis enables the determination of the conductivity components related to grain boundaries and the bulk phase.[2]

D. THE JONKER ANALYSIS

Simultaneous measurements of both σ and S in the gas/solid equilibrium for an amphoteric semiconductor (exhibiting an n–p transition) enables the determination of semiconducting properties and the related defect structure based on the Jonker-type analysis.[28] The Jonker analysis involves both S and σ measured as a function of equilibrium $p(O_2)$. However, the Jonker analysis does not require a knowledge of oxygen activity corresponding with the measured electrical properties. Accordingly, the Jonker analysis enables the elimination of any error in the determination of the oxygen activity.

The relationship between the thermopower and electrical conductivity can be expressed by the following equation:[26, 29]

$$S = \pm\ k/2e\left(E_g/kT + A_n + A_h\right)\left(1 - \sigma_{min}^2/\sigma^2\right)^{1/2} - \\ -(k/e)\ln\left(\sigma/\sigma_{min}\right)\left\{1 \pm \left[1 - \left(\sigma_{min}^2/\sigma^2\right)^{1/2}\right] + (k/2e)\ln(bd)\right. \tag{4.18}$$

where

$$d = \left(N_h/N_n\right)\exp\left(A_h - A_n\right) \tag{4.19}$$

$$b = \mu_h/\mu_n \tag{4.20}$$

σ_{min} in Equation (4.18) is the lowest value of σ which, in the first approximation, corresponds with an n–p transition (when $\mu_n = \mu_h$) and E_g is the band gap.

The principles of the Jonker analysis are illustrated in Figure 4.18. The analysis enables the determination of quantities such as (a) the band gap, (b) the ratio of mobility terms, and (c) the density of states of electronic carriers. The analysis involves the determination of the following quantities:

$$B = k/2e\ \left(E_g/kT + A_n + A_h\right) \tag{4.21}$$

and

$$D = k/2e\ \ln(bd) \tag{4.22}$$

Figure 4.19 shows a Jonker plot for undoped $BaTiO_3$ based on the data shown in Figures 4.14 and 4.17.

Details of the Jonker-type analysis for nonstoichiometric oxides, such as TiO_2 and $BaTiO_3$, are reported in Reference 30. These oxides both exhibit n- to p-type transitions.

The Jonker Equation (4.18) can be solved only for experimental data for both S and σ within the n–p transition. In many cases, however, these experimental data are available within the n- or the p-type regime. Then, a modified Jonker analysis enables the determination of semiconducting properties.[26,29]

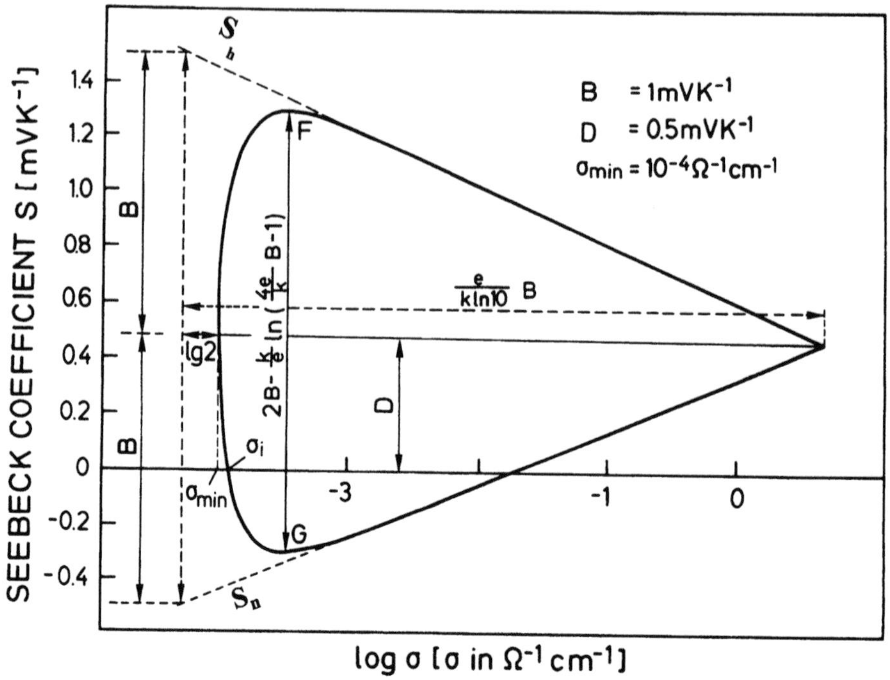

FIGURE 4.18. Schematic representation of the Jonker-type plot of thermopower, S, as a function of log σ. Symbols are explained in Section III.D.[26,27,29]

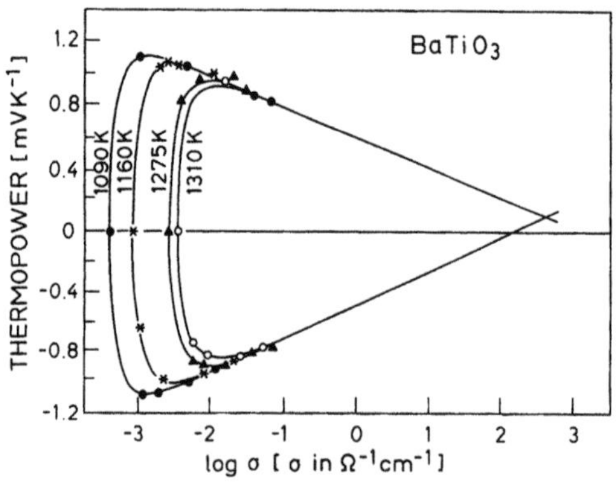

FIGURE 4.19. The Jonker-type plot for undoped $BaTiO_3$ in the temperature range 1090 to 1310 K (using the experimental data from Figures 4.14 and 4.17). (From Nowotny, J. and Rekas, M., *Ceram. Int.*, 1994, 20, 217. With permission.)

E. CONCLUSIONS

Electrical methods are very sensitive to defect-related properties of nonstoichiometric compounds and may be used for the determination of defect disorder and related (e.g., semiconducting) properties. Electrical methods are very useful in monitoring the processes involved in gas/solid equilibration.

Electrical properties, such as the WF, thermopower, and electrical conductivity, exhibit different sensitivities to interfaces. The WF is the electrical property most sensitive to surface properties and thus its measurement may be used in monitoring surface processes at elevated temperature. This property may be used in the determination of the defect structure of the outer surface layer as well as to monitor the processes that take place at the surface and within the outer surface layer.

IV. ELECTRICAL EFFECTS AT THE METAL OXIDE/OXYGEN INTERFACE

Interactions in gas/solid systems are important from the viewpoint of several applications, such as heterogeneous catalysis, high-temperature oxidation of metals, and sintering of ceramics. In this section, the mechanism and kinetics of processes taking place at the gas/solid interface will be considered for the metal oxide/oxygen system. The considerations will focus on the electrical effects accompanying interactions between gaseous oxygen and the surface of nonstoichiometric compounds.

Defects in the oxide lattice are formed at the surface as a result of interactions between the surface and gaseous oxygen. The mechanism of their formation initially involves adsorption of oxygen on the active adsorption sites, resulting in the formation of several ionized oxygen species, and then their incorporation into the surface layer. The process of incorporation involves the formation of defects within the near-surface layer. Then the defects are propagated into the bulk phase.

The mechanism of oxygen incorporation will be illustrated for a metal-deficient oxide involving cation vacancies and electron holes as the predominant lattice defects.

A. PHYSICAL ADSORPTION

Oxygen adsorption takes place on surface-active adsorption sites, such as surface metal ions, resulting in the formation of molecular or atomic species. The following equilibria may be considered for oxygen adsorption:

$$O_{2(gas)} \rightleftarrows O_{2(ads)} \tag{4.23}$$

$$O_{2(ads)} \rightleftarrows 2O_{(ads)} \tag{4.24}$$

The above forms of adsorption, which are predominant at lower temperatures, play an important role in catalytic oxidation processes.

The formation of electrically neutral adsorbed oxygen species does not involve charge transfer between the solid and the species. Thus, this kind of adsorption does not result in changes of electrical conductivity. However, changes in the WF may be expected if changes in dipole moment of the adsorbed species are involved. Major WF changes are observed during the formation of ionized surface oxygen species.

B. IONOSORPTION

Both molecular and atomic adsorbed oxygen species form acceptor centers which may be ionized at elevated temperatures. The degree of their ionization depends on the location of the centers in the band model in relation to the Fermi energy level of the crystal. The ionization degree of the surface centers is described by the Fermi–Dirac statistics.

The formation of chemisorbed oxygen species may be represented by the following equilibria:

$$O_{2(ads)} \rightleftarrows O_2^- + h^{\bullet} \tag{4.25}$$

$$O_2 \rightleftarrows O_2^{2-} + 2h^{\bullet} \tag{4.26}$$

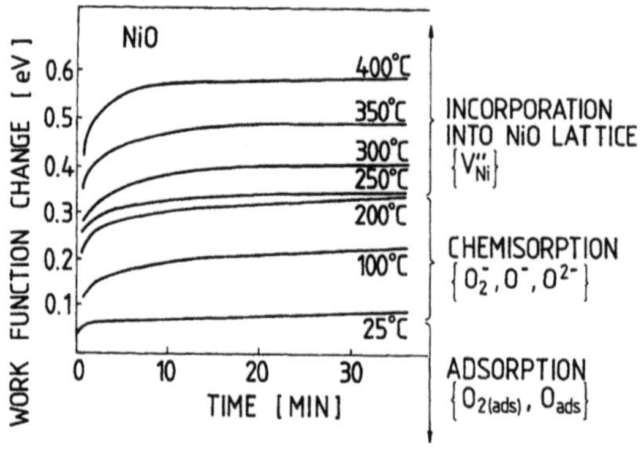

FIGURE 4.20. WF changes of undoped NiO during sorption of oxygen in the temperature range 20 to 400°C and related mechanism of oxygen interaction with the NiO lattice. (From Nowotny, J., in *Science of Ceramic Interfaces*, Nowotny, J., Ed., Elsevier, Amsterdam, 1991, 79–204. With permission.)

$$1/2\ O_{2(ads)} \rightleftarrows O^- + h^\bullet \tag{4.27}$$

$$O^- \rightleftarrows O^{2-} + h^\bullet \tag{4.28}$$

Figure 4.20 illustrates the WF changes of NiO during isothermal chemisorption of oxygen on an initially outgassed surface in the temperature range between 20 and 400°C. In all cases a sudden increase of the WF is observed followed by a slow change, approaching a stationary value that corresponds with chemisorption equilibrium at $p(O_2) = \text{const}$.[4,5] As seen, the extent of the electrical effect increases with temperature. This effect is determined by the surface coverage and ionization degree of the adsorbed species.

C. INCORPORATION OF OXYGEN INTO THE BOUNDARY LAYER

Reaction (4.28) results in the formation of doubly ionized oxygen species. These species, which already require stabilization in the crystal field, may be considered as cation vacancies in the outer surface layer. Using the Kröger–Vink notation,[31] the formation of metal vacancies in a metal oxide, MO, may be illustrated by the following equilibria:

$$1/2\ O_2 \rightleftarrows \left(V_M''\right)^* + 2h^\bullet + O_o \tag{4.29}$$

$$\left(V_M''\right)^* + h^\bullet \rightleftarrows \left(V_M'\right)^* \tag{4.30}$$

$$\left(V_M'\right)^* + h^\bullet \rightleftarrows \left(V_M\right)^* \tag{4.31}$$

where $(V_M)^*$, $(V_M')^*$ and $(V_M'')^*$ denote, respectively, neutral, singly ionized, and doubly ionized metal vacancies. The asterisk indicates the location of the defect in the near-surface layer, where due to specific interactions, defects should be distinguished from those in the periodic lattice of the bulk phase.

The characteristics of the isothermal WF changes of an oxide crystal, after oxygen activity in the gas phase is suddenly increased (oxidation), depend on the oxidation procedure.

At $p(O_2) = \text{const}$, the WF initially sharply increases, followed by a constant WF, which corresponds with chemisorption equilibrium (Figure 4.21). This is possible when the lattice

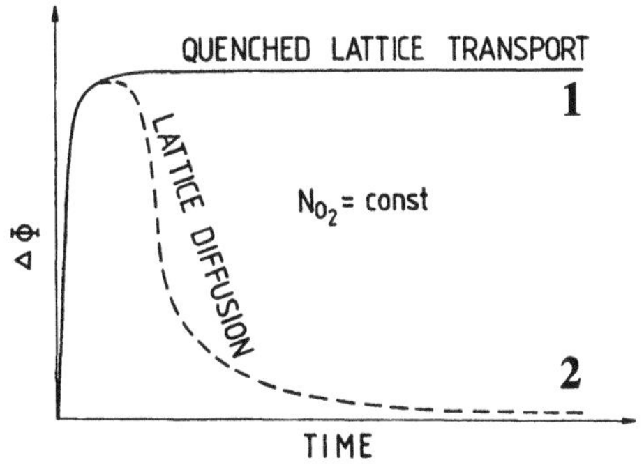

FIGURE 4.21. Schematic illustration of WF changes (1) during oxygen sorption at $p(O_2)$ = constant and (2) after sorption of a small oxygen dose comparable to a fraction of the monolayer coverage.

transport is quenched. Then the equilibrium value of $\Delta\phi$ is determined by (a) the concentration of the chemisorbed oxygen species in the adsorbed layer and (b) their ionization degree.

When this experimental procedure is performed with a constant amount of oxygen admitted to the reaction chamber, which is limited to a monolayer coverage, then the WF initially sharply increases, as a result of charging of the surface during chemisorption, and then decreases as a result of discharging of the surface due to incorporation of oxygen into the boundary layer (Figure 4.21). The rate of the discharging process is controlled by the transport of defects through the boundary layer.[32] Using a particular solution of Fick's second law, it is possible to determine the chemical diffusion coefficient, D, in the boundary layer:[33]

$$D = (1+z)\frac{\varepsilon L^2}{30\pi^2\left[h^\bullet\right]\tan^2\gamma} \tag{4.32}$$

where $\tan\gamma$ is the slope of the linear part of the relation $\Delta\phi/\phi_{max}$ vs. $1/t^{1/2}$, ϕ_{max} is the maximum WF value, ε is the dielectric constant, k is the Boltzmann constant, $[h^\bullet]_b$ is the concentration of electron holes in the bulk, and t is the time. Figures 4.22 and 4.23 show the experimental WF changes for NiO in the temperature range 175 to 425°C[30] and the normalized WF changes, $\Delta\phi/\phi_{max}$, as a function of $1/t^{1/2}$, respectively.[33]

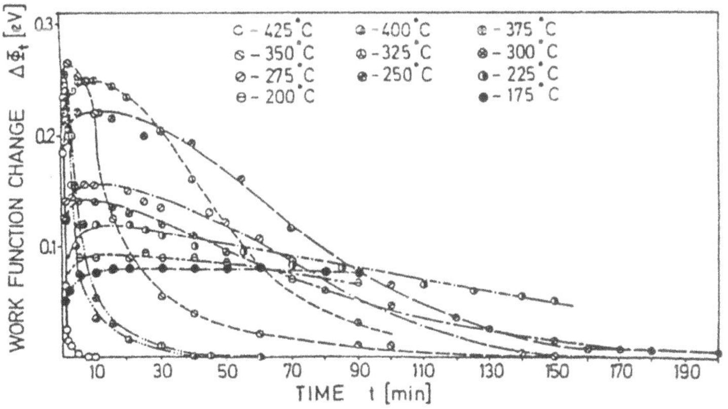

FIGURE 4.22. WF changes of undoped NiO during sorption of a small oxygen dose comparable to a fraction of the monolayer coverage in the temperature range 175 to 425°C. (From Nowotny, J., *J. Mater. Sci.*, 1977, 12, 1143–1160. With permission.)

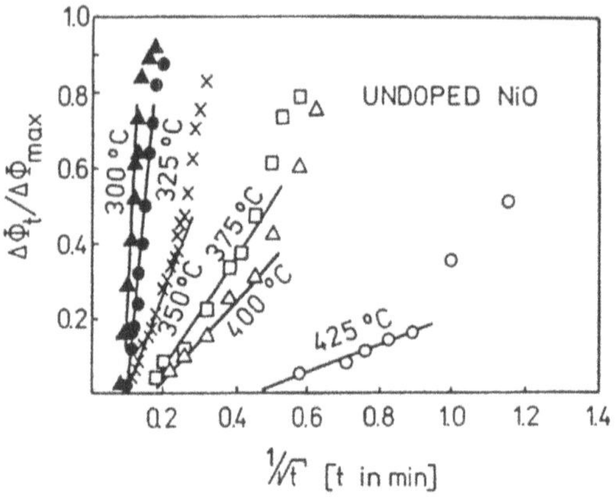

FIGURE 4.23. The normalized work function changes of undoped NiO (from Figure 4.22) as a function of $1/(t)^{1/2}$. (From Adamczyk, Z. and Nowotny, J., *J. Electrochem. Soc.*, 1980, 127, 1117–1120. With permission.)

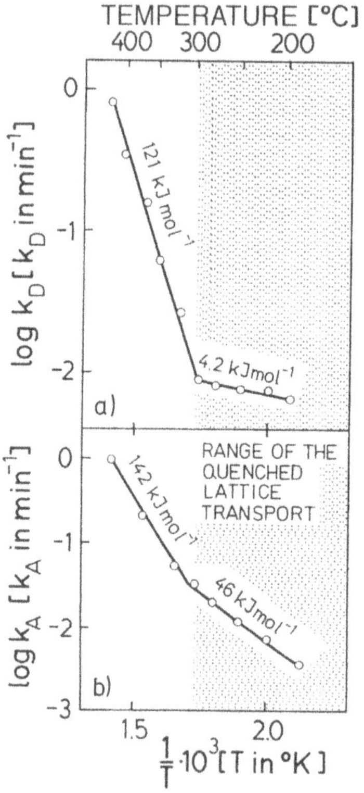

FIGURE 4.24. Arrhenius plot of rate constants corresponding with the discharging kinetics (k_D) and oxygen consumption (k_A). (Adapted from Nowotny, J., *J. Mater. Sci.*, 1977, 12, 1143–1160.)

The Arrhenius plot of the rate constant of the surface discharging (k_D) and the rate constant of oxygen sorption (k_A) indicate that the mechanism of interaction in the NiO/oxygen system exhibits a change at 300°C involving a diffusion-controlled process above 300°C and a surface diffusion regime below this temperature[32] (Figure 4.24).

D. BULK EQUILIBRATION KINETICS

Equilibrium in the metal oxide/oxygen system and related defect concentrations are determined by the oxygen activity in the gas phase and the temperature. When one of these is changed, then the system tends to achieve a new equilibrium state. The rate of the re-equilibration is determined by chemical diffusion in the oxide crystal.

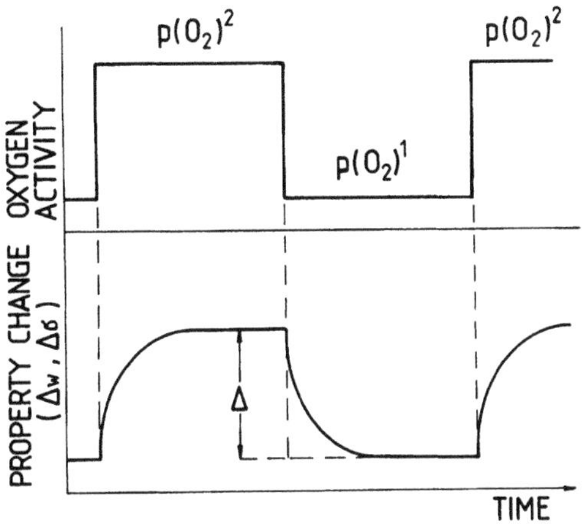

FIGURE 4.25. Schematic illustration of changes of a defect-related property (weight or electrical conductivity) imposed by changes of $p(O_2)$ for a nonstoichiometric oxide crystal, as a function of time.

Re-equilibration experiments may be performed either isothermally — where a new equilibrium is imposed by the suddenly changed $p(O_2)$ — or by changes of temperature at $p(O_2)$ = const.

When $p(O_2)$ is changed to a new value (within a single-phase region) over an initially equilibrated oxide crystal, then a new nonstoichiometry (and related concentration of defects) is imposed at the surface almost immediately. The new nonstoichiometry is then propagated into the crystalline bulk to regain new chemical potentials throughout the crystal. The rate of the propagation is determined by chemical diffusion. Therefore, the chemical diffusion coefficient may be determined from the re-equilibration kinetics. In the case of metal-deficient oxides, such as NiO and CoO, the mobile ionic species are metal ions, diffusing via cation vacancies while oxygen remains relatively immobile. Oxidation results in motion of metal ions to the surface (or motion of metal vacancies from the surface to the bulk); reduction involves the reverse transport.

The equilibration process in oxide systems can be monitored by changes in a bulk crystal property that is nonstoichiometry sensitive, such as weight, Δw, electrical conductivity, $\Delta\sigma$, or thermopower, ΔS. The kinetics of the re-equilibration process, provoked by isothermal changes of $p(O_2)$, are schematically illustrated in Figure 4.25. The rate of the equilibration is determined by chemical diffusion involving the transport of defects under a gradient of chemical potential (ambipolar diffusion). Chemical diffusion coefficients may be determined from the relevant solutions of Fick's second law for long and short times, respectively:[34,35]

$$D = \frac{0.934}{t\left(\frac{1}{a^2}+\frac{1}{b^2}+\frac{1}{c^2}\right)} \log \frac{0.533}{1-\Sigma} \tag{4.33}$$

where

$$\Sigma = \left(\Delta\sigma_t / \Delta\sigma_\infty\right) \tag{4.34}$$

a, b, and c are crystal dimensions, and

$$\Sigma^2 = \left\{\left(4\tilde{D}t/\pi\right)\left(q/v\right)^2\right\} \tag{4.35}$$

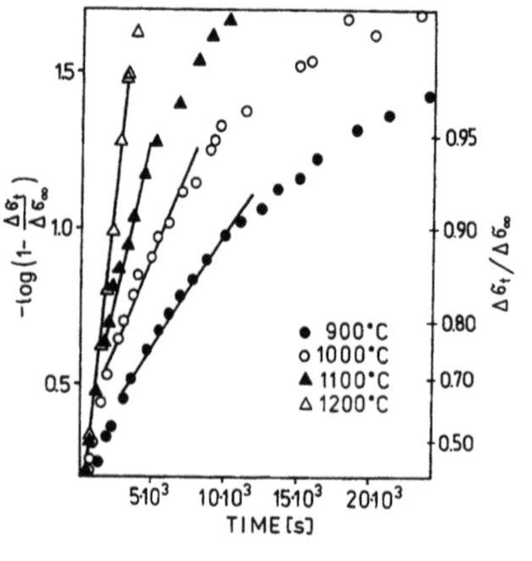

FIGURE 4.26. Normalized changes of the electrical conductivity of undoped NiO during the equilibration (logarithmic plot). (From Nowotny, J., Oblakowski, J., and Sadowski, A., *Bull. Pol. Acad. Sci. Chem.*, 1985, 33, 99–119. With permission.)

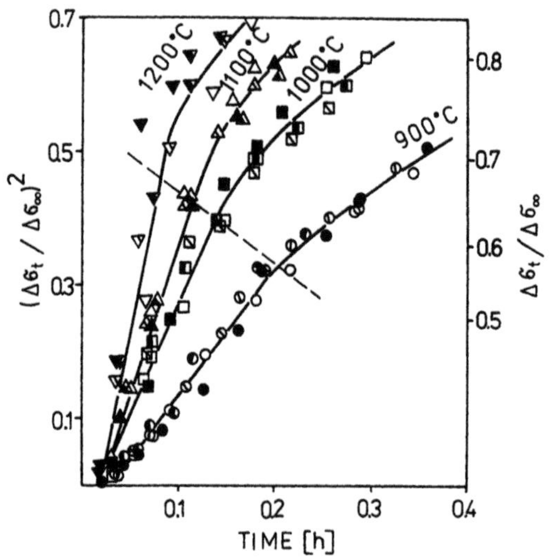

FIGURE 4.27. Normalized changes of the electrical conductivity of undoped NiO during equilibration (parabolic plot). (From Nowotny, J., Oblakowski, J., and Sadowski, A., *Bull. Pol. Acad. Sci. Chem.*, 1985, 33, 99–119. With permission.)

where $\Delta\sigma_t$ and $\Delta\sigma_\infty$ denote the change of σ after time t and after infinite time, respectively; q and v denote the surface area and the volume of the specimen. Equations (4.33) and (4.35) are valid for $\Sigma > 0.6$ and $\Sigma < 0.5$, respectively, assuming that in the equilibration range under study, only one type of ionized defect predominates and that the mobility term does not change with nonstoichiometry. Figures 4.26 and 4.27 show the respective plots for undoped NiO.[34] Both equations result in consistent data for the chemical diffusion coefficient within their validity range (Figure 4.28).

Equations (4.33) and (4.35) have been used widely in the determination of diffusion data of nonstoichiometric compounds assuming that the gas/solid equilibration kinetics are bulk diffusion controlled. It has been shown, however, that the rate of transport within strong electric fields, generated by segregation-induced enrichment of the boundary layer, may be controlled by diffusion through this layer even at elevated temperatures.[36] It has been documented that the gas/solid equilibration process may be considerably affected by the segregation-induced electrical potential barriers, resulting in the formation of a local diffusive resistance, even at high temperatures.[36] Accordingly, correct determination of bulk diffusion data requires evaluation of the segregation-induced diffusive resistance in the boundary layer, and

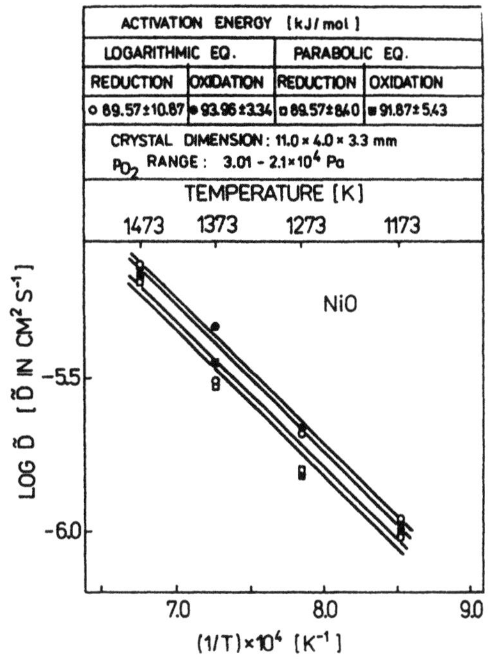

FIGURE 4.28. Chemical diffusion coefficient of undoped NiO determined using both logarithmic (Figure 4.27) and parabolic (Figure 4.28) plot. (From Nowotny, J., Oblakowski, J., and Sadowski, A., *Bull. Pol. Acad. Sci. Chem.*, 1985, 33, 99–119. With permission.)

its effect on the equilibration kinetics, if any. This effect will be considered in more detail in Section V.B.

The defect structure and related transport properties of binary metal oxides have been reported by Kofstad.[1]

E. CONCLUSIONS

Reactivity in metal oxide/oxygen systems depends on the temperature. Figures 4.29 and 4.30 illustrate the concentration of various species that are formed at various stages of the reaction between NiO and oxygen and the related effect on the surface potential, respectively. At lower temperatures, a physical form of oxygen adsorption predominates. As the temperature increases the adsorbed oxygen species become ionized, resulting in the formation of various chemisorbed species and, consequently, leading to charging of the surface. Above 300°C the NiO lattice becomes mobile, resulting in oxygen incorporation into the boundary layer and the formation of metal vacancies in this layer. At higher temperatures, the vacancies may be propagated into the bulk phase.

At elevated temperatures, corresponding with the gas/solid equilibrium, the surface properties are reproducible and the related surface charge, which is independent of the experimental procedure, depends only on temperature and $p(O_2)$. Below this temperature surface properties are essentially not reproducible.

V. SEGREGATION

A. EFFECT OF SEGREGATION ON INTERFACE COMPOSITION

1. Segregation of Foreign Elements

The chemical composition at interfaces may vary as a result of segregation of defects, resulting either in enrichment or in depletion. The direction and extent of segregation depend on the segregation driving force.[7,19]

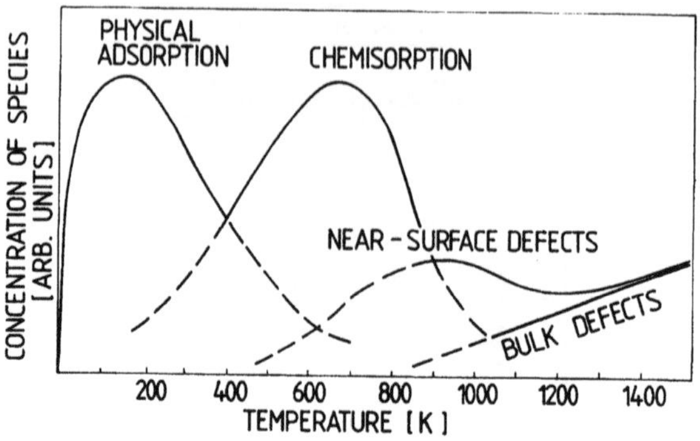

FIGURE 4.29. Schematic illustration of temperature ranges corresponding with different forms of oxygen species formed as a result of oxygen sorption on a nonstoichiometric oxide, such as NiO.

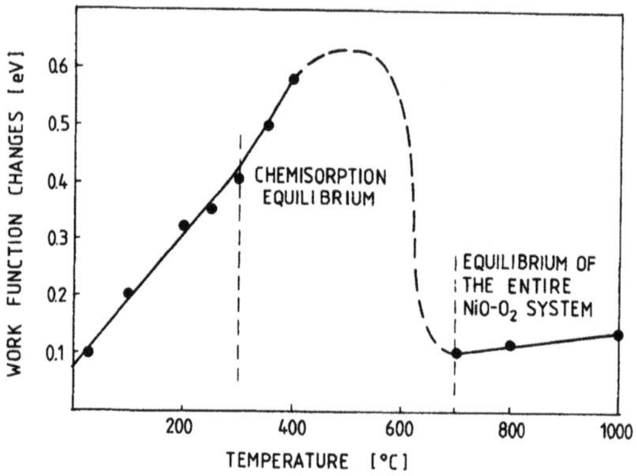

FIGURE 4.30. WF changes of undoped NiO after oxygen sorption at elevated temperatures.

Figure 4.31 shows the SIMS depth profiles of Cr in Cr-doped NiO after annealing under different gas-phase compositions.[37] As seen, Cr segregates to the surface of NiO, resulting in its enrichment. The enrichment depends on the composition of both the bulk phase[19] and the gas phase. There is a strong effect of the oxygen activity on Cr enrichment. If the annealing takes place in the gas/solid equilibrium, then the segregation enrichment data assume reproducible values for both oxidation (air) and reduction experiments performed successively.

It was documented that segregation also has a noticeable effect on the absolute value of the thermopower of polycrystalline materials.[38] An example is Cr segregation in NiO, and Figure 4.32 shows the Fermi energy as a function of bulk composition. As seen, good agreement between the experimental data of thermopower and the theoretical dependence (as determined from the mass action law) is observed only for very dilute solid solutions (below 0.1 at%) while, at higher Cr concentrations, there is a departure from the theoretical dependence owing to the increasing temperature gradient at the grain boundaries. This departure is related to the presence of a $NiCr_2O_4$-type bidimensional structure which is formed as a result of Cr segregation[19,38] as well as to a decrease in heat conduction at the necks between the grains. The difference between the bulk Fermi level and its value for the $NiCr_2O_4$ spinel phase indicates that the segregation-induced surface potential for Cr-doped NiO is about 0.5 V.

The surface model of undoped NiO, equilibrated at elevated temperatures, consists of a negatively charged outer layer, which is enriched in Ni vacancies, and the positive charge in the space charge layer formed of electron holes (Figure 4.33). The introduction of trivalent ions, such as Cr^{+3}, results in a change of the surface polarity, involving the formation of a

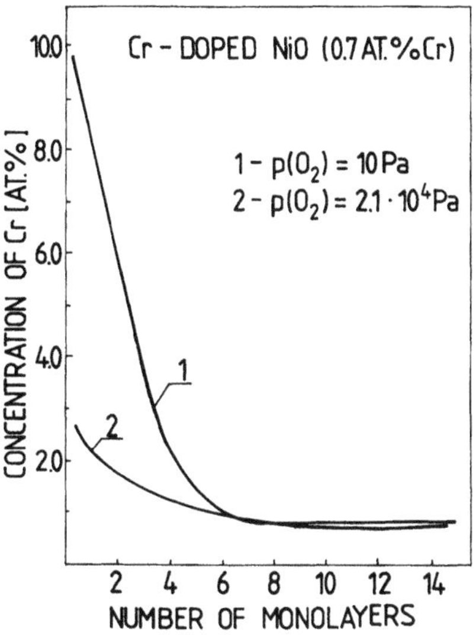

FIGURE 4.31. The SIMS depth profile of Cr for Cr-doped NiO annealed under reducing and oxidizing conditions (Ar and air, respectively). (Adapted from Hirschwald, W., Sikora, I., and Stolze, F., *Surf. Interface Anal.*, 1982, 18, 277–283.)

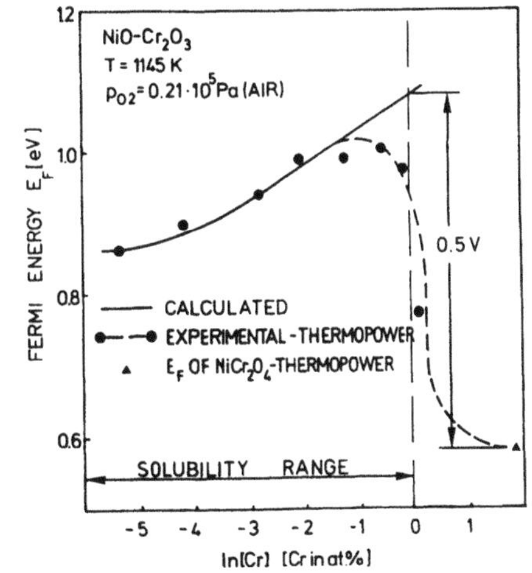

FIGURE 4.32. Fermi energy determined from the thermopower of Cr-doped NiO as a function of bulk Cr content. (From Nowotny, J. and Rekas, M., *Solid State Ionics*, 1984, 12, 253–261. With permission.)

positively charged outer layer, as a result of the formation of a spinel-type structure, and a negative space charge formed of Ni vacancies (Figure 4.34).

Cr also segregates to the CoO surface. As seen in Figure 4.35,[39] the segregation-enrichment coefficient strongly depends on the bulk composition, assuming very high values as the bulk concentration decreases. This tendency indicates that (a) the surface exhibits a tendency to assume a constant composition, which is independent of bulk Cr content; and (b) the segregation-induced surface enrichment may assume substantial values if the segregation driving force is high enough,[19] even if the bulk concentration of solute ions is at a negligible level. It has been shown that the surface enrichment coefficient (surface to bulk concentration) of Ca in yttria-doped ZrO_2 is about 10^5.[40]

Figure 4.36 shows the SIMS depth profile for the $CoCr_2O_4$ spinel phase, indicating that, in this case, Cr desegregates from the surface, resulting in the formation of a surface layer impoverished in Cr.[37]

Figures 4.37 to 4.39 show the effect of the gas phase composition on the SIMS segregation profiles of several impurities in the hematite phase.[41] As seen, the addition of a small amount of sulfur into the air results either in an increase of the segregation-induced enrichment (the

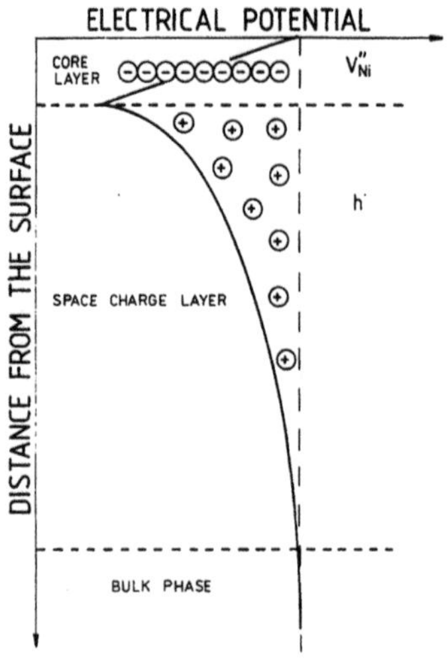

FIGURE 4.33. The surface model of undoped NiO.

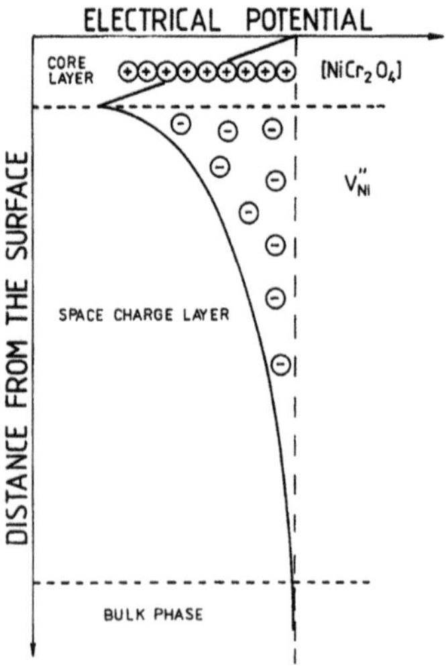

FIGURE 4.34. The surface model of Cr-doped NiO.

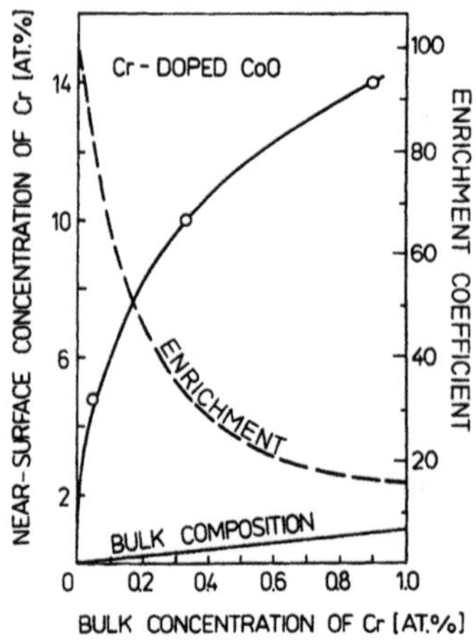

FIGURE 4.35. Surface vs. bulk composition of Cr-doped CoO. (From Haber, J., Nowotny, J., Sikora, I., and Stoch, J., *J. Appl. Phys.*, 1984, 17, 321–330. With permission.)

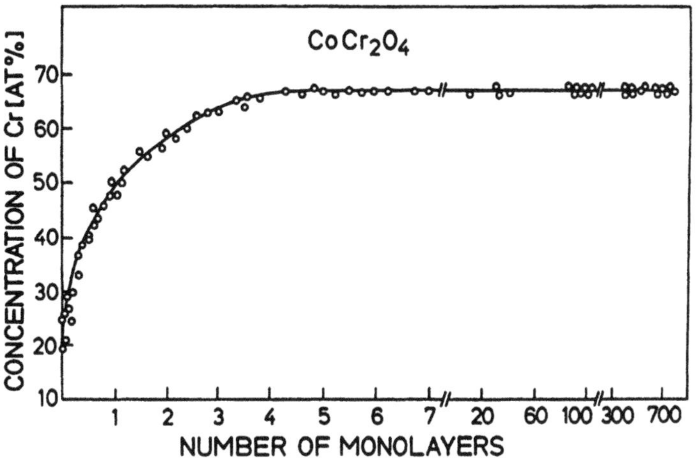

FIGURE 4.36. SIMS depth profile of a $CoCr_2O_4$ spinel phase. (Adapted from Hirschwald, W., Sikora, I., and Stolze, F., *Surf. Interface Anal.*, 1982, 18, 277–283.)

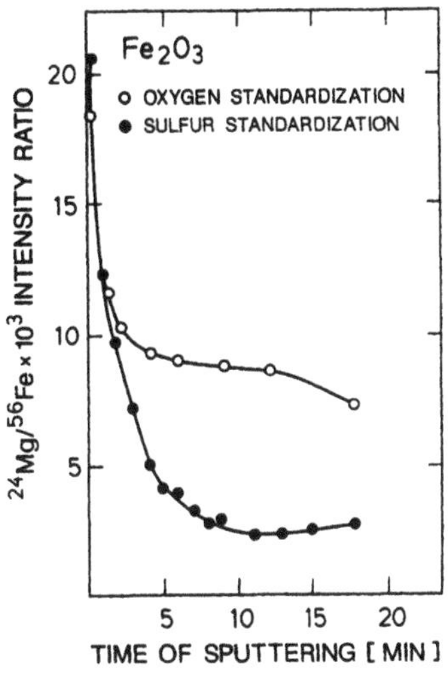

FIGURE 4.37. Effect of gas phase composition on depth profile of Mg in hematite [oxygen standardization: air; sulfur standardization: nitrogen (90%) $+H_2S$ (10%)]. (From Bernasik, A., Hirschwald, W., Janowski, J., Nowotny, J., and Stolze, F., *J. Mater. Sci.*, 1991, 26, 2527–2532. With permission.)

case of Mg — Figure 4.37) or in a change of the entire picture of segregation, as is the case for Si and Ca (Figures 4.38 and 4.39).

The experimental determination of the segregation of foreign lattice elements is a relatively easy matter, assuming they may reach surface concentrations which are within the detection limit of the applied surface technique. However, the determination of the effect of segregation on the surface nonstoichiometry of host lattice elements is subject to substantial difficulties.

2. Segregation of Host Elements

Besides the theoretical study of Duffy and Tasker,[42] who have shown that the surface of undoped NiO is enriched in Ni vacancies by a factor of about 40, little is known of the segregation of host lattice ions in compounds.

The recent study of Zhang et al.,[43] using angular dependent XPS, have shown that the Ti/Ba ratio at the surface of $BaTiO_3$ (analyzed at room temperature) depends on the gas phase composition during high-temperature annealing as well as on the cooling rate. The results indicate that the surface of $BaTiO_3$ is enriched in Ti, which is equivalent to an enrichment

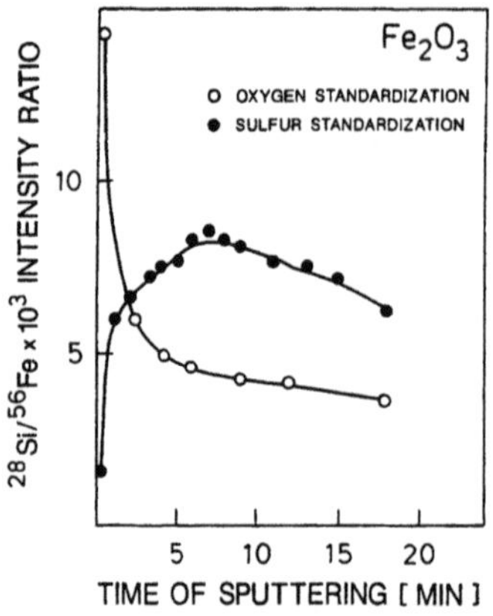

FIGURE 4.38. Effect of gas phase composition on depth profile of Si in hematite [oxygen standardization: air; sulfur standardization: nitrogen (90%) +H_2S (10%)]. (From Bernasik, A., Hirschwald, W., Janowski, J., Nowotny, J., and Stolze, F., *J. Mater. Sci.*, 1991, 26, 2527–2532. With permission.)

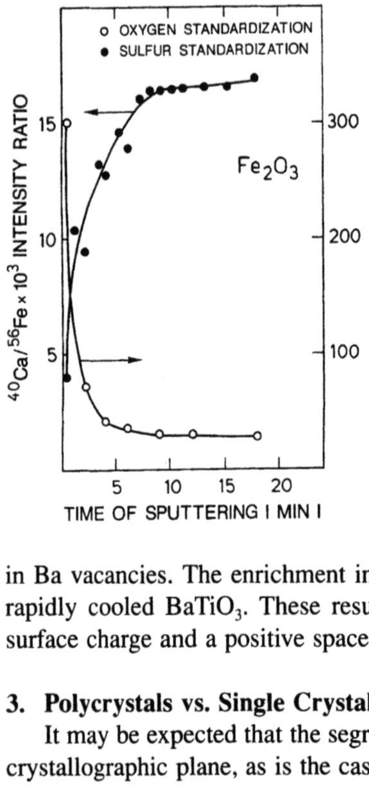

FIGURE 4.39. Effect of gas phase composition on depth profile of Ca in hematite [oxygen standardization: air; sulfur standardization: nitrogen (90%) + H_2S (10%)]. (From Bernasik, A., Hirschwald, W., Janowski, J., Nowotny, J., and Stolze, F., *J. Mater. Sci.*, 1991, 26, 2527–2532. With permission.)

in Ba vacancies. The enrichment in the oxidized specimen was higher for both slowly and rapidly cooled $BaTiO_3$. These results are consistent with the model involving a negative surface charge and a positive space charge (Figure 4.40).[43]

3. Polycrystals vs. Single Crystals

It may be expected that the segregation-induced enrichment substantially depends on the crystallographic plane, as is the case for metals. So far, however, segregation data for metal oxides are available mainly for polycrystalline materials. These data only illustrate the apparent segregation, which takes place on several planes. A better understanding of the picture of segregation in compounds requires the generation of experimental data on segregation for individual crystallographic planes of single crystals.

B. SEGREGATION-INDUCED BIDIMENSIONAL INTERFACE STRUCTURES

The outermost crystal layers, where the segregation-induced enrichment assumes the highest values, are the subject of structural deformations, thus resulting in the formation of bidimensional interface structures. These structures exhibit properties not displayed by the bulk phase.

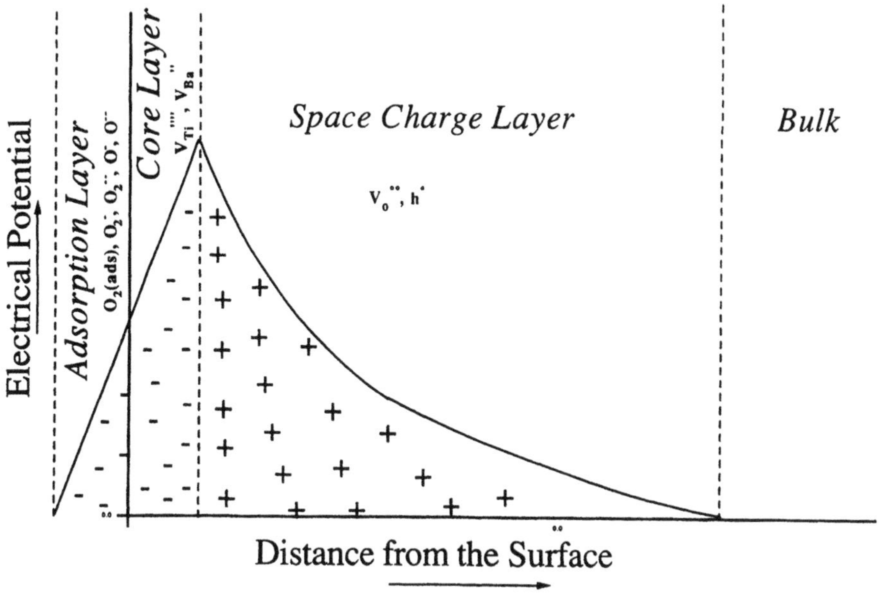

FIGURE 4.40. The surface model of undoped $BaTiO_3$. (From Zhang, Z., Pigram, P.J., Lamb, R.N., and Nowotny, J., *Proc. Int. Symp. Grain Boundary in Ceramics,* Kosuge, N., Ed., American Ceramic Society, Westerville, Ohio, in press. With permission.)

WF studies of undoped CoO[44] have shown that in the gas/solid equilibrium corresponding with the stability range of the CoO phase, grains of CoO are covered with a bidimensional layer of a spinel-type structure, Co_3O_4. These interface structures have been observed in other oxide systems, such as Cr-doped NiO and Cr-doped CoO.[19]

So far, little is known of these interface structures and their effect on materials properties. A better understanding of their local properties is very important for the preparation of advanced materials of improved properties. This is the reason why intensive studies are being undertaken in the characterization of low-dimensional systems, such as thin films, heterogeneous dispersion systems, and fine-grained ceramics (nanoparticle materials).[8]

C. EFFECT OF IMPURITIES ON THE SURFACE STATE IN EQUILIBRIUM

It has been a general assumption that the gas/solid equilibratium in nonstoichiometric oxides, after an oxygen potential gradient is imposed across a crystal, is determined by the bulk transport of the most rapid defects which may effectively remove the imposed chemical potential gradient. It has also been assumed that the surface layer assumes equilibrium state very rapidly. The latter assumption, however, is not correct when the crystal is contaminated with foreign ions which exhibit strong segregation. When their bulk content is very low, the time required to reach segregation equilibrium may be substantially longer than that required to remove oxygen potential gradient in the bulk phase. Then the changes in the concentration of the foreign ions in the bulk phase, due to segregation, are negligible. However, their changes at the surface are substantial and cannot be ignored.

Figure 4.41 shows changes in the CPD between zirconia (YSZ) and Pt at 780°C during reduction and oxidation runs[40] (the CPD changes are determined by the WF changes of the zirconia surface). Despite the fact that bulk equilibrium can be established after only a few minutes, a continuous change in the WF values during about 120 h can be observed. Figure 4.42 illustrates changes in the reciprocal of the $p(O_2)$ exponent, which varies between 4, as predicted by theory, to almost 2. The latter value is not consistent with the

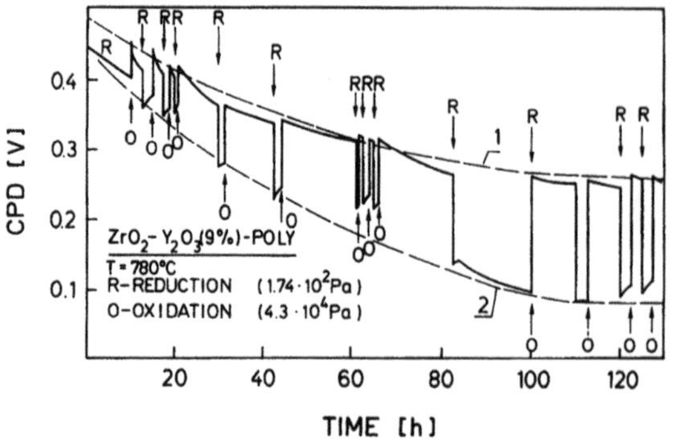

FIGURE 4.41. Contact potential difference of the zirconia–Pt system during prolonged annealing at 780°C. (From Nowotny, J., Sloma, M., and Weppner, W., *Solid State Ionics,* 1988, 28–30, 1445–1450. With permission.)

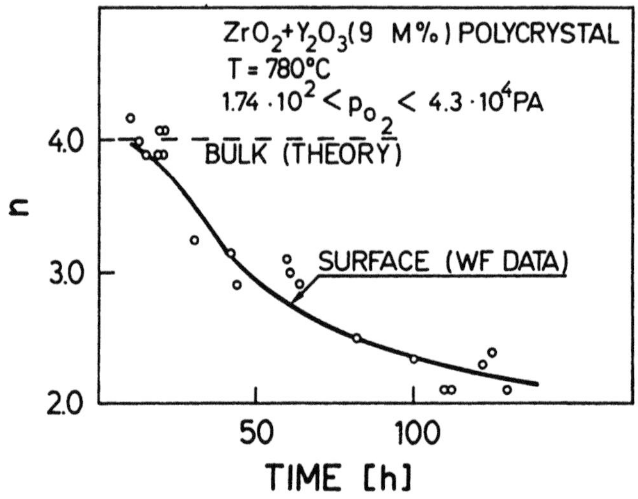

FIGURE 4.42. Changes of the reciprocal of oxygen exponent corresponding with changes of WF during prolonged annealing at 780°C. (From Nowotny, J., Sloma, M., and Weppner, W., *Solid State Ionics,* 1988, 28–30, 1445–1450. With permission.)

defect model of zirconia. The observed WF changes are caused by a slow process of defect segregation (impurities) which results in the formation of a low-dimensional surface layer. Its composition and properties are entirely different from those of the bulk phase. The proposed surface model of zirconia involves a "sandwich" composed of a surface layer enriched mainly in Ca, a subsurface layer enriched in yttria, and the bulk phase.

D. CONCLUSIONS

Segregation leads to an enrichment of interfaces in certain lattice defects and, consequently, results in the formation of concentration gradients (and associated electrical potential gradients) in the interface layer.

The segregation equilibrium, and related enrichment (or impoverishment) of the surface layer, are determined by the gas phase composition layer on one side and the bulk phase composition on the other side. While the equilibrium between the gas phase and the surface layer may be established very rapidly, the equilibratium between the surface layer and the bulk phase, of which the rate is controlled by the transport of impurities from the bulk to the surface, may be established after a long time even at high temperature.

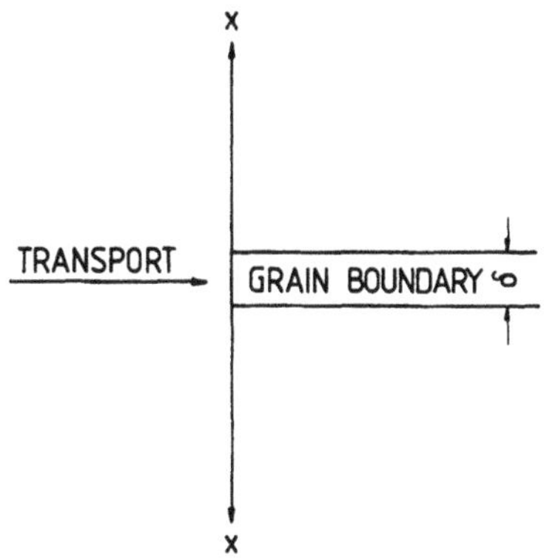

FIGURE 4.43. Schematic illustration of transport along grain boundary.

VI. EFFECT OF INTERFACES ON TRANSPORT

The mechanism and kinetics of transport along and across interfaces play an important role in the kinetics of gas/solid reactions. The mechanism of this transport is different from that in the bulk phase. This difference is caused by the structure and chemical composition of the interface layer, which are entirely different from those of the bulk.

A. TRANSPORT ALONG INTERFACES

Transport along interfaces, commonly considered as grain boundary diffusion in the literature, corresponds with diffusion parallel to interfaces, such as grain boundaries, and is limited to a thin grain boundary layer of thickness δ (Figure 4.43). This transport is usually much faster than the transport in the bulk phase. The enhancement effect is related to the microstructure of the interface region. The correct determination of the grain boundary diffusion coefficient of defects requires knowledge of the enrichment coefficient of the grain boundary by these defects.

Grain boundary transport is the subject of several fundamental publications of Kaur and Gust,[45] Atkinson and Taylor,[46] Deschamps and Barbier,[47] and Moya et al.[48]

B. TRANSPORT ACROSS INTERFACES

In contrast to the transport along interfaces, little is known about the transport across interfaces. This transport involves diffusion perpendicular to the interface, across the segregation-induced chemical potential gradients and related electric fields. This transport plays an important role in heterogeneous gas/solid and solid/solid processes.

It has been generally assumed that the rate of gas/solid re-equilibration at elevated temperatures is controlled by the bulk transport kinetics, while the processes that take place at the interface or within the interface layer are relatively fast. Recently, it has been shown that the rate of the gas/solid reactions may be controlled by a diffusive resistance, which is caused by segregation-induced electric fields within the interface layer.[36] The effect of these fields on the rate of transport of negatively charged defects is illustrated in Figure 4.44 in terms of $k^* [\tilde{D}/\delta]$ as a function of the normalized electric potential ($k^* [\tilde{D}/\delta]$ is a rate constant, $\tilde{D}$ is the chemical diffusion coefficient, δ the thickness of the diffusion layer, ψ the electrical potential, k the Boltzmann constant, and z the valency of the defects). With a positive migration effect the transport in the interface layer is accelerated and the process is controlled by bulk diffusion. However, when the migration effect assumes negative values, then the transport within the electric field in the boundary layer may control the entire gas/solid kinetics even at high temperatures.

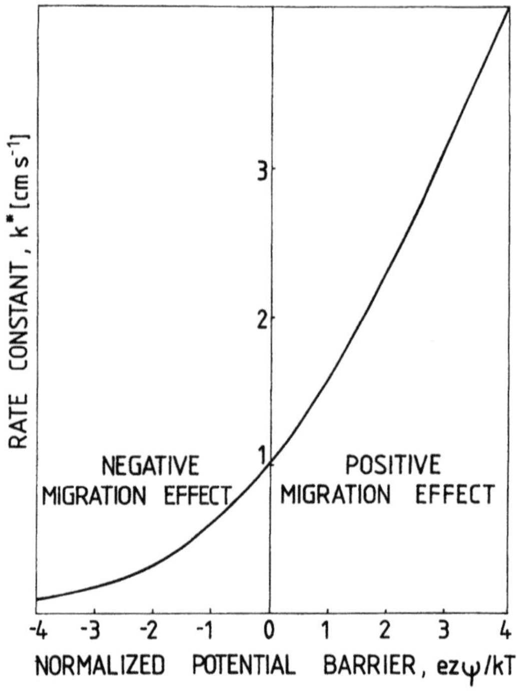

FIGURE 4.44. The effect of normalized electric potential barrier, ezψ/kT, on the diffusive transport kinetics in the boundary layer, $k^*\tilde{D}/\delta$. (From Adamczyk, Z. and Nowotny, J., *J. Phys. Chem. Solids*, 1986, 47, 11-27. With permission.)

Depending on the sign of the segregation-induced surface potential barrier and the charge of the diffusing species, the rate constant k* assumes different values, resulting in different solutions of the diffusion equation. When

$$k^* \gg \tilde{D}/\delta \tag{4.36}$$

then the rate of the gas/solid reaction is controlled by bulk diffusion. However, if

$$k^* \ll \tilde{D}/\delta \tag{4.37}$$

then the surface potential barrier assumes substantial values, resulting in the formation of a near-surface diffusive resistance, even at high temperature. When Condition (4.36) is applicable, then the general diffusion equation must be applied.[36]

Figure 4.45 shows the effect of the near-surface diffusive resistance on the apparent chemical diffusion coefficient of undoped NiO at different values of the segregation-induced potential barrier.[36] The effect of the surface barrier is seen to assume substantial values and to decrease with temperature.

C. CONCLUSIONS

Segregation-induced enrichment and related electric fields in the interface layer may have substantial effects on the heterogeneous gas/solid kinetics, even at high temperature. Thus a correct understanding of the gas/solid phenomena and resultant diffusion data require that the picture of segregation be well defined. Many diffusion data available in the literature were determined assuming that the bulk transport is rate controlling. In many cases this assumption is not valid. Thus diffusion data determined in such a way should be considered to be apparent

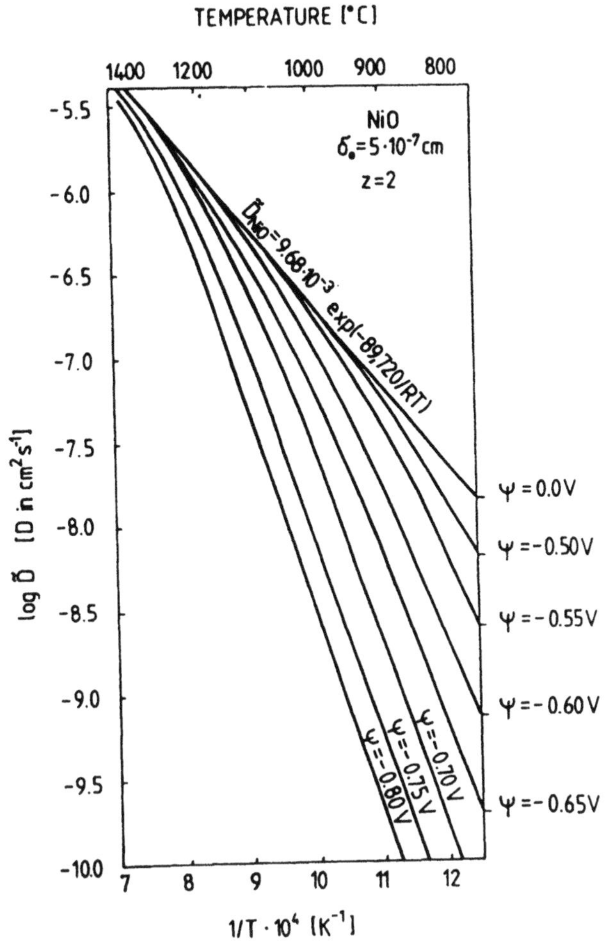

FIGURE 4.45. Arrhenius plot of the apparent chemical diffusion coefficient for undoped NiO at different surface potentials. (From Nowotny, J. and Sadowski, A., in *Transport in Non-stoichiometric Compounds*, Simkovich, G. and Stbican, V.S., Eds., Plenum Press, New York, 1985, 227–242. With permission.)

diffusion data corresponding to the near-surface diffusion resistance. The determination of these two components is required for a correct understanding of the physical sense of the reported diffusion data. Accordingly, the diffusion data for ionic solids available in the literature should be verified from the viewpoint of the effect of segregation on the local transport kinetics in the interface layer.

VII. APPLIED ASPECTS

Interfaces have a substantial impact on the properties (in particular, functional properties) of ceramics, such as varistors, dielectrics, sensor-type materials, catalysts, and superconductors. In many cases the properties of these materials are determined by the chemical composition and structure of the interfaces.

The PTCR may serve as a spectacular example of this effect, which is an exclusive property of polycrystalline materials (Figure 4.46). In other words, the PTCR effect can only be displayed by a crystal involving at least one interface (bicrystal). The extent of the PTCR effect can be modified by the chemical composition of the grain boundaries.

There is a growing awareness that the properties of gas sensors are determined by chemical composition of the gas/solid interface where the sensing signal is generated. Therefore, the

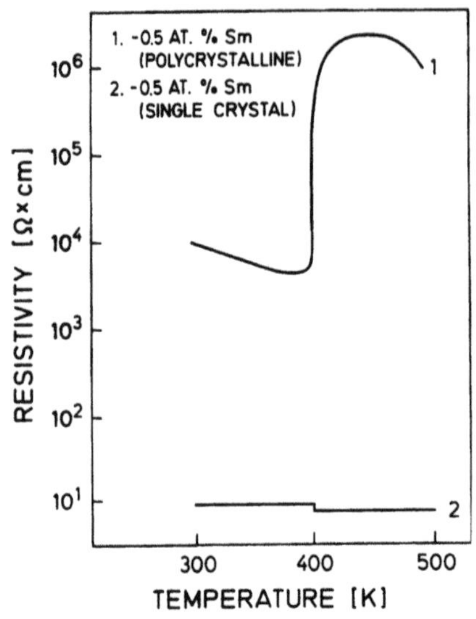

FIGURE 4.46. The PTCR of $BaTiO_3$ for both single crystal and polycrystalline specimen. (From Goodman, R., *J. Am. Ceram. Soc.*, 1963, 46, 48–51. With permission.)

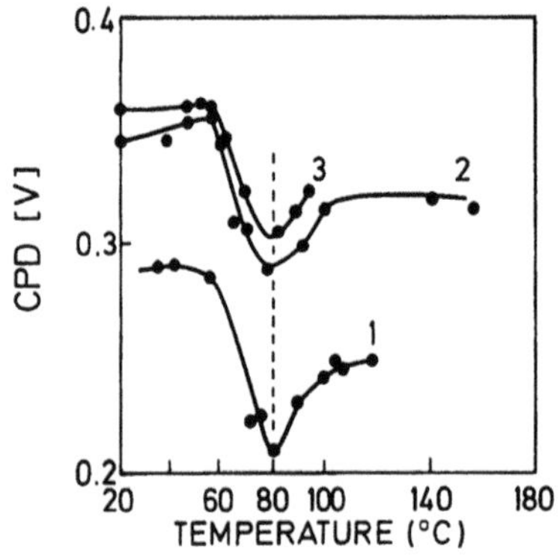

FIGURE 4.47. Changes of the CPD of a Cu_2O catalyst during consecutive heatings in the $CO+O_2$ gas mixture. (Adapted from Liashenko, V.I. and Stiepko, I.I., *Izv. Akad. Nauk SSSR, Ser. Fiz.* (in Russian), 1957, 21, 201–295.)

fabrication of a new generation of chemical gas sensors of improved sensitivity and selectivity may be achieved by appropriate engineering of its surface properties and, specifically, related gas/solid reactivity.

The catalytic properties of metal oxides are determined by active centers, which are formed at the surface during the catalytic process as a result of interactions between the gas phase and the catalyst. WF measurement is a very useful way of *in situ* monitoring of the surface during catalytic processes. These measurements provide information about electronic transitions that accompany the catalytic process. Figure 4.47 illustrates CPD changes (an increase in the CPD corresponds with an increase in the WF) during CO oxidation over Cu_2O for several consecutive runs. The appearance of the minimum indicates the beginning of the catalytic process and enables the determination of the direction of the charge transfer during the reaction.

Grain boundary barrier layer materials, based on $BaTiO_3$ and $SrTiO_3$, have wide applications as dielectrics. Their dielectric properties are determined by special processing resulting in the formation of an insulating grain boundary layer. Therefore, a better understanding of the grain boundary chemistry is of strategic importance to the preparation of dielectrics with improved properties.

The properties of high-T_c superconductors are determined by their grain boundary composition.[51] Intensive studies have been directed toward the elimination of the grain boundary weak links in oxide cuprates.

VIII. FUTURE PROSPECTS

It seems that future research on interfaces should focus on:

1. better understanding of the relationship between the local electrical properties and other properties such as chemical composition, structure, and nonstoichiometry and
2. engineering the interface properties in order to achieve desired properties of materials.

A. INTERFACE PROPERTIES

Still, very little is known about interface phenomena at the gas/solid and solid/solid interfaces at elevated temperatures and under controlled gas phase composition. Knowledge of these phenomena is important for better understanding the processing of materials. Electrical methods seem excellent for *in situ* monitoring the interface phenomena such as local transport kinetics, structural transitions, and chemisorption. There is an urgent need to develop new methods adequate to the complicated nature of interfaces, due to the limited dimensions of the interface layer and the gradients of properties within this layer.

B. ENGINEERING OF INTERFACES

Properties of polycrystalline materials are determined, or strongly influenced, by interfaces. Therefore, in order to produce materials with improved properties we have to be able to modify interface properties in a controlled manner. Thus we are observing the formation of a new scientific discipline: *interface engineering*. There is a need for progress in this discipline. The formation of basis of interface engineering depends on better understanding of interface properties of compounds and on the characterization of two-dimensional interface structures which are formed as a result of defect segregation. A strategy of modification of interface properties in a desired way should involve the formation of interface structures of controlled properties.

IX. SUMMARY AND FINAL CONCLUSIONS

It has been shown that the electrical properties of solids are very sensitive to both chemical composition and structure. The most commonly studied electrical properties, such as thermopower and electrical conductivity, are essentially bulk sensitive. However, their sensitivity to interfaces increases with a decrease in the grain size. Concordantly, most bulk-sensitive methods become sensitive to interfaces for fine-grained polycrystalline materials. The WF is selectively sensitive to the outermost surface layer. A great advantage of all these electrical methods involves a possibility of their application in studies of the high-temperature properties of solids and also in monitoring solid state processes at elevated temperatures.

A disadvantage of the electrical methods is the fact that they provide only indirect information on properties, such as chemical composition and structure. It is therefore important that studies using electrical methods are supplemented by other methods that are suitable for determining these properties directly.

The properties of many industrial materials are determined by the local properties of interfaces. The kinetics of gas/solid heterogeneous processes may be determined by the electric field that is generated as a result of segregation. This effect has many applied aspects. Also, the properties of materials that exhibit nonlinear characteristics are determined by interface composition and structure.

Further studies are needed to understand the picture of segregation and the related electrical effects at interfaces and their impact on properties of solids.

ACKNOWLEDGMENTS

This paper was reviewed by C. Ball, M. Rekas, and C. C. Sorrell. Their comments are sincerely appreciated. Thanks are also due to the editors for their remarks.

REFERENCES

1. Kofstad, P., *Electrical Conductivity, Nonstoichiometry and Diffusion in Binary Metal Oxides*, John Wiley & Sons, New York, 1972.
2. MacDonald, J.R. (Ed.), *Impedance Spectroscopy*, John Wiley & Sons, New York, 1987.
3. Bauerle, J.E., *J. Phys. Chem. Solids* 1969, *30*, 2657–2670.
4. Nowotny, J., *J. Chim. Phys. France*, 1978, *75*, 689–702.
5. Nowotny, J. and Sloma, M., in *Surface and Near-Surface Chemistry of Oxide Materials*, Nowotny, J. and Dufour, L. C., Eds., Elsevier, Amsterdam, 1988, 281–343.
6. Cabane, J. and Cabane, F., in *Interface Segregation and Related Processes in Materials*, Nowotny, J., Ed., Trans Tech Publications, Zurich, 1991, 1–159.
7. Nowotny, J., in *Surfaces and Interfaces of Ceramic Materials*, Dufour, L.C., Petot-Ervas, G., and Monty, C., Eds., Kluwer Academic Publishers, 1989, 205–239.
8. Saito, S., Soumura, T., and Maeda, T., *J. Vac. Sci. Technol.*, 1984, *A 2*, 1389–1382.
9. Nowotny, J., Sloma, M., and Weppner, W., *J. Am. Ceram. Soc.*, 1989, *72*, 569–570.
10. Lord Kelvin, *Philos. Mag.*, 1898, *46*, 82–120.
11. Zisman, W.A., *Rev. Sci. Instrum.*, 1932, *3*, 367–370.
12. Mignolet, J.C.P., *Discuss. Faraday Soc.*, 1950, *8*, 326–337.
13. Chrusciel, R., Deren, J., and Nowotny, J., *Exp. Technol. Phys.*, 1966, *14*, 127–133.
14. Besocke K. and Berger, S., *Rev. Sci. Instr.*, 1976, *47*, 840–842.
15. Haas, G.A., Thomas, R.E., Shin, A., and Marrian, C.R.K., *Ultramicroscopy*, 1983, *11*, 199–206.
16. Gordy, W. and Thomas, W., *J. Chem. Phys.*, 1956, *24*, 439–446.
17. Yamamoto, S., Susa, K., and Kawabe, U., *J. Chem. Phys.*, 1974, *60*, 4076–4080.
18. Nowotny, J. and Sikora, I., *J. Electrochem. Soc.*, 1978, *125*, 781–786.
19. Nowotny, J., in *Science of Ceramic Interfaces*, Nowotny, J., Ed., Elsevier, Amsterdam, 1991, 79–204.
20. Odier, P., Riflet, J.C., and Loup, J.P., in *Reactivity of Solids*, Haber, J., Dyrek, K., and Nowotny, J., Eds., Elsevier, Amsterdam, 1982, 458–466.
21. Joffe, A.F., *Physics of Semiconductors*, Academic Press, New York, 1960.
22. Bosman, A.J. and Van Daal, H.J., *Adv. Phys.*, 1970, *19*, 1–117.
23. Chaikin, P.M. and Beni, G., *Phys. Rev. B*, 1976, *13*, 647–51.
24. Hikes, R.R., in *Thermoelectricity*, Hikes, R.R. and Ure, R., Eds., Interscience, New York, 1961, chap. 4.
25. Austin, I.G. and Mott, N.F., *Adv. Phys.*, 1969, *18*, 41–102.
26. Nowotny, J. and Rekas, M., *Ceram. Int.*, 1994, *20*, 217.
27. Nowotny, J., Rekas, M., and Sikora, I., *J. Electrochem. Soc.*, 1984, *131*, 94–100.
28. Jonker, G.H., *Philips Res. Reps.*, 1968, *23*, 131.
29. Nowotny, J. and Rekas, M., *J. Am. Ceram. Soc.*, 1989, *72*, 1207–1214.
30. Nowotny, J., Radecka, M., and Rekas, M., *J. Phys. Chem. Solids*, submitted.
31. Kröger, F.A., *The Chemistry of Imperfect Crystals*, North Holland, Amsterdam, 1974.
32. Nowotny, J., *J. Mater. Sci.*, 1977, *12*, 1143–1160.
33. Adamczyk, Z. and Nowotny, J., *J. Electrochem. Soc.*, 1980, *127*, 1112–1120.
34. Nowotny, J., Oblakowski, J., and Sadowski A., *Bull. Pol. Acad. Sci. Chem.*, 1985, *33*, 99–119.
35. Nowotny, J. and Sadowski, A., in *Transport in Nonstoichiometric Compounds*, Simkovich, G. and Stbican, V.S., Eds., Plenum Press, New York, 1985, 227–242.
36. Adamczyk, Z. and Nowotny, J., *J. Phys. Chem. Solids*, 1986, *47*, 11–27.
37. Hirschwald, W., Sikora, I., and Stolze, F., *Surf. Interface Anal.*, 1982, *18*, 277–283.
38. Nowotny, J. and Rekas, M., *Solid State Ionics*, 1984, *12*, 253–261.
39. Haber, J., Nowotny, J., Sikora, I., and Stoch, J., *J. Appl. Phys.*, 1984, *17*, 324–330.
40. Nowotny, J., Sloma, M., and Weppner, W., *Solid State Ionics*, 1988, *28–30*, 1445–1450.
41. Bernasik, A., Hirschwald, W., Janowski, J., Nowotny, J., and Stolze, F., *J. Mater. Sci.*, 1991, *26*, 2527–2532.
42. Duffy, D.M. and Tasker, P.W., *Philos. Mag.*, 1984, *50*, 143–154.
43. Zhang, Z., Pigram, P.J., Lamb, R.N., and Nowotny, J., *Proc. Int. Symp. on Grain Boundary in Ceramics*, Kosuge, N., Ed., American Ceramic Society, Westerville, Ohio, 1994, in press.
44. Nowotny, J., Sloma, M., and Weppner, W., in *Nonstoichiometric Compounds*, Nowotny, J. and Weppner, W., Eds., Kluwer, Dordrecht, 1989, 265–277.
45. Kaur, I. and Gust, W., *Fundamentals of Grain and Interphase Boundary Diffusion*, Ziegler Press, Stuttgart, 1988.
46. Atkinson, A. and Taylor, R.I., *Philos. Mag.*, 1981, *43*, 979–998.
47. Deschamps, M. and Barbier, F., in *Science of Ceramic Interfaces*, Nowotny, J., Ed., Elsevier, Amsterdam, 1991, 323–369.

48. Moya, E.G., Moya, F., and Nowotny, J., in *Interface Segregation and Related Processes in Materials*, Nowotny, J., Ed., Trans Tech Publications, Zurich, 1991, 239–283.
49. Goodman, R., *J. Am. Ceram. Soc.*, 1963, *46*, 48–51.
50. Liashenko, V.I. and Stiepko, I.I., *Izv. Akad. Nauk SSSR, Ser. Fiz.* (in Russian), 1957, *21*, 201–295.
51. Nowotny, J., Rekas, M., Sarma, D.D., and Weppner, W., in *Surface and Near-Surface Chemistry of Oxide Materials*, Nowotny, J. and Dufour, L.C., Eds., Elsevier, Amsterdam, 1988, 669-699.

Chapter 5

DEFECT CHEMISTRY IN SOLID STATE ELECTROCHEMISTRY

Joop Schoonman

CONTENTS

I. INTRODUCTION

Solid electrolytes and mixed ionic–electronic conductors (MIECs) have been known for quite some time, going back at least to Faraday's observations in the 1830s that lead fluoride when heated to red hot conducts electricity similar to platinum. Some 50 years later, Warburg described the migration of sodium through a glass and its precipitation on the glass surface when a direct current flowed through the glass. At the turn of the century, Nernst[1] discovered that the high ionic conductivity of doped zirconia was due to the transport of oxide ions,[1] while the unusually high ionic conductivity of the α-phase of silver iodide which exists above 420 K, and which is comparable with the best conducting liquid electrolytes, was found in 1914.[2]

These observations concerning ionic transport can be explained on the basis of defect chemistry and crystal structure of the solid materials. The ideal crystal is in fact an abstract concept that is used in crystallographic descriptions. The lattice of a real crystal always contains imperfections. A suitable classification of crystalline defects can be achieved by first considering point defects and then proceeding to one- and higher-dimensional defects. Point defects are atomic defects whose effects are limited only to their immediate surroundings. They exist in a state of complete thermodynamic equilibrium. Examples are ionic vacancies in the regular crystal lattice, or interstitial atoms or ions.

Besides, crystals often contain extended defects. Among these are dislocations, which are classified as linear or one-dimensional defects. Grain boundaries, chemical twinning, stacking faults, crystallographic shear planes, intergrowth structures, and surfaces are two-dimensional

0-8493-8956-9/97/$0.00+$.50

TABLE 5.1
Solid Electrolytes

Point defect type		Positional disorder Cation-disordered sublattice		Orientational disorder	
Dilute	**Concentrated**	**Type I**	**Type II**		
		Low T		$NaNO_3$	Li_2SO_4
AgCl	ZrO_2-Y_2O_3	α-AgI	$LiAl_5O_8$	KNO_3	Na_2SO_4
$PbBr_2$	CaF_2-LaF_3	$RbAg_4I_5$		$NaBF_4$	K_2SO_4
LaF_3	$BaCeO_3$-Re	β-alumina			
		High T			
		$LiSO_4$			

defects. Finally, block structures, pentagonal column structures, infinitely adaptive structures, extended defect clusters, inclusions, or precipitates in the crystal lattice can be considered as three-dimensional defects. The concentrations of one- and higher-dimensional defects are not determined by thermodynamic equilibria. For the different kinds of defects, see also Chapter 3 of this handbook.

While certain preliminary ideas about the origin of ionic conduction in crystalline solids were derived from the crystal structure investigations of α-AgI by Strock,[3] and relate to a molten sublattice, i.e., no distinction can be made between a regular Ag lattice site and an Ag on an interstitial site, our knowledge of point defects is based mainly on the work of Frenkel,[4] Wagner and Schottky,[5] Schottky,[6] and Jost,[7] and was developed primarily by studying electrical and optical properties.

A possible classification of solids where ionic conductivity plays an important role is given in Table 5.1. In contrast to the situation in solutions, ionic transport in solids is accompanied by an electronic counterpart. For solid electrolytes the ionic contribution to the total electrical conductivity is predominant. Other cases represent MIECs.

A similar classification can be made for MIECs. MIECs are dealt with extensively by Riess in Chapter 7 of this book.

Type I positional disorder occurs because the lattice offers a large excess of available lattice positions for the mobile ions. These sites need not all be positions of the same energy, and indeed, generally they are not. Consequently, in the disordered state of the crystal the distribution of the ions in question among the sites to which they have access is not necessarily completely random. The ordered state of such a compound may have essentially the same structure, but with an ordered arrangement of occupied and unoccupied sites. Alternatively, it may have a different structure in which the number of sites now equals the number of ions.

Type II positional disorder is encountered in compounds in which some of the ions occupy quite definite positions, with the disorder confined to the positions of some or all of the remaining ions. The randomness involves one or more sublattices rather than all the lattice sites. Thus, type II disorder can occur in spinels of formula AB_2O_4, in which the oxide ions occupy fixed lattice positions, and any disorder concerns the whereabouts of the A and B ions. It is possible for a crystal to be simultaneously disordered in both ways, an instance of this being the high-temperature phase of the compound Ag_3SI.

Concentrated defect compounds show high lattice disorder and usually strong interactions between the moving ions, and thus resemble cation-disordered sublattice conductors. However, a distinction between normal and interstitial positions in a disordered sublattice is meaningless from a defect chemical point of view.

Orientational disorder in ionic solids arises because a diatomic or polyatomic ion has available to it two or more distinguishable orientations in the crystal lattice. This kind of disorder is a fairly common occurrence, especially when the polyatomic ions are sufficiently symmetrical. If these ions are associated with monoatomic ions of opposite charge, the situation simplifies in that the diatomic or polyatomic ions are to some extent protected from interference from ions of the same kind by the intervening shell of the monoatomic ions. Switching of diatomic or polyatomic ions from one orientation to another may induce local stress and hence facilitate the displacement of the monoatomic ion. A crystal can have positional disorder and at the same time orientational disorder of a polyatomic ion or molecule. Examples of this behavior are provided by lithium iodide monohydrate, and the high-temperature form of lithium sulfate.

While ionic conduction is mainly related to crystal structure, electronic conduction is determined by the electronic band gap, which depends more on the individual properties of the constituent ions. Thus in a series of compounds with comparable ionic conductivities, electronic conductivities can vary from virtually zero to quasimetallic. The first kind may find application as the solid electrolyte in a variety of solid state electrochemical devices, while the electrodes for these devices may be selected from the other.

The materials developed to date, for instance, for rechargeable solid state batteries, high-temperature fuel cells and electrolyzers, smart windows, and environmental gas sensors are numerous, and while many exhibit positional or orientational disorder, an overwhelming number conducts by virtue of point defects, and the defect chemistry of these materials is the focus of this chapter.

II. DEFECT CHEMISTRY OF BINARY AND TERNARY COMPOUNDS

Since the pioneering work of Frenkel, Schottky, Wagner, and Jost, a great number of studies, textbooks, and monographs have appeared which describe defect chemistry, dealing with equilibrium disorder in mainly binary ionic compounds.[8-23] In several of these references the one- and higher-dimensional defects mentioned before are being discussed in detail, while an excellent survey of positional and orientational disorder has been published by Parsonage and Staveley.[24]

The defect chemistry of ternary compounds has also attracted widespread attention, and the more fundamental concepts were developed mainly in the 1960s and 1970s.[8-10,17,25-34]

Since the field of solid state ionics was defined by Takahashi in the early 1960s, and the very first international meeting on the ever growing field of solid state ionics was held in Belgirate in 1972,[35] a great number of studies have appeared in which structural concepts and concepts of the defect chemistry of binary and ternary compounds have been applied in order to not only improve the electrical properties of existing materials, but also to guide solid state chemists and physicists and materials scientists in developing new materials. A selection of the literature is presented in References 36 to 59.

A. DEFECTS AND NONSTOICHIOMETRY IN BINARY COMPOUNDS

Point defects fall into two main categories: *intrinsic defects,* which are internal to the crystal in question, and *extrinsic defects,* which are created when an impurity atom or ion is inserted into the lattice. In, for instance, metal oxides containing transition metal ions, usually a component-dependent extrinsic disorder predominates.

The intrinsic defects fall into two main categories, i.e., Schottky disorder and Frenkel disorder. As these point defects do not change the overall composition, they are also referred to as *stoichiometric defects.* Their thermal generation will be exemplified for a metal oxide MO using the Kröger–Vink[8] notation, and assuming that activities of point defects are equal to their concentrations. Hence, the law of mass action is applicable to these equilibria.

Schottky disorder:

$$M_M^x + O_O^x \rightleftarrows V_M'' + V_O^{\bullet\bullet} + (MO)_{defect} \tag{5.1}$$

with equilibrium constant

$$K_s = \left[V_M''\right]\left[V_O^{\bullet\bullet}\right] \tag{5.2}$$

Here, $(MO)_{defect}$ denotes a lattice unit at the surface of the crystal.

Frenkel disorder:

$$M_M^x + V_i^x \rightleftarrows M_i^{\bullet\bullet} + V_M'' \tag{5.3}$$

The equilibrium constant is

$$K_F = \left[M_i^{\bullet\bullet}\right]\left[V_M''\right] \tag{5.4}$$

Frenkel disorder usually occurs in the cation sublattice. It is less common to observe Frenkel disorder in the anion sublattice *(anti-Frenkel disorder)*, and this is because anions are commonly larger than cations. An important exception to this generalization lies in the occurrence of anti-Frenkel disorder in fluorite-structured compounds, like alkaline earth halides (CaF_2, SrF_2, $SrCl_2$, BaF_2), lead fluoride (PbF_2), and thorium, uranium, and zirconium oxides (ThO_2, UO_2, ZrO_2). One reason for this is that the anions have a lower electrical charge than the cations, while the other reason lies in the nature of the open structure of the fluorite lattice.

Anti-Frenkel disorder is represented by:

$$O_O^x + V_i^x \rightleftarrows O_i'' + V_O^{\bullet\bullet} \tag{5.5}$$

with equilibrium constant

$$K_{aF} = \left[O_i''\right]\left[V_O^{\bullet\bullet}\right] \tag{5.6}$$

The temperature dependence of the number of Schottky defects is given by

$$\left[V_M''\right] = \left[V_O^{\bullet\bullet}\right] = n_s = N\, exp\left(-\Delta H_f/2kT\right) \tag{5.7}$$

where n_s is the number of one of the intrinsic Schottky defects per cubic meter at temperature T in a crystal with N cations and N anions per cubic meter. ΔH_f is the enthalpy required to form a set of vacancies. Equation (5.7) has been derived taking into account only variations in the configurational entropy. In the simplest case of the Einstein approximation for the limiting case of Dulong–Petit behavior, a crystal with N lattice atoms is considered to be a system of 3 N oscillators that all vibrate with the same frequency ν_o. From the partition function of this ensemble of oscillators, that part of the Gibbs free energy which arises from the crystal vibrations can be used to calculate the change in the vibrational frequencies, accompanying the introduction of point defects. The corresponding entropy change appears as a pre-exponential factor in Equation (5.7) and does not influence the temperature dependence of the concentration of point defects.

For the alkali and lead halides, a simple empirical relation exists between the formation enthalpy of Schottky defects and the melting temperature T_m,[13] i.e.,

$$\Delta H_f(eV) = 2.14 \times 10^{-3} T_m(K) \tag{5.8}$$

For oxides, no such simple empirical relation has been reported.

The concentration of the intrinsic point defects can be influenced by aliovalent impurity doping. Assuming that only ionic point defect concentrations are affected, the following lattice reactions exemplify the formation of extrinsic disorder by substitution in MO, e.g., MF_2 in MO

$$MF_2 \rightarrow M_M^x + 2F_O^{\bullet} + V_M'' \tag{5.9}$$

and Me_2O in MO

$$Me_2O \rightarrow 2Me_M^{\bullet} + O_O^x + V_O^{\bullet\bullet} \tag{5.10}$$

In general, in compounds with k components at constant temperature and pressure, the concentrations of point defects present in thermodynamic equilibrium are fixed by k-1 component activities or the related chemical potentials.

For the simple metal halides and metal oxides, at P and T fixed, usually the halogen or oxygen partial pressure is used, because this parameter can easily be varied. The dependence of the defect concentrations on component activities is determined by the predominating intrinsic defect structure and is analyzed using point defect thermodynamics.

1. Metal Halides

In the metal halides to be discussed here, Frenkel disorder predominates. Equilibrium with an X_2 containing ambient can be established according to

$$\tfrac{1}{2}X_2(g) \overset{K_a}{\rightleftarrows} X_X^x + V_M' + h^{\bullet} \tag{5.11}$$

and

$$X_X^x + M_M^x + V_i^x \overset{K_b}{\rightleftarrows} M_i^{\bullet} + \tfrac{1}{2}X_2(g) + e' \tag{5.12}$$

with

$$K_a = \frac{[V_M'][h^{\bullet}]}{p_{X_2}^{\frac{1}{2}}} \tag{5.13}$$

and

$$K_b = [M_i^{\bullet}][e']P_{X_2}^{\frac{1}{2}} \tag{5.14}$$

the respective equilibrium constants. The intrinsic ionic and electronic disorder reactions are

$$M_M^x + V_i^x \overset{K_F}{\rightleftarrows} M_i^{\bullet} + V_M' \tag{5.15}$$

and

$$O \underset{\rightleftarrows}{K_i} e' + h^{\bullet} \tag{5.16}$$

where O denotes the perfect lattice.

The equilibrium constants are

$$K_F = [M_i^{\bullet}]\,[V_M'] \tag{5.17}$$

and

$$K_i = [e']\,[h^{\bullet}] \tag{5.18}$$

Heyne[60] has shown that the electronic band gap of solid electrolytes should always be larger than $\frac{T}{300}$ (eV).

For $M_{1+x}X$ the mass balance equation reads

$$\begin{gathered}[M_M^x]+[M_i^{\bullet}]+[e']=1+x \\ (e' \equiv M_M')\end{gathered} \tag{5.19}$$

while for $M_{1-x}X$ the equation is

$$\begin{gathered}[X_X^x]+[X_i']+[h^{\bullet}]=1 \\ (h^{\bullet} \equiv X_x^{\bullet})\end{gathered} \tag{5.20}$$

The site balance equations are

$$[M_M^x]+[V_M']+[e']=1 \tag{5.21}$$

and

$$[X_x^x]+[h^{\bullet}]=1 \tag{5.22}$$

From Equations (5.19) and (5.21) one obtains for x

$$x=[M_i^{\bullet}]-[V_M'] \tag{5.23}$$

If P_{X_2} (0) represents the partial halogen pressure in equilibrium with stoichiometric MX, the following composition conditions can be distinguished:

$$(i)\quad x>0,\ [M_i^{\bullet}]>[V_M'] \ for\ P_{X_2}<P_{X_2}(0)$$

$$(ii)\quad x=0,\ [M_i^{\bullet}]=[V_M'] \ for\ P_{X_2}=P_{X_2}(0)$$

$$(iii)\quad x<0,\ [M_i^{\bullet}]<[V_M'] \ for\ P_{X_2}<P_{X_2}(0)$$

From the equations for K_a, K_F and x, one obtains for the relation between the relative partial halogen pressure $P_{X_2}/P_{X_2}(0)$ and the deviation from stoichiometry the relation

$$\left(\frac{P_{X_2}}{P_{X_2}(0)}\right)^{\frac{1}{2}}=\left[-\frac{x}{2K_F^{\frac{1}{2}}}+\left(\frac{x^2}{4K_F}+1\right)^{\frac{1}{2}}\right]\frac{[h^{\bullet}]}{[h^{\bullet}(0)]} \tag{5.24}$$

With $[h^{\bullet}(0)] = K_i^{1/2}$, K_i, and $[e'] - [h^{\bullet}] = x$, as well as the conditions $K_F^{1/2} \gg K_i^{1/2}$ and $x \ll K_F^{1/2}$, Equation (5.24) reduces to

$$log\frac{P_{X_2}}{P_{X_2}(0)}=2\ log/x/-log\ K_i \tag{5.25}$$

for $x < 0$ and $|x| \gg K_i^{1/2}$.

For given P_{X_2} the deviation from stoichiometry increases with increasing values for K_i. The relation between

$$\log\frac{P_{X_2}}{P_{X_2}(0)}$$

and the deviation from stoichiometry is schematically presented in Figure 5.1.

For the composition $M_{1-x}X$ the electroneutrality condition reads

$$[e']+[V_M']=[h^{\bullet}]+[M_i^{\bullet}] \tag{5.26}$$

With K_F, K_a, K_i and the condition $K_i \ll K_a$ $K_i \ll K_a P_{X_2}^{\frac{1}{2}}\left(=[V_M'][h^{\bullet}]\right)$ one obtains

$$[h^{\bullet}]^2=\frac{K_a^2P_{X_2}}{K_F+K_aP_{X_2}^{\frac{1}{2}}} \tag{5.27}$$

For small deviations from stoichiometry, i.e., $K_F \gg K_aP_{X_2}^{\frac{1}{2}}$ the concentration of the electron holes is given by

$$[h^{\bullet}]=K_aK_F^{-\frac{1}{2}}P_{X_2}^{\frac{1}{2}} \tag{5.28}$$

For large deviations from stoichiometry, i.e., $K_F \ll K_aP_{X_2}^{\frac{1}{2}}$ one obtains

$$[h^{\bullet}]=[V_M']=K_a^{\frac{1}{2}}P_{X_2}^{\frac{1}{4}} \tag{5.29}$$

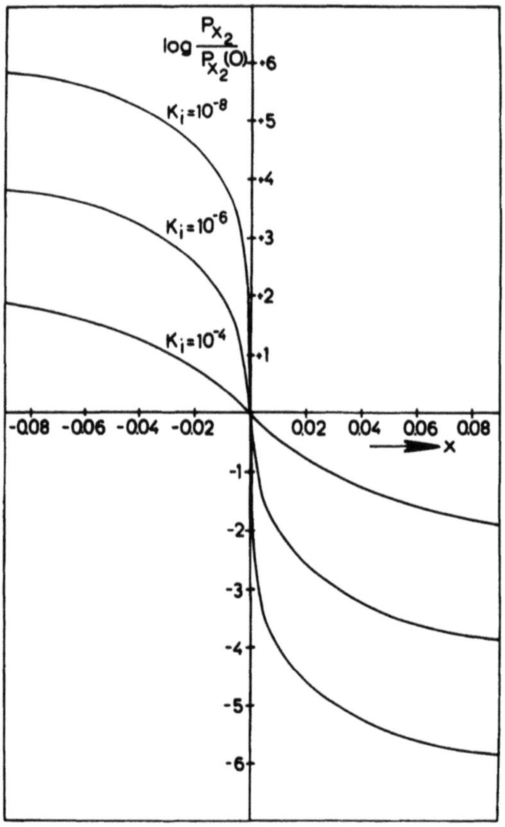

FIGURE 5.1. Schematic graph of the relative partial X_2 pressure as a function of the deviation x from the stoichiometric composition.

For the composition $M_{1+x}X$ it can be shown that for small deviations from the stoichiometry the electron concentration reads

$$[e'] = K_b K_F^{-\frac{1}{2}} P_{X_2}^{-\frac{1}{2}} \left(K_F \gg K_b P_{X_2}^{-\frac{1}{2}}\right) \quad (5.30)$$

and for large deviations

$$[e'] = [M_i^{\cdot}] = K_b^{\frac{1}{2}} P_{X_2}^{-\frac{1}{4}} \qquad \left(K_F \ll K_b P_{X_2}^{-\frac{1}{2}}\right) \quad (5.31)$$

Thus, in general, the concentration of the defects can be represented by

$$[i] = P_{X_2}^{n_i/2} \ \Pi_r K_r^{S_i} \quad (5.32)$$

Here, n_i and s_i are characteristic exponents (simple rational numbers). K_r is the equilibrium constant of the r-th defect reaction.

The results are commonly represented by the Kröger–Vink or Brouwer diagrams, i.e., plots of log defect concentration vs. log component activity, which quantitatively express the variation of defect chemistry within a given phase.[10,61]

The partial pressure dependence, according to

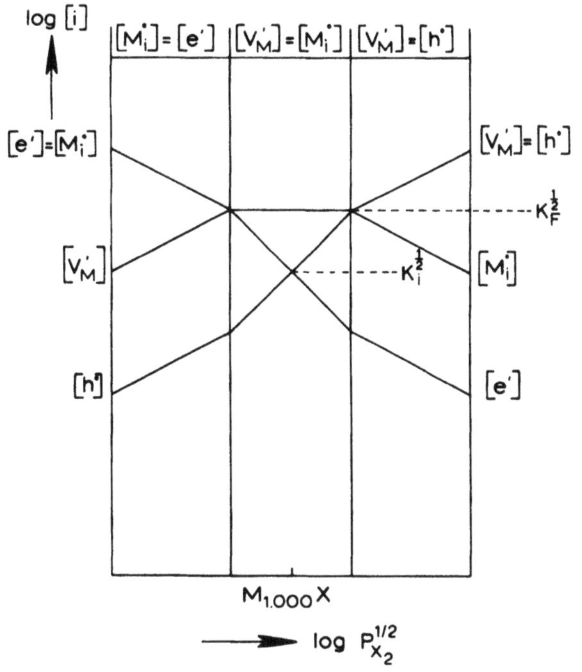

FIGURE 5.2. Kröger–Vink diagram of the model material MX with Frenkel disorder.

$$\left(\frac{d\ \log[i]}{d\ \log P_{X_2}^{\frac{1}{2}}}\right)_{P,T} = n_i, \tag{5.33}$$

is represented by straight lines in the Kröger–Vink diagram, as shown in Figure 5.2.

If aliovalent impurities are present, these have to be included into Equation (5.26) because they influence the point defect concentrations. A great number of studies, including extrinsic disorder in nonstoichiometric compounds, have appeared in the literature, and the concepts have been extended to ternary and multinary compounds. Before discussing these compounds, the concepts will be applied to transition metal oxides, because these represent an important technological class of materials.

2. Transition Metal Oxides

If a transition metal oxide MO exhibits Frenkel disorder, the following gas–solid equilibria can be established, thus forming nonstoichiometric $M_{1\pm x}O$,

$$M(g) + V_i^x \overset{K_1}{\rightleftarrows} M_i^{\bullet\bullet} + 2e' \tag{5.34}$$

$$M(g) + V_M'' \overset{K_2}{\rightleftarrows} M_M^x + 2e' \tag{5.35}$$

$$\tfrac{1}{2}O_2(g) \overset{K_3}{\rightleftarrows} O_O^x + V_M'' + 2h^{\bullet} \tag{5.36}$$

$$\tfrac{1}{2}O_2(g)+M_i^{\bullet\bullet} \overset{K_4}{\rightleftarrows} (MO)_{defect}+V_i^x+2h^{\bullet} \tag{5.37}$$

The equilibrium constants are, respectively,

$$K_1=\frac{\left[M_i^{\bullet\bullet}\right]\left[e'\right]^2}{a_M} \tag{5.38}$$

and

$$K_2=\frac{\left[e'\right]^2}{a_M\left[V_M''\right]} \tag{5.39}$$

With Equation (5.3) one obtains

$$K_1=K_2K_F \tag{5.40}$$

For the anion-excess oxide, the equilibrium constants are, respectively,

$$K_3=\frac{\left[h^{\bullet}\right]^2\left[V_M''\right]}{P_{O_2}^{\frac{1}{2}}} \tag{5.41}$$

and

$$K_4=\frac{\left[h^{\bullet}\right]^2}{\left[M_i^{\bullet\bullet}\right]P_{O_2}^{\frac{1}{2}}} \tag{5.42}$$

Here one obtains with Equation (5.3)

$$K_3=K_4K_F \tag{5.43}$$

The thermal generation of intrinsic electronic disorder is given by

$$O \rightleftarrows e'+h^{\bullet} \tag{5.44}$$

with

$$K_i=\left[e'\right]\left[h^{\bullet}\right] \tag{5.45}$$

Here we assume the defects to be fully ionized.

For the thermodynamic treatment of the point defect equilibria, one has to take into account the electroneutrality condition

$$\left[e'\right]+2\left[V_M''\right]=\left[h^{\bullet}\right]+2\left[M_i^{\bullet\bullet}\right] \tag{5.46}$$

For the determination of the dependence of the point defect concentrations on the oxygen partial pressure, one again has to distinguish between the regions of small and large deviations from stoichiometry, that is, for

$M_{1+x}O$:

$$[e'] + 2[V_M''] = 2[M_i^{\bullet\bullet}] \tag{5.47}$$

and

$$[e'] \approx 2[M_i^{\bullet\bullet}] \tag{5.48}$$

and for

$M_{1-x}O$:

$$2[V_M''] = [h^\bullet] + 2[M_i^{\bullet\bullet}] \tag{5.49}$$

and

$$2[V_M''] \approx [h^\bullet] \tag{5.50}$$

Electroneutrality Equations (5.47) and (5.49) hold for small deviations from stoichiometry, while Equations (5.48) and (5.50) hold for large deviations from stoichiometry.

The dependence of the point defect concentrations on the oxygen partial pressure can be derived using the expressions for K_1–K_4 and K_i, and the reduced electroneutrality conditions (5.47) to (5.50). For the regions of small deviations from stoichiometry, one obtains for

$M_{1+x}O$:

$$[e'] = K_i K_3^{-\frac{1}{2}} K_F^{\frac{1}{4}} P_{O_2}^{-\frac{1}{4}} \tag{5.51}$$

and for

$M_{1-x}O$:

$$[h^\bullet] = K_3^{\frac{1}{2}} K_F^{-\frac{1}{4}} P_{O_2}^{+\frac{1}{4}} \tag{5.52}$$

For large deviations from stoichiometry, the dependences are

$M_{1+x}O$:

$$[e'] = (2K_F)^{\frac{1}{3}} K_3^{-\frac{1}{3}} P_{O_2}^{-\frac{1}{6}} \tag{5.53}$$

and

$M_{1-x}O$:

$$[h^\bullet] = \left(\frac{K_4 K_i}{2}\right)^{\frac{1}{3}} P_{O_2}^{+\frac{1}{6}} \tag{5.54}$$

In the case of large deviations from stoichiometry, simple associates or more extended defect clusters can be formed. One example is the Koch–Cohen defect cluster in nonstoichiometric wüstite (FeO). This defect cluster bears a strong resemblance to the structure of Fe_3O_4. One can think of nonstoichiometric FeO as fragments of Fe_2O_3 intergrown in the rock salt structure of FeO.[22] Another well-known cluster in the oxide-interstitial defect cluster is anion-excess fluorite-structured UO_{2+x}. This cluster, comprising an O_i'' and two relaxed oxide ions, thus leading to 2 $V_{O,r}^{\bullet\bullet}$ and 2 $O_{i,r}''$, in addition to the regular interstitial, has been used to explain extended defect clustering in fluorite-type anion-excess solid solutions in the systems MeF_2-ReF_3 and MeF_2-UF_4. Here, Me stands for an alkaline earth metal and Re for a rare earth metal.

The defect chemistry of nonstoichiometric transition metal oxides has been studied extensively, and many of these studies[62,63] have been reviewed.[15,19,54,64-68]

Schematic Kröger–Vink diagrams of metal oxides have been reported,[15,68] and these resemble the defect diagram in Figure 5.2 as expected.

B. GENERALIZED APPROACH TO THE DEFECT CHEMISTRY OF TERNARY COMPOUNDS

For the simple binary compounds discussed in Sections II.A.1 and II.A.2, at given P and T, usually the partial pressure of the electronegative element is used. In ternary compounds a second component activity has to be determined, leading to a three-dimensional representation for the relationships between defect concentrations and component activities.

In ternary compounds of the type (A,B)X, the point defect concentrations and hence the deviation from stoichiometry depend on the concentrations of the two cations. Since in general the equilibrium constants of the different point defect equilibria in ternary compounds are composition dependent, their thermodynamic treatment is more difficult, and usually much more experimental data are required than in the case of the binary compounds. The complexity of the ternary compounds is the reason why point defect chemistry has been worked out more or less quantitatively only for some oxide and fluoride systems, using the concepts of Kröger and Vink,[8,10] Kröger,[10] Schmalzried and Wagner,[25] and Schmalzried.[26]

Schmalzried[26] has pointed out that the theories of the defect chemistry of spinels can be generalized and be applied to all ternary compounds. Groenink[34] and Groenink and Binsma[69] made this generalization for ternary compounds comprising two cations and one anion, i.e., $A_aB_bX_c$ with the cations $A^{\alpha+}$ and $B^{\beta+}$, and the anion $X^{\gamma-}$. The compound is formed from the binary compounds A_pX_q and B_rX_s. They choose $\beta \geq \alpha$. In this section the generalization reported by Groenink and Binsma[69] will be summarized.

For the ternary compound the following independent variables are chosen, i.e., total pressure P, temperature T, partial pressure P_{X_2}, and the activity $a_{A_pX_q}$ of the binary compound A_pX_q, hereafter referred to as a_0. The following defects can be present in $A_aB_bX_c$:

$$A_i^{(\alpha-j)\bullet}, \qquad j = 0,1,\ldots\ldots\alpha$$

$$B_i^{(\beta-k)\bullet}, \qquad k = 0,1,\ldots\ldots\beta$$

$$X_i^{(\gamma-l)'}, \qquad l = 0,1,\ldots\ldots\gamma$$

$$V_A^{(\alpha-j')'}, \qquad j' = 0,1,\ldots\ldots\alpha$$

$$V_B^{(\beta-k')'}, \qquad k' = 0,1,\ldots\ldots\beta$$

$$V_x^{(\gamma-l')\bullet}, \qquad l' = 0,1,\ldots\ldots\gamma$$

$$A_B^{(\beta-\alpha-m)'}, \quad m = 0,1,\ldots\ldots\beta-\alpha$$

$$B_A^{(\beta-\alpha-n)^\bullet}, \quad n = 0,1,\ldots\ldots\beta-\alpha$$

$$e'$$

$$h^\bullet$$

Here, ionization of the defects is assumed to occur. The total electroneutrality condition is, taking into account the possible defects,

$$\begin{aligned}
&[e'] + \sum_{j'=0}^{\alpha-1}(\alpha-j')\left[V_A^{(\alpha-j')'}\right] + \sum_{k'=0}^{\beta-1}(\beta-k')\left[V_B^{(\beta-k')'}\right] + \sum_{l=0}^{\gamma-1}(\gamma-1)\left[X_i^{(\gamma-1)^\bullet}\right] \\
&+\sum_{m=0}^{\beta-\alpha-1}(\beta-\alpha-m)\left[A_B^{(\beta-\alpha-m)'}\right] = [h^\bullet] + \sum_{j=0}^{\alpha-1}(\alpha-j)\left[A_i^{(\alpha-j)^\bullet}\right] \\
&+\sum_{k=0}^{\beta-1}(\beta-k)\left[B_i^{(\beta-k)^\bullet}\right] + \sum_{l'=0}^{\gamma-1}(\gamma-1')\left[V_X^{(\gamma-l')^\bullet}\right] \\
&+\sum_{n=0}^{\beta-\alpha-1}(\beta-\alpha-n)\left[B_A^{(\beta-\alpha-n)^\bullet}\right]
\end{aligned} \tag{5.55}$$

Usually, for a given combination of thermodynamic variables, the concentrations of two defects of opposite sign are much higher than the concentrations of all other defects. Using this approximation, Equation (5.55) reduces to a form for this majority defect pair, i.e., $[V_A^{\alpha'}] = [A_i^{\alpha\bullet}]$ for the majority defect pair $V_A^{\alpha'}$, $A_i^{\alpha\bullet}$.

Table 5.2 presents a number of defect equilibria and equilibria with the gas phase.

1. Variation of Defect Concentrations with a_0 and P_{X_2}

If the variation of the concentration of a defect j as a result of a change in a_0 and P_{X_2} occurs between equilibrium states in a reversible way, one can write at fixed P and T,

$$d[j] = \left(\frac{\partial[j]}{\partial_{P_{X_2}}}\right)_{P,T,a_0} dP_{X_2} + \left(\frac{\partial[j]}{\partial a_0}\right)_{P,T,p_{X_2}} da_0 \tag{5.68}$$

It is more convenient, however, to consider the change of the logarithm of the defect concentration, because this is a simple function of α, β, γ, a, b, and p, whereas d[j] still contains unknown defect concentrations. For the calculation of

$$\left(\frac{\partial log[j]}{\partial log P_{X_2}}\right)_{P,T,a_0} \text{ and } \left(\frac{\partial log[j]}{\partial log a_0}\right)_{P,T,P_{X_2}}$$

four equilibria are required, viz., Equations (5.64) to (5.67).

TABLE 5.2
Some Internal Equilibria and Equilibria with the Gas Phase

$$A_A^x + V_i^x \rightleftarrows A_i^\alpha + V_A^{\alpha'} \qquad K_{AF} = \left[A_i^{\alpha\cdot}\right]\left(V_A^{\alpha'}\right) \tag{5.56}$$

$$X_X^x + V_i^x \rightleftarrows X_i^{\gamma'} + V_X^{\gamma\cdot} \qquad K_{aF} = \left[X_i^{\gamma'}\right]\left[V_x^{\gamma\cdot}\right] \tag{5.57}$$

$$B_A^{(\beta-\alpha)\cdot} + V_i^x \rightleftarrows B_i^{\beta\cdot} + V_A^{\alpha'} \qquad K_4 = \frac{\left[B_i^{\beta\cdot}\right]\left[V_A^{\alpha'}\right]}{\left[B_A^{(\beta-\alpha)\cdot}\right]} \tag{5.58}$$

$$O \rightleftarrows aV_A^{\alpha'} + bV_B^{\beta'} + cV_X^{\gamma\cdot} \qquad K_S = \left[V_A^{\alpha'}\right]^a \left[V_B^{\beta'}\right]^b \left[V_X^{\gamma\cdot}\right]^c \tag{5.59}$$

$$O \rightleftarrows e' + h^\cdot \qquad K_i = [e']\left[h^\cdot\right] \tag{5.60}$$

$$X_i^{\gamma'} + h^\cdot \rightleftarrows X_i^{(\gamma-l)'} \qquad K_5 = \frac{\left[X_i^{(\gamma-1)'}\right]}{\left[X_i^{\gamma'}\right]\left[h^\cdot\right]^1} \tag{5.61}$$

$$A_pX_q(g) + pV_A^{\alpha'} + qV_i^x \rightleftarrows pA_A^x + qX_i^{\gamma'} \qquad K_6 = \frac{\left[A_A^x\right]^p \left[X_i^{\gamma'}\right]^q}{a_{A_pX_q}\cdot\left[V_A^{\alpha'}\right]^p} \tag{5.62}$$

$$2A_pX_q(g) + 2pV_A^{\alpha'} \rightleftarrows 2pA_A^x + 2p\alpha e' + qX_2(g) \qquad K_7 = \frac{[e']^{2p\alpha}\cdot P_{X_2}^q}{a_{A_pX_q}^2\cdot\left[V_A^{\alpha'}\right]^{2p}} \tag{5.63}$$

$$A_pX_q(g) + pV_A^{\alpha'} + qV_X^{\gamma\cdot} \rightleftarrows pA_A^x + qX_X^x \qquad K_8 = a_{A_pX_q}\cdot\left[V_A^{\alpha'}\right]^p\left[V_X^{\gamma\cdot}\right]^q \tag{5.64}$$

$$B_rX_s(g) + rV_B^{\beta'} + sV_X^{\gamma\cdot} \rightleftarrows rB_B^x + sX_X^x \qquad K_9 = a_{B_rX_s}\left[V_B^{\beta'}\right]^r\left[V_X^{\gamma\cdot}\right]^s \tag{5.65}$$

$$raA_pX_q(g) + pbB_rX_x(g) \rightleftarrows prA_aB_bX_c \qquad K_{10} = a_{A_pX_q}^{ra}\cdot a_{B_rX_s}^{pb} \tag{5.66}$$

$$2X_X^x \rightleftarrows 2V_X^{\gamma\cdot} + 2\gamma e' + X_2(g) \qquad K_{11} = \left[V_X^{\gamma\cdot}\right]^2 [e']^{2\gamma} P_{X_2} \tag{5.67}$$

Defining

$$\left(\frac{\partial log\left[\mathrm{V_A^{\alpha'}}\right]}{\partial log\mathrm{a_0}}\right)_{\mathrm{P,T,p_{X_2}}} = \xi_1,$$

$$\left(\frac{\partial log\left[\mathrm{V_B^{\beta'}}\right]}{\partial log\mathrm{a_0}}\right)_{\mathrm{P,T,p_{X_2}}} = \xi_2,$$

$$\left(\frac{\partial log\left[V_X^{\gamma\cdot}\right]}{\partial loga_0}\right)_{P,T,p_{X_2}} = \xi_3,$$

and

$$\left(\frac{\partial log[e']}{\partial loga_0}\right)_{P,T,p_{X_2}} = \xi_4,$$

and differentiating the logarithm of Equations (5.64) to (5.67) with respect to log a_0, one obtains

$$p\xi_1 + q\xi_3 = -1 \tag{5.69}$$

$$r\xi_2 + s\xi_3 = \frac{ar}{bp} \tag{5.70}$$

$$\xi_3 + \gamma\xi_4 = 0 \tag{5.71}$$

A fourth relation between ξ_1, ξ_2, ξ_3, and ξ_4 can be found by differentiating the logarithm of the electroneutrality condition for the majority defect pair considered with respect to loga_0. This electroneutrality condition can then be rewritten in the form:

$$x_1\xi_1 + x_2\xi_2 + x_3\xi_3 + x_4\xi_4 = 0 \tag{5.72}$$

with the aid of the following relation,

$$\frac{\partial log[j]}{\partial log x} = a_1 \frac{\partial log\left[V_A^{\alpha'}\right]}{\partial log x} + a_2 \frac{\partial log\left[V_B^{\beta'}\right]}{\partial log x} + a_3 \frac{\partial log\left[V_X^{\gamma'}\right]}{\partial log x} + a_4 \frac{\partial log[e']}{\partial log x} \tag{5.73}$$

in which x denotes either a_0 or P_{X_2}.

The numbers x_1, x_2, x_3, and x_4 in Equation (5.72) are characteristic for the given majority defect pair. Solving the four linear equations in ξ_i yields the values presented in Table 5.3.

The values for a_1, a_2, a_3, and a_4 are given in Table 5.4.

2. Composition and Defect Chemistry

The compound $A_aB_bX_c$ is made up of $L{\cdot}A_pX_q + M{\cdot}B_rX_s + NX_2$. L, M, and N are integers, L and M strictly positive, and N positive, zero, or negative. The total concentrations of A, B, and X are

$$[A]_{tot} = L \cdot p \tag{5.74}$$

$$[B]_{tot} = M \cdot r \tag{5.75}$$

$$[X]_{tot} = L \cdot q + M \cdot s + 2N \tag{5.76}$$

TABLE 5.3
The Dependence of the Logarithm of Some Defects on the Logarithm of a_0 and P_{X_2}

[j]	$\left(\frac{\partial log[j]}{\partial log a_0}\right)_{P,T,p_{X_2}}$	$\left(\frac{\partial log[j]}{\partial log P_{X_2}}\right)_{P,T,a_0}$
$V_A^{\alpha'}$	$\frac{-(a\alpha+\beta b)x_2-b(-\gamma x_3+x_4)}{C}$	$\frac{\alpha x_4}{B}$
$V_B^{\beta'}$	$\frac{(\alpha a+\beta b)x_2+a(-\gamma x_3+x_4)}{C}$	$\frac{\beta x_4}{B}$
$V_X^{\gamma\bullet}$	$\frac{-b\gamma x_1+a\gamma x_2}{C}$	$\frac{-\gamma x_4}{B}$
e′	$\frac{bx_1-ax_2}{C}$	$\frac{-ax_1-\beta x_2+\gamma x_3}{B}$
	$C=pb(\alpha x_1+\beta x_2-\gamma x_3+x_4)$	$B=2\gamma(\alpha x_1+\beta x_2-\gamma x_3+x_4)$

The composition of the compound is known if the ratios

$$\frac{[A]_{tot}}{[B]_{tot}} \text{ and } \frac{[X]_{tot}}{[A]_{tot}}$$

are known. For the ideal compound composition these ratios read

$$\frac{[A]_{tot}}{[B]_{tot}}=\frac{L\cdot p}{M\cdot r}=\frac{a}{b} \tag{5.77}$$

and

$$\frac{[X]_{tot}}{[A]_{tot}}=\frac{L\cdot p+M\cdot s+2N}{L\cdot p}=\frac{c}{a} \tag{5.78}$$

The deviation from the ideal composition can be described by two parameters, i.e., Δx, the deviation from molecularity, and Δy, the deviation from stoichiometry. An excess A_pX_q gives $\Delta x > 0$, and an excess B_rX_s $\Delta x < 0$. Δy is the parameter which determines, at a fixed ratio $\frac{L}{M}$, whether there is an excess $X_2(\Delta y > 0)$ or a deficiency $X_2(\Delta y < 0)$. The compound has the ideal composition for $\Delta x = \Delta y = 0$. Defined in terms of Equations (5.77) and (5.78), these parameters are

TABLE 5.4
Coefficients a_1, a_2, a_3, and a_4 in Equation 5.73 for All Defects Considered

[j]	a_1	a_2	a_3	a_4
$A_i^{(\alpha-j)\bullet}$	−1	0	0	+j
$B_i^{(\beta-k)\bullet}$	0	−1	0	+k
$X_i^{(\gamma-l)'}$	0	0	−1	+1
$V_A^{(\alpha-j')'}$	+1	0	0	−j′
$V_B^{(\beta-k')'}$	0	+1	0	−k′
$V_x^{(\gamma-l')\bullet}$	0	0	+1	−l′
$A_B^{(\beta-\alpha-m)'}$	−1	+1	0	−m
$B_A^{(\beta-\alpha-n)\bullet}$	+1	−1	0	+n
e′	0	0	0	1
h$^\bullet$	0	0	0	−1

$$\Delta x = \frac{r}{p}\left[\frac{[A]_{tot}}{[B]_{tot}} - \frac{a}{b}\right] \tag{5.79}$$

and

$$\Delta y = \frac{\gamma[X]_{tot}}{\alpha[A]_{tot} + \beta[B]_{tot}} - 1 \tag{5.80}$$

Groenink[34] has demonstrated that a given majority defect pair can only exist in a compound if certain conditions with regard to Δx and Δy are fulfilled. The conditions are specific for the majority defect pair. For Δx and Δy, general relations including all the possible point defects were derived,[34] and the condition for a given majority defect pair are then found by neglecting all the other point defect concentrations in the general relations for Δx and Δy. A complete survey of the conditions for all the possible majority defect pairs has been reported by Groenink[34] and Groenink and Binsma.[69] For given values of Δx and Δy, only a limited number of majority defect pairs can exist in a compound, and which of these pairs occurs in practice depends upon temperature, pressure, and formation energy of the various defects.

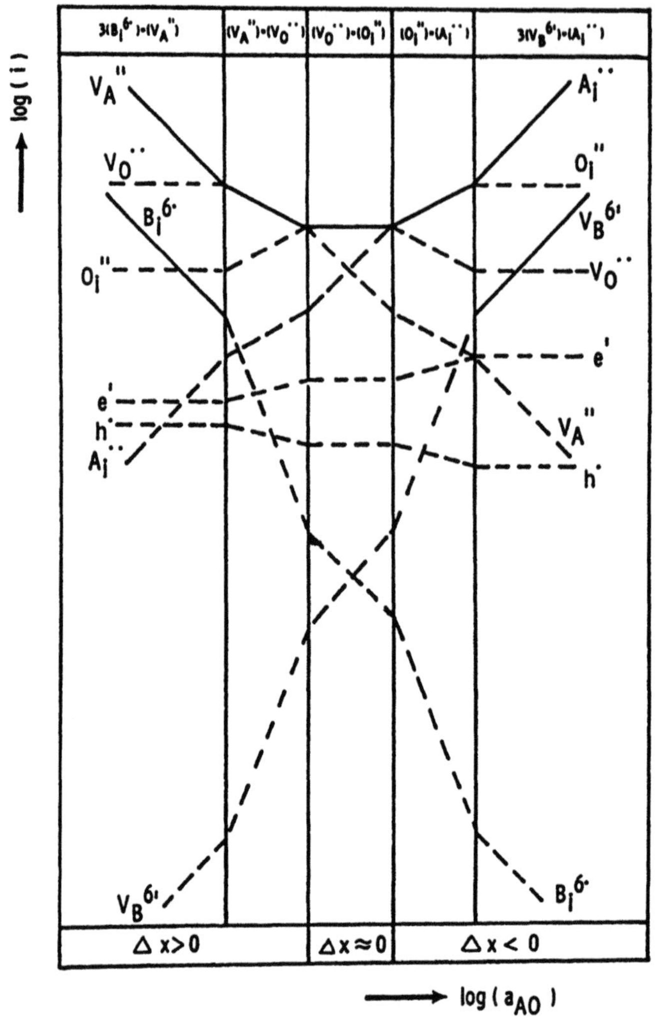

FIGURE 5.3. Kröger–Vink diagram of the model material A^{2+} B^{6+} O_4^{2-}, exhibiting deviations from molecularity.

3. Defect Diagrams

The Kröger–Vink diagrams can be constructed using Tables 5.3 and 5.4. It is evident that a_0 and P_{X_2} may vary only within the range of homogeneity of $A_aB_bX_x$. Figure 5.3 is the Kröger–Vink diagram of a model material ABO_4, in which A, B, and O are ions with a charge +2, +6, and –2, respectively. In this diagram Δx varies and Δy = 0. For variations in Δy at fixed a_{AO}, the diagram closely resembles that in Figure 5.2.

The Kröger–Vink diagrams of compounds ABX_4, in which A, B, and X are ions with charge +1, +3, and –1 are presented in Figures 5.4 and 5.5. It is assumed that the model material exhibits anti-Frenkel disorder.

Figure 5.4 presents the relations between the point defect concentrations and a deviation from stoichiometry, while in Figure 5.5 these relations are plotted in the case of deviations from the stoichiometry.[27]

The quantitative relations between the point defect concentrations and the compound activities are very useful in interpreting electrical properties of solid electrolytes and MIECs. The point defect–composition relations also define the electrolytic domain of a solid electrolyte, and hence determine experimental conditions to be fulfilled in order for the materials to be applicable in solid state electrochemical devices.

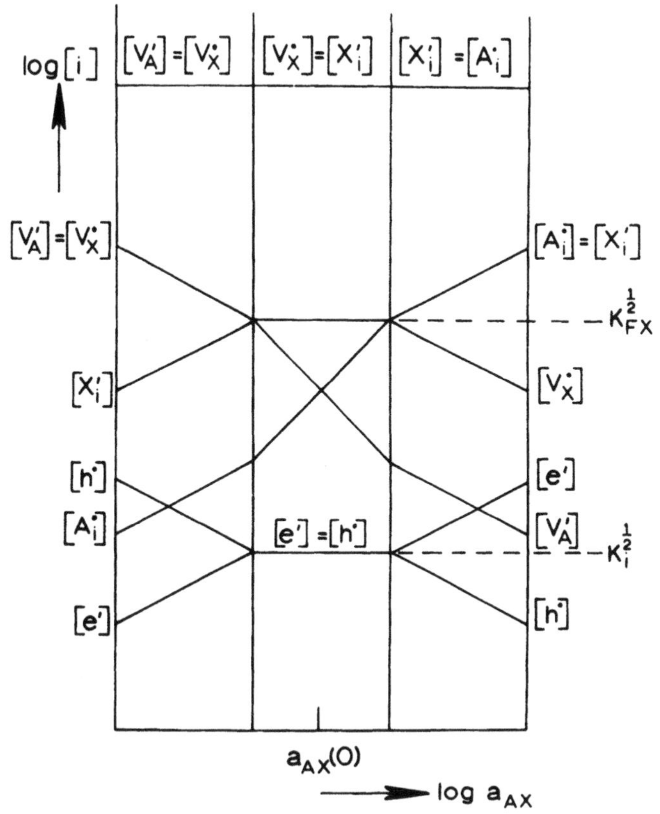

FIGURE 5.4. Kröger–Vink diagram of the model material $A^{+}\ B^{3+}\ X_4^{-}$, exhibiting deviations from molecularity.

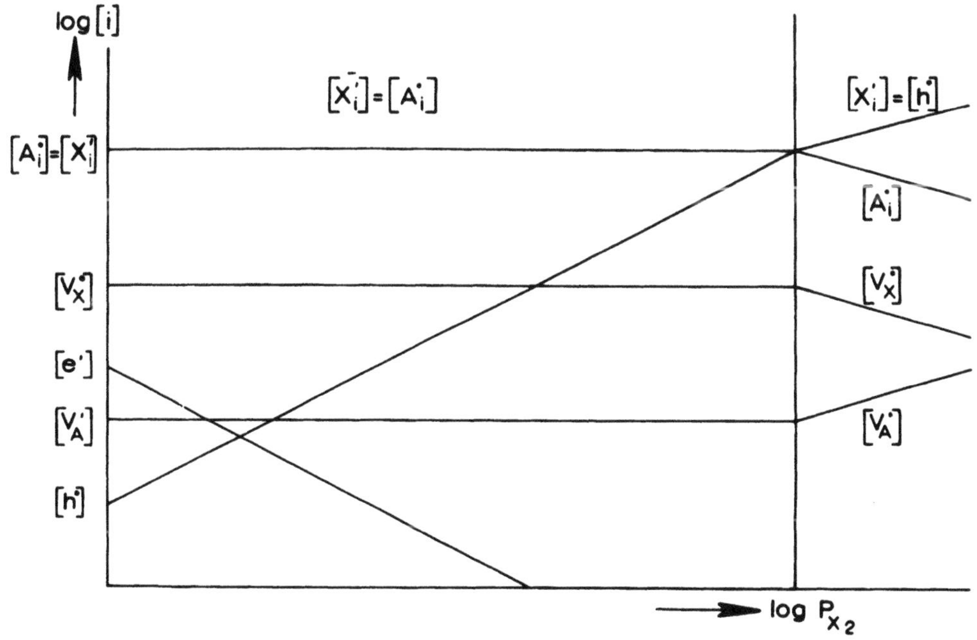

FIGURE 5.5. Kröger–Vink diagram of the model material $A^{+}\ B^{3+}\ X_4^{-}$, exhibiting deviations from stoichiometry for $\Delta x > 0$.

It should be borne in mind, though, that the traditional mass action laws which govern the point defect concentrations are based on the assumptions of negligible interactions and random distributions. They are not applicable in high-defect concentration regimes. A general defect chemistry formulation based on statistical thermodynamic consideration has recently been developed by Ling.[70] His statistical thermodynamic formulation explicitly incorporates long-range Coulombic interactions and the generalized exclusion effects. For solid solutions in the system CeO_2–CaO, Ling found good agreement between calculated and experimental conductivity data. The reaction enthalpies are found to be strongly influenced by defect interactions. Only at very high defect concentrations do the effects of defect exclusions become significant.

4. Application to Several Ternary Compounds

While intrinsic disorder of the Schottky, Frenkel, or anti-Frenkel type frequently occurs in binary metal oxides and metal halides, i.e., Equations (5.1), (5.3), and (5.5), Schottky disorder is seldomly encountered in ternary compounds. However, in several studies Schottky disorder has been proposed to occur in perovskite oxides. Cation and anion vacancies or interstitials can occur in ternary compounds, but such defect structures are usually to be related with deviations from molecularity (viz. Sections II.B.2 and II.B.3), which in fact represent extrinsic disorder and not intrinsic Schottky disorder. From Figures 5.3 and 5.4 it is apparent that deviations from molecularity always influence ionic point defect concentrations, while deviations from stoichiometry always lead to combinations of ionic and electronic point defects, as can be seen from Figures 5.2 and 5.5.

A material which has attracted widespread attention is the perovskite oxide $LaCrO_3$, and especially the $LaCrO_3$-based solid solutions. It is the state-of-the-art interconnection material in the solid oxide fuel cell (SOFC). $LaCrO_3$ is known to be a p-type conductor, owing to the presence of electron holes in the 3d-band of the Cr ions, i.e., $Cr^{\bullet}_{Cr}$. Conduction occurs by the thermally activated hopping of localized charge carriers ($Cr^{\bullet}_{Cr}$).[71] Doping with relatively small amounts of Mg^{2+}, Sr^{2+}, or Ca^{2+} enhances the electronic conductivity. Under oxidizing conditions, charge compensation takes place by a Cr^{3+} to Cr^{4+} transition, thereby increasing the electronic charge carrier density. Under reducing conditions, charge compensation takes place under the formation of oxygen ion vacancies, and no increased electronic conductivity is expected. The interconnection material in an SOFC operates under these extreme conditions. For Mg^{2+}-doped $LaCrO_3$ the defect Mg'_{Cr} is formed. The acceptor is electrically compensated by the Cr^{3+} to Cr^{4+} transition[71,72] or by $V^{\bullet\bullet}_O$. The electroneutrality condition reads

$$[Mg'_{Cr}] = [Cr^{\bullet}_{Cr}] + 2[V^{\bullet\bullet}_O] \tag{5.81}$$

Here, anti-Frenkel disorder is assumed to occur. The corresponding defect equilibrium can be written as

$$O^x_O + 2Cr^{\bullet}_{Cr} \rightleftarrows 2Cr^x_{Cr} + V^{\bullet\bullet}_O + \tfrac{1}{2}O_2(g) \tag{5.82}$$

For this material other defect models have been proposed.[73-76]

Flandemeyer et al.[73-74] and Anderson et al.[75] used a defect model based on Schottky disorder, i.e.,

$$O \rightleftarrows V'''_{La} + V'''_{Cr} + 3V^{\bullet\bullet}_O \tag{5.83}$$

To determine the influence of the oxygen partial pressure on the oxygen ion vacancy concentration, the following defect equilibrium was assumed:

$$V_{La}''' + V_{Cr}''' + 6Cr_{Cr}^{\bullet} + 3O_O^x \rightleftarrows 6Cr_{Cr}^x + \tfrac{3}{2}O_2(g) \tag{5.84}$$

However, it is unlikely that this defect equilibrium determines the oxygen ion vacancy concentration in Mg-doped $LaCrO_3$, as the presence of trivalent cation vacancies in ternary metal oxides is not favored thermodynamically.[10]

In the proposed defect models it is assumed that all defects are distributed randomly over the lattice. Singhal et al.,[76] Van Roosmalen and Cordfunke,[77] and Van Roosmalen[78] have proposed defect structures including defect clusters comprising $V_O^{\bullet\bullet}$ next to regular Cr_{Cr}^x ions, and $(Mg_{Cr}' - V_O^{\bullet\bullet} - Mg_{Cr}')^x$ clusters. While no rationale was presented for the trapping of oxygen ion vacancies at regular Cr_{Cr}^x sites, the $(2\,Mg_{Cr} \cdot V_O)^x$ cluster model seems to describe the experimentally determined oxygen ion vacancy concentrations as a function of P_{O_2} by Flandemeyer et al.[73,74] and Anderson et al.[75] The corresponding defect equilibrium including these clusters can be represented by

$$\begin{aligned} &(La_{La}^x)\,(Cr_{Cr}^{\bullet})_x\,(Cr_{Cr}^x)_{1-2x}\,(Mg_{Cr}')_x\,(O_O^x)_3 \\ &\rightleftarrows (La_{La}^x)\,(Cr_{Cr}^{\bullet})_{x-2\delta}\,(Cr_{Cr}^x)_{1-2x+2\delta} \\ &(Mg_{Cr}')_{x-2\delta}\left((Mg_{Cr}' - V_O^{\bullet\bullet} - Mg_{Cr}')^x\right)_\delta (O_O^x)_{3-\delta} \\ &+ \tfrac{\delta}{2}O_2(g) \end{aligned} \tag{5.85}$$

Usually, the interconnection material Mg-doped $LaCrO_3$ is produced using electrochemical vapor deposition (EVD). EVD is the key technology developed by Westinghouse for the production of gas tight layers of SOFC solid electrolyte and interconnection materials on porous electrode structures. Recently, Van Dieten et al.[79-81] used anti-Frenkel disorder in $LaCrO_3$ along with Equations (5.81) and (5.82) to quantitatively model the EVD growth of Mg-doped $LaCrO_3$.[81]

The clustering concept was not employed in the modeling of the EVD growth, because the EVD process is carried out at temperatures in the range of 1400 to 1600 K. At these temperatures these defect clusters are assumed to be not stable.

The presented concepts for Δx and Δy have been used in other studies on SOFC electrode and interconnection materials exhibiting mixed ionic (via $V_O^{\bullet\bullet}$) and electronic conductivity.[82-84]

The perovskites $La_{1-x}M_xB\,O_{3-\delta}$ (M = Sr, B = Co, Cr, Fe, Mn) have been reviewed by Alcock et al.[85] in terms of their stability in temperature and oxygen partial pressure of operation as active sensor material, nonstoichiometry, conductivity, and catalytic properties, and the effect of the strontium content, x.

The deviations from stoichiometry in these perovskites are accounted for using Equation (5.67). The defect models used for the MIECs discussed so far are all concordant with the defect model used by Van Dieten et al.[79-81] for the modeling of the EVD growth of the SOFC interconnection material.

Another important class of ternary oxides with mixed ionic–electronic conductivity is based on the pyrochlore system, $A_2B_2O_7$. Especially, Tuller and co-workers[86-89] have studied the defect chemistry of nominally pure, and donor- and acceptor-doped $Gd_2Ti_2O_7$ (GT). They have demonstrated that mixed ionic and electronic conductivity can be controlled over wide limits in the pyrochlore oxides described by compositions $Gd_2(Ti_{1-x}Zr_x)_2O_7$ (GZT) and $Y_2(Ti_{1-x}Zr_x)_2O_7$ (YZT) by control of both the parameter x and by doping. If GT is doped with Ca[86] or by selecting $x > 0.4$ in GZT,[87,88] an excellent solid electrolyte is obtained. Reduced or donor-doped GT exhibits predominantly a P_{O_2}-dependent semiconducting behavior.[90]

The ionic conductivity increases in GZT as Zr substitutes for Ti due to increasing anion disorder, and this appears to be related to the decrease in the average B to A cation radius ratio, which, in the limit, leads to full disordering to the defect fluorite structure $(A,B)_4O_7$. Also, the doping with multivalent donors and acceptors has been studied. An in-depth review has been reported by Spears and Tuller,[91] in which the application of defect modeling of the pyrochlores to materials design is discussed.

The electrical properties of the titanate-based pyrochlores can be described by point defect models in which the acceptor (A) and donor (D) impurities are compensated by oxide ion vacancies, or oxide ion interstitials, respectively.[92] The principal defect reactions include the redox reaction, Equation (5.67), the Frenkel disorder, Equation (5.57), dopant ionization, intrinsic electronic disorder, Equation (5.60), and the electroneutrality relation. For these compounds the total electroneutrality condition given by Equation (5.55) is, on the one hand, reduced, taking Equation (5.57) into account, and is, on the other hand, extended to include acceptor and donor impurities. In addition, defect association is included explicitly.

The electroneutrality relation thus reads,

$$2\left[O_i''\right]+\left[O_i'\right]+\left[e'\right]+\left[A'\right]+\left[\left(D\cdot O_i\right)'\right]= \\ 2\left[V_O^{\bullet\bullet}\right]+\left[V_O^{\bullet}\right]+\left[h^{\bullet}\right]+\left[D^{\bullet}\right]+\left[\left(A\cdot V_O\right)^{\bullet}\right] \tag{5.86}$$

Now the more complex defect equilibria are treated with the aid of numerical methods. Spears and Tuller[91] show how this approach can be used, not only to extract principal defect thermodynamic and kinetic data, but also to assist in the design and optimization of materials. Using all possible defect reactions and defect equilibria, combined with Equation (5.86), one arrives at an eight-order polynomial in, for instance, $[e']$, which can only be solved numerically. In a simplified approach for the nominally pure material, expressions for $[e']$ and $[h^{\bullet}]$ are obtained, which are concordant to Equations (5.51) and (5.52). The simple Kröger–Vink diagram for such a defect situation has already been presented by Kofstad.[93]

For the more complex extrinsic defect structure, the Newton–Raphson method was used to numerically solve the high-order polynomials, which come about during attempts to solve the simultaneous defect equations, using the complete electroneutrality condition. For acceptor-doped $Gd_2Ti_2O_7$ with strong acceptor-oxide ion vacancy association, Spears and Tuller[91] report the Kröger–Vink diagram for the log P_{O_2} (atm) regime –50 to +50. Using this approach, mixed conduction in $Gd_2(Zr_{0.3}Ti_{0.7})_2 O_7$ was modeled in terms of the P_{O_2} dependence of the point defect concentrations $[V_O^{\bullet\bullet}]$, $[O_i']$, and $[e']$ for log P_{O_2} (atm) in the region –25 to +5. For log $P_{O_2} < -15$ the reduced electroneutrality condition

$$2\left[V_O^{\bullet\bullet}\right]=\left[e'\right] \tag{5.87}$$

holds, while for larger P_{O_2} values the condition reads,

$$2\left[V_O^{\bullet\bullet}\right]=\left[e'\right]+2\left[O_i''\right] \tag{5.88}$$

In GZT, with x = 0,3, and doped with an acceptor of 1% substitution for Gd, i.e., A'_{Gd}, the oxide ion vacancy concentration is increased at the expense of $[O_i'']$, viz. Equation (5.57). Likewise, the electron hole conductivity is increased at the expense of the electron conduction.

The calculated P_{O_2} dependencies at 1373 K of the predominant defect concentrations in $Gd_2(Zr_{0.3}Ti_{0.7})_2O_7$ reveal that a transition from ionic disorder, $[V_O^{\bullet\bullet}] = [O_i'']$, at high P_{O_2} to reduction control $[e'] = 2\,[V_O^{\cdot\cdot}]$ occurs at $P_{O_2} \simeq 10^{-15}$ atm. For $P_{O_2} \simeq 10$ atm a p–n junction occurs.[91] GZT doped with 1 at. percent acceptor, A'_{Gd}, reveals a narrowing of the defect regime defined by $[e'] = 2\,[V_O^{\cdot\cdot}]$.

Also, in acceptor-doped perovskites $(Ba_xSr_{1-x})TiO_3$ the principal defects and their dependence on P_{O_2}, dopant concentrations, and temperature could be obtained.[91,94] The approach has recently also been applied successfully to the pyrochlore system $(Gd_{1-x}Ca_x)_2Sn_2O_{7\pm\delta}$.[95]

For cost and stability reasons, many materials studies for SOFCs are directed toward lowering the operating temperature from 1000°C to below 800°C. Since the pioneering work of Iwahara and co-workers,[96,97] who discovered high-temperature proton conductivity in the perovskite oxides $SrCeO_3$ and $BaCeO_3$, cerate-based perovskite-type oxide ceramics have attracted widespread attention. Doping with trivalent cations, which substitute for cerium, is essential for the occurrence of proton conduction in hydrogen-containing ambients at high temperatures.

$SrCe_{0.95}Y_{0.05}O_{3-\alpha}$, $SrCe_{0.95}Yb_{0.05}O_{3-\alpha}$, and $BaCe_{0.9}Nd_{0.1}O_{3-\alpha}$ are examples of this class of proton-conducting oxide ceramics. The state of the art has been reviewed by Iwahara.[98,99] Besides application in SOFCs, these solid electrolytes have been explored for use in high-temperature steam electrolyzers, hydrogen sensors and pumps, and electroceramic reactors for (de-)hydrogenation processes.[98-105] A major disadvantage of this class of materials is their instability in CO_2-containing atmospheres.[106,107]

The defect chemistry of these oxides has been studied in relation to the formation of protons in equilibrium with different ambients. The general formula for these oxides can be represented by

$$\left(M_M^x\right)\left(Ce_{Ce}^x\right)_{1-x}\left(Me'_{Ce}\right)_x\left(O_O^x\right)_{3-\frac{1}{2}x}\left(V_O^{\bullet\bullet}\right)_{\frac{x}{2}} \tag{5.89}$$

with electroneutrality condition.

$$\left[Me'_{Ce}\right]_x = 2\left[V_O^{\bullet\bullet}\right]$$

Here, M represents Sr, or Ba, and Me a rare earth element. Usually x is smaller than 0.1.

In dry oxygen atmosphere the formation of electron holes is described by

$$\tfrac{1}{2}O_2(g) + V_O^{\bullet\bullet} \overset{K_h}{\rightleftarrows} O_O^x + 2h^{\bullet} \tag{5.90}$$

In a wet oxygen atmosphere both protons and electron holes are generated according to

$$H_2O(g) + V_O^{\bullet\bullet} \overset{K_{H,1}}{\rightleftarrows} O_O^x + 2H_i^{\bullet} \tag{5.91}$$

and

$$H_2O(g) + 2h^{\bullet} \overset{K_{H,2}}{\rightleftarrows} 2H_i^{\bullet} + \tfrac{1}{2}O_2(g) \tag{5.92}$$

The equilibrium state of the oxides can be described by any two of these equilibrium reactions, as

$$K_{H,1} = K_h K_{H,2} \tag{5.93}$$

In a dry hydrogen-containing atmosphere the equilibrium

$$H_2(g)+2h^{\bullet} \rightleftarrows 2H_i^{\bullet} \tag{5.94}$$

is established.

Hence the electrical properties of the $MCeO_3$-based perovskite-type oxides are quite dependent on the ambients, and can be summarized as follows:[98,99]

- $MCe_{1-x}Me_xO_{3-½x}$ exhibits electron hole conductivity, when sintered in air, and measured in a dry, hydrogen-free atmosphere.
- In atmospheres containing water vapor or hydrogen, protonic conduction increases at the expense of electronic conductivity.
- In pure hydrogen the transference number of the protons becomes unity at high temperatures.
- Mixed oxide and proton conduction occurs in the $BaCeO_3$-based solid electrolytes, when utilized in a SOFC above 800°C.

The proton defect that appears in the gas–solid Equilibria (5.91), (5.92), and (5.94) is represented by $H_i^{\bullet}$. It is generally accepted that the protons are bonded to the oxide ions at regular lattice sites, thus forming a hydroxide ion, $OH_O^{\bullet}$, which is often referred to as $H_i^{\bullet}$.

Equilibrium (5.94) can also be represented by

$$H_2\ (g)+2O_O^x+2h^{\bullet} \rightleftarrows 2OH_O^{\bullet} \tag{5.95}$$

Likewise, equilibria (5.91) and (5.92) can be rewritten to express the formation of $OH_O^{\bullet}$ defects.

The introduction of protonic defects in oxides in hydrogen-containing ambients has been reported frequently, and has recently been reviewed by Colomban and Novak[108] and Strelkov et al.[108]

For a doped oxide M_2O_3 exhibiting anti-Frenkel disorder, Colomban and Novak[108] present a schematic Kröger–Vink diagram of the extrinsic and intrinsic point defects as a function of the partial water pressure. With regard to electrical properties, the proton conductivity in the binary metal oxides is usually much lower than in the perovskite-type oxides.[108]

C. MULTINARY COMPOUNDS

While in many of the fundamental studies of extrinsic and intrinsic disorder simple binary and ternary compounds have been investigated, the search for new materials for application in solid state batteries, high-temperature fuel cells and steam electrolyzers, electrocatalytic reactors, gas-separation membranes, smart windows, and chemical gas sensors has provided the solid state community with a wide variety of multinary materials. Prime examples are the β- and β″-aluminas, sodium ion conductors with the NASICON framework structure, lithium ion conductors with the LISICON structures, oxide Brownmillerite systems, MIECs in the doped ABO_3-$ABO_{2.5}$ series, layered perovskite structures, proton-conducting hydrated oxides, and the ceramic high T_c superconductors. In addition, solid solutions based on transition metal oxides represent interesting multinary compounds for solid state lithium batteries.

Many studies on multinary compounds have been reported and reviewed in the past decade.[109-118] Excellent surveys of crystalline multinary solid electrolytes have been reported by West[119] and Goodenough.[120]

These authors have reviewed positional disorder in stoichiometric compounds in relation to crystal structure, materials exhibiting first-order phase transformations, and diffuse, so-called Faraday, transitions, extrinsic disorder, doping strategies, and solid solutions.

In addition to these surveys, the crystal chemistry and electrochemical properties of a wide variety of multinary oxide solid electrolytes and MIECs are described in the studies reported in Reference 90. While defect chemistry has been successfully employed to model point defect generation and conduction mechanisms in nominally pure and doped binary and ternary compounds, complex defect equilibria occur in concentrated solid solutions based on ternary compounds and in multinary compounds. Usually, these complex defect equilibria are treated with the aid of numerical methods. Examples are the pyrochlores GT, GZT, and YZT as discussed in Section II.B.4.

Since the discovery of the high T_c ceramic superconductors,[121] various properties of these oxide ceramics have been studied, and this is especially true for $YBa_2Cu_3O_{7-x}$.

The critical temperature of the high T_c ceramic superconductors is intimately connected to the electronic charge carrier concentration, but at temperatures beyond the critical temperature in the nonsuperconducting state, mixed ionic–electronic conduction prevails, especially at the operation temperatures of the solid state electrochemical devices mentioned in the previous sections. In this regime a number of properties are determined by deviations from stoichiometry, which can be related to the high diffusivity of oxide ions in the structure of the high T_c ceramic superconductors. In particular, mixed ionic–electronic conductivity in $(La,Sr)_2\,CuO_{4-x}$, $YBa_2Cu_3O_{7-x}$, $Bi_2Sr_2Cu_6O_{6+x}$, and $Bi_2Sr_2CaCu_2O_{8+x}$ has attracted much attention, and has initiated defect chemical studies.[114,115,118,122-126]

For $YBa_2Cu_3O_{7-x}$ it has been established that a transition from electron hole to electron conduction occurs for $7-x < 6.3$. For $x < 0.7$ oxygen is incorporated according to

$$\tfrac{1}{2}O_2(g)+V_i^{x} \rightleftarrows O_i''+2h^{\bullet} \tag{5.96}$$

The oxide interstitials can trap an electron hole to form $O_i^{\bullet}$ defects.

$$O_i''+h^{\bullet} \rightleftarrows O_i' \tag{5.97}$$

Here, O_i' means $(O_i''\cdot h^{\bullet})'$ or $(O_i''\,Cu_{Cu}^{\bullet})'$. For $x < 0.7$ the isothermal conductivity reveals a $P_{O_2}^{1/2}$ dependence, indeed indicating the presence of O_i' defects.[122,123] The actual incorporation equilibrium reaction is therefore

$$\tfrac{1}{2}O_2(g)+V_i^{x} \rightleftarrows O_i'+h^{\bullet} \tag{5.98}$$

The preference of O_i' may be related to the fact that in the Cu plane of $YBa_2Cu_3O_{7-x}$, a Cu ion belongs to one structurally vacant oxygen site ($x = 1$). The incorporation of further oxygen should mainly lead to the formation of O_i^{x},

$$O_i'+h^{\bullet} \rightleftarrows O_i^{x} \tag{5.99}$$

In a different approach, the deviation from stoichiometry of $YBa_2Cu_3O_{7-x}$ ($0 \le x \le 0.7$) can be described using oxide ion vacancies. The gas–solid equilibrium is then described by

$$\tfrac{1}{2}O_2(g)+V_O^{\bullet\bullet}\left(or\ V_O^{\bullet}\right) \rightleftarrows O_O^{x}+2h^{\bullet}\left(or\ h^{\bullet}\right) \tag{5.100}$$

For nominally pure La_2CuO_4 the electron hole conductivity has been found to follow a 1/6 power law dependence on oxygen partial pressure, indicating the gas–solid equilibrium 5.96 to occur. For Sr-doped La_2CuO_4 incorporation reactions have been presented by Opila et al.[126] to account for an increase in $[h^{\bullet}]$ and $[V_O^{\bullet\bullet}]$. The total electroneutrality condition reads

$$2[O_i''] + [Sr_{La}'] + [e'] = 2[V_O^{\bullet\bullet}] + [h^{\bullet}] \qquad (5.101)$$

The Kröger–Vink diagrams have been calculated for $(La,Sr)_2\ CuO_4$,[124-126] assuming fixed doping concentration and temperature. With increasing oxygen partial pressure, the following reduced electroneutrality conditions hold:

$$[Sr_{La}'] = 2[V_O^{\bullet\bullet}],\ [Sr_{La}'] = [h^{\bullet}],\ and\ [h^{\bullet}] = [O_i'] \qquad (5.102)$$

At fixed oxygen partial pressure and temperature, an increase in dopant concentration reveals the following majority defect conditions:

$$[h^{\bullet}] = 2[O_i''],\ [Sr_{La}'] = [h^{\bullet}],\ [Sr_{La}'],\ = 2[V_O^{\bullet\bullet}] \qquad (5.103)$$

For $YBa_2Cu_3O_{7-x}$ tentative Kröger–Vink diagrams have been constructed under the assumption of excess Y, and in the case of anti-Frenkel disorder.[124-126] In the first case the electroneutrality conditions are with increasing P_{O_2}:

$$[Y_{Ba}'] = [O_i'],\ [O_i'] = [h^{\bullet}] \qquad (5.104)$$

In the case of Frenkel disorder the majority defect pairs are

$$[O_i'] = [V_O^{\bullet}],\ [O_i'] = [h^{\bullet}] \qquad (5.105)$$

In the high P_{O_2} regimes, i.e., $[O_i^{\bullet}] = [h^{\bullet}]$ the concentration of O_i^x exceeds that of the charged point defects.

In order to obtain quantitative information on the oxide ion conductivity of high T_c ceramic superconductors, solid state electrochemical cells of the type

$$Pt(air)/YSZ/YBa_2\,Cu_3O_{7-x}/YSZ/Pt(air) \qquad (5.106)$$

were studied by impedance spectroscopy.[127,128] Here, YSZ represents yttria-stabilized zirconia. Similar cell configurations were used to study other high T_c ceramic superconductors.[124,129]

A detailed discussion of oxide ion conductivity as obtained using the solid state electrochemical techniques is beyond the scope of this chapter.

Recent high-temperature applications of these ceramic superconductors are

- a novel chromatography detector[130]
- heterogeneous catalysis,[131-133] and
- Taguchi-type gas sensors.[134-137]

McNamara et al.[130] used $YBa_2Cu_3O_{7-x}$ as a catalyst for redox reactions between organic molecules and NO_2 in a chromatography detector, based on chemoluminescence. NO_2 is converted to NO, which is subsequently detected via a chemoluminescence reaction with ozone, i.e.,

$$NO + O_3 \rightarrow NO_2 + O_2 + h\nu \tag{5.107}$$

Hansen et al.[131] have studied the oxydation of hydrocarbons to CO_2 over $YBa_2Cu_3O_{7-x}$ as a catalyst. Besides full oxydation to CO_2, process conditions can be manipulated such as to partially oxidize hydrocarbons to a variety of industrial chemicals.

In addition to deviations from stoichiometry in the anion sublattice, $YBa_2Cu_3O_{7-x}$ has been shown to absorb nitrogen, argon, hydrogen, and nitrogen oxide. The solubility of these gases is dependent on the oxide stoichiometry, but does not affect the perovskite structure. At 300°C, 3 mol% of NO can be dissolved into $YBa_2Cu_3O_{7-x}$. Desorption occurs at 500°C. The amount of absorbed NO increases with the parameter x. The NO molecules occupy the empty oxide ion sites. For high concentrations of NO, decomposition occurs according to

$$2NO + 2YBa_2Cu_3O_{7-x} \rightarrow 3CuO + Y_2BaCuO_5 + Ba(NO_2)_2 + 2BaCuO_2 \tag{5.108}$$

For simplicity, x has been assumed to be zero here.

Besides absorption of NO, $YBa_2Cu_3O_{7-x}$ has been shown to be very effective in the catalytic decomposition of NO to N_2 and O_2. This decomposition is related to the amount of absorbed NO, and hence to the deviation from stoichiometry. This is, in fact, one of the first examples of the direct decomposition of NO in N_2 and O_2.[133]

The high T_c ceramic superconductors have been tested in Taguchi-type gas sensors.[134-137] The Taguchi gas sensor[138] is based on a semiconducting metal oxide, which exhibits a resistance change when gas molecules adsorb at the surface and influence the charge carrier concentration in the metal oxide. SnO_2, In_2O_3, and solid solutions based on these oxides have frequently been studied as active material of the Taguchi sensor.

The sensitivity to NO_x and CO_x has been investigated for $YBa_2Cu_3O_{7-x}$, $Ba_{1.5}La_{1.5}Cu_3O_x$, $La_{1.85}Sr_{0.15}CuO_x$, $Nd_{1.85}Ce_{0.15}CuO_{4-x}$, $Bi_2Sr_2CuO_{6+x}$, $Bi_2Sr_2CaCu_2O_{8+x}$, and $Bi_{1.8}Pb_{0.2}Sr_2Ca_2Cu_3O_{10+x}$. As expected, $YBa_2Cu_3O_{7-x}$ reacts with NO and decomposes. $Ba_{1.5}La_{1.5}Cu_3O_x$ and $La_{1.85}Sr_{0.15}CuO_x$ did not show any sensitivity or selectivity of practical interest, while $Nd_{1.85}Ce_{0.15}CuO_{4+x}$ has no selectivity for NO against CO. However, the Bi-family materials are good candidates for the development of Taguchi-type NO_x gas sensors. $Bi_2Sr_2CuO_{6+x}$, $Bi_2Sr_2CaCu_2O_{8+x}$, and $Bi_{1.8}Pb_{0.2}Sr_2Ca_2Cu_3O_{10+x}$ exhibit good stability, reproducibility, and high selectivity against CO and CO_2. $Bi_2Sr_2CaCu_2O_{8+x}$ exhibits the best response behavior. Composites of $Bi_2Sr_2CaCu_2O_{8+x}$ with NiO or Fe_2O_3 show improved selectivity to NO against CO. The addition of Al_2O_3 does not affect the selectivity, but does enhance the response rate. Lithium intercalation also improved the selectivity to NO against CO. The detection mechanisms have been reported by Huang et al.,[134-137] and they are based on reactions (5.96) and (5.100).

D. MULTICOMPONENT MATERIALS

Since the initial observations that the dispersion of small particles of nonconducting alumina in lithium iodide leads to a dramatic increase of the lithium ion conductivity,[139,140] a number of studies have been devoted to the electrical properties of these multicomponent systems.[141-147]

In general, it is found that the increase in isothermal ionic conductivity is greater the smaller the size of the dispersed inert particle, and that the conductivity–composition curve exhibits a maximum between 10 and 40 mol% of inert particles. It is generally accepted that the conductivity enhancement can be ascribed to an increase in the defect concentration responsible for extrinsic conduction in a narrow space charge layer in the conducting compound at the grain boundary interface. This "composite effect" occurs mainly at low temperatures.[148] Surprisingly, practically all the studies on this composite effect have dealt with cation conducting compounds. A few studies have been reported on YSZ–alumina composites,[149-151] but the results have not led to firm conclusions. In a more systematic study, Filal et al.[152] have investigated single crystalline, polycrystalline, and composites of YSZ. For composites comprising YSZ (9.9 mol%) and Mg-doped α-alumina of submicron size, a maximum is observed for 2 mol% α-Al_2O_3. At 250°C the grain boundary conductivity is increased by a factor of 3.7 and at 600°C by a factor of 2. While the composite effect has been established for stabilized zirconia, a mechanism is lacking. For CaF_2-Al_2O_3 composites the composite effect has been observed by Fujitsu et al.[153] Here, a complication may occur, as the ionic radii of oxide and fluoride ions are comparable, and CaF_2 exhibits anti-Frenkel disorder. As a consequence, CaF_2, or more general the alkali earth fluorides, are easily contaminated with oxide ions to form the defects $O_F' + V_F^{\cdot}$. If at the interface CaF_2-Al_2O_3 this doping effect occurs, it may mask space charge effects.

Maier[154,155] has calculated the concentration profiles of minority charge carriers in ideal space charge regions. In addition, the solid state electrochemical Hebb–Wagner polarization technique was extended to the case of multicomponent materials.[154] The behavior of the minority carriers in the space charge regions is determined by the profiles of the majority carriers. By applying the theoretical model on the composite AgCl-γ-Al_2O_3, Maier[155] found that both the electron and the electron hole conductivity increased, compared to single-phase AgCl. In contrast, a shift in charge carrier type is observed on going from pure LiI (electron holes)[139] to LiI–Al_2O_3 (electrons).[145] Similar observations were reported for β-AgI-alumina composites. Pure β-AgI exhibits electron conduction,[146] while the composite has electron hole conduction. As has been pointed out by Maier,[154] the extra space charge contribution due to the minority charge carriers depends sensitively on impurity levels, and these are not specified in the studies on LiI- and β-AgI-based composites.

In order to shed more light on these phenomena, Maier[156] has calculated the defect concentrations for ionic and electronic defects in space charge regions as a function of temperature and component activities.

Using the model compound, of which the bulk defect chemistry is treated in Section II.A.1 of this chapter, Maier constructed a Kröger–Vink diagram of the model compound MX for bulk and boundary layer. His results are schematically included in Figure 5.2, and presented in Figure 5.6.

In this example it has been assumed that the bulk structure extends up to the top surface (interface) layer, in which the excess charges are accommodated and in which the potentials and defect concentrations have discretely different values. For the top layer a linear geometry is assumed. Reaction (5.12) holds for the top layer. As long as one or both ionic point defects determine the charge density alone, both ionic point defect concentrations can be handled as a constant (cf. Reaction (5.15)), as is done for the bulk. In the Kröger–Vink diagram the same slopes arise for the ionic point defects, i.e., 0, and ±½ for the electronic defects, i.e., Equations (5.28) and (5.30). Due to the space charge effect the ranges of validity are different. In the present example a positive space charge potential has been assumed, which implies accumulation of V_M' and e′, and depletion of $M_i^{\cdot}$ and $h^{\cdot}$.

The model has been used to calculate the point defect profiles as a function of component activity and temperature for the multicomponent system AgCl-Al_2O_3.

For an acceptor-doped metal oxide, Maier presents a qualitative Kröger–Vink diagram assuming a positive space charge potential. Here, the extrinsic regime is increased due to depletion of e′ and accumulation of $V_O^{\cdot\cdot}$.[156] At low P_{O_2} values the curves are approaching the bulk behavior. This particular situation is observed for SnO_2 in O_2 ambients, and has been used to discuss the sensing behavior of SnO_2 in a Taguchi sensor.

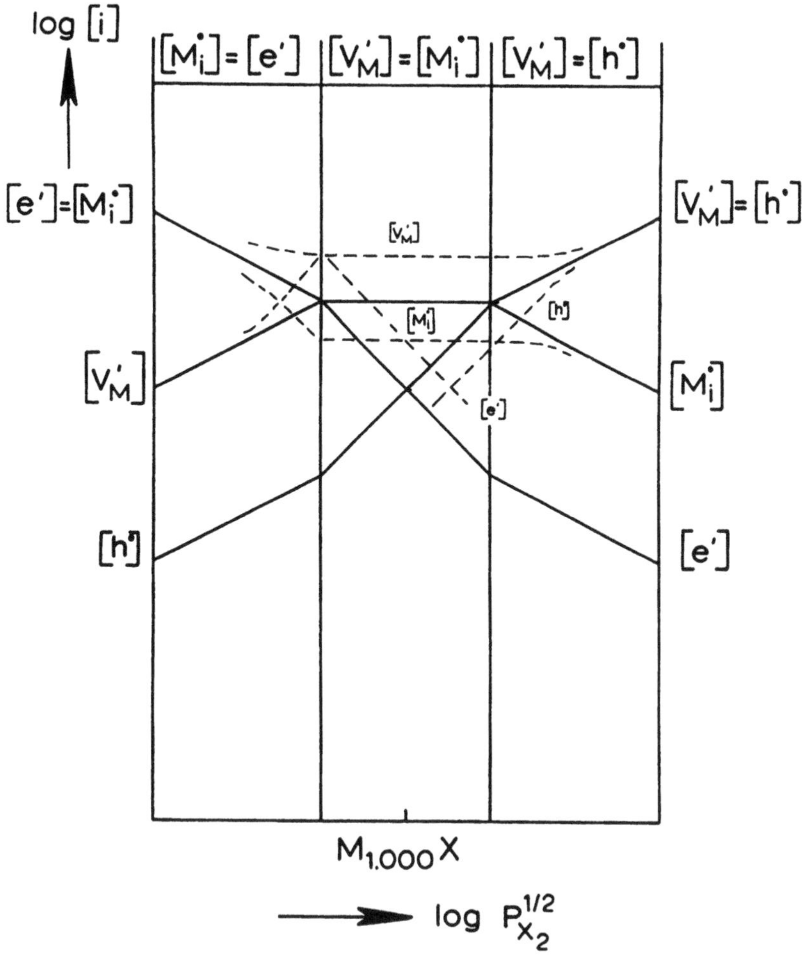

FIGURE 5.6. Kröger–Vink diagrams of MX with Frenkel disorder (——— = bulk, --- = boundary layer).

A large number of problems in solid state electrochemistry are determined by the defect chemistry in boundary regions rather than in the bulk. Taking the defect chemistry in space charge layers into account, unusual partial pressure dependences of electronic conductivity, which cannot be understood in terms of bulk defect chemistry, can be made plausible. Göpel and Lampe[157] observed for acceptor-doped ZnO, with a defect chemistry quite similar to SnO_2, an accumulation layer for e′ and as exponents for the P_{O_2} dependence of the conductivity −0.25 (bulk) and −0.15 ± 0.03 (space charge layer). The bulk value is easily explained by a constant ionic point defect concentration in the acceptor regime.

If a strong depletion of oxide ion vacancies occurs in the space charge layer, electrons should form the majority carriers there. Maier[156] found under these conditions and under the assumptions that no frozen-in profiles exist and a low mobility for the acceptor impurities, for the P_{O_2} dependence of the space charge conductivity theoretically the exponent −⅛, which is concordant with the experimental value.

III. CONCLUDING REMARKS

Defect chemistry is a chemistry within the solid state that is analogous to the long-familiar chemistry in the liquid phase, and arises from departures from the ideal crystal structure which are thermodynamically unavoidable, the point defects. While defect chemistry foundations were established over 60 years ago, this area of chemistry enables one to date to describe in a unified manner a wide variety of phenomena, based on the dynamical behavior of ionic and electronic species, in electroceramics, MIECs, chemically active ceramics, optoceramics, photographic materials, scale growth on metals, and many more.

It has been demonstrated that the classical equilibrium defect chemical concepts derived for binary compounds can be applied to ternary and multinary compounds. In the case of multicomponent materials, the space charge effects will become very important in cases in which the dimensions are no longer large compared with the thickness of the space charge layers, as in extremely thin films or in structural and functional ceramics with crystallites of nanometer dimensions. The formation of latent images in silver halide photography represents a prelude to effects of point defects in nanostructured materials, and is related to enlarged concentrations of point defects in boundary layers.

To date the development of nanostructured materials is receiving widespread attention, and it is believed that classical defect chemical concepts will contribute to the pace of advance in our understanding of improved or new properties of either single-phase or multiphase materials. With regard to multiphase materials the ceramic–metal composites (cermets) with percolation-type conductivity behavior like YSZ-Pd or $BaCeO_3$-Pd, which exhibit enhanced diffusion of oxide ions and protons, respectively, represent a novel class of gas-separation membranes. While mixed oxide, proton, and electronic conductivity may occur in these cermets, the mechanisms of mass and charge transport will largely be governed by the dynamical behavior of ionic point defects. Hence, classical defect chemistry is invaluable for the field of solid state electrochemistry.

REFERENCES

1. Nernst, W., *Z. Elektrochem.*, 1899, *6*, 41.
2. Tubandt, C. and Lorentz, E., *Z. Phys. Chem.*, 1914, *87*, 513.
3. Strock, L.W., *Z. Phys. Chem.*, 1934, *B25*, 441.
4. Frenkel, J., *Z. Physik*, 1926, *35*, 652.
5. Wagner, C. and Schottky, W., *Z. Phys. Chem.*, 1930, *B11*, 163.
6. Schottky, W., *Z. Phys. Chem.*, 1935, *B29*, 335.
7. Jost, W., *J. Chem. Phys.*, 1933, *1*, 466.
8. Kröger, F.A. and Vink, H.J., in *Solid State Physics: Advances in Research and Applications*, Vol. 3, Seitz, F. and Turnbull, D., Eds., Academic Press, New York, 1956, 307.
9. van Bueren, H.G., *Imperfections in Crystals*, North-Holland, Amsterdam, 1960.
10. Kröger, F.A., *The Chemistry of Imperfect Crystals*, North-Holland, Amsterdam, 1964.
11. Hauffe, K., *Reaktionen in und an festen Stoffen*, Springer-Verlag, Berlin, 1966.
12. Fong, F.K., in *Progress in Solid State Chemistry*, Vol. 3, Reiss, H., Ed., Pergamon Press, Oxford, 1967, 135.
13. Barr, L.W. and Lidiard, A.B., in *Physical Chemistry. An Advanced Treatise*, Vol. X, Jost, W., Ed., Academic Press, New York, 1970, 151.
14. Schoonman, J., *J. Solid State Chem.*, 1972, *4*, 466.
15. Rickert, H., *Elektrochemie fester Stoffe*, Springer-Verlag, Berlin, 1973.
16. Hayes, W., Ed., *Crystals with the Fluorite Structure*, Clarendon Press, Oxford, 1974.
17. Schmalzried, H., *Solid State Reactions*, Verlag Chemie, Weinheim, 1974.
18. Geller, S., Ed., *Solid Electrolytes*, Springer-Verlag, Berlin, 1977.
19. Kofstad, P., *Nonstoichiometry, Diffusion, and Electrical Conductivity in Binary Metal Halides*, Krieger Publishing, Malabar, India, 1983.
20. West, A.R., *Solid State Chemistry and its Applications*, John Wiley & Sons, Chichester, 1985.
21. Tilley, R.J.D., *Defect Crystal Chemistry*, Blackie, London, 1987.
22. Smart, L. and Moore, E., *Solid State Chemistry*, Chapman & Hall, London, 1992.
23. Deportes, C., Duclot, M., Fabry, P., Fouletier, J., Hammou, A., Kleitz, M., Siebert, E., Souquet, J.-L., *Electrochimie des Solides*, Presses Universitaires de Grenoble, Grenoble, 1994.
24. Parsonage, N.G. and Staveley, L.A.K., *Disorder in Crystals*, Clarendon Press, Oxford, 1978.
25. Schmalzried, H. and Wagner, C., *Z. Phys. Chem. N.F.*, 1962, *31*, 198.

26. Schmalzried, H., in *Progress in Solid State Chemistry,* Riess, H., Ed., Pergamon Press, Oxford, 1965.
27. Schoonman, J., Huggins, R.A., in *Solid Electrolyte Materials,* Tech. Rep. No. 3, Stanford University, 1974, 62.
28. George, W.L. and Grace, R.E., *J. Phys. Chem. Solids,* 1968, *30,* 881.
29. Long, S.A. and Blumenthal, R.N., *J. Am. Ceram. Soc.,* 1971, *54,* 515, 577.
30. Seuter, A.M.J.H., *Philips Res. Rep. Suppl. 3,* 1974.
31. Jarzebski, Z.M., *Mater. Res. Bull., 9,* 1974.
32. Smyth, D.M., *J. Solid State Chem., 16,* 1976.
33. Smyth, D.M., *J. Solid State Chem.,* 1977, *20,* 359.
34. Groenink, J.A., The Electrical Conductivity and Luminescence of $PbMoO_4$ and $PbWO_4$. Ph.D. thesis. Utrecht University, The Netherlands, 1979.
35. van Gool, W., Ed., *Fast Ion Transport in Solids. Solid State Batteries and Devices,* American Elsevier, New York, 1973.
36. Wapenaar, K.E.D., Ionic Conduction and Dielectric Loss in Fluorite-Type Solid Solutions, Ph.D. thesis, Utrecht University, The Netherlands, 1981.
37. Wapenaar, K.E.D., van Koesveld, J.L., and Schoonman, J., *Solid State Ionics,* 1981, *2,* 145.
38. Metselaar, R., Heijligers, H.J.M., and Schoonman, *J. Solid State Chemistry 1982. Studies in Inorganic Chemistry 3,* Elsevier, Amsterdam, 1983.
39. Roos, A., Electrical and Dielectrical Phenomena of Solid Solutions $La_{1-x}Ba_xF_{3-x}$. Ph.D. thesis, Utrecht University, The Netherlands, 1983.
40. Iwahara, H., Uchida, H., and Tanaka, S., *Solid State Ionics,* 1983, *9–10,* 1021.
41. Chowdary, B.V.R., Radhakrishna S., Eds., *Materials for Solid State Batteries,* World Scientific, Singapore, 1986.
42. Takahashi, T., Ed., *High Conductivity Solid Ionic Conductors. Recent Trends and Applications,* World Scientific, Singapore, 1989.
43. Nazri, G.-A., Huggins, R.A., and Shriver, D.F., Eds., *Solid State Ionics, MRS Symposium Proceedings,* Vol. 135, MRS, Pittsburgh, 1989.
44. Nazri, G.-A., Shriver, D.F., Huggins, R.A., and Balkanski, M., Eds., *Solid State Ionics II, MRS Symposium Proceedings,* Vol. 210, MRS, Pittsburgh, 1991.
45. Colomban, P., Ed., *Proton Conductors. Chemistry of Solid State Materials 2,* Cambridge University Press, Cambridge, UK, 1992.
46. Chowdary, B.V.R., Chandra, S., and Shri Singh, Eds., *Solid State Ionics. Materials and Applications,* World Scientific, Singapore, 1992.
47. Balkanski, M., Takahashi, T., and Tuller, H.L., *Solid State Ionics,* Proc. Symp. A2 of ICAM 1991, North-Holland, Amsterdam, 1992.
48. Scrosati, B., Magistris, A., Mari, C.M., and Mariotto, G., Eds., *Fast Ion Transport in Solids, Proc. NATO Advanced Research Workshop, Series E: Applied Sciences,* Vol. 250, Kluwer Academic, Dordrecht, 1993.
49. Chowdary, B.V.R., Yahaya, M., Talib, I.A., and Salleh, M.M., Eds., *Solid State Ionic Materials,* World Scientific & Global Publishing Services, Singapore, 1994.
50. Hagenmuller, P. and van Gool, W., *Solid Electrolytes. General Principles, Characterization, Materials, Applications,* Academic Press, New York, 1987.
51. Vashishta, P., Mundy, J.N., and Shenoy, G.K., *Fast Ion Transport in Solids. Electrodes and Electrolytes,* North-Holland, New York, 1979.
52. Bates, J.B. and Farrington, G.C., *Fast Ion Transport in Solids,* North-Holland, Amsterdam, 1981.
53. Kleitz, M., Sapoval, B., and Chabre, Y., *Solid State Ionics-83,* Parts I & II, North-Holland, Amsterdam, 1983.
54. Petot-Ervas, G., Matzke, H.J., and Monty, C., *Transport in Nonstoichiometric Compounds,* North-Holland, Amsterdam, 1984.
55. Boyce, J.B., DeJonghe, L.C., and Huggins, R.A., *Solid State Ionics-85,* Parts I & II, North-Holland, Amsterdam, 1986.
56. Weppner, W. and Schulz, H., *Solid State Ionics-87,* Parts I & II, North-Holland, Amsterdam, 1988.
57. Hoshino, S., Ishigame, M., Iwahara, H., Iwase, M., Kudo, T., Minami, T., Okazaki, H., Yamamoto, O., Yamamoto, T., and Yoshimura, M., *Solid State Ionics-89,* Parts I & II, North-Holland, Amsterdam, 1990.
58. Nicholson, P.S., Whittingham, M.S., Farrington, G.C., Smeltzer, W.W., and Thomas, J., *Solid State Ionics-91,* North-Holland, Amsterdam, 1992.
59. Boukamp, B.A., Bouwmeester, H.J.M., Burggraaf, P.J., van der Put, P.J., and Schoonman, J., *Solid State Ionics-93,* Parts I & II, North-Holland, Amsterdam, 1994.
60. Heyne, L., *Fast Ion Transport in Solids. Solid State Batteries and Devices,* American Elsevier, New York, 1973, 123.
61. Brouwer, G., *Philips Res. Rep.,* 1954, *9,* 366.
62. Darken, L.S. and Gurry, R.W., *J. Amer. Chem. Soc.,* 1945, *67,* 1398.
63. Fender, B.E.F. and Riley, F.D., *J. Phys. Chem. Solids,* 1969, *30,* 793.
64. Alcock, C.B., in *Annual Review of Materials Science,* Vol. 1, Huggins, R.A., Bube, R.H., and Roberts, R.W., Eds., Annual Reviews Co., Palo Alto, CA, 1971, 219.

65. Hasiguti, R.R., in *Annual Review of Materials Science*, Vol. 2, Huggins, R.A., Bube, R.H., and Roberts, R.W., Eds., Annual Reviews Co., Palo Alto, CA, 1972, 69.
66. Petersen, N.L. and Chen, W.K., in *Annual Review of Materials Science*, Vol. 3, Huggins, R.A., Bube, R.H., and Roberts, R.W., Eds., Annual Reviews Co., Palo Alto, CA, 1973, 75.
67. Greenwood, N.N., *Ionic Crystals, Lattice Defects and Nonstoichiometry*, Butterworths, London, 1970.
68. Dieckmann, R., *Solid State Ionics*, 1984, *12*, 1.
69. Groenink, J.A. and Binsma, H., *J. Solid State Chem.*, 1979, 29, 227.
70. Ling, S., *Solid State Ionics*, 1994, *70-71*, 686.
71. Karim, D.P. and Aldred, A.T., *Phys. Rev. B*, 1979, *20*, 2255.
72. Weber, W.J., Griffin, C.W., and Bates, J.L., *J. Am. Ceram. Soc.*, 1987, *70*, 265.
73. Flandemeyer, B.K., Nasrallah, M.M., Agrawal, A.K., and Anderson, H.U., *J. Am. Ceram. Soc.*, 1984, *67*, 195.
74. Flandemeyer, B.K., Nasrallah, M.M., Sparlin, D.M., and Andersen, H.N., *High Temp. Sci.*, 1985, *20*, 259.
75. Anderson, H.U., Nasrallah, M.M., Flandemeyer, B.K., and Agrawal, A.K., *J. Solid State Chem.*, 1985, *56*, 325.
76. Singhal, S.C., Ruka, R.J., and Sinharoy, S., Interconnection Materials Development for Solid Oxide Fuel Cells. DOE-MC-21184, Final Report, Department of Energy, 1985.
77. van Roosmalen, J.A.M. and Cordfunke, E.H.P., *J. Solid State Chem.*, 1991, *93*, 212.
78. van Roosmalen, J.A.M., Some Thermochemical Properties of (La, Sr) $MnO_{3+\delta}$ as a Cathode Material for Solid Oxide Fuel Cells, Ph.D. thesis, University of Amsterdam, Amsterdam, 1993.
79. van Dieten, V.E.J., Dekker, J.P., van Zomeren, A.A., and Schoonman, J., in *Fast Ion Transport in Solids, Proc. NATO Advanced Research Workshop, Series E: Applied Sciences*, Vol. 250, Kluwer Academic, Dordrecht, 1993, 231.
80. van Dieten, V.E.J. and Schoonman, J., in *Solid Oxide Fuel Cells IV*, Dokiya, M., Yamamoto, O., Tagawa, H., and Singhal, S.C., *The Electrochemical Society*, Pennington, New Jersey, 1995, 960.
81. van Dieten, V.E.J., Electrochemical Vapour Deposition of SOFC Interconnection Materials. Ph.D. thesis, Delft University of Technology, Delft, 1995.
82. Stevenson, J.W., Armstrong, T.R., and Weber, W.J., in *Solid Oxide Fuel Cells IV*, Dokiya, M., Yamamoto, O., Tagawa, H., and Singhal, S.C., *The Electrochemical Society*, Pennington, New Jersey, 1995, 454.
83. Armstrong, T.R., Stevenson, J.W., Pederson, L.R., and Raney, P.E., in *Solid Oxide Fuel Cells IV*, Dokiya, M., Yamamoto, O., Tagawa, H., and Singhal, S.C., *The Electrochemical Society*, Pennington, New Jersey, 1995, 944.
84. Yokokawa, H., Horita, T., Sakai, N., Kawada, T., Dokiya, M., Nishiyama, H., and Aizawa, M., in *Solid Oxide Fuel Cells IV*, Dokiya, M., Yamamoto, O., Tagawa, H., and Singhal, S.C., *The Electrochemical Society*, Pennington, New Jersey, 1995, 975.
85. Alcock, C.B., Doshi, R.C., and Shen, Y., *Solid State Ionics*, 1992, *51*, 281.
86. Kramer, S., Spears, M., and Tuller, H.L., *Solid State Ionics*, 1994, *72*, 59.
87. Moon, P.K. and Tuller, H.L., in *Solid State Ionics, MRS Symposium Proceedings*, Vol. 135, Nazri, G.-A., Huggins, R.A., and Shriver, D.F., Eds., MRS, Pittsburgh, 1989, 149.
88. Moon, P.K. and Tuller, H.L., *Solid State Ionics*, 1988, *28-30*, 470.
89. Tuller, H.L., Kramer, S., and Spears, M.A., in *High Temperature Electrochemical Behaviour of Fast Ion and Mixed Conductors*, Poulsen, F.W., Bentzen, J.J., Jacobsen, T., Skou, E., and Ostegard, M.J.L., Eds., Risø National Laboratory, Roskilde, 1993, 151.
90. Kosacki, I. and Tuller, H.L., in *Solid State Ionics IV, MRS Symposium Proceedings*, Vol. 369, Nazri, G.A., Tarascon, J.M., and Schreiber, M., Eds., 1995, 703.
91. Spears, M.A. and Tuller, H.L., in *Solid State Ionics IV, MRS Symposium Proceedings*, Vol. 369, Nazri, G.A., Tarascon, J.M., and Schreiber, M., Eds., 1995, 271.
92. Tuller, H.L. and Moon, P.K., *Mater. Sci. Eng.*, 1988, B1, 171.
93. Kofstad, P., *Nonstoichiometry, Diffusion and Electrical Conductivity in Binary Oxides*, John Wiley & Sons, New York, 1977, 40.
94. Choi, G.M. and Tuller, H.L., *J. Am. Ceram. Soc.*, 1988, *71*, 201.
95. Yu, T.-H. and Tuller, H.L., in *Solid State Ionics IV, MRS Symposium Proceedings*, Vol. 369, Nazri, G.A., Tarascon, J.M., and Schreiber, M., Eds., MRS, Pittsburgh, 1995, 371.
96. Iwahara, H., Esaka, T., Uchida, H., and Maeda, N., *Solid State Ionics*, 1981, *3/4*, 359.
97. Iwahara, H., Uchida, H., Ono, K., and Ogaki, K., *J. Electrochem. Soc.*, 1988, *135*, 529.
98. Iwahara, H., in *Proton Conductors. Chemistry of Solid State Materials 2*, Colomban, P., Ed., Cambridge University Press, Cambridge, UK, 1992, 122.
99. Iwahara, H., in *Solid State Ionics*, Proc. Symp. A2 of ICAM 1991, North-Holland, Amsterdam, 1992, 575.
100. Hammou, A., *Adv. Electrochem. Sci. and Eng.*, 1992, *2*, 87.
101. Vangrunderbeek, J., Luyten, J., de Schutter, F., van Landschoot, R., Schram, J., and Schoonman, J., in *Solid State Ionics*, Proc. Symp. A2 of ICAM 1991, North-Holland, Amsterdam, 1992, 611.
102. Luyten, J., de Schutter, F., Schram, J., and Schoonman, J., *Solid State Ionics*, 1991, *46*, 117.
103. de Schutter, F., Vangrunderbeek, J., Luyten, J., Kosacki, I., van Landschoot, R., Schram, J., and Schoonman, J., *Solid State Ionics*, *57*, 1992, 77.

104. Kosacki, I., Schoonman, J., and Balkanski, M., *Solid State Ionics*, 1992, *57*, 345.
105. Kosacki, I., Becht, J.G.M., van Landschoot, R., and Schoonman, J., *Solid State Ionics*, 1993, *59*, 287.
106. Scholten, M.J., Schoonman, J., van Miltenburg, J.C., and Oonk, H.A.J., *Solid State Ionics*, 1993, *61*, 83.
107. Scholten, M.J., Schoonman, J., van Miltenburg, J.C., and Oonk, H.A.J., in *SOFC-III*, Singhal, S.C. and Iwahara, H., Eds., Vol. 93-4, The Electrochemical Society, Pennington, New Jersey, 1993, 146.
108. Colomban, P. and Novak, A., (p. 61), Strelkov, A.V., Kaul, A.R., and Tretyakov, Yu.D., in *Proton Conductors. Chemistry of Solid State Materials 2*, Cambridge University Press, Cambridge, UK, 1992, 605.
109. Pistoria, G., Ed., *Lithium Batteries, Industrial Chemistry Library*, Vol. 5, Elsevier, Amsterdam, 1994.
110. Julien, C. and Nazri, G.A., *Solid State Batteries: Materials Design and Optimization*, Kluwer, Boston, 1994.
111. Bruce, P.G., Ed., *Solid State Electrochemistry*, Cambridge University Press, Cambridge, UK, 1995.
112. Megahed, S., Barnett, B.M., and Xie, L., *Rechargeable Lithium and Lithium-Ion Batteries*, Proc. Vol. 94-28, The Electrochemical Society, Pennington, New Jersey, 1995.
113. Steele, B.C.H., in *Solid State Ionics*, Proc. Symp. A2 of ICAM 1991, North-Holland, Amsterdam, 1992, 17.
114. Jorgensen, J.D., Kitazawa, K., Terascon, J.M., Thompson, M.S., and Torrance, J.B., Eds., *High Temperature Superconductors: Relationships between Properties, Structure, and Solid State Chemistry*, MRS Symp. Proc., Vol. 156, Materials Research Society, Pittsburgh, 1989.
115. Narlikar, A., Ed., *Studies of High Temperature Superconductors*, Vols. 1 and 2, Nova Sci. Publ., Commack, 1989.
116. Davies, P.K. and Roth, R.S., *Chemistry of Electronic Ceramic Materials*, Technomic Publ. Co. Inc., Lancaster, 1990.
117. Aucouturier, J.-L., Cauhapé, J.-S., Destrian, M., Hagenmuller, P., Lucat, C., Ménil, F., Portier, J., and Salardenne, J., Proc. 2nd. International Meeting on Chemical Sensors, Bordeaux, 1986.
118. Schoonman, J., High-temperature applications of ceramic superconductors, in *Supergeleiding*, Stuivinga, M. and van Woerkens, E.C.C., Eds., SCME, Delft, 1990 (in Dutch).
119. West, A.R., in *Solid State Electrochemistry*, Bruce, P.G., Ed., Cambridge University Press, Cambridge, UK, 1995, 7.
120. Goodenough, J.B., in *Solid State Electrochemistry*, Bruce, P.G., Ed., Cambridge University Press, Cambridge, UK, 1995, 43.
121. Bednorz, J.G. and Müller, K.A., *Z. Phys.*, 1986, *B64*, 189.
122. Maier, J., Murugaraj, P., Pfundtner, G., and Sitte, W., *Ber. Bunsenges. Phys. Chem.*, 1989, *93*, 1350.
123. Maier, J., Murugaraj, P., and Pfundtner, G., *Solid State Ionics*, 1990, *40/41*, 802.
124. Maier, J., Pfundtner, G., Tuller, H.L., Opila, E.J., and Wuensch, B.J., *High Temperature Superconductors*, Vincenzini, P., Ed., Elsevier Science, Amsterdam, 1991, 423.
125. Maier, J.G. and Pfundtner, G., *Adv. Mater.*, 1991, *3*, 292.
126. Opila, E.J., Pfundtner, G., Maier, J., Tuller, H.L., and Wuensch, B.J., in *Solid State Ionics*, Proc. Symp. A2 of ICAM 1991, North-Holland, Amsterdam, 1992, 553.
127. Vischjager, D.J., van der Put, P.J., Schram, J., and Schoonman, J., *Solid State Ionics*, 1988, *27*, 199.
128. Vischjager, D.J., van Zomeren, A.A., and Schoonman, J., *Solid State Ionics*, 1990, *40/41*, 810.
129. Zhu, W. and Nicholson, P.S., *J. Electrochem. Soc.*, 1995, *142*, 513.
130. McNamara, E.A., Montzka, S.A., Barkley, R.M., and Sievers, R.E., *J. Chromatogr.*, 1988, *452*, 75.
131. Hansen, S., Otamiri, J., Borin, J.-O., and Andersson, A., *Nature*, 1988, *334*, 143.
132. Arakawa, T. and Adachi, G., *Mater. Res. Bull.*, 1989, *24*, 529.
133. Tabata, K., *J. Mater. Sci. Lett.*, 1988, *7*, 147.
134. Huang, X.J., High T_c Ceramic Superconductors in Chemical Devices, Ph.D. thesis, Delft University of Technology, Delft, 1993.
135. Huang, X.J., Schoonman, J., and Chen, L.Q., *Sensors Actuators*, 1995, *B22*, 211, 219.
136. Huang, X.J., Schoonman, J., and Chen, L.Q., *Sensors Actuators*, 1995, *B22*, 227.
137. Huang, X.J. and Schoonman, J., *Solid State Ionics*, 1994, *72*, 338.
138. Taguchi, N., Japanese Patent 45-38200, 1962.
139. Liang, C.C., *J. Electrochem. Soc.*, 1973, *120*, 1289.
140. Liang, C.C., Joshi, A.V., and Hamilton, N.E., *J. Appl. Electrochem.*, 1978, *8*, 445.
141. Shahi, K., Wagner, Jr., J.B., and Owens, B., in *Lithium Batteries*, Gabano, J.P., Ed., Academic Press, London, 1980, 407.
142. Phipps, J.B., Johnson, D.L., and Whitmore, D.H., *Solid State Ionics*, 1981, *5*, 393.
143. Jow, T. and Wagner, Jr., J.B., *J. Electrochem. Soc.*, 1979, *126*, 1963.
144. Shahi, K. and Wagner, Jr., J.B., *J. Electrochem. Soc.*, 1981, *128*, 6.
145. Poulsen, F.W., Andersen, N.H., Kindl, B., and Schoonman, J., *Solid State Ionics*, 1983, *9&10*, 119.
146. Mazumdar, D., Govindacharyulu, P.A., and Bose, D.N., *J. Phys. Chem. Solids*, 1982, *43*, 933.
147. Wagner, Jr., J.B., in *Transport in Nonstoichiometric Compounds*, Simkovich, G. and Stubican, V.S., Eds., Plenum Press, New York, 1985, 3.
148. Shukla, A.K., Manoharan, R., and Goodenough, J.B., *Solid State Ionics*, 1988, *26*, 5.
149. Verkerk, M.J., Middelhuis, B.J., and Burggraaf, A.J., *Solid State Ionics*, 1982, *6*, 159.

150. Buchanan, R.P. and Wilson, D.M., *Adv. Ceram.*, 1984, *10*, 526.
151. Mori, M., Yoshikawa, M., Itoh, H., and Abe, T., *J. Am. Cer. Soc.*, 1994, *77*, 2217.
152. Filal, M., Petot, C., Mokchah, M., Chateau, C., and Carpentier, J.L., *Solid State Ionics*, 1995, *80*, 27.
153. Fujitsu, S., Miyayama, M., Konmoto, K., Yanagida, H., and Kanazawa, T., *J. Mater. Sci.*, 1985, *20*, 2103.
154. Maier, J., *Ber. Bunsenges. Phys. Chem.*, 1989, *93*, 1468.
155. Maier, J., *Ber. Bunsenges. Phys. Chem.*, 1989, *93*, 1474.
156. Maier, J., *Solid State Ionics*, 1989, *32/33*, 727.
157. Göpel, W. and Lampe, U., *Phys. Rev.*, 1980, *B22*, 6447.

Chapter 6

SURVEY OF TYPES OF SOLID ELECTROLYTES

Tetsuichi Kudo

CONTENTS

I. INTRODUCTION

Solid electrolytes are a class of materials exhibiting high ionic conductivity like electrolyte solutions, though they are in the solid state. Their history dates back to 1834 when Michael Faraday observed high electrical conductivity of PbF_2 at high temperature. In the early part of this century, cation conduction in silver halides has been extensively studied, contributing to the establishment of solid state electrochemistry. Since the sodium sulfur battery employing Na^+-conductive β-alumina ceramics was proposed in the 1960s, the practical importance of solid ionic conductors has been widely noticed. Nowadays, as also emerges from other chapters in this handbook, these materials are indispensable for many kinds of electrochemical devices such as sensors, high-temperature fuel cells, etc. In this chapter a survey is given of different types of solid electrolytes.

II. OXIDE ION CONDUCTORS

Up to their melting points CeO_2 and ThO_2 adopt the cubic fluorite (CaF_2) type of structure which can be derived from the CsCl structure by removing its cations alternately. Thus the packing of oxygens in these compounds is not close and there are 8-coordinated cubic interstices. Probably due to such structural looseness, they easily form solid solutions with alkaline earth or rare earth oxides in an unusually wide range of composition. In the case of the CeO_2-Gd_2O_3 system, for example, the range for a single fluorite-type phase extends to a composition of Gd/(Ce + Gd) = 0.5 (in molar ratio). Pure ZrO_2 has three polymorphs as a function of temperature: a monoclinic form (<1100°C), a tetragonal form with a distorted

0-8493-8956-9/97/$0.00+$.50

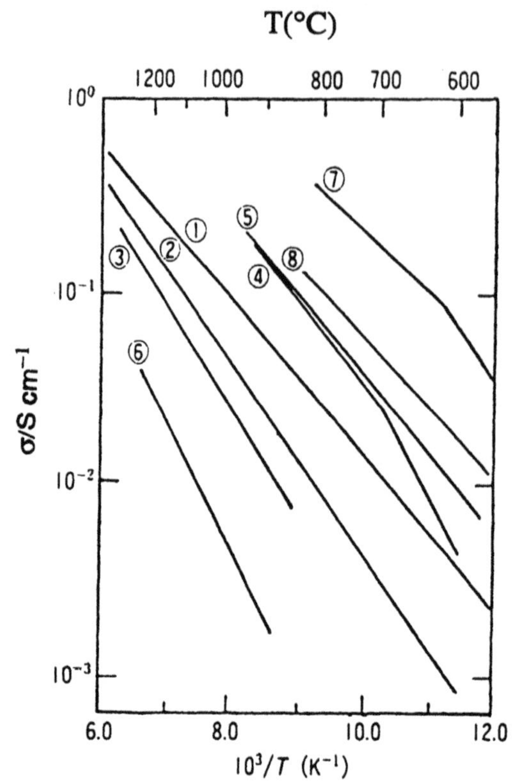

FIGURE 6.1 Conductivity (σ) of some oxide ion conductors plotted against 1/T. (Adapted from Arai, H., *Bull. Ceram. Soc. Jpn.*, 1992, *27*, 100. With permission.)

1. $(ZrO_2)_{0.9}(Y_2O_3)_{0.1}$
2. $(ZrO_2)_{0.87}(CaO)_{0.13}$
3. $(ThO_2)_{0.93}(Y_2O_3)_{0.07}$
4. $(CeO_2)_{0.8}(GdO_{1.5})_{0.2}$
5. $(CeO_2)_{0.8}(SmO_{1.5})_{0.2}$
6. $(HfO_2)_{0.88}(CaO)_{0.12}$
7. $(Bi_2O_3)_{0.75}(Y_2O_3)_{0.25}$
8. $(Bi_2O_3)_{0.75}(WO_3)_{0.25}$

fluorite structure (<2400°C), and a cubic form at higher temperature. However, incorporation of 15 to 25 mol% of CaO gives a cubic fluorite-type solid solution which is structurally stable up to its melting point and is thus called stabilized zirconia. Similar stabilized cubic phases are also formed with Y_2O_3 or other rare earth oxides.

In these solid solutions aliovalent guest cations such as Ca^{2+} or Y^{3+} reside on the sites for host cations (Zr^{4+}), generating oxide ion vacancies ($V_O^{\cdot\cdot}$) for electrical neutrality. Therefore, they can be formulated, for instance, as $Zr_{1-x}Ca_xO_{2-x}$ and $Zr_{1-x}Y_xO_{2-x/2}$, containing x and x/2 mole of $V_O^{\cdot\cdot}$ per formula unit, respectively. High oxide ion conductivity is found at high temperature, as shown in Figure 6.1.[1] The latter is due to a highly defective anion sublattice and, in addition, its loose structure is far from close packed. As shown in Figure 6.2,[2] the conductivity is accordingly increased with increasing the concentration of aliovalent guest cations at the initial stage of doping. However, further doping decreases conductivity after a relatively sharp maximum. This phenomenon, i.e., leveling off of the ionic conductivity, is often explained as due to the formation of defect pairs according to

$$V_O^{\bullet\bullet} + Ca_{Zr}'' = \left(Ca_{Zr} - V_O\right)^x \qquad (6.1)$$

which cannot contribute to electric conduction because they are effectively neutral. A drastic decrease in the heavily doped region may be interpreted as caused by ordering of vacancies or formation of a superstructure in the anion sublattice, equivalent to the arrangement in the

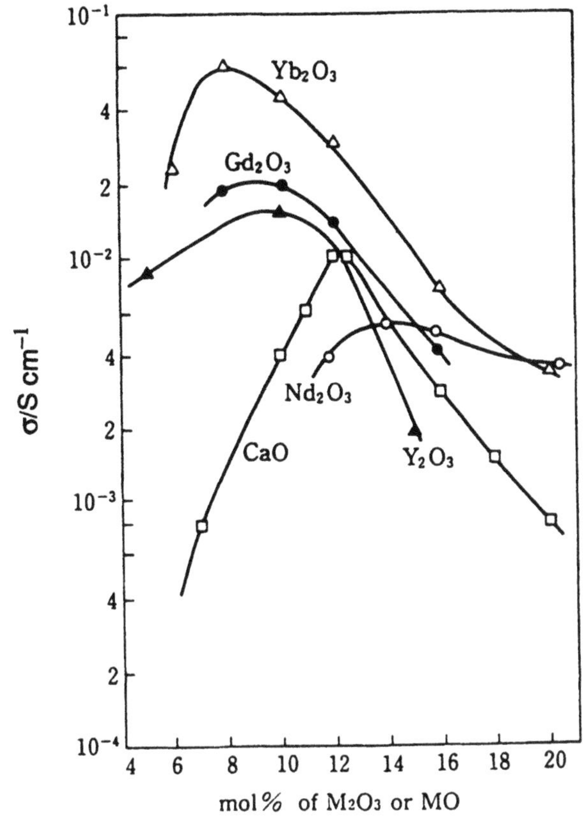

FIGURE 6.2 Composition dependency of conductivity (800°C) of solid solutions in the ZrO_2-M_2O_3(or -MO) system. (Adapted from Tannenberger, H., *Proc. J. Int. Etude Piles Combust.*, 1965, 19.)

C-type rare earth oxide like Y_2O_3. The conductivity of Y_2O_3 is orders of magnitude lower than that of the stabilized zirconia phases. In fact, there is a compound $CaZr_2O_5$, resembling the C-type oxide in structure, and calcia-stabilized zirconia (CSZ) tends to disproportionate very slowly into a ZrO_2 phase, and this compound at high temperature.[3]

The ionic conductivity of CeO_2-based solid solutions such as $Ce_{1-x}Gd_xO_{2-x/2}$ is remarkably higher than that of CSZ and yttria-stabilized zirconia (YSZ), as shown in Figure 6.1. However, their ionic transference number $t_{O^{2-}}$ starts to decrease at relatively high oxygen partial pressure (P_{O_2}), because Ce^{4+} is reduced much more easily than Zr^{4+}, generating conduction electrons by the following defect reaction:

$$O_O^x = \tfrac{1}{2}\, O_2 + V_O^{\bullet\bullet} + 2e' \tag{6.2}$$

Applying the law of mass action to this reaction, the electron concentration in equilibrium is given by

$$[e'] = k(T)\, P_{O_2}^{-1/4} \tag{6.3}$$

where k(T) is a parameter depending solely on the temperature, because $[V_O^{\bullet\bullet}]$ can be regarded as substantially unchanged with reduction for doped CeO_2 in which a great number of vacancies are already generated by the dopant addition. In such a case, the ionic conductivity is independent of P_{O_2} for the same reason, and the ionic transference number $t_{O^{2-}}$ (= $\sigma_i/(\sigma_i + \sigma_e)$) may be expressed in a form,

TABLE 6.1
Po_2^* (Pressure at which $t_{O^{2-}} = 1/2$) of Mixed Conductors Based on CeO_2

Composition	Po_2^* (atm) 500°C	700°C	850°C	1,000°C
$Ce_{0.9}Ca_{0.1}O_{1.9}$	10^{-20}	-10^{-13}	10^{-9}	—
$Ce_{0.905}Y_{0.095}O_{1.95}$	10^{-26}	10^{-17}	10^{-13}	10^{-10}
$Ce_{0.9}Gd_{0.1}O_{1.95}$	—	1.2×10^{-19}	1.7×10^{-15}	—
$Ce_{0.5}Gd_{0.5}O_{1.75}$	—	1.5×10^{-16}	6.5×10^{-14}	—
(CSZ)		-10^{-34}	-10^{-30}	-10^{-26}

$$t_{O^{2-}} = \left[1 + \left(P_{O_2} / P^*_{O_2}\right)^{-1/4}\right]^{-1} \tag{6.4}$$

where $P^*_{O_2}$ is a parameter associated with k(T). As is evident from Equation (6.4), $P^*_{O_2}$ at a given temperature corresponds to the oxygen partial pressure at which $t_{O^{2-}} = 1/2$. As given in Table 6.1,[4] the value of $P^*_{O_2}$ for CeO_2-based solid solutions at 850°C is higher than 10^{-15} atm, while that for CSZ is estimated to be as low as 10^{-30} atm. This is the reason why CSZ and YSZ are more useful for fuel cell application than the more conductive CeO_2-based solid electrolytes. The latter are useful, for example, as electrocatalysts, in which mixed conduction plays an essential role.[5]

Bismuth sesquioxide, Bi_2O_3, exhibits a high oxide ion conductivity at high temperature without doping of aliovalent cations. The oxide transforms from the monoclinic α form to the cubic δ form at 729°C. This transition is accompanied by a jump of conductivity from ~10^{-3} to ~1 S cm^{-1}. The δ phase has a structure related to fluorite, but with a complicated disorder in the oxygen sublattice. It is a kind of statistical distribution of oxygen vacancies which gives rise to the high ionic conductivity of δ-Bi_2O_3.[6] This highly conductive phase can be stabilized to lower temperatures by incorporation with Y_2O_3, Gd_2O_3, WO_3, etc., as shown in Figure 6.1. Of these $(Bi_2O_3)_{0.75}(Y_2O_3)_{0.25}$ exhibits an oxide ion conductivity as high as ~10^{-2} S cm^{-1} at a temperature as low as 500°C. Unfortunately, these materials undergo reduction as easily as CeO_2-based solid electrolytes, and their ionic transference number is decreased rapidly in reducing atmospheres. An oxygen pressure higher than 10^{-13} atm is needed for $(Bi_2O_3)_{0.75}(Y_2O_3)_{0.25}$ to maintain $t_{O^{2-}} > 0.99$ at 600°C.

Some ABO_3 compounds with the perovskite-type structure exhibit relatively high oxide ion conductivity or diffusivity due to their oxygen deficient nonstoichiometry. They are built of a ReO_3-type framework, in which a larger A ion is accommodated in an interstice coordinated by 12 oxygens while a smaller B ion resides on a 6-coordinated Re-site. This arrangement is structurally very stable so that a large part of the A and/or B ions can be replaced by various aliovalent ions without a change in the basic structure, and thus a large number of oxygen vacancies can be generated.

The ionic conductivity of perovskite ($CaTiO_3$) itself has been reported[7] to be about 10^{-3} S cm^{-1} at 1000°C, which is lower than that of fluorite-type solid solutions like YSZ, but much higher than oxides based on close-packed oxygen arrays. A nonstoichiometric, but isostructural compound, $CaTi_{0.7}Al_{0.3}O_{2.85}$, formed by substituting Al for Ti in the mother perovskite, shows a high conductivity almost comparable to CSZ.[8] As reported previously by Takahashi and Iwahara,[9] a series of perovskite-type compounds based on $LaAlO_3$, in which oxygen vacancies are generated by the replacement of its A ion (La) with an aliovalent cation like Ba, also show a high ionic conductivity. This was probably the first report about perovskite-type oxide ion conductors. Recently, it has been found that a similar type of compound, $Nd_{0.9}Ca_{0.1}Al_{0.5}Ga_{0.5}O_3$, exhibits a conductivity as high as 0.1 S cm^{-1} at 1000°C.[10] It is interesting to note that compounds like $BaCe_{0.9}Nd_{0.1}O_{3-\delta}$, known as a proton conductor, turn into an oxide ion conductor under a certain condition at elevated temperature (~1000°C).

Some perovskite-type oxides composed of lanthanide elements (as an A-site cation) and transition metals (B-site) show relatively fast oxide ion transport properties, though their electronic conductivities are very high (often even metallic). Electrochemically measured chemical diffusion coefficients of oxide ions in $Nd_{0.8}Sr_{0.2}CoO_3$ and $La_{0.8}Sr_{0.2}CoO_3$ are ~10^{-11} cm^2 s^{-1} at 25°C[11] and ~10^{-13} at 76°C,[12] respectively. It has also been reported that $La_{0.79}Sr_{0.2}MnO_3$, which is used as an interconnector material for solid oxide fuel cells, shows a chemical diffusion coefficient in the range of 10^{-8} to 10^{-6} cm^2 s^{-1} between 700 and 860°C.[13] For that purpose, such a high diffusivity of oxide ions is considered to be a problem. Mixed conduction of these perovskites is discussed in more detail in Chapters 7 and 14 of this handbook.

The crystal structure of brownmillerite (Ca_2AlFeO_5) is closely related to that of perovskite, being an oxygen-deficient perovskite structure with vacancies ordered along the [101] direction. Some compounds with this structure have been investigated as an oxide ion conductor. However, they do not show a high conductivity until they experience a transition from a vacancy ordered state to a statistically disordered state. For example, $Ba_2In_2O_5$ shows a conductivity jump of about 2 orders of magnitude at the transition temperature (930°C).[14] The ionic conductivity after transition is as large as 0.1 S cm–1. On the other hand, the conductivity of $Sr_2Fe_2O_5$ at 1000°C is as low as that of stoichiometric perovskite compounds, because its oxygen sublattice is not completely disordered up to this temperature.[7] It is interesting to note that $BaBi_4Ti_3MO_{14.5}$ (M = Ga, In, and Sc) compounds that consist of intergrowths between the Aurivillius phase ($Bi_4Ti_3O_{12}$) and brown millerite layers of $BaMO_{2.5}$ show a high conductivity (10^{-2} to 10^{-1} S cm–1 at 900°C) after a similar order–disorder transition.[15]

III. FLUORIDE ION CONDUCTORS

In general, the fluoride ion is thought to be more mobile in the solid state than the oxide ion because it is monovalent and its radius is smaller. Indeed, a fluorite-type solid solution of fluorite (CaF_2) itself with 1 mol% NaF shows a F^- conductivity of 2×10^{-4} S cm^{-1} at 350°C,[16] which is far higher than the O^{2-} conductivity of CSZ at the same temperature. It is noted that, as shown in Figure 6.3,[17] even pure CaF_2 exhibits considerable conductivity.

The β form of PbF_2, also having the fluorite structure and consisting of cations (Pb) with high polarizability, exhibits a much higher conductivity (~1 S cm^{-1} at 400°C) than CaF_2. As shown in Figure 6.4,[18] β-PbF_2 undergoes a diffuse transition from a low- to a high-conductive phase at around 300°C, the conductivity of the latter being almost the same as that of its molten phase. During this transition its basic crystal structure is unchanged. According to neutron diffraction,[19] there is a large number of anti-Frenkel pairs in β-PbF_2 at low temperature. It is therefore believed that the conduction in the low-conductive phase is due to F^- hopping via such defects. On the other hand, the "super ionic" behavior in the high-conductive phase is interpreted as due to a semifused state of the F^- sublattice associated with large anharmonic thermal vibrations along the [111] direction.[20] In fact, a specific heat anomaly accompanying the transition has also been observed.

A variety of mixed fluorides with fluorite-type or related structures have been derived from PbF_2. Of these, $PbSnF_4$ is the best F^- conductor to date in a relatively low-temperature region. As shown in Figure 6.3, its conductivity shows a kink around 80°C, which was previously thought to be due to a structural transition. Recently, however, it has been attributed to the transition in the conduction mechanism as in the case of PbF_2.

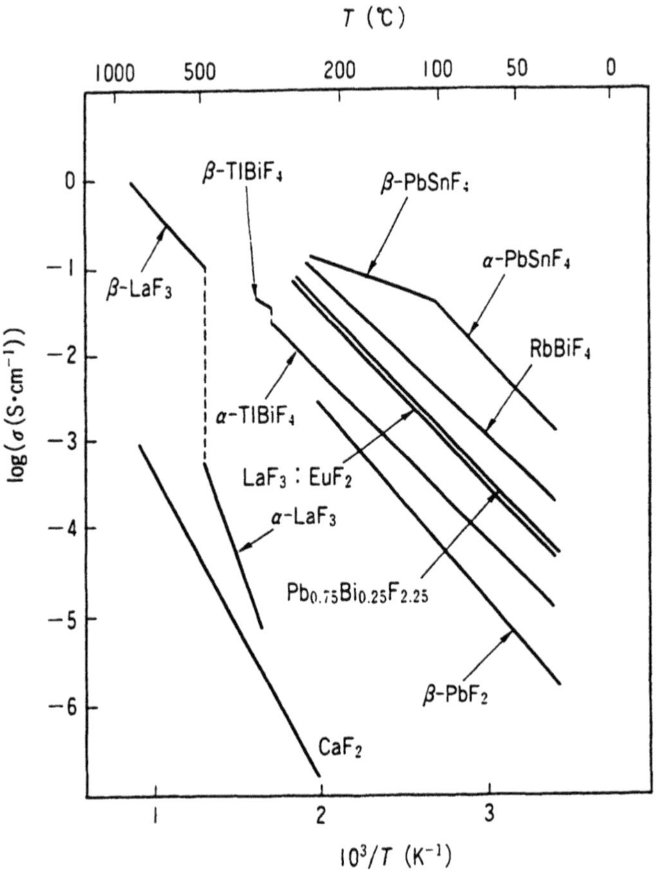

FIGURE 6.3 Conductivity of some fluoride ion conductors as a function of 1/T. (Adapted from Kawamoto, Y., *Bull. Ceram. Soc. Jpn.*, 1992, *27*, 105. With permission.)

In contrast with their oxide counterparts, CaF_2 and PbF_2 can form fluorite-type solid solutions with fluorides of higher-valency metals such as YF_3, BiF_3, UF_4, etc. They are expressed, for example, by $Ca_{1-x}Y_xF_{2+x}$, in which the excess F^- ions are accommodated in the interstitial sites, forming various kinds of clusters with $V_F^{\bullet}$, depending on the concentration of F_i'. They are usually good F^- solid electrolytes, and the conductivity of $Pb_{1-x}Bi_xF_{2+x}$ (x = 0.25) is shown in Figure 6.3 as an example.

Another class of crystalline F^- conductors consists of the rare earth trifluorides with the tysonite-type structure as represented by LaF_3, which exhibits a conductivity of 3×10^{-6} S cm^{-1} already at room temperature.[21] In its trigonal structure, there are three crystallographically nonequivalent F^- sites in a ratio of 12:4:2.[22] However, according to ^{19}F NMR,[23] two of these sites are equivalent for ionic motion; consequently two kinds of sites exist in a ratio of 12:6. At low temperature, F^- diffusion takes place via Schottky-type defects generated in minority sites. At elevated temperatures, however, hopping between the two sites also takes place. Tysonite-type solid solutions are formed with divalent metal fluorides such as BaF_2 or EuF_2; $La_{1-x}Eu_xF_{3-x}$, for example, has anion vacancies. The conductivity of these phases is higher than that of pure LaF_3 as shown in Figure 6.3.

Apart from crystalline phases, various glassy solid F^- conductors have also been developed. A glass in the ZrF_4-BaF_2 system, the first example of such conductors, exhibits a conductivity of ~10^{-6} S cm^{-1} at 200°C.[24] Extensive studies have been performed with this system, including short-range structural analyses, ^{19}F NMR, simulation based on molecular dynamics, etc. According to the results, there are basically two kinds of fluorine ions, i.e., a

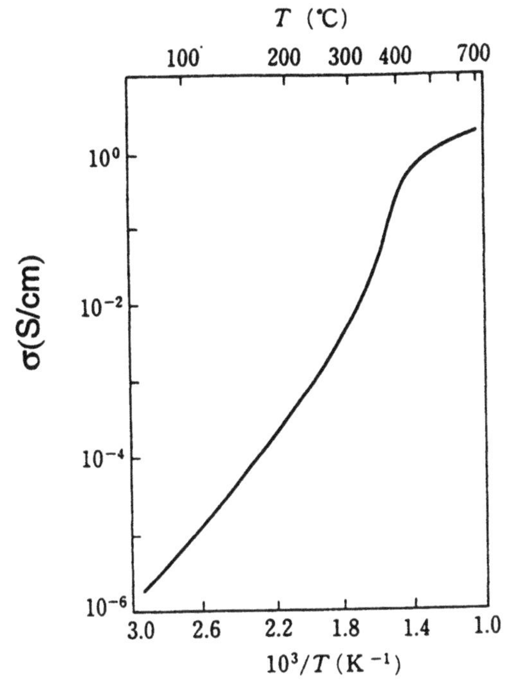

FIGURE 6.4 Conductivity of β-PbF_2 as a function of 1/T. (Adapted from Benz, R., *Z. Phys. Chem.*, N.F., 1975, *95*, 25. With permission.)

bridging fluorine (forming a bonding Zr-F-Zr) and a nonbridging one (Zr-F-Ba). The positional exchange takes place much easier between the latter fluorine-ions.[25,26] It is therefore believed that F^- can migrate in the glassy structure through channels connecting positions of the nonbridging fluorine.

Recently, high F^- conductivities have been reported for glass phases based on GaF_3, FeF_3, and InF_3. Of these, $35InF_3{\cdot}30SnF_2{\cdot}35PbF_2$ shows the highest conductivity to date, ~10^{-3} S cm^{-1} at 200°C, which is higher than that of β-PbF_2 at the same temperature.[27] Some oxyfluoride glasses, for example, the composition $45PbF_2{\cdot}45MnF_2{\cdot}10SiO_2$, also show good conductivity of ~10^{-6} S cm^{-1} at 200°C.[28,29]

IV. SILVER AND COPPER ION CONDUCTORS

Sodium chloride-type crystals of AgCl and AgBr show notable conductivity at relatively low temperature due to transport of Ag^+ via Frenkel defects. In a pure sample, conduction takes place mainly by migration or diffusion of interstitial Ag^+ ($Ag_i^{\bullet}$), because its mobility is much higher than that of vacancies (V_{Ag}'). According to measurements of the correlation factor f (= D_T/D),[30] this is not by the direct interstitial mechanism, but the indirect one (interstitialcy mechanism) involving collinear jumps of interstitial and lattice silver ions. When doped with a small amount of aliovalent halides such as $CdBr_2$, they decrease in conductivity because of the resulting shift (decrease) of the equilibrium concentration of $Ag_i^{\bullet}$. Heavier doping increases the conductivity as the vacancy diffusion mechanism becomes more important than the interstitialcy mechanism.

Silver iodide, AgI, adopts the wurtzite-type structure at room temperature. This β form of AgI transforms into the α form (bcc) at 147°C with a jump of conductivity over more than 3 orders of magnitude, as given in Figure 6.5. The silver ion conductivity after the transformation is ~1 S cm^{-1}, which is almost the same as that of the molten phase (m.p. 552°C) and even comparable to an H_2SO_4 solution. Cuprous iodide, CuI, undergoes a similar transition

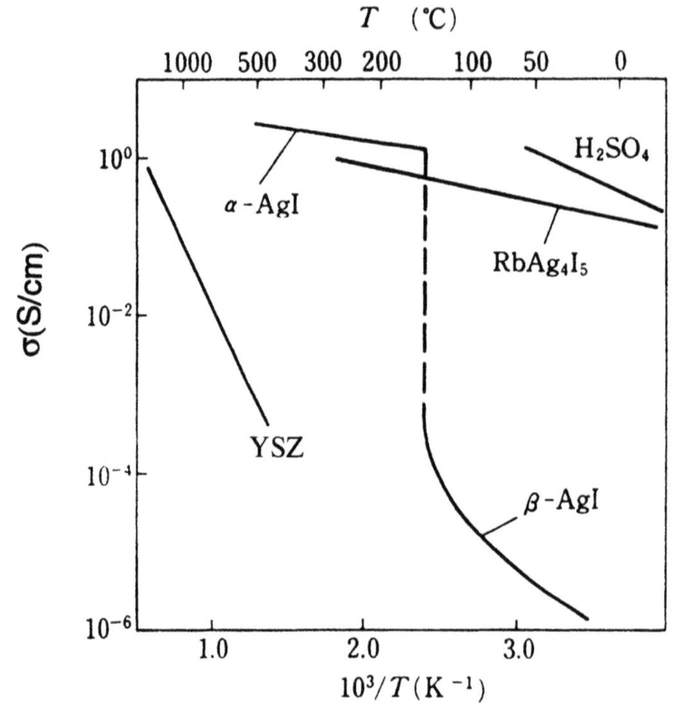

FIGURE 6.5 Temperature dependency of conductivity for typical silver ion conductors, AgI and $RbAg_4I_5$, as compared with those for CSZ (O^{2-}) and an aqueous solution of H_2SO_4 (H^+).

at 430°C and turns into a very good Cu^+ conductor. A common feature of both compounds is that they are composed of extremely polarizable ions.

The ionic conduction of α-AgI (or CuI) is quite different from that of AgBr or of CSZ. In the latter phases conduction is generated by point defects, whereas in the former it is due to partial occupation of crystallographic sites, which is sometimes called "structural disorder." Thus ionic conductivity is enabled without use of aliovalent dopants. Strock's[31] model of α-AgI proposed in 1934 is that the unit cubic cell (a = 5.0 Å, z = 2, space group: Im3m) is constructed by the bcc arrangement of the I^- sublattice and two Ag^+ ions are statistically distributed over 42 sites (i.e., 6b, 12d, and 24h) among anions (average structure model). Recent studies revealed that Ag^+ ions are normally situated on the tetrahedrally coordinated 12d sites with a large amplitude anharmonic vibration, tending to move toward the adjacent 12d position via the 24h intermediate sites. Such a structural situation in α-AgI could be taken as a "half-fused" state where only the cation sublattice loses its crystallographic order. Indeed, AgI and CuI show extraordinarily large entropy changes when they transform from the β- into the α form, as shown in Table 6.2, which are comparable to or even larger than those for melting.[32] Very high cationic conductivity in the α (high-temperature) forms of Ag_2S, Ag_2Se, Ag_2HgI_4, CuBr, etc., is also explained as due to the average structure of the cation sublattice.

TABLE 6.2
Entropy Changes for Phase Transition and Fusion (ΔS_t and ΔS_m) for AgI and CuBr

Compound	Transition temp. (T_t/K)	ΔS_t/JK^{-1}mol^{-1}	Melting point (T_m/K)	ΔS_m/JK^{-1}mol^{-1}
AgI	419	14.5	830	11.3
CuBr	664	9.0	761	12.6

Data from O'Keefe, M. and Hyde, B.G., *Philos. Mag.*, 1976, *33*, 219.

In 1961, Ag_3SI was found to have a very high ionic conductivity (0.01 S cm^{-1}) at room temperature.[33] After this finding, various room-temperature silver and copper ion conductors have been developed based on either AgI or CuI. A typical example is $RbAg_4I_5$ (RbI·4AgI),[34] which shows a conductivity as high as diluted H_2SO_4, as shown in Figure 6.5. According to single-crystal X-ray analysis, the compound has a complex structure called the β-Mn type (cubic, a = 11.2 Å, z = 4), in which 20 I^- ions form 56 tetrahedra sharing each face and in which 16 Ag^+ ions are statistically distributed.[35] Silver ions can migrate easily to adjacent vacant tetrahedral sites through a common face of the tetrahedra. This situation is therefore similar to α-AgI in that the structure is composed of face-sharing coordination tetrahedra and the tetrahedral sites thus formed are only partially occupied by Ag^+ ions. Thermodynamically, $RbAg_4I_5$ is metastable below 27°C, tending to disproportionate into Rb_2AgI_3 and AgI, though the rate is very low. A copper version of this compound has been synthesized by substituting partially Cl for I, i.e., $Rb_4Cu_{16}I_7Cl_{13}$,[36] exhibiting a conductivity higher than $RbAg_4I_5$ above ~50°C, as shown in Figure 6.6. Room-temperature conductivities of various silver and copper ion conductors are summarized in Table 6.3.[37]

TABLE 6.3
Room Temperature Conductivity of Silver and Copper Ion Conductors

Compound	Conductivity (S/cm, 25°C)	Ref.
Ag_3SI	0.01	33
$RbAg_4I_5$	0.26	38
$[(CH_3)_4N]_2Ag_{13}L_{15}$	0.04	39
$Ag_{15}I_{15}P_2O_7$	0.09	40
$Ag_7I_4PO_4$	0.019	40
$Ag_2Hg_{0.25}S_{0.5}I_{1.5}$	0.07	41
$Ag_6I_4WO_4$	0.047	42
7 $CuBr{\cdot}C_6H_{12}N_4CH_4Br$	0.017	43
47 $CuBr{\cdot}3\ C_6H_{12}N_2{\cdot}2\ CH_3Br$	0.035	44
$RbCu_3Cl_4$	0.0023	45
$Rb_3Cu_7Cl_{10}$	0.004	46
$Rb_4Cu_{16}I_7Cl_{13}$	0.34	36
$NH_4Cu_4Cl_3\ (I_{2-x}Cl_x)$	0.21	47

Some glass phases based on AgI also show a high Ag^+ conduction at room temperature. For example, a glass with the composition $3AgI{\cdot}Ag_2MoO_4$ obtained by quenching a molten mixture of AgI and Ag_2MoO_4 exhibits a conductivity of 0.02 S cm^{-1} at 25°C with the Ag^+ transference number being unity.[48] Although details of the mechanism are not completely understood, the glassy structure itself is held responsible for the high conductivity because most of these compounds show a decrease in conductivity by 3 orders of magnitude when crystallized at high temperature.[49] Conductivities of glassy silver conductors are listed in Table 6.4.[50]

V. SODIUM AND POTASSIUM ION CONDUCTORS

The β-alumina family, a series of compounds in the Na_2O-Al_2O_3 system, is one of the most important groups of solid electrolytes not only because it is practically useful for advanced batteries, but also because it is a typical two-dimensional ionic conductor, from which or by analogy with which, various kinds of solid electrolyte materials can be derived.

The principal members are the β and β″ phase, the ideal compositions of which are $Na_2O{\cdot}11Al_2O_3$ ($NaAl_{11}O_{17}$) and $Na_2O{\cdot}5{\cdot}33Al_2O_3$, though they are usually nonstoichiometric.

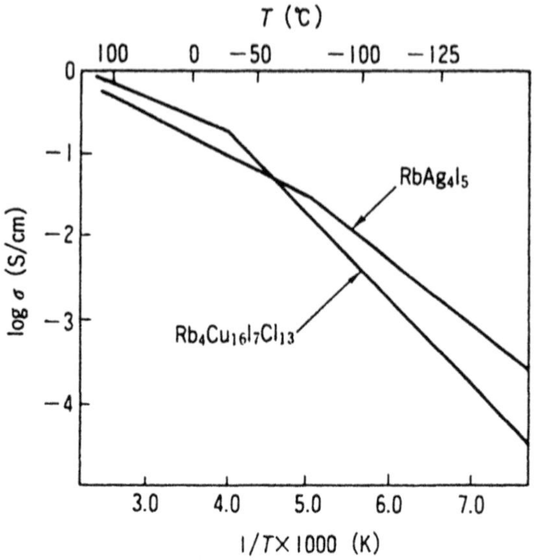

FIGURE 6.6 Conductivity of $RbAg_4I_5$ and $Rb_4Cu_{16}I_7Cl_{13}$ as a function of 1/T. (Adapted from Yamamoto, O., *Bull. Ceram. Soc. Jpn.*, 1992, *27*, 128. With permission.)

TABLE 6.4
Conductivity and Transference Number of Typical Ag^+ Conductive Glasses

Composition	σ_{Ag^+} (25°C)/Ω^{-1} cm^{-1}	t_{ion}	Ref.
4 AgI-$Ag_2Cr_2O_7$	2.2×10^{-2}	1.0	49
3 AgI-$Ag_4P_2O_7$	4×10^{-3}	~1.0	51
3 AgI-Ag_2MoO_4	2.1×10^{-2}	1.00	40
3 AgI-Ag_2WO_4	3.6×10^{-2}	1.00	40
3 AgI-Ag_4SiO_4	5.4×10^{-3}	—	40
6 AgI-3 Ag_2O-B_2O_3	5.5×10^{-3}	—	40
6 AgI-3 Ag_2O-B_2O_3	(8.5×10^{-3})	1.00	52

For β-alumina this is generally represented by $Na_{1+x}Al_{11}O_{17+x/2}$, where x is typically about 0.2. On the other hand, the β″ phase in the binary Na_2O-Al_2O_3 system is unstable, and incorporation of aliovalent cations such as Mg^{2+} is needed for stabilization; then, the composition is represented, for example, by $Na_{1+x}Mg_xAl_{11-x}O_{17}$, where x is typically 2/3.

The crystal structure of β-alumina was solved in 1937,[53] 30 years before the discovery of its high Na^+ conductivity in 1967.[54,55] It is based on a hexagonal unit cell (a = 5.58 Å, c = 22.5 Å, space group: $P6_3/mmc$), where the $Al_{11}O_{16}$ blocks in a spinel-like arrangement (spinel block) and the NaO^+ layers are stacked alternately along the c-axis. The β″ structure (a = 5.59 Å, c = 33.95 Å, z = 3, space group: R3m) is constructed of a similar alternate stacking of the spinel blocks and NaO layers, but the sequences of the oxygen layers in the spinel block are different.[56,57] In both phases, migration of Na^+ takes place in the NaO layers, which is often expressed by a honeycomb arrangement, as shown in Figure 6.7. The sites marked with BR and a-BR are nonequivalent for the β phase due to the difference in their distances from oxygens in the spinel block, and its sodium ions are normally situated on the former. However, the excess Na^+ ions usually present in this phase are located on the mO (mid-oxygen) sites with pairing with a Na^+ relocated from a BR site. It is believed that diffusion of Na^+ takes place in a manner such that a mO-Na^+ moves to the adjacent mO site via a-BR,

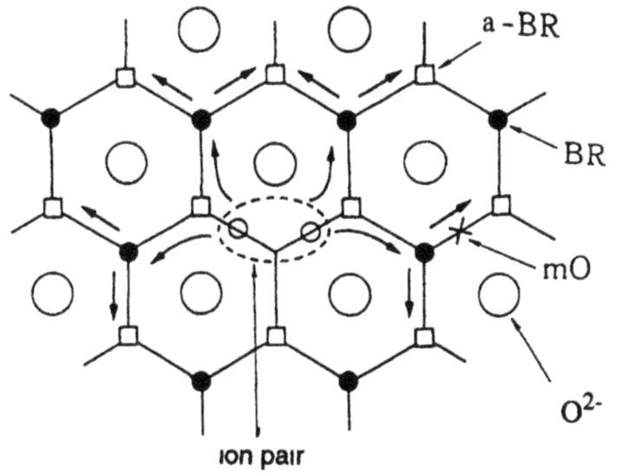

FIGURE 6.7 Structure of the (NaO) layer in β-alumina and migration paths for sodium ions. Normally, Na^+ is situated on the BR site.

to generate a new pair with a Na^+ pulled out from the BR. This is a kind of interstitialcy diffusion, the correlation factor for which has been calculated to be 3/5, agreeing with the observed value of $f = 0.61$ for Ag-substituted β-alumina.[58] The existence of ion pairs has also been confirmed by quasielastic light-scattering measurement.[59] In contrast with the β phase, the two kinds of sites (BR and a-BR) in the β″ phase are equivalent and Na^+ ions statistically occupy these sites. Diffusion for this case is believed to be due to the vacancy mechanism.[60] Nevertheless, unusually fast ionic motion in this phase cannot be interpreted by a simple hopping model, as in the case of CSZ.

Conductivity of a typical single crystal of β-alumina is as high as ~0.01 S cm^{-1} at 25°C along the conduction plane (perpendicular to the c-axis).[61] It is anisotropic because of the two-dimensional nature of the structure and several orders of magnitude lower along the c-axis. The β″ phase exhibits a conductivity 1 order of magnitude higher than the β phase. Sintered polycrystalline β″ phase, often used for the Na/S batteries, shows a poorer conductivity at room temperature, but its conductivity reaches a level high enough for practical use at high temperature (~300°C), as shown in Figure 6.8.[62]

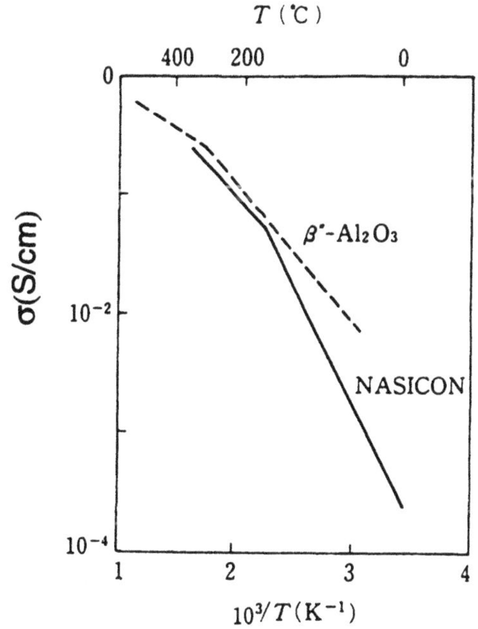

FIGURE 6.8 Conductivity of NASICON and sintered β″-alumina as a function of 1/T. (Adapted from Clearfield, A., Subramanian, M.A., Wang, W., and Jerus, P., *Solid State Ionics*, 1983, *9/10*, 895. With permission.)

TABLE 6.5
Conductivity of Various β (or β″)-Al_2O_3 Single Crystals (in conduction plane)

Incorporated cation	Ionic radius of cation(Å)	$\sigma_i(\Omega^{-1}\ cm^{-1}$, 25°C or 200°C)	Ref.
Na^+	0.95	1.4×10^{-2} (25°C)	61
Li^+	0.60	1.3×10^{-4} (25°C)	63
K^+	1.33	6.5×10^{-5} (25°C)	63
Rb^+	1.48	(1.2×10^{-6}) (25°C)	63
Ag^+	1.26	0.67×10^{-2} (25°C)	61
Tl^+	1.49	2×10^{-6} (25°C)	63
H_3O^+	1.40	5×10^{-3} (25°C)	64
Ba^{2+}	1.35	2×10^{-5} (200°C)	65
Pb^{2+}	1.32	6×10^{-2} (200°C)	66
Gd^{3+}	1.11	4×10^{-9} (200°C)	65

Sodium ions in either the β or the β″ phase can be easily exchanged for various cations because of fast ionic diffusion in the layer. Ion exchange is usually carried out by immersing sample crystals in a fused salt of the cation to replace Na^+. For example, Na^+ in β-alumina is totally replaced by K^+ in fused KNO_3 at 300 to 350°C; K-β-alumina thus obtained shows a K^+ conductivity of 6.5×10^{-4} S cm^{-1} at 25°C. A variety of ionic conductors in the β-alumina family have been similarly obtained, as listed in Table 6.5. It is interesting to note that the conductivities of the mother compound and its Ag derivative (Ag-β-alumina) are much higher than those of other derivatives of monovalent cations. This is explained by the barrier for diffusion having a minimum at an ionic radius near that of Na^+ or Ag^+.[60] A positive pressure effect on the conductivity has been observed for K-β-alumina,[67] as the radius of K is too large for the spacing of the conduction layer.

In the β″ phase, trivalent cations like Eu^{3+} can also be substituted for Na^+. For instance, when a slab of this crystal is heated in $EuCl_3$ powder for 24 h, almost all the Na is replaced by Eu.[68] This vapor-phase ion exchange utilizing fast ion transport is a typical example of materials synthesis by "soft chemistry." Unfortunately, rare earth β″-aluminas are not good ionic conductors, but they are promising candidates as crystals for solid state lasers.[69]

As Ga_2O_3 and α-Fe_2O_3 are isostructural with Al_2O_3, they form Ga- (or Fe-) versions of β- and β″-alumina. The ionic conductivity of Na-β″-Ga_2O_3 is reported to be 6×10^{-2} S cm^{-1} at 25°C,[70] which is somewhat higher than that of its Al counterpart. However, compounds with β-Fe_2O_3 are mixed conducting.[71] Utilizing the volatile nature of Ga_2O_3-based compounds, a thin film form of Na-β-Ga_2O_3 has been obtained by sputtering a mixture of Na_2O-$4Ga_2O_3$ and $NaGaO_2$ under an Ar/O_2 atmosphere and subsequent annealing at 1100°C, exhibiting a Na^+ conductivity of ~10^{-2} S cm^{-1} at 300°C.[72] In addition, β-Ga_2O_3 is useful as a host structure for proton conductors (see Section VII).

The NASICON (Na super ionic conductor) family is another important series of Na^+ conductors. Original NASICON, discovered in 1976 through careful structural examination of oxyacid salt frameworks with three-dimensional tunnels,[73] is a solid solution in the $NaZr_2(PO_4)_3$-$Na_4Zr_2(SiO_4)_3$ system and can be represented by $(1-x/3)NaZr_2(PO_4)_3\cdot(x/3)Na_4Zr_2(SiO_4)_3$ or $Na_{1+x}Zr_2P_{3-x}Si_xO_{12}$. A great variety of modified NASICONs have been synthesized by replacing Zr with, for example, Co, Ti, V, and/or by substituting As or Ge for P or Si.

One of the end members, $NaZr_2(PO_4)_3$, itself has a framework structure provided with three-dimensional channels suitable for ionic conduction. However, its conductivity is not excellent (~10^{-4} S cm^{-1} at 300°C) because one of the two kinds of Na sites (Na1 and Na2) along the conduction channel is preferentially occupied with Na^+ ions and all are fixed at Na1. The structure of NASICON, rhombohedral at high temperature and monoclinic at low temperature, consists of a similar framework based on corner sharing of a ZrO_6 octahedron with a PO_4 or SiO_4 tetrahedron, in which two kinds of Na sites exist, the Na2 sites being separated into two nonequivalent sites in the monoclinic form. As x is increased, the Na^+

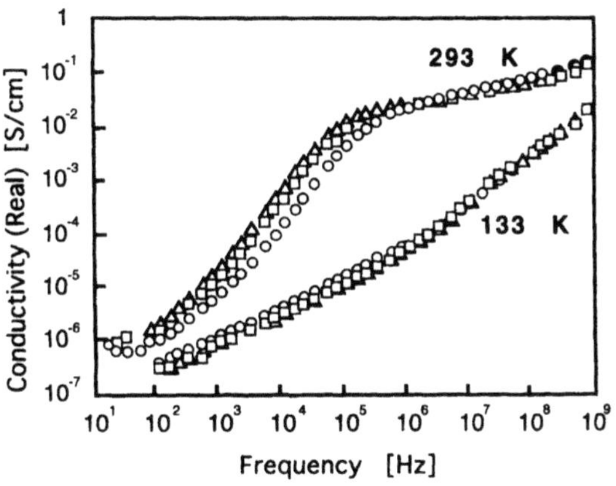

FIGURE 6.9 Frequency dependency of the real part of the complex conductivity measured for some priderite compounds, $K_x(B, Ti)_8O_{16}$. (○): B = Mg; (□): Al; (△): 0.4Mg + 0.6Al. (Adapted from Doyama, Y., Funatani, H., Matsui, N., Yoshikado, S., and Taniguchi, I., *Proc. 18th Symp. on Solid State Ionics in Japan,* Fukuoka, 1992, 114. With permission.)

ions also partially occupy Na2 sites, giving the compound a better conductivity. It is believed that diffusion takes place due to exchange of Na^+ between Na1 and Na2 sites. The maximum conductivity is obtained at x = 2, at which composition the activation energy for ionic conduction shows a minimum. Hydrothermally synthesized NASICON, with approximately this composition, shows a conductivity comparable with that of β″-alumina,[74] as shown in Figure 6.8. A kink in the conductivity is due to the transition from monoclinic to rhombohedral. Though NASICON shows excellent conductivity, there is the problem that they are unstable with respect to liquid sodium.

Compounds in the hollandite family also serve as Na^+ or K^+ conductors. Hollandite is a mineral with a composition $Ba_xMn_8O_{16}$ ($x \leq 2$), in which Ba^{2+} ions are located in one-dimensional tunnels formed along the c-axis of a pseudotetragonal framework structure composed of MnO_6 octahedra. Mixed titanates, $A_x(B,Ti)_8O_{16}$ (priderite, A: alkaline metal, B: Mg, Al, Ga, etc.), are isostructural and exhibit one-dimensional ionic conduction along the tunnel. For example, a single crystal sample of $K_x(Mg,Ti)_8O_{16}$ (KMTO) shows K^+ conductivity as high as 0.1 S cm^{-1} at 293 K along the c-axis in the very high-frequency region, though it rapidly decreases as the measuring ac frequency is lowered, as shown in Figure 6.9.[75] High-frequency conductivity may represent short-range motion of K^+ between bottlenecks in the tunnel. It is believed that such bottlenecks are generated due to large impurity ions, like Rb, and/or excess oxygens occupying the tunnel sites. Moreover, it is noted that the shortest diameter of the tunnel is about 2.4 Å, which is not large enough for K^+ migration. Analogous compounds, such as $K_xGa_8Ga_{8+x}Ti_{16-x}O_{56}$, show a better conductivity because their tunnel is larger in diameter.[76]

Glassy Na^+ conductors have also been investigated extensively. They are described in the following section together with Li^+ conductive glasses.

VI. LITHIUM ION CONDUCTORS

From a practical point of view, Li^+-conducting solid electrolytes are especially important because they enable the development of a solid state lithium battery with high energy density; thus, a variety of materials as represented by those shown in Figure 6.10[77] have so far been investigated as a Li^+ conductor.

The simplest example is LiI, which shows not an excellent, but still considerable conductivity of $\sim 5 \times 10^{-7}$ S cm^{-1} at room temperature, while other lithium halides are almost insulators, probably because of the less polarizable nature of their halogen ions. It is

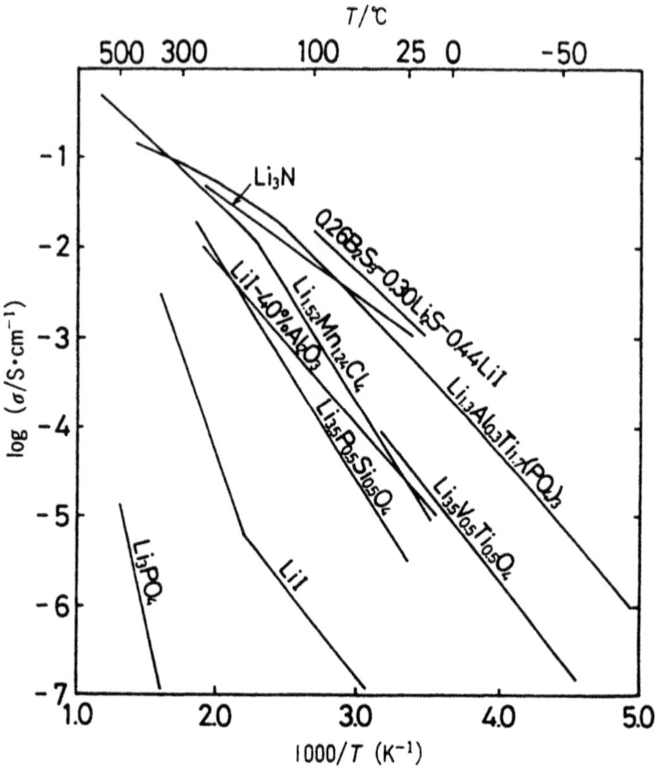

FIGURE 6.10 Conductivity of some lithium ion conductors as a function of 1/T. (Adapted from Adachi, G. and Aono, H., *Bull. Ceram. Soc. Jpn.*, 1992, *27*, 117. With permission.)

noteworthy that LiI was the first practical solid electrolyte used in a Li/I_2-complex cell in 1972, in which a layer of LiI was formed by contact of the negative and the positive electrodes. The layer was so thin that the resistance was small enough for low current uses, such as in a cardiac pacemaker.

Doping LiI with aliovalent impurities such as CaI_2 increases its conductivity for the same reason as described before, the conductivity of LiI-1mol% CaI_2 being ~10^{-5} S cm^{-1} at 25°C.[78] Besides such a classical approach, it was found in 1973 that addition of a fine insulator (Al_2O_3) particles into the LiI matrix as a dispersed second phase enhanced conductivity much more.[79] Since then, much research has been directed toward this phenomenon. Such a dispersed-type solid electrolyte (often called "composite electrolyte") is prepared, for example, by heating a well-blended mixture of anhydrous LiI and activated Al_2O_3 powder with high specific surface areas (>100 m^2 g^{-1}) in a very dry atmosphere at elevated temperature (~600°C). The cooled product is ground and pressed into a green pellet, which is used as a solid electrolyte sample. Typical Li^+ conductivities of LiI-Al_2O_3 measured as a function of Al_2O_3 content are shown in Figure 6.11.[80] The conductivity increases with increasing Al_2O_3 content up to about 40 mol%, at which it reaches a maximum value more than 2 orders of magnitude higher than that of pure LiI at 25°C. The beneficial effect of Al_2O_3 addition becomes smaller at higher temperatures. It has been pointed out in this connection that a plot of conductivity vs. the volume fraction of Al_2O_3 instead shows a linear relationship before showing a maximum at 30%.[81] Decreases in conductivity seen in the higher content region are due to the blocking effect of insulating Al_2O_3.

This conductivity enhancement is not due to the defect generation through classical homogeneous doping (such as $Al_2O_3 = 2Al_{Li}^{\cdot\cdot} + V_{Li}' + 3O_i'$), since the effect of Al_2O_3 extends far beyond its solubility limit in LiI. It is instead believed that defects (V_{Li}' and/or $Li_i^{\cdot}$)

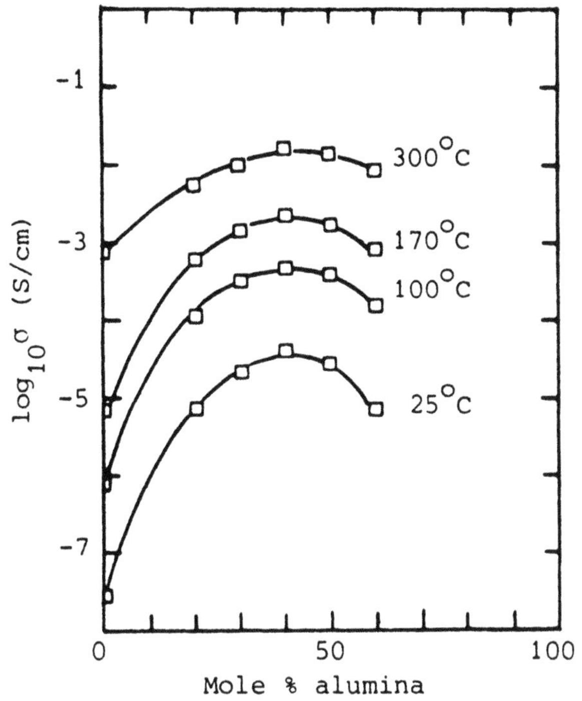

FIGURE 6.11 Conductivity isotherms for the LiI-Al_2O_3 composite system. (Adapted from Poulsen, F., Andersen, N.H., Kindl, B., and Schoonman, J., *Solid State Ionics*, 1983, *9/10*, 119. With permission.)

generated by space charge effects at the ionic conductor (LiI)/insulator (Al_2O_3) interface are responsible. Therefore, the finer the particle size of the dispersed insulator, the greater the enhancement effect.[82] In addition, the activation energy for conduction of every LiI-Al_2O_3 composite is usually very close to that of bulk LiI, excluding the possibility that enhanced conduction takes place using a newly generated migrating path at the interface. Smaller effects at higher temperature may be explained as being due to a reduction of the thickness of the space charge layer (the Debye length). Similar phenomena are also observed with AgCl, AgBr, CuCl, etc., in contact with insulating compounds, serving as a generally applicable method to develop new solid electrolytes. The mechanism of the phenomenon is discussed in more detail in Reference 83.

Another Li^+ conductor binary compound is Li_3N, the crystal structure of which is built up of Li_2N layers in hexagonal arrangement stacked along the c-axis with the rest of the Li in between. Intralayer migration of the latter Li^+ (Li2) is responsible for ionic conduction. Thus its single crystal shows highly anisotropic conductivity, i.e., 1.2×10^{-3} S cm^{-1} at 25°C along the layers, while it is about 2 orders of magnitude lower along the c-axis.[84] The conductivity of polycrystalline samples is typically 7×10^{-4} S cm^{-1} at room temperature. However, the reported conductivities of this compound depend on the synthetic method or sample history, suggesting a structure sensitivity, i.e., involvement of defects in the Li2 conducting layer. In fact, Li_3N intentionally doped with hydrogen exhibits a higher conductivity due to defects of Li2 introduced by partial reduction of N^{3-} to NH^{2-}. It has also been reported that Li_3N so carefully prepared that it is not contaminated shows a conductivity as low as 10^{-5} S cm^{-1} regardless of the crystallographic direction.[85]

Though Li_3N has an excellent conductivity, there is a disadvantage from a practical point of view, in that its decomposition voltage is as low as ~0.45 V, which means that it is thermodynamically impossible to construct a battery with emf exceeding this value. To overcome this problem, a number of derivatives of Li_3N have been synthesized and tested as

Li^+ conductors. The first example is a compound in the Li_3N-LiCl system,[86] which shows a high decomposition voltage (~2.5 V), but its conductivity is much poorer than that of the mother compound. A more successful example is Li_3N-LiI-LiOH (1:2:0.77) with the decomposition voltage between 1.6 to 1.8 V, which, at the same time, exhibits a high conductivity of ~0.01 S cm^{-1} as Li_3N with the electronic transference number being smaller than 10^{-5}.[87] The high conductivity of this compound probably originates from its structure based on the bcc arrangement of the anion (N^{3-} and OH^-) sublattice, which is the same as that of α-AgI.

More complex lithium compounds such as Li_2CdCl_4 in the spinel phase also show remarkable Li^+ conductivity at high temperature.[88,89] Their spinel structure is the inverse one where half of the lithium ions are accommodated at the tetrahedral interstices of the cubic close packing (ccp) array of anions and the other half are distributed statistically over the octahedral sites together with divalent cations (Cd, Mg, Mn, etc.). Nonstoichiometric compounds based on these chloride spinels, expressed as $Li_{2-2x}M_{1+x}Cl_4$, are also formed and show higher conductivity than the stoichiometric ones, as shown in Figure 6.12.[90] It is noted that the conductivity of $Li_{2-2x}Cd_{1+x}Cl_4$ (x = 0.05) has a record of 0.35 S cm^{-1} at 400°C. Kinks seen in the conductivity curves are due to the phase transition from spinel to a defective NaCl-type structure, in which the anion sublattice remains unchanged.

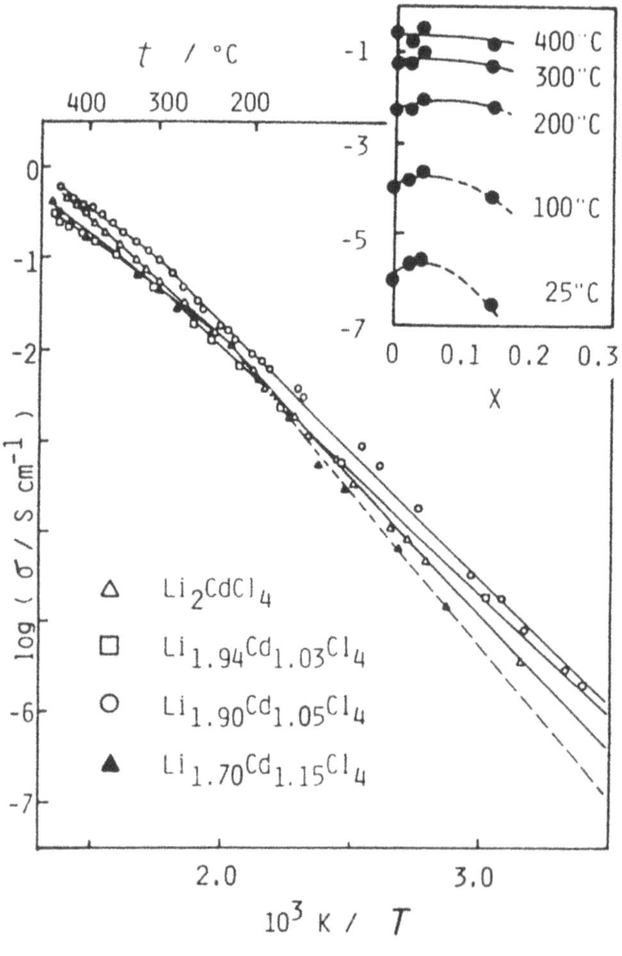

FIGURE 6.12 Temperature dependence of conductivity for spinel-type $Li_{2-2x}Cd_{1+x}Cl_4$. (Adapted from Kanno, R., Takeda, Y., and Yamamoto, O., *J. Electrochem. Soc.*, 1984, *131*, 469. With permission.)

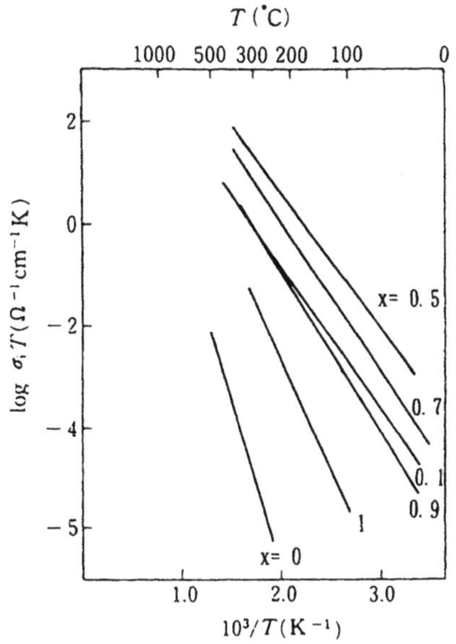

FIGURE 6.13 Plot of σT vs. 1/T where σ is the conductivity of solid solutions in the $(1{-}x)Li_3PO_4{\cdot}Li_4SiO_4$. (Adapted from the Hu, Y.-W., Raistrick, I.D., and Huggins, R.A., *J. Electrochem. Soc.*, 1977, *124*, 1240. With permission.)

Now let us survey oxide lithium ion conductors. Most of them are based on oxyacid salts of lithium such as Li_3PO_4 or Li_4SiO_4. The γ_{II} type of Li_3PO_4 itself shows a small but not negligible conductivity at high temperature, as shown by the line for x = 0 in Figure 6.13.[91] This structure belongs to the orthorhombic system (pseudohexagonal), in which the oxygens are packed into a hexagonal close packing (hcp) array, and both Li and P occupy its tetrahedral interstices, to form a framework with PO_4 and LiO_4 tetrahedra connected with each other. All Li^+ ions are used for building the structure such that they are only slightly mobile in pure Li_3PO_4. On incorporation of other oxyacid anions like Li_4SiO_4, however, the conductivity of the γII phase is remarkably increased as shown in Figure 6.13, because excess mobile Li^+ ions are introduced into the octahedral interstices as a result of solid solution formation $(1{-}x)Li_3PO_4{\cdot}xLi_4SiO_4$ or $Li_{3+x}P_{1-x}Si_xO_4$. The conductivity has a maximum at x = 0.5, where half of the octahedral site in the conduction plane is filled. Replacement of P with V or As and/or of Si with Ge or Ti gives analogous solid solutions.[92] Of these, the As-Ti and V-Ti systems show a higher conductivity of ~5×10^{-5} S cm^{-1} at room temperature, probably because their interstices for migration are enlarged by the larger cations. A solid electrolyte known as LISICON is also based on the γ_{II}-Li_3PO_4 structure.

The compounds $LiZr_2(PO_4)_3$ and $LiTi_2(PO_4)_3$ are isostructural with $NaZr_2(PO_4)_3$, the mother compound of NASICON mentioned earlier. They form NASICON-like solid solutions with $Li_3M(PO_4)_3$, which are expressed, for example, by $(1{-}x/2)LiTi_2(PO_4)_3{\cdot}(x/2)Li_3M(PO_4)_3$ or $Li_{1+x}M_xTi_{2-x}(PO_4)_3$, where M is a trivalent cation such as Sc, Al, In, etc.[93,94] Most of these exhibit a much higher Li^+ conductivity than the end members, as shown in Figure 6.14,[77] for the same reason, at least partly, as explained for NASICON. Especially, the conductivity of $Li_{1+x}Al_xTi_{2-x}(PO_4)_3$ (x = 0.3) reaches a value of 7×10^{-4} S cm^{-1} at room temperature, being almost equal to that of Li_3N. It has been pointed out that improved intergranular contact in these solid solutions is also responsible for their higher conductivity.

Oxide and sulfide glasses are extensively being investigated as Li^+ conductors. Simple binary glasses in the Li_2O-SiO_2 and Li_2O-Bi_2O_3 system show a conductivity of ~10^{-6} S cm^{-1} at room temperature, while the phosphate system is somewhat less conductive at the same

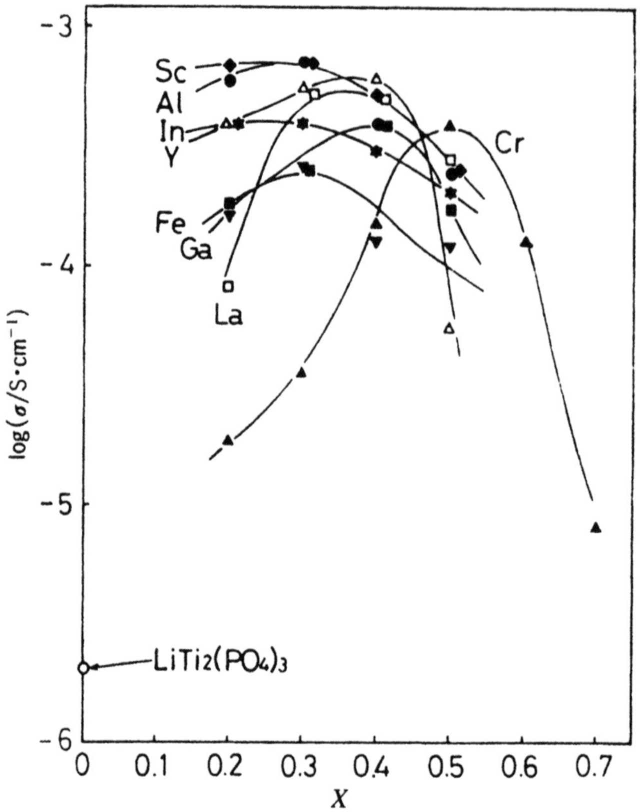

FIGURE 6.14 Composition dependence of conductivity (25°C) of solid solutions in the $(1-x/2)LiTi_2(PO_4)_3 \cdot (x/2)Li_3M(PO_4)_3$ system. (Adapted from Adachi, G. and Aono, H., *Bull. Ceram. Soc. Jpn.*, 1992, *27*, 117. With permission.)

Li_2O content. At the meta composition (1:1, i.e., $LiPO_3$, $LiBO_2$ and Li_2SiO_3), their structures are, in principle, formed by long chains of repeat units, for example, $-O(PO_2)-$ (i.e., corner-sharing PO_4 tetrahedra). It is suggested that the phosphate glasses exhibit a lower conductivity than others because the fraction of oxygens with a full negative charge is the smallest of these three systems.[95] Adding Na_2O as a network modifier, replacing Li_2O, gives Na^+-conductive glasses, though their conductivities are usually lower, as shown in Figure 6.15[96] for the Na_2O-B_2O_3 system (as an end composition of the system Li_2O-Na_2O-B_2O_3).

As appears from Figure 6.15 the partial replacement of Li_2O with Na_2O gives rise to a dramatic decrease in conductivity. Whereas the total cation concentration in the system is kept constant, conductivity at the midcomposition is 3 orders of magnitude lower than that expected from the Vegard-type additive rule. This phenomenon is not special for this ternary system and is called the "mixed alkaline effect". In contrast, addition of a co-glass former often enhances conductivity, and this is called the "mixed former (or anion) effect". For example, glasses in the Li_4SiO_4-Li_3BO_3 system show a maximum conductivity at the composition 1:1, which is about 1 order of magnitude higher than that expected from the Vegard interpolation.[97] A similar effect is induced by partial replacement of a modifier by another containing a different anion. The conductivity of LiCl-Li_2O-B_2O_3 spans the range 10^{-2} to 10^{-3} S cm^{-1} at 300°C, which is higher than that of the respective binary systems. The origin of the "mixed alkaline effect" as well as of the "mixed anion effect" is still under debate.

Binary sulfide glasses in the system Li_2S-B_2S_3 or Li_2S-SiS_2, for example, show a higher conductivity than their oxide counterparts because S^{2-} is larger and more polarizable. That for the former exhibits a conductivity as high as 10^{-4} S cm^{-1} at room temperature.[98] Doping

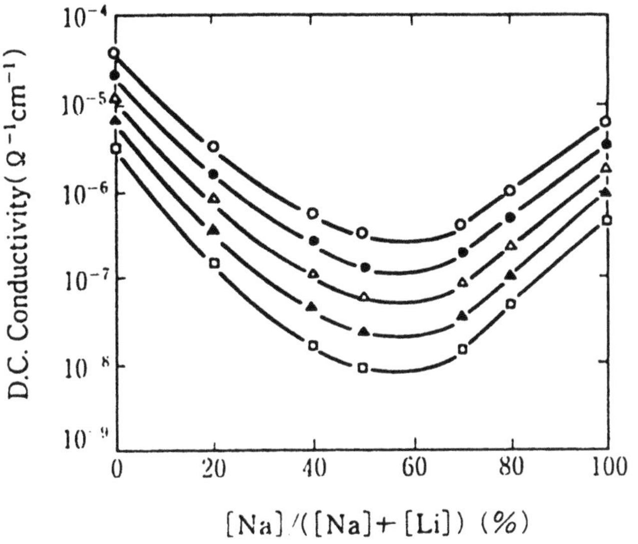

FIGURE 6.15 Conductivity isotherms for glasses in the Li_2O-Na_2O-B_2O_3 system in which the total alkaline content (Na + Li) is kept constant at 30%. (○): 300°C; (●): 280°C; (△): 260°C; (▲): 240°C; (■): 220°C. (Adapted from Kawamura, J., Mishima, S., Sato, R., and Shimoji, M., *Proc. 13th Symp. on Solid State Ionics in Japan*, Tokyo, 1986, p. 21. With permission.)

with LiI enhances their conductivity probably due to the same mixed anion effect as in the case of the oxides. Room temperature conductivities of some ternary sulfide glasses in the M_2Sn-Li_2S-LiI (M = Si, P, etc.) systems are shown in Figure 6.16.[95]

Li^+ conductors in noncrystalline polymeric phases are discussed in Section VIII.

VII. PROTON CONDUCTORS

Solid proton conductors are of great importance in relation to the development of fuel cells, sensors, and electrochromic devices. As shown in Figure 6.17,[99] these can be divided into two groups according to the temperature range in which they serve as a solid electrolyte. Members of the low- or moderate-temperature–type group are some solid state acids and the family of β-aluminas ion exchanged with protons. The high-temperature group consists of oxides belonging to the perovskite family.

Hydrated solid state inorganic acids more or less conduct protons in their crystal structure. Of these, $H_3[PMo_{12}O_{40}]\cdot nH_2O$ (n ~ 29), a kind of heteropolyacid, shows a remarkably high conductivity (~0.1 S cm^{-1}) at room temperature.[100] This compound has a cubic crystal structure, in which two diamond-type sublattices, $[PMo_{12}O_{40}]^{3-}$ (a Keggin-type polyanion) and $[H_3\cdot 29H_2O]^{3+}$ (a cationic cluster), are interpenetrated.[101] The latter cluster is formed by a complex hydrogen bonding network, which is connected to the network of the adjacent cluster. Thus the migration path of a proton utilizing such a hydrogen bonding network spreads over an entire section of the crystal. It is believed that proton conduction in this case takes place according to the Grotthus-type mechanism of proton transfer from a H_3O^+ molecule to an adjacent H_2O molecule by the tunneling effect in the hydrogen bonding, followed by the rotation of molecules for the next transfer. Proton conduction by this type of mechanism is characterized by a small activation energy (~10 kJ mol^{-1}). As shown in Figure 6.17, the conductivity of $H_3[PMo_{12}O_{40}]\cdot nH_2O$ (n ~ 29) depends very slightly on temperature, though the measuring range is narrow. At higher temperature or under dry conditions, its conductivity is drastically reduced because of loss of water. Some other acidic crystals such as $Sb_2O_5\cdot nH_2O$ and $HUO_2PO_4\cdot 4H_2O$ show considerable conductivity in humid atmospheres.[102]

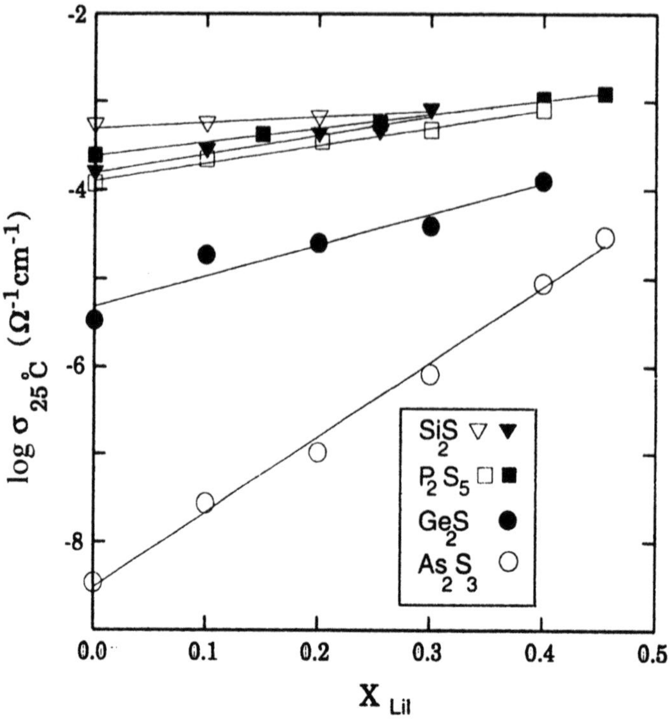

FIGURE 6.16 Ionic conductivity (25°C) vs. LiI content (X) in some ternary glass systems M_2Sn-Li_2S-LiI (M = Si, P, Ge, As, and n = 2, 3). (Adapted from Magistris, A., *Fast Ion Transport in Solids*, Scrosati, B. et al., Eds., NATI ASI Series, Kluwer Academic Publishers, Dordrecht, 1993, 213. With permission.)

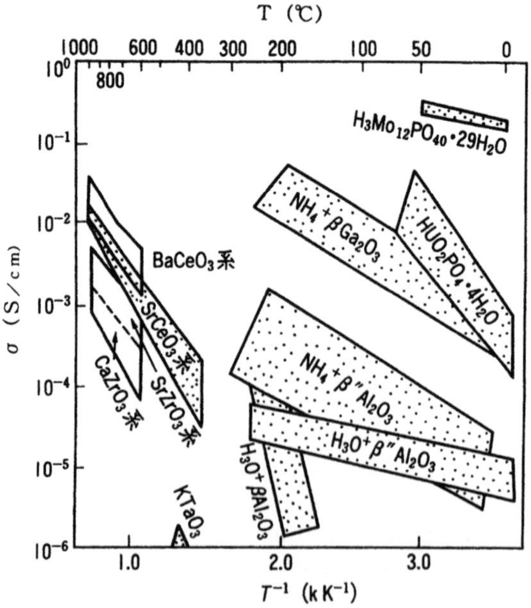

FIGURE 6.17 Conductivity of some proton conductors as a function of 1/T. The overlapping area between $BaCeO_3$ and $SrCeO_3$ indicates solid solutions based on these compounds. (Adapted from Iwahara, H., *Bull. Ceram. Soc. Jpn.*, 1992, *27*, 112. With permission.)

Compounds in the β-alumina family turn into proton conductors when their Na^+ ions are ion exchanged with H_3O^+ and/or NH_4^+. Derivatives from the β″-phase usually show a higher conductivity than those from the β phase, as shown in Figure 6.17. The highest conductivity of 10^{-4} S cm^{-1} has been reported for the compound $(NH_4)(H_3O)_{2/3}Mg_{2/3}Al_{31/3}O_{17}$, which is synthesized by soaking β″-alumina in fused ammonium salt for a long time.[103] Ionic conduction in this class of compounds, of course, occurs within layers between the spinel blocks, but it is still unclear whether protons are transported by a Grotthus-type mechanism or by a "vehicular" mechanism in which they are assumed to migrate as multiatomic ions such as H_3O^+ or NH_4^+. Gallium analogs of β- or β″-alumina are ion exchanged more smoothly, and NH_4-β-gallates thus formed show higher conductivity than their alumina counterparts, probably because the spacing of the conduction plane is wider for these compounds.[104]

A thin film form of proton conductors attracts much attention from a practical point of view, especially as an ion-conducting layer for electrochromic devices. One of the promising candidates is an amorphous film of $Ta_2O_5{\cdot}nH_2O$, which is formed on a substrate by spin coating of a peroxo-polytantalate solution.[105] The structure of this noncrystalline compound consists of anionic clusters fragmented out of L-Ta_2O_5 with cationic species like $H(H_2O)^+$ linking them. Its conductivity measured after treatment at 80°C is reported to be 4×10^{-5} S cm^{-1} at 25°C with a small activation energy (~12 kJ mol^{-1}), suggesting the Grotthus mechanism being applicable. Moreover, the conductivity remains almost constant down to such a low humidity as corresponds with a dew point of –60°C. Feasibility of an electrochromic cell

$$WO_3|Ta_2O_5 \cdot nH_2O|HIrO_2$$

has also been demonstrated.

Since high proton conductivity of perovskite-type oxides based on $SrCeO_3$ was discovered in 1981,[106] this class of high-temperature–type proton conductors has been extensively investigated. Mother compounds like $SrCeO_3$ or $BaCeO_3$ are not good conductors in themselves unless they are doped with aliovalent cations, to form a solid solution $SrCe_{1-x}M_xO_{3-\delta}$ where M = Sc, Y, Yb, etc. The solid solutions are merely p-type semiconductors without a hydrogen source like H_2 or H_2O; the electron hole conductivity of $SrCe_{1-x}Yb_xO_{3-\delta}$ (x = 0.05), for example, in dry air is 0.01 S cm^{-1} at 800°C. When they are contacted with hydrogen sources, protons are introduced into their structure by the reaction,

$$H_2 + 2h^{\bullet} = 2H^+ \tag{6.5}$$

or

$$V_O^{\bullet\bullet} + H_2O = O_O^x + 2H^+ \tag{6.6}$$

generating proton conduction much superior to that due to electron holes. It is noted that the former reaction reduces the concentration of electron holes at the same time. The conductivity of $SrCe_{1-x}Yb_xO_{3-\delta}$ (x = 0.05) in contact with H_2 at 800°C is about 0.02 S cm^{-1}, which has been confirmed as being almost solely due to proton migration by the fact that, on passing current through a electrochemical cell of type

$$Pt(H_2)|SrCe_{1-x}Yb_xO_{3-\delta}|Pt$$

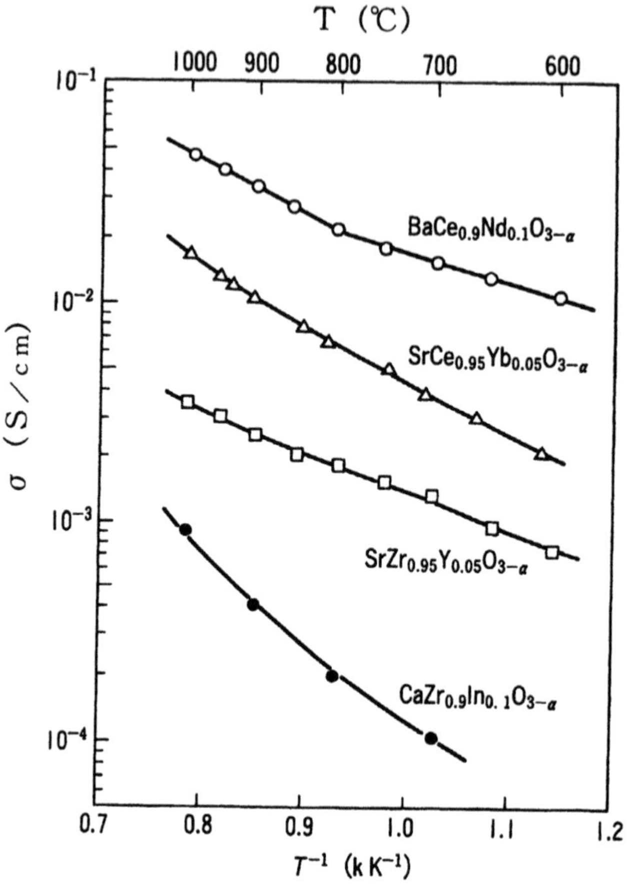

FIGURE 6.18 Conductivity of perovskite-type solid solutions as a function of 1/T. (Adapted from Iwahara, H., *Bull. Ceram. Soc. Jpn.*, 1992, *27*, 112. With permission.)

hydrogen evolves at its anode in obedience to Faraday's law.[107] Similar experiments have also shown that protons are not transported as H_3O^+, but as H^+ itself because no formation of water is observed at the anode. The conductivity of electron holes under these conditions has been reported to be 2 orders of magnitude lower than that of protons.[108] In Figure 6.18,[99] conductivities of compounds in this family measured in a hydrogen atmosphere are shown as a function of temperature. It is noted that at high temperatures (~1000°C), a contribution from oxide ion conduction becomes considerable, especially in the case of compounds in the $BaCeO_3$ series.

The detailed mechanism of proton conduction in these perovskite-type compounds is still controversial. Recent infrared spectroscopy studies have shown that protons exist in their structure as an OH species, though the O to H distance is considerably longer than usual, suggesting a contribution of hydrogen bonding.[109] It is of interest to compare them with H_xReO_3, which is built up of a structurally equivalent framework and exhibits very fast proton transport, though electronic conduction is overwhelming. According to neutron diffraction, its hydrogen atoms are statistically located at 72 equivalent sites that are on a sphere about 1 Å apart from each oxygen.[110] It is thus reasonable to think that protons in a ReO_3- or perovskite-type framework are transported by hopping from an OH group to one of the nearest oxygens.

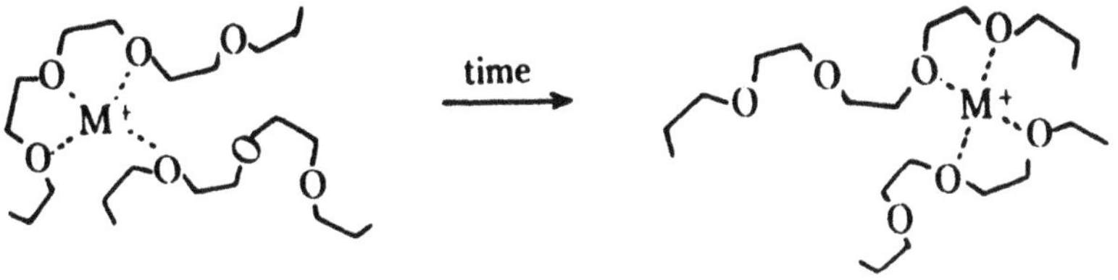

FIGURE 6.19 Scheme of ionic transport in an amorphous polymer matrix. (Adapted from Ogata, N., *Dodensei Kobunshi (Electrically Conductive Polymers)*, Ogata, N., Ed., Kodansha, 1990, 137. With permission.)

VIII. POLYMER SOLID ELECTROLYTES

Some vinyl fluoride-based polymers with side chains of perfluorosulfonic acid (the Nafion family) are important ion-exchange membrane materials used in practice for electrolysis of NaCl and in certain fuel cells. They show a proton conductivity of 0.01 S cm^{-1} at room temperature. However, such fast ionic transport occurs only when they are swollen with water. It is therefore not appropriate to call them solid electrolytes in the true sense of the word. It was in 1970 that anionic conductivity, though not high, was reported for crown ether complexes such as dibenzo-18-crown-6:KSCN, in which cations are trapped by the ligand.[111] A few years later much higher cationic (instead of anionic) conduction was found in complexes of a chain-like polyether such as PEO or PPO with alkaline salts; here, PEO stands for poly(ethyleneoxide), $(CH_2CH_2\text{-}O)_n$, and PPO for poly(propyleneoxide).[112,113] These were the first examples of "true" polymer solid electrolytes and were followed by a great number of studies. Polymeric electrolytes are advantageous in practice because they are easily processed and formed into flexible films.

Polymer electrolytes such as $(PEO)_n{\cdot}LiCF_3SO_3$ are obtained as a film, for example, by casting a mixed solution of PEO and $LiCF_3SO_3$ and then evaporating the solvent completely. Films thus formed are usually composed of three regions: a crystalline phase of a PEO-MX(alkaline salt) complex, a crystalline phase of PEO itself, and a noncrystalline solid solution of PEO with MX. Previously it was assumed that the first complex phase was responsible for ionic conduction, because X-ray studies showed a structure in which cations were accommodated in a helical tunnel of PEO seemingly suitable for ionic conduction. More recently, however, it is believed that transport of ions takes place mainly in the noncrystalline phase, in which an alkaline salt like LiX is electrolytically dissociated as if it were dissolved in a polar liquid and Li^+ ions are coordinated (or solvated) by oxygens in the ether chain. Cation transport occurs as a result of a sequence of association and deassociation steps of Li–O accompanied by local thermal motion of the polymer chains, as shown in Figure 6.19.[114]

The conductivity of $(PEO)_x{\cdot}LiCF_3SO_3$ (x = 12) and $(PEO)_x{\cdot}LiClO_4$ (x = 20) is very high at high temperatures (i.e., 3 to 5×10^{-3} S cm^{-1} at 125°C), but is drastically decreased as the temperature is lowered (~10^{-8} S cm^{-1} at 25°C).[115] The unexpectedly low conductivity at low temperature is mainly due to crystallization of the polymer chains. To prevent this, various crosslinked derivatives of PEO or PPO have been investigated as an alternative polymer matrix. A typical example is a PEO network prepared by a crosslinking reaction of triol-type PEO with tolylene-2,4-diisocyanate. Composites of this PEO derivative and $LiClO_4$ show high conductivity even at room temperature, as shown in Figure 6.20.[116] It is noted that solid electrolytes based on a crosslinked polymer are advantageous for applications also because their films have improved mechanical properties and thermal stability.

As shown in Figure 6.20, the plots of conductivity vs. 1/T do not obey the Arrhenius-type dependence. The observed convex dependence is characteristic of noncrystalline phases showing conductivity according to the foregoing mechanism. The conductivity for this case

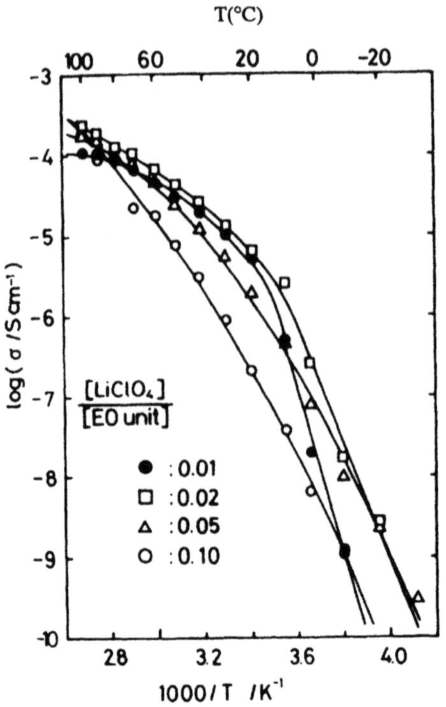

FIGURE 6.20 Temperature dependence of ionic conductivity for PEO-$LiClO_4$ complexes. (Adapted from Watanabe, M., Nagano, S., Sanui, K., and Ogata, N., *Solid State Ionics*, 1986, *18/19*, 338. With permission.)

is usually well fitted by a function derived from the free volume theory, called the Williams–Landel–Ferry's (WLF) relationship,

$$\log\left[\sigma(T)/\sigma(T_g)\right] = C_1\,(T - T_g)/\left[C_2 + (T - T_g)\right] \quad (6.7)$$

where σ(T) is the conductivity at temperature T, T_g the glass transition temperature, and C_1, C_2 are constants.[117] Sometimes the VTF relationship[118] obtained from a similar free volume model is also used for curve fitting. Successful fitting with these types of relationship suggests that ionic transport in polymers and liquid phases is similar.

Besides PEO and PPO, many other macromolecules, even including inorganic polymers like poly(phosphazene), $(\text{-N} = PCl_2\text{–})_n$, can serve as a base material for this class of solid electrolyte. A compound, poly(bis-(methoxyethoxyethoxide)phosphazene), synthesized through the reaction of polyphosphazene with a Na salt of 2-(2methoxyethoxy)ethanol, forms amorphous complexes with $LiBF_4$, $AgCF_3SO_3$, etc., each of which shows very high conductivity for a polymer electrolyte, as shown in Figure 6.21,[119] probably because of its very low T_g.

Very recently, new solid ionic conductors, so-called "polymer in-salt" materials,[120] have been reported, in which lithium salts are mixed with small quantities of the polymers PEO and PPO, while conventional polymer electrolytes ("salt-in-polymer") contain only one Li per about ten repeat units of ether. The reported conductivity in the $AlCl_3$-LiBr-$LiClO_4$-PPO system is as high as 0.02 S cm^{-1} at room temperature.

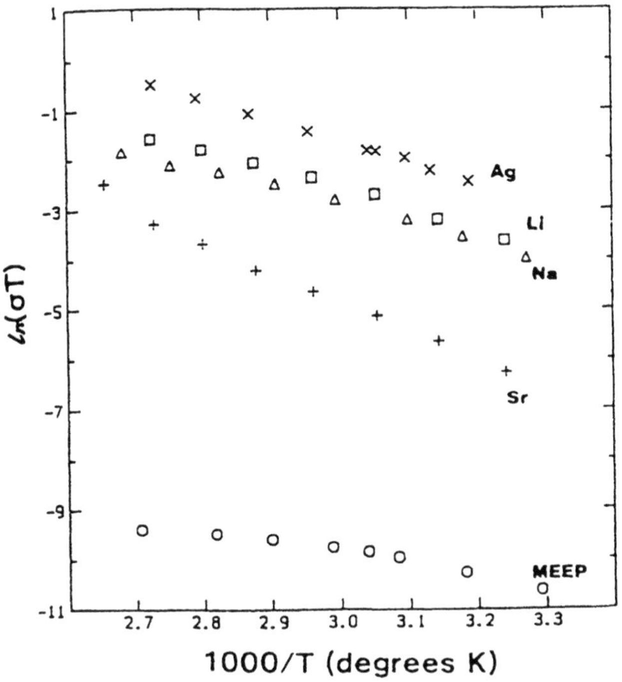

FIGURE 6.21 Ionic conductivity (S cm^{-1}) as plotted in ln (σT) vs. 1/T(K) for $[M(CF_3SO_3)_x]_{0.25}$·MEEP complexes. MEEP = $[NP(OC_2H_4OC_2H_4OCH_3)_2]_n$ and M = Ag, Li, Na, and Sr. (Adapted from Blonsky, P.M. and Shriver, D.F., *Solid State Ionics*, 1986, *18/19*, 258. With permission.)

REFERENCES

1. Arai, H., *Bull. Ceram. Soc. Jpn.*, 100 *27*, 1992.
2. Tannenberger, H., *Proc. J. Int'l Etude Piles Combust*, 1965, 19.
3. Alpress, J.G. and Rossel, H.J., *J. Solid State Chem.*, 1975, *15*, 68.
4. Kudo, T. and Fueki, K., *Solid State Ionics*, VCH and Kodansha, 1990, 133.
5. Takahashi, T., Iwahara, H., and Suzuki, Y., *Proc. Int. Etude Piles Combust.*, 1969, *3*, 113.
6. Wells, A.F., *Structural Inorganic Chemistry*, 5th ed., Oxford University Press, Oxford, 1982, 890.
7. Grenier, J.-C., Ea, N., Pouchard, M., and Hagenmuller, P., *J. Solid State Chem.*, 1985, *58*, 243.
8. Takahashi, T., in *Superionic Conductors*, Mahan, G.B. and Doth, N.L., Eds., Plenum Press, New York, 1976.
9. Takahashi, T. and Iwahara, H., *Energy Convers.*, 1971, *11*, 105.
10. Ishihara, T., Matsuda, H., Mizuhara, Y., and Takita, Y., Proc. 18th Symp. Solid State Ionics in Japan, Fukuoka, 1992, p. 75.
11. Kudo, T., Obayashi, H., and Gejo, T., *J. Electrochem. Soc.*, 1975, *122*, 159.
12. Kobussen, A.G.C. et al., *J. Electroanal. Chem.*, 1978, *87*, 389.
13. Belzner, A., Guer, T.M., and Huggins, R.A., *Solid State Ionics*, 1992, *57*, 327.
14. Goodenough, J.B., Ruiz-Diaz, J.E., and Zhen, Y.S., *Solid State Ionics*, 1990, *44*, 21.
15. Thomas, J.K., Kendall, K.R., and zur Loye, H.-C., *Solid State Ionics*, 1994, *70/71*, 225.
16. Chandra, S., *Super Ionic Solid*, North-Holland, Amsterdam, 1981, 120.
17. Kawamoto, Y., *Bull. Ceram. Soc. Jpn.*, 1992, *27*, 105.
18. Benz, R., *Z. Phys. Chem.*, N.F. 1975, *95*, 25.
19. Dickens, M.H., Hayes, W., Hutchings, M.T., and Smith, C., *J. Phys.*, 1982, *C15*, 4043.
20. Beckmann, R. and Schultz, H., *Solid State Ionics*, 1983, *9/10*, 521.
21. Roos, A., Aalders, A.F., Schoonman, J., Arts, A.F.M., and deWijn, H.W., *Solid State Ionics*, 1983, *9/10*, 571.
22. Cheetham, A.K., Fender, B.E.F., Fuess, H., and Wright, A.F., *Acta Crystallogr.*, 1976, *B32*, 94.

23. Goldman, M. and Shen, L., *Phys. Rev.*, 1966, *144*, 321.
24. Leroy, D., Lucas, J., Poulain, M., and Ravaine, D., *Mater. Res. Bull.*, 1978, *13*, 1125.
25. Kawamoto, Y. and Fujiwara, J., *Phys. Chem. Glasses*, 1990, *31*, 117.
26. Inoue, H. and Yasui, I., Extended Abs. 4th Int'l Symp. on Halide Glasses, 1987, 106.
27. Kawamoto, Y., Nohara, I., Fujiwara, J., and Umetani, Y., *Solid State Ionics*, 1987, *24*, 327.
28. Soga, N., Hirano, K., and Tsujimura, A., *MRS Int. Meet. on Advanced Mat.*, 1989, *3*, 535.
29. Schultz, P.C. and Mizzoni, M.S., *J. Am. Ceram. Soc.*, 1973, *56*, 65.
30. Compton, W.D. and Maurer, R.J., *J. Phys. Chem. Solids*, 1956, *1*, 191.
31. Strock, L.W., *Z. Phys. Chem.*, 1934, *B25*, 441.
32. O'Keefe, M. and Hyde, B.G., *Philos. Mag.*, 1976, *33*, 219.
33. Reuter, B. and Hardel, K., *Naturwisserschaften*, 1961, *48*, 161.
34. Bradley, J.N. and Green, P.D., *Trans. Faraday Soc.*, 1967, *63*, 424.
35. Geller, S., *Science*, 1967, *157*, 310.
36. Takahashi, T., Yamamoto, O., Yamada, S., and Hayashi, S., *J. Electrochem. Soc.*, 1979, *126*, 1654.
37. Yamamoto, O., *Bull. Ceram. Soc. Jpn.*, 1992, *27*, 128.
38. Owens, B.B. and Argue, A.G., *Science*, 1967, *157*, 308.
39. Owens, B.B., *Advances in Electrochemistry and Electrochemical Engineering*, Vol. 8, Delahay, P. and Tobias, C.W., Eds., John Wiley & Sons, 1971, 1.
40. Takahashi, T., Ikeda, S., and Yamamoto, O., *J. Electrochem. Soc.*, 1972, *119*, 477.
41. Takahashi, T., Kuwabara, K., and Yamamoto, O., *J. Electrochem. Soc.*, 1973, *120*, 1607.
42. Takahashi, T., Ikeda, S., and Yamamoto, O., *J. Electrochem. Soc.*, 1973, *120*, 647.
43. Takahashi, T., Yamamoto, O., and Ikeda, S., *J. Electrochem. Soc.*, 1973, *120*, 1431.
44. Takahashi, T. and Yamamoto, O., *J. Electrochem. Soc.*, 1975, *122*, 83.
45. Matsui, T. and Wagner, J.B., *J. Electrochem. Soc.*, 1977, *124*, 941.
46. Kanno, R., Takeda, Y., Matsuyama, F., Yamamoto, O., and Takahashi, T., *Solid State Ionics*, 1983, *11*, 221.
47. Geller, S., Akridge, J.R., and Wilber, S.A., *J. Electrochem. Soc.*, 1980, *127*, 251.
48. Kuwano, J., Isoda, T., and Kato, M., *Denki Kagaku, (J. Electrochem. Soc. Jpn.)*, 1978, *46*, 353.
49. Kuwano, J., Isoda, T., and Kato, M., *Denki Kagaku, (J. Electrochem. Soc. Jpn.)*, 1975, *43*, 734.
50. Kudo, T. and Fueki, K., *Solid State Ionics*, VCH and Kodansha, 1990, p. 87.
51. Kuwano, J., Isoda, T., and Kato, M., *Denki Kagaku*, 1977, *45*, 104.
52. Minami, T., Shimizu, T., and Tanaka, M., *Solid State Ionics*, 1983, *9/10*, 577.
53. Beevers, C.A. and Ross, M.A., *Z. Kristallogr.*, 1937, *97*, 59.
54. Weber, N. and Kummer, J.T., *Proc. Ann. Power Sources Conf.*, 1967, *21*, 37.
55. Yao, Y.F. and Kummer, J.T., *J. Inorg. Nucl. Chem.*, 1967, *29*, 2453.
56. Yamaguchi, G. and Suzuki, K., *Bull. Chem. Soc. Jpn.*, 1968, *41*, 93.
57. Bettman, M. and Peters, M., *J. Phys. Chem.*, 1969, *73*, 1774.
58. Whittingham, M.S. and Huggins, R.A., *J. Electrochem. Soc.*, 1971, *118*, 1.
59. Suemoto, T., Takeda, T., and Ishigame, M., *Solid State Commun.*, 1988, *68*, 581.
60. Wang, J.C., Gaffari, M., and Choi, S., *J. Chem. Phys.*, 1975, *63*, 772.
61. Whittingham, M.S. and Huggins, R.A., *J. Chem. Phys.*, 1971, *54*, 414.
62. Kudo, T. and Fueki, K., *Solid State Ionics*, VCH and Kodansha, 1990, p. 104.
63. Whittingham, M.S. and Huggins, R.A., National Bureau of Standards (NBS) Spec. Publ. 364, *Solid State Chem.*, 1972, 139.
64. S. Olejhik, et al., *Spec. Disc. Faraday Soc.*, 1971, *1*, 188.
65. Dunn, B. and Farrington, G.C., *Solid State Ionics*, 1983, *9/10*, 223.
66. Thomas, J.O., Alden, M., and Farrington, G.C., *Solid State Ionics*, 1983, *9/10*, 301.
67. Radzilowski, R.H. and Kummer, J.T., *J. Electrochem. Soc.*, 1971, *118*, 714.
68. Farrington, G.C. and Dunn, B., *Solid State Ionics*, 1982, *7*, 267.
69. J.D. Barrie, et al., *J. Luminescence*, 1987, *37*, 303.
70. Chandrashekher, G.V. Foster, L.M., *J. Electrochem. Soc.*, 1977, *124*, 434.
71. Takahashi, T., Kuwabara, K., and Kase, K., *Denki Kagaku, (J. Electrochem. Soc. Jpn.)*, 1975, *43*, 273.
72. Miyauchi, K., Kudo, T., and Suganuma, T., *Appl. Phys. Lett.*, 1980, *37*, 799.
73. Hong, H.Y.-P., *Mater. Res. Bull.*, 1976, *11*, 173.
74. Clearfield, A., Subramanian, M.A., Wang, W., and Jerus, P., *Solid State Ionics*, 1983, *9/10*, 895.
75. Doyama, Y., Funatani, H., Matsui, N., Yoshikado, S., and Taniguchi, I., *Proc. 18th Symp. on Solid State Ionics, in Japan*, Fukuoka, 1992, p. 114.
76. Yoshikado, S., Ohachi, T., Taniguchi, I., Watanabe, M., Onoda, Y., and Fujiki, R., Proc. 16th Symp. on *Solid State Ionics*, in Japan, Zao, 1990, p. 137.
77. Adachi, G. and Aono, H., *Bull. Ceram. Soc. Jpn.*, 1992, *27*, 117.
78. Schlaikjer, C.R. and Liang, C.C., *J. Electrochem. Soc.*, 1971, *118*, 1447.

79. Liang, C.C., *J. Electrochem. Soc.*, 1973, *120*, 1289.
80. Poulsen, F., Andersen, N.H., Kindl, B., Schoonman, J., *Solid State Ionics*, 1983, *9/10*, 119.
81. Maier, J., *Ber. Bunsenges. Phys. Chem.*, 1984, *88*, 1057.
82. Jow, T. and Wagner, Jr., J.B., *J. Electrochem. Soc.*, 1979, *126*, 1963.
83. Maier, J., *Science and Technology of Fast Ion Conductors*, Tuller, H.L. and Balkanski, M., Eds., Plenum Press, New York, 1989, 89.
84. Rabenau, R., *Solid State Ionics*, 1981, *6*, 267.
85. Bell, M.F., Breitschwerdt, A., and von Alpen, O., *Mater. Res. Bull.*, 1981, *16*, 267.
86. Hartwig, P., Weppner, W., and Wickelhaus, W., *Fast Ion Transport in Solid Electrodes and Electrolytes*, Vashsta et al., Eds., North-Holland, Amsterdam, 1979, 487.
87. Obayashi, H., Nagai, R., Goto, A., Mochizuki, S., and Kudo, T., *Mater. Res. Bull.*, 1981, *16*, 587.
88. Kanno, R., Takeda, Y., and Yamamoto, O., *Mater. Res. Bull.*, 1981, *16*, 999.
89. Lutz, H., Schmidt, W., Haeusler, H., *Z. Anorg. Allg. Chem.*, 1979, *453*, 121.
90. Kanno, R., Takeda, Y., and Yamamoto, O., *J. Electrochem. Soc.*, 1984, *131*, 469.
91. Hu, Y.-W., Raistrick, I.D., and Huggins, R.A., *J. Electrochem. Soc.*, 1977, *124*, 1240.
92. Rodger, A.R., Kuwano, J., and West, A.R., *Solid State Ionics*, 1985, *15*, 185.
93. Chun, L.S. and Xiang, L.Z., *Solid State Ionics*, 1983, *9/10*, 835.
94. Aono, H., Sugimoto, E., Sadaoka, Y., Imanaka, N., and Adachi, G., *J. Electrochem. Soc.*, 1990, *137*, 1023.
95. Magistris, A., *Fast Ion Transport in Solids*, Scrosati, B. et al., Eds., NATO ASI Series, Kluwer Academic Publishers, Dordrecht, 1993, 213.
96. Kawamura, J., Mishima, S., Sato, R., and Shimoji, M., *Proc. 13th Symp. on Solid State Ionics in Japan*, Tokyo, 1986, p. 21.
97. Machida, S., Tatsumisago, M., and Minami, T., *Proc. 13th Symp. on Solid State Ionics in Japan*, Tokyo, 1986, p. 21.
98. Levasseur, A., Olazcuaga, R., Kbala, M., Zahir, M., and Hagenmuller, P., *Synthese et Propriete Electriques de Nouveaux Verres Soufres de Conductivite Ionique Elevee*, C. R. Acad. Sci., 1981, p. 563.
99. Iwahara, H., *Bull. Ceram. Soc. Jpn.*, 1992, *27*, 112.
100. Nakamura, O., Kodama, T., Ogino, I., and Miyake, Y., *Chem. Lett.*, 1979, 17.
101. Bradley, A.J. and Illingworth, J.W., *Proc. R. Soc.*, 1936, *A57*, 113.
102. Miura, N., Ozawa, Y., and Yamazoe, N., *Nippon Kagaku Kaishi*, 1988, *12*, 1959.
103. Farrington, G.C. and Briant, J.L., *Science*, 1979, *204*, 1371.
104. Ikawa, H., Tsurumi, T., Urabe, K., and Udagawa, S., *Solid State Ionics*, 1986, *20*, 1.
105. Sone, Y., Kishimoto, A., and Kudo, T., *Solid State Ionics*, 1993, *66*, 53.
106. Iwahara, H., Esaka, T., Uchida, H., and Maeda, M., *Solid State Ionics*, 1981, *3/4*, 359.
107. Iwahara, H., Uchida, H., and Yamazaki, I., *Int. Hydrogen Energy*, 1987, *12*, 73.
108. Iwahara, H., Esaka, T., Uchida, H., Yamauchi, Y., and Ogaki, K., *Solid State Ionics*, 1986, *18/19*, 1003.
109. Shin, S., Huang, H.H., Ishigame, M., and Iwahara, H., *Solid State Ionics*, 1990, *40*, 910.
110. Dickens, P.G. and Weller, M.T., *J. Solid State Chem.*, 1983, *48*, 407.
111. Owens, B.B., *J. Electrochem. Soc.*, 1970, *117*, 1576.
112. Fenton, D.E., Parker, J.M., and Wright, P.V., *Polymer*, 1973, *14*, 589.
113. Wright, P.V., *J. Polym. Sci., Polym. Phys.* ed. 1976, *14*, 955.
114. Ogata, N., *Dodensei Kobunshi (Electrically Conductive Polymers)*, Ogata, N., Ed., Kodansha, 1990, 137.
115. Armand, M.B., *Ann. Rev. Mat. Sci.*, 1986, *16*, 245.
116. Watanabe, M., Nagano, S., Sanui, K., and Ogata, N., *Solid State Ionics*, 1986, *18/19*, 338.
117. Williams, M.L., Landel, R.F., and Ferry, J.D., *J. Am. Chem. Soc.*, 1955, *77*, 3701.
118. Fulture, G.S., *J. Am. Ceram. Soc.*, 1925, *8*, 339.
119. Blonsky, P.M. Shriver, D.F., *Solid State Ionics*, 1986, *18/19*, 258.
120. Angell, C.A., Liu, C., Sanchez, E., Nature, 1993, *362*, 137.

Chapter 7

ELECTROCHEMISTRY OF MIXED IONIC–ELECTRONIC CONDUCTORS

Ilan Riess

Key words: Mixed ionic electronic conductors, ionic conductivity, electronic conductivity, ambipolar diffusion, I–V relations, defect distributions, applications of mixed conductors.

CONTENTS

0-8493-8956-9/97/$0.00+$.50

LIST OF SYMBOLS

A	concentration of fixed positive defects
a_M	activity of the chemical (neutral) component M
c_M	concentration of species M, irrespective of charge
d	thickness
$D_{Tr}, D_k, \tilde{D}$	diffusion coefficient: tracer, component, and chemical, respectively
I, I_e, I_h, I_{el}, I_i	current: general, electrons, holes, electrons + holes, and ions, respectively
$j_t, j_e, j_h, j_{el}, j_i, j_{Mi}, j_{Vi}$	current density: total, electrons, holes, electrons + holes, ionic defects, $M_i^{\cdot}$ and V_M', respectively*
k	Boltzmann constant
k_1	electrode diameter
L	length of MIEC

* The charge is relative to the perfect crystal as defined in the Kröger–Vink notation. See also Chapter 1.

MIEC/MIECs	mixed ionic–electronic conductor(s)
MVIEC/MVIECs	mixed-valence ionic–electronic conductor(s)
$M, M_i^{\cdot}\ M_i^{\cdot\cdot}$	chemical (neutral) component, singly charged cation interstitial, and doubly charged cation interstitial, respectively*
$n, n(x), n_0, n_L$	quasi-free electron concentration, its spatial distribution, $n_0 = n(0)$ and $n_L = n(L)$
n_i	quasi-free electron concentration for the intrinsic composition
$N_i, N_i(x), N_i(0)$	mobile ion concentration
O_i''	doubly charged oxygen interstitial*
$p, p(x)$	hole concentration
$P(O_2)$	oxygen partial pressure
$P(O_2, -), P(O_2, +)$	oxygen partial pressure for which $\sigma_e = \sigma_i$, and $\sigma_h = \sigma_i$, respectively
q	elementary charge
r	radius of circular MIECs
R	resistance of amperometer
R_{el}, R_i	resistance of MIEC to the flow of electrons + holes and ions, respectively
S	cross-sectional area of MIEC
SE/SEs	solid electrolyte(s)
t	time
$t_i, \bar{t}_i, t_e$	transference number: ionic, average ionic, and of electrons, respectively
T	temperature
V	voltage
V_a	applied voltage
V_{th}	theoretical cell EMF (Nernst voltage)
V_{oc}	open circuit voltage
$V_O^{\cdot\cdot}, V_M'$	vacancy of oxygen ion and of monovalent cation M, respectively*
W	thermodynamic factor
x	(I) position, or (II) composition when it refers to a chemical formula
z	valence of mobile atomic defect, relative to the perfect crystal*
β	$1/kT$
δ	deviation from stoichiometric composition
Δ	difference
$\mu, \mu_M, \mu_{M1}, \mu_i, \mu_e, \mu_h$	chemical potential: general, chemical (neutral) component M, $M_i^{\cdot}$ defects, ions, electrons, and holes, respectively
$\tilde{\mu}, \tilde{\mu}_i, \tilde{\mu}_{M1}, \tilde{\mu}_e, \tilde{\mu}_h$	electrochemical potential: general, ions, $M_i^{\cdot}$ defects, electrons, and holes, respectively
∇	gradient
ν, ν_e, ν_h, ν_i	mobility: general, electrons, holes, and ions, respectively
ρ	space charge density
$\sigma, \sigma_e, \sigma_h, \sigma_{el}, \sigma_i, \sigma_t$	conductivity: general, electrons, holes, electrons + holes, ions, and total, respectively
φ	internal electric potential
□	defect concentration

I. INTRODUCTION

Solid mixed ionic–electronic conductors (MIECs) exhibit both ionic and electronic (electron/hole) conductivity. Naturally, in any material there are in principle nonzero electronic and ionic conductivities (σ_{el}, σ_i). It is customary to limit the use of the name MIEC to those materials in which σ_i, and σ_{el} do not differ by more than 2 orders of magnitude. It is also customary to use the term MIEC if σ_i and σ_{el} are not too low (σ_i, $\sigma_{el} \geq 10^{-5}\ \Omega^{-1}\ cm^{-1}$). Obviously, these are not strict rules. There are processes where the minority carriers play an important role despite the fact that σ_i/σ_{el} exceeds those limits and σ_i, $\sigma_{el} < 10^{-5}\ \Omega^{-1}\ cm^{-1}$. For example, the small electronic conductivity in a "purely" ionic conductor ($\sigma_i \gg \sigma_{el}$), i.e., in a solid electrolyte (SE), is a necessary condition for ion permeation through the SE and therefore, e.g., shortens the lifetime of a battery based on this SE. On the other hand, a small ionic conductivity in an electronic conductor is a necessary condition for permeation of ions through the electronic conductor, which may be of advantage in some applications (e.g., as electrode material).

Interest in MIECs has grown in recent years as possible applications of MIECs become apparent. This was accompanied by progress in materials preparation methods and the development of new materials. Deeper understanding was also gained of the defect chemistry of MIECs as well as the understanding of the I–V relations of electrochemical cells based on MIECs. In particular, the I–V relations were calculated for a wide range of boundary conditions, removing the limitation on the applied voltage to one of the following: open circuit, polarization, or short circuiting.

There are many books and reviews on solid electrolytes (see, e.g., Hladik),[1] but only a few of them pay attention to mixed conduction. Some of the relevant books on material properties of MIECs are by Kröger,[2] Hauffe,[3] Kofstad,[4] and Jarzebski.[5] Reactions in solids, including those enabled by the coupled motion of ions and electrons/holes, are discussed by Schmalzried,[6] Kofstad,[4] and Martin.[7] An excellent review of the electrochemistry of MIECs (before the year 1981) can be found in the book by Rickert.[8] Reviews on the topics were written by Wagner,[9–10] Heyne,[12] Gurevich and Ivanov–Schitz,[13] Tuller[14] (on mixed conducting oxides), Riess and Tannhauser[15] (on I–V relations), Weppner and Huggins,[16] and Dudley and Steele[17] (on experimental methods used to investigate MIECs).

An outstanding contribution to the field before the year 1977 was made by C. Wagner. This includes important contributions to the theoretical understanding of the electrochemistry of MIECs as well as (together with co-workers) important experimental work in the field.

The present review discusses MIEC materials, the I–V relations of electrochemical cells based on MIECs, methods for determining the partial conductivities and the chemical (ambipolar) diffusion coefficient, the defect distributions in MIECs under electrical and chemical gradients, and finally, applications.

II. MIXED CONDUCTOR MATERIALS

A. GENERAL

The materials of main interest in the field of electrochemistry of mixed conductors are ionic compounds. The existence of ions is then assured by definition. However, metals and covalent bonded semiconductors can also be considered as long as atomic defects exist which have a fixed effective electric charge (including zero), and move under an electrochemical gradient according to the transport equations discussed in Section III, which hold for MIECs.

Certain crystallographic and glass structures allow rather easy motion of ions, in both MIECs and SEs.[1,18] For ionic conduction in solids to occur, ions have to move through a rather dense matrix (whether crystalline or amorphous) consisting of ionic species of comparable size. To enable this, three conditions must be fulfilled: (a) an empty site exists in the "forward" direction, into which a conducting ion can move; (b) the propagation of the ion

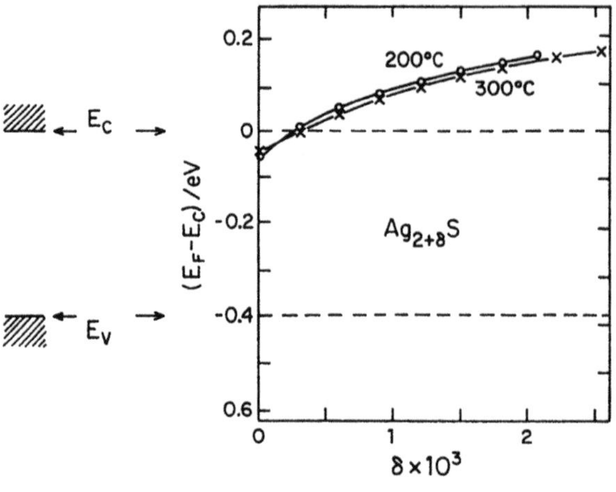

FIGURE 7.1. Fermi energy as a function of δ in $Ag_{2+\delta}S$. (From Rickert, H., *Electrochemistry of Solids*, Springer-Verlag, Berlin, 1982. With permission.)

from site to site is not impeded; and (c) there is a continuous path of sites, extending from one side of the sample to the other side which fulfill conditions (a) and (b).[19] Ion transport occurs normally via interstitial sites or by hopping into a vacant site (vacancy motion) or a more complex combination based on interstitial and vacant sites.[20] These ionic defects (interstitials and vacancies) can be formed in three ways: (a) by thermal excitation, (b) by change of stoichiometry, and (c) by doping.[15] Thermal excitation forms, e.g., Frenkel pairs of an interstitial and a corresponding vacancy.[2] Change of stoichiometry changes the composition, e.g., as in CeO_{2-x}, which introduces oxygen vacancies into ceria.[21–30] Doping may introduce, for example, mobile interstitial donors as in Li-doped Ge and Li-doped Si.[31] Doping may also introduce mobile vacancies as in Gd_2O_3-doped CeO_2, where each pair of Gd cations substituting for a pair of Ce cations introduces an oxygen vacancy on the anion sublattice.[32,33] These vacancies are mobile at elevated temperatures.

Electronic (electron/hole) conductivity occurs via delocalized states in the conduction/valence band or via localized states by a thermally assisted hopping mechanism.[34] The electronic conductivity is generated in three ways: (a) thermal excitation, (b) deviation from stoichiometry, and (c) doping[15] (except in stoichiometric metals where free electrons are present anyway). Though, formally, these are the same three ways as are used to generate mobile ionic defects, the mechanisms are quite different. Thermal excitation generates an electron hole pair across the band gap. Deviation from stoichiometry introduces native defects that may act as donors or acceptors. For example, an oxygen vacancy acts as a donor, an oxygen interstitial as an acceptor. Quasi-free electrons and holes contributed by deviation from stoichiometry may depend also on thermal excitation if the native donor (acceptor) electronic levels are not degenerate with or above the conduction band (or: degenerate with or below the valence band). An example is $\alpha - Ag_{2+\delta}\,S$ at $T > 177°C$ (see Figure 7.1). For a small deviation from stoichiometry $0 \le \delta \le 10^{-6}$ it is practically an intrinsic semiconductor (and an ionic conductor) with electron hole pairs excited thermally across a 0.4-eV gap.[8,35,36] For intermediate values $\delta \le 10^{-3}$ it is an *n*-type semiconductor since the excess silver is incorporated as interstitials that act as donors. For large values $10^{-3} < \delta \le 2.5 \times 10^{-3}$ it is a metal as the Fermi level is degenerate with the conduction band. Another example is $YBa_2Cu_3O_x$, which is an intrinsic semiconductor for $x \approx 6.0$, a *p*-type semiconductor for $6.0 < x \le 6.5$ and a metal for $6.5 \le x \le 7.0$.[37] Doping may introduce donors or acceptors. Electronic conduction by doping may require also thermal excitation if the donor (acceptor) electronic levels are not degenerate with or above the conduction band (or: degenerate with or below the valence band). An example where thermal excitation is not required is U-doped CeO_2, where the uranium donor level lies above the bottom of the conduction band of ceria.[38,39]

MIEC materials can follow complex defect models. The complexity increases with the type number of mobile defects and traps. However, the more common and well-known MIECs can be classified using the following models:

a. Highly disordered MIECs with one type of mobile ionic defects in which the concentration of the mobile ionic defects, N_i, is higher than the concentration of both the conduction electrons, n, and holes, p, i.e., $n,p \ll N_i$. The ionic defect concentration is so high that its change in response to changes in stoichiometry, due to changes in the chemical potential of the mobile component, can be neglected.
b. The concentration of conduction electrons or holes is in proportion to the mobile ionic defect concentration, i.e., $n = zN_i$ or $p = -zN_i$ where z is the effective charge of the mobile ionic species, in the sense of Kröger–Vink relative defect notation.[2]
c. High-electronic concentration MIECs. The electron or hole concentration is high and fixed, e.g., by doping of a semiconductor or in a metal MIEC. Then $n \gg N_i$ or $p \gg N_i$.

There exist MIECs which exhibit different p, n, N_i relations, depending on the experimental conditions. Some MIECs are undefined, as measurements have not yet revealed their n, p, N_i relations.

Because of the variety of defect models and in order to be able to describe them accurately, we use the ratio of the different defects to denote each model rather than a short name that is therefore inherently vague.

We also group materials of special interest, though they fall under the models "a" to "c": the defect concentrations in many insertion compounds[40] fall under model "b". YBCO ($YBa_2Cu_3O_x$), a material which represents the high-temperature superconductors, draws much attention and its MIEC properties are intensively investigated. Its classification "a" to "c" changes with oxygen composition, x.[37,41–43]

B. HIGH-DISORDER MIXED CONDUCTORS

For MIECs of high disorder the concentration of mobile ions is large. It is assumed that the concentration of electrons and holes is small. The defect model is therefore denoted by "$p, n \ll N_i$". In this class fall the SEs. However, a number of dominantly electronic conductors may be found there as well. The reason is that a large electronic conductivity need not reflect a large electronic (electron/hole) concentration. When the mobility of the electron/hole is many orders of magnitude higher than that of the ions it is possible to find $\sigma_{el} \gg \sigma_i$ in spite of the fact that $n, p \ll N_i$.

AgBr and AgCl are SEs. They have been intensively investigated primarily because of their use in photographic emulsions.[8,44–49] The ionic defects are Frenkel pairs on the cation sublattice. These defects facilitate the ionic motion. σ_i of AgBr is of the order of ~0.5 S/cm ($S = \Omega^{-1}$) at $T \sim 400°C$.[50] Under high silver activity, these halides are n-type MIECs and under low silver activity, p-type MIECs. Mizusaki and Fueki[48] have pointed out the difficulty in measuring σ_{el} due to rapid decomposition of the halides. In their experimental setup decomposition was suppressed. Solid solutions of $(AgBr)_{1-x}(AgCl)_x$ exhibit similar behavior.[44] Hellstrom and Huggins[51] have reported that $AgGaS_2$ and $AgAlS_2$ are predominant ionic conductors, i.e., $\sigma_i \gg \sigma_{el}$. For $AgGaS_2$, $\sigma_{Ag^+} = 5.3 \times 10^{-7}$ S/cm at 200°C. Ag_9AlS_6 and Ag_9GaS_6 are MIECs with high silver ion conductivity. For Ag_9GaS_6, $\sigma_{Ag^+} = 0.53$ S/cm at 200°C. The spinel phase $AgAl_5S_8$ is a MIEC with a low ionic and electronic conductivity ($\sigma_{Ag^+} = 2.3 \times 10^{-7}$ S/cm at 200°C).

Copper halides are good ionic conductors with values of $\sigma_i \sim 0.1 - 1$ S/cm and $\sigma_{el} \ll \sigma_i$ at ~400°C.[52–55] The ionic conductivity of copper is facilitated by Frenkel disorder on the cation sublattice. It was believed for many years, on the basis of the two-point Hebb–Wagner (H–W) polarization measurements[56,57] (discussed in Section IV), that CuBr is a p-type MIEC.[52] However, it was recently shown that CuBr decomposes quite rapidly and therefore the H–W method used to measure σ_{el} yields false results.[56,57] Other measurements[58] show that CuBr is a weak *n*-type MIEC. CuI is a predominantly ionic conductor in the high-temperature α phase ($405 < T < 605°C$) and β phase ($369 < T < 405°C$), and it gradually becomes a predominantly electronic conductor in the low-temperature γ phase ($T < 369°C$) as T is lowered below ~200°C.[52,59]

Cu_2S exhibits mixed conductivity, with $\sigma_{Cu+} \sim 0.2$ S/cm at 420°C.[60] The electronic conductivity is contributed by electrons and holes. For Cu_2S equilibrated with copper, $\sigma_{el} = 0.16$ S/cm.[61] As the mobility of the electrons is expected to be at least an order of magnitude larger than that of the ions, we conclude that $n, p \ll N_i$. Yokota[62] also found that this class ("$n, p \ll N_i$") fits the experimental data. However, for $T < 100°C$, Allen and Buhks[63] find that the class $p = N_i$ fits their experimental data on Cu_2S. This indicates that at elevated temperatures thermally excited ionic defects dominate. However, thermal excitation of defect pairs is not effective at low temperatures ($T < 100°C$), and one kind of ionic defect (copper vacancies) is formed by deviation from stoichiometry being accompanied by electronic defects (holes). Mixed conductivity is observed also in $Cu_{2-x}Se$.[64] Direct measurement of p and N_i ($n < p$) shows that $p \ll N_i$. Copper phosphates with the NASICON or alluaudite type structure exhibit mixed conductivity with a wide range of ratios σ_i/σ_{el}.[65]

$PbBr_2$, $PbCl_2$, and $PbCl_{2-x}Br_x$ are anion conductors. Their electronic conductivity is low.[66,67] Attempts to measure the electronic conductivity yield results which are not consistent with existing theories, and it is believed that decomposition of the sample affects the measurements.[68,69] PbF_2 is a fast anion conductor[70] for F^-, in particular above the Faraday transition temperature[19,71] where $\sigma_i \sim 1$ S/cm^{-1}. For PbF_2 in equilibrium with Pb the electronic (hole) conductivity is 10^{-10} S/cm^{-1} at 300°C.[72]

Orthorhombic and tetragonal PbO are mixed conductors conducting oxygen via charged oxygen interstitials.[73,74] For orthorhombic PbO, σ_i is about 2×10^{-6} S/cm at ~500°C, while σ_{el} is slightly larger. Both electron and hole conductivities are observed.

Stabilized zirconia is a well-known SE with σ_i approximately 10^{-2} S/cm at 1000°C (depending on the cation used for stabilization). The ionic current is carried by oxygen vacancies introduced by doping ZrO_2 with lower valent cations.[8] Under high oxygen partial pressures, $P(O_2) \sim 1$ atm, some *p*-type electronic conductivity is detected. Under reducing conditions an *n*-type electronic conductivity is observed. For 8 mol% Y_2O_3 in ZrO_2 the hole conductivity σ_h at $P(O_2) = 1$ atm and $T = 1000°C$ is about 0.3×10^{-4} S/cm, and the electronic transference number $t_{el} = \sigma_h/(\sigma_i + \sigma_h)$ is about 10^{-2}. At $P(O_2) = 10^{-17}$ atm, $T = 1000°C$, the electron conductivity σ_e is 10^{-4} S/cm, and $t_{el} = \sigma_e/(\sigma_i + \sigma_e)$ is also about 10^{-2}.[75] For yttria-stabilized ZrO_2 doped also with CeO_2: $(ZrO_2)_{0.87}(Y_2O_3)_{0.12}(CeO_2)_{0.01} \cong Zr_{0.777}Y_{0.214}Ce_{0.009}O_{1.893}$ mixed conduction is observed under reducing conditions. Changing the oxygen composition from 1.893 to 1.893 – x, the ionic transference number $t_i = \sigma_i/(\sigma_i + \sigma_{el})$ is reduced (at 1186 K) to $t_i = 0.71$ at $x = 0.0048$ and $t_i = 0.16$ at $x = 0.0140$.[76] Yttria-stabilized zirconia containing 5 to 10 mol% TiO_2 exhibits enhanced mixed oxygen ion and electronic conduction under reducing atmospheres at elevated temperatures.[77,78] A small electronic conductivity was also observed in partially stabilized tetragonal ZrO_2.[79]

Cerium cations can be present in two forms: Ce^{3+} and Ce^{4+}. It is therefore possible to reduce pure as well as doped CeO_2. CeO_{2-x} corresponds to the class "$n = zN$" and will be discussed later. However, CeO_2 doped with cations of lower valency is an oxygen ion conductor,[32,33] as is stabilized zirconia. When reduced it also conducts electrons. At 700°C $Ce_{0.8}Gd_{0.2}O_{1.2-x}$ exhibits an ionic transference number t_i about 0.5 for $P(O_2) = 10^{-19}$ atm.[32] Also, $(CeO_2)_{0.9}(Y_2O_2)_{0.1}$ becomes a predominantly electronic conductor at elevated temperature under oxygen partial pressures of $P(O_2)$ below 10^{-10} atm at 1000°C and 10^{-5} at 1400°C.[80] A similar trend is observed for $(CeO_2)_{0.95}(Y_2O_2)_{0.05}$[81] and for SrO-doped CeO_2.[82]

$Gd_2(Zr_{0.3}Ti_{0.7})\,O_7$ doped with 0.5 to 2 mol% Ca or 1 mol% Ta was shown to exhibit mixed conductivity for $800 \leq T \leq 1100°C$.[83] The ionic conductivity in the pyrochlore structure is predominantly due to anion Frenkel defects for undoped samples and to oxygen vacancies, controlled by impurities, for doped samples. Both *n*-type and *p*-type electronic conductivity is observed with $t_e > t_i$ for most $P(O_2)$ values.[83]

ThO_2 doped with lower valent cations is a SE which, at $P(O_2) \geq 10^{-5}$ atm, is a predominant *p*-type semiconductor at elevated temperature.[8] The effect of doping on ThO_2 with variable valent Ce(+3,+4) was examined by Fujimoto and Tuller.[84] Electrical measurements at relatively low temperatures on doped ThO_2 $165 \leq T \leq 500$°C are reported by Näfe.[85]

$La_{1-x}Ca_xAlO_{3-\delta}$ is an oxygen ion conductor[86] with the perovskite-type structure. At relatively high pressures ($P(O_2)$ above 10^{-7} atm at ~1400°C) it exhibits also electronic (hole) conductivity which predominates at higher $P(O_2)$.

$LiNbO_3$ is a Li^+ conductor via Li vacancies with an electronic *n*-type conductivity that depends on $P(O_2)$. The ionic transference number t_i deviates significantly from unity for $P(O_2) < 10^{-3}$ atm at 1000°C.[87]

C. COMPARABLE CONCENTRATIONS OF MOBILE IONS AND ELECTRONS OR HOLES

In this model the mobile ionic defects (of charge z or $-z$) are compensated by quasi-free electrons or holes to maintain charge neutrality. This model is therefore denoted by "$n = zN_i$ or $p = -zN_i$." CeO_{2-x} is a well-investigated[21–30] example for this model. Oxygen vacancies are introduced by deviation from stoichiometry under reducing conditions. Electrons are then thermally excited from the V_O^x donor levels to the conduction band. Analysis[23] of conductivity data shows that $V_O^{\bullet\bullet}$ are dominant at small x ($x < 10^{-3}$). This changes toward singly charged oxygen vacancies, $V_O^{\bullet}$, at larger deviation from stoichiometry, x. For $x \geq 10^{-2}$ significant defect interaction is observed.

Li in Ge and Li in Si are mobile even close to room temperature. The defects formed by doping are ionized Li interstitials and electrons.[31] This corresponds to the model "$n = N_i$". At elevated temperatures many other dopants exhibit relatively high mobility in Si, e.g., Cu, H, Ni, and Fe.[88,89] However, they yield deep donor and/or acceptor levels, and most of the mobile impurities are not ionized. A suitable model description would then have to consider electrons and holes and both the mobile charged ionic defects and the neutral ones. Relatively fast diffusion of impurities is also observed in III–V compounds, e.g., both Zn and Be in GaAs and in InP.[88,90] Zn and Be are acceptors in the III–V compound which conform to the "$p = -zN_i$" or "$p = -zN_i$"[91] model.

Intercalation compounds usually comply with the model "$n = zN_i$" or "$p = -zN_i$." Intercalation is the reversible insertion of large concentrations of mobile guest species into a host solid in which structural entities of the initial solid are maintained.[40,91] The intercalation compound usually exhibits sufficient ionic and electronic conductivity to allow for a reasonable rate of diffusion of the guest atoms in the host material. The materials to be listed below are ionic ones. However, alloying[92] and hydriding[93] also fall under the above definition. The host solids have in many cases a layered structure, such as graphite. However, three-dimensional and one-dimensional structures allowing intercalation also exist. These solids comply with the model "$n = zN_i$" or "$p = -zN_i$" when the guest atom is ionized and contributes free electrons or holes. While this is generally the case, insertion may also result in electrons and holes freed from the guest atom, but localized on the host ions.[94] These localized electrons and holes can, however, be excited at elevated temperatures and contribute to the electronic conductivity. Some of the intercalated compounds exhibit interesting optical properties and are considered for light control in "smart windows".[95] The change in absorption and reflection in Li_xWO_3 and H_xWO_3 was shown to originate from changes in the concentration of conduction electrons.[96] As the plasma frequency is in the UV region (~4eV for x ~ 0.1), a high electron concentration typical of metals must exist.[96] The solids most investigated are metal chalcogenides with layered structure such as MoS_2, $MoSe_2$, $TiSe_2$, $TiTe_2$, InSe, In_2Se_3, GaSe, VSe_2, VS_2, Bi_2Se_3, Bi_2S_3, $HfTe_2$, and the oxides WO_3, WO_2, ReO_3, MoO_3, MoO_2, Ir_2O_3, Nb_2O_3, V_2O_5, CoO_2, NiO, RuO_2, OsO_2, and IrO_2. The intercalated guest atoms are usually alkali metal ions, in particular Li and Na, alkaline earth atoms, copper atoms, or protons. Graphite allows a variety of atoms, cations and anions, as well as groups of atoms to intercalate into it.

D. HIGH ELECTRON OR HOLE CONCENTRATION

In this defect model the concentration of electrons or holes is larger than that of the mobile ions. It is therefore denoted by "n or $p \gg N_i$". It is not difficult to prepare doped semiconductors that comply with the model requirements $n \gg N_i$ or $p \gg N_i$. This can be achieved by doping an n-type or p-type semiconductor (containing fixed donors or acceptors) with a relative small concentration of mobile ions. The same effect can be achieved by varying the stoichiometry of a compound. For example, $La_{1-x}Ca_xCrO_{3+\delta}$ is a p-type electronic conductor at $P(O_2)$ above 10^{-8} atm and T about 1000°C.[97] For $P(O_2)$ below 10^{-8} atm the electronic conductivity becomes $P(O_2)$ dependent. This indicates that ionic defects $V_O^{\bullet\bullet}$ are generated and that these defects are mobile as they can diffuse from the surface into the bulk. At low $P(O_2)$ ($<10^{-8}$ atm) the model "$p = N_i$" holds, but for high $P(O_2)$ the concentration of $V_O^{\bullet\bullet}$ is small and the concentration of holes is higher and is determined by the acceptor concentration $[Ca'_{La}]$ so that the model "$p \gg N_i$" holds. Indium–tin–oxide exhibits one of the highest electronic conductivities among oxides.[98] It depends on $P(O_2)$. A very small chemical diffusion coefficient for oxygen is measured $\tilde{D} \sim 0.1 \exp(-18000/T)$ cm²/sec.[99] The oxygen defect concentration is of the order of 10^{18} to 10^{19} cm^{-3} and appears to be smaller than p.[99]

In U-doped CeO_2 the conduction electron concentration is fixed by the U concentration.[38] Ionic transport of oxygen depends on the deviation from stoichiometry. As long as this deviation is small, the concentration of mobile defects $N_i = [O_i'']$ or $N_i = [V_O^{\bullet\bullet}]$ is small and $n \gg N_i$.

E. METALLIC MIXED CONDUCTORS

Some MIECs exhibit metallic properties. These materials can be found under all three models, "$p, n \ll N_i$", "p or $n \sim N_i$", and "p or $n \gg N_i$". $Ag_{2+\delta}S$, $\delta \approx 2.5 \times 10^{-3}$, $T > 177$°C, is an example of a metallic MIEC that conducts electrons and silver ions and corresponds to the model of high disorder "$p, n \ll N_i$".[35,36] Ag_2Se is similar to $Ag_{2+\delta}S$, and the high-temperature α phase exhibits metallic conductivity.[100] Li_xWO_3 with $x \geq 0.1$ exhibits metallic conductivity and represents the second model "$n = N_i$".[95,96] The metallic hydrides conform to the model "$n = zN_i$" where z need not be an integer, since the hydrogen need not be fully ionized.[93] At low hydrogen content they conform to the "$n \gg N_i$" defect model. An alloy of a low concentration of Li in Al[92] can be viewed as a MIEC metal with high electron concentration corresponding to the model "$n \gg N_i$".

F. MISCELLANEOUS MIXED CONDUCTORS

Many of the MIECs are characterized experimentally by measuring their electronic and ionic conductivities, σ_{el}, σ_i, without determining also the concentrations n, p, N_i. Therefore, although mixed conductivity can be detected, it is not always known which concentration model describes a MIEC best. The model is of interest as it will be shown in Section III that the current voltage, I–V, relations depend on the model for the defect concentration ratio. We list here materials for which mixed conductivity has been established in recent years, but the defect concentration ratio has not yet been determined.

$La_{0.6}A_{0.4}Co_{0.8}Fe_{0.2}O_{3-\delta}$ (A=La,Ca,Sr) and $La_{0.6}Sr_{0.4}Co_{0.8}B_{0.2}O_{3-\delta}$ (B=Fe,Co,Ni,Cu) are excellent MIECs exhibiting high metallic conductivity. The oxygen ionic conductivity, though a few orders of magnitude lower, is also high compared with other ionic conductors.[101,102] The ionic conductivity is of the order of that of yttria-stabilized zirconia. It is reasonable to assume that these compounds fit the model "$n = zN_i$" as the concentration of quasi-free electrons and of $V_O^{\bullet\bullet}$ are expected to be of the order of unity per formula and therefore of the same order of magnitude. The electronic conductivity is then much higher than the ionic one because of a large difference in the mobilities, with $\nu_e \gg \nu_i$.

Tm_2O_3 is a *p*-type MIEC at $P(O_2) \geq 10^{-6}$ atm.[103] The p–$P(O_2)$ relations suggest that the mobile ionic defects are excess oxygen, O_i'', entering the quasi-vacant sites of the c-structure. If this is indeed the case, then the defect model that holds for this oxide is $p = 2N_i$. β-Ta_2O_5 and Nb_2O_5-doped β-Ta_2O_5 exhibit ionic conduction via $V_O^{\bullet\bullet}$.[104,105] For high $P(O_2)$ (e.g., $P(O_2) \geq 10^{-5}$ atm at $T \sim 1000°C$) $N_i = [V_O^{\bullet\bullet}]$ is fixed by residual impurities or the Nb_2O_5 dopant concentration, respectively. At lower $P(O_2)$ when the concentration of $V_O^{\bullet\bullet}$ exceeds that of the impurities, the electroneutrality condition become $n = 2N_i$.

For ABC_2 (A = Cu,Ag; B = Ba,In; C = Se,Te) analysis of conductivity vs. $P(O_2)$ and of $P(O_2)$ vs. *T* data can be understood either if the covalent bonding instead of the ionic bonding exists, or if anti-site defects exist, where cations occupy vacant anion sites.[106] In these materials as well as in $CuInS_2$,[107] *p*-type as well as *n*-type conductivity are observed. The mobile ions are Cu or Ag.

The SE Gd-doped CeO_2 exhibits mixed conductivity at elevated *T* and low $P(O_2)$ due to partial reduction. The model describing it is "$p \ll n \ll N_i$". The electrons are generated by excitation from oxygen vacancies. It was shown that the electron concentration can be reduced by trapping by adding a second dopant such as Pr. In this way σ_{el} could be reduced by 2 orders of magnitude.[108]

Amorphous or glassy materials have been investigated in particular as possible SEs. Some were found to be MIECs. $(40 - x)Fe_2O_3 - xNa_2O$-$60P_2O_5$ is a glassy MIEC conducting Na^+ and electrons.[109] The electrons propagate by hopping between localized states. The ionic conductivity for $x = 35$ is $\sim 10^{-10}$ S/cm at 25°C and 10^{-7} S/cm at 150°C. The sum of concentrations of conduction electrons and of mobile ions varies slowly with *x* as compared to the variation in the electron and ion concentrations. This forms a new (approximate) defect model "$n + N_i \approx$ const" which is different from the three models usually considered. Cu_xCS_2 and Ag_xCS_2 were reported to be amorphous MIECs conducting Cu^+ and Ag^+ ions, respectively, and electrons exhibiting $t_i \sim 1/2$, $\sigma_{Cu^+} \sim 10^{-2}$ S/cm, for $x = 3.60$ and $\sigma_{Ag^+} \sim 3 \times 10^{-2}$ S/cm, for $x = 3.60$ at room temperature.[110,111]

Cu_2O has been considered for the semiconductor industry before the Si era and in recent years for photovoltaic cells. It has therefore been intensively investigated. Yet there is still much disagreement on the exact defect nature of Cu_2O. The electronic conductivity is *p*-type. There is no agreement on the nature of atomic defects. It has been suggested that the following defects exist: V_{Cu}', V_{Cu}^x, associates of $(V_{Cu}' V_{Cu}^x)$, and O_i'', for relatively high $P(O_2)$ (e.g., $P(O_2) \simeq 10^{-5}$ atm at 1000°C).[112–118a] The presence of neutral defects together with charged ones is usually assumed. V_{Cu}^x as well as V_{Cu}' are claimed to be mobile.[117] When all three defects V_{Cu}^x, V_{Cu}', and *p* are mobile, then mixed valence ionic–electronic conduction (MVIEC) occurs. This is a different defect model; the I–V characteristics are quite distinct from those for the other models, as will be explained in Section III.H. Park and Natesan[115] suggest different dominant mobile defects: O_i'' and *p* with $p = 2[O_i'']$, which corresponds to the defect model "$p = -zN_i$".

The high-temperature superconductor oxides, in particular $YBa_2Cu_3O_x$ (YBCO), have drawn much attention in recent years. YBCO is *p*-type metal for $6.5 \leq x \leq 7$. It exhibits superconductivity at reduced temperatures.[119] It is a *p*-type semiconductor for $6 < x \leq 6.5$, showing a transition to *n*-type conductivity for $x \sim 6$.[119–123] Ionic conductivity is difficult to establish. Diffusion of oxygen through YBCO is observed at elevated *T* ($T \geq 800$ K) using different experimental methods.[124–128] However, it is not clear whether the oxygen diffuses as a neutral species, say O_i^x, or as a charged defect, say O_i''.

Measurements aimed at directly measuring the ionic conductivity in YBCO using ion-selective electrodes yield conflicting results.[129,130] It is questionable if the various methods can distinguish between motion of charged and neutral atomic defects in an electronic conductor, as will be shown in Section IV. It was independently shown that not only O_i'', but also a high concentration of O_i^x exists in YBCO, and it was suggested that both kinds of defects might be mobile.[129,131] In conclusion, true MIEC in YBCO has not yet been established.

G. *pn* JUNCTIONS IN MIXED CONDUCTORS

MIECs may be made nonuniform to the extent that they become *n*-type on one side and *p*-type on the other side, thus forming *pn* or *pin* (i = intrinsic) junctions. This nonuniformity can be introduced by doping and by deviations from stoichiometry under a chemical gradient (fixed by the electrode compositions), with or without an applied electric potential gradient.

TiO formed anodically on Ti becomes *n*-type on the metal side and *p*-type on the liquid electrolyte side due to a gradient in the Ti to O concentration ratio.[132,133] This forms a *pn* junction in the TiO film. In a Ta_2O_5 film grown anodically on Ta, a *pin* junction is formed.[134]

The formation of *p* regions and *n* regions in MIECs under applied voltage has been predicted theoretically for different defect models.[135–139] The electric field either pushes mobile ions to one side, as in Li-doped Si and when using ion-blocking electrodes, or it can induce or modify the gradient in deviation from stoichiometry in MIECs that interact with their surroundings. Also, *pn* junctions were formed by applying a voltage to ZrO_2 + 10 mol% Y_2O_3,[140] doped $BaTiO_3$,[141] doped $SrTiO_3$,[142] $CuInSe_2$,[143,144] and $Hg_{0.3}Cd_{0.7}Te$.[145] In the latter case the applied voltage must be large enough to yield a graded *n*, graded *p*, *pn*, or *pin* junction. While forming a *p*- or *n*-rich region may result in degradation of an electronic device, it may be beneficial in other cases. The fabrication of *pn* junctions in this manner has been suggested.[137,138,144,145]

III. I–V RELATIONS

A. GENERAL

The I–V relations of solid electrochemical cells based on MIECs are of primary interest in this review. They describe the performance of these cells and are also the theoretical basis that allows one to extract the partial ionic and electronic conductivities from I–V measurements. The geometry of the cell considered here is mainly one dimensional (taken to be in the x direction) as shown in Figure 7.2. Reference will also be made to the Van der Pauw configuration, where the MIECs have a flat, singly connected shape and the electrodes are applied on the periphery of the MIEC. We refer in particular to circular flat samples with electrodes equally spaced as shown in Figure 7.3.

The analysis is normally done for the current density j instead of for the current I, as the first is independent of the cross-sectional area of the MIEC.

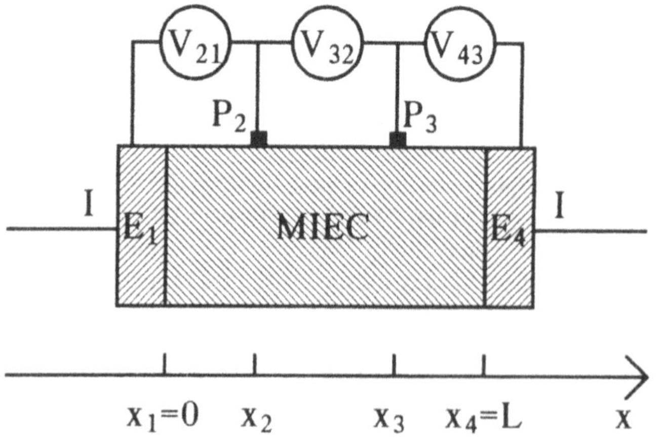

FIGURE 7.2. Schematic of a linear cell with four electrodes and an MIEC. The current-carrying electrodes E_1 and E_4 are reversible except when stated otherwise. The probes P_2 and P_3 present an option, being selective either for electrons or for ions. Reversible probes should be implanted into the MIEC as shown in Figure 7.6

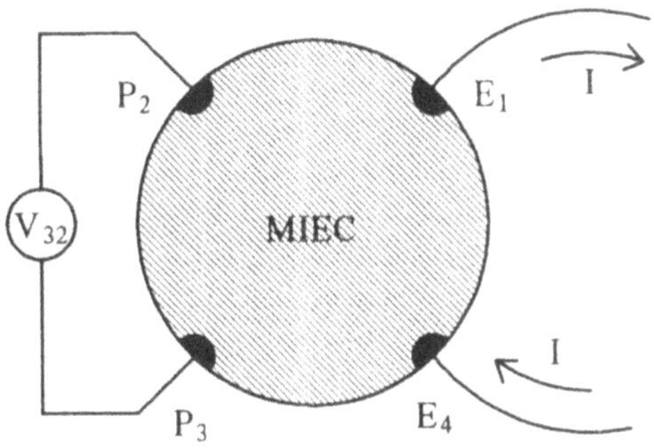

FIGURE 7.3. Schematics of a Van der Pauw, circular, configuration with four equally spaced electrodes. The current-carrying electrodes E_1 and E_4 are reversible except when stated otherwise. The probes, P_2 and P_3, are optional, being either reversible or selective for electrons or for ions.

An MIEC, say with the composition MX, has at least one mobile ionic species, say $M_i^{\cdot}$, and at least one electronic defect, say e'. The electrical current densities for $M_i^{\cdot}$ and e' are

$$j_{M1} = -\frac{\sigma_{M1}}{q}\nabla\tilde{\mu}_{M1} \tag{7.1}$$

where M_1 indicates $M_i^{\cdot}$, and

$$j_e = \frac{\sigma_e}{q}\nabla\tilde{\mu}_e, \tag{7.2}$$

where q is the elementary charge, $\sigma = q\nu[]$ is the conductivity, ν the mobility, [] the concentration of the relevant mobile species, and $\tilde{\mu}$ the corresponding electrochemical potential. The latter is related to the chemical potential, μ, and to the internal electric potential, φ, by

$$\tilde{\mu} = \mu + qz\varphi \tag{7.3}$$

where z is the valency of the defect. The use of chemical and electrochemical potentials for the defects, as structure elements, is not obvious.[6] A problem may arise as these defects are not independent of each other, and the change in concentration of one defect necessarily induces a change in the concentration of another defect. Thus, for example, forming $M_i^{\cdot}$ from $M_M^{\times}$ is accompanied by the formation of V_M'. However, no real problem arises with the use of the concept of chemical potential of structure elements as long as one uses the minimum number of species required to describe the MIEC, taking care of material, site, and charge balance.[6,146,147] Thus, in an oxide conducting material with movement of oxygen via oxygen vacancies, one can represent the oxygen flux either by the flux of $V_O^{\cdot\cdot}$ or the flux of $O_O^{\times}$ ($\cong O^{2-}$), but not by both. Similarly, the electronic current should only be represented by the contribution from the electrons in the conduction band e' and holes in the valence band $h^{\cdot}$ and not also by the current of the electrons present in the valence band $e^{\times}$ and the holes present in the conduction band, $h^{\times}$.

When the MIEC also conducts metal ion vacancies and holes, one has to consider also the electrical current density equations:

$$j_{V1} = \frac{\sigma_{V1}}{q}\nabla\tilde{\mu}_{V1}, \quad V1 \equiv V_M', \tag{7.4}$$

and

$$j_h = -\frac{\sigma_h}{q}\nabla\tilde{\mu}_h. \tag{7.5}$$

Due to the creation/annihilation reaction of Frenkel pairs, Equations (7.1) and (7.4) can be combined to yield the total ionic current density, j_i:

$$j_i = j_{M1} + j_{V1} = -\frac{\sigma_i}{q}\nabla\tilde{\mu}_{M1}, \quad \sigma_i = \sigma_{V1} + \sigma_{M1}. \tag{7.6}$$

Similarly, due to the creation/annihilation reaction of electron-hole pairs, Equations (7.2) and (7.5) can be combined to yield the total electronic current density, j_{el}:

$$j_{el} = j_e + j_h = \frac{\sigma_{el}}{q}\nabla\tilde{\mu}_e, \quad \sigma_{el} = \sigma_e + \sigma_h. \tag{7.7}$$

Equations (7.6) and (7.7) can be combined to yield[62,148]

$$j_{el} = \frac{\sigma_{el}}{\sigma_t} j_t + \frac{\sigma_{el}\sigma_i}{q\sigma_t}\nabla\mu_M, \tag{7.8}$$

and

$$j_i = \frac{\sigma_i}{\sigma_t} j_t - \frac{\sigma_{el}\sigma_i}{q\sigma_t}\nabla\mu_M, \tag{7.9}$$

where

$$\sigma_t = \sigma_{el} + \sigma_i, \quad j_t = j_{el} + j_i, \tag{7.10}$$

and

$$\mu_M = \tilde{\mu}_{M1} + \tilde{\mu}_e = \mu_{M1} + \mu_e, \tag{7.11}$$

is the chemical potential of the neutral species *M*. Equations (7.8) and (7.9) are quite general, the only limitation being that $|z|$ is constant (in the example $|z| = 1$), and therefore hold for many defect models, for both neutral and space charge-controlled systems and for time-dependent problems. They shall be used in the analysis of diffusion problems under open circuit conditions where $j_t = 0$.

In the steady state the concentrations of all defects do not change with time. For the one-dimensional configuration and for $|z|$ having only one value, j_i and j_{el} must be uniform. Equations (7.6) and (7.7) can then be integrated to yield the general dependence of the ionic current $I_i = j_i S$ and the electronic current $I_{el} = j_{el} S$ (S is the cross-sectional area of the MIEC) on the voltage drop, V, across the MIEC and the chemical potential difference, $\Delta\mu_M$, on the MIEC as fixed by the electrodes:[149–151]

$$I_i = \frac{V_{th} - V}{R_i}, \quad R_i = \frac{1}{S}\int_0^L \frac{dx}{\sigma_i}, \tag{7.12}$$

$$I_{el} = -\frac{V}{R_{el}}, \quad R_{el} = \frac{1}{S}\int_0^L \frac{dx}{\sigma_{el}}, \tag{7.13}$$

where L is the length of the MIEC and

$$V_{th} = -\frac{\Delta\mu_M}{q}, \tag{7.14}$$

is the Nernst voltage. The cell current, I, being measured in the external circuit equals the sum $I_i + I_{el}$. Equation (7.12) shows that $I_i = 0$ corresponds with $V = V_{th}$. This is denoted as "polarization condition", a name given to reflect the fact that a gradient in the chemical potential is induced in the MIEC. Equation (7.13) shows that $I_{el} = 0$ corresponds with $V = 0$. This is denoted as "short-circuit condition". Under open-circuit conditions, $I = 0$, and $I_i = -I_{el} \neq 0$ are given by Equations (7.12) and (7.13) with V equal to the open-circuit voltage V_{oc}, which is

$$V_{oc} = \frac{R_{el}}{R_i + R_{el}} V_{th} = \bar{t}_i V_{th} \tag{7.15}$$

where $\bar{t}_i$ is the average ionic transference number under open-circuit conditions. It is important to notice that in general, R_i and R_{el} are not constant and may depend on V and V_{th}. When this is the case, any attempt to substitute an equivalent circuit with fixed resistors for the electrochemical cell and using it to analyze the cell performance over a wide range of I and V yields misleading results. It is only under conditions of small chemical potential gradients in the MIEC ($qV_{th} \ll kT$, where k is the Boltzmann constant, T is the temperature) that R_i and R_{el} can be considered to be approximately constant and an equivalent resistor circuit can be used.[152]

Equations (7.12), (7.13), and (7.15) hold for all defect models discussed in Section II, whether local neutrality inside the MIEC holds or not. This would not be true if the MIEC conducts ions with different values of $|z|$, as is explained in Section III.H when discussing MVIECs.

Further discussion of the I–V relations requires specification of the defect model relevant for the MIEC under investigation.

The I–V relations discussed here hold whether the electrodes are reversible or not, since V and V_{th} are by definition the values actually existing on the MIEC. When the electrodes are reversible with respect to both the electronic current and the ionic current, then V equals the voltage applied to the electrodes and $-qV_{th} = \Delta\mu_M$ is the chemical potential difference of M in the two current carrying electrodes.

B. I–V RELATIONS WHEN HIGH DISORDER PREVAILS

1. General Relations

The relevant defect model for the MIEC is "$p, n \ll N_i$", where $p \ll n$, $p \sim n$, and $p \gg n$ are possible. A special case is "$p \ll n \ll N_i$" (or "$n \ll p \ll N_i$"). This case was most intensively investigated, but mainly under ion blocking or open circuit conditions. *I–V* relations for a wide range of applied voltages were derived under the conditions:

a. (approximate) local neutrality,
b. low electron and hole concentrations so that they follow Boltzmann statistics,
c. steady state; we notice that true steady state requires also that the rate of decomposition of the MIEC is negligible.

The I–V relations were evaluated in parametric form by Choudhury and Patterson[153] and by Tannhauser.[154] An explicit analytic solution was then derived by Riess.[149,150] These calculations rely on the fact that due to the high concentration of ionic defects, i.e., high disorder,

$$\nabla \mu_i = 0, \quad \nabla \sigma_i = 0. \tag{7.16}$$

The explicit I–V relations found are[149,150]

$$j_e = -\sigma_e(0)\frac{V_{th} - V}{L}\frac{e^{-\beta q(V_{th}-V)} - e^{-\beta q V_{th}}}{1 - e^{-\beta q(V_{th}-V)}}, \tag{7.17}$$

and

$$j_i = \sigma_i \frac{V_{th} - V}{L}, \tag{7.18}$$

where $\sigma_e(0)$ is the electron conductivity of the MIEC close to the contact at $x = 0$, and $\beta = 1/kT$.

The sign convention for the linear cell is $V > 0$, $V_{th} > 0$ when the potential is higher on the right-hand side, and $j > 0$ when the current flows to the right, i.e., in the + x direction.

2. Polarization Conditions

By polarization conditions we mean the case where j_i vanishes. Then by Equation (7.12) $V = V_{th}$. Substituting $V = V_{th}$ into the general $I_e - V$ relations of Equation (7.17) yields for the polarization conditions:

$$j_e = -\sigma_e(0)\frac{kT}{qL}\left(1 - e^{-\beta q V}\right). \quad V = V_{th}, \quad j_i = 0. \tag{7.19}$$

The ionic current can be suppressed using ion-blocking electrodes. This is the Hebb–Wagner polarization method for determining $\sigma_e(0)$.[9,155,156] Alternatively, j_i can be eliminated by applying a voltage equal to a fixed V_{th}.[151] Obviously this is not trivial to do. When using an ion-blocking electrode the equality $V = V_{th}$ is readily obtained, since V_{th} is not fixed and in the steady state it adjusts itself automatically to the value V imposed on the MIEC.

When a space charge exists near the MIEC edge due to redistribution of the mobile ions (e.g., near a contact with another phase), this affects the I–V relations. For currents drawn parallel to this edge the current density changes with the distance from the edge. The $\sigma_e(0)$ determined is an average over the space charge region.[157] These corrections, however, are expected to be negligible for samples of sizes much larger than the space charge region.

3. Short-Circuit Conditions

In the limit $j_{el} = 0$, $V = 0$ Equation (7.18) yields

$$j_i = \sigma_i \frac{V_{th}}{L}, \quad V = 0, \quad j_{el} = 0. \tag{7.20}$$

The electronic current can be suppressed using electron-blocking electrodes, i.e., SEs.[8,16] Alternatively, one can eliminate j_{el} by short circuiting the MIEC imposing $V = 0$, hence $j_{el} = 0$. This is the "short-circuit" method for determining σ_i.[151]

4. Open-Circuit Conditions

Under open-circuit conditions, $j_t = j_l + j_i = 0$ and,

$$-j_e = j_i = \sigma_i \frac{V_{th} - V_{oc}}{L}, \tag{7.21}$$

where

$$V_{oc} = V_{th} + \frac{kT}{q} \ln \frac{\sigma_e(L) + \sigma_i}{\sigma_e(0) + \sigma_i} = V_{th} + \frac{kT}{q} \ln \frac{t_i(0)}{t_i(L)}, \tag{7.22}$$

$\sigma_e(L)$ is the electron conductivity in the MIEC near the contact at $x = L$, $t_i(0)$, $t_i(L)$ with $[t_i = \sigma_i/(\sigma_i + \sigma_e)]$ are the local ionic transference numbers in the MIEC near $x = 0$ and $x = L$, respectively. It is customary to define $\bar{t}_i$ using the equations:[9]

$$V_{oc} \equiv \bar{t}_i V_{th} \Rightarrow \bar{t}_i = 1 + \frac{kT}{qV_{th}} \ln \frac{t_i(0)}{t_i(L)}. \tag{7.23}$$

Equation (7.22) can also be obtained by a different mathematical approach, suitable for open-circuit conditions, which is based on the parametric solution of Choudhury and Patterson[153] as was shown by Crouch–Baker.[158]

When the MIEC is in equilibrium with the gas phase, say oxygen, then the electronic conductivity is related to the oxygen partial pressure $P(O_2)$ in the gas, and $\bar{t}_i$ can be expressed as[6,12,159]

$$\bar{t}_i = 1 + \frac{kT}{qV_{th}} \ln \frac{P(O_{2,}-)^{-1/4} + P(O_{2,}L)^{-1/4}}{P(O_{2,}-)^{-1/4} + P(O_{2,}0)^{-1/4}}, \quad V_{th} = \frac{kT}{4q} \ln \frac{P(O_{2,}L)}{P(O_{2,}0)}, \tag{7.24}$$

where $P(O_2,L)$ and $P(O_2,0)$ are the values of $P(O_2)$ at $x = 0$ and L, respectively. When the electrodes are reversible, then $P(O_2,L)$ and $P(O_2,0)$ are equal to the oxygen partial pressure in the gas phase near the corresponding electrodes. $P(O_2,-)$ is the value of $P(O_2)$ for which $\sigma_e = \sigma_i$.

C. I–V RELATIONS FOR HIGH DISORDER IN THE PRESENCE OF BOTH ELECTRONS AND HOLES

1. General Relations

We discuss now the more general high-disorder defect model "$p,n \ll N_i$", i.e., without restriction on the p/n ratio. The transport equations describing the cell are now Equation (7.1) and Equation (7.7). The analysis of the general I–V relations is done under the same condition (*a*) through (*c*) of Section III.B.1 and Equation (7.16).[138] The ionic current density is then also given by Equation (7.18). The j_{el}–V relations are obtained in implicit form:

$$JI(\bar{n}_L) = \beta qV, \tag{7.25a}$$

where,

$$J = \frac{j_{el}L}{qv_e n_0 (V_{th} - V)}, \tag{7.25b}$$

$$\bar{n}_L = \frac{n_L}{n_0} \tag{7.25c}$$

n_0 is the electron concentration in the MIEC near $x = 0$, and

$$I(\bar{n}) = \begin{cases} \frac{1}{\sqrt{\Delta}} \ln\left|\frac{2\bar{n} - J - \sqrt{\Delta}}{2\bar{n} - J + \sqrt{\Delta}} \times \frac{2 - J + \sqrt{\Delta}}{2 - J - \sqrt{\Delta}}\right|, & \Delta > 0 \\ \frac{2}{J - 2\bar{n}} - \frac{2}{J - 2}, & \Delta = 0 \\ \frac{2}{\sqrt{-\Delta}}\left[\arctan\frac{2\bar{n} - J}{\sqrt{-\Delta}} - \arctan\frac{2 - J}{\sqrt{-\Delta}}\right], & \Delta < 0 \end{cases} \tag{7.25d}$$

where

$$\bar{n} = n/n_0 \tag{7.25e}$$

and

$$\Delta = J^2 - 4n_i^2 v_h / (v_e n_0^2) \tag{7.25f}$$

and n_i is the intrinsic electron concentration. For reversible electrodes and for a given MIEC, n_0 and V_{th} are determined by the electrode compositions. The j_{el}–V relations for a fixed V_{th}, calculated numerically from Equations (7.25a) to (7.25e), are shown in Figure 7.4. It should be noticed that the asymptotic j_{el}–V relations are linear (not exponential). The reason is that there exists no space charge, and the change in slope reflects the fact that the electronic resistance is changed (see Section VI).

2. Polarization Conditions

In this limit $j_i = 0$ ($V = V_{th}$) and explicit j_{el}–V relations can be obtained. These relations, derived by different methods,[150,156] are

$$j_{el} = -\frac{kT}{qL}\left(\sigma_e(0)\left(1 - e^{-\beta qV}\right) + \sigma_h(0)\left(e^{\beta qV} - 1\right)\right), \quad V = V_{th} \tag{7.26}$$

where $\sigma_h(0)$ is the hole conductivity in the MIEC near $x = 0$. Figure 7.5 shows these j_{el}–V relations (with $V_{th} = V$) for a cation conductor, MX, using M as a reversible electrode and assuming $\sigma_e(0) \gg \sigma_h(0)$. As the applied voltage increases, the concentration of holes near $x = L$ increases. Eventually for large V the term in Equation (7.26) containing $\sigma_h(0)$ becomes dominant. This is reflected by the exponential increase of j_{el} at high V.

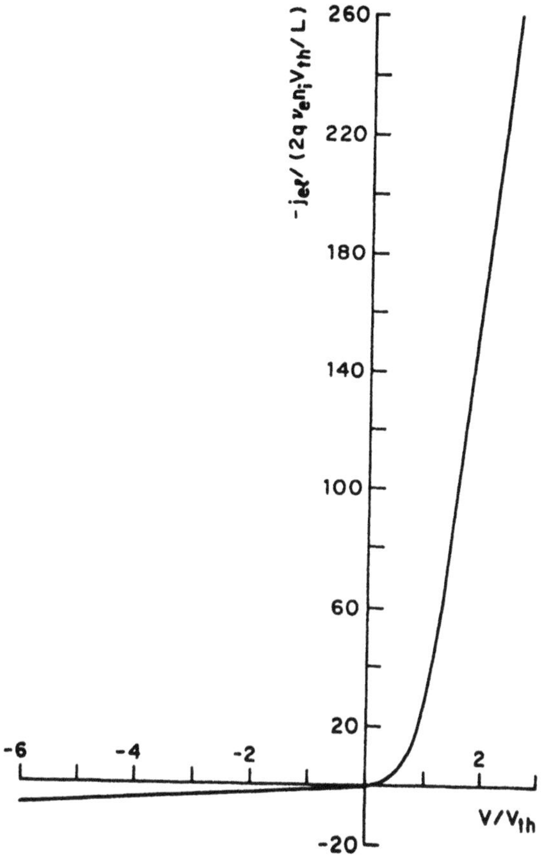

FIGURE 7.4. $j_{el} - V$ relations for MIECs with the defect model "$p, n \ll N_i$", using reversible electrodes. (From Riess, I., *Phys. Rev.*, B35, 1987, 5740. With permission.)

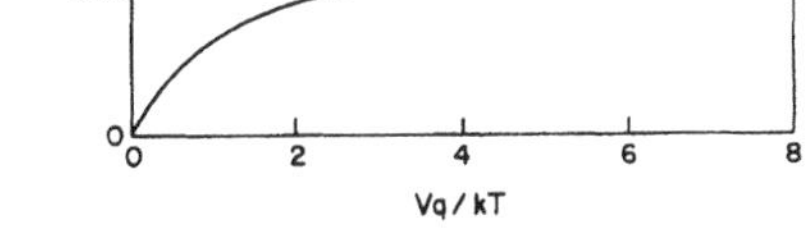

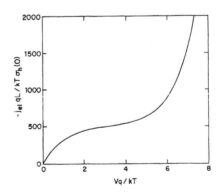

FIGURE 7.5. $j_{el} - V$ relations for MIECs with the defect model "$p, n \ll N_i$" under Hebb–Wagner polarization conditions (j_i = 0). ($\sigma_e(0)/\sigma_h(0)$ = 500.) (From Riess, I., *J. Phys. Chem. Solids*, 47, 1986, 129. With permission.)

3. Short-Circuit Conditions

In the limit $j_{el} = 0$ ($V = 0$), j_i is given by Equation (7.20). j_{el} can be eliminated either by using a SE as an electrode or by short circuiting.

4. Open-Circuit Conditions

An explicit I–V relation can be obtained also for this limit (where $j_t = j_{el} + j_i = 0$),

$$-j_{el} = j_i = \sigma_i \frac{V_{th} - V_{oc}}{L} \tag{7.27}$$

where[158]

$$V_{oc} = \frac{kT}{q}\frac{\sigma_i}{\sigma^*}\ln\left\{\frac{\left[2\sigma_h(0)e^{\beta qV_{th}}+\sigma_i-\sigma^*\right]\left[2\sigma_h(0)+\sigma_i+\sigma^*\right]}{\left[2\sigma_h(0)e^{\beta qV_{th}}+\sigma_i+\sigma^*\right]\left[2\sigma_h(0)+\sigma_i-\sigma^*\right]}\right\} \tag{7.28a}$$

and

$$\sigma^* = \sqrt{\sigma_i^2 - 4\sigma_h(0)\sigma_e(0)}\ . \tag{7.28b}$$

For an MIEC in equilibrium with the gas phase, say oxygen, the open-circuit voltage can be expressed in terms of $P(O_2)$ (for $P(O_2,+) \gg P(O_2,-)$), as:[6,12,159]

$$V_{oc} = \frac{kT}{q}\left[\ln\frac{P(O_{2,}-)^{1/4}+P(O_{2,}L)^{1/4}}{P(O_{2,}-)^{1/4}+P(O_{2,}0)^{1/4}}+\ln\frac{P(O_{2,}+)^{1/4}+P(O_{2,}0)^{1/4}}{P(O_{2,}+)^{1/4}+P(O_{2,}L)^{1/4}}\right] \tag{7.29}$$

where $P(O_2, +)$ is the value of $P(O_2)$ for which $\sigma_h = \sigma_i$ and $P(O_2, -)$ is the value of $P(O_2)$ for which $\sigma_e = \sigma_i$.

D. I–V RELATIONS WHEN THE CONCENTRATIONS OF MOBILE IONS AND ELECTRONS OR HOLES ARE COMPARABLE

1. General Relations

The I–V relations for the model "$p \ll n = zN_i$" (or "$n \ll p = -zN_t$") are obtained from Equations (7.1) and (7.2) (or Equations (7.4) and (7.5), respectively), under the conditions:

a. $n = N_i$ (or $p = N_i$),
b. low electron, hole, and mobile ion concentrations so that Boltzmann statistics hold for these defects
c. steady state

The I–V relations are[150]

$$j_e = -2\sigma_e(0)\frac{kT}{qL}\left(1-e^{-\beta qV_{th}/2}\right)\frac{V}{V_{th}}, \tag{7.30}$$

and

$$j_i = 2\sigma_i(0)\frac{kT}{qL}\left(1-e^{-\beta qV_{th}/2}\right)\frac{V_{th}-V}{V_{th}}, \tag{7.31}$$

where $\sigma_i(0)$, $\sigma_e(0)$ are the ion and electron concentrations in the MIEC near $x = 0$. In this model the local ionic and electronic transference numbers are uniform throughout the MIEC:

$$t_i = \frac{v_i}{v_e + v_i}, \quad t_e = \frac{v_e}{v_i + v_e} \tag{7.32}$$

hence $\bar{t}_i = t_i$.

2. Polarization Conditions

In this limit $j_i = 0$ ($V = V_{th}$) and Equation (7.30) reduces to:

$$j_e = -2\sigma_e(0)\frac{kT}{qL}\left(1 - e^{-\beta qV/2}\right), \quad V = V_{th}, \quad (j_i = 0) \tag{7.33}$$

The ionic current can be eliminated as mentioned before using an ion-blocking electrode or by adjusting V to V_{th}. The j_e–V relations under polarization for the model "$p \ll n \ll N_i$" (and "$p \ll n \ll N_i$") are different from those for the present model. Comparing Equations (7.19) and (7.33), one notices that (a) Equation (7.33) depends exponentially on $\beta qV/2$ rather than on βqV and (b) Equation (7.33) contains a factor 2 which does not appear in Equation (7.19).

3. Short-Circuit Conditions

In this limit $j_{el} = 0$ ($V = 0$) and Equation (7.31) reduces to:

$$j_i = 2\sigma_i(0)\frac{kT}{qL}\left(1 - e^{-\beta qV_{th}/2}\right), \quad V = 0, \left(j_{el} = 0\right). \tag{7.34}$$

The electronic current can be eliminated as mentioned before using an electron-blocking electrode or by short circuiting.

4. Open-Circuit Conditions

Under open-circuit conditions ($j_t = j_e + j_i = 0$):

$$-j_e = j_i = 2t_e\sigma_i(0)\frac{kT}{qL}\left(1 - e^{-\beta qV_{th}/2}\right), \tag{7.35a}$$

and

$$V_{oc} = t_i V_{th}, \tag{7.35b}$$

where t_i,t_e are given by Equation (7.32).

E. I–V RELATIONS WHEN THE ELECTRON OR HOLE CONCENTRATION IS HIGH

The I–V relations for the model "$n \gg N_i \gg p$" (or "$p \gg N_i \gg n$") are derived from Equations (7.1) and (7.2) under the assumptions

$$\nabla\mu_e = 0, \quad \nabla\sigma_e = 0, \tag{7.36}$$

in the steady state, assuming (approximate) local neutrality. Then:

$$j_i = -\frac{V}{L}\sigma_i(0)\frac{\left(e^{-\beta qV} - e^{-\beta qV_{th}}\right)}{\left(e^{-\beta qV} - 1\right)}, \tag{7.37a}$$

and

$$j_e = -\sigma_e\frac{V}{L}. \tag{7.37b}$$

Normally, due to the high electronic conductivity, V is kept low, much smaller than 1V and V_{th} is of the order of 1V. Then $V \ll V_{th}$ and Equation (7.37a) reduces to

$$j_i = \frac{kT}{qL}\sigma_i(0)\left(1 - e^{-\beta q V_{th}}\right). \qquad (7.37c)$$

where $\sigma_i(0)$ is the ionic conductivity in the MIEC at $x = 0$.

F. I–V RELATIONS WHEN THE TOTAL CONCENTRATION OF MOBILE DEFECTS IS FIXED

The model "$n + N_i$ = const." (or "$p + N_i$ = const.") has been considered under blocking conditions where either the electronic charge carriers or ions are blocked, and under the additional conditions: (a) local electroneutrality, (b) low defect concentrations, and (c) steady state.[160]

Materials complying with this model are not often encountered. $(40 - x)Fe_2O_3 - xNa_2O–60P_2O_5$ glass[109] and Cu_3VS_4[198] can serve as examples. This model allows an analytical evaluation of the space charge for ion-blocking conditions and is therefore of interest.

For $n + N_i = A$, where N_i now represents the concentration of negative mobile ions V_M' and A is constant, Equations (7.2) and (7.4) yield under ion-blocking conditions:

$$j_e = \sigma_A \frac{kT}{qL} \ln \frac{\left(A - N_i(0)\right)e^{-\beta q V} + N_i(0)}{A}, \quad j_i = 0, \quad \left(V = V_{th}\right), \qquad (7.38)$$

where $\sigma_A = q\nu_e A$ is the conductivity as if both types of defects e', and V_M' would have the mobility ν_e, and $N_i(0)$ is the concentration of the mobile ionic defects in the MIEC near $x = 0$. When the electrode near $x = 0$ is reversible, $N_i(0)$ is fixed by the electrode chemical potential and the measured I–V relations are fully described by Equation (7.38).

G. I–V RELATIONS IN FOUR POINT MEASUREMENTS ON MIXED CONDUCTORS

1. General Relations

The application of four electrodes on a MIEC is not a trivial modification of the two-electrode configuration. The application of the additional two voltage probes may affect the I–V relations even if the impedance of the voltmeter is extremely high. The reason is that a high-impedance voltmeter (or open circuit) eliminates the total current through the probes, but not the electronic and ionic components of the current, i.e., at each probe $I_{el} + I_i = 0$, but I_{el} and I_i separately are not necessarily zero. In other words, the exchange of charge and matter between the voltage probes and the MIEC reflects a chemical reaction which affects the I–V relations and defect distributions. These effects can be taken into consideration in the Van der Pauw configuration, e.g., the one shown in Figure 7.3.

The steady-state I–V relations for the Van der Pauw configuration, for MIECs with electron and ion conductivity, with quite arbitrary ratios of the electron and ion concentrations, have been derived for small applied chemical potential differences $|\Delta\mu_M| \ll kT$.[161] If the current-carrying electrodes are not reversible, the applied voltage must also be limited so that $|qV| < kT$, to avoid polarization resulting in large chemical potential gradients. The I–V relations of interest are $V_{41} - I$, $V_{21} - I$, $V_{32} - I$, and $V_{43} - I$, where I is the total current through the MIEC. The current I is measured in the external circuit. Other relations of interest are those between the aforementioned voltage drops and the electronic component of the current. We quote here the relation $V_{32} - I$,[161] with geometrical parameters relevant to a circular sample as shown in Figure 7.3:[162]

$$V_{32} = -\frac{\ln 2}{\pi d \sigma_t} I - \frac{\sigma_i}{\sigma_t} \frac{\mu_M(3) - \mu_M(2)}{q}, \tag{7.39}$$

where $\sigma_t = \sigma_e + \sigma_i$, d is the thickness of the MIEC, and μ_M (3), $\mu_M(2)$ are the chemical potentials of M in the MIEC at the probes 3 and 2. By definition $I > 0$ when it enters through electrode 4.

In the derivation of Equation (7.39) finite electronic and ionic currents through the probes were allowed and taken into consideration. It was only assumed that the sum of these currents through each probe vanishes. When the probes are reversible with respect to the electronic and ionic currents, $\mu_M(3)$, and $\mu_M(2)$ are equal to the chemical potentials of M in the probes.

2. Polarization Conditions

To block the ionic current, three of the electrodes must be blocking to ions. The fourth should be a reversible one, thus fixing a well-defined composition of the MIEC at the point of contact. If the fourth electrode is also ion blocking, then the MIEC composition is not well defined and depends on the sample history. We choose electrodes 2, 3, 4 or 1, 3, 4 as ion-blocking ones and electrode 1 or 2, respectively, as the reversible electrode.

The I_e–V were derived for small applied voltages by Riess and Tannhauser[161] for an arbitrary ratio of the electron and ion concentrations. The limitation of small applied voltages was then removed in analyzing a MIEC for which the model "$n \ll p \ll N_i$" (or "$p \ll n \ll N_i$") holds (together with Equation (7.16)).[163] All $V_{ij} - I$, $i,j = 1,...4$ relations have been evaluated in the steady state for the circular sample of Figure 7.3.[163] We quote here the $V_{32} - I$ relation: first that of interest when the electrode 1 is the reversible one,

$$V_{32} = -\frac{kT}{q} \ln\left(1 - \frac{A_e \ln 2}{1 - A_e \ln(2r/k_1)}\right), \quad A_e = \frac{qI}{\pi d k T \sigma_e(1)}, \tag{7.40a}$$

and then that of interest when probe 2 is the reversible electrode

$$V_{32} = -\frac{kT}{q} \ln(1 - B_e \ln 2), \quad B_e = \frac{qI}{\pi d k T \sigma_e(2)}, \tag{7.40b}$$

where $\sigma_e(1)$ and $\sigma_e(2)$ are the local conductivities in the MIEC in the vicinity of electrodes 1 and 2, respectively, k_1 is the diameter of the reversible electrode 1, and r is the radius of the MIEC.

For the one-dimensional configuration the models "$n \ll p \ll N_i$" (or "$p \ll n \ll N_i$") as well as the model "$n,p \ll N_i$" under polarization conditions ($j_i = 0$) have also been analyzed.[163] We quote here the $V_{32} - I$ relation

$$V_{32} = -\frac{kT}{q} \ln\left(\frac{kT\sigma_e(1) + x_3 q j_e}{kT\sigma_e(1) + x_2 q j_e}\right) = -\frac{kT}{q} \ln\left(1 + \frac{q j_e}{kT\sigma_e(2)}(x_3 - x_2)\right), \tag{7.41}$$

where x_3 and x_2 are defined in Figure 7.2.

The contact of the ion-blocking probes to the MIEC sample can be established either directly or via MIEC contacts.[164]

3. Short-Circuit Conditions

To suppress the electronic current, at least three of the electrodes must be electron blocking. Then the voltage of the MIEC proper vanishes, as under short-circuit conditions. It is advantageous to make the fourth electrode reversible for fixing a known composition in the MIEC near that electrode. The V_{32}–I relations for small signals, and for the configuration of Figure 7.3, are,[161] (where v_{32} is the voltage measured on the electron-blocking probes, each composed of a SE backed by a reversible contact)

$$V_{32} = -\frac{\ln 2}{\pi d \sigma_i} I, \quad I_{el} = 0, \tag{7.42}$$

The contact of the electron-blocking probes to the MIEC sample can be established either directly or via MIEC contacts.[164]

4. One Ion-Blocking Electrode and Three Reversible Electrodes

The interaction of the reversible probes with the MIEC enables one to manipulate the I–V relations. In particular, when a single blocking electrode is used as a current carrying one and the other three electrodes are reversible, the I–V relations in different parts of the MIEC are governed by different conductivities σ_{el} or σ_i. This situation, under steady state, has been analyzed[165] for the models "$p \ll n \ll N_i$" ("$n \ll p \ll N_i$") and "$p,n \ll N_i$", assuming in addition that $\sigma_{el} \ll \sigma_i$, i.e., that the MIEC is a SE exhibiting a relatively small electronic conductivity.

The linear configuration is shown in Figure 7.6. The reversible probes are inserted into the sample to assure the one-dimensional character of the configuration. The I–V relations (for the model "$n,p \ll N_i$") are

$$I = -\frac{S}{x_3 - x_2} \sigma_i V_{32}, \tag{7.43}$$

and

$$I = \frac{kT}{q} \frac{S}{L - x_3} \left[\sigma_e(3)\left(1 - e^{-\beta q V_{43}}\right) + \sigma_h(3)\left(e^{\beta q V_{43}} - 1\right) \right], \tag{7.44}$$

where $\sigma_e(3)$, $\sigma_h(3)$ are the electron and hole conductivities in the MIEC near probe 3.

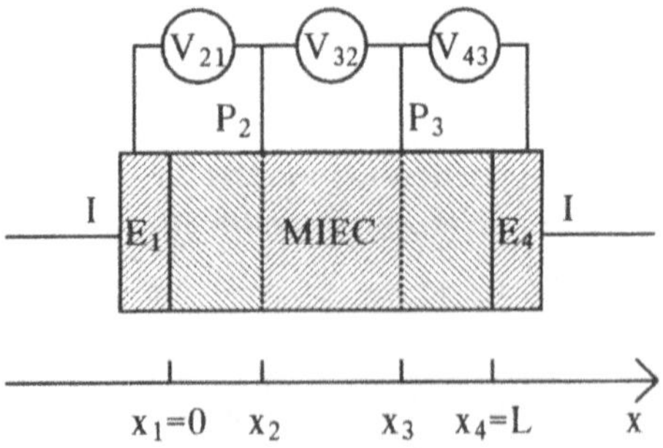

FIGURE 7.6. Schematics of a linear cell with three reversible electrodes (E_1, P_2, and P_3) and one blocking for ions (E_4). The reversible voltage probes (P_2, P_3) are implanted as grids in the MIEC to preserve the one-dimensional symmetry.

The *I–V* relations for the Van der Pauw configuration with three reversible electrodes (numbers 1, 2, and 3, respectively) and one ion-blocking electrode (number 4) for circular samples as shown in Figure 7.3 have been evaluated for the same model "$p \ll n \ll N_i$" (and "$n \ll p \ll N_i$") and $\sigma_e \ll \sigma_i$:

$$V_{32} = \frac{\ln 2}{\pi d \sigma_i} I \tag{7.45}$$

and

$$1 - e^{-\beta q V_{43}} = \frac{q}{kT} \frac{I}{\pi d \sigma_e(3)} \left[\frac{4 \ln(r/r_1) \ln(2r/r_1)}{3 \ln(r/r_1) + \ln 2} + \ln(r_1/r_4) \right], \tag{7.46}$$

where r_1 is the radius of electrodes 1, 2, and 3 (assumed to have equal radii) and r_4 is the radius of electrode 4.

H. I-V RELATIONS FOR MIXED VALENCE IONIC–ELECTRONIC CONDUCTORS

When the MIEC conducts ionic species having different values of $|z|$ (absolute value of the relative charge) the *I–V* relations become more complicated. Solutions have been obtained only for limiting cases, open-circuit conditions, and for suppressed material transport.[166–168] That the *I–V* relations are quite different from those for MIECs with mobile ionic defects having a single $|z|$ can be seen from the following consideration: let us assume that the mobile species are $M_i^{\cdot}$ and $M_i^{\cdot\cdot}$, and that the electrodes impose a chemical potential difference $\Delta\mu_M$ across the MIEC. The ionic current density $j(M_i^{\cdot})$ of the $M_i^{\cdot}$ species, vanishes when the voltage across the MIEC equals the Nernst voltage $v_{th,1} = -\Delta\mu_M/q$. The ionic current density $j(M_i^{\cdot\cdot})$ of the $M_i^{\cdot\cdot}$ species vanishes when the voltage on the MIEC equals the Nernst voltage $V_{th,2} = -\Delta\mu_M/2q = 0.5V_{th,1}$. When the electronic current can be neglected then, under open-circuit conditions, the total electrical current must vanish

$$j_i = j(M_i^{\cdot}) + j(M_i^{\cdot\cdot}) = 0 \,. \tag{7.47}$$

The open-circuit voltage cannot equal both $V_{th,1}$ and $V_{th,2}$. It turns out to be between $V_{th,1}$ and $V_{th,2}$, so that $j(M_i^{\cdot}) = -j(M_i^{\cdot\cdot}) \neq 0$. As a result, there is a nonzero material transport given by the flux:

$$J = j(M_i^{\cdot})/q + j(M_i^{\cdot\cdot})/2q \,. \tag{7.48}$$

When the electronic current density, j_e, cannot be neglected, then $j(M_i^{\cdot})$, $j(M_i^{\cdot\cdot})$, and j_e are all nonzero under open-circuit conditions.

Attempts to block the ionic current by blocking material transport fail in these MIECs. Let us consider, for example, graphite as the quasi-ion–blocking electrode. This electrode is capable of blocking the material flux, i.e., set $J = 0$. However, as appears from Equation (7.48), $J = 0$ does not imply that j_i (given by Equation [7.47]) vanishes. Thus the current in the external circuit is equal to the sum $j_i + j_e$ (with $j_i \neq 0$) despite the use of an inert graphite electrode.

I. CROSS TERMS, L_{ij}, BETWEEN ELECTRONIC AND IONIC CURRENTS

In metals the high flux of electrons may transfer momentum to mobile defects, forcing them to move in the same direction as the electrons.[169] This so-called electron wind is not present in MIECs, which usually exhibit lower electronic conductivities typical of semiconductors or insulators.

One can, however, consider quasi-coupling terms, L_{ij}, between ionic and electronic currents in MIECs which conduct ionic defects having different $|z|$ values. Let us consider a MIEC that conducts $M_i^{\cdot}$, $M_i^{\cdot\cdot}$, and e'. The coupling terms arise if one attempts to describe the three particle systems in terms of two only.[11] The coupling terms then depend on which of the particles is used to describe the system. Thus, if these are $M_i^{\cdot\cdot}$ and e', then $M_i^{\cdot}$ is presented as $M_i^{\cdot\cdot} + e'$. Then, eliminating $j(M_i^{\cdot})$ from the current equations causes the appearance of coupling terms between the remaining current densities $J(M_i^{\cdot\cdot})$ and j_e.[170]

Coupling constants also arise when a system containing one type of mobile ionic defects, say $M_i^{\cdot\cdot}$, is described in terms of the driving forces and currents of defects with a different $|z|$, say, $M_i^{\cdot}$.

The fact that coupling constants in MIECs can sometimes be measured[17,171–175] shows that there are mobile ionic defects with different values of $|z|$ in the MIECs or that the defects used in the theoretical analysis have a different $|z|$ value than the mobile ionic defects.

IV. METHODS FOR MEASURING THE PARTIAL IONIC AND ELECTRONIC CONDUCTIVITIES

A. HEBB–WAGNER POLARIZATION METHOD FOR DETERMINING σ_e AND σ_h

The H–W polarization method[9,155,156] is aimed at measuring the electron and hole conductivities of MIECs. It is based on blocking the ionic current through the MIEC using an ion-blocking electrode. Ion blocking is achieved when, e.g., an inert electrode such as graphite or Pt is used in the steady state under correct polarity (applied to drive the mobile ions away from the blocking electrode). For example, to measure σ_{el} in AgBr the cell configuration would be: (–)Ag/AgBr/Pt(+). The nonblocking electrode is preferably a reversible one so that the composition of the MIEC near this electrode is well defined. Using this method σ_{el} was determined among others in: AgBr,[45,49,176,177] AgI,[176] AgCl,[176] AgTe,[178] $RbAg_4I_5$,[179] tetragonal zirconia polycrystals stabilized with 3 mol% Y_2O_3,[79] CuBr,[52,55,56,58,165] CuI,[52,59] CuCl,[52] Cu_2S,[61] PbF_2,[72] and $PbBr_2$.[68,69]

Originally[9,156] the I–V relations were derived for the "$p, n \ll N_i$" defect model MIEC, in the linear configuration using two electrodes. The I–V relations were later derived also for the "$n = zN_i$" (or "$p = -zN_i$") defect model MIEC,[150,180] and for MIECs whose defect model changes, near the blocking electrode, from the "$p, n \ll N_i$" model to the "$n = zN_i$" (or "$p = -zN_i$") model under a high applied voltage.[180,181]

The derivations of σ_{el} from the measured I–V relations can be made independent of the defect model (with the limitation that the mobile ion defects must have a single absolute value of valency $|z|$). This is obtained at the expense of additional experimental work. One repeats the measurements at another applied voltage close the original value. Then σ_{el} at L near the blocking electrode can be obtained from the derivative,

$$\frac{\partial I}{\partial V} = -\frac{S}{L}\sigma_{el}(L). \tag{7.49}$$

If geometry is extended to a Van der Pauw one, then the I–V relations for a four-electrode configuration can be derived for quite general MIEC provided qV is much smaller than kT[161] and for the "$n,p \ll N_i$" defect model MIEC for arbitrary V.[163]

The H–W method has serious experimental limitations which may lead to misinterpretation of the measurements and to wrong conclusions:[182]

a. charge transfer may take place over the free surface at a rate which is not negligible;
b. when there are mobile ionic defects with different absolute valence values $|z|$, then the inert electrode is unable to block the ionic current in the MIEC (see Section III.H);
c. the resistance of the blocking electrode to ionic current must be higher than the resistance of the MIEC to electronic current, otherwise the MIEC becomes a blocking electrode for electrons, and one may be measuring the ionic conductivity of the inert metal blocking electrodes if material can be supplied from a contact or from the gas phase;[183]
d. decomposition of the sample is a severe problem with some MIECs such as the halides;[56–58,68,69]
e. the electrodes may not be reversible with respect to the electronic current so that the electrode overpotential is not negligible.

The use of the four-electrode H–W method[163] enables the detection of deviations from ideal behavior due to the problems mentioned before. For example, the four-probe method was used to prove that CuBr decomposes rapidly in a H–W experiment.[56–58]

B. THE ANALOG OF THE HEBB–WAGNER METHOD FOR DETERMINING σ_i

A method analogous to the H–W method is used to determine the ionic conductivity σ_i. It is based on blocking the electronic current through the MIEC in steady-state conductivity measurements. Such a blocking electrode is a SE backed by a reversible contact. For determining $\sigma(M_i^{\cdot})$ in a MIEC the cell configuration would be, e.g.,

$$(-)M/\mathrm{SE}(M_i^{\cdot})/\mathrm{MIEC}(M_i^{\cdot}, e', h^{\cdot})/M(+)$$

where the electrode M is either the pure metal or an alloy containing M.[8,183] Then

$$R_i = \frac{V}{I} \tag{7.50}$$

where V is the applied voltage and I the measured current through the cell. When σ_i is not uniform, Equation (7.50) can only yield an average value for σ_i. Using this method σ_i was determined, among others, in: Ag_2S,[8] $YBa_2Cu_3O_{7-x}$,[129,130] and $La_{0.8}Sr_{0.2}Co_{0.8}Fe_{0.2}O_{3-x}$.[184]

A local value of σ_i in the MIEC near the boundary with the SE, $\sigma_i(b)$, can be obtained by repeating the measurement at a voltage close to V. Then

$$\frac{\partial I}{\partial V} = \frac{S}{L}\sigma_i(b)\,. \tag{7.51}$$

The method suffers from problems similar to those of the H–W method:[182]

a. the rate of material transfer across the free interface should be negligible or be blocked;
b. $|z|$ must have a single value, otherwise the SEs used are unable to block the electronic current;
c. the resistance of the SE to electronic current must be larger than the resistance of the MIEC to ionic current,[183] otherwise the MIEC is blocking the ionic current and one measures the resistance of the SE to electronic current;
d. the decomposition rate of the MIEC must be negligible;
e. the electrode overpotential should be small or a four-electrode arrangement should be used.[8,183,184]

Instead of using one SE as the blocking electrode, two SEs are sometimes used on opposite sides of the MIEC.[130,184] This may introduce another severe problem. Material is being pumped electrochemically through the second (superfluous) SE and may accumulate at the SE/MIEC interface. When the applied voltage exceeds a few millivolts the chemical potential of the material accumulated may be so high that it precipitates. When it is a gas, large pressures may build up and the gas is lost. In both cases the measurements are then incorrect.[182]

When low-voltage ac measurements are used with the cell M/SE/MIEC/SE/M, the signal frequency f must be low to obtain (approximate) steady-state conditions. This means that $f^{-1} \gg L^2/\tilde{D}$ where L is the length of the MIEC and $\tilde{D}$ the chemical diffusion coefficient of the mobile species. For most MIECs with $\tilde{D} \ll 1$ cm²/sec and $L \sim 1$ cm f must be quite low ($f \ll 1$ Hz) and it is questionable if an ac signal is of any use at all for determining σ_{el} and σ_i in MIECs by selective blocking. However, one can use low-voltage ac measurements to determine the chemical diffusion coefficient $\tilde{D}$ (see Section V.G) from which partial information can be obtained concerning the partial conductivities (see Section IV.H).

C. SELECTIVE PROBES

For selective measurements of the driving force for electrons ($\nabla\tilde{\mu}_e$) one uses electron conductors which are blocking for ions. These can be inert metals or semiconductors. For selective measurements of the driving force for ions ($\nabla\tilde{\mu}_i$) one uses SEs which are blocking for electrons. A small leak of the other charge type can be tolerated as long as it is small compared to the currents in the cell.[183] This leads to the suggestion that even so-called reversible probes may sometimes be used as blocking electrodes if they are small enough.[185] Thus microprobes ($\phi = 10$ μm) of Ag on AgBr were used as metallic probes, arguing that the ionic current due to introduction of Ag into AgBr can be neglected.

D. MEASURING THE IONIC CONDUCTIVITY BY THE "SHORT-CIRCUITING" METHOD

It is possible to eliminate the electronic current in a MIEC by short circuiting the MIEC.[151] This sets the difference in $\tilde{\mu}_e$ across the MIEC to zero ($\Delta\tilde{\mu}_e = 0$). If $|z|$ is single, then also locally $\nabla\tilde{\mu}_e = 0$ and the driving force for electron and hole motion vanishes. The short circuiting is done through a low-impedance amperometer. One can then derive the resistance of the MIEC to ionic current R_i,

$$R_i = \frac{V_{th}}{I} \tag{7.52}$$

where I is the short-circuit current measured by the amperometer and V_{th} is the Nernst voltage determined by the compositions of the reversible electrodes. The value of I is then equal to the purely ionic current inside the MIEC.

Polarization at the electrode can introduce an error in the value of V_{th}.[184,186] This difficulty can be overcome by using a four-electrode arrangement.[151] For a MIEC with a low electronic resistance, R_{el}, the resistance R of the amperometer may exceed R_{el}. Then no short circuiting may be assumed. For these extreme cases an auxiliary SE should be used in series with the MIEC (between the two short-circuited electrodes) to increase the effective R_{el} of the cell. The use of a SE serves also to suppress the electronic current generated by uncontrolled temperature gradients.[151] When measuring σ_i in SEs, on the other hand, due to their high electronic resistance, they can be short circuited using high resistances, R, in the $M\Omega$ range[187] as long as $R_{el} \gg R$.

E. SIMULTANEOUS MEASUREMENT OF THE ELECTRONIC AND IONIC CONDUCTIVITIES

A four-electrode arrangement was proposed which allows the simultaneous measurement of σ_e or σ_h and σ_i in SEs,[165] making use of three reversible electrodes and one ion-blocking electrode.

A seven-electrode arrangement also enables the simultaneous measurement of σ_{el} and σ_i in MIECs. The arrangement includes one reversible electrode, one ion-blocking and one electron-blocking electrode, one pair of ion-blocking probes, and one pair of electron-blocking probes.[182]

F. MEASUREMENT OF THE IONIC CONDUCTIVITY IN SOLID ELECTROLYTES

For SEs one can measure the total conductivity σ_t, and use the results for σ_i as $\sigma_i \sim \sigma_t$. Accurate measurements of σ_t are obtained by using four reversible electrodes.[55]

G. THE TUBANDT OR HITTORF METHOD

A mixed ionic electronic current I is sent through the cell which includes the MIEC and two reversible electrodes.[8] The change in electrode mass is measured, which reflects the ionic charge transported, and this yields I_i. The electronic component of the current is then obtained from $I_{el} = I - I_i$. From I_i, I_{el}, the Nernst voltage, V_{th}, and the applied voltage V one obtains:[151]

$$R_i = \frac{V_{th} - V}{I_i} \tag{7.53}$$

and

$$R_e = -\frac{V}{I_{el}} \tag{7.54}$$

H. DETERMINING PARTIAL CONDUCTIVITY BY PERMEATION MEASUREMENTS

The ambipolar diffusion of ions and electronic charge carriers, under chemical potential gradients, yields a material transport through the MIEC. The rate is given by[12] (use Equation [7.9] with $j_t = 0$),

$$j_i = -\frac{1}{q}\frac{\sigma_{el}\sigma_i}{\sigma_{el} + \sigma_i}\nabla\mu_M . \tag{7.55}$$

This rate can be measured in various ways, e.g., by pumping through an auxiliary SE.[75,188] When the driving force $\nabla\mu_M$ is known the conductivity factor $\sigma_{el}\sigma_i/(\sigma_{el} + \sigma_i)$ can be determined. When $\sigma_{el} \ll \sigma_i$ or $\sigma_i \ll \sigma_{el}$ the smaller conductivity is obtained. This method is referred to as the permeation method. It is very useful for determining very small minority charge carrier conductivities, provided leaks of material can be eliminated.

This was used, for example, in determining the low electronic conductivity, σ_{el} in ZrO_2 – 8%mol Y_2O_3[75] and in determining the low ionic conductivity σ_i in the electronic conductor $YBa_2Cu_3O_{6+x}$.[188]

The method fails to distinguish between ambipolar motion and diffusion of neutral defects. Thus in the case of $YBa_2Cu_3O_{6+x}$[188] it is not certain that the measurement determines the ionic conductivity. It can very well be that one measures the rate of diffusion of neutral oxygen interstitials. It was shown that $O_i^{\times}$ is present in large concentrations in $YBa_2Cu_3O_{6+x}$, and diffusion of $O_i^{\times}$ cannot be excluded.[42,129,131]

The conductivity factor can be determined also by measuring the (ambipolar) chemical diffusion coefficient $\tilde{D}$. However, $\tilde{D}$ includes the thermodynamic factor W that must be determined separately, as

$$\tilde{D} = \frac{1}{q^2}\frac{\sigma_i \sigma_{el}}{\sigma_i + \sigma_{el}}\frac{1}{c_M}\frac{\partial \ln a_M}{\partial \ln c_M} \equiv \frac{1}{q^2}\frac{\sigma_i \sigma_{el}}{\sigma_i + \sigma_{el}}\frac{W}{c_M} \tag{7.56}$$

where a_M is the activity of the chemical component M which contributes mobile species and c_M is the concentration of M (regardless of its electrical charge).[8,16,167,189,190]

I. DETERMINING THE ELECTRONIC AND IONIC CONDUCTIVITIES FROM THE ACTIVITY DEPENDENCE OF THE TOTAL CONDUCTIVITY

It is possible to determine σ_e, σ_h, and σ_i by making use of their different dependence on the activity of the compounds that interact with the MIEC.[14] In particular, let us consider oxides and the dependence of σ_e, σ_h, and σ_i on the oxygen partial pressure $P(O_2)$. One measures the total conductivity as a function of $P(O_2)$ and T. Under favorable conditions, σ_e, σ_h, or σ_i dominate $\sigma_t = \sigma_e + \sigma_h + \sigma_i$ at different $P(O_2)$, T ranges. Then σ_t yields the corresponding partial conductivity. Furthermore, the details of the $\sigma_t - P(O_2)$, T dependence allow one to determine the nature of the charged point defects.

This method has been applied, for instance, to Y_2O_3-doped CeO_2,[80,81] SrO-doped CeO_2,[82] CeO_2-doped ThO_2,[84] $Gd_2Ti_{2-x}Zr_xO_7$, and $La_{1-x}Ca_x\ AlO_{3-\delta}$.[86]

J. DETERMINING THE AVERAGE IONIC TRANSFERENCE NUMBER BY EMF MEASUREMENTS

The open-circuit voltage measured across a MIEC subject to a chemical potential difference $\Delta\mu_M$ is $V_{oc} = \bar{t}_i V_{th}$. The average ionic transference number, $\bar{t}_i$, can be equal to the local t_i, as is the case for the "$n = zN_i$" ("$p = -zN_i$") defect model MIECs.[150] On the other hand, the average $\bar{t}_i$ can be quite different from the local one when the latter varies over a wide range, as may be the case in the "$p \ll n \ll N_i$" defect model MIECs where $\bar{t}_i = 1 + (kT/qV_{th})\ln(t_i(0)/t_i(L))$ (Equation [7.23]). One can determine $\bar{t}_i$ as long as it is not too small, i.e., $\bar{t}_i \geq 0.1$. Then also $\bar{t}_e = 1 - \bar{t}_i$ can be obtained. Alternatively, both $\bar{t}_i$ and $\bar{t}_e$ can be measured using four probes (two pairs of selective probes).[191]

$\bar{t}_i$ has been measured, for instance, in yttria-doped ceria,[191] gadolinium-doped ceria,[32] and PbO, and TiO_2-doped PbO.[73]

K. COPING WITH ELECTRODE OVERPOTENTIAL

Two electrode measurements used to determine σ_e, σ_h, or σ_i of a bulk MIEC may suffer from errors due to the electrode overpotential. Methods used for eliminating this adverse effect in electronic conductors or SEs cannot be simply applied to MIECs. The use of a four-probe arrangement requires a detailed analysis of the I–V relations which depend on the defect nature of the MIEC. The result is a nontrivial extension of the two point I–V relations, as the I–V relations may not be linear, but exponential. For the "$p, n \ll N_i$" defect model MIECs the nonlinear $I_{el} - V$ relations were derived for both the linear and Van der Pauw configurations.[163] The $I_i - V$ relations, being linear, can be obtained from geometrical considerations. For the "$n = zN_i$" (or "$p = -zN_i$") defect model MIECs the $I_{el} - V$ and $I_i - V$ relations for four-point measurements can readily be derived from the general I–V relations given in the literature.[150]

For predominantly ionic or electronic conductors, four reversible electrodes rather than selective ones can be used when measuring the dominant partial conductivity.[55]

ac measurements, usually used for either ionic or electronic conductors to separate the electrode contribution to the total impedance from the bulk contribution, may be questionable when used on MIECs as mentioned before (Section IV.B). In predominantly ionic or electronic conductors the ac technique can be used for determining the dominant partial conductivity.

The MIEC contribution to the overall resistance can be identified by varying the length[192] or weight[193] of the MIEC. This is true provided the electrode impedance can be assumed not to vary from one sample to the other.

V. MEASURING THE CHEMICAL DIFFUSION COEFFICIENTS IN MIXED CONDUCTORS

A. GENERAL

Two diffusion coefficients are of interest in MIECs: the component diffusion coefficient, D_k, and the chemical diffusion coefficient, $\tilde{D}$. The component diffusion coefficient reflects the random walk of a chemical component. It is therefore equal to the tracer diffusion coefficient, D_{Tr}, except for a correlation factor which is of the order of unity. It is also proportional to the component mobility as given by the Nernst–Einstein relations.[8] The chemical diffusion coefficient, $\tilde{D}$, reflects the transport of neutral mass under chemical potential gradients. In MIECs mass is carried by ions, and transport of neutral mass occurs via ambipolar motion of ions and electrons or holes so that the total electric current vanishes.[190,194]

$\tilde{D}$ can be determined from steady-state permeation measurements,[127] as mentioned in Section IV.H. However, $\tilde{D}$ is usually determined from the time dependence of a response to a step change in a parameter, e.g., the applied current. Alternatively, $\tilde{D}$ is determined from the response to an ac signal applied to the MIEC.

B. VOLTAGE RESPONSE TO A STEP CHANGE IN THE APPLIED CURRENT

Yokota[62] has analyzed the voltage response to a step change in a small dc current applied to the MIEC, assuming that deviation from local neutrality and displacement current can be neglected. The current can be applied either through ion- or electron-blocking electrodes. Two selective probes, either ion blocking or electron blocking, are used to follow the voltage change after the step change in the current. This is done with either type of current carrying electrodes (i.e., four combinations). The method has been applied to determine $\tilde{D}$(Ag) in αAg_2Te.[62] Miyatani[148] extended Yokota's analysis to larger applied current densities and determined $\tilde{D}$(Ag) in αAg_2Te and $\tilde{D}$(Cu) in βCu_2S. This method has been reviewed and extended to two-dimensional diffusion problems.[195,196] Millot and de Mierry[197] have applied the method to determine $\tilde{D}$(O) in CeO_{2-x}. The configuration seems different, but is equivalent to that of Yokata[62] with ion-blocking electrodes and electron-blocking probes. However, the measured signals are different and yield $\Delta\mu_M$ rather than $\Delta\tilde{\mu}_i$. Jin and Rosso[198] used a three-electrode configuration to determine $\tilde{D}$(Cu) in Cu_3VS_4. Using a nonlinear configuration with a point electrode,[199] $\tilde{D}$(Cu) was determined in $CuInS_2$ and $CuInSe_2$.[200] Weppner[140,201] has considered the limiting case of SEs ($\sigma_{el} \ll \sigma_i$) and determined $\tilde{D}$(O) of oxygen in yttria-stabilized zirconia using one reversible and one ion-blocking electrode.

C. CURRENT RESPONSE TO A STEP CHANGE IN THE APPLIED VOLTAGE

$\tilde{D}$ can be determined from the change of the current through the MIEC as a result of a step change in the applied voltage.[200,202] This method has been used, for instance, to determine $\tilde{D}$(O) in $YBa_2Cu_3O_{7-x}$.[125,203]

D. RESPONSE OF THE VOLTAGE ON PART OF THE MIXED CONDUCTOR TO A STEP CHANGE IN THE APPLIED VOLTAGE

A change in the applied voltage induces a change in the composition of the MIEC. This gradual change can be followed by measuring the voltage at a point in the MIEC with respect to a reversible electrode. This has been used to determine $\tilde{D}$(Ag) in *n*-type AgBr.[204]

E. EMITTER–COLLECTOR METHOD

Miyatani and Tabuchi[205] extended the Haynes–Shockley method for determining $\tilde{D}$ and D_k (or drift mobility) in a MIEC. A pulse of material is introduced into the MIEC through a SE. The response is measured at a remote site on the MIEC using another SE and a reference electrode. A voltage applied to the MIEC forces the excess material to drift. The diffusion broadens the measured signal. Both the ambipolar drift and ambipolar diffusion coefficients can be determined. This method was applied to αAg_2S.

F. RESPONSE TO A CHANGE IN COMPOSITION THROUGH INTERACTION WITH A GAS

1. General

The composition of a MIEC can be conveniently altered suddenly by inducing a step change in the composition of the atmosphere which interacts with the MIEC. $\tilde{D}$ can then be determined from the time dependence of a physical parameter that changes with the composition, provided the reaction rate is not limited by the reaction process at the free surface.

2. Change of Weight or Length

Oxidation of metals results in an increase of weight and a change in dimension. It is straightforward to follow the weight change as a response to a step change in $P(O_2)$ or other gases. This was used to determine the effect of doping on $\tilde{D}$ of the excess component in NiO + Li, NiP + Cr, and CoO + Li,[206] and $\tilde{D}$(Fe) in $Fe_{1.12}Te$ in the presence of a gas containing Te.[207]

3. Change of Resistance

On reduction or oxidation the electrical resistance changes. The time dependence of the resistance change of a MIEC can be used to determine $\tilde{D}$. This method was used for determining $\tilde{D}$ in NiO + Li, NiO + Cr, CoO–Li,[206] $\tilde{D}$(Cu) in Cu_2O,[208,209] and $\tilde{D}$(O) in $YBa_2Cu_3O_{6+x}$.[124]

4. Change in Optical Properties

Reduced stabilized zirconia is black. As it oxidizes it become transparent. Following the propagation of the front between the dark and transparent part in a single crystal $\tilde{D}$(O) has been determined.[210] Instead of following the front, one can follow the integrated intensity changes due to diffusion of defects. This has been used to determine $\tilde{D}$(O) in $SrTiO_3$ doped with Fe impurities by following the absorption lines of the Fe^{4+} ions.[211] The concentration of the optically active Fe^{4+} ions increases at the expense of the Fe^{3+} ions as the MIEC oxidizes.

G. DETERMINING $\tilde{D}$ FROM LOW-AMPLITUDE ac IMPEDANCE MEASUREMENTS

Yokota and Miyatani[215] have analyzed the ac impedance of a MIEC for zero bias, low ac signals at low frequencies so that the displacement current and deviation from local electroneutrality can be neglected. It was further assumed that $|z|$ has a single value and that selective electrodes (for ions or electrons) are used. The analysis yields the corresponding voltage on selective probes applied on the MIEC, i.e., for the four possible combinations of selective electrodes and selective probes. The authors use this method to determine $\tilde{D}$(Cu) in Cu_2S and $\tilde{D}$(Ag) in Ag_2Te.

Macdonald and Franceschetti[216] made an analysis of small-signal ac impedance measurement which can be applied also to MIECs. They allow for high frequencies and therefore for the existence of a nonnegligible displacement current and deviation from local electroneutrality. They also consider electrodes with partial blocking and internal reaction between defects. The model is limited to dilute concentrations and zero dc bias.

H. GALVANOSTATIC AND POTENTIOSTATIC INTERMITTENT TITRATION TECHNIQUE

The composition of a MIEC can be changed by electrochemically pumping material through a SE into the MIEC, which serves as electrode. One can follow the time dependence of the change in the chemical potential, μ, of a chemical component in the MIEC after a step change in the pumping dc current (galvanostatic intermittent titration technique, GITT). $\tilde{D}$ is then determined from the time dependence of $\mu(t)$.[16,189] Alternatively, one can follow the pumping current through the galvanic cell after a step change in the applied voltage (potentiostatic intermittent titration technique).[16,217] An improved GITT method was later suggested to overcome the adverse effect of the overpotential at the SE/MIEC interface on the measurements. In the modified method a second SE is used to monitor the chemical potential $\mu'(t)$ at the remote end of the MIEC where no current flows. One then determines $\tilde{D}$ from the time dependence of $\mu'(t)$.[218]

I. MISCELLANEOUS METHODS FOR DETERMINING $\tilde{D}$

1. NMR and ESR Imaging

Nuclear magnetic resonance (NMR) is used for determining the component diffusion coefficient D_k, from the relaxation times T_1 and T_2, where T_1 is the relaxation time of magnetic polarization induced parallel to a magnetic field and T_2 is the relaxation time of the polarization induced perpendicular to the direction of the magnetic field.[212] On the other hand, NMR and ESR imaging (tomography) can be used to follow the chemical diffusion of a chemical component and therefore to determine $\tilde{D}$.[212,213]

2. $\tilde{D}$ by Creep Measurements

$\tilde{D}$ in CoO and Cu_2O was determined by creep measurements.[214] As the applied load is varied the defect concentration changes. This relaxation process can be followed by following the change in sample length under load. The method seems to be less accurate at this stage than those mentioned before.

3. Use of Work Function Measurements

It has been proposed that $\tilde{D}$ in the near to surface layers in MIEC oxides can be determined by following the change in the work function after a step change in $P(O_2)$ has occurred.[219]

VI. DEFECT DISTRIBUTION IN MIXED CONDUCTORS UNDER ELECTRICAL AND CHEMICAL POTENTIAL GRADIENTS

A. GENERAL

The defect distributions in MIECs, in linear galvanic cells, under steady-state conditions[23,138,150] have been evaluated for the main defect models ("$p,n \ll N_i$", $p \ll n = zN_i$" and "$n \gg N_i \gg p$"). The possibility of solving for the defect distribution enabled also the evaluation of the I–V relations for arbitrary voltage, V (discussed in Section III). The defect distributions as well as the I–V relations have been evaluated under local neutrality condition. Criteria for local neutrality are discussed at the end of this section.

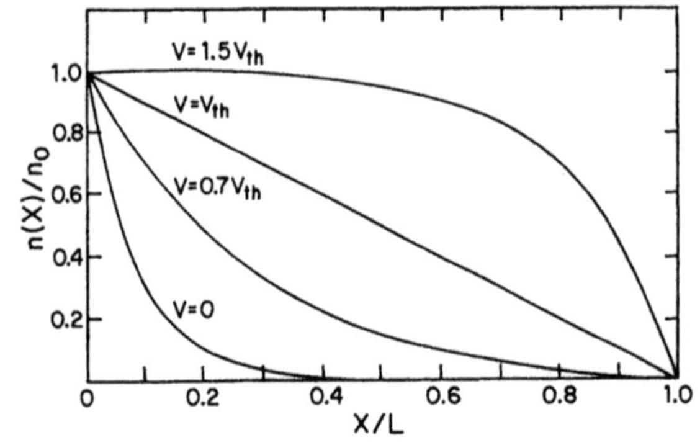

FIGURE 7.7. Electron distribution *n(x)* for the defect model "$p \ll n \ll N_i$". [V_{th} = 1 V, T = 1000 K, $n_L/n_0 = 10^{-5}$.] (From Riess, I., *J. Phys. Chem. Solids*, 47, 1986, 129. With permission.)

B. ELECTRON DISTRIBUTION WHEN HIGH DISORDER PREVAILS

In MIECs with high ionic disorder corresponding to the defect model "$p \ll n \ll N_i$", the ion concentration is approximately uniform, i.e., deviations δN_i from the average value N_i are negligible ($\delta N_i \ll N_i$). On the other hand, the concentration of electrons (and holes) is not uniform and depends on the position x, the voltage V, and the chemical potentials of the electrodes through n_0 and $-qV_{th}$:[23,150]

$$n(x)=n_0\left[1-\frac{1-e^{-\beta qV_{th}}}{1-e^{-\beta q(V_{th}-V)}}\left(1-e^{-\beta q(V_{th}-V)x/L}\right)\right], \tag{7.57}$$

where $n_0 = n(x = 0)$. The value of $n(x)$ is shown in Figure 7.7. Three limiting conditions are of prime interest: $j_i = 0$, $j_e = 0$, and $j_t = j_e + j_i = 0$.

For the polarization condition:

$$n(x)=n_0\left[1-\left(1-e^{-\beta qV_{th}}\right)\frac{x}{L}\right],\quad j_i=0,\quad V=V_{th}\,. \tag{7.58}$$

For the short-circuiting condition:

$$n(x)=n_0e^{-\beta aV_{th}x/L},\quad j_e=0\,,\quad V=0\,. \tag{7.59}$$

For the open-circuit condition:[150,158]

$$n(x)=n_0\left[\frac{\sigma_e(0)+\sigma_i}{\sigma_e(0)}\left(\frac{\sigma_e(0)e^{-\beta qV_{th}}+\sigma_i}{\sigma_e(0)+\sigma_i}\right)^{x/L}-\frac{\sigma_i}{\sigma_e(0)}\right], \tag{7.60}$$

$$j_e+j_i=0\,,\quad V=V_{oc}\,,$$

where $\sigma_e(0)$ is the local electron conductivity in the MIEC near the electrode at $x = 0$.

C. ELECTRON AND HOLE DISTRIBUTIONS WHEN HIGH DISORDER PREVAILS

Implicit equations have been obtained for *n(x), p(x)* in MIECs corresponding to the defect model "$p,n \ll N_i$".[138] Figure 7.8 exhibits *n(x)* and *p(x)* for different values of the applied voltage. Explicit expressions are obtained for the limits $j_i = 0$ and $j_{el} = 0$.

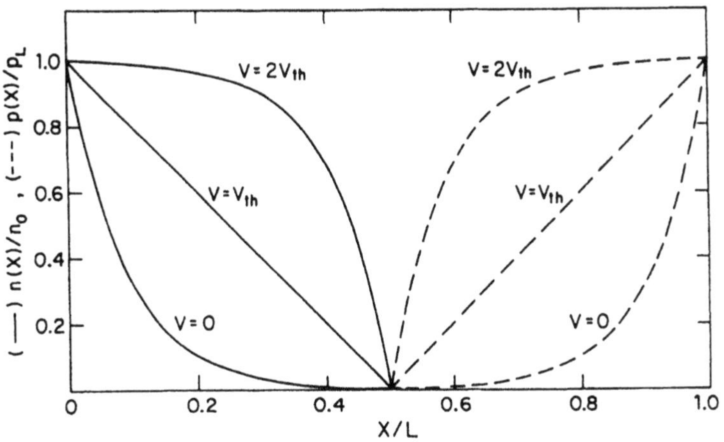

FIGURE 7.8. Electron and hole distributions, $n(x)$ and $p(x)$, respectively, for the defect model "$p, n \ll N_i$". [V_{th} = 1 V, T = 1000 K, $n_L/n_0 = 10^{-5}$, $p_0 = n_L$, $p_L = n_0$, $\nu_e = \nu_h$.] (From Riess, I., *Phys. Rev.*, B35, 1987, 5740. With permission.)

For the polarization condition:[150]

$$n(x) = \frac{D + \left(D^2 + 4v_e v_h n_i^2\right)^{\frac{1}{2}}}{2v_e}, \quad j_i = 0, \quad V = V_{th} \tag{7.61}$$

where

$$D = \left(v_e n_L - v_h p_L\right)\frac{x}{L} + \left(v_e n_0 - v_h p_0\right)\left(1 - \frac{x}{L}\right). \tag{7.62}$$

The value of $p(x)$ is also known, since $p(x) = n_i^2/n(x)$, where n_i is the intrinsic concentration of electrons (and holes).

For the short-circuiting condition:

$$n(x) = n_0 e^{-\beta q V_{th} x/L}, \quad j_{el} = 0, \quad V = 0 \tag{7.63}$$

D. ELECTRON DISTRIBUTION WHEN THE CONCENTRATIONS OF ELECTRONS AND MOBILE IONS ARE COMPARABLE

For the defect model "$p \ll n = zN_i$" the concentrations of the mobile ions and of the electrons vary linearly with x:[150]

$$N_i(x) = n(x) = n_0\left[1 - \left(1 - e^{-\beta q V_{th}}\right)\frac{x}{L}\right]. \tag{7.64}$$

The defect distributions are independent of V. When electron-blocking electrodes are used V_{th} is fixed by V: $V_{th} = V$.

E. ION DISTRIBUTION WHEN THE ELECTRON OR HOLE CONCENTRATION IS HIGH

For the defect model "$n \gg N_i \gg p$" the concentration of electrons is large and hardly affected by the electrode compositions and applied voltage. Therefore, n is uniform and $N_i(x)$ varies along the MIEC:

$$N_i(x) = N_i(0)\left[e^{-\beta qVx/L} + \frac{e^{-\beta qV_{th}} - e^{-\beta qV}}{1 - e^{-\beta qV}}\left(1 - e^{-\beta qVx/L}\right)\right], \tag{7.65}$$

where $N_i(0) = N_i(x = 0)$.

F. CRITERIA FOR NEGLECTING THE SPACE CHARGE

The I–V relations, defect distributions, and diffusion experiments are usually evaluated under the assumption of local neutrality. However, one should, in principle, use the Poisson equation $\nabla^2\varphi = -\rho/\varepsilon$ (where φ is the electrical potential, ρ the space charge density, and ε the dielectric constant). The question is, when is the assumption $\rho \to 0$ valid? This question has been answered for MIECs in galvanic cells operated under steady-state conditions.[220] Local neutrality can be assumed as long as gradients in the concentration of the charged majority defects are small. These gradients may be of different origin. They can be caused by contact between different materials, nonuniform doping, applied voltage, applied chemical potential difference, $\Delta\mu$, and a combination of these factors. As a rule, for a single-phase MIEC with uniform doping under a given nonzero $\Delta\mu$, the gradients in defect concentration increase with voltage for high applied voltages. At large enough applied voltages local neutrality may not be assumed.

VII. HETEROGENEOUS MIXED CONDUCTING SYSTEMS

A. GALVANIC CELLS WITH DIFFERENT MIECS CONNECTED IN SERIES

In a MIEC placed under an electrical and/or chemical potential gradient, the distribution of at least one kind of defect is not uniform. Key examples have been discussed in Section VI. In all cases the changes in the defect distribution and in the chemical potential distribution are monotonic. This may not be the case in galvanic cells containing different MIECs placed in series.

Let us consider two MIECs conducting $M_i^{\cdot}$ ionic defects and electrons, having different ionic transference numbers t_{i1} and t_{i2}. The galvanic cell is (+)E_1|MIEC$_1$|MIEC$_2$|E_2(–). We assume for the sake of simplicity that the electrodes E_1 and E_2 have the same composition, with a chemical potential μ_M *(E)*. It turns out that the value of μ_M *(i.f.)* at the interface between the two MIECs can be either higher or lower than μ_M*(E)*.[221-223] This means that the gradients $\nabla\mu_M$ in the two MIECs have different signs, i.e., there is no monotonic change in μ_M along the galvanic cell. When μ_M *(i.f.)* is large enough precipitation of M at the interface occurs and the cell turns into E_1|MIEC$_1$|M|MIEC$_2$|E_2. When μ_M represents gaseous species, highly pressurized gas bubbles can be formed at the interface (or internally, where t_i is discontinuous).[223]

The changes of μ_M *(i.f.)* under steady state can be obtained from the general I–V relations of Equations (7.12) and (7.13) applied to each MIEC separately. Under steady state and when M does not precipitate at the interface, I_i and I_{el} are uniform, i.e., the same in both MIECs. The value of μ_M *(i.f.)* is then given by:

$$V_{th,2} = V_2 - V_a\frac{R_{i1}}{R_{i1} + R_{i2}} \tag{7.66}$$

where $V_{th,2} = -(\mu_M\,(E) - \mu_M\,(i.f.))/q$, V_a is the applied voltage, V_2 the voltage across MIEC$_2$, and R_{i1} and R_{i2} the resistance of the corresponding MIEC to ionic current. For instance, for MIECs with $R_{i1} \sim R_{i2}$ and $t_{e1} \ll t_{e2}$ (i.e., $V_2 \ll V_a$):

$$V_{th,2} \approx -\frac{1}{2}V_a\,. \tag{7.67}$$

B. MODIFICATION OF THE CONCENTRATIONS OF MOBILE DEFECTS

A different property of heterogeneous systems is the space charge that may be formed near the interface. In the space charge region the concentration of mobile ionic and electronic defects may be increased (or decreased).[157,224,225] This has an effect, in particular, on the conductivity parallel to the interface. For *a priori* poor conductors the overall conductivity might be significantly enhanced by adding the second phase. For that purpose the second phase can also be an insulator. For some examples and a further discussion, see Chapter 6, Section V of this handbook.

VIII. MAGNETIC MEASUREMENTS ON MIXED CONDUCTORS

The application of a magnetic field on a MIEC placed in a galvanic cell affects the current density equation which now contains also the effect of the Lorentz force.[196] The use of a magnetic field should enable the determination of the Hall coefficient and magneto resistance. The interpretation of the results is straightforward and can be done in terms of the two-band semiconductor model[226] when the composition of the MIEC is uniform and there is one dominant mobile ionic species and one dominant electronic species. Uniformity holds when the two electrodes have the same composition, are reversible, and a steady state has been reached.

Hall coefficient measurements on MIECs usually reveal the properties of the more mobile electronic (electron/hole) defects. These measurements can be done for different compositions. The composition can be conveniently altered by coulometric titration or via interaction with the gas phase. Hall coefficient measurements yielding the electron or hole concentration have been reported for MnO,[227] Ag_2S,[36,228] Ag_2Se,[229] and Ag_2Te.[230,231]

Hall measurements on the ionic charge carrier can be done only in SEs, i.e., when $\sigma_{el} \ll \sigma_i$ so that the signal is not masked by the electronic charge carriers. These measurements are difficult because of the relatively low mobility of the ionic defects. Funke and Hackenberg[232] have measured the ion Hall coefficient in α-AgI.

IX. THERMOELECTRIC POWER OF CELLS WITH MIXED CONDUCTORS

The galvanic cells considered so far were assumed to have a uniform temperature. Temperature gradients exert driving forces on both the electronic and the ionic defects. The thermoelectric power, TEP, is the open-circuit voltage measured on the MIEC under a temperature gradient. Due to the possible motion of ions as well as electrons and holes and because of possible interaction with the electrode material or the gas phase, care must be taken to control or identify the conditions of the experiments. A detailed analysis was given by Wagner.[10] The following experimental conditions are considered by Wagner (with $I_t = 0$ in all cases):

a. A uniform composition. The measurement is done rapidly, after a temperature gradient is applied to a uniform MIEC, to avoid large changes in the composition.
b. Steady-state gradient in composition (Soret effect). At least one electrode is ion blocking so that both $I_{el} = 0$ and $I_i = 0$.
c. Uniform activity of the metal. This is achieved, e.g., by using the pure metal as the reversible electrode material.
d. Uniform partial pressure of the cation or anion material in the surrounding atmosphere. The activity of the metal in the MIEC is then, in general, not uniform.

The interpretation of the TEP in terms of Q_k^*, the heat transfer of species *k,* and $\bar{S}_k$, the partial molar entropy of species *k,* is different under the different experimental conditions.

References to TEP measurements on MIEC can be found in Reference 10. More recent examples are $UO_{2.03}$[233] and yttria-stabilized zirconia.[234,235]

X. APPLICATIONS OF MIXED CONDUCTORS

A. GENERAL

The applications of SEs have been repeatedly reviewed. We limit our discussion here to application of MIECs for which the ionic transference number, t_i, is not close to unity as in SEs. Most applications are then specific to MIECs and cannot be found with SEs except for two, mentioned at the end, which refer to unconventional uses of MIECs in applications believed, so far, to require SEs.

B. MAIN APPLICATIONS IN THE FIELD

1. Electrodes in Fuel Cells

The electrodes in fuel cells, fuel $|SE(O^{=})|$air, are usually porous ones made of inert metals or semiconducting oxides.[236] The electrode reaction is limited to the so-called "triple-phase boundary", i.e., the narrow area along the edge of the pores in the electrodes where the three phases, electrode, SE, and gas, meet. Using MIEC as electrodes was suggested, thus turning the whole nominal area of the electrode into an active electrode area where the electrode reaction can take place.[237-239] This might reduce the electrode impedance by orders of magnitude if other factors (such as the bulk resistance of the MIEC layer or space charge at the SE/MIEC interface) do not severely increase the impedance. (See Chapters 8 and 12 of this handbook for a more detailed discussion.)

2. Insertion Electrodes

A different use of MIECs is as electrode material in batteries, where MIECs are used to react with and store the species mobile in the SE.[40,91] These MIECs are usually layered compounds, such as graphite and MoS_2, which can accommodate large concentrations of foreign atoms with little change in the chemical potential of these atoms. (See Chapter 11 of this handbook for a more detailed discussion.)

3. "Smart Windows"

Some MIECs change color as their composition changes. Using the MIEC Li_xWO_3 as an electrode in an electrochemical cell of the form $Li|SE(Li^+)|Li_xWO_3$, the Li_xWO_3 oxide becomes dark blue as the concentration *x* increases, and recovers its transparency when Li is removed from the oxide. This cell can therefore serve to shade a room from the sunlight by means of applying an electric signal to a window covered with layers forming such electrochemical cells.[95]

An alternative method suggested is to apply a dc voltage to such MIECs via blocking electrodes, thus forcing the defects to redistribute.[137] Under favorable conditions the defect concentration on one side of the MIEC is sufficient to yield both the color and the intensity as needed. A necessary condition for the latter method to work is that the light is affected by conduction electrons and not by localized absorption centers. The high internal electric field that exists in this case in the MIEC enhances the migration and allows for fast switching of the color.[137] (See Chapter 16 of this handbook for a more detailed discussion.)

4. Selective Membranes

We consider now an application based on a MIEC where neither SE nor electrodes are used. Let a MIEC that conducts $O^{=}$ ions and say electrons separate two compartments, one open to the ambient air and the other connected to a mechanical pump. When the pump is operating, the pressure in the corresponding compartment drops and oxygen permeates selectively through the $O^{=}$ conducting MIEC from the high $P(O_2)$ side to the low $P(O_2)$ side.[12]

Very low values of $P(O_2)$ can be achieved by reaction with a fuel. Then it is possible to remove traces of oxygen, e.g., from molten metals placed in the other compartment. (See Chapter 14 of this handbook for a more detailed discussion.)

5. Sensors Based on the EMF Method

MIECs can be used in concentration cells to measure differences in the chemical potential of the corresponding components. From Equation (7.15) the open circuit is $V_{oc} = \bar{t}_i V_{th}$. By calibrating the average ionic transference number, $\bar{t}_i$, one can use a sensor based on a MIEC. However, this, usually, is possible only over a relatively narrow range of chemical potential values, where $\bar{t}_i \geq 0.1$. (See Chapter 10 of this handbook for a more detailed discussion.)

6. Sensors Based on Changes in Stoichiometry

The change in stoichiometry introduces a change in the charge carrier concentration and therefore in the electrical conductivity. Therefore, for example, for MIEC oxides one can determine $P(O_2)_x$ in the gas phase by measuring the conductivity σ once a calibration between σ and $P(O_2)_x$ has been done. Thus it was recently suggested to use $YBa_2Cu_3O_{6+x}$ in the tetragonal phase as an oxygen sensor at $T \geq 300°C$.[240] This material exhibits a high sensitivity of the electrical conductivity to variations in the oxygen content.

This method of sensing, e.g., oxygen, is different from another method used to sense different gases by modifying the concentration of oxygen only near the surface of oxides such as tin oxide and zinc oxide. The first method is an equilibrium one and may require heating the sensor to high temperatures to achieve equilibrium rapidly. The other method is a nonequilibrium one and relies on large changes in the surface conductivity of the oxide.[241,242] (See Chapter 10 of this handbook for a more detailed discussion.)

7. Catalysis

The change of stoichiometry may affect the catalytic properties of a MIEC. For instance, in the galvanic cell $Ag|AgI|Ag_2S$, the activity of silver, a_{Ag} and of sulfur, a_S, and the Fermi level of the electrons in the MIEC Ag_2S are governed by the applied voltage. Thus the voltage may affect the catalytic activity of the MIEC.[243-245] (See Chapter 13 of this handbook for a more detailed discussion.)

C. OTHER APPLICATIONS

1. Solid Lubricants

Solid lubricants are layered materials such as graphite and MoS_2. Intercalation of atomic or molecular species can substantially improve the lubricating properties of these solids. The reason is that the intercalated species may widen the van der Waals gap.[246]

2. Variable Resistance

A change of stoichiometry of a MIEC results in a change of electrical resistance. For example, the galvanic cell $Ag|AgI|Ag_2S$ allows one to modify the composition and resistance of the MIEC Ag_2S.

3. The Photographic Process

In the photographic process silver ions and electrons migrate to traps to recombine there and form small nuclei for precipitates of metallic silver.[6,44,247]

4. Gettering

The fast ambipolar motion possible in MIECs allows one to use them for buffering the gas phase. Thus reduced ZrO_2 readily reacts with oxygen and can serve as a getter.

5. Transmission of the Electrochemical Potentials of Electrons and Ions

MIECs can be used to transmit the value of the electrochemical potential of electrons and ions, simultaneously.[164] This holds provided both the electronic and ionic components of the current through the MIEC vanish. The MIEC then acts analogous to a metal wire that transmits $\tilde{\mu}_e$ when the electric current in the metal vanishes.

6. Device Failure Due to Ion Migration

It has been shown that the properties of solid state devices, that operate under dc bias for a length of time, may change due to migration of ionic species, even at room temperature. This eventually results in device failure. Such a migration was observed in ZnO varistors,[248] in Cu_2S/CdS solar cells,[63] and in TiO_2 electrodes.[247]

D. NONCONVENTIONAL USES OF MIXED CONDUCTORS

1. The Use of Mixed Conductors Instead of Solid Electrolytes in Fuel Cells

The design of fuel cells is commonly based on the use of a SE as the ion-conducting membrane. It was at first believed that MIECs were excluded due to the internal electronic leak. However, a detailed analysis showed that MIECs can be used as long as $\bar{t}_i > 0.5$.[149,250] Since $\bar{t}_i$ is an average ionic transference number, the local value of t_i in part of the MIEC can be much lower (see Equation [7.23]). This widens the list of materials that can be used as membranes in fuel cells, and in particular allows one to consider doped CeO_2 as the electrolyte in high-temperature fuel cells.[149,154,250,251]

2. The Use of Mixed Conductors Instead of Solid Electrolytes in EMF Sensors

EMF sensors are constructed using SEs. Then the open-circuit voltage measured is $V_{oc} = V_{th}$. We have mentioned before the use of MIECs instead of SEs where $V_{oc} = \bar{t}_i V_{th}$ provided $\bar{t}_i$ is quite close to unity. When this is not the case the cell can be constructed differently, containing two MIECs in series and a power supply that applies a voltage V to one of them.[252] The power supply is used to apply the voltage needed to suppress the ionic current through the sensor, while the second MIEC is used to generate the control signal which detects when this condition is achieved. When the ionic current vanishes the voltage of the power supply equals the measured V_{th}. The design of this type of sensor is more complex than that of a normal EMF sensor. However, it allows extension of the measurements to conditions under which the known SEs turn into MIECs and to build sensors for detecting materials for which only MIECs are known, but no SEs.

3. Electronic Device Fabrication

It was shown in Section IV that the concentration profiles of electrons, holes, and ionic defects can be altered by an applied voltage for given material properties, temperature, and electrode compositions. One can then control the shape of *pn* junctions at elevated temperatures, where both the ions and electronic charge carries are mobile.[138] Such profiles can be quenched.[137,138] Quenching is of interest if the ionic conductivity vanishes while that of the electrons and holes stays finite.

Instead of heating and quenching, one can apply a large transient electric field at room temperature on MIECs (e.g., on p-$CuInSe_2$ and on doped $Hg_{0.7}Cd_{0.3}Te$) to obtain *pn* or *pnp* junctions.[144,145] However, in this method one loses the freedom to choose the shape of the *pn* junction.

A combination of a transient applied field and quenching is used in γ radiation detectors[31] where controlled *pn* junctions are formed in Li-doped Si.[135]

XI. CONCLUDING REMARKS

MIECs are not of a single type, but are characterized by diverse relations between the concentrations of the mobile ionic defects, the electrons, the holes, and the immobile charged defects having fixed or variable charge. The I-V relations of solid electrochemical cells based on MIECs are qualitatively different for MIECs in which the relation between the defect concentrations is different. Similarly, the defect distributions are also qualitatively different for different defect concentration relations. So far, the I-V relations and defect distributions were evaluated only for that part of the defect models which are more frequently encountered at the present time.

The applications of MIECs began to emerge in recent years. In the past the extremes (the solid electrolytes on the one hand and the semiconductors on the other) found a wide range of applications. It has been demonstrated that MIECs are important for use in a wide range of applications. It is expected that more applications for MIECs will be found as we get to know them better and as we discover and prepare new MIEC materials.

The MIECs are mainly characterized by the concentration of the different defects, their relations, and by the conductivities of the mobile species. To determine defect concentrations one can rely in part on classical methods as used in other materials. However, new and special methods have been developed to obtain the partial conductivities and to obtain the defect distributions from macroscopic I-V relations measurements.

ACKNOWLEDGMENT

The author thanks the United States–Israel Binational Science Foundation (BSF), the Basic Research Foundation, administered by the Israeli Academy of Science and Humanities, and The Fund for Promotion of Joint Research, Niedersachsen–Israel for supporting the research programs that have contributed in great measure to the development of the material in this chapter. This chapter has profited considerably from discussions with H. Schmalzried, D.S. Tannhauser, and H.L. Tuller.

REFERENCES

1. *Physics of Electrolytes*, Vols. 1&2; Hladik, J., Ed., Academic Press, New York, 1972.
2. Kröger, F.A., *The Chemistry of Imperfect Crystals*, Vol. II, North-Holland, Amsterdam, 1974. (See also Chapter 3 of this handbook.)
3. Hauffe, K., *Reaktionen in und an Festen Stoffen*, Springer-Verlag, Berlin, 1966.
4. Kofstad, P., *Nonstoichiometry, Diffusion and Electrical Conductivity in Binary Metal Oxides*, Robert E. Krieger, Malabar, FL, 1983.
5. Jarzebski, Z.M., *Oxide Semiconductors*, Pergamon Press, Oxford, 1973.
6. Schmalzried, H., *Solid State Reactions*, Verlag Chemie, Weinheim, 1981.
7. Martin, M., *Mater. Sci. Eng. Rep.*, 1990, *7(1&2)*, 1.
8. Rickert, H., *Electrochemistry of Solids*, Springer-Verlag, Berlin, 1982.
9. Wagner, C., *Proc. Int. Comm. Electrochem. Thermodyn. Kinetics (CITCE)*, 1957, *7*, 361.
10. Wagner, C., *Progr. Solid State Chem.*, 1972, *7*, 1.
11. Wagner, C., *Progr. Solid State Chem.*, 1975, *10*, 3.
12. Heyne, L., Electrochemistry of mixed ionic-electronic conductors, in *Solid Electrolytes*, Geller, S., Ed., Springer-Verlag, Berlin, 1977, 169–221.
13. Gurevich, Yu.Ya. and Ivanov-Schits, A.K., Semiconductor properties of superionic materials, in *III-V Compound Semiconductors and Semiconductor Properties: Semiconductors and Semimetals*, Willardson, R.K., Ed., Academic Press, New York, 1988, 229–372.

14. Tuller, H.L., Mixed conduction, in *Nonstoichiometric Oxides,* Toft-Sørensen, O., Ed., Academic Press, New York, 1981, 271–335.
15. Riess, I. and Tannhauser, D.S., Mixed ionic electronic conductors, in *High Conductivity Solid Ionic Conductors,* Takahashi, T., Ed., World Scientific, Singapore, 1989, 478–512.
16. Weppner, W. and Huggins, R.A., *Annu. Rev. Mater. Sci.*, 1978, *8,* 269.
17. Dudley, G.J. and Steele, B.C.H., *J. Solid State Chem.*, 1980, *31,* 233.
18. *Ionic Solids at High Temperatures,* Stoneham, A.M., Ed., World Scientific, Singapore, 1989.
19. Riess, I., Crystalline anionic fast ion conduction, in *Science and Technology of Fast Ion Conductors,* Tuller, H.L. and Balkanski, M., Eds., Plenum Press, New York, 1989, 23–50.
20. Manning, J.R., *Diffusion Kinetics for Atoms in Crystals,* D van Nostrand, Princeton, NJ, 1968, 2–9.
21. Vinokurov, I.V., Zonn, Z.N., and Ioffe, V.A., *Sov. Phys. Solid State,* 1968, *9,* 2659.
22. Ban, Y. and Nowick, A.S., *National Bureau of Standards Special Publication 364, Solid State Chemistry, Proceedings of 5th Materials Research Symposium,* July 1972, 353–365.
23. Tuller, H.L. and Nowick, A.S., *J. Electrochem. Soc.*, 1979, *126,* 209.
24. Tuller, H.L. and Nowick, A.S., *J. Phys. Chem. Solids,* 1977, *38,* 859.
25. Riess, I., Janczicowski, H., and Nölting, J., *J. Appl. Phys.*, 1987, *61,* 4931.
26. Naik, I.K. and Tien, T.Y., *J. Phys. Chem. Solids,* 1978, *39,* 311.
27. Lol, H.B. and Pratap, V., *Indian J. Pure Appl. Phys.*, 1978, *16,* 519.
28. Blumenthal, R.N. and Hofmaier, R.L., *J. Electrochem. Soc.*, 1974, *121,* 126.
29. Dawicke, J.W. and Blumenthal, R.N., *J. Electrochem. Soc.*, 1986, *133,* 904.
30. Chang, E.K. and Blumenthal, R.N., *J. Solid State Chem.*, 1988, *72,* 330.
31. Knoll, G.F., *Radiation Detection and Measurement,* John Wiley & Sons, New York, 1979, 414.
32. Kudo, T. and Obayashi, H., *J. Electrochem. Soc.*, 1976, *123,* 415.
33. Riess, I., Braunshtein, D., and Tannhauser, D.S., *J. Am. Ceram. Soc.*, 1981, *64,* 479.
34. Mott, N.F. and Davis, E.A., *Electronic Processes in Non-crystalline Materials,* Clarendon Press, Oxford, 1971, 102–120.
35. Rickert, H., Sattler, V., and Wedde, Ch., *Z. Phys. Chem. NF,* 1975, *98,* 339.
36. Sohége, J. and Funke, K., *Ber. Bunsenges. Phys. Chem.*, 1984, *88,* 657.
37. Choi, G.M., Tuller, H.L., and Tsai, M.-J., in *Nonstoichiometric Compounds,* Nowotny, J. and Weppner, W., Eds., Kluwer Academic, Dordrecht, 1989, 451; *Mater. Res. Soc. Symp. Proc.*, 1988, *99,* 141.
38. Stratton, T.S. and Tuller, H.L., *J. Chem. Soc. Faraday Trans. II,* 1987, *83,* 1143.
39. Tuller, H.L., *Solid State Ionics,* 1992, *52,* 135.
40. West, K., Intercalation compounds: metal ions in chalcogenide and oxide hosts, in *High Conductivity, Solid Ionic Conductors,* Takahashi, T., Ed., World Scientific, Singapore, 1989, 447–477.
41. Metha, A. and Smyth, D.M., Nonstoichiometry and defect equilibria in $YBa_2Cu_3O_x$, in *Non-Stoichiometric Compounds,* Nowotny, J. and Weppner, W., Eds., Kluwer Academic, Dordrecht, 1989, 509–520.
42. Maier, J., Murugaraj, P., Pfundtner, G., and Sitte, W., *Ber. Bunsenges. Phys. Chem.*, 1989, *93,* 1350.
43. Nowotny, J., Rakas, M., and Weppner, W., *J. Am. Ceram. Soc.*, 1990, *73,* 1040.
44. Marchetti, A.P. and Eachus, R.S., *Adv. Photochem.*, 1992, *17,* 145.
45. Wagner, C., *Z. Elektrochem.*, 1959, *63,* 1027.
46. Weiss, K., *Z. Phys. Chem. NF,* 1968, *59,* 242.
47. Raleigh, D.O., *J. Phys. Chem. Solids,* 1965, *26,* 329.
48. Mizusaki, J. and Fueki, K., *Rev. Chim. Miner.*, 1980, *17,* 356.
49. Mizusaki, J., Fueki, K., and Mukaibo, T., *Bull. Chem. Soc. Jpn.*, 1975, *48,* 428.
50. Miller, A.S. and Maurer, R.J., *J. Phys. Chem. Solids,* 1958, *4,* 196.
51. Hellstrom, E.E. and Huggins, R.A., *J. Solid State Chem.*, 1980, *35,* 207.
52. Wagner, J.B. and Wagner, C., *J. Chem. Phys.*, 1957, *26,* 1597.
53. Bierman, W. and Oel, H.J., *Z. Phys. Chem. NF,* 1958, *17,* 163.
54. Viswarath, A.K. and Radhakrishna, S., Copper ion conductors, in *High Conductivity Solid Ionic Conductors,* Takahashi, T., Ed., World Scientific, Singapore, 1989, 280–326.
55. Safadi, R., Riess, I., and Tuller, H.L., *Solid State Ionics,* 1992, *57,* 125.
56. Riess, I. and Safadi, R., *Solid State Ionics,* 1993, *59,* 99.
57. Riess, I., Safadi, R., and Tuller, H.L., *Solid State Ionics,* 1994, *72,* 3.
58. Riess, I., Safadi, R., and Tuller, H.L., *Solid State Ionics,* 1993, *59,* 279.
59. Jow, T. and Wagner, J.B., *J. Electrochem. Soc.*, 1978, *125,* 613.
60. Hirahara, E., *J. Phys. Soc. Jpn.*, 1951, *6,* 422.
61. Wagner, J.B. and Wagner, C., *J. Chem. Phys.*, 1957, *26,* 1602.
62. Yokota, I., *J. Phys. Soc. Jpn.*, 1961, *16,* 2213.
63. Allen, L.H. and Buhks, E., *J. Appl. Phys.*, 1984, *56,* 327.
64. Ema, Y., *Jpn. J. Appl. Phys.*, 1990, *29,* 2098.
65. Warner, T.E., Milius, W., and Maier, J., *Ber. Bunsenges, Phys. Chem.*, 1992, *96,* 1607.
66. Lumbreras, M., Protas, J., Jebbari, S., Dirksen, G.J., and Schoonman, J., *Solid State Ionics,* 1985, *16,* 195.

67. Lumbreras, M., Protas, J., Jebbari, S., Dirksen, G.J., and Schoonman, J., *Solid State Ionics*, 1986, *18&19*, 1179.
68. Schoonman, J. and Macke, A.J.H., *J. Solid State Chem.*, 1972, *5*, 105.
69. Schoonman, J. and Dijkman, F.G., *J. Solid State Chem.*, 1972, *5*, 111.
70. Koto, K., Schulz, H., and Huggins, R.A., *Solid State Ionics*, 1980, *1*, 355.
71. Sato, H., Some theoretical aspects of solid electrolytes, in *Solid Electrolytes*, Geller, S., Ed., Springer-Verlag, Berlin, 1977, 5.
72. Joshi, A.V. and Liang, C.C., *J. Phys. Chem. Solids*, 1975, *36*, 927.
73. Maier, J. and Schwitzgebel, G., *Mater. Res. Bull.*, 1982, *17*, 1061.
74. Maier, J. and Schwitzgebel, G., *Mater. Res. Bull.*, 1983, *18*, 601.
75. Park, J.-H. and Blumenthal, R.N., *J. Electrochem. Soc.*, 1989, *136*, 2868.
76. Fouletier, J. and Kleitz, M., *J. Electrochem. Soc.*, 1978, *125*, 751.
77. Liou, S.S. and Worrell, W.L., *Appl. Phys.*, 1989, *A49*, 25.
78. Liou, S.S. and Worrell, W.L., in *Solid Oxide Fuel Cells, First Int. Symp.*, Vol. 89-11, Singhal, S.C., Ed., The Electrochemical Society, Pennington, NJ, 1990, 81.
79. Kopp, A., Näfe, H., and Weppner, W., *Solid State Ionics*, 1992, *53–56*, 853.
80. Calès, B. and Baumard, J.F., *J. Phys. Chem. Solids*, 1984, *45*, 929.
81. Tuller, H.L. and Nowick, A.S. *J. Electrochem. Soc.*, 1975, *122*, 255.
82. Blumenthal, R.N. and Garnier, J.E., *J. Solid State Chem.*, 1976, *16*, 21.
83. Moon, P.K. and Tuller, H.L., *Solid State Ionics*, 1988, *28–30*, 470.
84. Fujimoto, H.H. and Tuller, H.L., Mixed ionic electronic transport in thoria electrolytes, in *Fast Ion Transport in Solids*, Vashishta, P., Mundy, J.N., and Shenoy, G.K., Eds., Elsevier/North-Holland, Amsterdam, 1978, 649–652.
85. Näfe, H., Low temperature ion conductivity of a solid oxide electrolyte: the role of electrode polarization, in *Solid State Ionics*, Balkanski, M., Takahashi, T., and Tuller, H.L., Eds., North-Holland, Amsterdam, 1992, 253–258.
86. Mizusaki, J., Yasuda, I., Shimoyama, J.-I., Yamauchi, S., and Fueki, K., *J. Electrochem. Soc.*, 1993, *140*, 467.
87. Metha, A., Chang, E.K., and Smyth, D.M., *J. Mater. Res.*, 1991, *6*, 851.
88. Gösele, U.M., *Annu. Rev. Mater. Sci.*, 1988, *18*, 257.
89. *VLSI Technology*, Sze, S.M., Ed., McGraw-Hill, 1983, 208.
90. Yarbrough, D.W., *Solid State Technol.*, Nov. 1968, 23.
91. Whittingham, M.S., The fundamental science underlying intercalation electrode for batteries, in *Solid State Ionics*, Chowdari, B.V.R. and Radhakrishna, S., Eds., World Scientific, Singapore, 1988, 55–74.
92. Huggins, R.A., *J. Power Sources*, 1988, *22*, 341.
93. Libowitz, G.G., *The Solid-State Chemistry of Binary Metal Hydrides*, W.A. Benjamin, New York, 1965.
94. Schöllhorn, R., *Solid State Ionics*, 1989, *32/33(I)*, 23.
95. Goldner, R.B., Electrochromic smart window glass, in *Solid State Ionics*, Chowdari, B.V.R. and Radhakrishna, S., Eds., World Scientific, Singapore, 1988, 379–390.
96. Goldner, R.B., Norton, P., Wong, K., Foley, G., Goldner, E.L., Seward, G., and Chapman, R., *Appl. Phys. Lett.*, 1985, *47*, 536.
97. Yosuda, I. and Hikita, T., *J. Electrochem. Soc.*, 1993, *140*, 1699.
98. Frank, G. and Köstlin, H., *Appl. Phys. A*, 1982, *27*, 197.
99. Berger, J., Riess, I., and Tannhauser, D.S., *Solid State Ionics*, 1985, *15*, 225.
100. Shukla, A.K., Vasan, H.N., and Rao, C.N.R., *Proc. Roy. Soc. London, A*, 1981, *376A*, 619.
101. Teraoka, Y., Zhang, H.M., Okamoto, K., and Yamazoe, N., *Mater. Res. Bull.*, 1988, *23*, 51.
102. Teraoka, Y., Nobunaga, T., Okamoto, K., Mirua, N., and Yamazoe, N., *Solid State Ionics*, 1991, *48*, 207.
103. Choi, J.S., Kim, K.H.K., and Chung, W.Y., *J. Phys. Chem. Solids*, 1986, *47*, 117.
104. Meltale, A.F. and Tuller, H.L., *J. Am. Ceram. Soc.*, 1985, *68*, 646.
105. McHale, A.E. and Tuller, H.L., *J. Am. Ceram. Soc.*, 1985, *68*, 651.
106. Neuman, H., *Crystal Res. Technol.*, 1983, *18*, 483.
107. Kleinfeld, M. and Wiemhöfer, H.-D., *Ber. Bunsenges. Phys. Chem.*, 1986, *90*, 711.
108. Maricle, D.L., Swarr, T.E., and Karavolis, S., *Solid State Ionics*, 1992, *52*, 173.
109. Tsuchiya, T. and Yoshimura, N., *J. Mater. Sci.*, 1989, *24*, 493.
110. Kuo, H.-J., Cappelletti, R.L., and Pechan, M.J., *Solid State Ionics*, 1987, *24*, 315.
111. Lemaire, P.K., Hunt, E.R., and Cappelletti, R.L., *Solid State Ionics*, 1989, *34*, 69.
112. Zirin, M.H. and Trivich, D.., *J. Chem. Phys.*, 1963, *39*, 870.
113. Malueda, J., Farhi, R., and Petot-Ervas, G., *J. Phys. Chem. Solids*, 1981, *42*, 911.
114. Toth, R.S., Kilkson, R., and Trivich, D., *Phys. Rev.*, 1961, *122*, 482.
115. Park, J.-H. and Natesan, K., *Oxid. Met.*, 1993, *39*, 411.
116. Tretyakov, Y.D., Komarov, V.F., Prosvirnina, N.A., and Kutsenok, I.B., *J. Solid State Chem.*, 1972, *5*, 157.
117. Mrowec, S., Stoklosa, A., and Godlewski, K., *Cryst. Lattice Defects*, 1974, *5*, 239.
118. Xue, J. and Dieckmann, R., *J. Phys. Chem. Solids*, 1990, *51*, 1263.

118a. Porat, O. and Riess, I., *Solid State Ionics*, 1995, 81, 29.
119. Cava, R.J., Batlogg, B., Chen, C.H., Rietman, E.A., Zahurak, S.M., and Werder, D., *Phys. Rev. B.*, 1987, *36*, 5719.
120. Shafer, M.W., Ponney, T., Olson, B.L., Greene, R.L., and Koch, R.H., *Phys. Rev. B.*, 1989, *39*, 2914.
121. Yoo, H.-I. and Choi, H.-S., *J. Am. Ceram. Soc.*, 1992, *75*, 2707.
122. Fisher, B., Genossar, J., Lelong, I.O., Kessel, A., and Ashkenazi, J., *Physica C.*, 1988, *153–155*, 1348.
123. Choi, G.M., Tuller, H.L., and Tsai, M.-J., in *Nonstoichiometric Compounds*, Nowotny, J. and Weppner, W., Eds., Kluwer Academic, Dordrecht, 1989, 451.
124. LaGraff, J.R. and Payne, D.A., *Phys. Rev. B.*, 1993, *47*, 3380.
125. MacManus, J., Fray, D., and Evetts, J., *Physica C.*, 1990, *169*, 193.
126. Jantsch, S., Ihringer, J., Maichle, J.K., Prendl, W., Kemmler-Sack, S., Kiemel, R., Lösch, S., Schäfer, W., Schlichenmaier, M., and Hewat, A.W., *J. Less. Comm. Met.*, 1989, *150*, 167.
127. Patrakeev, M.V., Leonidov, I.A., Kozhevnikov, V.L., Tsidilkovskii, V.I., and Demin, A.K., *Physica C.*, 1993, *210*, 213.
128. Gür, T.M. and Huggins, R.A., *J. Electrochem. Soc.*, 1993, *140*, 1990.
129. Maier, J., Murugaraj, P., and Pfundtner, G., *Solid State Ionics*, 1990, *40/41*, 802.
130. Vischjager, D.J., van Zomeren, A.A., and Schoonman, J., *Solid State Ionics*, 1990, *40/41*, 810.
131. Porat, O., Riess, I., and Tuller, H.L., *Physica C.*, 1992, *192*, 60; *J. Superconduct.*, 1993, *6*, 313.
132. Huber, F. and Rottersman, M., *J. Appl. Phys.*, 1962, *33*, 3385.
133. Huber, F., *J. Electrochem. Soc.*, 1963, *110*, 846.
134. Sasaki, Y., *J. Phys. Chem. Solids*, 1960, *13*, 177.
135. Pell, E.M., *J. Appl. Phys.*, 1960, *31*, 291.
136. Kharkats, Y.I., *Sov. Electrochem.*, 1984, *20*, 234.
137. Weppner, W., *Solid State Ionics*, 1986, *18&19*, 873.
138. Riess, I., *Phys. Rev. B*, 1987, *35*, 5740.
139. Baiatu, T., Waser, R., and Härdtl, K.-H., *J. Am. Ceram. Soc.*, 1990, *73*, 1663.
140. Weppner, W., *J. Solid State Chem.*, 1977, *20*, 305.
141. Desu, S.B. and Yoo, I.K., *J. Electrochem. Soc.*, 1993, *140*, L133.
142. Waser, R., Baiatu, T., and Härdtl, K.-H., *J. Am. Ceram. Soc.*, 1990, *73*, 1654.
143. Tomlinson, R.D., Elliot, E., Parkes, J., and Hampshire, M.J., *Appl. Phys. Lett.*, 1975, *26*, 383.
144. Jakubowicz, A., Dagan, G., Schmitz, C., and Cahen, D., *Adv. Mater.*, 1992, *4*, 741.
145. Gartsman, K., Chernyak, L., Gilet, J.M., Cahen, D., and Triboulet, R., *Appl. Phys. Lett.*, 1992, *61*, 2429.
146. Kröger, F.A., *J. Phys. Chem. Solids*, 1980, *41*, 741.
147. Yoo, H.-I., Schmalzried, H., Martin, M., and Janek, J., *J. Phys. Chem. NF*, 1990, *168*, 129.
148. Miyatani, S., *J. Phys. Soc. Jpn.*, 1981, *50*, 1595.
149. Riess, I., *J. Electrochem. Soc.*, 1981, *128*, 2077. [Please note the following main typographic errors in this reference: (a) v_i is missing in the right-hand side of Equation (14); (b) replace (–) by (+) in the denominator in Equation (28); (c) 6th line beyond Equation (33): change the inequality sign to read $r > 0.9$; (d) V is missing in Equation (40) that should read $P = (S\sigma_i/L)V(V_{th} - V)r = \ldots$].
150. Riess, I., *J. Phys. Chem. Solids*, 1986, *47*, 129.
151. Riess, I., *Solid State Ionics*, 1991, *44*, 207.
152. Patterson, J.W., Ionic and electronic conduction in nonmetallic phases, in *ACS Symposium Series No. 89, Corrosion Chemistry*, Brubaker, G.R., Beverley, P., and Phipps, P.P., Eds., American Chemical Society, Washington, D.C., 1979, 96–125.
153. Choudhury, N.S. and Patterson, J.W., *J. Electrochem. Soc.*, 1971, *118*, 1398.
154. Tannhauser, D.S., *J. Electrochem. Soc.*, 1978, *125*, 1277.
155. Hebb, M.H., *J. Chem. Phys.*, 1952, *20*, 185.
156. Wagner, C., *Z. Electrochem.*, 1956, *60*, 4.
157. Maier, J., *Ber. Bunsenges. Phys. Chem.*, 1989, *93*, 1468.
158. Crouch-Baker, S., *Z. Phys. Chem.*, 1991, *174*, 63.
159. Schmalzried, H., *Ber. Bunsenges. Phys. Chem.*, 1962, *66*, 572.
160. Gurevich, Yu. and Kharkats, Yu.I., *Solid State Ionics*, 1990, *38*, 241.
161. Riess, I. and Tannhauser, D.S., *Solid State Ionics*, 1982, *7*, 307.
162. Riess, I., *J. Appl. Phys.*, 1992, *71*, 4079.
163. Riess, I., *Solid State Ionics*, 1992, *51*, 219.
164. Riess, I., *J. Electrochem. Soc.*, 1992, *139*, 2250.
165. Safadi, R. and Riess, I., *Solid State Ionics*, 1992, *58*, 139.
166. Maier, J. and Schwitzgebel, G., *J. Phys. Stat. Sol. (b)*, 1982, *113*, 535.
167. Maier, J., *Z. Phys. Chem. NF*, 1984, *140*, 191.
168. Maier, J., *J. Am. Ceram. Soc.*, 1993, *76*, 1218.
169. Ho, P.S. and Kwok, T., *Rep. Prog. Phys.*, 1989, *52*, 301.
170. Wiemhöfer, H.-D., *Solid State Ionics*, 1990, *40/41*, 530.

171. Dudley, G.J. and Steele, B.C.H., *J. Solid State Chem.*, 1977, *21*, 1.
172. Miyatani, S., *Solid State Commun.*, 1981, *38*, 257.
173. Yokota, I. and Miyatani, S., *Solid State Ionics*, 1981, *3/4*, 17.
174. Millot, F. and Gerdanian, P., *J. Phys. Chem. Solids*, 1982, *43*, 507.
175. Yoo, H.-I., Lee, J.-H., Martin, M., Janek, J., and Schmalzried, H., *Solid State Ionics*, 1994, *67*, 317.
176. Ilschner, B., *J. Chem. Phys.*, 1958, *28*, 1109.
177. Mizusaki, J., Sasaki, J., Yamauchi, S., and Fueki, K., *Solid State Ionics*, 1982, *7*, 323.
178. Grientschnig, D. and Sitte, W., *Z. Phys. Chem. NF*, 1990, *168*, 143.
179. Boris, A.V. and Bredkhin, S.I., *Solid State Ionics*, 1990, *40/41*, 269.
180. Crouch-Baker, S., *Solid State Ionics*, 1991, *45*, 101.
181. Riess, I., *Solid State Ionics*, 1993, *66*, 331.
182. Riess, I., Critical review of methods for measuring partial conductivities in mixed ionic electronic conductors, in *Ionic and Mixed Conducting Ceramics*, (Proceedings volume 94-12), Ramanarayanan, T.A., Worrell, W.L., and Tuller, H.L., Eds., The Electrochemical Society, Pennington, NJ, 1994, 286–306.
183. Riess, I., *Solid State Ionics*, 1991, *44*, 199.
184. Worrell, W.L., *Solid State Ionics*, 1992, *52*, 147.
185. Grosse, V.T. and Schmalzried, H., *Z. Phys. Chem.*, 1991, *172*, 197.
186. Riess, I., Kramer, S., and Tuller, H.L., Measurement of ionic conductivity in mixed conducting pyrochlores by the short circuit method, in *Solid State Ionics*, Balkanski, M., Takahashi, T., and Tuller, H.L., Eds., North-Holland, Amsterdam, 1992, 499–505.
187. Näfe, H., *Z. Phys. Chem. NF*, 1991, *172*, 69.
188. Patrakeev, M.V., Leonidov, I.A., Kozhevnikov, V.L., Tsidilkovskii, V.I., Demin, A.K., and Nikolaev, A.V., *Solid State Ionics*, 1993, *66*, 61.
189. Weppner, W. and Huggins, R.A., *J. Electrochem. Soc.*, 1977, *124*, 1572.
190. Maier, J., *J. Am. Ceram. Soc.*, 1993, *76*, 1212.
191. Dudney, N.J., Coble, R.L., and Tuller, H.L., *J. Am. Ceram. Soc.*, 1981, *64*, 621.
192. Liu, M. and Joshi, A., *Proc. Electrochem. Soc. 91-12, Proc. Int. Symp. Ionic and Mixed Conduc. Ceram.*, 1991, 231–246.
193. Yushina, L.D., Tarasov, A.Ya., and Karpachev, S.V., *Electrochim. Acta*, 1977, *22*, 797.
194. Chebotin, V.N., *Russ. Chem. Rev.*, 1986, *55*, 495.
195. Miyatani, S., *J. Phys. Soc. Jpn.*, 1984, *53*, 4284.
196. Miyatani, S., *J. Phys. Soc. Jpn.*, 1985, *54*, 639.
197. Millot, F. and de Mierry, P., *J. Phys. Chem. Solids*, 1985, *46*, 797.
198. Jin, Y. and Rosso, M., *Solid State Ionics*, 1988, *26*, 237.
199. Rickert, H. and Wiemhöfer, H.-D., *Solid State Ionics*, 1983, *11*, 257.
200. Kleinfeld, M. and Wiemhöfer, H.-D., *Solid State Ionics*, 1988, *23–30*, 1111.
201. Weppner, W., *Z. Naturforsch. A*, 1976, 31A, 1336.
202. Buck, R.P., *J. Phys. Chem.*, 1989, *93*, 6212.
203. Scolnik, Y., Sabatani, E., and Cahen, D., *Physica C*, 1991, *174*, 273.
204. Sasaki, J., Mizusaki, J., Yamauchi, S., and Fueki, K., *Bull. Chem. Soc. Jpn.*, 1981, *54*, 2444.
205. Miyatani, S. and Tabuchi, A., *J. Phys. Soc. Jpn.*, 1981, *50*, 2050.
206. Gesmundo, F., *J. Phys. Chem. Solids*, 1985, *46*, 201.
207. Magara, M., Tsuji, T., and Naito, K., *Solid State Ionics*, 1990, *40/41*, 284.
208. Maluenda, J., Farhi, R., and Petot-Ervas, G., *J. Phys. Chem. Solids*, 1981, *42*, 697.
209. Ochin, P., Petot-Ervas, G., and Petot, C., *J. Phys. Chem. Solids*, 1985, *46*, 695.
210. Ben-Michael, R. and Tannhauser, D.S., *Appl. Phys. A*, 1991, *53*, 185.
211. Bieger, T., Maier, J., and Waser, R., *Solid State Ionics*, 1992, *53–56*, 578.
212. Brinkmann, D., *Magn. Reson. Rev.*, 1989, *14*, 101.
213. Suits, B.H. and White, D., *Solid State Commun.*, 1984, *50*, 291.
214. Clauss, C., Dominguez-Rodriguez, A., and Castaing, J., *Rev. Phys. Appl.*, 1986, *21*, 343.
215. Yokota, I. and Miyatani, S., *Jpn. J. Appl. Phys.*, 1962, *1*, 144.
216. Macdonald, J.R. and Franceschetti, D.R., *J. Chem. Phys.*, 1978, *68*, 1614.
217. Belzner, A., Gür, T.M., and Huggins, R.A., *Solid State Ionics*, 1990, *40/41*, 535.
218. Becker, K.D., Schmalzried, H., and von Wurmb, V., *Solid State Ionics*, 1983, *11*, 213.
219. Adamczyk, Z. and Nowotny, J., *J. Electrochem. Soc.*, 1980, *127*, 1112.
220. Riess, I., *Solid State Ionics*, 1994, *69*, 43.
221. De Jonghe, L.C., *J. Electrochem. Soc.*, 1982, *129*, 752.
222. Virkar, A.V., *J. Mater. Sci.*, 1985, *20*, 552.
223. Virkar, A.V., Nachlas, J., Joshi, A.V., and Diamond, J., *J. Am. Ceram. Soc.*, 1990, *73*, 3382.
224. Maier, J., *J. Electrochem. Soc.*, 1987, *134*, 1524.
225. Maier, J., *Ber. Bunsenges. Phys. Chem.*, 1989, *93*, 1474.

226. Ziman, J.M., *Principles of the Theory of Solids*, Cambridge University Press, Cambridge, 1964, 215–216.
227. Gvishi, M., Tallan, N.M., and Tannhauser, D.S., *Solid State Commun.*, 1968, *6*, 135.
228. Sohége, J. and Funke, K., *Ber. Bunsenges. Phys. Chem.*, 1989, *93*, 122.
229. Sohége, J. and Funke, K., *Ber. Bunsenges. Phys. Chem.*, 1989, *93*, 115.
230. Kellers, J., Hock, S., and Funke, K., *Ber. Bunsenges. Phys. Chem.*, 1991, *95*, 180.
231. Miyatani, S. and Yokota, I., *J. Phys. Soc. Jpn.*, 1959, *14*, 750.
232. Funke, K. and Hackenberg, R., *Ber. Bunsenges. Phys. Chem.*, 1972, *76*, 883.
233. Millot, F. and Gerdanian, P., *J. Nucl. Mater.*, 1980, 92, 257.
234. Yoo, H.-I. and Hwang, J.-H., *J. Phys. Chem. Solids*, 1992, *53*, 973.
235. Ratkje, S.K. and Tomii, Y., *J. Electrochem. Soc.*, 1993, *140*, 59.
236. Brown, J.T., *Energy (Oxford)*, 1986, *11*, 209.
237. Worrell, W.L., *Solid State Ionics*, 1988, *28–30*, 1215.
238. Van Dijk, M.P., De Vries, K.J., and Burggraaf, A.J., *Solid State Ionics*, 1986, *21*, 73.
239. Burggraaf, A.J., De Vries, K.J., and Van Dijk, M.P., *Solid State Ionics*, 1986, *18/19*, 807.
240. Park, J.-H., Kostic, P., Sreckovic, T., Kovaćevic, M., and Ristick, M.M., *Microelectron. J.*, 1992, *23*, 665.
241. Moseley, P.T. and Williams, D.E., Oxygen surface species on sermiconducting oxides, in *Techniques and Mechanisms in Gas Sensing*, Moseley, P.T., Norris, J.O.W., and Williams, D.E., Eds., Adam Hilger, Bristol, 1991, 46–60.
242. Goto, K.S., *Solid State Electrochemistry and Its Applications to Sensors and Electronic Devices*, Elsevier, Amsterdam, 1988, 333–371.
243. Kobayashi, H., and Wagner, C., *J. Chem. Phys.*, 1957, *26*, 1609.
244. Vayenas, C.G., Bebelis, S., Yentekakis, I.V., and Lintz, H.-G., *Catal. Today*, 1992, *11*, 303.
245. Gellings, P.J. and Bouwmeester, H.J.M., *Catal. Today*, 1992, *12*, 1.
246. Sutor, P., *MRS Bull.*, May 1991, p. 24.
247. Hamilton, J.F., *Adv. Phys.*, 1988, *37*, 359.
248. Chiang, Y.M., Kingery, W.D., and Levinson, C.M., *J. Appl. Phys.*, 1982, *53*, 9080.
249. Kiwiet, N.J. and Fox, M.A., *J. Electrochem. Soc.*, 1990, *137*, 561.
250. Riess, I., *Solid State Ionics*, 1992, *52*, 127.
251. Ross, B.N. and Benjamin, T.G., *Power Sources*, 1976/77, *1*, 311.
252. Riess, I., *Solid State Ionics*, 1992, *51*, 109.

Chapter 8

ELECTRODICS

Ilan Riess and Joop Schoonman

CONTENTS

0-8493-8956-9/97/$0.00+$.50

LIST OF SYMBOLS

$Ag_i^{\cdot}$	positively charged silver interstitial in the Kröger–Vink notation
c_o	concentration
C	ion-blocking electrode. In Equation (8.16) C is capacitance
C_{dl}	double-layer capacitance
C_e	capacitance of the electrodes together with the sample forming a two-plate capacitor
C_1,C_2	constants defined in Equations (8.22) and (8.23), determined by diffusion rates
E	electrode
E^{ref}	reference electrode
f	frequency
GITT	galvanostatic intermittent titration technique
I	current
I_{el}	electronic (electron/hole) current
I_i	ionic current
I_t	total current. $I_t = I_{el} + I_i$
i_f	interface
k	Boltzmann constant
L	length of sample SE or MIEC
LSC	$La_{1-\alpha}Sr_{\alpha}CoO_3$
LSM	$La_{1-\alpha}Sr_{\alpha}MnO_3$
MGFC	mixed-gas fuel cell
MIEC	mixed ionic–electronic conductor
M^{z+}	mobile ion with charge z+
$O''_{surface}$	doubly charged oxygen ion, adsorbed on the surface, in the Kröger–Vink notation
$P(O_2)$	oxygen partial pressure
$P(O_2)^{ext}$	$P(O_2)$ of the external atmosphere surrounding an electrode
q	elementary charge
R	resistance
R_a,R_c	dc resistance of the anode and cathode in a SOFC, respectively
ref	reference
Ref. A, Ref. C	auxiliary electrodes for measuring the overpotential of the cathodes and anodes in SOFCs
S	cross-sectional area

SCO	Sm_2O_3-doped CeO_2
SE	solid electrolyte
SOFC/SOFCs	solid oxide fuel cell(s)
T	temperature
tpb	triple-phase boundary
V	voltage
V_a	applied voltage
V'_{Ag}	negatively charged silver vacancy, in the Kröger–Vink notation
V_e	voltage measured to yield $\Delta\tilde{\mu}_e$
V_i	voltage measured to yield $\Delta\tilde{\mu}_i$
V_O^x	neutral oxygen vacancy in the Kröger–Vink notation
x	position
YSZ	Y_2O_3-stabilized ZrO_2
z	valence of mobile ion
$Z(\omega)$	impedance, $Z(\omega) = ReZ + iImZ$
$Z_C(\omega)$	impedance of a capacitor
$Z_{RC}(\omega)$	impedance of a parallel RC element
$Z_W(\omega)$	impedance of a Warburg diffusion element
$Z_D(\omega)$	impedance of a finite path length diffusion element
Δ	length of finite diffusion path
η	overpotential
μ_e	chemical potential of electrons
$\tilde{\mu}_e$	electrochemical potential of electrons
μ_i	chemical potential of ions
$\tilde{\mu}_i$	electrochemical potential of ions
$\mu(O_2)$	chemical potential of oxygen
$\mu(O^{2-})$	chemical potential of oxygen ions
$\tilde{\mu}(O^{2-})$	electrochemical potential of oxygen ions
ω	$2\pi f$
[ ]	concentration

ABSTRACT

This review discusses the processes at electrodes and grain boundaries, the applications of electrodes, and the characterization of electrodes and grain boundaries. The electrode processes considered are charge transfer and diffusion limitation. The applications of electrodes are presented for both current-carrying contacts and voltage probes. The nature of grain boundaries and the asymmetry in their conductivity for currents flowing parallel and perpendicular to the boundary are discussed. Experimental methods for characterization of electrodes and grain boundaries, mainly electrochemical ones, are reviewed.

I. INTRODUCTION

This review is concerned with the characteristics of interfaces and the methods for characterization of interfaces.

Interfaces form naturally between two phases, but can also occur within a single phase. The interfaces of interest here are between solids (including glasses) and between solids and gases. The interfaces within a single phase are those between grains at the grain boundaries, where a discontinuity exists in the lattice structure of two adjacent grains. The two grains are single crystals of the same material and the discontinuity is due to different crystal plain orientations. In solid state electrochemistry the interfaces of interest are found at electrodes and at grain boundaries in ceramic materials.

Electrodes can serve as current-carrying contacts or as voltage probes where the total current vanishes. Current-carrying electrodes introduce into or remove from a sample, in general, both electric charge and material. When the material is supplied from the gas phase an auxiliary conducting material must also be present on the surface to supply or remove electrons. (An exception is when the surface is electronically conducting.) For example, the introduction of oxygen into the solid electrolyte (SE), Y_2O_3–doped ZrO_2 (YSZ, or yttrium-stabilized zirconia), in the form of ions can be achieved by using, as electrodes, porous platinum layers applied onto the YSZ. One speaks then of an electrode involving three phases and the electrode process taking place close to the three-phase boundary (tpb). However, the electrode processes, when analyzed in terms of elementary steps, occur, each at an interface between two phases only, or in a single phase.

The transfer of material and charge at the electrodes can consist of a series of elementary steps. Each elementary step may include the motion of neutral species (molecules, atoms, or neutral point defects), or the motion of charged species (ions, charged point defects, electrons, or holes). The path for the motion of the species in a single elementary step can be either within a single phase (e.g., diffusion of O_2 molecules in the gas phase toward the tpb of a Pt electrode on YSZ), parallel to a phase boundary (e.g., diffusion of adsorbed oxygen along the surface of a Pt grain), and across an interface. In particular, the transfer of an electron to an adsorbed oxygen atom on a Pt grain across the metal boundary is

$$O_{ad} + e^- \rightarrow O_{ad}^- \tag{8.1}$$

and the introduction of an oxygen ion into YSZ is

$$O_{ad}^{2-} \rightarrow O^{2-}(YSZ) \tag{8.2a}$$

This reaction is written in more detail using the Kröger–Vink notation:[1]

$$O''_{surface} + V_O^{\bullet\bullet} \rightarrow O_O^x + V_{O,surface}^x \tag{8.2b}$$

where $V_{O,surface}^x$ denotes an empty adsorption site for oxygen on the surface of YSZ.

The motion of species in a single phase as part of the electrode reaction should be noticed. In addition to the example of O_2 diffusion in the small pores of the Pt layer, other examples are the motion of electrons in the electron-conductive material (e.g., Pt) and the diffusion of possibly neutral oxygen in silver when used in an oxygen electrode, and the diffusion of oxygen ions in the electrode material $La_{0.9}Sr_{0.1}CoO_3$ which is reported to be a mixed ionic–electronic conductor (MIEC), conducting electron/holes and oxygen ions.[2]

In voltage probes the total current vanishes. However, the partial ionic and electronic current there need not vanish, it is only their sum that vanishes. Thus matter and charge may flow across the interface probe/sample, when the sample is a MIEC.

When selective voltage probes are used which can exchange only electrons or only ions, the net current vanishes (provided a high-impedance voltmeter is used). However, this is a dynamic equilibrium where the mobile species are transported across the probe/sample interface with a net zero flux. The probes, then, are used to transfer information, namely, the electrochemical potential of ions or electrons.

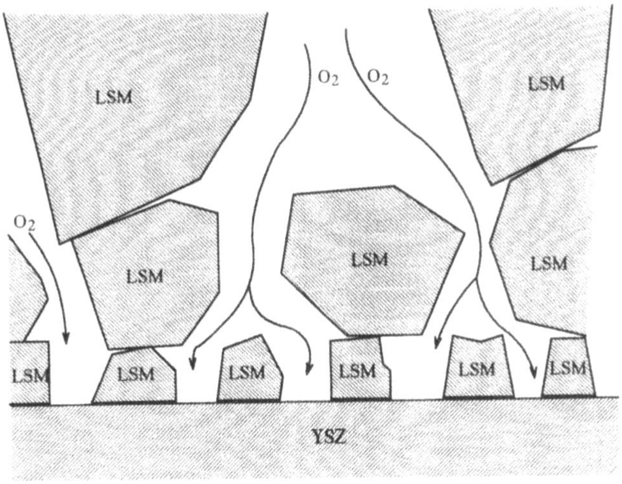

FIGURE 8.1. Cathode for SOFCs, morphology, and diffusion path for molecular oxygen.

Electrodes can be used to fix certain chemical potentials. They then serve as reference electrodes. Reference electrodes can be current-carrying ones or voltage probes with zero total current through them.

In some arrangements the current-carrying electrodes are made blocking to the flow of ions or electrons. This is done in order to allow only the other type of charge carrier to flow across the interface. This is used mainly in experiments for determining the partial electronic or ionic conductivity in MIECs.[3-5]

The catalytic properties of electrodes are of interest, in general, as they control the reaction rates at the electrodes. There are, however, two arrangements where the catalytic properties play a key role. First it has been shown that fuel cells (FCs) can be operated on uniform mixtures of fuel with air applied, with the same composition, to both electrodes.[6] The symmetry is broken by using two electrodes having different catalytic properties.[7] The second arrangement is one in which the rate of reaction of gases is changed by changing the voltage across a SE, thus changing the Fermi level of the electrons in the electrodes exposed to the reacting gases.[8]

We shall discuss electrodes, grain boundaries, and methods for their characterization.

II. ELECTRODES

A. CURRENT-CARRYING ELECTRODES

1. Cathodes in Solid Oxide Fuel Cells

The processes believed to prevail in cathodes of solid oxide fuel cells (SOFCs) are not unique and depend, for given electrolytes, on the electrode materials and in particular whether the electrodes are electronic conductors or MIECs. In all cases oxygen has to diffuse from air toward the SE near the tpb via the pores in the electrode structure, as shown schematically in Figure 8.1. This diffusion is, in modern electrodes, not rate limiting. To reduce the impedance to gas diffusion the electrode is made of few layers with coarse grains and large pores in the outer layer, and layers with finer grains and pores closer to the SE.[9] The possible elementary steps inside an electrode are summarized in Figure 8.2.

I. In the process denoted as I the oxygen molecules diffuse all the way to the SE where they are adsorbed, decompose into two adsorbed atoms, diffuse on the SE toward the metal, i.e., toward the tpb, where they combine with two electrons while entering the SE. This mechanism was suggested for Pt and Au electrodes on Er_2O_3-doped Bi_2O_3[10] and for Au electrodes on Sm_2O_3-doped CeO_2.[11]

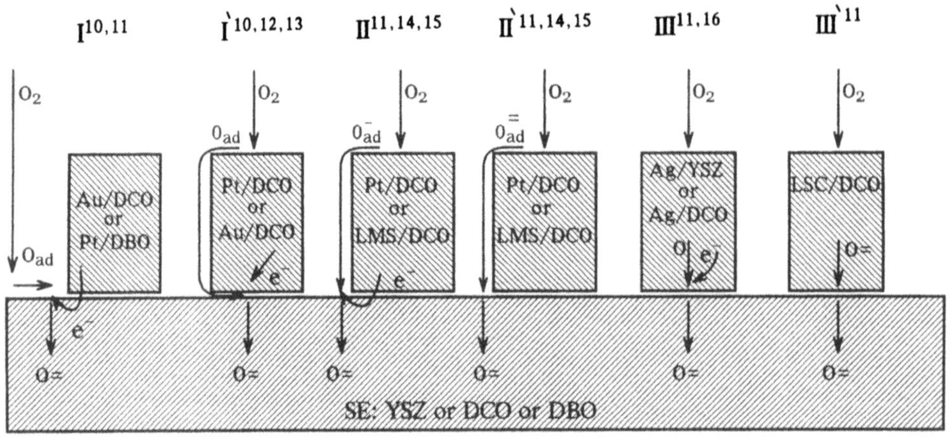

FIGURE 8.2. Series of elementary steps comprising different possible overall-electrode reactions. SE: solid electrolyte; YSZ: Y_2O_3-stabilized ZrO_2; DCO: CaO-, Y_2O_3-, Sm_2O_3, or Gd_2O_3–doped CeO_2; DBO: stabilized Bi_2O_3; LSM: $La_{0.8}Sr_{0.2}MnO_3$; LSC: $La_{0.8}Sr_{0.2}CoO_3$. Superscripts indicate reference numbers.

I′. In this process the oxygen is adsorbed on the metal; the molecule dissociates there into two adsorbed atoms which diffuse along the metal and along the metal/SE interface, where they are charged and enter the SE. This mechanism was suggested for Ag electrodes on CaO and Y_2O_3-doped CeO_2,[12] for Pt on YSZ,[10] and for Pt on Gd_2O_3-doped CeO_2.[13] However, it was also suggested[11,14,15] that the latter metal/SE pair behaves according to model II discussed next.

II. Oxygen molecules are adsorbed on the metal and dissociate to form two adsorbed atoms. A charge transfer occurs on the metal, where electrons from the metal are transferred to the adsorbed atoms to form singly ionized ions. The ions diffuse along the metal to the SE. Another electron is transferred to the ion via the SE to form doubly ionized oxygen ions that enter the SE. This mechanism was suggested for Pt on doped CeO_2[11,14,15] and for $La_{0.8}Sr_{0.2}MnO_3$ (LSM) on doped CeO_2.[11]

II′. In this case two electrons are transferred to the adsorbed oxygen atom on the metal. This ion then diffuses to the SE and enters the SE. This mechanism was suggested as an alternative one for Pt on doped CeO_2 by Wang and Nowick[14,15] and for LSM on doped CeO_2.[11]

III. This mechanism can take place only at metallic electrodes that allow diffusion of oxygen through the metal. Oxygen is adsorbed and dissociates on the metal, diffuses through the metal as a neutral species and at the interface with the SE it reacts with two electrons and then enters the SE. This was suggested to be the mechanism for Ag electrodes on YSZ[16] and Ag on doped CeO_2.[11] However, it was also suggested that Ag on doped CeO_2 follows model I′.[12]

IV. This model can be applied only to electrodes that are MIECs. The oxygen molecule dissociates on the MIEC. The adsorbed atoms are ionized there, then diffuse as ions through the MIEC and are transferred directly into the SE at the MIEC/SE interface. This mechanism was suggested for the MIEC $La_{0.8}Sr_{0.2}CoO_3$ (LSC) on doped CeO_2.[11]

The difference in description, by different authors, of the electrode mechanism for seemingly similar electrode/SE pairs may arise from the fact that the electrodes are not identical, but the data do not reveal this. The rate of the diffusion steps in the gas phase or along the grain boundaries depends on the morphology that might be different in different experiments, and the charge transfer reaction rate depends on the composition of the electrode, probably through the Fermi level, while small amounts of impurities may have a dramatic effect on these rates.[18]

When the electrode material allows fast diffusion of oxygen through it, as in Ag and LSC, the electrode can be applied as a continuous thin layer backed by a coarse current collector, thereby increasing the active area of the electrode. The continuous layer must be thin in order to reduce its resistance to the diffusion of oxygen ions. The coarse current collector is required in order to reduce the sheet resistance of the electrode to the flow of electrons without impeding the diffusion of oxygen molecules toward the thin layer.

A novel planar SOFC design proposed by Jaspers et al.[19] makes use of continuous MIEC electrodes. It is based on the loose stacking of gas-tight MIEC electrode elements, conventional plate-type electrolytes, and interconnectors. At the operating temperature oxygen is reduced at the gas/cathode two-phase contact zone, and transport of the resulting oxygen ions and electronic charge carriers takes place through the dense mixed conducting electrode layer toward the SE. The oxygen ions flow through the SE (which need not be gas tight) and enter a mixed conducting anode. Making the anode hollow (yet gas tight) allows the oxygen to react with the fuel gas in the cavity.

The series of elementary reactions taking place at the electrodes, shown in Figure 8.2, can occur simultaneously. By singling out a particular series, one points at the fastest process believed to dominate. This can change with temperature as the temperature dependence of the various elementary reactions is different and because the nonstoichiometry of the SE near the interface as well as of the MIECs used as electrode material can change with temperature. For example, LSC exhibiting an ionic transference number $t_i \sim 0.5$ at 800°C is practically an electronic conductor ($t_i \ll 1$) at 600°C.[20]

2. Anodes in SOFCs

The reactions at the anode of SOFCs based on oxygen ion conductors are more complex than the reactions at the cathode. Therefore, the uncertainty in determining the series of elementary reactions is higher for the anode process. While for the cathode, one material, oxygen, has to be transferred as ions into the SE, on the anode side, the fuel, oxygen ions emerging from the SE, and the products of the reaction have to be considered. For example, for the "simple" fuel hydrogen one has to consider H_2, H_2O, and their adsorbed species and ions, O^{2-}(SE), and maybe O_2 and reaction products, e.g., OH^- or NiO, a possible reaction product for a Ni-CSZ cermet electrode. The reaction between hydrogen and oxygen at the anode can proceed in different ways:

$$H_2(g) \rightarrow 2\,H_{ad,SE} \tag{8.3a}$$

$$2\mathrm{H}_{\mathrm{ad,SE}} + \mathrm{O}^{2-}_{\mathrm{ad,SE}} \rightarrow \mathrm{H_2O} + 2\mathrm{e}^-_{\mathrm{SE}} \tag{8.3b}$$

$$e^-_{SE} \rightarrow e^-_{anode} \tag{8.3c}$$

or

$$2O^{2-}_{ad,SE} \rightarrow O_2(g) + 4e^-_{SE} \tag{8.4a}$$

$$2H_2(g) + O_2(g) \rightarrow 2H_2O(g) \tag{8.4b}$$

$$e^-_{SE} \rightarrow e^-_{anode} \tag{8.4c}$$

or[21]

$$H_2(g) \rightarrow 2H^+_{ad,anode} + 2e^-_{anode} \tag{8.5a}$$

$$H^+_{ad,anode} \text{ diffuses toward the SE to yield } H^+_{ad,SE} \tag{8.5b}$$

$$2H^+_{ad,SE} + 2O^{2-}_{ad,SE} \rightarrow 2OH^-_{ad,SE} \tag{8.5c}$$

$$2OH^-_{ad,SE} \rightarrow H_2O(g) + O^{2-}_{ad,SE} \tag{8.5d}$$

The best performance, i.e., the lowest electrode impedance, for the cathodes is found for electrodes made of MIECs.[11] It is expected that the same will be true for anodes. However, no suitable MIEC was found that can be used as anode, having a low resistance to electronic as well as ionic current, being stable under the reducing conditions prevailing there, not reacting with the SE, and having a thermal expansion coefficient similar to that of the SE. A quasi-MIEC electrode can, however, be prepared by mixing fine powders of the SE material and the metal Ni (for YSZ- or CeO_2-based SE).

3. Insertion Electrodes

Insertion electrodes are MIECs that provide a source or sink for material as well as a conductive path for electric charge. For example, TiS_2 serves as a cathode in Li batteries. It allows the intercalation of Li ions that arrive through a lithium-conducting SE.[22] One expects that the charge transfer process across the SE/MIEC interface will exhibit a Butler–Volmer-type current–overpotential relation and that the diffusion in the MIEC electrode will yield a diffusion-limited current at high current densities. A detailed analysis confirms this mechanism.[23]

4. Solid/Solid Interfaces at Electrodes

In the discussion of electrodes in SOFCs, the gas/solid interface played an important role. However, examining the possible elementary reactions shown in Figure 8.2 it is evident that reactions at solid/solid (electrode/SE) interfaces may take place. This is true for electrode materials that allow diffusion of both electrons and ions (cases III and IV in Figure 8.2), and also for metals and semiconductors that allow only the diffusion of electrons (or holes) if electrons have to cross the electrode/SE interface. A solid/solid interface arises also when insertion electrodes are applied on SEs and when a parent metal is used as an electrode on a compound, e.g., Ag on Ag_2S.

The transfer of material from one ionic conductor to the other occurs via direct transfer of the ions from one solid to the other. This has been established for two contacting YSZ solids[24] and has been suggested also for a contact between a MIEC and a SE.[25] Schmalzried[26] has classified the interfaces as coherent, semicoherent, or incoherent, depending on the similarity or dissimilarity of the sublattices in the solids. The different types of interfaces should exhibit different exchange rates of ions. Another contribution to the exchange rate is the defect concentration at the interface. While this has been shown for a gas/solid interface,[27] it is plausible to assume that also the exchange rate at the solid/solid interface depends on the defect concentration. The exchange rate may depend on the concentration of active sites and the mechanical pressure pressing the two solids together. Careful measurements on the Ag/Ag_2S interface show that the overpotential for transfer of Ag^+ ions from Ag_2S into bulk Ag is 2 orders of magnitude larger than for the transfer of Ag^+ ions from Ag_2S into Ag whiskers.[28] The latter could sustain a current density of ~9 A/cm^2 with an overpotential $\eta <$ 5 mV at 300°C. The explanation suggested is that there are active sites in the Ag_2S surface where the whiskers tend to grow, e.g., at dislocations, and that the reaction rate there is much higher. However, the explanation can also be that the actual contact area of two contacting solid blocks pressed together is rather small and that, therefore, the real current density through the area of contact is large. This is in accord with the fact that the reaction rate increases with mechanical pressure on the cell. The dependence of the ion current across the interface on exchange current, defect concentration, and electrical potential barrier is described by the Butler–Volmer equation[29] discussed in Section IV.D.

The discussion in the previous paragraph mentions causes which could affect the exchange current across the interface. The electrical potential barrier, though, depends on the difference in the Fermi level of the two solids, and the difference in the unperturbed chemical potential of the mobile ions for the two solids before contact is established. The contact introduces changes in the distribution of the point defects in the space charge and the corresponding internal electrical field.[30-33] This will be discussed in more detail in Section III on grain boundaries.

5. Reversible Electrodes

Electrodes are defined as reversible ones if they transfer electrons and ions with negligible impedance. Then also under current, the electrochemical potential of electrons, $\tilde{\mu}_e$, and ions, $\tilde{\mu}_i$, do not change across the different interfaces which may exist in an electrode (see Figure 8.2). As a result, the chemical potential of the neutral species is also constant across the interface. For example, for an oxygen electrode comprising a metal on an oxide, $\tilde{\mu}_e$, $\tilde{\mu}(O^{2-})$, and $\mu(O_2)$ in the oxide edge near the electrode are equal to the corresponding parameters in the electrode, i.e., to $\tilde{\mu}_e$ in the metal, to $\tilde{\mu}(O^{2-})$ of oxygen ions adsorbed on the metal and on the SE, and to $\mu(O_2)$ in the gas phase, respectively.

6. Ion-Blocking Electrodes

Plates of inert metals such as platinum and graphite serve in solid state electrochemical cells to supply electrons to the sample, but block the passage of material and ions for voltages below the thermodynamic decomposition voltage of the SE. The ionic current is blocked since the metals do not contain the required material that can provide the ions. In addition, the metal blocks are applied in a way that blocks the exchange of material with, e.g., the surrounding atmosphere, as well. Not only inert metals can serve for that purpose, but also any electron conductor, whether metal or semiconductor, that stays chemically intact and is impermeable to the material flow either in the form of ions or neutral species.

One important use of ion-blocking electrodes is in selective measurements of the partial electronic (electron and hole) conductivities of MIECs. This is done using the Hebb–Wagner method.[3,4] A schematic of the cell used is shown in Figure 8.3. The blocking electrode is denoted as C. Let us consider a MIEC that is an oxide and conducts oxygen ions, electrons, and holes. The blocking electrode isolates the oxygen gas phase from the oxide MIEC. $\mu(O_2)_{if}$ in the MIEC at the interface with C is therefore not determined by the chemical potential of oxygen in the surrounding of electrode C. In the case that the MIEC conducts only ions with a single absolute value of charge, $|z|$, the ionic current is blocked.[5,34] Then $\mu(O_2)_{if}$ is determined, in the steady state, by the oxygen chemical potential at the reversible reference electrode, $\mu(O_2)_{ref}$ and the cell voltage,

$$\mu(O_2)_{if} = \mu(O_2)_{ref} - 4qV \tag{8.6}$$

where q is the elementary charge. As an electronic current is flowing through the cell, a possible overpotential at the blocking electrode η_c (and one at the reference electrode, η_{ref}) may exist. Then the voltage V in Equation (8.6) does not equal the applied voltage V_a, but should read $V_a - \eta_c$.

7. Electron-Blocking Electrodes

Electrodes that transmit ions and block electrons are SEs. In the ideal case they exhibit an infinitely high resistance to electronic (electron/hole) current. Then the electronic current

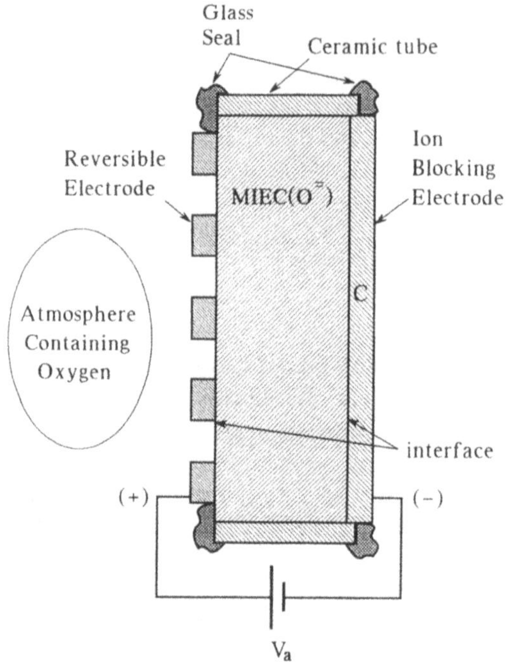

FIGURE 8.3. Cell arrangement for the measurement of the partial electronic conductivity according to the Hebb–Wagner method. Specific example: the MIEC conducts O^{2-} ions and electrons/holes. The polarity of the applied voltage is fixed accordingly.

through the SE vanishes. Contrary to common belief, the voltage drop on the blocking electrode need not be equal, in the general case, to the applied voltage.[35]

One important use of electron-blocking electrodes is in determining the partial ionic conductivity of MIECs. This is done using a method analogous to the Hebb–Wagner one whereby the electronic current is blocked.[36] The cell is shown schematically in Figure 8.4. This works only as long as $|z|$ is the same in the MIEC and the blocking electrode.[5,34,35] Even so, since in reality the SEs do not exhibit infinite resistance to electronic current, a careful examination of the resistances of the MIEC and the blocking electrode in the cell is required in order to be certain that the electronic current is indeed blocked and that the resistance measured is the ionic resistance of the tested sample.[5,37]

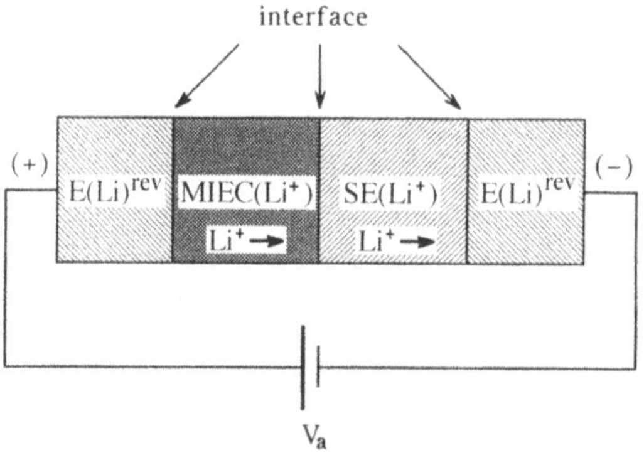

FIGURE 8.4. Cell arrangement for the measurement of the partial ionic conductivity using an SE as an electron-blocking electrode. Specific example: the MIEC conducts Li^+ ions and electrons/holes. The polarity of the applied voltage is fixed accordingly. $E(Li)^{rev}$: reversible electrode with a fixed μ_{Li}.

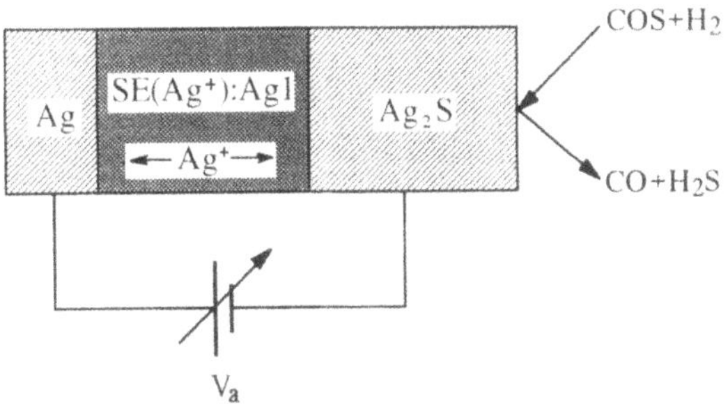

FIGURE 8.5. Cell for catalyzing the reaction $COS + H_2 = CO + H_2S$ using Ag_2S as the catalyst. The catalytic properties of the Ag_2S electrode on the SE(Ag^+) AgI is modified by the applied voltage. (Drawn from data of Kobayashi, H. and Wagner, C., *J. Chem. Phys.*, 1957, *26*, 1609.)

8. Reference Electrodes

Reference electrodes are used to fix a certain chemical potential in the sample near the interface with the electrode. As such they are expected to buffer, i.e., maintain their chemical potential even if material is removed or introduced into them. Examples are metal blocks of the elements, e.g., a Ag electrode that maintains a constant chemical potential of silver irrespective of mass added or removed from the electrode, pressed mixtures of coexisting oxides, e.g., a Cu_2O/CuO mixture which has a fixed oxygen (and copper) chemical potential irrespective of changes in overall oxygen content and a P_t electrode in air.

A reference electrode can carry a current, as in the Hebb–Wagner method (Figures 8.3 and 8.4), or can be a voltage probe where the total current, I_t, vanishes. It should be noticed that $I_t = 0$ does not imply that the partial ionic, I_i, nor the partial electronic, I_{el}, current through the probe, vanishes. For example, using in SOFC research a reference probe made of LSC on the MIEC oxide Sm_2O_3-doped CeO_2 to impose the surrounding $\mu(O_2)$ at the LSC/MIEC interface, the O^{2-} and e^- partial currents, through the LSC probe, do not vanish.

9. Electrodes as Catalysts

The rate of reaction between gases can be affected in the presence of a solid which serves then as a catalyst. The catalyst does not participate in the overall reaction, though it can take part in intermediate reaction steps. The catalytic properties of the solid can be altered by shifting the Fermi level (electrochemical potential of the electrons).[8] This shift affects the rate of transfer of electrons to an adsorbed atom. It can also be altered by changing the point defect concentration when associates between adsorbed species and point defects are important in intermediate reaction steps. The morphology is also of importance, as the active reaction sites may depend on the local structure. For a given morphology, both the Fermi level and the defect concentration can be modified in the catalyst when it serves as an electrode in a galvanic cell. This is done by changing the applied voltage on the cell.

An example of such a galvanic cell is shown in Figure 8.5. Ag_2S is used to catalyze the reaction $COS + H_2 = CO + H_2S$. The Ag electrode serves as a reference electrode. The chemical potential of Ag in the Ag_2S electrode is governed by the applied voltage and as long as the current of the Ag^+ ions is small:

$$\mu_{Ag}(Ag_2S) = \mu_{Ag}(Ag) - qV \tag{8.7}$$

The Fermi level difference between the catalyst Ag_2S and the Ag reference electrode is qV. This cell was used by Kobayashi and Wagner[38] to demonstrate that the catalytic properties of the Ag_2S electrode can be changed by the applied voltage V. Under reaction the chemical potential and concentration of Ag and S in Ag_2S are fixed by both the applied electric field and the reacting species. Under these conditions the Ag^+ ionic current through the cell does

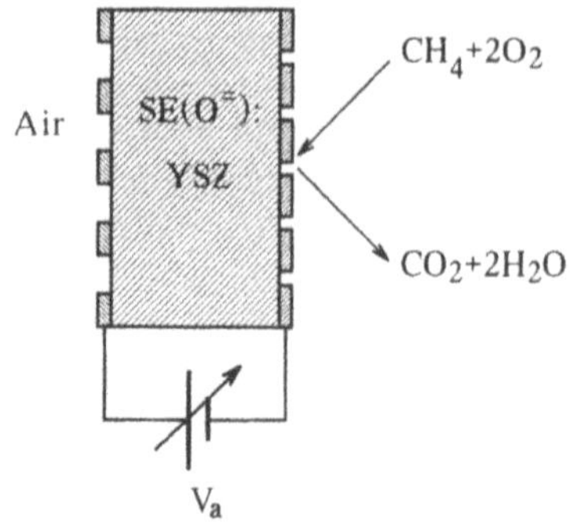

FIGURE 8.6. Schematic of the cell used to catalyze reactions of gases with O_2. Specific example: $CH_4 + 2O_2 \rightarrow CO_2 + 2H_2O$. (Drawn from data of Vayenas, C.G., Bebelis, S., Yentekakis, I.V., and Lintz, H.-G., *Catal. Today.*, 1992, *11*, 303.)

not vanish. When the applied voltage is changed, a transient current arises as the composition of the Ag_2S electrode changes to a new steady-state value.

A similar arrangement has been investigated extensively in recent years by Vayenas and co-workers.[8] Many of the experiments done by this group use YSZ as the SE and Pt slightly porous layers as the catalyst electrode, as shown in Figure 8.6. The applied voltage changes the Fermi level of the electrons and the chemical potential of oxygen in the right hand side electrode with respect to the left hand side reference electrode. The rate of oxidation of CH_4 depends on V and can be changed by up to a factor of 70.[8]

An ionic current is generated in the steady state due to deviation of the oxygen chemical potential at the electrode from the value calculated in analogy to Equation (8.7). This current depends on the resistance of the cell which is determined by the resistance of the SE and the electrodes. Vayenas et al.[8] found ways to reduce the current by using fine-grain porous, but rather dense Pt electrodes which exhibit a high impedance to gas diffusion. The result is that the current of oxygen through the YSZ, though not zero, is much smaller than the flow of oxygen taking part in the reaction in the gas phase. Because of that they refer to the process as non-Faradic electrochemical modification of catalytic activity (NEMCA).

10. Using Differences in Catalytic Properties of Electrodes to Drive Fuel Cells

It has been demonstrated[6,39] that a cell composed of $E_1/SE/E_2$ (E_1,E_2-electrodes) placed in a uniform mixture of fuel with oxygen, can generate power. To make it clear, the fuel is supplied mixed with air or pure oxygen and the same mixture is supplied to both electrodes. We therefore refer to the cell as "mixed-gas fuel cell" (MGFC). The symmetry is broken by using different electrode materials, one that catalyzes the adsorption of oxygen to generate oxygen ions, and the other catalyzes the reaction of fuel adsorption and reaction with the oxygen ions.[7] The fuel used should not exhibit fast direct reaction with oxygen in the gas phase at the given temperature.

B. VOLTAGE PROBES

1. General

In solid state electrochemistry it is of interest to measure the driving forces acting on the electrons and on the mobile ions. One therefore seeks means to determine differences in the electrochemical potential of the electrons ($\Delta\tilde{\mu}_e$) and of the mobile ions ($\Delta\tilde{\mu}_i$). There is no complete analogy between these two cases, because the driving forces are of different origin while the measuring device stays the same, i.e., a voltmeter which measures a voltage V. $\Delta\tilde{\mu}_e/q$ equals V while $\Delta\tilde{\mu}_i/q$ equals the corresponding Nernst voltage minus V.[35] The voltmeter responds therefore to $\Delta\tilde{\mu}_e$. To be able to measure $\Delta\tilde{\mu}_i$ using a voltmeter, a suitable analog translator must be used. This is discussed in Section II.B.3.

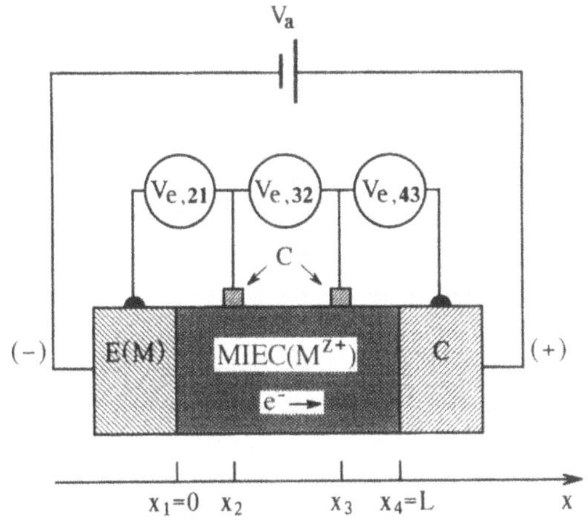

FIGURE 8.7. Linear cell configuration for measuring differences $\Delta\tilde{\mu}_e$ in MIECs and electrode overpotentials, using four electrodes. The arrangement, which includes three ion-blocking electrodes, can also serve to determine the partial electronic conductivity of the MIEC. Specific example: the MIEC conducts M^{z+} ions. The polarity of the applied voltage is fixed accordingly. E(M): electrode with fixed μ_M.

2. Probes for Measuring $\Delta\tilde{\mu}_e$

A typical cell configuration for measuring the electronic conductivity in MIECs is shown in Figure 8.7.[40] The voltage probes are No. 2 and 3. The voltmeter $V_{e,32}$ reads,

$$V_{e,32} = -\left(\tilde{\mu}_e(3) - \tilde{\mu}_e(2)\right)/q \equiv -\Delta\tilde{\mu}_{e,32}/q \tag{8.8}$$

It should be noticed that Equation (8.8) holds for the points where the metallic leads from the voltmeter contact the probes. For $V_{e,32}$ to reflect $\Delta\tilde{\mu}_{e,32}$ in the MIEC sample just under the probes, the probe material must be either a pure electronic conductor (i.e., a metal or semiconductor) or a MIEC.[41] Reference electrodes cannot be used as probes in the linear configuration. The reason is that they alter the composition of the sample MIEC and thus alter the experimental conditions. In particular, the linear symmetry is broken. One can tolerate reference electrodes as probes in certain Van der Pauw configurations, as there their effect on the potential distributions can be taken into consideration, for small applied chemical gradients.[42]

SEs cannot serve as probes in this case. SEs will serve as probes for measuring $\Delta\tilde{\mu}_i$, as shown next.

The use of blocking electrodes is experimentally not trivial, and care must be taken to make sure that blocking is achieved.[5]

3. Probes for Measuring $\Delta\tilde{\mu}_i$

A typical cell configuration for measuring the ionic conductivity in MIECs is shown in Figure 8.8. The voltmeter connected to the probes reads a difference in $\Delta\tilde{\mu}_e$. However, it is there to read a difference in $\Delta\tilde{\mu}_i$. To achieve this the probes are made of SE material and both are backed by reference electrodes E(M). When using identical reference electrodes in the two probes, then for mobile ions with charge z,

$$zqV_{i,32} = \tilde{\mu}_i(3) - \tilde{\mu}_i(2) \equiv \Delta\tilde{\mu}_{i,32} \tag{8.9}$$

and $\Delta\tilde{\mu}_{i,32}$ is determined from the measured voltage $V_{i,32}$.

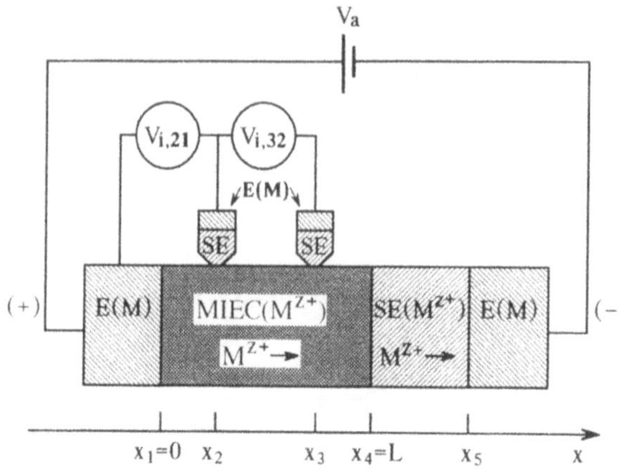

FIGURE 8.8. Linear cell configuration for measuring differences $\Delta\tilde{\mu}_i$ in MIECs and electrode overpotentials, using four electrodes. The arrangement, which includes three SEs, can also serve to determine the partial ionic conductivity of the MIEC. Specific example: the MIEC conducts M^{z+} ions. E(M): electrode with fixed μ_M.

4. Use of MIECs as Intermediate Contacts

MIEC contacts can be used to mediate the information $\tilde{\mu}_e$ and $\tilde{\mu}_i$ from the same contact points with the sample.[41] These MIEC contacts must then be backed by selective probes and two voltmeters, as described in Sections II.B.2 and II.B.3 above for measuring $\Delta\tilde{\mu}_e$ and $\Delta\tilde{\mu}_i$.

III. GRAIN BOUNDARIES

A. INTRODUCTION, SINGLE CRYSTALS

In a single crystal of a chemical compound, in the state of equilibrium, the chemical potentials of the components are all uniform. In ionic crystals in equilibrium, the electrochemical potentials of the ions are also uniform and so is the electrochemical potential of the electrons. For instance, in AgCl μ_{Ag}, μ_{Cl}, $\tilde{\mu}_{Ag^+}$, $\tilde{\mu}_{Cl^-}$, and $\tilde{\mu}_e$ are all uniform in the state of equilibrium. This, however, does not necessarily imply that either the chemical potential or the concentration of the ions are uniform. If the chemical potential is not uniform, an internal electric field must exist, governed by the Poisson equation through the local space charge. The result can be an accumulation of one type of ions on the free surface, a counter charge spread out in the bulk forming the so-called space charge region, and a double layer region at the interface.[33]

The physical reason for the nonuniform distribution of the ions in a single crystal is the fact that the forces acting on an ion close to the surface are different from those acting deep inside the bulk.

The different concentrations of ionic defects near the free surface and the electric field in the space charge region have the following effects:[33] (I) The conductivity parallel to the surface is different from the bulk. This is because of the difference in the defect concentrations. This holds for both the ionic and the electronic conductivity. (II) Likewise, the conductivity perpendicular to the surface is different from the bulk because of both the different charge carrier concentrations and the local electric field. It exhibits a capacitance due to changes in the space charge under an applied electric potential.

Experimentally, the effect of the free surface can usually be neglected. It is only when the surface-to-volume ratio is high that surface effects become important, i.e., for very small crystals which are rarely of interest as single crystals.

In nano-sized grains, the bulk and the grain boundary regions coincide and the differentiation between grain boundary region resistance and bulk resistance is not valid anymore.[30]

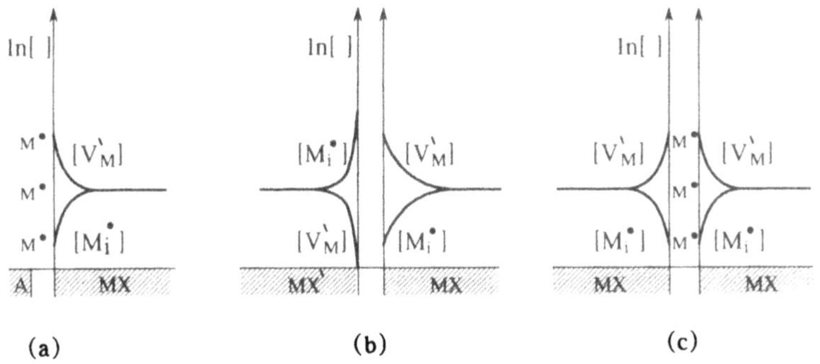

FIGURE 8.9. Distribution of the majority, Frenkel pair, defects near the grain boundaries between an SE (or an MIEC in which the majority charged defects are ionic), and a second phase. Specific example: assumes $M_i^{\bullet}$ to be a mobile ion. (a) A(insulator)/MX(MIEC or SE). (b) Two different MIECs or SEs with a common mobile ion $M_i^{\bullet}$. (c) Two grains of the same MIEC or SE material. (Adapted from Maier, J., in *Recent Trends in Superionic Solids and Solid Electrolytes,* Chandra, S. and Laskar, A., Eds., Academic Press, New York, 1989, 137.)

B. SINGLE-PHASE POLYCRYSTALLINE MATERIALS

A ceramic material of single phase is an assembly of contacting single crystals. The grain boundary is characterized by an abrupt change in the crystal planes orientation. The latter can be viewed at, formally, as a large change in the orientation. The grain boundary between two grains is then analogous to the free surface of the single crystal. The surface region, now the grain boundary region, maintains its peculiar properties though, quantitatively, the concentration of the ions drawn to the surface and the nonuniformity in the defect concentration can be different.

Experimentally since ceramics can be made with small (<1 μm) grains the effect of grain boundary on the parallel and perpendicular resistance can be significant. This has been demonstrated for AgCl and for AgBr.[30] (See Figure 8.9c.)

C. TWO-PHASE MATERIALS

The contact between two grains of different phases yields qualitatively similar results to those discussed before. Now the boundary between the two grains is the interface of interest. The ions removed from one phase need not concentrate on the boundary, but can spread into the other phase, if it conducts the ions.

A clear demonstration of the effect of a contact between grains of two different phases is given by the case where an insulator is dispersed into a SE and the result is an enhancement of the isothermal ionic conductivity of the SE. For example, by adding fine powder of the insulator γ-Al_2O_3 to fine powder of the ionic material LiI, which is a SE for Li ions, the ionic conductivity is enhanced by a factor of 50.[43] Similarly, adding Al_2O_3 to AgCl enhances in AgCl the ionic conductivity parallel to the grain boundary.[30] The explanation is that some Ag^+ ions are extracted to the interface between the insulator Al_2O_3 and the SE, Ag^+ being depleted in the grain boundary region inside the AgCl (Figure 8.10). As a result, the ionic conductivity of AgCl which is dominated by $\sigma(V'_{Ag})$ increases dramatically in the grain boundary region to such an extent that it affects the overall resistance of the two-phase material and can therefore be measured.

Figure 8.9 summarizes the different configurations that can be visualized for grain boundaries between a SE or a MIEC and another phase. This holds also for MIECs, which have small concentrations of electrons or holes compared to the mobile ion concentration.[30]

D. EFFECT OF ELECTRIC FIELDS APPLIED PERPENDICULAR TO THE GRAIN BOUNDARY

With the electric field applied perpendicular to the grain boundary, the defect distribution and the local space charge are altered. A general analytical solution has not been derived for

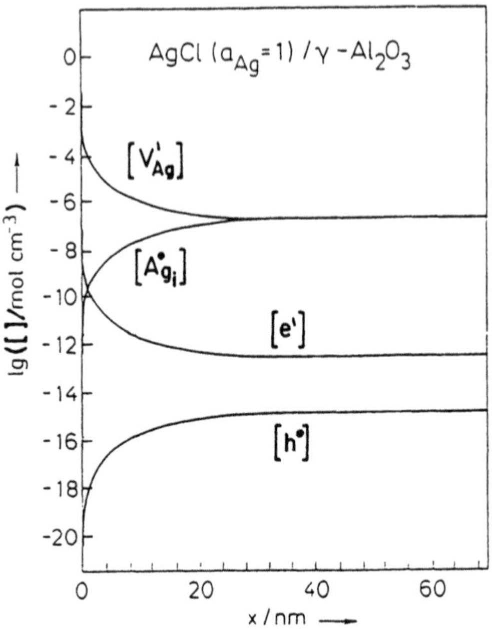

FIGURE 8.10. Defect concentration distribution in AgCl (with $a_{Ag} = 1$) in contact with $\gamma-Al_2O_3$. Majority defects: $Ag_i^{\bullet}$ and V'_{Ag}, minority defects: e' and $h^{\bullet}$. (From Maier, J., in *Recent Trends in Superionic Solids and Solid Electrolytes,* Chandra, S. and Laskar, A., Eds., Academic Press, New York, 1989, 137. With permission.)

an arbitrary applied electric field. However, an analysis for small applied signals has been given.[31-33] It turns out that the change of the charged defect concentration is linear in the applied potential for small signals, i.e., exhibits a capacitance-like behavior. Since a dc current can also be driven through the grain boundary region, this capacitance is coupled in parallel to a grain boundary resistance.

IV. EXPERIMENTAL METHODS FOR CHARACTERIZING ELECTRODES AND GRAIN BOUNDARIES

A. GENERAL

We concentrate here on electrochemical methods for characterizing electrodics. In these methods one characterizes the electrodics by analyzing the relations between the electrical and chemical driving forces on the one hand, and the electronic and ionic currents through interfaces on the other hand. For full characterization one has also to use other methods aimed at determining the chemical composition of the species in the various elementary steps taking place at an electrode. Similarly, one has to examine the chemical composition of the sample and the electrodes, or of the sample at the grain boundary region. The overall chemistry of the electrode reaction can be followed by monitoring the reaction products, e.g., by gas chromatography.[6,8] *In situ* optical measurements coupled with mass spectroscopy can reveal information on the chemical species also at intermediate steps.[44]

Let us examine the electrochemical methods used, with special emphasis on those used to investigate electrodes in SOFCs based on SEs which conduct O^{2-} ions. The chemical potential applied to the test electrode is fixed by the oxygen partial pressure, $P(O_2)$, there.

The electrical methods used are based on the current voltage, I-V, relations obtained under the following conditions:

I. Small signal, alternating currents (ac impedance).
II. Small signal, step change in the applied voltage. This method is equivalent to the former small signal ac one. The information obtained by the step method in the time domain can be transformed into the ac one in the frequency domain by using the Laplace transform. The ac impedance method seems to be experimentally more convenient at relatively high frequencies (above ~0.1 Hz). The reason is that a few cycles are required to determine the impedance at each frequency. At lower frequencies the step function method is more convenient. The full frequency range can best be covered by a combination of the two methods.[45]
III. dc voltage–current relations up to high voltages. (The high voltage limit is characteristic of the specific electrode investigated. It is, however, always $\gg kT/q$).
IV. Large current, interruption method, where the voltage is measured as a function of time.

B. DIRECT MEASUREMENT OF THE ELECTRODE OVERPOTENTIAL

1. General

The current–voltage, I-V, relations of a galvanic cell are determined by the bulk and electrode impedances. We shall show below that by scanning the I-V relations over a wide frequency range using an ac signal, one can identify and separate the impedance contributed by the electrodes and by the bulk (solid and grain boundaries).

The different contributions to the impedance can also be identified from dc I-V measurements over a wide current range. The reason is that the voltage drop on the bulk is to a good approximation linear in the current while the electrode current–overpotential relations are not linear, as will be shown in Section IV.D.

There are other ways to separate and measure the electrode overpotential only, i.e., without the voltage drop in the bulk. These methods will be discussed now. First, methods suitable for purely ionic or purely electronic conductors are discussed and, subsequently, methods which are suitable also for MIECs.

2. Four-Point Method

The four-point method utilizes four electrodes; two are current-carrying electrodes and two are voltage probes. The configuration is similar to that shown in Figure 8.7. For a purely electronic conductor, this configuration is well known and allows determination of the electrical conductivity of the sample without the adverse effect of the electrode impedance. This method can also be used to determine the electrode overpotential. The notation of Figure 8.7 is used next. If the electrodes were reversible, the voltage drop on the electrode would vanish and the voltage drop across the sample, V_{41}, would be ohmic only. For the linear configuration,

$$\frac{V_{41}}{V_{32}} = \frac{L}{x_3 - x_2} \tag{8.10}$$

If an overpotential η_a does exits at the anode, and η_c at the cathode, the applied voltage, V_a, will not equal V_{41}, but instead,

$$V_a = V_{41} + \eta_a + \eta_c \tag{8.11}$$

Then,

$$\frac{V_a}{V_{32}} = \frac{V_{41}}{V_{32}} + \frac{\eta_a + \eta_c}{V_{32}} = \frac{L}{x_3 - x_2} + \frac{\eta_a + \eta_b}{V_{32}} \tag{8.12}$$

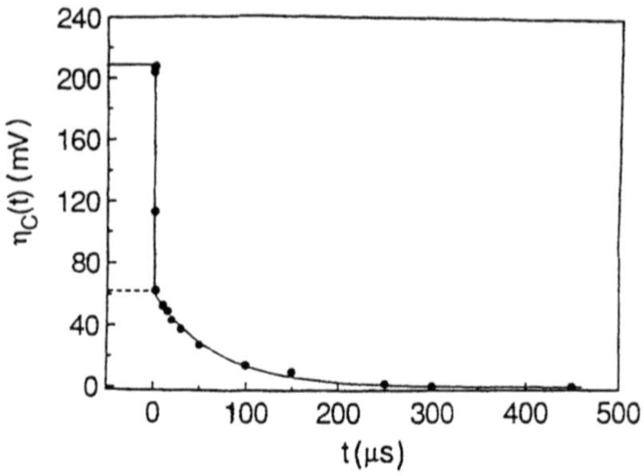

FIGURE 8.11. Current interruption results for the MIEC cathode LSC, on the MIEC electrolyte Sm_2O_3–doped CeO_2, at 700°C. (From Riess, I., Gödickemeier, M., and Gauckler, L.J., *Solid State Ionics,* in press.)

$\eta_a + \eta_c$ can be determined using Equation (8.12) as all other parameters in the equation are measurable. In order to determine η_a and η_c separately, one notices that,

$$V_{21,meas} = V_{21} + \eta_a, \quad V_{43,meas} = V_{43} + \eta_c \tag{8.13}$$

where,

$$\frac{V_{21}}{V_{32}} = \frac{x_2}{x_3 - x_2}, \quad \frac{V_{43}}{V_{32}} = \frac{L - x_3}{x_3 - x_2} \tag{8.14}$$

Using Equations (8.13) and (8.14) allows determination of η_a and η_c by measuring the voltage drops V_{32}, $V_{43,meas}$ and $V_{21,meas}$ and the geometrical parameters x_2, x_3, and L.

Instead of a linear configuration one can also use a Van der Pauw configuration with planar samples.[46] It is most convenient to use circular samples with four equally spaced electrodes applied near the edge.[47]

3. Current Interruption Method

A different method uses current interruption. It is based on the experimental findings that when the current is set to zero the fastest response in, say SOFCs, is that of the ohmic IR voltage drop which follows the rate of decay of the current. Only then starts a slower decay of the overpotential on the electrodes (which originates from adsorption, charge transfer, and diffusion limitations). Figure 8.11 exhibits the voltage drop on the cathode and part of the bulk for a CeO_2-based SOFC as a function of time before and after current interruption.[48] The voltage just after the IR jump is the cathode overpotential due to causes not associated with an IR drop.

When only two electrodes are used, the method yields the sum of the overpotential at the cathode and anode (except for the IR drop there): i.e., $\eta_a + \eta_c - I(R_a + R_c)$, where R_a, R_c is the ohmic resistance of the anode and cathode, respectively. In order to separate the contributions of the anode and cathode, one uses a four-electrode configuration, as shown schematically, e.g., in Figure 8.7, and follows $V_{21,meas}$ and $V_{43,meas}$ as a function of time before and after current interruption. This was the method used with the arrangement shown in Figure 8.12 to obtain the results shown in Figure 8.11.

4. Use of High dc Voltages

The next method for directly measuring the electrode overpotential uses rather high dc voltages (from approximately 1 V to a few tens of volts). It is therefore applied preferably

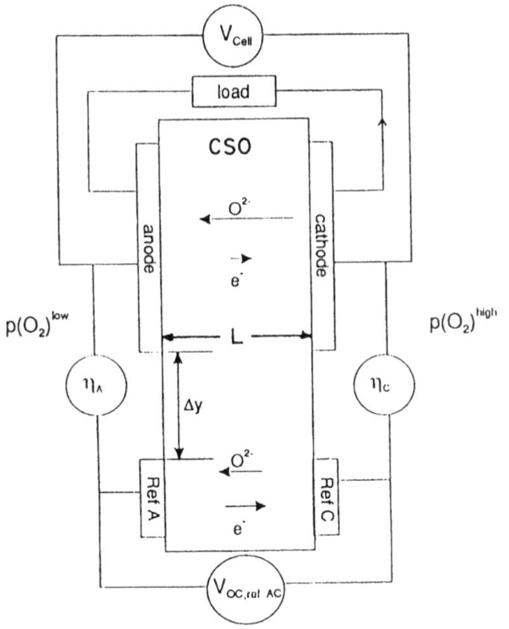

FIGURE 8.12. Four-electrode arrangement for determining the electrode overpotential in a SOFC, using non-blocking voltage probes (Ref. A and Ref. C), with $\Delta y \gg L$. Specific example: the MIEC Sm_2O_3–doped CeO_2 (CSO) as the solid electrolyte. (From Riess, I., Gödickemeier, M., and Gauckler, L.J., *Solid State Ionics,* in press.)

to high-resistance samples. It is based on the fact the I-V relations of the ohmic drop and the electrode processes are different. When the current I increases more slowly than linear with the overpotential at the electrode, the $I - \eta_a$, or $I - \eta_c$ relations dominate the I-V relations at high voltages, i.e., η_a or $\eta_c \sim V_a$. The method is useful also when the current I increases faster than linearly with η_a and η_c.[49] One measures the I-V relations up to high values of the applied voltage where the I-V relations become linear. This is an indication that the ohmic, IR voltage drop becomes the dominant term and one can determine R, the ohmic resistance of the bulk plus the electrodes, from the I-V relations. Now the IR drop can be subtracted from the measured total voltage V_a to yield information on the electrode overpotential for the entire measured range of the I-V relations,

$$V_a - IR = \eta_a + \eta_c - I(R_a + R_c) \tag{8.15}$$

In many cases $R_a + R_c \ll R$ and R equals the resistance of the bulk.

To measure η_a and η_c separately, one can use a four-electrode arrangement and follow $V_{21,meas}$ and $V_{43,meas}$ as a function of current to high current values until the I-V relations become either $I - \eta_a$ or $I - \eta_c$ dominated in one case, or become linear in the other case. The configuration need not be linear. It can be a Van der Pauw one. When the ohmic resistance of the electrodes can be neglected, the shape of the sample is not important and the four-electrode method allows determination of η_a and η_c separately.

Methods of measurement to be applied to MIECs should distinguish between the overpotential for electron/hole current and the overpotential for ion current.

5. Overpotential at Ion-Blocking Electrodes

When the sample is a MIEC the current through the bulk and electrode consists, in general, of an electronic component and an ionic one unless special care is taken to block one of the partial currents, e.g., by using blocking electrodes.[3-5]

A configuration using three ion-blocking electrodes and the fourth electrode a nonblocking one is shown in Figure 8.7. The current through the cell is purely electronic (when the MIEC sample conducts defects having a single $|z|$ value).[5] The overpotential at the electrodes is therefore due to the electron/hole current only. The I-V relations for this galvanic cell are not linear. However, when the nonlinear relations are taken into consideration the voltage drops, on the MIEC, i.e., V_{21} and V_{43} can be calculated from the measured value of V_{32} and the geometry, and compared with the measured values of $V_{21,meas}$ and $V_{43,meas}$. The difference between the measured and the calculated values yields the overpotentials η_a and η_c.[40,42] One then attributes the overpotential of the electrodes to electronic current only as the ionic current is blocked.

6. Overpotential at Electron-Blocking Electrodes

Figure 8.8 exhibits an arrangement that utilizes three-electron blocking electrodes (SEs) to block the electronic partial current through the cell under certain conditions.[5] The overpotential at the interface between the MIEC sample and the electrode E(M) is therefore due to the ionic current. The driving force is a gradient in $\tilde{\mu}_i$ and the overpotential is due to a difference $\Delta\tilde{\mu}_i/zq$ at the electrode. If the same material M is used also to back the probe at x_2, then $zqV_{i,21} = \Delta\tilde{\mu}_{i,21}$. The overpotential is determined then by measuring $V_{21,meas}$ and by calculating the value of V_{21}. The difference between the calculated value and the corresponding measured value yields the overpotential $\Delta\tilde{\mu}_i/zq$ at electrode M.

7. Overpotential at Nonblocking Electrodes

If all four electrodes are nonblocking, one has to take into consideration that both ionic and electronic currents are flowing through the bulk and the four electrodes. The measurements can be analyzed for an arbitrary MIEC in the Van der Pauw configuration, when small signals are used.[42] This enables separation of the potential drops on the bulk from the overpotential at the electrodes. In other cases an analysis, suitable to the specific defect model and geometry, has to be developed that predicts the I-V relations. The latter has been done for CeO_2-based SOFCs having the shape of thin disks, as shown in Figure 8.12.[48]

C. AC IMPEDANCE MEASUREMENTS

Small-amplitude ($V < kT/q$) ac impedance measurements are used to determine the bulk, and both the electrode and the grain boundary characteristics. As mentioned before, the grain boundaries respond to a signal applied perpendicular to the boundary, as a capacitor and a resistor in parallel. For electrodes, the charge transfer process under small ac signals also exhibits a response that can be described by a capacitor connected with a resistor in parallel. The corresponding impedance is

$$Z_{RC}(\omega) = \frac{R - i\omega R^2 C}{1 + \omega^2 R^2 C^2}, \quad i = \sqrt{-1} \tag{8.16}$$

where $\omega = 2\pi f$, f is the signal frequency, $0 \le \mathrm{Re}Z_{RC} \le R$, and $0 \le -\mathrm{Im}Z_{RC} \le R/2$, with

$$(\mathrm{Re}\,Z_{RC} - R/2)^2 + (\mathrm{Im}\,Z_{RC})^2 = (R/2)^2 \tag{8.17}$$

Equation (8.17) shows that drawing $-\mathrm{Im}Z_{RC}$ vs. $\mathrm{Re}Z_{RC}$ forms a semicircle with a radius R/2, centered around R/2 on the real axis. In combination with other elements which contribute to the impedance, the center is shifted from that point.

For ion-blocking electrodes applied on a SE, the dc ($\omega \to 0$) resistance of the electrodes is infinite. The reason is that charge can be introduced or removed from a SE only via ions which require electrodes that allow material and charge transport. However, for ion-blocking electrodes this is not possible. The electrode and the nearby surface of the SE behave as a double-layer capacitor, C_{dl}, with an impedance,

$$Z_C(\omega) = \frac{-i}{\omega C_{dl}} \tag{8.18}$$

The transport of mass when limited by diffusion exhibits a Warburg-type impedance,[13,50]

$$Z_W(\omega) = \frac{B(1-i)}{\omega^{1/2}}, \qquad B = \frac{kT}{z^2q^2}\frac{1}{Sc_0\sqrt{2D}} \tag{8.19}$$

where S is the cross-sectional area of the diffusion path, D the chemical diffusion coefficient, c_o the equilibrium concentration of the diffusing species, and zq the charge of the mobile ions inside the MIEC formed from or equal to the diffusing species. Equation (8.19) holds for an infinitely long diffusion path. For a finite diffusion path Δ the impedance is[13,50]

$$Z_D(\omega) = Z_W(\omega) \times \tanh\left((1+i)\Delta\sqrt{\omega/2D}\right) \tag{8.20a}$$

where for $\omega \to 0$ $Z_D(0)$ is real and finite,

$$Z_D(0) = \frac{kT}{z^2q^2}\frac{\Delta}{Sc_0D} \tag{8.20b}$$

The electrochemical cell exhibits an impedance that is, usually, a combination of the elements mentioned before. They can be coupled in series and in parallel. A simple Cole–Cole plot (–ImZ vs. ReZ) is shown in Figure 8.13, assuming diffusion limitation, D, in the gas phase in series with a charge transfer impedance at the electrodes, contributed by C_{dl} and R_{CT}, and in series with the bulk resistance R_B which shunts a capacitance C_e of the dielectric between the plate electrodes. In drawing Figure 8.13 it was assumed that the characteristic time constants of the three circuit elements differ so much that no overlap occurs in the corresponding impedance plots of Figure 8.13b. The equivalent circuit yielding a given Cole–Cole plot is not unique, as demonstrated in Figure 8.13.

When the impedance plot is sufficiently detailed and the various contributions resolved, one can determine the various resistances, capacitance, and diffusion process parameters. Care must be taken in analyzing the data as different equivalent circuits can yield the same or very similar impedance plots.[50] It may therefore be meaningless to use sophisticated programs for fitting complicated equivalent circuits comprising many elements and to look for best fit if the data are not sufficiently accurate, which is usually the case. It is then advisable to guess an as simple equivalent circuit as possible that can be justified electrochemically for describing the supposed processes in the cell and fit its parameters.

To assist in identifying the phenomena which determine the ac impedance spectrum one can vary parameters that affect the relaxation processes differently. The typical parameters are temperature, oxygen partial pressure, dc voltage applied on top of the ac one, length of MIEC, and grain size. Raising the temperature will enhance the contribution of the process with the higher activation energy. Lowering the oxygen partial pressure, $P(O_2)$, will promote oxygen diffusion limitation and the corresponding impedance. In some cases a variation in $P(O_2)$ changes the stoichiometry of the electrode and/or sample to the extent that the electrode parameters are significantly changed. Applying a dc voltage does not affect the impedance

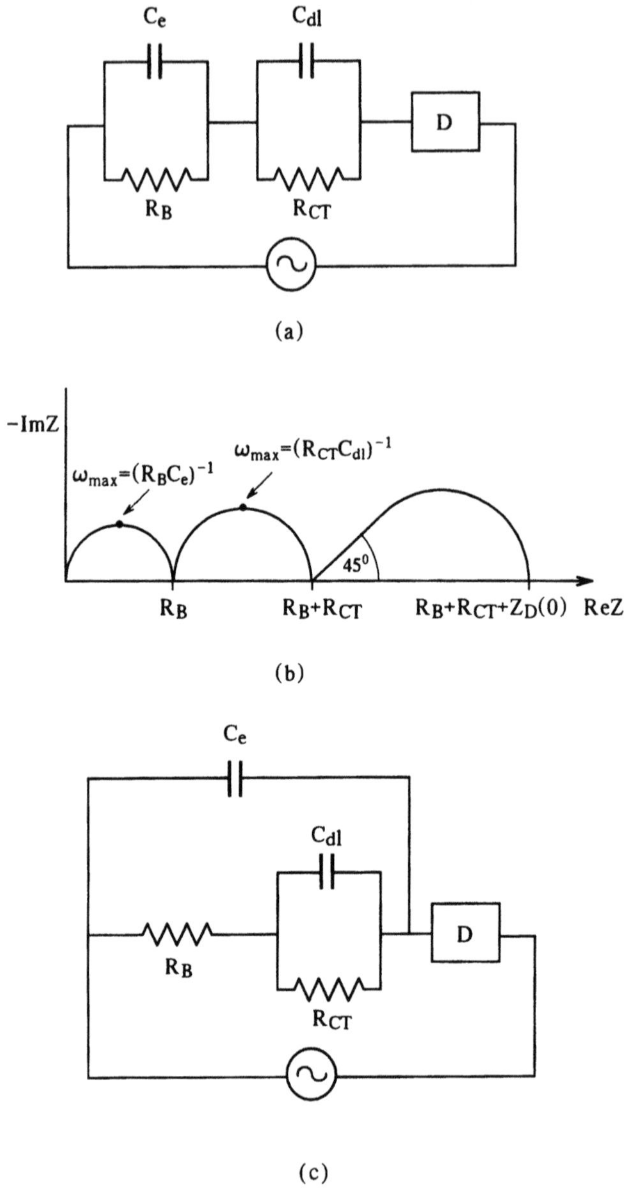

FIGURE 8.13. ac impedance of a typical equivalent circuit of an electrochemical cell. (a) The equivalent circuit, where the element D represents a diffusion path with length Δ (Equation [8.20a]). (b) The corresponding –ImZ vs. ReZ plot. (c) A different equivalent circuit that can result in the same –Imz vs. ReZ plot.

of components which are ohmic ($V \alpha I$), but changes the impedance of the nonohmic components. In particular, when the current depends exponentially on the voltage, as in the Butler–Volmer relations,[29] the dc bias reduces the ac impedance.[49] For a diffusion-limited process the dc bias increases the ac impedance. Increasing the length of the MIEC, keeping the nature of the electrodes the same, should increase the bulk impedance linearly with length, but should not change the electrode impedance. Increasing the grain size reduces the total grain boundary surface area per unit volume and should therefore reduce the impedance associated with the grain boundaries.

To assist further the identification of the processes that determine the ac impedance spectrum, it is sometimes necessary to replace components of the galvanic cell by different ones. This enables verification that the attribution of certain parts of the ac impedance spectrum to a specific process is correct.[51]

The electrode contribution can be removed by using a four-electrode ac measurement.[49] However, this may not be convenient as most ac impedance measuring devices do not enable true four-point measurements.

The shape of the ac impedance plots may deviate from that expected for the simple RC and Warburg elements. There are different reasons for deviations.[50] Typical reasons are rough surfaces, constriction resistance, and distribution of elements with different characteristic parameters, mainly in the bulk. The constriction resistance is due to a smaller contact area of the electrode than the nominal electrode area. At low frequencies the capacitance reflects the actual contact area, while at high frequencies the capacitance reflects the area of the electrode material which may be larger. Thus the contact cannot be described by a single capacitance.[52] It has also been shown[53] that for a MIEC electrode the impedance of transfer of oxygen from the gas phase into the MIEC and the impedance of diffusion inside the MIEC, though coupled in series, do not yield separated parts in the Cole–Cole plot.

D. I-V RELATIONS

Much work has been done on determining the overpotential of electrodes in SOFCs. It is observed that usually there is one rate-determining step at low current densities and a different one at high current densities. At low ones the rate-determining step is a charge transfer one.[48] The corresponding I-η relation can be approximated by the equation

$$I = 2I_0 \sinh\left(\frac{\alpha q \eta}{kT}\right) \tag{8.21}$$

Equation (8.21) can be applied to different charge transfer processes both at the cathode and the anode by fitting different parameter values for α and I_0 to these processes. The approximation in Equation (8.21) for the Butler–Volmer equation[29] is much better than the Tafel approximation.[29,48] The reason is that it is exact both for large currents (in one of the directions) and for zero current. The equation can be used as an approximation for the Butler–Volmer equation in both current directions if a different pair of parameters values α and I_0 are used for each direction.

The rate-determining step at high current densities in SOFCs is usually impeded diffusion. The I-η relations for molecular oxygen diffusion are[13,48]

$$\eta = \frac{kT}{4q}\ln\left(\frac{P(O_2)^{ext}}{P(O_2)^{ext} - C_1 I_i}\right) = -\frac{kT}{4q}\ln\left(1 - \frac{I_i}{P(O_2)^{ext}/C_1}\right) \tag{8.22}$$

and for monoatomic diffusion

$$\eta = \frac{kT}{2q}\ln\left(\frac{\left(P(O_2)^{ext}\right)^{1/2}}{\left(P(O_2)ext\right)^{1/2} - C_2 I_i}\right) = -\frac{kT}{2q}\ln\left(1 - \frac{I_i}{\left(P(O_2)^{ext}\right)^{1/2}/C_2}\right) \tag{8.23}$$

where I_i is the ionic current though the MIEC and C_1, C_2 are constants determined by the diffusion rate. The monoatomic species can be a neutral oxygen atom or an oxygen ion diffusing along a conducting surface of constant electrical potential. Equations (8.22) and (8.23) exhibit limiting current behavior under high η values with the limits $P(O_2)^{ext}/C_1$ and

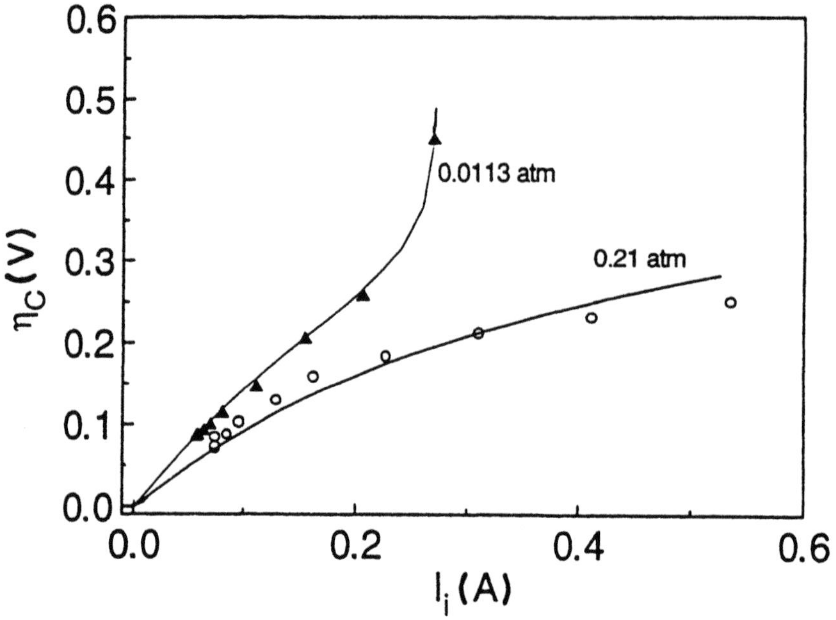

FIGURE 8.14. Overpotential–current relations for the MIEC LSC, on the MIEC electrolyte Sm_2O_3–doped CeO_2, at T = 700°C, $P(O_2)^{ext}$ = 0.21 and 0.0113 atm. (From Riess, I., Gödickemeier, M., and Gauckler, L.J., *Solid State Ionics,* in press.)

$(P(O_2)^{ext})^{1/2}/C_2$, respectively. The limitation on the current is due to limitation on the driving force, i.e., the concentration gradient which reaches its maximum when the low concentration can be neglected while the high concentration maintains a fixed value. Normally, the limiting current in SOFC cathodes exhibits a linear dependence on $(P(O_2)^{ext})^{1/2}$. However, in the past it has been reported that the diffusion limited current was linear in $P(O_2)^{ext}$ rather than in $(P(O_2)^{ext})^{1/2}$. This can be ascribed to excessive diffusion impedance to the gas in the sputtered and probably quite dense Pt electrode used.[54]

A characteristic I-η relation measured on a cathode of an SOFC is shown in Figure 8.14. For $P(O_2)^{ext}$ = 0.21 atm the overpotential is determined by charge transfer limitation only. For $P(O_2)^{ext}$ = 0.0113 atm a limiting current is observed at high current densities. By examining the $P(O_2)$ dependence of the limiting current one can determine whether the diffusion is limited by molecular diffusion or monoatomic diffusion. The diffusion limitation in Figure 8.14 is of monoatomic species. In general, the main contribution to the cathode impedance in modern SOFCs comes from a Butler–Volmer-type relation, most probably a charge transfer process. Only at high current densities, above 1 A/cm^2, can diffusion limitation be detected.[11,48]

In addition to charge transfer and diffusion limitation, chemical reactions can also take place at intermediate reaction steps. For instance, in Ni/CeO_2 cermet anodes, a possible intermediate reaction product can be NiO. These processes seem not to be rate limiting, and they have not been uniquely identified and thoroughly investigated.

E. DIFFUSION IN INTERCALATION ELECTRODES

It has been shown that due to the thin structure of the cathodes in SOFCs it is possible to reduce the impedance of diffusion of the O_2 molecules to the extent that it can be neglected, and that the impedance to diffusion of atomic oxygen is quite small. In solid state batteries using large insertion electrodes the diffusion path may be quite long and the diffusion impedance no longer negligible. It is then of interest to determine the chemical diffusion coefficient of the diffusing species in the insertion electrode. This is done most conveniently using the galvanostatic intermittent titration technique.[55] In this method, a pulse of material is pumped from a reference electrode through a SE into the insertion electrode. The rate of diffusion of the material from the interface region is monitored by the EMF of the cell while pumping and after current interrupt. Polarization overpotential can be generated at the SE/electrode interface by the current. This introduces an error in the EMF measured. To avoid this the method is modified and the EMF is measured on the other end of the insertion electrode following the rate at which the material reaches that side.[56] No overpotential exists there, as the current there vanishes.

F. DETERMINING THE GRAIN BOUNDARY PROPERTIES

The electrochemical methods discussed before for electrode characterization can be applied in principle also for characterizing grain boundaries. We have to remember that one has to consider the impedance for current perpendicular to the boundary and the impedance for current parallel to the boundary. These impedances are different. For the determination of the first one the ac impedance method is usually used, yielding the capacitance and resistance of the interface. The interpretation is simple when the boundary properties are rather uniform. The resistance of a current parallel to the surface can be determined when it is much smaller than the bulk resistance in parallel with it. Then the overall resistance in dominated by the grain boundary region low resistance.[30]

We have considered grain boundaries at quasi-equilibrium. In reacting solids the boundary between the phases moves. Also, in a single phase under sintering conditions the grain boundaries move. The characterization in these cases are usually done by structural observations as a function of time. The phenomena associated with reactions in solids are reviewed extensively in two books by Schmalzried,[26,57] and sintering in a single phase is reviewed in many good monographs. These topics will therefore not be discussed here.

V. SUMMARY

This review concentrates on the properties of electrodes and grain boundaries, the applications of electrodes, and the characterization of electrodes and grain boundaries mainly by solid state electrochemical methods.

ACKNOWLEDGMENT

One of the authors (I.R.) thanks the German–Israel Foundation for Research and Development (GIF) for supporting a research program which has contributed in great measure to the development of the material in this chapter. This chapter has profited considerably from discussions with L.J. Gauckler, M. Gödickemeier, and J. Maier.

REFERENCES

1. Kröger, F.A., *The Chemistry of Imperfect Crystals,* Vol. II, North-Holland, Amsterdam, 1974, 14.
2. van Hassel, B.A., Kawada, T., Sakai, N., Yokokawa, H., Dokiya, M., and Bouwmeester, H.J.M., *Solid State Ionics,* 1993, *66,* 295.
3. Hebb, M.H., *J. Chem. Phys.,* 1952, *20,* 185.
4. Wagner, C., *Z. Elektrochem.,* 1956, *60,* 4.
5. Riess, I., Critical review of methods for measuring partial conductivities in mixed ionic electronic conductors, in *Ionic and Mixed Conducting Ceramics,* 2nd Intl. Symp., Ramanarayanan, T.A., Worrell, W.L., and Tuller, H.L., Eds., PV 94-12, The Electrochemical Society, Pennington, NJ, 1994, 286–306.
6. Asano, K., Hibino, T., and Iwahara, H., *J. Electrochem. Soc.,* 1995, *142,* 3241.

7. Riess, I., van der Put, P., and Schoonman, J., *Solid State Ionic,* 1995, *82,* 1.
8. Vayenas, C.G., Bebelis, S., Yentekakis, I.V., and Lintz, H.-G., *Catal. Today,* 1992, *11,* 303.
9. Sasaki, K., Wurth, J.P., Gschwend, R., Gödickemeier, M., and Gauckler, L.J., *J. Electrochem. Soc.,* 1996, *143,* 530.
10. Verkerk, M.J., Hammink, M.W.J., and Burggraaf, A.J., *J. Electrochem. Soc.,* 1983, *130,* 70.
11. Gödickemeier, M., Gauckler, L.J., and Riess, I., *J. Electrochem. Soc.,* in press.
12. Wang, D.Y. and Nowick, A.S., *J. Electrochem. Soc.,* 1981, *128,* 55.
13. Braunshtein, D., Tannhauser, D.S., and Riess, I., *J. Electrochem. Soc.,* 1981, *128,* 82.
14. Wang, D.Y. and Nowick, A.S., *J. Electrochem. Soc.,* 1979, *126,* 1155.
15. Wang, D.Y. and Nowick, A.S., *J. Electrochem. Soc.,* 1979, *126,* 1166.
16. van Herle, J. and McEvoy, A.J., *J. Phys. Chem. Solids,* 1994, *55,* 339.
17. van Herle, J., McEvoy, A.J., and Ravindranathan Thampi, K., *Electrochim. Acta,* 1994, *39,* 1675.
18. Olmer, L.J. and Isaacs, H.S., *J. Electrochem. Soc.,* 1982, *129,* 345.
19. Jaspers, B.C., van Dongen, B.A.M., Monaster, G.A., van Roosmalen, J.A.M., Plaisier, K.H., and Schoonman, J., Proc. 4th Int. Symp. on Solid Oxide Fuel Cells, SOFC-IV, Japan, 1995.
20. Gödickemeier, M., Ph.D. thesis, ETH, Zurich, 1995.
21. Møgensen, M. and Lindegaard, T., The Kinetic of Hydrogen Oxidation on a Ni-YSZ SOFC Electrode at 1000°C, in *Proc. 3rd Intl. Symp. SOFC,* Singhal, S.C. and Iwahara, H., Eds., *The Electrochemical Society,* Pennington, NJ, Proc. Vol. 93-4, 1993, 484–493.
22. Julien, C. and Nazri, G.-A., *Solid State Batteries: Materials Design and Optimization,* Kluwer Academic Publishers, Dordrecht, 1994, 369–511.
23. Honders, A. and Broers, G.H.J., *Solid State Ionics,* 1985, *15,* 173.
24. Fabry, P., Schouler, E., and Kleitz, M., *Electrochim. Acta,* 1978, *23,* 539.
25. Hammouche, A., Siebert, E., Hammou, A., and Kleitz, M., *J. Electrochem. Soc.,* 1991, *138,* 1212; Siebert, E., Hammou, A., and Kleitz, M., *Electrochim. Acta,* 1995, *40,* 1741.
26. Schmalzried, H., *Solid State Reactions,* Verlag Chemie, Weinheim, 1981, 118, 155.
27. Kilner, J.A., Isotopic exchange in mixed and ionically conducting oxides, in *Ionic and Mixed Conducting Ceramics,* 2nd Intl. Symp., Ramanarayanan, T.A., Worrell, W.L., and Tuller, H.L., Eds., The Electrochemical Society, Pennington, NJ, 1994, 174–190.
28. Cornish, J. and Warde, C.J., *Ber. Bunsenges. Phys. Chem.,* 1978, *82,* 282.
29. Bard, A.J. and Faulkner, L.R., *Electrochemical Methods, Funadmentals and Applications,* John Wiley & Sons, New York, 1980, 101–107.
30. Maier, J., in *Recent Trends in Superionic Solids and Solid Electrolytes,* Chandra, S. and Laskar, A., Eds., Academic Press, New York, 1989, 137; *Ber. Bunsenges. Phys. Chem.,* 1989, *93,* 1474.
31. Macdonald, J.R., Franceschetti, D.R., and Lehnen, A.P., *J. Chem. Phys.,* 1980, *73,* 5272.
32. Liu, S.H. and Kaplan, T., *Solid State Ionics,* 1986, *18&19,* 65.
33. Jamnick, J., Pejovnik, S., and Maier, J., *Electrochim. Acta,* 1993, *14,* 1975.
34. Maier, J., *J. Am. Ceram. Soc.,* 1993, *76,* 1218.
35. Riess, I., *Solid State Ionics,* 1995, *80,* 129.
36. Rickert, H., *Electrochemistry of Solids,* Springer-Verlag, Berlin, 1982, 96–100.
37. Riess, I., *Solid State Ionics,* 1991, *44,* 199.
38. Kobayashi, H. and Wagner, C., *J. Chem. Phys.,* 1957, *26,* 1609.
39. Dyer, C.K., *Nature,* 1990, *343,* 547.
40. Riess, I., *Solid State Ionics,* 1992, *51,* 219.
41. Riess, I., *J. Electrochem. Soc.,* 1992, *139,* 2250.
42. Riess, I. and Tannhauser, D.S., *Solid State Ionics,* 1982, *7,* 307.
43. Liang, C.C., *J. Electrochem. Soc.,* 1973, *120,* 1289.
44. Wiemhöfer, H.D. and Göpel, W., *Fresenius J. Anal. Chem.,* 1991, *341,* 106; Wiemhöfer, H.D., Vohrer, U., and Göpel, W., *Mater. Sci. Forum.,* 1991, *76,* 265.
45. Lee, J.-S. and Yoo, H.-I., *J. Electrochem. Soc.,* 1995, *142,* 1169.
46. van der Pauw, L.J., *Philips Res. Rep.,* 1958, *13,* 1.
47. Riess, I., *J. Appl. Phys.,* 1992, *71,* 4079.
48. Riess, I., Gödickemeier, M., and Gauckler, L.J., *Solid State Ionics,* in press.
49. Cohen, Y., Davidovich, A., and Riess, I., *Appl. Phys. Lett.,* 1995, *66,* 706.
50. Raistrick, I.D., Macdonald, J.R., and Franceschetti, D.R., in *Impedance Spectroscopy,* Macdonald, J.R., Ed., John Wiley & Sons, New York, 1987.
51. Jak, M., Riess, I., and Schoonman, J., Proc. 10th Intl. Conference on Solid State Ionics, SSI-10, Singapore, Dec. 3–8, 1995.
52. Fleig, J. and Maier, J., Proc. 3rd Intl. Symp. on Electrochemical Impedance Spectroscopy, Belgium, May 1995.
53. Adler, S.B., Lane, J.A., and Steele, B.C.H., *J. Electrochem. Soc.,* in press.
54. Gür, T.M., Raistrick, I.D., and Huggins, R.A., *J. Electrochem. Soc.,* 1980, *127,* 2620.
55. Weppner, W. and Huggins, R.A., *J. Electrochem. Soc.,* 1977, *124,* 1569; *Annu. Rev. Mater. Sci.,* 1978, *8,* 269.
56. Becker, K.D., Schmalzried, H., and von Wurmb, V., *Solid State Ionics,* 1983, *11,* 213.
57. Schmalzried, H., *Chemical Kinetics of Solids,* VCH Publishers, Weinheim, Germany, 1995.

Chapter 9

PRINCIPLES OF MAIN EXPERIMENTAL METHODS

Werner Weppner

CONTENTS

I. INTRODUCTION

Ionic conductors may act as transducers for a large variety of fundamental thermodynamic and kinetic quantities. The most important and very attractive features are

- thermodynamic and kinetic properties are directly — without any intermediate step — converted into electrical quantities;
- the precision is very high compared to other techniques, since voltages, electrical currents, and time are readily measurable with high precision; some types of information otherwise not accessible, or obtained only with great difficulty, may be acquired readily;

0-8493-8956-9/97/$0.00+$.50

- the composition may be controlled precisely *in situ,* allowing in this way the collection of data as a function of even very small changes of stoichiometry and the drastic reduction of the number of samples to be employed;
- the experimental arrangement is commonly very simple; however, the requirement of chemical compatibility of the various galvanic cell materials, which are in contact with each other, and skills of the experimentalist are required;
- solid state electrochemical methods provide, furthermore, the advantage of a large applicable temperature range, especially the application of high temperatures when thermodynamic equilibria are more readily achieved.

Some fundamental principles underlying all electrochemical methods will be briefly discussed first. A convention followed throughout this chapter is to treat quantities such as the chemical potential, electrochemical potential, and the activity as properties of a *single* particle. Voltages and currents are assumed to be positive if the right-hand side is positively charged relative to the left-hand side, and positive species move to the right (or negative ones to the left), respectively (Stockholm convention).[1]

II. FUNDAMENTAL ASPECTS

The equilibrium voltage of a galvanic cell is determined by thermodynamic quantities, whereas the electrical current is related to dynamic processes of the mobile ionic and electronic species in one or more parts of the galvanic cell. Despite this clear separation between kinetic and thermodynamic information, many experimental methods involve the determination and/or control of both types of electrical quantities.

A. FORMATION OF GALVANIC CELL VOLTAGES

The application of a voltmeter even with very high input impedance (“electrometer”) requires a small current. A chemical reaction will occur in the galvanic cell by the transfer of the ions. The chemical energy of this process at constant total pressure and temperature, the Gibbs energy of reaction, ΔG_r, corresponds to the electrical work that accompanies the electrical current:

$$\Delta G_r = -nqE, \tag{9.1}$$

where n, q, and E are the number of charges that are transferred for the reaction, the elementary charge, and open-circuit cell voltage (emf), respectively. The change of the Gibbs energy with the number of atoms in the electrodes corresponds to the chemical potential μ of this component. Accordingly, the transfer of ions A^{z+} corresponds to the cell voltage

$$E = \frac{1}{zq}\left(\mu_A^l - \mu_A^r\right). \tag{9.2}$$

l and r stand for the left- and right-hand electrodes, respectively. The chemical potential μ may be written in terms of the activity a or the equilibrium partial pressure p_A,

$$\mu_A = \mu_A^0 + kT \ \ln a_A = \mu_A^0 + kT \ \ln p_A . \tag{9.3}$$

μ^0, k, and T are the chemical potential in the standard state, Boltzmann's constant, and absolute temperature, respectively. Substitution of Equation (9.3) into (9.2) results in

$$E = \frac{kT}{zq} \ln \frac{a_A^l}{a_A^r} = \frac{kT}{zq} \ln \frac{p_A^l}{p_A^r}. \tag{9.4}$$

The voltage is determined by the reaction that really occurs by the application of the voltmeter. This does not always agree with the thermodynamically most favorable reaction.

B. GALVANIC CELL CURRENTS

If an electric current is forced or permitted to pass through the galvanic cell, the partial current density i_n of any species n (ions A^{z+}, electrons e, holes h) is, in general, proportional to the overall transport force acting upon that species. If the transport is isothermal and involves no volume change, we can neglect all but chemical and electrostatic forces. Correlations with the fluxes of other particles may also often be neglected.[2] The proportionality constant between the current and the electrochemical potential gradient is given by the partial electrical conductivity σ_n,

$$i_n = -\frac{\sigma_n}{z_n q} \frac{\partial \eta_n}{\partial x}, \tag{9.5}$$

where η_n is the electrochemical potential of species n which is composed of an electrostatic potential ϕ and a chemical potential μ_n contribution according to the relation

$$\eta_n = \mu_n + z_n q \phi. \tag{9.6}$$

The conductivity σ_n may be written in terms of the product of the concentration c_n and the diffusivity D_n of the mobile species n:[3]

$$\sigma_n = \frac{c_n D_n z_n^2 q^2}{kT} \tag{9.7}$$

where D_n is the diffusivity (which is also sometimes called "component diffusion coefficient"), which is related to the electrical mobility u_n (drift velocity per unit electrical field) or the general mobility b_n (drift velocity per unit general force) by the Nernst–Einstein relation[3] regardless of whether the material may be considered to be an ideal or a nonideal solution:

$$D_n = u_n \frac{kT}{|z_n| q} = b_n kT. \tag{9.8}$$

This quantity describes the random motion of species n in the absence of concentration gradients. It is related to the tracer diffusion coefficient $D_{Tr,n}$ (or self-diffusion coefficient $D_{s,n}$), which is determined by the use of radioactive isotopic tracers according to $D_{Tr,n} = f\, D_n$,[3] where f is the correlation factor or, more generally, the Haven ratio.[4]

With the help of Equations (9.3), (9.6), and (9.7), Equation (9.5) can be cast into the form

$$i_n = -\sigma_n \frac{\partial \phi}{\partial x} - z_n q D_n \frac{\partial \ln a_n}{\partial \ln c_n} \frac{\partial c_n}{\partial x}. \tag{9.9}$$

The first term of the right-hand side is Ohm's law for the migration of species n under the influence of an electrical field. The second term is Fick's first law for diffusion under the influence of a concentration gradient. Equation (9.9) holds in general for isothermal conditions and is valid at any position in a galvanic cell, assuming that there is no volume change and no other forces are involved in the transport, e.g., correlation effects.

Equation (9.9) holds both in electrolyte phases, in which the transport of charge is primarily due to the motion of ionic species, and in mixed ionic and electronic conducting electrode phases. In most cases of interest one may assume that practically useful ionic conductors in galvanic cells are characterized by a very large concentration of mobile ionic defects. As a result, the chemical potential of the mobile ions (which is proportional to the logarithm of the activity or the concentration in a first approach) may be regarded as being essentially constant within the material. Thus any ionic transference in such phases must be predominantly due to the influence of an internal electrostatic potential gradient:

$$i_{\text{ion (electrolyte)}} = -\sigma_{\text{ion}} \frac{\partial \phi}{\partial x} . \tag{9.10}$$

On the other hand, for the electronic species (electrons and holes), which may be considered to be comparatively dilute in an electrolyte, the concentration gradient is the more important driving force.

A different situation holds in the case of electrodes that transport both electrons and ions, but predominantly electronic species. In that case the chemical potentials of the electrons or holes may be regarded as practically independent of location within the solid, due to their high concentrations. Thus, their transport is primarily due to the effect of an internal electric field in such materials. Because of the large concentrations as well as the high mobilities, this field will be very small.

If an ionic flux occurs across an electrode/electrolyte interface in connection with the current flow by the motion of ions in the electrolyte, the ionic transport within the electrode material must be predominantly determined by the presence of a local ionic concentration gradient:[3]

$$i_{\text{ion (electrode)}} = -z_{\text{ion}} q D_{\text{ion}} \frac{\partial \ln a_{\text{ion}}}{\partial \ln c_{\text{ion}}} \frac{\partial c_{\text{ion}}}{\partial x} \tag{9.11}$$

The "thermodynamic" factor $\partial \ln a_{ion}/\partial \ln c_{ion}$ (which is a special case of the Wagner factor, as described later) is sometimes very large and enhances the ionic flux above that which would be expected from the concentration gradient alone.[5] In a predominantly electronic conductor in which the concentrations of electrons or holes are very large, i.e., the chemical potentials of the electronic species are essentially uniform throughout the material, the gradients of the chemical potentials of neutral atoms and their respective ions are identical. The transport of ions may be considered to be the same as the net transport of neutral species in this case.

If an electrode material is not overwhelmingly an electron or hole conductor, the internal electric field may not be completely neglected with regard to the movement of ions. The transport of ionic species occurs under the combined influence of electrostatic and chemical potential gradients. This situation can be expressed by an equation similar to Equation (9.11) by replacing $\partial \ln a_{ion}/\partial \ln c_{ion}$ by the more general Wagner ("enhancement") factor W:[6]

$$i_{\text{ion (mixed conductor)}} = -z_{\text{ion}} q D_{\text{ion}} W \frac{\partial c_{\text{ion}}}{\partial x} \tag{9.12}$$

TABLE 9.1
Wagner Factor which Relates the Chemical Diffusion Coefficient $\tilde{D}$ to the Diffusivity D ($\tilde{D}$ =WD) for Various Conditions.

Transference number $t_{i,e}$	Concentrations $c_{i,e}$	Activity coefficients $\gamma_{i,e}$	Wagner factor W
General	General	General	$t_e[\partial \ln a_i/\partial \ln c_i + z_i\ \partial \ln a_e/\partial \ln c_i]$ $= t_e[\partial \ln a_i^*/\partial \ln c_i^*]$
General	General	$\gamma_{i,e}$ = constant	$t_e[1 + z_i^2\, c_i/c_e]$ or $t_e + t_i\, D_e/D_i$
$t_i \ll t_e$	General	γ_i = general, γ_e = constant	$t_e\, \partial \ln a_i/\partial \ln c_i + z_i^2\ c_i/c_e$
$t_i \ll t_e$	General	$\gamma_{i,e}$ = constant	$1 + z_i^2\, c_i/c_e$
$t_i \ll t_e$	$c_i \ll c_e$	$\gamma_{i,e}$ = constant	1
$t_i \ll t_e$	$\lvert z_i \rvert c_i = c_e$	$\gamma_{i,e}$ = constant	$1 + \lvert z_i \rvert$
$t_i \ll t_e$	$c_i \gg c_e$	$\gamma_{i,e}$ = constant	$z_i^2\, c_i/c_e$
$t_i \gg t_e$	General	General	$\ll[\partial \ln a_i^*/\partial \ln c_i^*]$

Note: Ionic species i and electrons e are the predominantly mobile species. a, c, and z are the activity, concentration, and charge number, respectively. Asterisks symbolize neutral species.

The values of the Wagner factor under various conditions are summarized in Table 9.1. This term of thermodynamic quantities can be interpreted kinetically as the generation of internal electric fields by two differently mobile species. If one kind, e.g., electrons, has a significantly higher mobility than the other kind, e.g., ions, the more mobile species will tend to move ahead of the other ones. The requirement for overall charge flux neutrality causes the more mobile species to be slowed down and the less mobile ones to be speeded up.[6]

The total charge transport within a solid (especially in an electrode) may be composed of several partial currents whose relative contributions vary with position. In the external circuit only the total current that is passed through the overall cell may be observed. However, an electrolyte serves as an ion-pass filter and only ionic species can cross phase boundaries between electrodes and such an electrolyte. Since under steady-state conditions the ionic current must be continuous, the total externally measurable current must be equal to the partial ionic current, as expressed in Equation (9.12), within the electrode material just at the interface.

It is possible that the most mobile ions within the electrolyte are not those that are predominantly transferred in the electrode. In such a case, the overall cell current is given by Equation (9.12) when it is written in terms of the ionic species that are most mobile in the electrode phase.

III. THERMODYNAMIC PROPERTIES

A. DETERMINATION OF PHASE EQUILIBRIA

The determination of phase equilibria is based on coulometric titration which was originally developed for the variation of the stoichiometry of binary compounds.[7] A charge transport $\int I dt$ that is passed through the galvanic cell

$$\left.\begin{matrix}\text{counter-}\\ \text{reference-}\end{matrix}\text{electrode}\,\right|\begin{matrix}\text{electrolyte}\\ \left(A^{z+}\text{ ions}\right)\end{matrix}\left|\,\text{sample}\right. \qquad \text{(I)}$$

changes the mass m_A of species A in the sample according to Faraday's law:

$$\Delta m_A = \frac{M_A}{zF}\int I\,dt. \qquad (9.13)$$

M_A and F are the atomic weight of A and Faraday's constant, respectively. This corresponds to a change in the stoichiometric number x of the component A in the sample by

$$\Delta x_A = \frac{1}{mzF}\sum_{i=1}^{k} x_i M_i\,. \qquad (9.14)$$

The summation runs over all components of the sample. m is the total mass of the sample with the starting composition.

Two other important aspects are the Gibbs-Duhem relation

$$\sum_{i=1}^{k} \mu_i dc_i = 0 \qquad (9.15)$$

and the decrease of the number of independently variable chemical potentials in case of equilibrium between an increasing number of phases as given by Gibbs' phase rule at constant total pressure and temperature,

$$f = k - p + 2 \qquad (9.16)$$

which relates the thermodynamic degree of freedom f to the number of phases p for a system with k components.

The current is interrupted from time to time, and the equilibrium cell voltage (emf) is measured as a function of the composition. Since the emf is an indication of the chemical potential of the electroactive component according to Equation (9.2), a constant composition-independent value is observed if the number of coexisting phases corresponds to the number of components. As soon as a smaller number of phases is present, a variation of the emf with composition must occur because of the increased number of degrees of freedom. The cell voltage has to change monotonically since the equilibrium condition of the minimum Gibbs energy for any given composition may not be fulfilled otherwise.

Figure 9.1 shows as an example the voltage drops at the stoichiometries of the single phases Li_3Sb and Li_3Bi of the binary systems Li-Sb and Li-Bi.[8] By using elemental antimony and bismuth as starting materials, small voltage steps are observed at the compositions Li_2Sb

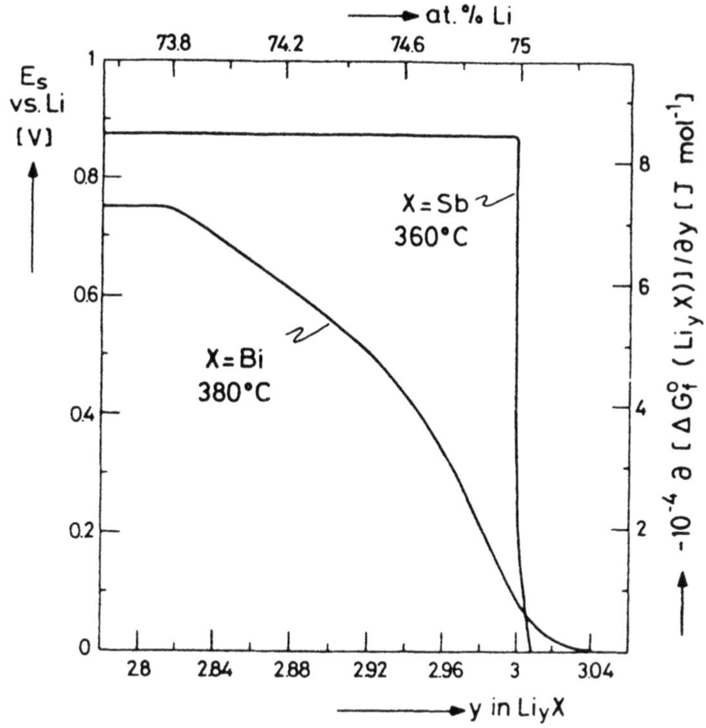

FIGURE 9.1. Coulometric titration curves for the systems Li-Sb and Li-Bi using a molten salt electrolyte for lithium ions. The equilibrium cell voltage is plotted with reference to elemental lithium as a function of the lithium concentration.

and LiBi. The voltage is otherwise independent of the composition within the two-phase regions ("plateaus").

The coulometric titration technique (CTT) requires only a small number of sample preparations. Only one sample has been sufficient to study the phase diagram of the system Li-Sb or Li-Bi. A ternary system requires only that many samples as required commonly for studying a binary system. Since the cell voltage also provides thermodynamic information (as will be described later), this may be taken into consideration and the number of sample preparations may be even further reduced.

The determination of ternary phase diagrams is shown in Figures 9.2 and 9.3 for the example of the system Cu-Ge-O. The composition of the ternary sample follows a straight line in the Gibbs triangle which connects the starting composition with the corner of the electroactive component which has been oxygen in the present case by using zirconia solid electrolytes.[9] Several voltage plateaus are observed which correspond to three-phase equilibria of the ternary system. These plateaus are separated by small two- (or one-) phase regions. Data are shown for four different Cu:Ge ratios (4:1, 3:2, 1:1, and 1:3).

Plateaus of the same cell voltage belong to the same three-phase region for all samples with different Cu:Ge ratios. In Figure 9.3, these results are transduced into Gibbs' phase diagram ("triangle"). By connecting the borders of the three-phase voltage plateau regions, one ternary phase, $CuGeO_3$, is found at the experimental temperature of 900°C. Three-phase regions which have one side in common with one of the binary legs of the electroactive component (Cu-O or Ge-O) show the same emf as the corresponding two-phase regions of the binary system. The third phase has no (or only little) influence, since it does not become involved in the cell reaction (with the exception of some dissolution of the third component in the binary phases). Only the relative ratios of the binary compounds along the binary legs are changed by the coulometric titration.

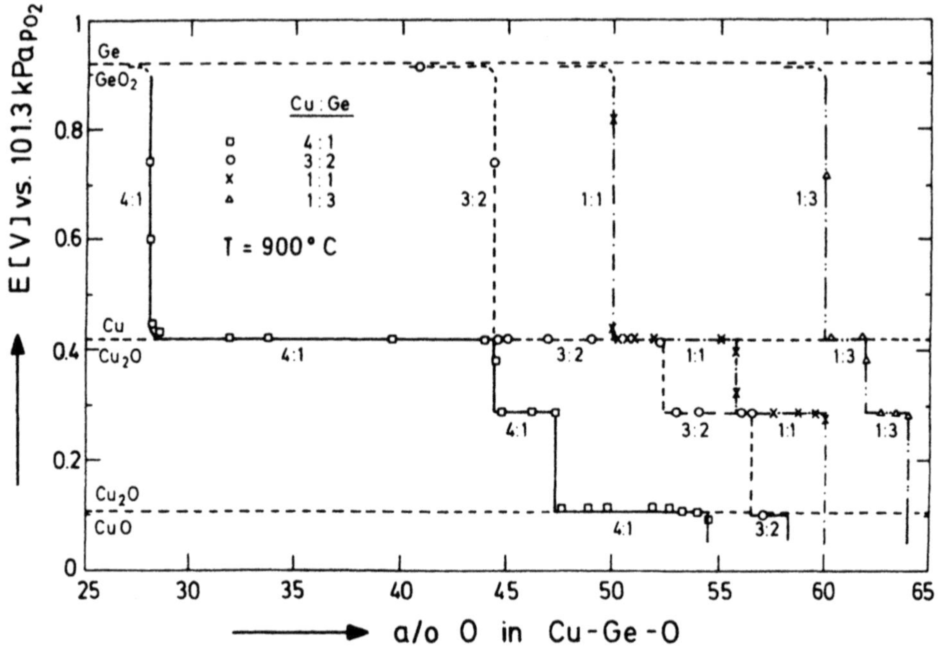

FIGURE 9.2. Coulometric titration curves for the ternary system Cu-Ge-O for samples with different Cu-Ge ratios using a solid zirconia oxygen ion conductor. The voltages related to the equilibria of the binary oxide systems are shown as horizontal broken lines.

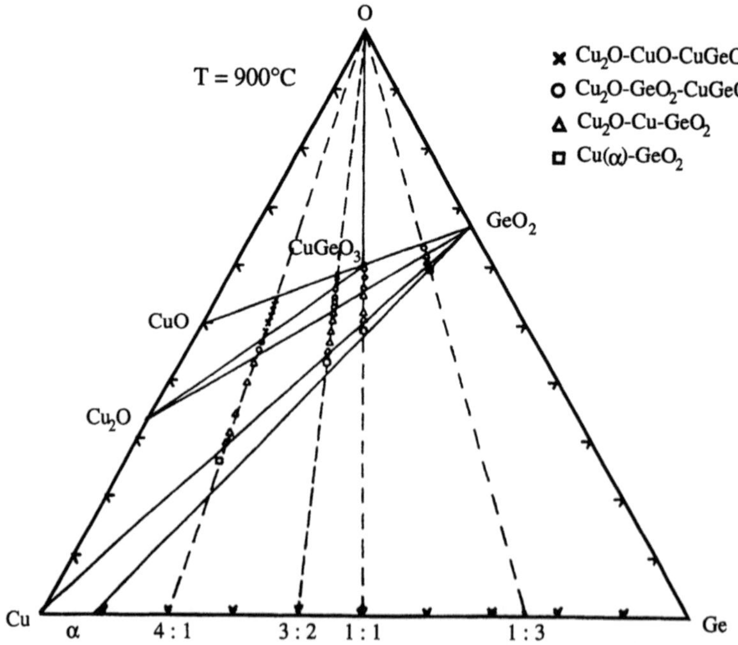

FIGURE 9.3. Phase diagram (Gibbs triangle) of the ternary system Cu-Ge-O. The ranges of plateau voltages in Figure 9.2 are translated into this diagram, and regions of identical voltages for different Cu:Ge ratios are combined to three-phase regimes indicated by the same symbols.

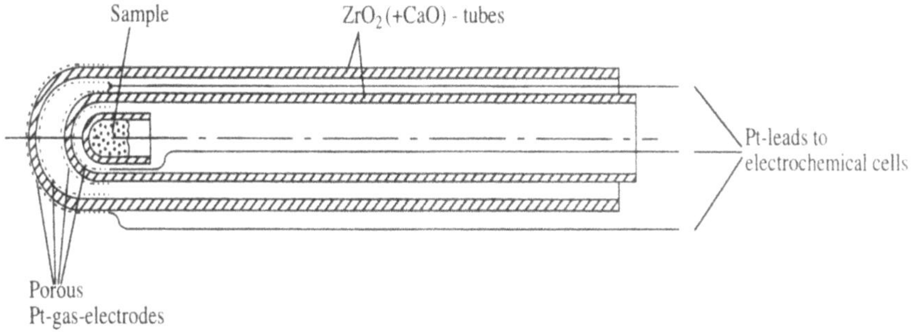

FIGURE 9.4. Experimental arrangement for the coulometric titration and emf measurements of the ternary systems Y-Fe-O and Dy-Fe-O. A double electrolyte cell is employed to avoid oxygen permeation from the outside air to the sample. The oxygen partial pressure in the intermediate gas gap is automatically controlled by a potentiostat to have the same pressure as over the sample. A current is passed to that purpose through the outer electrolyte.

For the construction of the phase diagram it should be verified that the voltage along any straight line through the corner of the electroactive component of the Gibbs triangle changes monotonically. This may be helpful to exclude possibly formed metastable states. Very useful also is the comparison of the titration results (i) in both directions, (ii) at longer annealing times, and (iii) as a result of temperature variations.

The stoichiometric resolution for the various regions of the phase diagram is extraordinarily high as a result of the very sensitive measurement of currents and time. An experimentally easily controllable current of 1 μA for the period of 10 s corresponds to a mass change of the sample which may not be achievable for any balance, e.g., 8.3×10^{-10} g for oxygen or 7.2×10^{-10} g for lithium. Compounds which are commonly called line phases may be readily resolved with regard to the stoichiometric width. As an example, the stoichiometric width for lithium in $LiAlCl_4$ was found to be 2×10^{-6}.[10]

The CTT does not require any quenching of the samples and may therefore also be applied to such system which may not be conventionally analyzed because of phase transitions during quenching. The materials are investigated *in situ* and the sample does not need to be destroyed. The equilibration process is always visible by viewing the time dependence of the emf. There is no need to prepare identical samples which are successively analyzed after different periods of time of annealing. Also, annealing for unnecessarily long periods of time is avoided.

The observed emf during the titration provides fundamental information on the thermodynamics of the system in addition to the phase equilibria. The knowledge of the emf as a function of stoichiometry allows the determination of thermodynamic data with high resolution as a function of the composition of the sample.

Other systems may be investigated in a similar way. Experimentally, many differences are observed, however. In the case of the systems Y-Fe-O and Dy-Fe-O, the long equilibration times have required drastic reduction in the oxygen leakage rate of the electrolyte by the simultaneous motion of ions and electrons or holes. A double electrolyte cell with a gas space in between has been used in this case, as shown in Figure 9.4. The oxygen partial pressure in the space between the tubes has been kept at the same value as in the inner compartment which contains the sample. Accordingly, no driving force for oxygen permeation through the inner electrolyte occurs and the leakage is suppressed. Experimentally, a current flux through the outer electrolyte removed the oxygen that has permeated through this electrolyte which has been controlled automatically by employing a potentiostat.[11]

Depending on the temperature, quite different phase equilibria were found (Figures 9.5 to 9.7). It is seen that the phase YFe_2O_{4-x} occurs only above 1010°C. Above 1078°C a change of the coexistence of the phases Fe-$YFeO_3$ occurs toward the equilibrium between Y_2O_3 and YFe_2O_{4-x}.[11] Instead of carrying out coulometric titrations at different temperatures, it is more

(9.5)

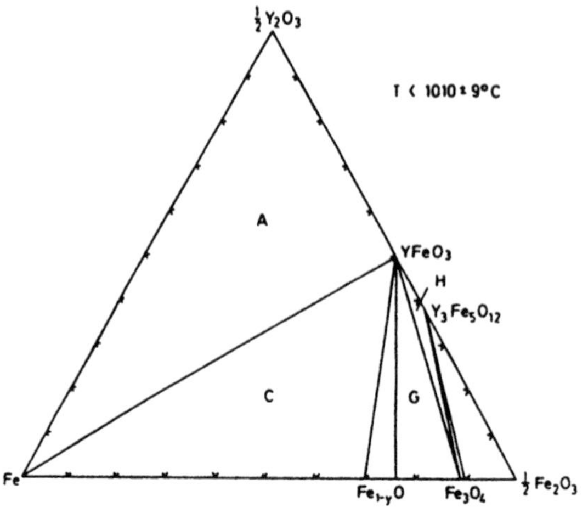

(9.6)

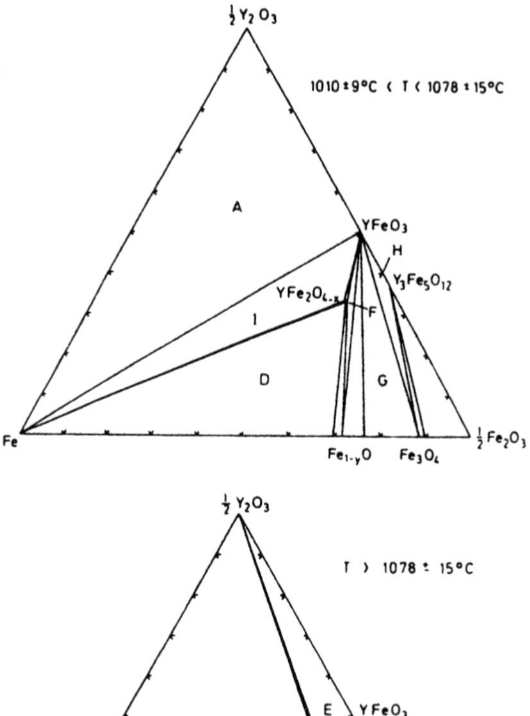

(9.7)

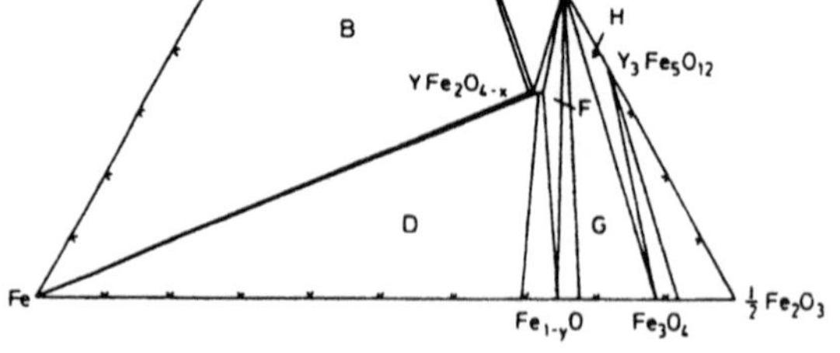

FIGURES 9.5.–9.7. Partial Gibbs triangle of the system Y-Fe-O for different temperature regimes. The phase YFe_2O_{4-x} occurs only above 1010°C. The coexistence of the phases Fe-$YFeO_3$ changes at 1078°C in favor of the equilibrium between Y_2O_3 and YFe_2O_{4-x}. The letters in the various three-phase regions refer to the thermodynamic data reported in Reference 11.

convenient to find the transition temperatures from the changes of slope, intersection, and branching of the straight lines of the emfs for compositions of the various three-phase regions when the measurement is performed as a function of temperature.

So far, mixed ionic–electronic conducting systems were considered and employed as electrodes. Phase equilibria of predominantly ionically conducting compounds may also be easily studied electrochemically by employing these materials themselves as ionic conductors in galvanic cells.[10] The phase equilibria of the ternary system Li-Al-Cl were verified by equilibration of the lithium ion conductor $LiAlCl_4$ at one side with an LiCl, Al mixture and at the other side with an $AlCl_3$, Al mixture. The measured emf allows the determination of the Gibbs energy of formation of $LiAlCl_4$ from LiCl and $AlCl_3$:

$$E = \frac{1}{q}\left\{\Delta G_f\left(LiAlCl_4\right) - \Delta G_f\left(LiCl\right) - \Delta G_f\left(AlCl_3\right)\right\}. \tag{9.17}$$

With this value the thermodynamically most favorable equilibria may be determined (see Reference 11). The calculated activities of all components should show the highest possible value for any composition for any hypothetically assumed three-phase equilibrium.

The ternary systems Li-N-Cl, Li-N-Br, and Li-N-I show several ternary fast ion conductors along the quasibinary sections Li_3N–lithium halide. The Gibbs energies of formation were determined by electrochemical decomposition of the electrolyte and forming the equilibrium electrode phases *in situ*. The decomposition voltages were then used to calculate the thermodynamically most favorable phase relations.[12]

The discussed techniques may be extended to the determination of phase diagrams with any number of components. A general treatment is described in the literature.[13] The determination of thermodynamic data from the coulometric titration curves will be illustrated in the next chapter.

B. DETERMINATION OF GIBBS ENERGIES OF FORMATION

The emf multiplied by the charge $z\delta q$ of a small current of δ ions through the galvanic cell is equivalent to the Gibbs energy $\Delta\delta G$ of the corresponding chemical reaction. Accordingly, the Gibbs energy of formation for any composition may be obtained by integration of the emf along the path of coulometric titration:[13]

$$\Delta G = zq\int_{x_0}^{x} E(x)\,dx + \Delta G_0. \tag{9.18}$$

The integration runs from the stoichiometry of the starting composition (x_o) to the stoichiometry of the sample under consideration (x). ΔG_o is an integration constant corresponding to the Gibbs energy of formation of the material with the starting composition. This relation allows one to determine the Gibbs energy of formation as a function of stoichiometry.

In case of thermodynamic equilibrium of k phases of a system of k components, the cell voltage is related to the Gibbs energies of formation of the k phases, independent of their relative amounts at constant temperature and total pressure. The "determinant formula" turned out to be very convenient in this regard to calculate the emf E,[13] especially in the case of multicomponent systems and complicated stoichiometric ratios:

$$E = -\frac{1}{zqd}\sum_{i=1}^{k}(-1)^i\, d_{i1}\Delta G_f^0\left(A_{x_i}B_{y_i}\ldots\right), \tag{9.19}$$

assuming the application of an ionic conductor for A^{z+} ions. The emf is given in this equation with reference to the pure (elemental) electroactive component under standard conditions. d is the determinant formed by the k stoichiometric numbers x_i, y_i,... of the k compounds which are in equilibrium with each other, e.g., in the case of a ternary system,

$$d = \begin{vmatrix} x_1 & y_1 & z_1 \\ x_2 & y_2 & z_2 \\ x_3 & x_3 & z_3 \end{vmatrix}. \tag{9.20}$$

d_{i1} is the minor which is formed by elimination of the i-th row and first column which lists the stoichiometric numbers of the conducting component, i.e., in the case of the ternary system

$$d_{11} = \begin{vmatrix} y_2 & z_2 \\ y_3 & z_3 \end{vmatrix}, \; d_{21} = \begin{vmatrix} y_1 & z_1 \\ y_3 & z_3 \end{vmatrix}, \; d_{31} = \begin{vmatrix} y_1 & z_1 \\ y_2 & z_2 \end{vmatrix} \tag{9.21}$$

Measurements of the emfs of the regions of equilibrium of k phases allow in reverse to determine the Gibbs energies of formation of all phases, at least in the case of small stoichiometric widths of all phases, i.e., approximately the same stoichiometries and accordingly Gibbs energies of the compounds for all k-phase equilibria in which the compound is involved.

The temperature dependence of the Gibbs energy of formation allows one to determine the entropy of formation according to the Gibbs–Helmholtz relation $\Delta S = -(\partial \Delta G / \partial T)$, i.e., the temperature dependence of the cell voltage relates to the entropy of the cell reaction, $\Delta S = nq(\partial E/\partial T)$. The enthalpies of formation and cell reaction follow according to $\Delta H = \Delta G + T\Delta S$. Knowledge of these fundamental thermodynamic quantities allows the comprehensive determination of the thermodynamic properties of the system.

The experimental technique of electrochemical determination of thermodynamic data will be illustrated by a few examples. The standard Gibbs energy of formation for any composition along the binary system Li-Sb is shown in Figure 9.8 as determined from the integration of the coulometric titration curve. The enlargement of the region around the composition Li_3Sb shows the high resolution of the technique.

The variation of the Gibbs energy of formation of a phase with an even narrower stoichiometric width may be determined by integration of the coulometric titration curve as shown in Figure 9.9 for $LiAlCl_4$. The data are given relative to the ideal stoichiometry. The resolution is well below 1 J/mol and is hard to be matched by other techniques.

The technique allows one to obtain the Gibbs energy of formation for any composition, not only for those along the titration path toward the corner of the electroactive component of the Gibbs triangle. Results for the quasibinary Section InSb-Li_3Sb of the ternary system Li-In-Sb are shown in Figure 9.10.[14]

Another technique of determination of thermodynamic data is the measurement of the emfs of the k-phase equilibria of a k-component system as a function of temperature as shown in Figure 9.11 for the example of the ternary system Y- Fe-O. The emf is converted into the oxygen partial pressure according to Equation (9.4). These data made it possible to calculate the Gibbs energies of formation of all ternary phases in that system.[11]

It may be possible that metastable phases are formed in the course of the coulometric titrations because it is necessary to employ not negligibly small current densities. Since the emf corresponds to the Gibbs energy of the reaction which really takes place, this information may be employed in this case for the determination of the Gibbs energy of the metastable phase. The formation of the thermodynamically stable phases has to provide, of course, the lower Gibbs energy of reaction and therefore the smaller emf. Figure 9.12 shows the example of the electrochemical formation of "Li_2S_2" at 400°C by the reaction of FeS_2 with Li. The

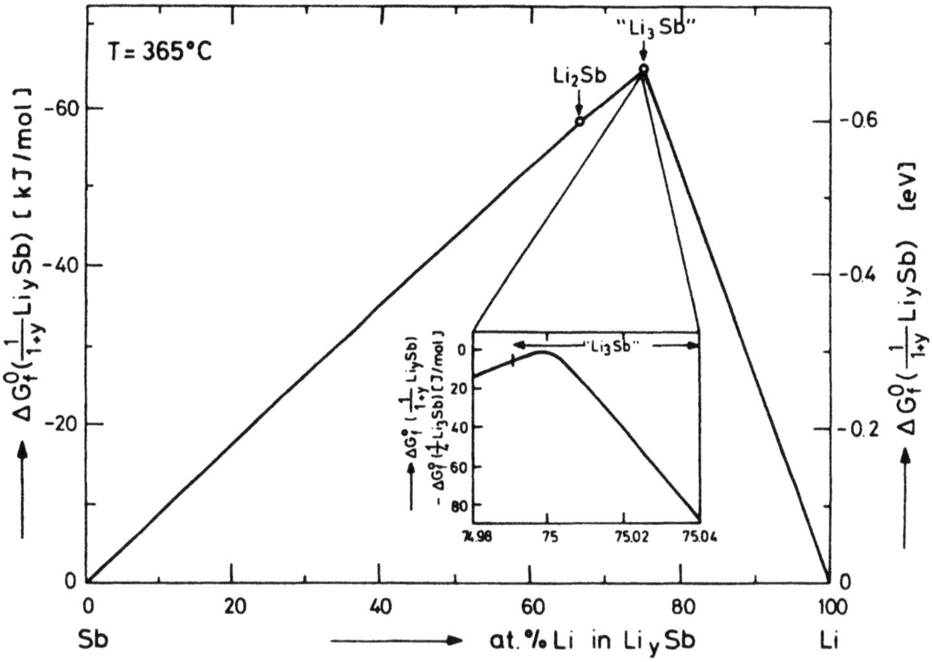

FIGURE 9.8. Standard Gibbs energy of formation as a function of composition for the binary system Li-Sb.

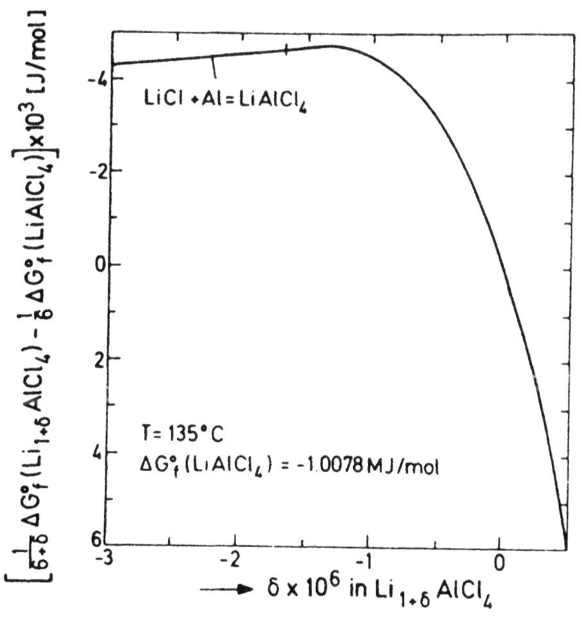

FIGURE 9.9. Gibbs energy of formation of $Li_{1-\delta}AlCl_4$ from the three elements under standard conditions as a function of the deviation from the ideal stoichiometry. The change of the Gibbs energy of formation is plotted compared to the value for the ideal stoichiometric compound.

formation of the metastable phase shows a voltage which is 79 mV smaller than the equilibrium value which is observed after a sufficiently long period of equilibration. The thermodynamic reaction of the formation of Li_2S and S is more stable than the formation of "Li_2S_2" by about 119 kJ/mol.[15]

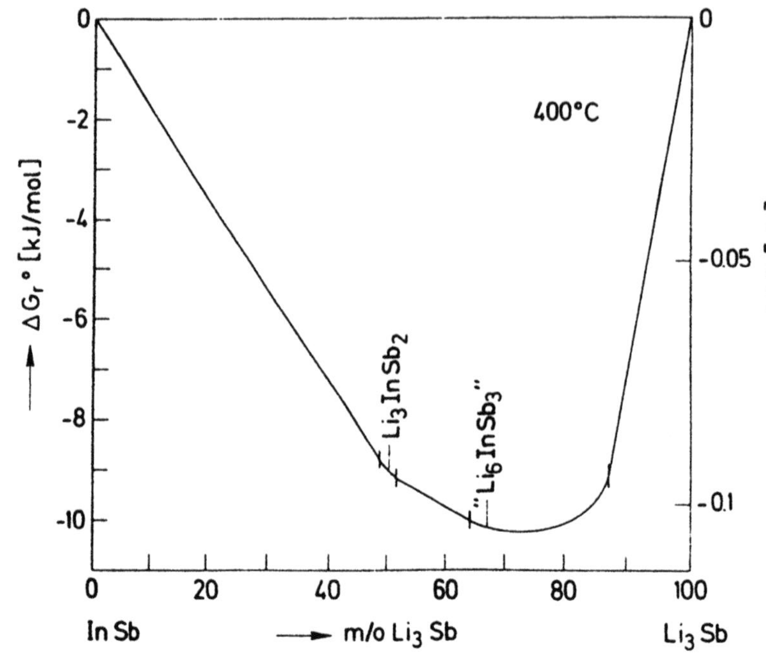

FIGURE 9.10. Standard Gibbs energies of formation along the quasibinary section InSb-Li_3Sb as calculated from the phase diagram Li-In-Sb and the emfs of the three-phase regions.

Of special interest are those systems which include a gaseous component. The emf depends in this case on the partial pressure of the gas. The galvanic cell acts in this way as a gas sensor. The gaseous component does not necessarily have to agree with the electroactive component of the electrolyte. As an example, the voltage across the lithium ion conductor $LiAlCl_4$ to which LiCl or $AlCl_3$ was added at the interface that is exposed to the gas shows an emf depending on the chlorine partial pressure.[10]

So far, nearly exclusively, the properties of solids in thermodynamic equilibrium or under metastable conditions have been analyzed. In the following, techniques for determining kinetic properties will be considered.

IV. KINETIC PROPERTIES

A. SOLID IONIC CONDUCTORS

1. Conductivity of Majority Charge Carriers (Ions)

According to Equation (9.10), the ionic conductivity in an electrolyte with negligible electronic conduction ($i_{total} \cong i_{ion}$) may be determined by using Ohm's law, provided that nonpolarizable reversible electrodes are employed which allow unimpeded delivery of the mobile ions on one side and their removal on the other side. To overcome the limitation, separate voltage probes, in the form of identical electronic leads connected to the electrolyte at positions separated by a distance L, are employed. Under these conditions the ionic conductivity is given by

$$\sigma_{ion} = -\frac{iL}{E}. \qquad (9.22)$$

This technique is called the four-point dc technique.[16] An alternative that may be used to overcome polarization problems is the use of alternating current techniques using a wide

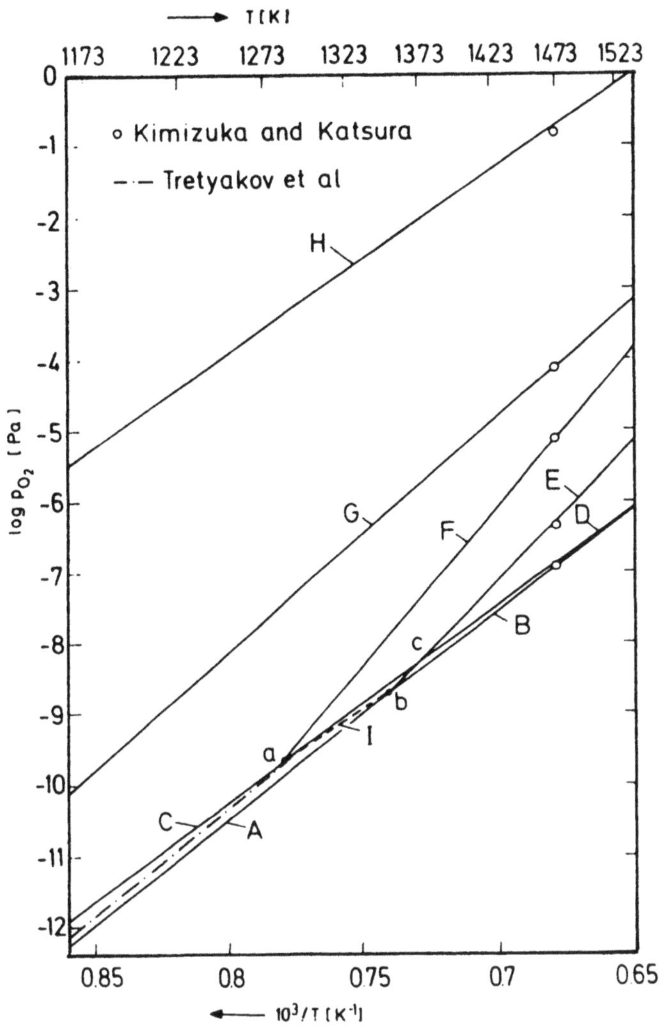

FIGURE 9.11. Temperature dependence of the oxygen partial pressure of the various three-phase equilibria of the system Y-Fe-O (cf. Figures 9.5–9.7). The branching shows the formation of new phase equilibria at the corresponding temperature.

range of frequencies, commonly from 10^{-2} to 10^{+6} Hz. This technique ("impedance spectroscopy") is discussed in detail elsewhere.[17]

The identity of the mobile ionic species may be determined in simple cases from the changes in the mass of the ionic sink and source electrodes after a defined charge-flux through the cell. Electrodes must be used that are able to exchange all potential ionic species involved in the transport within the solid ionic conductor. If one species dominates the total charge transport, Faraday's law can be used to relate the mass change $|\Delta m|$ of either one of the electrodes per unit time and unit current by

$$\frac{|\Delta m|}{\int I dt} = \frac{A}{|z|F}, \tag{9.23}$$

where A is the atomic weight of the particles transferred as ions of charge z through the electrolyte. The atomic weight divided by the charge number relates to the mobile species.

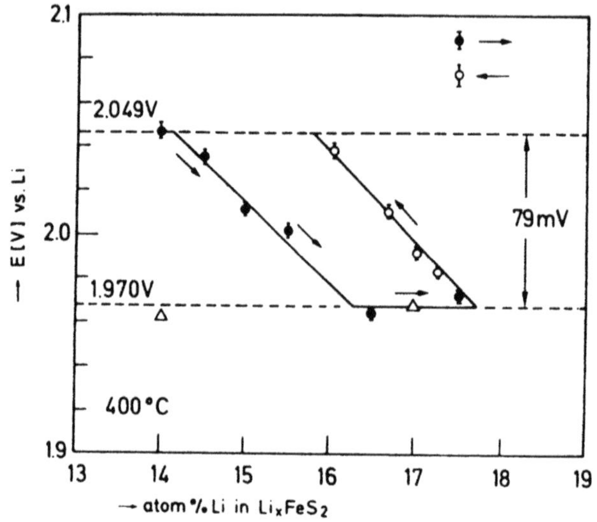

FIGURE 9.12. Thermodynamic equilibrium voltages for the titration of lithium into FeS_2 (upper broken line) and metastable apparent equilibrium voltage as a result of the formation of the metastable compound "Li_2S_2" from Li and FeS_2 (lower broken line).

2. Conductivity of Minority Charge Carriers (Electrons, Holes)

In spite of the small electronic conductivity in a practically useful solid electrolyte, electrons and holes are responsible for many materials properties. For example, electronic conductivity acts as an internal short circuit within a galvanic cell, thus reducing the energy available to do useful work in an external circuit. In addition, electronic conductivity reduces the cell voltage. Furthermore, the transport of electrons or holes controls the rate of equilibration of the composition within an electrolyte, since the condition of local charge flux neutrality requires equivalent amounts of different charged species to be transported within the solid.

The minority electronic properties of solid electrolytes may vary considerably with changes in composition. It is therefore often necessary to study the minority charge carrier transport as a function of the activity of the components.

Materials are commonly either predominantly ionic or electronic conductors. This information may be obtained from transference measurements. More precise values may be obtained from the so-called emf technique.[18] This approach makes use of the effect of the partial internal short circuit of the electrolyte by the movement of electrons or holes, which results in a decrease of the externally measurable voltage of a galvanic cell. In contrast to the transference experiment, the electrolyte is here sandwiched between two electrodes that have different, but precisely known, chemical potentials for the electroactive species.

Because no current is allowed to pass through the external electric circuit, the charge neutrality condition demands that the current densities i_n of the different kinds of species n, i.e., all ions, electrons, and holes, are electrically compensated by each other:

$$\sum_n i_n = 0 \tag{9.24}$$

Substitution of Equation (9.5) for the current densities and making use of the equilibrium relation between ions, neutral atoms, and electrons, $\mu_A = \eta_{A^{z+}} + z\eta_{e^-}$, integration over the length of the electrolyte yields the following equation for the difference in the electrochemical potentials of the electrons between the two electrodes:

$$E = \frac{1}{q} \sum_{\text{ions}} \frac{1}{z_n} \int_{\text{electrolyte}} t_n \, d\mu_n , \tag{9.25}$$

where μ_n is the chemical potential of the neutral species n.

In general, one has to assume that the ionic transference number is strongly dependent on the chemical potential of the components. The transference numbers of the species n may be determined as a function of their activities from the change in the cell voltage with the variation of the chemical potential of the neutral n-th component at one electrode while all other chemical potentials are held constant. Differentiation of Equation (9.25) with regard to the upper limit of the integral, μ_n^r, yields

$$t_n\left(\mu_n^r\right) = -z_n q \frac{\partial E}{\partial \mu_n^r} . \tag{9.26}$$

While in the experiments described so far both ionic and electronic species are transferred, it is also possible to determine the partial electronic conductivity by using ionically blocking electrodes to suppress the ionic transport so that only electrons and holes can pass. This technique is known as the asymmetric polarization or Hebb–Wagner technique.[19] By using a chemically inert electronic conducting material, no ions will be delivered to the electrolyte when a voltage is applied with such a polarization that the mobile ions tend to be depleted at the inert electrode. An electrode used on the other side fixes the chemical potential of the mobile component of the electrolyte by the applied voltage at that phase boundary.

If the electrolyte has large ionic disorder and since the blocked ionic current i_{ion} is zero, Equations (9.5) and (9.6) indicate that no gradient can exist in the electrostatic potential within the sample. Then the steady-state transport of electrons and holes occurs only due to diffusion under the influence of gradients in their concentrations. These gradients must be uniform if the diffusion coefficient does not depend markedly on the concentration. From Equations (9.2) and (9.3) and the ionization equilibrium, the cell voltage determines the ratio of the activities of the electronic species at both sides of the electrolyte:[28]

$$E = \frac{kT}{q} \ln \frac{a_e^r}{a_e^l} = -\frac{kT}{q} \ln \frac{a_h^r}{a_h^l} . \tag{9.27}$$

Integration of the diffusion part of the general transport (Equation [9.5]) over the length L of the sample, differentiation of the result with regard to the lower limit of the integral, ln a_e^r, and consideration of Equation (9.27) yields

$$\frac{di}{dE} = -\frac{1}{L}\left(\sigma_e^r + \sigma_h^r\right) . \tag{9.28}$$

This equation allows the determination of the total electronic conductivity $\sigma_e^r + \sigma_h^r$ within the electrolyte adjacent to the interface with the inert electrode. The corresponding chemical potential of the electroactive component at that location is given by the cell voltage according to Nernst's (Equation [9.4]). Thus, by varying the applied voltage, $\sigma_e + \sigma_h$ may be obtained as a function of the component activities or the composition of the sample. Usually, the electronic conductivity depends significantly upon the cell voltage because of changes in the defect concentration within the material. In certain potential ranges either the hole or the electron conductivity may be dominant. Then σ_e or σ_h may be studied separately as a function of the composition.

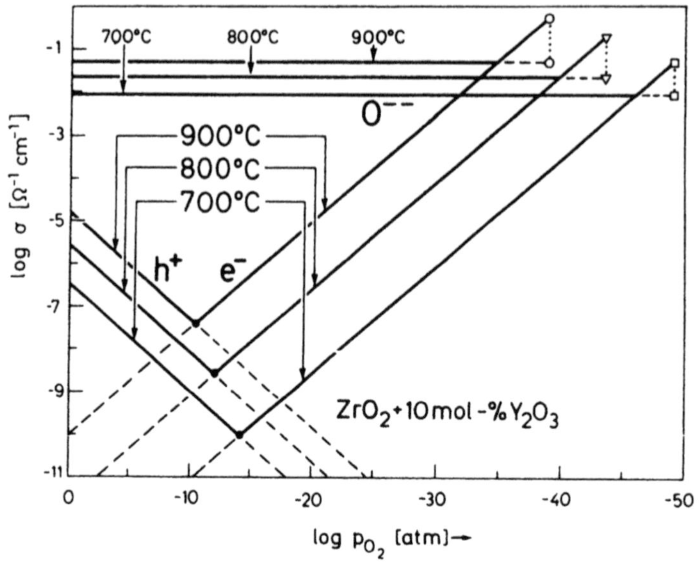

FIGURE 9.13. Oxygen ion, electron, and hole conductivity of ZrO_2 (+ 10 m/o Y_2O_3) as a function of the oxygen partial pressure at three temperatures.

The relation between the current and the voltage in a dc polarization experiment may be obtained by assuming that ratios of the activities of electrons and holes can be replaced by ratios of their concentrations. Such an assumption is usually valid up to concentrations of about 10^{19} particles per cubic centimeter. Then, integration of Equation (9.28) with the help of Equation (9.27) (assuming that the diffusion coefficients of electrons and holes are concentration independent) yields

$$i = \frac{kT}{qL}\left\{\sigma_e^{rev.}\left[1 - \exp\left(\frac{Eq}{kT}\right)\right] + \sigma_h^{rev.}\left[\exp\left(-\frac{Eq}{kT}\right) - 1\right]\right\}. \tag{9.29}$$

Equation (9.29) includes one term that approaches a plateau of the current and another term that shows an exponential increase of the current with increasing voltage. The relative contributions depend on the magnitudes of the prefactors which are the partial conductivities at the activity imposed by the reversible electrode. The shape of the log (current) vs. voltage curves allows one to determine the type of minority charge carriers and to derive their conductivities from the plateau value and the intersection with the log (current) axis of the extrapolated straight line.

By fitting the experimental results to Equation (9.29) both σ_e and σ_h may be determined at the chemical potentials defined by equilibrium with the reversible electrode.

The dc asymmetric polarization technique has been used quite extensively to evaluate the partial conductivity of the electronic minority charge carriers in many solid electrolytes, e.g., doped zirconia,[20] silver halides,[21] copper halides,[22] and solid lithium conductors.[23]

Figure 9.13 shows results obtained for ZrO_2 doped with 10 mol% Y_2O_3. σ changes with P_{O_2} with the fourth power. This dependence corresponds to divalent oxygen vacancies as the predominant type of disorder. Under very low oxygen partial pressures the material becomes mainly electronically conducting before it is decomposed.[20]

3. Mobility and Concentration of Minority Charge Carriers (Electrons, Holes)

The steady-state measurement of the electronic conductivity allows one to obtain only the product of the concentration and either the mobility or diffusion coefficient, according

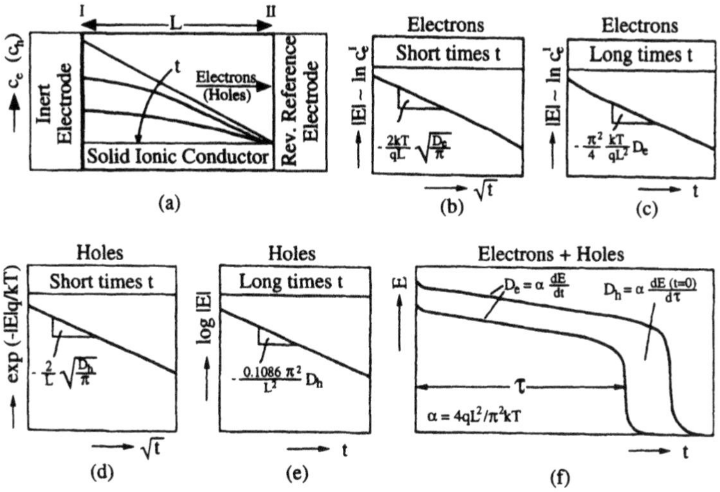

FIGURE 9.14. Voltage relaxation technique for the determination of the diffusion coefficients (or mobilities) of electrons and holes within predominantly ionically conducting solids. The various possibilities for calculating the diffusion coefficients D_e and D_h from short- ($t \ll L^2/D_{e,h}$) and longtime ($t \gg L^2/10\ D_{e,h}$) behavior are indicated. $c_{e,h}$: concentration of electrons, holes; q: elementary charge; k: Boltzmann's constant; T: absolute temperature.

to Equation (9.7). In the following, transient phenomena will be considered which are similar to many technical processes and natural phenomena. Most important for these processes are the transport properties of the two predominantly mobile species, which are in most cases one type of ions and electrons or holes. The motions of these species are coupled by local electroneutrality conditions. Rate determining is generally the type of species which provides the second largest contribution to the transport process. It will be shown that electrochemical techniques allow one to obtain very elegantly a large amount of fundamental information on the kinetic materials properties.

a. *Voltage Relaxation Method*[24]

A galvanic cell identical to that used for dc polarization measurements is polarized until a steady-state current is reached. Assuming that the diffusion coefficients are concentration independent and that concentrations are sufficiently small to make the thermodynamic factor of the electrons unity, linear concentration profiles will be established for the electronic species. For charge compensation, a corresponding gradient in the chemical composition of the compound must, of course, also be present. If the applied voltage is then switched off, a relaxation of the cell voltage is observed, as schematically shown in Figure 9.14. Microscopically, this occurs by diffusion of electrons and holes, coupled with an equivalent transport of the mobile ions (in order to maintain the local electrical neutrality) until a homogeneous composition is attained across the whole sample.

The transport of species under the influence of composition gradients is usually called chemical diffusion. In the case of good ionic conductors in which electrons and holes are the minority charge carriers, the movement of these electronic species is the rate-determining process, and their diffusion coefficient may thus be obtained.

As under steady-state polarization conditions, the electrons and holes drift predominantly under the influence of their activity gradients in good electrolytes. The comparatively large conductivity of the ions means that a significant internal electric field may not be present, i.e., in the electrical neutrality condition $\left(\sum_n i_n = 0\right)$ the electronic migration terms $\sigma_{e,h}$ grad ϕ may be neglected compared to σ_{ion} grad ϕ. Thus, the electronic concentration profiles may be calculated as a function of time, assuming only concentration gradient-driven diffusion by using appropriate boundary conditions. If open-circuit measurements are made, no exchange of either ions or electrons occurs across the phase boundary between the electrolyte and the inert electrode, and the electronic concentrations at the interface with the reversible electrode remain unchanged. The change in the concentrations of the electronic species at the inert electrode may be observed by monitoring the cell voltage as a function of time, in accordance with Equation (9.27). Such data are analyzed by comparison with the theoretical solution of the diffusion problem. The various possibilities for the evaluation of the diffusion coefficients from either short-time or long-time results, assuming that the concentration and conductivity of either the electrons or holes are dominant, are illustrated in Figure 9.14b to e. If both electrons and holes are initially present as the dominant electronic species in different parts of the sample, i.e., if a p–n junction exists within the sample, a complex superposition of the relaxation of both species is observed, as shown in Figure 9.14f. In that case a positive voltage was applied to a galvanic cell employing an yttria-doped zirconia electrolyte with an air reference electrode. Despite this apparent complication, it is possible to calculate both the electron and hole diffusion coefficients by proper analysis of different portions of the curve, as indicated in Figure 9.14f. The mobilities of electrons may be calculated by using the Nernst–Einstein relation (Equation [9.8]).

***b. Charge Transfer Technique*[25]**

While the voltage relaxation method allows the direct determination of the diffusion coefficients of the electronic species, these quantities may also be obtained from measurements of the conductivities as described above and the species concentrations. The charge transfer technique is an electrochemical method for the determination of concentrations of electronic species. As in the case of the dc polarization and voltage relaxation methods, an asymmetric cell with an ionically blocking electrode and a reversible reference electrode is used. The voltage is again applied in such a direction that no steady-state ionic current flows. A concentration profile of the electronic species is gradually established with time. The corresponding number of electrons and holes passed into the ionic conductor may be evaluated from the total charge transfer during the transient period. Since part of the charge is also used to change the charge of the double layers, this amount may be eliminated by carrying out similar experiments with samples of different lengths (Figure [9.15]). The difference in the charge transfers in the two experiments, ΔQ, is given by the difference in the areas of the two triangles according to Equation (9.27) (assuming ideal behavior for the electronic species):

$$\Delta Q = \frac{q}{2}\left(V_2 - V_1\right)c^{r}_{e,h}\left[\exp\left(\frac{E_2 q}{kT}\right) - \exp\left(\frac{E_1 q}{kT}\right)\right]. \tag{9.30}$$

$c^{r}_{e,h}$ is the concentration of the predominant electronic species in the sample at the reversible electrode side; V_1 and V_2 are the two sample volumes; the voltage is changed from E_1 to E_2. The technique has been applied to study the hole concentration in CuCl.

The voltage relaxation and the charge transfer technique are generally applicable for obtaining information about the two contributions to the electronic conductivity. However, under some conditions one technique is more favorable than the other. The voltage relaxation method is especially applicable in the case of low diffusion coefficients while the charge transfer technique is preferred for highly mobile electronic species.

B. SOLID MIXED CONDUCTORS

Mixed conductors are considered to have predominantly electronic conductivity, but also show an appreciable ionic conductivity. In most kinetic studies it is necessary to use auxiliary solid or liquid electrolytes.

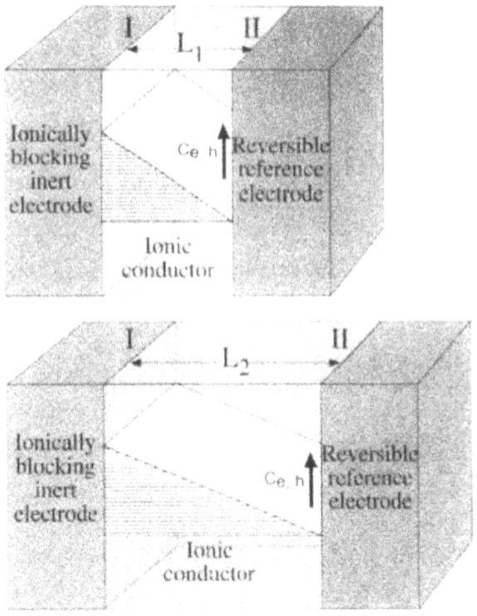

FIGURE 9.15. Charge transfer technique for the measurement of the concentration of electrons or holes in the predominantly ionic conductors; two samples of different lengths are polarized by the same voltage.

1. Partial Ionic Conductivity

Similar to the dc polarization technique for the determination of the partial electronic conductivity in ionic conductors, the electronic majority charge carriers may be blocked by introducing a pure ionic conductor that is permeable to the ions under consideration.[19] In order to provide defined thermodynamic conditions, reversible reference electrodes, which can act as a source or sink for the ions, are employed at the other side of the sample (Figure 9.16).

The ionic current that is passed through the sample is measured by the electric current in the external circuit. Since no electronic species are transferred under steady-state conditions, their electrochemical potentials must be zero. At high concentrations, the chemical potential of the predominant electronic charge carriers is approximately constant, and no electric field may be built up accordingly. Concentration gradients form the driving force for the ionic minority charge carriers. The flux is given by Fick's first law,

$$\sigma_{ion} = -\frac{iL}{E}, \tag{9.31}$$

assuming that the partial ionic conductivity is independent of the chemical potential of the mobile component by sufficiently high disorder. Electrode polarizations should be negligible, or separate probes of ionic conductors and reference electrodes have to be employed.

Rather than by a reversible electronically conducting ion sink or source, the activity of the mobile component is fixed at the right hand side by another electrolyte and their applied voltage. The electronically conducting sample is in this way sandwiched between two ionic conductors which fix the activities at both interfaces. The resulting (identical) currents in both circuits are measured and evaluated for the conductivity according to Equation (9.31). Again, possible polarization voltages by the current fluxes across both electrolyte/sample interfaces have to be subtracted, e.g., by using separate voltage probes or by briefly switching off the applied voltages. By varying the voltages, the partial ionic conductivity may be determined for different composition ranges.

Another, and often more convenient, way of determining the ionic conductivity in predominantly electronically conducting mixed conductors is the calculation from the diffusion coefficient as obtained from transient measurements which are described below.[26] Such a

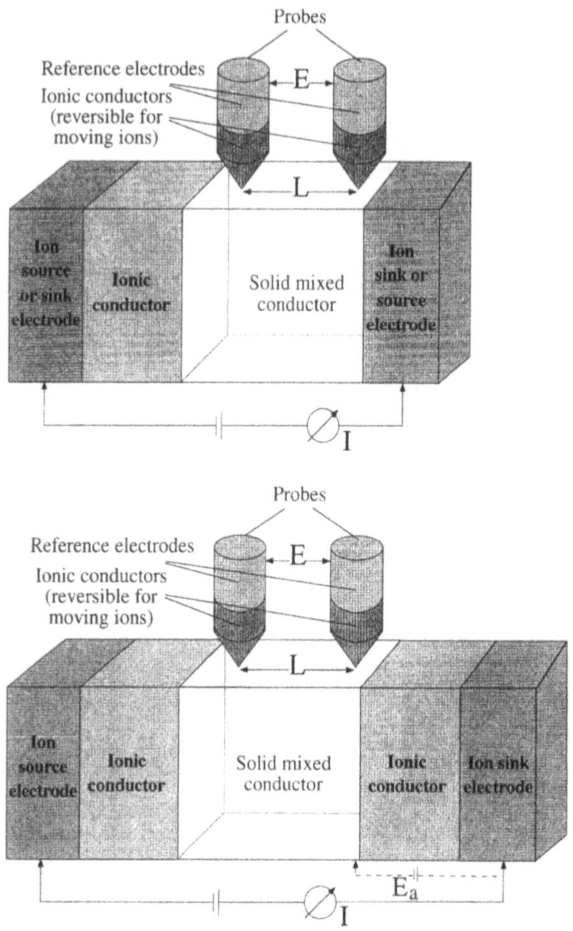

FIGURE 9.16. Dc polarization technique for the determination of the partial ionic conductivity of mixed conductors by blocking the electronic current with an auxiliary electrolyte. Probes are used to measure the gradient of the electrochemical potential of the mobile ions. The stoichiometry of the mixed conductor is controlled by a reversible electrode (or by applying another auxiliary electrolyte with a counter electrode and the application of a voltage) that defines the activities at the right-hand side.

technique offers the advantage of the determination of a large number of important kinetic and thermodynamic parameters as a function of stoichiometry in a single set of experiments. For the galvanostatic intermittent titration technique (GITT), the appropriate formulas are presented in Table 9.2.

2. Partial Electronic Conductivity

The electronic conductivity of good electronic conductors is often sufficiently close to the total conductivity. For dc or ac measurements one must be sure to use chemically compatible electrodes that do not react with the sample to form electronically insulating phases. In contrast to predominantly ionically conducting materials, one should observe Ohm's law instead of Fick's law in dc polarization measurements. Also, emf measurements may be employed in the same way as in the case of ionic conductors. The mixed conducting samples are brought into equilibrium with two different activities of the ionically mobile component.

TABLE 9.2
Kinetic and Thermodynamic Quantities that can be Determined by the Galvanostatic Intermittent Titration Technique (GITT). Species A of the component A_yB are considered to be mobile.

Quantity	Formula	Conditions
Chemical diffusion coefficient $\tilde{D}$	$\frac{4}{\pi\tau}\left(\frac{m_B V_M}{M_B S z_A}\right)^2 \frac{\Delta E_s}{\Delta E_t}$	$\tau \ll L^2/\tilde{D}$
Partial ionic conductivity σ_A	$-\frac{4 m_B V_M I_0}{\pi M_B S^2}\frac{\Delta E_s}{(\Delta E_t)^2}$	$\tau \ll L^2/\tilde{D}$
Diffusivity D_A	$-\frac{4kT\, m_B V_M I_0}{\pi c_A z_A^2 q^2 M_B S^2}\frac{\Delta E_s}{(\Delta E_t)^2}$	$\tau \ll L^2/\tilde{D}$
Electrical ionic mobility u_A	$-\frac{4 m_B V_M I_0}{\pi c_A \lvert z_A\rvert q\, M_B S^2}\frac{\Delta E_s}{(\Delta E_t)^2}$	$\tau \ll L^2/\tilde{D}$
General ionic mobility b_A	$-\frac{4 m_B V_M I_0}{\pi c_A z_A^2 q^2\, M_B S^2}\frac{\Delta E_s}{(\Delta E_t)^2}$	$\tau \ll L^2/\tilde{D}$
Parabolic tarnish; rate constant k_t	$\frac{1}{y}\left\lvert\int_0^L \tilde{D}\, dy\right\rvert$	$\tau \ll L^2/\tilde{D}$
Thermodynamic enhancement factor $\partial \ln a_A/\partial \ln c_A$	$-\frac{z_A q c_A V_M}{kT\, N_A}\frac{dE}{dy}$	$\tau \ll L^2/\tilde{D}$
Gibbs free energy of formation ΔG_f^0	$-z_A q\int_{y_0}^{y} E_s\, dy + DG_f^0\left(A_{y_0}B\right)$	$\tau \ll L^2/\tilde{D}$

Note: $\Delta E_{s,t}$: change of steady-state equilibrium and transient voltage; c_A: concentration of species A; I_0: galvanostatic current; m_B: mass of species B; N_A: Avogadro's number; q: elementary charge; k: Boltzmann's constant; S: contact area between sample and electrolyte; T: absolute temperature; V_M: molar volume; y: stoichiometric number; z_A: charge number; τ: duration of the galvanostatic pulse.

3. Chemical Diffusion

The chemical diffusion coefficient describes the relaxation of compositional gradients to achieve a homogeneous composition. This quantity is important in the case of many phenomena of practical interest, e.g., for corrosion processes or the performance of electrodes in batteries. The chemical diffusion is commonly determined by the simultaneous diffusion of (at least) two different charge carriers in order to maintain local electroneutrality. In most cases it is the diffusion of one type of ions and electrons or holes.

For the measurement of the chemical diffusion coefficient $\tilde{D}$, an auxiliary electrolyte is used both to measure the flux of ions into and out of the sample and to determine the concentration of the mobile species at the interface with the electrolyte. The rate-determining factor is assumed to be the diffusion of species within the mixed conducting solid and not the transport of ions across the interface. The change in the double layer charge at the interface must also be small compared to the charge transported into or out of the bulk of the sample. In order to verify this assumption, typical values for the double layer capacities may be assumed and, better, samples with different lengths should be employed.

The experimental arrangement is shown in Figure 9.17. It is preferable to use a separate reference electrode for potential measurements in addition to the current-supplying counter

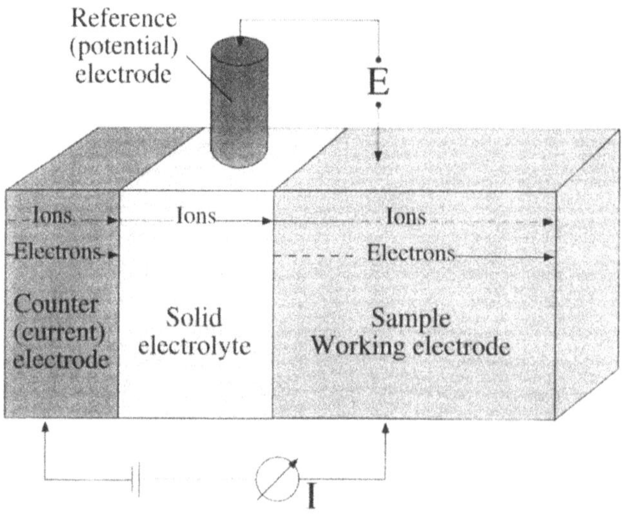

FIGURE 9.17. Experimental arrangement for the determination of the chemical diffusion coefficient and other kinetic quantities by transient techniques. The separate reference electrode allows one to determine the activity of the electroactive species at the interface between the sample and the electrolyte.

electrode. The reference electrode should be attached to the electrolyte near the sample in order to avoid large contributions from ohmic polarization losses. (Under certain experimental conditions, e.g., in the case of constant currents, this is unimportant, however).

In the case of diffusion of foreign atoms in small concentrations into a host lattice, e. g., a solid or liquid metal, this situation is simplified because of the applicability of the laws of ideal diluted solid solutions.[27] The concentration of the electroactive species at the interface with the electrolyte is given directly by Nernst's law. In case of nonideal behavior, further information concerning the activity–concentration relation of the diffusing species is necessary. This may be acquired from coulometric titration measurements.

In principle, any transient transport measurement method may be employed to determine the chemical diffusion coefficient, if Fick's second law may be properly solved. The mathematical solutions are known for a number of initial and boundary conditions. Among those are potentiostatic, galvanostatic, current pulse, and ac conditions. Figure 9.18 shows the concentration profiles of the diffusing species for various times in the sample for linear and cylindrical geometries for potentiostatic, galvanostatic, and short pulse modes. It is assumed that no transport of species occurs across the phase boundary at the opposite side from the electrolyte in the linear geometry or across the axis in the cylindrical geometry. The observed currents are shown at the right-hand side against linearizing time scales for short and long intervals after the start of the experiment.

a. *Galvanostatic Intermittent Titration Technique (GITT)*[6,28]

In the galvanostatic mode, a constant current is applied which implies a time-independent concentration gradient of the mobile species in the sample just inside the interface with the electrolyte according to Fick's first law. The change in the cell voltage with time that results from this condition is measured.

An important experimental advantage is the fact that the IR drop of the cell arrangement is time independent and is merely added to the cell voltage as a constant. This does not change the shape of the voltage–time curve. Also, the location of the reference electrode is not important for the same reason. This is in contrast to the potential step mode in which an infinitely large current should pass through the cell initially in order to fulfill the boundary conditions precisely. Thus, a considerable time-dependent IR drop is superimposed upon the

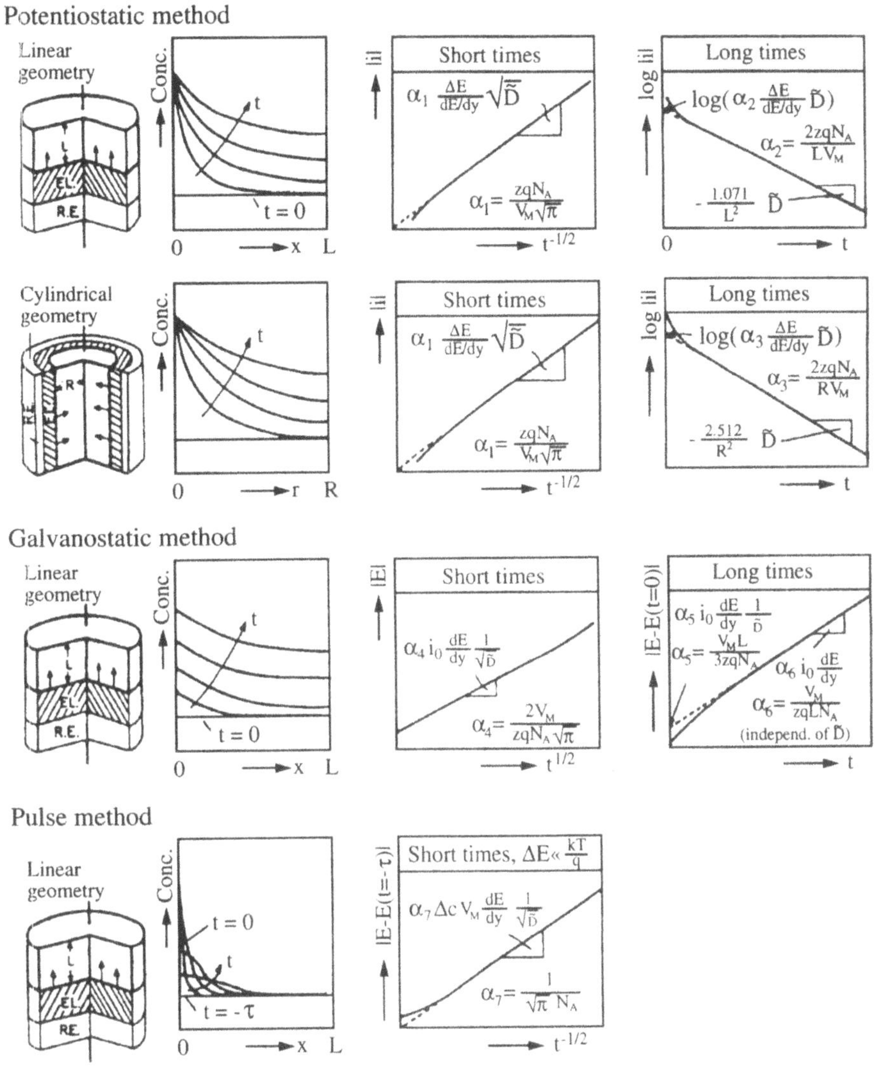

FIGURE 9.18. Compilation of various procedures for measuring chemical diffusion coefficients. The application of several signals and geometries is shown: voltage steps for linear and cylindrical geometrical arrangements, constant currents, and short galvanostatic or potentiostatic pulses. The concentrations of the mobile ions as a function of location are shown in the second row for several times. Linearized plots of the current or voltage response are given in the third and fourth rows for short ($t \ll L^2/\tilde{D}$ or $R^2/\tilde{D}$) and long times ($t \gg L^2/10\,\tilde{D}$ or $R^2/10\,\tilde{D}$). The various possibilities for evaluating $\tilde{D}$ are indicated. E: cell voltage; i: current density; i_0: galvanostatically controlled current density; t: time; y: stoichiometric number; q: elementary charge; z: charge number of mobile species; Δc: total change of the overall concentration during the pulse; N_A: Avogadro's number; V_M: molar volume of the sample.

potentiostatically applied potential. This reduces the effective voltage and changes the boundary conditions, which is observed by deviations from the theoretical current–time curve. On the other hand, the galvanostatic mode does not allow the determination of the chemical diffusion coefficient from the slope of the concentration for long times, since the diffusion coefficient does not influence the measured concentration change at the interface. Another advantage of the galvanostatic mode is that the charge, and therefore the overall concentration change of the sample is easily measured from the current–time product.

The relationship between the voltage and the composition is obtained from the charge transfer across the galvanic cell and the change in the steady-state voltage (coulometric titration). This relationship between the changes in the cell voltage and the concentration of the mobile species varies as a function of stoichiometry. Assuming that the lattice of the other species remains unchanged in a first approximation, we have

$$\frac{dE}{d\ \ln c_m} = y\frac{dE}{dy} \tag{9.32}$$

where c_m is the concentration of the mobile species and y is the stoichiometric number of species m in the sample. The relation between the charge ΔQ and the concentration change Δc_m is

$$\Delta c_m = \frac{\Delta Q}{zqV} \tag{9.33}$$

where V is the sample volume and z the charge number of the mobile species. Nernst's law and Equation (9.32) can be used to give the kinetically important thermodynamic or Wagner factor by which the diffusion of ions is enhanced when a composition gradient is present in the case of predominantly electronically conducting mixed conductors:

$$\left|\frac{\partial\ \ln a_m}{\partial\ \ln c_m}\right| = -\frac{zqy}{kT}\frac{dE}{dy} \tag{9.34}$$

where a_m stands for the conductivity of the mobile species, which is assumed to be the electroactive component. For the case that the mobile ionic species is not identical to the electroactive one, the Gibbs–Duhem equation may be used to relate the activity of the electroactive species to that of the mobile one.

The GITT combines both transient and equilibrium measurements as illustrated in Figure 9.19. Galvanostatic currents are applied to the cell for a time interval τ in order to study the voltage as a function of time for the evaluation of $\tilde{D}$. This produces a change in the stoichiometry. After such a titration pulse, a new equilibrium voltage will be established. If it is found that the voltage transient follows a square root of time law, it is sufficient to measure only the total change of the transient voltage (reduced by the IR drop) and the change of the steady-state equilibrium voltage in order to determine the chemical diffusion coefficient. The applicable equations are given in Table 9.2. Subsequently, the procedure may be repeated and $\tilde{D}$ may be measured as a function of stoichiometry.

The starting stoichiometry is known from the preparation or may be established from the shape of the coulometric titration curve, which exhibits an inflection point for small concentrations of defects.[29]

An advantage of the technique is also the possibility to determine many other kinetic and thermodynamic properties as a function of stoichiometry. Some of these are included in Table 9.2 and are discussed below.

b. Potential Step Technique[30]

In this case, the activity of the electroactive species and therefore the concentration of the mobile species is suddenly changed from a homogeneous distribution to one in which a new time-independent value is imposed at the phase boundary in contact with the electrolyte. A time-dependent current flows when transport occurs across the interface as species diffuse into (or out of) the bulk of the sample in response to the internal concentration gradient until a new homogeneous composition is reached according to the applied voltage. The relation between the cell voltage and the concentration is obtained from separate coulometric titration

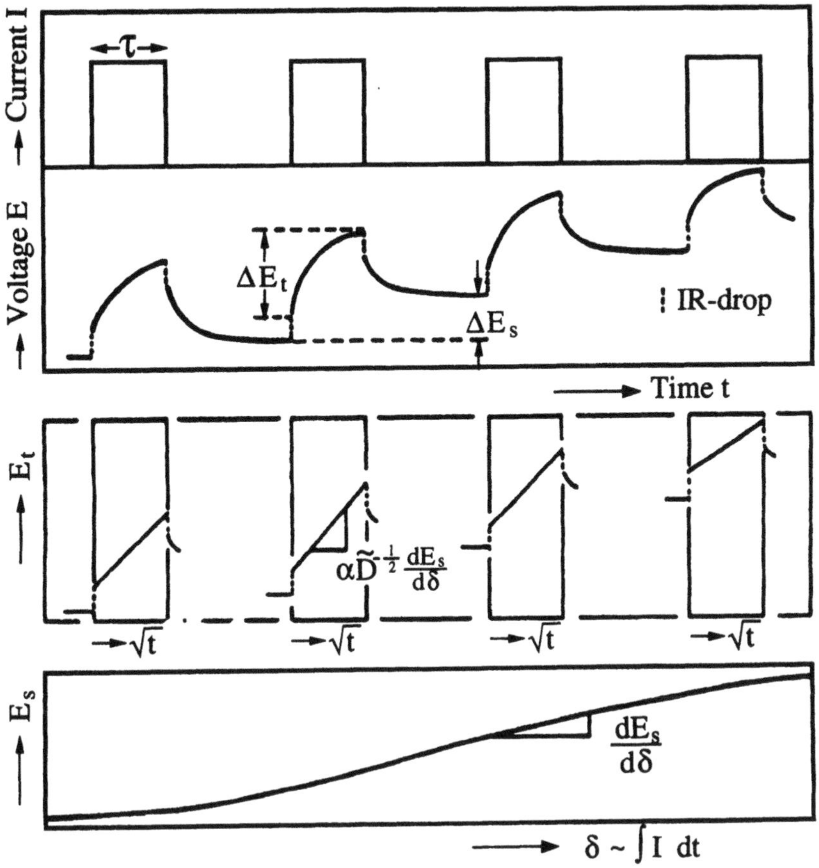

FIGURE 9.19. Schematic representation of the galvanostatic intermittent titration technique (GITT). During galvanostatic current pulses, the transient voltage E_t is observed as a function of time t and analyzed to evaluate the chemical diffusion coefficient $\tilde{D}$. The steady-state equilibrium voltage is studied as a function of the composition (y).

measurements or from measuring the total amount of charge passed through the galvanic cell during the full transient period until the new equilibrium is reached in the same experimental run. It has to be taken into consideration that the current changes by several orders of magnitude.

The time dependence of the electric current may be analyzed in various ways in order to evaluate the chemical diffusion coefficient. It is especially valuable to consider the short- and long-time behavior separately. The appropriate formulas for determining $\tilde{D}$ are given in Figure 9.18. For the solutions at long times in the cases of both linear and cylindrical geometry, it is not necessary to know the value of Δc_m. This is caused by the finite length (or radius) of the sample which influences the concentration profile, so that the transport kinetics are no longer dependent upon the original homogeneous concentration of species in the sample. The short-time solutions are identical to those for samples with infinite length.

c. *Short Pulse Technique*[31]

In this case, a short galvanostatic current pulse is applied. The concentration of the mobile species is changed in a narrow region at the sample–electrolyte interface and the time-dependent relaxation of the voltage is then analyzed. In order to meet the boundary conditions for this solution to the diffusion equation, the pulse time has to be $\ll L^2/2\,\tilde{D}$. The diffusion problem is similar to that used in radiotracer diffusion experiments. An advantage of the technique is that any external rate-limiting processes, such as the transport of ions across the electrolyte or the electrolyte–sample interface, are eliminated. The electrolyte does not act in this case simultaneously as both a transmitter of ions and a transducer to measure the concentration of the mobile species. However, large local surface concentrations are sometimes applied which may produce large local and time-dependent variations in the enhancement factor and in the value of the chemical diffusion coefficient within the sample.

d. Ionic Diffusivity, Mobility, and Conductivity

The diffusivity D_m of the mobile ions in a predominantly electronic conducting sample may be calculated from the chemical diffusion coefficient and the slope of the coulometric titration curve according to the relation

$$D_m = \frac{\tilde{D}}{\partial \ln a_m / \partial \ln c_m} \tag{9.35}$$

The Wagner enhancement factor $\partial \ln a_m / \partial \ln c_m$ may be determined from the coulometric titration curve according to Equation (9.32). When using GITT, all the necessary information is already included in the quantities ΔE_t, ΔE_s and the stoichiometry of the sample. Therefore, D may be directly obtained from the formula given in Table 9.2.

The diffusivity is related to the radiotracer diffusion coefficient by the correlation factor or, more generally, the Haven ratio. If one knows or can assume a value for this quantity, the tracer diffusion coefficient may be determined. Alternatively, if tracer data are available, information about the diffusivity provides a value of the Haven ratio, which is related to the mechanism of the ion jumps.

With the help of Equation (9.8) the electrical and general mobilities of the ions may be determined from the diffusivity, since all three quantities describe the same microscopic effect of the random motion of ions in the lattice in the absence of concentration gradients.

The partial ionic conductivity of the mixed conductor is determined by the product of the concentration and the diffusivity of the mobile ionic species according to Equation (9.7). With the knowledge of the concentration of mobile ions (or the concentration of mobile ionic defects), and the measured diffusivity of the mobile ions (or mobile ionic defects) the partial ionic conductivity may be determined.

C. KINETICS OF SOLID STATE REACTIONS

In addition to single-phase samples, electrochemical methods may also be employed to investigate reactions of mixed conducting materials with other phases. If the reaction rates are determined by bulk diffusion, the parabolic tarnishing rate constant k_t results from a comparison of the parabolic tarnishing rate law and Fick's first law. By integration of this equation over the thickness of the sample L, i.e., the stoichiometric range over which the sample exists, one obtains

$$k_t = \frac{1}{\bar{y}} \left| \int_O^L \tilde{D}(y)\, dy \right| \tag{9.36}$$

The integral may be calculated from the stoichiometry dependence of the chemical diffusion coefficient. $\bar{y}$ is the average stoichiometric number of the compound A_yB.[32] Equation (9.36) allows the determination of the tarnishing rate constant as a function of all combinations of activities of the components at both sides of the sample.

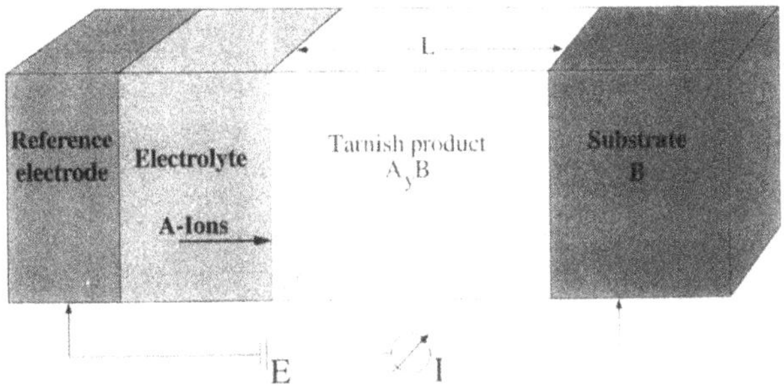

FIGURE 9.20. Galvanic cell for studying the formation of tarnish product A_yB from the components B and A. The latter one is delivered by the electrolyte. The growth rate and conditions are controlled by the voltage E and the current i.

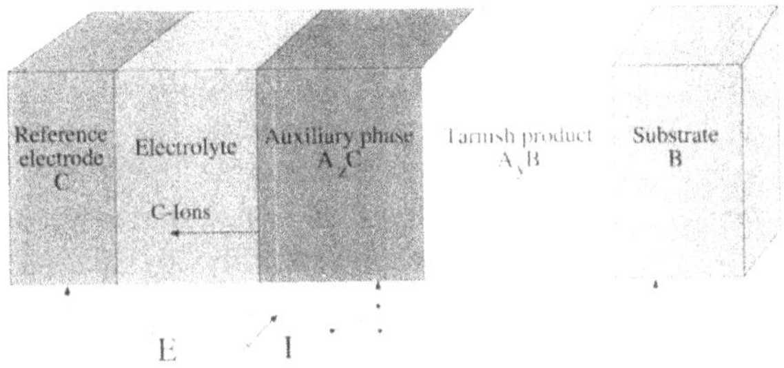

FIGURE 9.21. Galvanic cell for studying the formation of the compound A_yB from B and A. The latter is made available by the auxiliary phase A_zC due to the electrolytic removal of C ions from this phase.

Another way to study the kinetics of solid state reactions is the control or measurement of the flow of reactants to or from the product phase by means of an ionic conductor.[33] This is shown schematically in Figure 9.20. A separate potential electrode must be used if polarization at the reference electrode is not negligibly small. The current i through the electrolyte reflects the growth of a product phase A_yB by the reaction of species A delivered by the electrolyte with species B from the substrate. Either A or B (or both) must diffuse through the product phase. So the resolution is very high, since currents may be measured very precisely. The parabolic tarnishing rate constant is determined according to

$$k_t = \frac{i V_t L}{y z_A F} \tag{9.37}$$

where V_t, L, z_A, and F are the molar volume of the product A_yB, the thickness of the layer, the charge number of the A ions, and Faraday's constant, respectively.

In cases where no suitable ionic conductor is available for either of the reactants, the electrochemical technique may nevertheless be employed by using auxiliary phases. By the transport of C ions across the electrolyte, the activity of A is increased in the auxiliary phase (or this phase may even be decomposed). Instead of the electrolyte, the auxiliary phase provides the species A for the growth of the tarnish product phase and Equation (9.37) applies as well. The switch in Figure 9.21 permits a check on whether or not the transport of electrons in the product phase has an influence on the reaction rate. If the voltage is applied to the auxiliary phase, both A ions (or B ions) and electronic species have to move through the tarnish layer, whereas only ions move in steady state if the electronic lead is connected to the substrate.

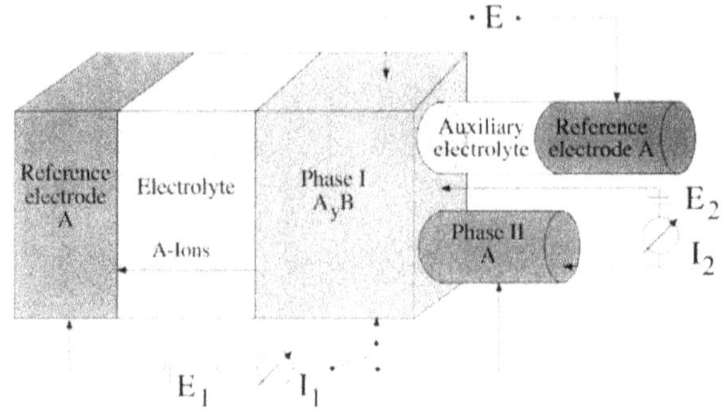

FIGURE 9.22. Arrangement for investigating the transport of A ions, electrons, and both of them in equivalent amounts across the interface between phases I and II. In the lower switch position only ions pass the interface, whereas both ions and electrons are moving across the interface in the upper switch position. The current i_2 is only due to the movement of electrons. The auxiliary galvanic cell at the upper right-hand side is used to determine the activity of A at the interface.

Electrochemical methods may also be used to investigate the transport of species across interfaces. By use of different locations for the electronic leads, and with the help of auxiliary galvanic cells, the transport of electrons, ions, and neutral species may be investigated separately. Such an arrangement is shown schematically in Figure 9.22. The voltage E determines the activity of A in phase I at the interface with phase II, and deviations from equilibrium with phase II are indicated by values different from 0. If voltage E_2 is applied, the transport of electrons across the interface is studied. If, when applying E_1, the switch is in the lower position, only A ions move across the phase boundary, whereas neutral A atoms (i.e., both ions and electrons) in equivalent amounts pass if the switch is in the upper position. Experiments of this type have been carried out to study the transport of silver across the interface Ag|Ag_2X (X = S, Te, Se) using solid silver ion conductors and silver as a reference electrode.[34]

In a similar arrangement to that shown in Figure 9.22, sublimation processes may also be studied electrochemically. Phase II is replaced by vacuum or inert gas. A steady-state current through the cell is only possible by decomposition of A_yB by sublimation of species B. The rate of this process is measured by the cell current. The voltage of the auxiliary galvanic cell controls or measures the activities of the components in the sample at the interface with the vapor phase. Investigations of this type have been recorded for the sublimation of iodine from CuI using CuBr electrolytes[35] and of sulfur from Ag_2S using an AgI electrolyte.[36] The latter case has also been combined with the Knudsen cell.[37]

If a gas that reacts with phase I is employed rather than a vacuum or inert gas, the cell current reflects the reaction rate at the phase boundary phase I/gas. The activity of the mobile component is fixed by the applied voltage. Likewise, changes in the cell potential may be monitored to obtain information about nucleation processes. This technique has been applied to study the reaction of Ag_2S with H_2, H_2S mixtures.[38]

V. SUMMARY

Fast ionic conductors have the unique property of transducing kinetic and thermodynamic quantities into readily measurable or controllable electric currents and voltages. Table 9.3 provides a summary of the techniques.

TABLE 9.3
Summary of Main Electrochemical Techniques for Determining Thermodynamic and Kinetic Properties in Solid Ionic and Mixed Conductors

Predominantly moving species	Quantity	Technique	Remarks
Ions	Gibbs energy of formation	Emf measurements	Phase equilibria with adjacent phases of the phase diagram at both electrodes
		Electrochemical decomposition	Reversible and inert electrode
	Ionic conductivity	Four-point dc technique	Reversible electrodes, electronic probes
		Ac technique	Inert or reversible electrodes
		Transference measurements	Reversible electrodes, electrons unblocked
		Emf measurements	Reversible electrodes with known activities, fast screening tool
	Electronic conductivity	Dc polarization (Hebb–Wagner technique)	One ionic blocking and one reversible electrode, high precision
		Transference measurements	Reversible electrodes, low precision
		Emf measurements	Reversible electrodes with known activities
	Electronic diffusion (-mobility, -concentration)	Voltage relaxation method	One ionic blocking and one reversible electrode, transient process
		Charge transfer technique	One ionic blocking and one reversible electrode, transient process
	Reaction rate	Dc polarization	Reversible electrode, reactant phase
Electrons or holes	Gibbs energy of formation	Emf measurements	Auxiliary electrolyte for one component; n-phase equilibrium for n phases
	Nonstoichiometry, phase diagram	Coulometric titration	Auxiliary electrolyte for stoichiometrically variable component
	Partial ionic conductivity	Dc polarization	Electrons blocked by auxiliary electrolyte, reversible electrode
		Transference experiments	Reversible electrode, small weight changes
		Transient experiments	Auxiliary electrolyte, general technique
	Electronic conductivity	Dc polarization	Ions blocked, reversible electrode
		Transference experiments	Reversible electrode, practically total conductivity
		Emf measurements	Auxiliary electrolytes
	Chemical diffusion	Transient technique combined with coulometric titration	Auxiliary electrolyte, various signals
	Ionic diffusivity (-mobility)	Transient experiments with coulometric titration	Auxiliary electrolyte, very general procedure
	Reaction rate	Transient experiments with coulometric titration	Auxiliary electrolyte, stoichiometric dependence
		Electrochemical formation	Formation controlled by auxiliary electrolyte

REFERENCES

1. de Bethune, A.J., *J. Electrochem. Soc.*, 1955, *102,* 288C–92C.
2. Wagner, C., *Prog. Solid State Chem.*, 1975, *10,* 3–16.
3. Rickert, H., *Solid State Electrochemistry, an Introduction,* Springer-Verlag, Berlin, 1982.
4. Le Claire, A.D., *Physical Chemistry,* Jost, W., Ed., Academic Press, New York, 1970, *10,* 261–330.
5. Darken, L.S., *Trans. AIME,* 1948, *175,* 184–201.
   Wagner, C., *Atom Movements,* Am. Soc. Met.: Cleveland, 1951, 153–173.
6. Weppner, W. and Huggins, R.A., *Electrochem. Soc.*, 1977, *124,* 1569–1578.
7. Wagner, C.J., *Chem. Phys.*, 1953, *21,* 1819.
8. Weppner, W. and Huggins, R.A., *J. Electrochem. Soc.*, 1978, *125,* 7–14.
   Weppner, W. and Huggins, R.A., *J. Solid State Chem.*, 1977, *22,* 297–308.
9. Li-chuan, Chen and Weppner, W., *Naturwisserschaften,* 1978, *65,* 595.
   Weppner, W., Li-chuan, Chen, and Rabenau, A., *J. Solid State Chem.*, 1980, *31,* 257.
10. Weppner, W. and Huggins R.A., *Fast Ion Transport in Solids; Electrodes and Electrolytes,* Vashishta, P., Mundy, J.N., and Shenoy, G.K., Eds., Elsevier/North-Holland, New York, 1979, 475.
    Weppner, W. and Huggins, R.A., *Solid State Ionics,* 1980, *1,* 3.
11. Piekarczyk, W., Weppner, W., and Rabenau, A., *Z. Naturforsch.*, 1979, *34a,* 430–436.
12. Hartwig, P., Weppner, W., and Wichelhaus, W., *Fast Ion Transport in Solids: Electrodes and Electrolytes,* Vashishta, P., Mundy, J.N., Shenoy, G.K., Eds., Elsevier/North-Holland, New York, 1979, 487–490.
    Hartwig, P., Weppner, W., Wichelhaus, W., and Rabenau, A., *Angew. Chem.*, 1980, *92,* 72; *Angew. Chem. Int. Ed. Engl.*, 1980, *19,* 74.
13. Weppner, W., Li-chuan, Chen, and Piekarczyk, W., *Z. Naturforsch.*, 1980, *35a,* 381–388.
14. Sitte, W. and Weppner, W., *Appl. Phys.*, 1985, *38,* 31–36.
    Sitte, W. and Weppner, W., *Z. Naturforsch.*, 1987, *42a,* 1–6.
15. Schmidt, J.A. and Weppner, W., Phase equilibria and thermodynamics of Li-Fe-S, in *Proc. 7th Int. Conf. on Solid Compounds of Transition Elements, Grenoble, France,* 1982, pp. IB 13 a–d.
16. Kohlrauch, F., *Praktische Physik,* Vol. 2, Teubner, Stuttgart, 1968, 284.
17. Macdonald, J.R., *Impedance Spectroscopy,* John Wiley & Sons, New York, 1987.
18. Wagner, C., *Z. Phys. Chem.*, 1933, *21,* 25–41, 42–47.
    Schmalzried, H., *Z. Phys. Chem. Frankfurt,* 1963, *38,* 87–102.
19. Hebb, M.H., *J. Chem. Phys.*, 1952, *20,* 185–190.
    Wagner, C., *Proc. 7th Meet. Int. Comm. on Electrochem. Thermodynam. Kinet., Lindau, 1955,* Butterworths, London, 1957, 361–377.
    Wagner, C., *Z. Elektrochem.*, 1956, *60,* 4–7.
20. Friedmann, L.M., Oberg, K.E., Boorstein, W.M., and Rapp, R.A., *Metall. Trans.*, 1973, *4,* 69–74.
    Patterson, J.W., Bogren, E.C., and Rapp, R.A., *J. Electrochem. Soc.*, 1967, *114,* 752–758.
    Burke, L.D., Rickert, H., and Steiner, R., *Z. Phys. Chem. Frankfurt,* 1971, *74,* 146–167.
    Weppner, W., *J. Solid State Chem.*, 1977, *20,* 305–314.
21. Ilschner, B., *J. Chem. Phys.*, 1958, *28,* 1109–1112.
22. Wagner, J.B. and Wagner, C., *J. Chem. Phys.*, 1957, *26,* 1597–1601.
23. Weppner, W. and Huggins, R.A., *J. Electrochem. Soc.*, 1977, *124,* 35–38.
    Weppner, W., Huggins, R.A., *Phys. Lett.*, 1976, *58A,* 245–248.
24. Weppner, W., *Z. Naturforsch,* 1976, *31a,* 1336–1343.
    Weppner, W., *Electrochim. Acta,* 1977, *2,* 721–727.
    Weiss, K., *Z. Phys. Chem. Frankfurt,* 1968, *59,* 242–250.
    Weiss, K., *Electrochim. Acta,* 1971, *16,* 201–221.
25. Raleigh, D.O., *Z. Phys. Chem. Frankfurt,* 1969, *63,* 319–322.
    Joshi, A.V., *Ph.D. dissertation,* Northwestern University, Evanston, IL, 1972, 110.
    Joshi, A.V. and Wagner, J.B., *J. Phys. Chem. Solids,* 1972, *33,* 205–210.
    Joshi, A.V., *Fast Ion Transport in Solids — Solid State Batteries and Devices,* van Gool, W., Ed., North-Holland, Amsterdam, 1973, 173–180.
    Wagner, J.B., *Fast Ion Transport in Solids — Solid State Batteries and Devices,* van Gool, W., Ed., North-Holland, Amsterdam, 1973, 489–502.
    Joshi, A.V. and Wagner, J.B., *J. Electrochem. Soc.*, 1975, *122,* 1071–1080.
26. Weppner, W. and Huggins, R.A., *Electrode Materials and Processes for Energy Conversion and Storage,* Electrochemical Society, Princeton, NJ, 1977, 833–846.
27. Rickert, H., *Electromotive Force Measurements in High Temperature Systems,* Alcock, C.B., Ed., Institute of Mining and Metallurgy: London, 1968, 59–90.
    Mijatani, S. *J. Phys. Soc. Jpn.*, 1955, *10,* 786–793.
    Wagner, C. *J. Chem. Phys.*, 1953, *21,* 1819–1827.
    Rickert, H. and Steiner, R., *Naturwissenschaften,* 1965, *52,* 451–452.

Rickert, H. and Steiner, R., *Z. Phys. Chem. Frankfurt,* 1966, *49,* 127–137.
Raleigh, D.O., *J. Electrochem. Soc.,* 1967, *114,* 493–494.
Rickert, H. and El Miligy, A., *Z. Metallk.,* 1968, *59,* 635–641.
Bazan, J.C., *Electrochim. Acta,* 1968, *13,* 1883–1885.
Pastorek, R.L. and Rapp, R.A., *Trans. Met. Soc. AIME,* 1969, *245,* 1711–1720.
Raleigh, D.O. and Crowe, H.R., *J. Electrochem. Soc.,* 1969, *116,* 40–48.
Oldham, K.B. and Raleigh, D.O., *J. Electrochem. Soc.,* 1971, *118,* 252–255.
Ramanarayanan, T.A. and Rapp, R.A., *Met. Trans AIME,* 1972, *3,* 3239–3246.
28. Weppner, W. and Huggins, R.A. *J. Solid State Chem.,* 1977, *22,* 297–308.
Liebert, B.E., Weppner, W., and Huggins, R.A., *Electrode Materials and Processes for Energy Conversion and Storage,* Electrochemical Society, Princeton, NJ, 1977, 821–832.
29. Wagner, C., *Progr. Solid State Chem.,* 1971, *6,* 1–15.
30. Chu, W. F., Rickert, H., and Weppner, W., *Fast Ion Transport in Solids — Solid State Batteries and Devices,* van Gool, W., Ed., North-Holland, Amsterdam, 1973, 181–191.
Rickert, H. and Weppner, W., *Z. Naturforsch.,* 1974, *29a,* 1849–1859.
Steele, B.C.H. and Ricardi, C.C., *Metallurgical Chemistry,* Kubaschewski, O., Ed., Her Majesty's Stationery Office, London, 1972, 125–135.
Steele, B.C.H., *Mass Transport Phenomena in Ceramics,* Cooper, A.R. and Heuer, A.H., Eds., Plenum Press, New York, 1975, 269–283.
Steele, B.C.H., *Mater. Sci. Res.,* 1975, *9,* 269–283.
31. Winn, D.A. and Steele, B.C.H., *Mater. Res. Bull.,* 1976, *11,* 559–566.
32. Weppner, W. and Huggins, R.A., *Z. Physikal. Chem. N. F. (Frankfurt),* 1977, *108,* 105–122.
33. Rickert, H., in *Reactivity of Solids, 5th Int. Symp., 1964,* Schwab, G.M., Ed., Elsevier, Amsterdam, 1965, 214–226.
Bartkowicz, I. and Mrowec, S., *Zesz. Nauk. Akad. Gorn.-Hutn. Cracow, Mat. Fiz. Chem.,* 1973, *10,* 41–63.
34. Rickert, H. and Wagner, C., *Z. Phys. Chem. Frankfurt,* 1962, *31,* 32–39.
Rickert, H. and O'Brian, C.D., *Z. Phys. Chem. Frankfurt,* 1962, *31,* 71–86.
Contreras, L. and Rickert, H., *Fast Ion Transport in Solids — Solid State Batteries and Devices,* van Gool, W., Ed., North-Holland, Amsterdam, 1972, 523–531.
35. Mrowec, S. and Rickert, H., *Z. Elektrochem.,* 1962, *66,* 14–17.
36. Birks, N. and Rickert, H., *Ber. Bunsenges. Phys. Chem.,* 1963, *67,* 97–98.
37. Detry, D., Drowart, J., Goldfinger, P., Keller, H., and Rickert, H., *Z. Phys. Chem. Frankfurt,* 1967, *55,* 314–319.
38. Birks, N. and Rickert, H., *Ber. Bunsenges. Phys. Chem.,* 1963, *67,* 501–505.
Rickert, H. and Tostmann, K.H., *Werkst. Korros.,* 1970, *21,* 965–973.

Chapter 10

ELECTROCHEMICAL SENSORS

Pierre Fabry and Elisabeth Siebert

CONTENTS

0-8493-8956-9/97/$0.00+$.50

LIST OF ABBREVIATIONS

emf	electromotive force
FIA	flow injection analysis
ISE	ion-selective electrode
ISFET	ion-sensitive field effect transistor
NASICON	Na super ionic conductor
SIC	solid ionic conductor
YSZ	yttria-stabilized zirconia

LIST OF SYMBOLS

a_i	activity of species i
A	area
C_i	concentration of species i (general notation)
d	diameter
D_i	diffusion coefficient of species i
e^-	electron of metal (electrode)
e'	band (or free) electron (Kröger–Vink notation in solid state chemistry)
E	potentiometric sensor voltage
E°	standard voltage
E_{dc}	dc voltage
E_{homo}	homogeneous voltage (thermoelectric sensor)
$E_{I=0}$	voltage without current
E_{rev}	reversible voltage (thermoelectric sensor)
E_s	potentiometric signal amplitude (pump-gauge sensor)
E_{th}	thermodynamic voltage
F	Faraday constant
I	current
I_l	limiting current
J_i	flux of species i
K_{eq}	equilibrium constant (e.g., K_{ex}, K_e, K_i, K_O, etc.)
$K_{i,j}$	selectivity coefficient (i: analyzed species, j: interfering species)
K'_d	Knudsen coefficient
L	length

$M^{\cdot}_{Ag}$	metal M^{2+} on Ag^+ site (Kröger–Vink notation in solid state chemistry)
M_i	molar mass of species i
$O_O^{\times}$	oxygen on oxygen site (Kröger–Vink notation in solid state chemistry)
P_i	partial pressure of species i
R	gas constant
t	time
$\bar{t}_i$	average transport number of species i
T	absolute temperature
Ti'_{Ti}	Ti^{3+} on Ti^{4+} site (Kröger–Vink notation in solid state chemistry)
$Ti^{\times}_{Ti}$	Ti^{4+} on Ti^{4+} site (Kröger–Vink notation in solid state chemistry)
$\tilde{u}_i$	electrochemical mobility
U	total voltage of an electrochemical cell
V°	volume
V'_{Ag}	silver vacancy (Kröger–Vink notation in solid state chemistry)
$V_O^{\cdot\cdot}$	oxygen vacancy (Kröger–Vink notation in solid state chemistry)
W_a	activation energy or reaction enthalpy (other examples: W_e, W_i, W_{ae}, W_{aO})
z_i	charge number of species i
α_{conv}	convection coefficient
ΔG	change of free enthalpy
ΔG°	change of standard free enthalpy
η_a	anodic overvoltage
η_b	cathodic overvoltage
μ_i	chemical potential of species i
μ_i°	standard chemical potential of species i
$\tilde{\mu}_i$	electrochemical potential of species i
ϕ	internal electric potential
σ_i	conductivity of species i
σ_t	total conductivity
τ	time constant
τ_{eq}	equivalent time constant
ω	frequency of the ac signal
[i]	concentration of the point defect i in a SIC (solid state chemistry)

I. INTRODUCTION

Chemical sensors are increasingly used in numerous applications, such as analysis setups, chemical processes, food industry, environmental control (water, air, industrial wastes, combustion monitoring, etc.), and biomedical applications (analysis and monitoring). Other new fields of applications could arise in the near future; for instance, integrated home systems. Electrochemical sensors are particularly attractive because the chemical quantities which are measured are directly transduced in electrical signals. As for electronic components, two types can be distinguished: active sensors, which give a voltage (potentiometric sensors), and passive sensors, for which an electrical source is necessary to apply a signal, the response of which is analyzed afterwards (amperometric, coulometric, and conductometric sensors).

We will limit our description of electrochemical sensors to those using a solid state material with an ionic conductivity playing a major role in the detection mechanism. These sensors are not always assembled in "all-solid state" technology, but in most of the devices such an assembling technique could be considered.

A. BRIEF HISTORY

Basic electrochemical concepts were established in the 19th century by Volta (classification of redox couples), Faraday (relation between electricity and quantity of matter), Kohlrausch (conductivity measurements), and Nernst (thermodynamic approach), but the first sensors were developed at the end of the century and at the beginning of the 20th century. The pH electrode using a glass membrane was discovered by Cremer[1] in 1906. Industrial development was achieved only about 20 years later. After a first approach made by Nikolskii[2] in 1937, a theoretical model was established by Eisenman[3] to explain interfering phenomena in sensors using a glass membrane. The use of the pH electrode for P_{CO_2} measurement was proposed by Severinghaus and Bradley[4] in 1957. After the ion-selective electrode (ISE) for the Cl^- ion, introduced in 1937,[5] other ISEs based on silver salts or on a LaF_3 monocrystalline membrane have been developed in the 1960s.[6,7] Ceramics were first used for O^{2-} analysis in molten salts in 1960.[8]

Since the work of Kiukkola and Wagner,[9] solid oxide cells have been used for thermodynamic measurements. The first oxygen gauges were tested in 1961.[10,11] Similar sensors based on other solid electrolytes have been proposed for halogens,[12,13] sulfur oxides,[14] and hydrogen.[15]

Amperometric and coulometric oxygen sensors are more recent.[16,17] The first conductometric sensors using a TiO_x-based material were discovered in the 1970s.[18]

B. CURRENT TRENDS

At the present time, research tends toward the use of new ionic materials, amorphous[19] or crystallized, to improve the selectivity properties.[20] New systems coupling amperometric and potentiometric measurements are proposed.[21] In this field, digital electronic development is a tool which is well suited for sophisticated devices. The miniaturization techniques are actually developed for mass production. The basic idea is to adapt microelectronic technology to the fabrication of chemical sensors. This is the main trend, whatever the kind of sensor. Solid ionic conductors (SIC) are very suitable for these methods of fabrication. Thin- or thick-layer techniques, using physical or chemical deposition methods, can be used.[22-26,70,71] We can mention the first work of Velasco et al.[26] concerning oxygen sensors. Since the development of the ion-sensitive field effect transistor by Bergveld[27] in the 1970s, several ways have been proposed to deposit an ionic membrane on the gate of field effect transistors. The first device of this type, using a silver halide, was proposed by Buck and Hackerman.[28]

C. PREFATORY NOTE

This article cannot be an exhaustive review on all the solid state electrochemical sensors which have been developed up to now. The literature in this field is abundant and we in particular mention some recent reviews.[29-41] The goal of this paper is to explain, using solid state electrochemical concepts, the working principle of solid state electrochemical sensors and to show their limits of utilization. This theoretical treatment is illustrated by some examples. Field effect chemical transistors are not described here, and we refer to a recent review.[42]

II. DIFFERENT KINDS OF SENSORS

A. POTENTIOMETRIC SENSORS

1. Fundamental Principles

The fundamental principles of potentiometric sensors based on solid electrolytes have been reviewed in several papers.[37,43-50] Potentiometric sensors are based on the existence of an equilibrium at the interface between a SIC and the analyzed medium, by exchange of electrochemical species. This hypothesis is acceptable (after a transient response) if there is no net dc current. We classify different kinds of sensors based on the nature of the interfacial exchange, and therefore it is useful to give a description of elementary steps in an electrochemical sensor.

a. Ionic Junction of the First Kind

In this case, only a single electrochemical species (X^{z+}, mobile ion) is exchanged between the SIC and the analyzed solution (sol). Typical examples are LaF_3/F^-, stabilized zirconia/O^{2-}, $AgCl/Ag^+$, or Na^+-based glass/Na^+ interfaces in ISE. The fundamental hypothesis can be expressed as

$$X^{z+}_{SIC} \Leftrightarrow X^{z+}_{sol} \tag{10.1a}$$

The equilibrium condition is $\Delta\tilde{G} = 0$, e.g.,

$$\tilde{\mu}_{X^{z+}(SIC)} = \tilde{\mu}_{X^{z+}(sol)} \tag{10.1b}$$

where $\tilde{\mu}_{X^{z+}}$ is the electrochemical potential of X^{z+} which can be written in terms of the chemical potential μ and the electrical energy $zF\phi$ in the considered phase (see also Chapter 2 of this handbook), by the relation

$$\tilde{\mu}_{X^{z+}} = \mu_{X^{z+}} + zF\phi \tag{10.2}$$

with

$$\mu_{X^{z+}} = \mu^{o}_{X^{z+}} + RT \ln a\left(X^{z+}\right) \tag{10.3}$$

$\mu^o{}_Xz+$ being the standard chemical potential, $a(X^{z+})$ the activity of X^{z+}, T the absolute temperature, and R and F have their usual meaning. For dilute phases, concentration and activity can be assumed to be equal. For concentrated phases, interactions between species must be taken into account by introducing an activity coefficient which can be calculated, for example, from the Debye–Hückel theory (see some common books of electrochemistry).[51,52]

Equations (10.1b) to (10.3) give the Galvani potential between both phases (SIC and sol),

$$\phi_{SIC} - \phi_{sol} = \frac{1}{zF}\left(\mu^{o}{}_{X^{z+}(sol)} - \mu^{o}{}_{X^{z+}(SIC)}\right) + \frac{RT}{zF} \ln a\left(X^{z+}\right)_{sol} - \frac{RT}{zF} \ln a\left(X^{z+}\right)_{SIC} \tag{10.4}$$

If the concentration of the mobile ions in the SIC ($a(X^{z+})_{SIC}$) is high, its activity can be considered to be constant and so the Galvani potential is directly related to the activity of ions in the analyzed solution ($a(X^{z+})_{sol}$). Figure 10.1a shows a schematic diagram of ϕ, μ, and $\tilde{\mu}$ variations. The Galvani potential cannot be measured directly, and a longer thermodynamic chain is necessary up to both metal connections. A more detailed description is given later (see also Chapter 2 of this handbook).

b. Ionic Junction of the Second Kind

In this case, the mobile ion of the SIC reacts with the analyzed ion to give a complex or an insoluble chemical product at the interface

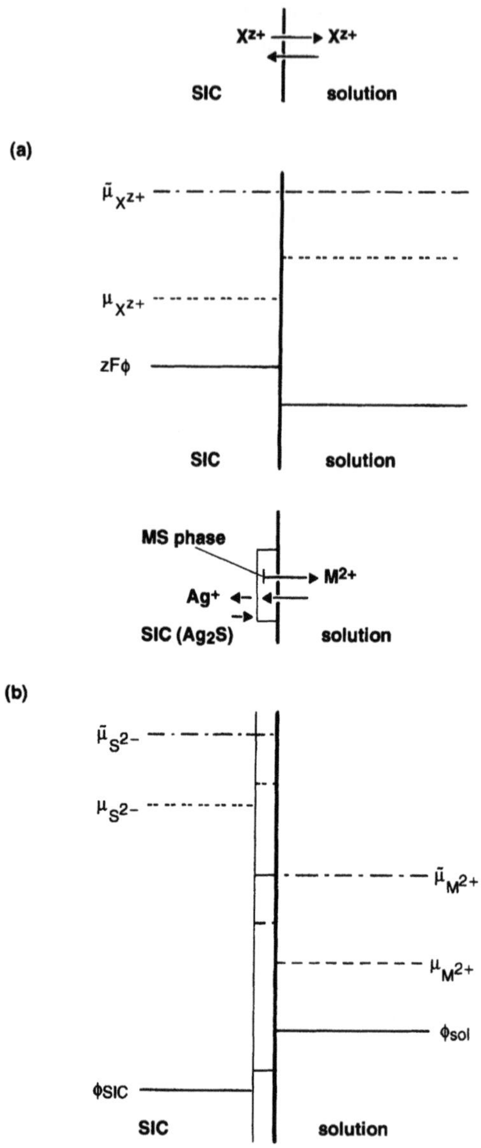

FIGURE 10.1. Schematic presentation of electrical (ϕ), chemical (μ), and electrochemical ($\tilde{\mu}$) potential variations in ionic junctions: (a) first kind, (b) second kind.

$$y X_{SIC}^{z+} + z B_{sol}^{y-} \Leftrightarrow X_y B_z \tag{10.5a}$$

and the thermodynamic relation is

$$\begin{aligned} &y\mu^{o}{}_{X^{z+}(SIC)} + yRT \ln a\left(X^{z+}\right)_{SIC} + yzF\phi_{SIC} + z\mu^{o}{}_{B^{y-}(sol)} \\ &+ zRT \ln a\left(B^{y-}\right)_{sol} - yzF\phi_{sol} = \mu_{X_y B_z} \end{aligned} \tag{10.5b}$$

If X_yB_z forms a separate phase, its chemical potential is equal to the standard potential $\mu°$. The Galvani potential is given by

$$\phi_{SIC} - \phi_{sol} = \frac{1}{yzF}\left(\mu^{o}{}_{X_yB_z} - y\mu^{o}{}_{X^{z+}(SIC)} - z\mu^{o}{}_{B^{y-}(sol)}\right) - \frac{RT}{zF}\ln a\left(X^{z+}\right)_{SIC} - \frac{RT}{yF}\ln a\left(B^{y-}\right)_{sol} \tag{10.5c}$$

As in case a, if the concentration of X^{z+} in the SIC is high, the voltage is proportional to the logarithm of the activity of the analyzed ion. Of course, the variation is a function of the numerical value of y (according to its sign). Typical examples are AgCl membranes and Ag_2S-AgCl mixtures to determine the activity of Cl^- ions. If AgCl is a separate phase, Equation (10.5c) can be applied and the Nernst equation is obeyed well. If AgCl is dissolved in another phase, the value of the chemical potential $\mu_{X_yB_z}$ may not be constant and Equation (10.5c) would no longer be valid.

The thermodynamic chain is more complex in the case of a ternary compound. To illustrate this, we consider the Ag_2S-MS/M^{2+} interface (for example, Cu^{2+}, Cd^{2+}, or Pb^{2+} ISEs will operate according to this mechanism). A schematic view is given in Figure 10.1b. In this case, the SIC is a conductor for Ag^+ ions. Equilibrium at the interface is ensured by the following exchange reactions, in a macroscopic description,

$$M^{2+}_{sol} + S^{2-}_{SIC} \Leftrightarrow MS \tag{10.6a}$$

and, in the bulk of the membrane, near the interface

$$2Ag^{+}_{SIC} + S^{2-}_{SIC} \Leftrightarrow Ag_2S \tag{10.6b}$$

The Galvani potential is now obtained as

$$\phi_{SIC} - \phi_{sol} = \frac{1}{2F}\left(\mu^{o}{}_{M^{2+}(sol)} - 2\mu^{o}{}_{Ag^{+}(SIC)} + \mu_{Ag_2S} - \mu_{MS}\right) - \frac{RT}{F}\ln a\left(Ag^{+}\right)_{SIC} + \frac{RT}{2F}\ln a\left(M^{2+}\right)_{sol} \tag{10.6c}$$

The voltage obeys the Nernst equation with respect to $a(M^{2+})_{sol}$ only if

- the concentration of Ag^+ in the SIC is sufficiently high to assume that $a(Ag^+)_{SIC}$ is constant,
- Ag_2S is the dominant phase and $\mu_{Ag_2S} = \mu^{o}{}_{Ag_2S}$
- μ_{MS} is constant, e.g., MS is a separate phase or is sufficiently concentrated in Ag_2S (or other solid solvent like a glass)

Similar conclusions are obtained if we use a microscopic model. The M^{2+} ion is assumed to be dissolved in a silver salt according to the following reaction, using the Kröger–Vink notation,

$$V'_{Ag} + M^{2+}_{sol} \Leftrightarrow M^{\bullet}_{Ag} \tag{10.7a}$$

The Galvani potential is then given by

$$\phi_{SIC} - \phi_{sol} = \frac{1}{2F}\left(\mu^{o}{}_{V'_{Ag}(SIC)} + \mu^{o}{}_{M^{2+}(sol)} - \mu^{o}{}_{M^{\bullet}_{Ag}(SIC)}\right)$$
$$- \frac{RT}{2F} \ln a\left(M^{\bullet}_{Ag}\right)_{SIC} + \frac{RT}{2F} \ln a\left(V'_{Ag}\right)_{SIC} + \frac{RT}{2F} \ln a\left(M^{2+}\right)_{sol} \tag{10.7b}$$

This description shows that

- the M^{2+} ion concentration in the solid matrix needs to be buffered, and
- the Ag^{+} vacancy concentration must be high.

c. *Electrode Reaction of the First Kind*

One material of the junction is an electronic conductor, generally a metal or a semiconductor which shows catalytic properties. This type of junction is the transducing part from the ionic conductor to the metallic connection in ISE devices. It is also the gas electrode in potentiometric gauges.

In most simple cases, the SIC and the metal have a common chemical species and the following equilibrium can be written:

$$M^{n+}_{SIC} + n\, e^{-}_{M} \Leftrightarrow M \tag{10.8a}$$

M has two functions: it is a neutral chemical species and acts as a donor (or acceptor) of electrons. The Galvani potential is then given by

$$\phi_{M} - \phi_{SIC} = \frac{1}{nF}\left(\mu^{o}{}_{M^{n+}(SIC)} - \mu^{o}{}_{M} + n\mu_{e}\right) + \frac{RT}{nF} \ln a\left(M^{n+}\right)_{SIC} \tag{10.8b}$$

The concentration of electrons is generally very high, so μ_e is constant. With the usual assumption that $a(M^{n+})_{SIC}$ is high, the Galvani potential is constant during the measurement. Equations (10.8b) and (10.4) or (10.5b), (10.5c), (10.6c), and (10.7b) give a Nernst equation for the metal potential as a function of the activity of the analyzed ion. Of course, if M is not a pure phase, for instance, an amalgam or a solid solution, $\mu^{o}{}_{M}$ must be replaced by μ_M in Equation (10.13) and the activity of M in this solution must be taken into account. This activity must be constant during the measurements (no chemical modifications may take place by faradaic effects due to parasitic currents).

In the case of a gas electrode, three phases are in contact (Figure 10.2a): inactive metal (M), gas (X_2), and SIC (for instance, conducting X^{z-}-ions). The equilibrium reaction is given by

$$X^{z-}_{SIC} \Leftrightarrow \frac{1}{2} X_2 + z\, e^{-} \tag{10.9a}$$

and the Galvani potential

$$\phi_{M} - \phi_{SIC} = \frac{1}{zF}\left(\frac{1}{2}\mu^{o}{}_{X_2} - \mu^{o}{}_{X^{z-}(SIC)} + z\mu_{e}\right) - \frac{RT}{zF} \ln a\left(X^{z-}\right)_{SIC}$$
$$+ \frac{RT}{2zF} \ln P_{X_2} \tag{10.9b}$$

which corresponds with the Nernst equation for X_2 analysis.

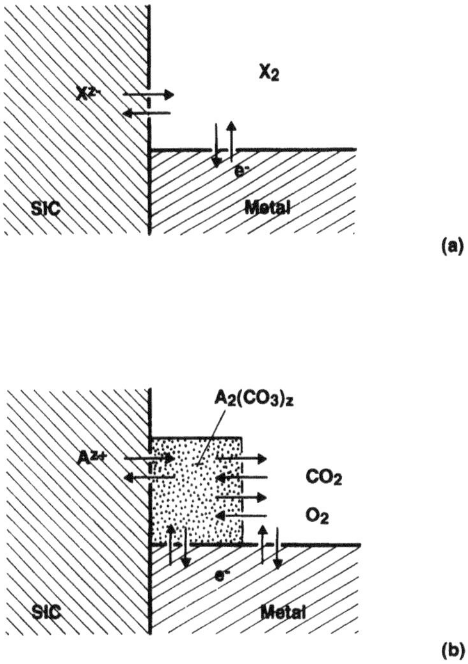

FIGURE 10.2. Schematics of electrode reactions: (a) first kind, (b) second kind.

d. *Electrode Reaction of the Second Kind*

As for an ionic junction of the second kind, the mobile ion of the SIC reacts with the analyzed species, forming an intermediate phase (separate phase or dissolved species). Simple examples are halide gauges (I_2, Cl_2, Br_2) using a cationic conductor. The equilibrium reaction is given by

$$\frac{z}{2}X_2 + nA^{z+}_{SIC} + nz\,e^- \Leftrightarrow A_nX_z \qquad (10.10a)$$

and the Galvani potential

$$\phi_M - \phi_{SIC} = \frac{1}{F}\left(\frac{1}{2n}\mu^o{}_{X_2} + \frac{1}{z}\mu^o{}_{A^{z+}(SIC)} + \mu_e - \frac{1}{nz}\mu_{A_nX_z}\right) + \frac{RT}{zF}\ln a\left(A^{z+}\right)_{SIC} + \frac{RT}{2nF}\ln P_{X_2} \qquad (10.10b)$$

As in the previous case, the same conclusions are obtained: the voltage obeys the Nernst equation if $a(A^{z+})_{SIC}$ and $a(A_nX_z)$ are constant during measurement. The second condition is justified if A_nX_z is a separate phase or if it is dissolved in a solid and its concentration is buffered.

In some sensors, two gases react simultaneously. A typical example is a CO_2 gauge in air using a cationic conductor to give a carbonate at the interface (Figure 10.2b). In such a case, the equilibrium reaction is more complex,

$$2ze^- + 2zA^{z+}_{SIC} + \frac{z}{2}O_2 + zCO_2 \Leftrightarrow A_2(CO_3)_z \qquad (10.11a)$$

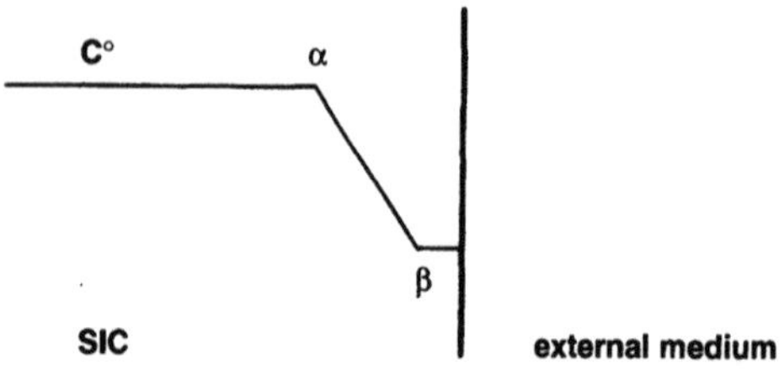

FIGURE 10.3. Concentration profile in a nonhomogeneous SIC (linear model).

and a similar derivation gives

$$\phi_M - \phi_{SIC} = \frac{1}{F}\left(\frac{1}{4}\mu^o{}_{O_2} + \frac{1}{z}\mu^o{}_{A^{z+}(SIC)} + \mu_e + \frac{1}{2}\mu^o{}_{CO_2} - \frac{1}{2z}\mu_{A_2(CO_3)_z}\right)$$
$$+\frac{RT}{zF}\ln a\left(A^{z+}\right)_{SIC} + \frac{RT}{4F}\ln P_{O_2} + \frac{RT}{2F}\ln P_{CO_2} \quad (10.11b)$$

The voltage obeys the Nernst equation for CO_2 if $A_2(CO_3)_z$ has a constant activity (when it is, e.g., a separate phase) and also if the oxygen pressure is fixed, for instance in air. In usual CO_2 sensors, the species A is currently silver or an alkaline ion and so z = 1.[95-98] Other sensors following this principle can be proposed.

e. Ionic Exchange at Interfaces

If a SIC is in contact with an active species, it can penetrate into the bulk by exchange followed by diffusion. It is then called an interfering species. The SIC becomes nonhomogeneous in composition due to the penetration of the interfering species. If the concentration of the mobile ion in the SIC is sufficiently high, it does not change much in use. A linear diffusion profile is assumed in Figure 10.3. At the interface between the SIC and the external medium, there is the following equilibrium:

$$I_{SIC} + K_{ext} \Leftrightarrow I_{ext} + K_{SIC} \quad (10.12)$$

where I is the normal species and K the interfering one. The subscript "ext" is used to indicate the external medium (solution or electrode material). This reaction is shown only qualitatively, and is written for neutral species. Stoichiometric coefficients would be necessary if I and K have different valences. The interfacial potential between the SIC (β, see Figure 10.3) and the external medium is fixed by the equilibrium

$$I_{SIC} \Leftrightarrow I_{ext} \quad (10.13)$$

The potential is given by

$$\phi_\beta - \phi_{ext} = E^o + \frac{RT}{z_i F}\ln\frac{C_i(ext)}{C_i(\beta)} \quad (10.14)$$

where E° includes the usual constants (μ°, etc.) and C_i are the concentrations (we suppose that the solid solution is not sufficiently disturbed to change the activity coefficient). A similar relation can be obtained with the interfering species.

Without a dc current, the sum of the partial current densities is equal to zero and the general relation is

$$\sum_j z_j F C_j \tilde{u}_j \frac{d\tilde{\mu}_j}{dx} = 0 \qquad (10.15)$$

where $\tilde{u}_j$ is the electrochemical mobility. Using Equation (10.2), the gradient of ϕ is given by

$$\frac{d\phi}{dx} = -\frac{RT}{F}\frac{\sum_j z_j \tilde{u}_j \frac{dC_j}{dx}}{\sum_j z_j^2 C_j \tilde{u}_j} \qquad (10.16)$$

To simplify the derivation, we consider the example for which there are only two ions (I and K). The electroneutrality relation then reads

$$z_i C_i + z_k C_k = z_i C^o \qquad (10.17)$$

where C° is the initial concentration of the ion I. In this case, the following differential relation holds

$$\frac{dC_k}{dx} = -\frac{z_i}{z_k}\frac{dC_i}{dx} \qquad (10.18)$$

By integration of Equation (10.16), taking into account Equations (10.17) and (10.18), the variation of ϕ between two points in the bulk of the SIC, in a linear model, we obtain

$$\phi_\beta - \phi_\alpha = \frac{RT}{F}\frac{(\tilde{u}_k - \tilde{u}_i)}{(z_i\tilde{u}_i - z_k\tilde{u}_k)} \ln \frac{C_i(\beta)[z_i\tilde{u}_i - z_k\tilde{u}_k] + z_k C^o \tilde{u}_k}{C_i(\alpha)[z_i\tilde{u}_i - z_k\tilde{u}_k] + z_k C^o \tilde{u}_k} \qquad (10.19)$$

For a potentiometric sensor, the measured potential between the SIC bulk (α) and the external medium can be calculated according to

$$\phi_{SIC} - \phi_{ext} = (\phi_\alpha - \phi_\beta) + (\phi_\beta - \phi_{ext}) \qquad (10.20)$$

in which the term $\phi_\beta - \phi_{ext}$ corresponds to the Galvani potential as defined in Equation (10.14).

f. Electrochemical Chain

Equations (10.14), (10.19), and (10.20) allow one to calculate the emf developed between two phases which are nonhomogeneous. In a potentiometric sensor, two metal electrodes are necessary, which are connected to a voltmeter. A schematic view is given in Figure 10.4. To obtain the SIC voltage, it is necessary to realize a reversible junction with a metal connection as described in Section II.A.1.c. A direct exchange is not always easy, and therefore a more complex chain must be elaborated by adding an intermediate system such as:

M(1) I ionic junction I SIC I solution I ref. M(2)
—> —> —>
e⁻ <— ion" ion' <— ion <—

FIGURE 10.4. Schematics of an electrochemical chain (ion-selective electrode).

- an ionic conducting material which can exchange common ions with the SIC and the connection material (exchange of the first or second kind),
- a mixed (ionic and electronic) conducting material which can exchange ions with the SIC and electrons with the metal.

In ionic sensors, such systems are called ionic bridge or internal reference. As the second electrode, an external reference is used; for instance, a calomel electrode with a liquid junction.

A fundamental assumption for evaluation of the voltage is that the SIC and other intermediate materials are equipotential. The electrochemical potential of the mobile ion is assumed to be constant throughout the material. The chemical potential can be considered to be constant if the mobile ion concentration in the SIC is high. From Equation (10.2) it follows that the electrochemical potential is constant when ϕ = constant.

For potentiometric devices, the sensor voltage is given by an algebraic relation including all phases and interface voltages. The interface voltages are given by Galvani potentials previously established and the difference of electrical potential in a phase is supposed to be negligible under steady-state conditions, except if it is a nonhomogeneous material. We will discuss the possible fluctuations of the electrical potential in such a material later.

2. Ionic Sensors

Table 10.1 summarizes the characteristics of common ISEs and a number of new sensors in this field. We have not included in this table the liquid or polymer membrane-based electrodes which are selective, but rather fragile (for more details on such membranes see References 58,59). ISEs of the first kind are not very numerous, e.g., F^--ISE (monocrystalline membrane based on LaF_3), Ag^+-ISE (silver salts), or Na^+-ISE (Na alumino-silicate glass or polycrystalline NASICON [Na super ionic conductor] membranes). Most of the ISEs are of the second kind and are based on insoluble silver salts; for example, halide ISEs (Cl^-, Br^-, I^-), Cd^{2+}, Pb^{2+}, Cu^{2+}, etc. Such ISEs use mixtures of insoluble salts based on silver sulfide or silver selenide. Recently, Vlasov et al.[60,61] and Neshkova[62] have proposed several glasses sensitive to transition metals. Typical ISE devices are shown in Figure 10.5. Thin-layer chemical sensors based on chalcogenide glasses have also been developed.[63]

a. Solid Internal Reference Systems

The usual internal systems are liquid solutions, because a salt can be dissolved to give an ionic junction. For instance, in a Na^+-ISE, a NaCl aqueous solution gives an equilibrium with the Na-SIC (Na^+ exchange) and an electrode reaction with an Ag/AgCl connection (Cl^- exchange). Such systems are advantageous because

- the NaCl concentration is high so that there is a good buffering effect (this point is discussed later), the electrochemical reversibility is improved (interface and electrode reaction), and the internal resistance is low; and
- the interfacial contact is very good.

Nevertheless, there are some disadvantages, such as

TABLE 10.1
Main Ion Selective Electrodes

Ion (x)	Membrane	Concentration range	Main interfering ions $K_{x,y} \geq 10^{-2}$	Ref.
Ag^+	Ag_2S	10^{-1}–10^{-7}	Cl^-, Br^-, SCN^-, CN^-, S^{2-}, Hg^{2+}	53
Br^-	$Ag_2S + AgBr$	1–10^{-5}	I^-, SCN^-, CN^-	53–55
	AgBr			
Cd^{2+}	$Ag_2S + CdS$	10^{-1}–10^{-7}	S^{2-}, Pb^{2+}, Ag^+,	53,54
Cl^-	$Ag_2S + AgCl$	10^{-1}–5 10^{-5}	CN^-, I^-, Br^-, Ag^+, S^{2-}, SCN^-	53,54
CN^-	$Ag_2S + AgI$	10^{-2}–10^{-6}	I^-, S^{2-}, Ag^+	49,53,55
Cu^{2+}	$Ag_2S + CuS$	10^{-1}–10^{-8}	Ag^+, Hg^{2+}, S^{2-}, Cl^-, Br^-, SCN^-, CN^-	49,55
F^-	LaF_3	1–10^{-6}	OH^-	53,54
H_3O^+	Glass	1–10^{-14}		49,53
Hg^{2+}	$Ag_2S + AgI$		Ag^+, Cl^-, Br^-, SCN^-, CN^-	56
I^-	$Ag_2S + AgI$	1–5 10^{-8}	S^{2-}, Ag^+, CN^-	53–55
	AgI			
Na^+	Glass	1–5 10^{-7}	H_30^+, K^+, Li^+,	53, 57
	NASICON	1–10^{-4}	H_30^+	57
Pb^{2+}	$Ag_2S + PbS$	1–10^{-7}	Ag^+, Hg^{2+}, Cu^{2+}, Cd^{2+}, Cl^-, Br^-, SCN^-, CN^-	53–55
S^{2-}	Ag_2S	1–10^{-7}	Ag^+, Hg^{2+}	53,54
SCN^-	$Ag_2S + AgSCN$	1–5 10^{-6}	I^-, Cl^-, OH^-, Br^-, CN^-, S^{2-}	49,59

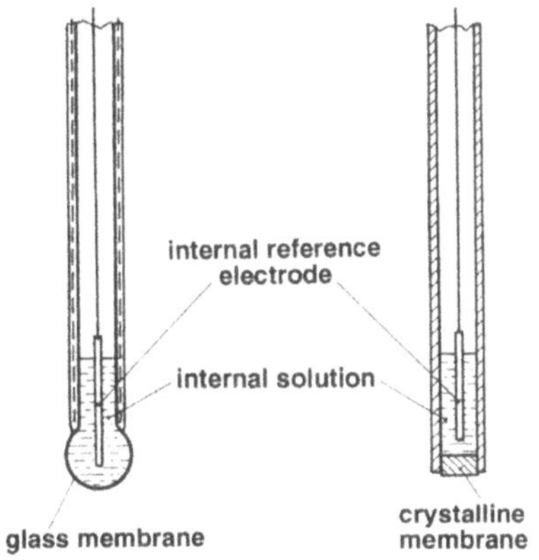

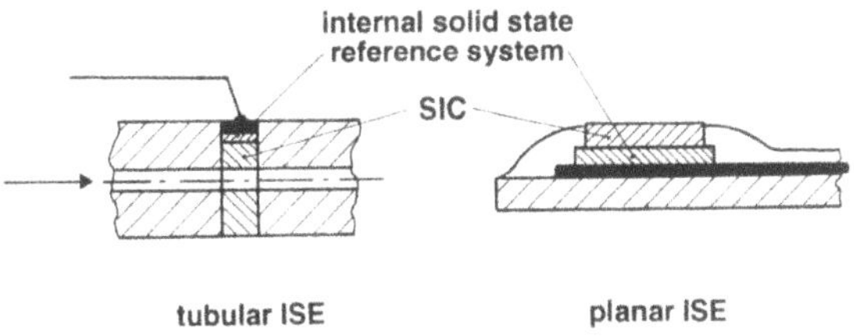

FIGURE 10.5. Typical ISE devices.

- there are limitations in the working temperature;
- there is a restriction in the positioning (it is necessary to put the sensitive part below to maintain the liquid phase in contact with the membrane); one can remark here that to increase the freedom in use, it is possible to replace the liquid solution by a gel; and
- the systems are not suitable for miniaturization.

For these reasons, several improvements have been suggested which are described in detail in the review paper of Nikolskii and Materova.[64] We can distinguish three classes of internal reference systems:

(i) metal having cations in common with the SIC. The calculation of the voltage is easy from Equations (10.8b) and (10.4) (or [10.5c] or [10.6c]). As an example, for a M^{2+}-ISE with silver as the internal connection, the electrochemical chain is

$$\textbf{connection : } e^-, \textbf{Ag/SIC : } Ag^+, S^{2-}, M^{2+}\textbf{/sol : } M^{2+}$$

We obtain,

$$\phi_M - \phi_{sol} = \frac{1}{F}\left(\mu_e - \mu^o_{Ag} + \frac{1}{2}\mu_{Ag_2S} - \frac{1}{2}\mu_{MS} + \frac{1}{2}\mu^o_{M^{2+}(sol)}\right) + \frac{RT}{2F}\ln a\left(M^{2+}\right)_{sol} \tag{10.21}$$

As ϕ_{sol} can be related to the reference electrode potential, one can write the usual equation

$$\phi_M - \phi_{sol} = E^o + \frac{RT}{2F}\ln a\left(M^{2+}\right)_{sol} \tag{10.22}$$

where E° includes the different constants. The aforementioned assumptions are supposed to be obeyed (essentially constant concentrations of Ag_2S and MS or a similar assumption in a microscopic description). A similar law is obtained for the electrochemical chain,

$$\textbf{connection (Me) : } e^-, \textbf{ A/A}_X\textbf{F/SIC : } F^-\textbf{/sol : } F^-$$

which is used in a F^--ISE.[65-67]

(ii) mixed conductor (intercalation compounds, glasses) which exchanges electrons with the connection and ions with the SIC.[68-71] As a typical example, we describe here an alkaline ISE (sensitive to A^+). The electrochemical chain is

$$\textbf{connection (Me) : } e^-\textbf{/mixed conductor : } e^-, \; A^+\textbf{/SIC : } A^+\textbf{/sol : } A^+$$

and the voltage is given by

$$\phi_M - \phi_{sol} = \frac{1}{F}\left(\mu_{e(Me)} - \mu_{e(int)} + \mu^o_{A^+(sol)} - \mu^o_{A^+(int)}\right) - \frac{RT}{F}\ln a\left(A^+\right)_{int} + \frac{RT}{F}\ln a\left(A^+\right)_{sol} \tag{10.23}$$

From this equation it clearly appears that the voltage response obeys the Nernst equation only if the concentrations of e^- and A^+ in the mixed conductor (int) are constant. Any parasitic dc current or the use of a voltmeter having too low an input impedance can change their values by a faradaic effect, which results in a drift of the cell voltage. Therefore, the initial amount of intercalated species must be high enough so that the changes in its concentration are small enough to avoid these problems.

(iii) solid solution having one ion in common with the SIC and another ion common with the internal connection (generally in an electrode reaction of the second kind), for instance a polymer or a silver halide solution.[72-74] As an example of such a solid solution, we consider the following chain

$$\textbf{connection : } e^-, \textbf{ M/int : } M^{n+}, \ A^{z+}\textbf{/SIC : } A^{z+}\textbf{/sol : } A^{z+}$$

which leads to

$$\phi_M - \phi_{sol} = \frac{1}{F}\mu_{e(M)} + \frac{1}{nF}\left(\mu^o{}_{M^{n+}(int)} - \mu^o{}_M\right) + \frac{1}{zF}\left(\mu^o{}_{A^{z+}(sol)} - \mu^o{}_{A^{z+}(int)}\right) \quad (10.24)$$

$$+\frac{RT}{nF}\ln a\left(M^{n+}\right)_{int} - \frac{RT}{zF}\ln a\left(A^{z+}\right)_{int} + \frac{RT}{zF}\ln a\left(A^{z+}\right)_{sol}$$

The same remark as that concerning the amount of intercalated compound can be made about the concentrations of common ions in the internal material. The solid solution AgI-NaI is not stable at room temperature, and a drifting voltage was observed with a Na^+-ISE using such an internal reference system.[74] The solid solutions AgCl-NaCl or AgBr-NaBr are more stable, but their conductivities are low.

b. Interfering Phenomena

When a SIC, conducting ions I, is in contact with a solution of ions J (interfering ions), an ionic exchange reaction takes place:

$$I_{SIC} + J_{sol} \Leftrightarrow I_{sol} + J_{SIC} \quad (10.25)$$

The solubility of J in the SIC is generally poor and limited to its surface. As a result of the ionic exchange reaction, the membrane also becomes sensitive to the ion J. Nikolskii[2] has first proposed the empirical equation

$$E = E° + \frac{RT}{z_i F}\ln\left(a_i + \sum_j K_{i,j} a_j^{z_i/z_j}\right) \quad (10.26)$$

where z_i and z_j are the charge numbers of ions I and J, a_i and a_j their activities in the solution, and $K_{i,j}$ the selectivity coefficients (sometimes also called interfering coefficients). The lower the values of $K_{i,j,}$ the higher the selectivity of the ISE. If a_j (or $K_{i,j}$) is very low, Equation (10.26) is equivalent to the Nernst equation which gives the voltage as a function of a_i. On the other hand, if the interfering ion J predominates, Equation (10.26) corresponds to the Nernst equation for the activity of a_j (E° is thus shifted).

Several models have been proposed to explain the form of Equation (10.26). For instance, in the case of insoluble membranes (silver salts), it is possible to relate the selectivity coefficient $K_{i,j}$ to the ratio of solubility constants of insoluble salts such as MI and MJ (see the example of AgCl and AgI described by Morf).[49]

In this review we only present a model based on the ion mobilities as developed, for instance, by Eisenman.[3-75] In this case, the ion J can penetrate into the SIC which is not doped initially with this species. To simplify the derivation, we consider the example of a Na^+-SIC in contact with a Li^+ containing solution. The SIC is considered to be nonhomogeneous near the interface (see Figure 10.3). Equation (10.19) in this case becomes

$$\phi_\beta - \phi_\alpha = \frac{RT}{F} \ln \frac{C_{Na}(\alpha)[\tilde{u}_{Na} - \tilde{u}_{Li}] + C^o\tilde{u}_{Li}}{C_{Na}(\beta)[\tilde{u}_{Na} - \tilde{u}_{Li}] + C^o\tilde{u}_{Li}} \qquad (10.27)$$

which is the junction potential between the bulk of the SIC and the interface. The Galvani potential between point α and the solution is obtained from Equation (10.4). With $C_{Na}(\alpha) = C^\circ$, the total sensor voltage can be written as

$$\phi_{SIC} - \phi_{sol} = E^o + \frac{RT}{F} \ln \frac{\left(C_{Na}(\beta)[\tilde{u}_{Na} - \tilde{u}_{Li}] + C^o\tilde{u}_{Li}\right)C_{Na}(sol)}{C^o\tilde{u}_{Na}C_{Na}(\beta)} \qquad (10.28)$$

The equilibrium constant of the reaction given by Equation (10.25) in this case reads

$$K_{ex} = \frac{C_{Li}(\beta)C_{Na}(sol)}{C_{Na}(\beta)C_{Li}(sol)} \qquad (10.29)$$

Combining Equations (10.29) and (10.17), we obtain

$$C_{Na}(\alpha) = \frac{C^o\,C_{Na}(sol)}{K_{ex}\,C_{Li}(sol) + C_{Na}(sol)} \qquad (10.30)$$

Equations (10.28) and (10.29) lead to

$$\phi_{SIC} - \phi_{sol} = E^{o'} + \frac{RT}{F} \ln \left(C_{Na}(sol) + K_{ex} \frac{\tilde{u}_{Li}}{\tilde{u}_{Na}} C_{Li}(sol) \right) \qquad (10.31)$$

This relation is similar to Equation (10.26) with a selectivity coefficient $K_{Na,\,Li}$ given by

$$K_{Na,Li} = K_{ex} \frac{\tilde{u}_{Li}}{\tilde{u}_{Na}} \qquad (10.32)$$

This corresponds to the conclusion drawn by Eisenman: the selectivity coefficient is a function of the exchange equilibrium constant and of the ratio of the mobility of the interfering ion to that of the normal ion. The relation becomes more complex in the case of multiply charged ions, but the results remain qualitatively the same. Since this model does not take into account any kinetic aspects in the mixed ionic exchange at the interface, it must be considered as a first approximation only (a complete model is not available in the literature).

The above description may be used to explain the improvement of selectivity coefficients obtained on a crystallized membrane, as compared with glass membranes. The size of the conduction sites is nonhomogeneous in such amorphous materials, while it is well defined in size in a crystalline material. The mobility of interfering ions is therefore lower in the latter type of structure, which will favor a higher selectivity (see Equation [10.32]). This is well illustrated by NASICON, where the selectivity coefficient for alkali ions is improved by a factor of at least ten.[57,74] The main interfering ions of common membranes are indicated in Table 10.1. The methods for the determination of the selectivity coefficients are described in specific papers.[36,49,76-78]

For H^+ interference, this model is not suitable, because protons do not penetrate into the membrane (their mobility is very low). The interfering phenomenon is a surface reaction with

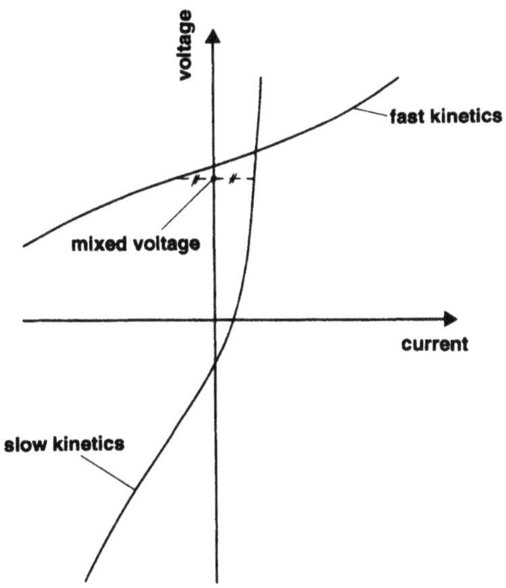

FIGURE 10.6. Mixed voltage in the case of two electrochemical reactions at the interface.

groups such as silanol groups, which show both acid and base reactions. Nevertheless, it is possible to model the interfering phenomenon in a similar way: the surface layer (gel) can be viewed as a thin protonic membrane. The selectivity coefficient $K_{Na,H}$ is about 10 to 100 for a Na^+ membrane and about 10^{13} for a pH membrane (i.e., $K_{H,Na} = 10^{-13}$).

c. *Influence of Electronic Conductivity*

If there is no redox couple in the analyzed solution, potentiometric measurements are not influenced by the existence of an electronic conductivity in the SIC. The voltage is fixed only by the ionic exchange reaction at the interface between the SIC and the analyzed solution. Typical examples are Ag_2S-based membranes.

On the other hand, if there are one or more redox couples in the solution, mobile ions and electrons both take part in the exchange, fixing the voltage. If the ion exchange is very fast, the voltage corresponds to that previously described in Section II.A.1.a. If the electron exchange is very fast, the Galvani potential is equal to

$$\phi_{SIC} - \phi_{sol} = \frac{1}{F}\mu_{e(SIC)} + \frac{1}{nF}\left(\mu^{\circ}_{OX(sol)} - \mu^{\circ}_{RED(sol)}\right) + \frac{RT}{nF}\ln\frac{a(OX)_{sol}}{a(RED)_{sol}} \quad (10.33)$$

Generally, the voltage takes an intermediate value. As for a mixed voltage in traditional electrochemistry, the value is close to the Galvani potential of the couple which exhibits the fastest exchange rates (Figure 10.6). To avoid any interfering phenomenon by redox couples in the solution (O_2/OH^-, etc.), it is better to use a membrane having a high ionic conductivity and a fast ionic exchange at the interface.

d. *Limit of Detection*

Interferences can restrict the measurement of ion concentrations, if the interfering ions are highly concentrated or if their selectivity coefficients are high (see Equation [10.26]). For instance, in the preceding case, $K_{Na,H}$ is about 100 for a glass membrane and if the pH of the analyzed solution is 7, the Na^+ detection limit (see Figure 10.7 for the definition of this parameter) is about 10^{-5} M. If the pH is 5, this limit is about 10^{-3} M.

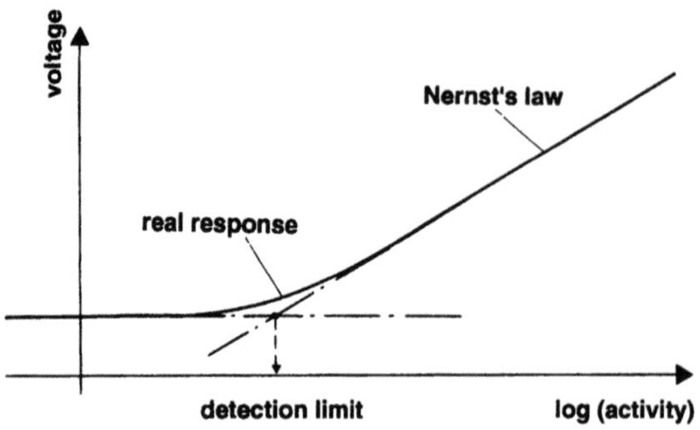

FIGURE 10.7. Definition of the detection limit for an ISE.

The solubility of the membrane in the solution is another limiting factor. The calculations are not specific to solid state electrochemistry; a good quantitative description is given, for example, in the case of AgX membranes.[49] The order of magnitude of the detection limit of X^- in the solution is about $\sqrt{K_s}$ (K_s is the solubility product of AgX). For the general case the relation is more complex, but the conclusion remains qualitatively the same.

3. Gas Sensors (Electrochemical Gauges)

The Severinghaus and Bradley[4] sensors use an ISE to measure an active gas pressure. The gas is supposed to be in equilibrium with an ion (typically protons) in an intermediate solution (carbonate solution for CO_2 measurement) which is in contact with the external medium, through a material permeable to the gas (TEFLON®). They are described in specialized literature; we do not treat them here because they do not belong to the field of solid state electrochemistry. Some of the existing ISEs could be used in this way: for instance, a pH electrode can be used for CO_2, NH_3, SO_2, and NO_2; S^{2-}-ISE for H_2S; Ag^+-ISE for HCN, F^--ISE for HF; and Cl^--ISE for Cl_2.[56]

The variety of "traditional" solid state potentiometric gas sensors is smaller than for ISEs. Table 10.2 summarizes some of the developed electrochemical gauges. Most of them use crystalline materials with a conductivity sufficiently high at moderate temperature (a few hundred degrees centigrade) or at high temperature (above 1000°C). Generally, these sensors operate according to a reaction of the first kind, e.g., O_2 using an oxide conductor; I_2, Cl_2, and Br_2 using the respective halide conductor; and H_2 using a protonic conductor. Examples of devices are shown in Figure 10.8. Exceptionally, they are of the second kind, e.g., CO_2 or NO_2 using a silver or sodium conductor.

a. Working Temperature

The choice of the working temperature depends on the conductivity of the SIC (impedance of the sensor) and on the electrode kinetics. With thin-film technologies, it is possible to reduce the SIC impedance strongly, but, unfortunately, the electrode kinetics is not affected. So, the working temperature of microsensors cannot be greatly improved with respect to that of macrosensors (traditional electrochemical gauges). The best way to increase the rate of the electrode reactions is to use catalytic materials. We may mention, for example, the use of such materials to lower the working temperature of oxygen gauges.[104-108] A surface treatment has also been proposed to improve the performance of a ZrO_2-based sensor at low temperature.[108,109] In the case of a material having electronic and ionic charge carriers (mixed conductors), the electrode scheme is given on Figure 10.9. The rate of reaction is a function of the ionic conductivities of both materials (SIC and catalyst): it is fixed by the material exhibiting the lowest rate of reaction. In this field, oxides other than stabilized zirconia are now being proposed, for example, based on bismuth oxide.[110,111] The redox stability domain of the SIC must be compatible with the usual pressure range. Generally, the lower pressure

TABLE 10.2
Examples of Electrochemical Gauges (Gas Potentiometric Sensors)

Analyzed species	Electrolyte	Operating temperature range (K)	Concentration range (ppm)	Ref.
O_2	ZrO_2–based ceramic	750–1100	0.1–10^6	79,85
	$PbSnF_4$	370–520	10^3–10^6	92,93
	LaF_3	300–373	10–10^6	94
Cl_2	$SrCl_2$-KCl	350–500	0.1–10^6	13
	$BaCl_2$-KCl	350–500	0.1–10^6	86
I_2	KAg_4I_5	313	10^3 -10^6	12
H_2	H.U.P.	300	>10^4	87
	β'-Al_2O_3	300	>10^4	91
	NASICON	300	>10^4	89
	$BaCeO_3$-based ceramic	770–1100	10^3–10^6	89,90
CO_2	Na_2CO_3/NASICON	570–1000	>10 in dry air	95,96,98
	Li_2CO_3/$LiTi_2(PO_4)_3$	620–1000	80–10^4	97,98
SO_2-SO_3	K_2SO_4			
	Ag_2SO_4 + Li_2SO_4			
	Na_2SO_4/β'-Al_2O_3	500–1200	>10 in air	99
	Na_2SO_4/NASICON			
	Li_2SO_4/ZrO_2-based ceramic			
NO_2	$NaNO_3$/β', β''-Al_2O_3			
		400	0.1–10^3 in air	100
(NO_x)	$NaNO_3$/NASICON			
AsH_3	Na^+ β-Al_2O_3	1000	50–2 10^3	
As_4O_8			1–190	101
H_2O	$SrCeO_3$-based ceramic	870–1270		103
	Ag^+-Glass	300		102

is fixed by the physicochemical properties of the mechanical part of the setup (adsorption, etc.), and so the typical limiting value is about 10^{-7} atm. (10^{-2} Pa).

The presence of temperature gradients is a crucial problem for the accuracy of measurements leading to errors due to the thermoelectric effect. Thermoelectric sensors have been made on this principle.[112,113] In this case, the cell can schematically be represented by

$$X_2, Me(T)/SIC/X_2, Me(T+dT)$$

The X_2 pressure is constant, and each electrode is assumed to be isothermic. The voltage is composed of two parts:

(i) The "reversible" emf which can be related to the thermodynamic properties of the electrode reaction. Its value is

$$E_{rev} = \frac{1}{2xF}\left(\frac{\partial \mu^\circ{}_{X_2}(T)}{\partial T} + R \ln P_{X_2}\right) dT \tag{10.34}$$

where x is the charge number of X and μ° the standard chemical potential of X_2 (function of T). For example, for O_2, the value of μ° was given by Goto et al.[114] The numerical expression of Equation (10.34) then is

$$E_{rev} = \left(\frac{R}{4F} \ln P_{O_2} - \frac{198.3}{4F} - \frac{0.0421}{4F} T\right) dT \tag{10.35}$$

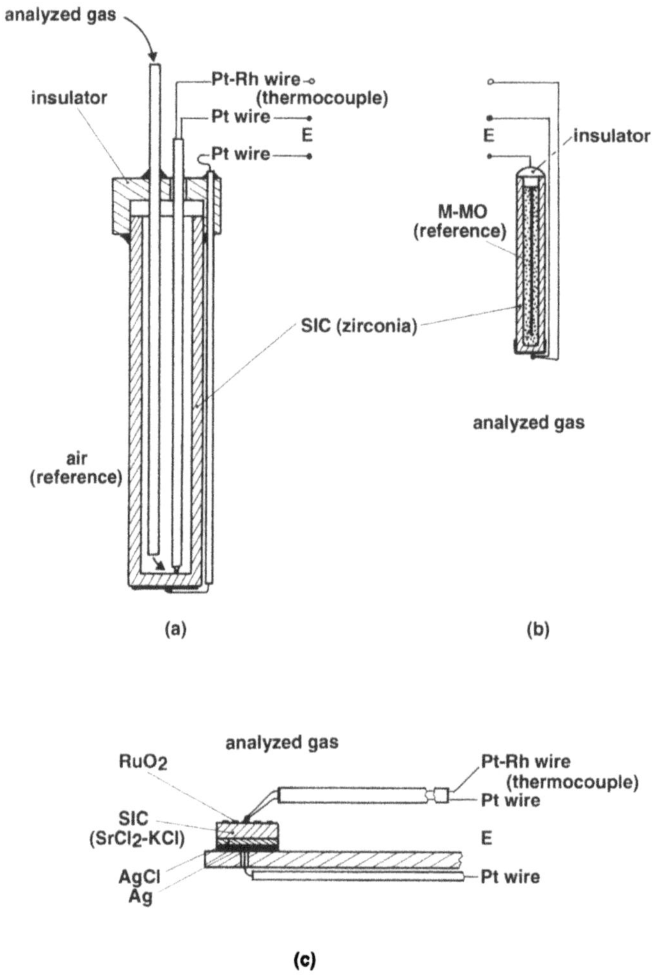

FIGURE 10.8. Examples of electrochemical gauges (potentiometric sensors): (a) laboratory oxygen sensor (air reference), (b) oxygen minisensor with a solid state internal reference (M-MO), (c) chlorine sensor with a solid state internal reference (Ag-AgCl).

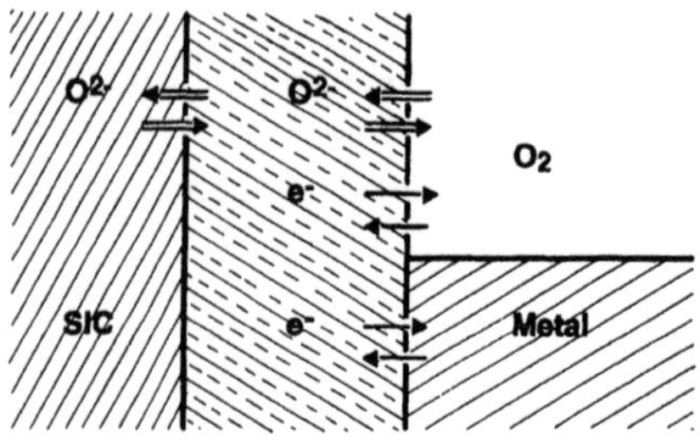

FIGURE 10.9. Electrode reaction scheme with electronic and ionic carriers in the electrode material.

(ii) The "homogeneous" emf which is a function of the conduction properties of the SIC. Its modeling is not easy and is not treated here (see, for example Reference 115). It is generally written as

$$E_{homo} = \alpha \ dT \tag{10.36}$$

The value of α can be determined experimentally when a sample of the SIC is located in a temperature gradient. For oxides, the value of α is of the order of 0.5 mV K^{-1}.[116-117] For dT = 10°C, the total voltage measured on stabilized zirconia is about 2 to 5 mV, depending on the oxygen pressure (1 to 10^{-6} atm.).

b. Reference Systems

As mentioned previously, two electrodes are required in potentiometric sensors. In the most simple case, the reference point is an electrode exposed to a gas with a well-known partial pressure. For example, in oxygen gauges, air can be used. Equation (10.9b) for each electrode equilibrium leads to the following relation,

$$E = \frac{RT}{4F} \ln \frac{P(O_2)}{P_{ref}} \tag{10.37}$$

where $P_{ref} = 0.209\ P_{atm}$, P_{atm} being the ambient pressure. Equation (10.37) holds at conditions close to equilibrium, i.e., the electrochemical potential of the O^{2-} ion must be constant throughout the SIC material.

Other gas systems can be used, e.g., CO-CO_2 or H_2-H_2O equilibrium, for buffering purposes to avoid any change due to a parasitic oxygen flux (permeability, semipermeability, etc.).

Solid state reference systems are often advisable, for instance, in miniaturized devices. M-MO_n (or $MO_{n'}$-$MO_{n''}$) can be used for oxygen gauges or M-MX_m for halide gauges. Generally, the metal corresponding with the cation of the SIC reduces the material and cannot be used directly (e.g., zirconia is reduced by metallic zirconium). The choice of the M-MO_n couple is a function of the stability domain of the SIC.

A first approach consists of writing the local equilibrium

$$M + \frac{n}{2} O_2 \Leftrightarrow MO_n \tag{10.38}$$

and an oxygen pressure corresponding to this equilibrium can be defined. The voltage is then given by an equation similar to Equation (10.37), with

$$\ln P_{ref} = \frac{2\ \Delta G^o}{nRT} \tag{10.39}$$

where $\Delta G°$ is the standard free enthalpy of formation of MO_n. Then the voltage can be written as

$$E = -\frac{\Delta G^o}{2nF} + \frac{RT}{4F} \ln P(O_2) \tag{10.40}$$

This approach is not satisfactory for the following two reasons:

(i) the number of molecules can be lower than one in a small reference volume, essentially for miniaturized sensors; and
(ii) it has been shown by several authors that the electrode reaction rate is a function of the oxygen pressure (for instance, the polarization resistance R is proportional to $P(O_2)^\alpha$, α varying with the nature of the rate-determining step);[118-120] therefore, for low oxygen partial pressure, the response time is expected to be very high; this is not the case.

Another more rigorous approach takes into account the equilibrium at the solid interface. In macroscopic description, we have

$$M + n\ O^{2-}_{SIC} \Leftrightarrow MO_n + 2n\ e^- \tag{10.41a}$$

or in the Kröger–Vink notation,

$$M + n\ O^{\times}_{O} \Leftrightarrow MO_n + nV^{\bullet\bullet}_{O} + 2n\ e^- \tag{10.41b}$$

where e^- are the electrons of the electrode material. Whatever the description, if the SIC is homogeneous and if the same metal is used in isothermal connections, the voltage of the oxygen gauge is given by

$$E = \frac{1}{2nF}\left(\mu^o{}_M - \mu^o{}_{MO_n} + \frac{n}{2}\mu^o{}_{O_2}\right) + \frac{RT}{4F}\ \ln P(O_2) \tag{10.42}$$

which is equivalent to Equation (10.40).

Common reference oxide systems are Pd-PdO, Ni–NiO, Cu-CuO, Fe-FeO, and Pb-PbO. Polarization effects can be observed due to oxidation mechanisms (a thin oxide layer does not give a good buffer effect). It was shown that, for instance, Ni–NiO is not a good reference system from this point of view.[121] The choice is a function of polarization effects due to a detrimental current through the gauge (see preceding discussion on ISE) or due to oxygen semipermeability as described in the following section. It is also a function of the $\Delta G°$ value: the corresponding value must be located in the electrolytic domain of the SIC. Cr-Cr_2O_3 is therefore a system which is not recommended, especially at high temperature. Diffusion of metal species into the SIC gives donor or acceptor levels, and this can lead to electronic conductivity (for example, Fe or Pb by a change of valency).

c. *Influence of Electronic Conductivity*

The first effect of the electronic conductivity on emf measurements was demonstrated by Wagner.[122] A short description is given in this section for an oxide SIC. Electrons and oxygen ions are assumed to be the predominant species. Under steady-state conditions, without dc current, Equations (10.15) or (10.16) can then be written

$$\text{grad}\ \phi = -\frac{\sigma_e}{F\sigma_t}\ \text{grad}\ \mu_e \tag{10.43}$$

with

$$\sigma_t = \sum_j \sigma_j \tag{10.44}$$

and

$$\sigma_j = z_j^2 F^2\ C_j \tilde{u}_j \tag{10.45}$$

where j stands for electrons or O^{2-} ions. Integration of Equation (10.45) gives

$$\phi_1 - \phi_2 = -\frac{1}{F}\int_2^1 \frac{\sigma_e}{\sigma_t}\, d\mu_e = \bar{t}_e\left(\frac{\mu_{e(2)} - \mu_{e(1)}}{F}\right) \tag{10.46}$$

where $\bar{t}_e$ is the average electronic transport number between the points 1 and 2 in the SIC. At both electrodes, the following equilibrium occurs:

$$\frac{1}{2}O_2 + V_O^{\bullet\bullet} + 2e' \Leftrightarrow O_O^{\times} \tag{10.47a}$$

with the corresponding equilibrium constant

$$K_e = \frac{[O_O^{\times}]}{[V_O^{\bullet\bullet}][e']^2\, P(O_2)^{1/2}} \tag{10.47b}$$

using the Kröger–Vink notation. For an SIC, the oxide ion and vacancy concentrations can be assumed to be constant. The free electron concentration is then given by

$$[e'] = K_e'\, P(O_2)^{-1/4} \tag{10.48}$$

and Equation (10.46) becomes

$$\phi_1 - \phi_2 = \bar{t}_e\, \frac{RT}{F}\, \ln \frac{P(O_2)_1}{P(O_2)_2} = \bar{t}_e\, E_{th} \tag{10.49}$$

In the derivation given in Section II.A3.b, ϕ_{SIC} was assumed to be constant (see Equations [10.9b] and [10.37]). In the present derivation, the voltage variation between the points 1 and 2 must be taken into account. The total measured value is then

$$E_m = E_{th} - (\phi_1 - \phi_2) = E_{th}(1 - \bar{t}_e) \tag{10.50}$$

This equation shows the error in the voltage measurement. Generally, the error is small (the relative error is of the order of a few thousandths). In fact, the main error is due to the concomitant semipermeability flux.[123] Because of the emf between both electrodes, there is a mass transfer of oxygen, without a dc current, because the ionic current is electrically balanced by the electronic current in accordance with

$$I = I_e + I_O = 0 \tag{10.51}$$

Two phenomena are observed:

(i) An oxygen flux passes through the SIC, from the oxygen-rich atmosphere to the oxygen poor. Then the measured partial pressure is modified. The electrode reaction can be represented by

$$\frac{1}{2}O_2 + V_O^{\bullet\bullet} + 2e' \rightarrow O_O^{\times} \tag{10.52}$$

and its reverse at the other electrode. From Equation (10.52), the oxygen flux is related to the free electron flux by

$$J(O_2) = 1/4\,J_{e'}\left(\text{or } 1/2\ J_{O^{2-}}\right) \tag{10.53}$$

and so this flux can be written as

$$J(O_2) = \frac{\tilde{u}_e[e']}{4}\ \text{grad}\ \tilde{\mu}_e \approx \frac{\tilde{u}_e[e']}{4}\ \text{grad}\ \mu_e \tag{10.54}$$

and using Equation (10.45), if μ_e is approximately linear, which is the case if t_i is close to unity,

$$J(O_2) = \frac{RT}{4F^2L}\left(\sigma''_e - \sigma'_e\right) \tag{10.55}$$

where L is the thickness of the SIC, and σ''_e and σ'_e are the electronic conductivities of the SIC at the location of the electrodes. In the general case, if one electrode is under a reducing atmosphere, so that the SIC shows n-type electronic conductivity, and the other under an oxidizing one where the SIC shows p-type electronic conductivity, the general formula giving the semipermeability flux is

$$J(O_2) = \frac{RT}{4F^2L}\left(\sigma''_h + \sigma'_e\right) \tag{10.56}$$

with, from Equation (10.48),

$$\sigma'_e = \sigma^\circ_e\ P'(O_2)^{-1/4} \tag{10.57}$$

and, from the equivalent relation for p-type equilibrium,

$$\sigma''_h = \sigma^\circ{}_h\ P''(O_2)^{1/4} \tag{10.58}$$

Of course, the modification of the partial pressure in the measured atmosphere is a function of the experimental conditions (closed chamber or gas flux).

(ii) The second phenomenon is the overvoltage effect at the electrodes. From an electrochemical point of view, it may seem paradoxical to have an overpotential if there is no dc current. In fact, this phenomenon is due to the ionic current which is not zero. The phenomenon can be described qualitatively by the oxygen flux; it comes from the high-pressure atmosphere, e.g., air used as reference, and goes to the lower pressure (i.e., the measured pressure). This flux increases the oxygen activity at the low-pressure electrode which may become polarized because of slow desorption of oxygen. The lower the measured pressure, the higher the overvoltage and so the higher the error. This phenomenon has been studied by Fouletier et al.[123] An example of the error observed on a platinum electrode is shown in Figure 10.10. At 1000°C the error can be higher than 1 order of magnitude, if the measured pressure is about 10^{-1} Pa (10^{-6} atm). For solid reference systems, the oxygen flux displaces the equilibrium. If this system is reversible, from a kinetic point of view, no serious problems will arise, only a decrease of the lifetime of the sensor. In cases without a buffering effect, for example, the Ni–NiO system, the discussed error can be very important.

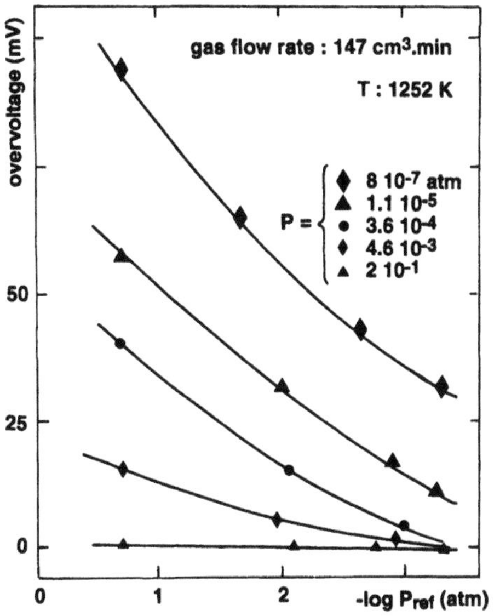

FIGURE 10.10. Semipermeability overvoltage of a platinum electrode used in an oxygen gauge. P and P_{ref} are the pressures of the analyzed gas and the reference gas, respectively. (Data from Fouletier, J., Fabry, P., and Kleitz, M., *J. Electrochem. Soc.*, 1976, *123*, 204–213. With permission.)

d. Interfering Phenomena

Interfering phenomena have not been systematically studied for gas sensors. There is no theoretical description available in the literature. As a first approximation, we propose to apply the same description as used for ISEs. For the sake of illustration we consider an X_2 gauge of the first kind (SIC conductor by X^-) in contact with Y_2 halogen gas ($z_X = z_Y = -1$). The derivation in the general case, where both species take different charge numbers, becomes very complex and is not presented here.

Assuming that local equilibrium,

$$\frac{1}{2}Y_2 + X^-_{SIC(\alpha)} \Leftrightarrow \frac{1}{2}X_2 + Y^-_{SIC(\alpha)} \tag{10.59a}$$

is established at the electrode, the equilibrium constant is given by

$$K_{eq} = \frac{P(X_2)^{1/2}\, C_Y(\alpha)}{P(Y_2)^{1/2}\, C_X(\alpha)} \tag{10.59b}$$

where α is the part of SIC near the surface (with Y^- ions) and β the bulk (without Y^- ions) (see Figure 10.3). Using the above notations, Equation (10.19) becomes

$$\phi_\beta - \phi_\alpha = \frac{RT}{F}\ln\frac{C^o\,\tilde{u}_X}{C_X(\alpha)\left[\tilde{u}_X - \tilde{u}_Y\right] + C^o\tilde{u}_Y} \tag{10.60}$$

The electrode reaction can be written with X_2 or Y_2, for example,

$$X_2 + 2e^- \Leftrightarrow 2X^-_{SIC(\alpha)} \tag{10.61a}$$

and the Galvani potential is

$$\phi_M - \phi_\alpha = E^o + \frac{RT}{2F} \ln P(X_2) - \frac{RT}{F} \ln C_X(\alpha) \tag{10.61b}$$

The electroneutrality condition ($C_X(\alpha) + C_Y(\alpha) = C^o$), Equations (10.59b), (10.60), and (10.61b), give the final voltage,

$$\phi_M - \phi_{SIC} = E'^o + \frac{RT}{F} \ln \left[P(X_2)^{1/2} + K_{eq} \frac{\tilde{u}_Y}{\tilde{u}_X} P(Y_2)^{1/2} \right] \tag{10.62}$$

which is similar to the equation derived for an ISE. Of course, in this model, kinetic phenomena have not been taken into account; for instance the catalytic effect of the electrode material. In such a case, the electrode equilibrium (Equation [10.59]) is shifted.

A similar description can be proposed for halide conductors used as oxygen-sensitive materials, e.g., $SrCl_2$-KCl,[124] $PbSnF_4$,[125] PbF_2, or LaF_3.[126-128] These electrolytes can be used only if there are no noticeable X_2 traces in the analyzed gas. The SIC can be doped by oxide ions from gas-phase oxygen, but in this case the C^o value, which is included in E'^o, must be constant. It is therefore better to dope the SIC chemically with an oxide to avoid any fluctuation of C^o.

If the electrode material shows a high catalytic activity toward gaseous X_2, the sensor is not sensitive to the oxygen pressure in the presence of X_2 (e.g., RuO_2 used as chloride electrode in the range 10^5 to 10^{-1} Pa of Cl_2 in air).[13]

B. AMPEROMETRIC SENSORS

1. Principle

Amperometric sensors are based on electrochemical reactions which are governed by the diffusion of the electroactive species through a barrier.[129] The barrier usually consists of a hole (see Figure 10.11a) or a porous neutral layer. The control of the gas inflow by an electrochemical method was proposed recently.[130] The voltage is fixed on the diffusion plateau of the I(U) curve (Figure 10.11b). For a reaction limited by the mass transport process, the general flux equation in a one-dimensional model is

$$J_i = -D_i \frac{\partial P_i}{\partial x} + \alpha_{conv} P_i \tag{10.63}$$

where P_i is the partial pressure of the gas i and α_{conv} the convection coefficient of the gas forming the ambient atmosphere.

Two types of diffusion control can be distinguished:

(i) If the hole diameter is very small (i.e., < about 1 μm), a Knudsen mechanism is prevalent and the flux is equal to

$$J_i = K'_d (P_i^o - P_{el}) \tag{10.64}$$

where P^o_i and P_{el} are the pressure of the component i in the analyzed gas and at the electrode, respectively, and K'_d is the Knudsen coefficient, defined by

$$K'_d = \frac{d}{3L\sqrt{\pi M_i RT}} \tag{10.65}$$

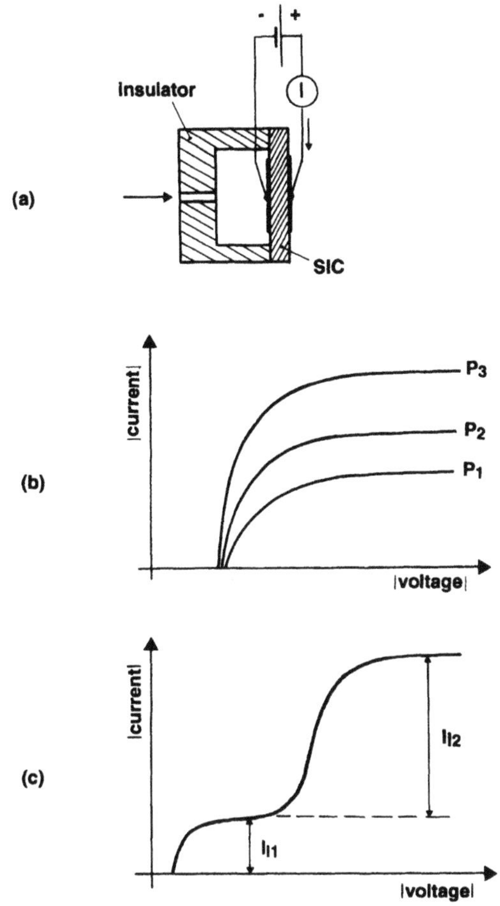

FIGURE 10.11. Amperometric gas sensors: (a) schematics of a typical device, (b) I(U) curves as a function of pressure, (c) I(U) curve with two electroactive gases.

where d and L are the diameter and the length of the hole, respectively, and M_i is the molar mass of the gas i. In the diffusion-limited regime, all the electroactive species are consumed at the electrode faradaically and so $P_{el} = 0$. Then the limiting current is given by

$$I_l = 2zK'_d F\pi d^2 P_i^o \tag{10.66}$$

z being the charge number of X^{z-} ions in the electrode reaction

$$X_2 + 2ze^- \rightarrow 2X^{z-} \tag{10.67}$$

As follows from Equations (10.65) and (10.66), the limiting current is proportional to $T^{-1/2}$.

(ii) If the hole diameter is larger than 10 μm, the limiting current equation is given by[123b]

$$I_l = \frac{2zFD^oT^{3/4}\pi d^2}{RL} \ln\left(1 - x_i\right) \tag{10.68}$$

where x_i is the molar fraction of species i in the ambient atmosphere (it is equal to the partial pressure if the total pressure is 1 atm) and D° the standard diffusion coefficient of i in the analyzed gas. In this case the variation of I as function of T is different.

As pointed out elsewhere, a concentration up to 100% can be determined in the Knudsen regime.[31]

2. Experimental Cells

To measure $P°_i$, it is necessary to have $P_{el} = 0$; the difficulty is then to fix the corresponding U value. If this is too high (in absolute value), other reactions can occur (reduction of CO_2, H_2O, or the SIC). The voltage of the working electrode therefore must be carefully controlled. The general U(I) equation is the following,

$$U = E_{(I=0)} + RI + \eta_a - \eta_c \tag{10.69}$$

where $E_{(I=0)}$ is the voltage without dc current, RI is the ohmic drop of the cell, and η_a and η_c are the anodic and cathodic overvoltages, respectively.

If a two-electrode cell is used, the counter electrode (anode) must be unpolarizable. For instance, on an oxygen electrode it has been shown that the higher the oxygen partial pressure, the lower the overvoltage. So, a high oxygen pressure P_i can be used in the counter electrode compartment. A counter electrode of large area is also desirable because the current density is then low and the overvoltage can be negligible.

Another possibility is to use a third electrode as reference, for instance, a M-MO_n mixture for oxygen sensors. Of course, such a device is more expensive, but also more reliable. In such a case electronic control by means of a potentiostat is required.

Other more sophisticated devices using pump and gauge parts will be described later.

3. Different Species Analyzed by Amperometry

Up to now, only gas sensors are based on amperometric principles, essentially oxygen sensors (there is no amperometric sensor for ion analysis using solid electrolytes). The SIC used for oxygen sensors are generally based on stabilized zirconia (e.g., YSZ), as for potentiometric sensors.[131] From a general point of view, the conductivity must be as high as possible to avoid an excessive ohmic drop (see Equation [10.69]). Another problem is the influence of the temperature on the diffusion plateaus of the I(U) curves. The diffusion is slow at low temperatures. On the other hand, the voltage might become too difficult to control at very low temperatures.[132] A working temperature of about 400°C is often a good compromise. An example of an amperometric oxygen sensor with an air reference is shown in Figure 10.12. Thin-film devices have also been proposed.[133,134]

Recently, Liaw and Weppner[135] have proposed tetragonal zirconia for measurements down to 250°C. Bi_2O_3-Er_2O_3 solid solutions are also good candidates for low-temperature sensors.[136] Amperometric oxygen sensors based on $PbSnF_4$ already operate at room temperature.[137]

Besides O_2, similar SICs can be used to analyze species such as CO_2 or H_2O by electrochemical reduction. The different electrode reactions are

$$O_2 + 4e^- \rightarrow 2O^{2-}$$

$$CO_2 + 2e^- \rightarrow CO + O^{2-}$$

$$H_2O + 2e^- \rightarrow H_2 + O^{2-}$$

For instance, in a neutral gas, two limiting currents can be observed in a I(U) curve corresponding to O_2 and H_2O.[138] These can be deduced from the total I(U) plot or measured for two corresponding voltages as depicted in Figure 10.11c.

To apply Faraday's law, the voltage at which the above reactions occur must be compatible with the electrolytic domain of the SIC. When this condition is not fulfilled, an electronic

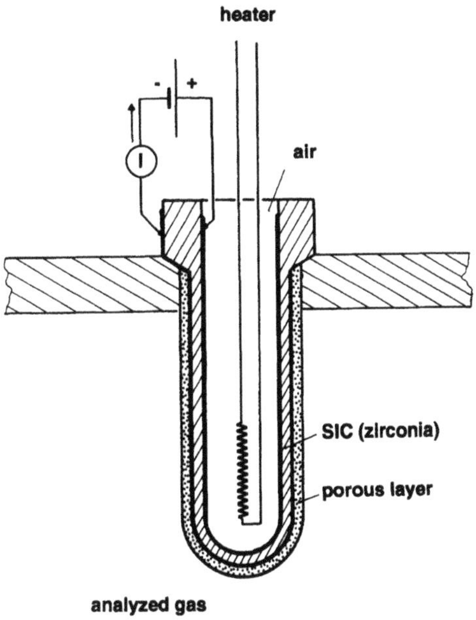

FIGURE 10.12. Example of amperometric device (oxygen sensor).

conductivity, either n or p type, will appear. Generally, with an electronic transport number of a few percent, the faradaic efficiency is close to one and the error is less disastrous than for potentiometric devices.

Very few amperometric gas sensors have been studied so far. A protonic conductor has been used for a H_2 amperometric sensor by Miura et al.[139] One may also mention the chlorine sensor proposed by Liu and Weppner,[140] which is based on β″-alumina as SIC with an AgCl layer. The electrode reaction is given by

$$Ag^+ + e^- + 1/2\,Cl_2 \rightarrow AgCl$$

The lifetime of this sensor could be limited by the formation of a covering AgCl layer at the electrode.

A reducing gas can also be analyzed indirectly by first oxidizing it, for instance, with an excess of oxygen produced by an oxygen pump (I = constant). The O_2 excess is then measured by an amperometric oxygen sensor.[141] A schematic representation is given in Figure 10.13.

C. COULOMETRIC SENSORS

1. Principle

Coulometric sensors are very similar to the amperometric devices. The gas species is dosed by the faradaic effect, but, in this case, the diffusion flux between the chamber and the analyzed atmosphere is negligible during the measurement. The current pumps the internal partial pressure (P_{in}) from the external value (P_{out}), which corresponds to equilibrium condition, to zero or from zero to P_{out}. The electroactive gas is supposed to be an ideal gas and so its partial pressure is equal to

$$P_{in} = \frac{\int_t^{t+\Delta t} RT\, I(t)\, dt}{zF\, V^o} \qquad (10.70)$$

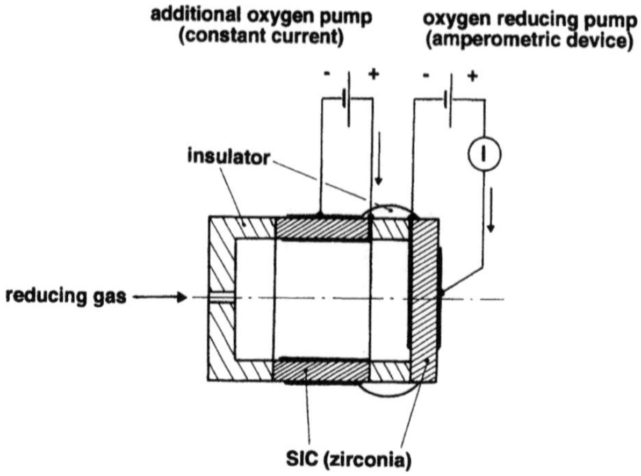

FIGURE 10.13. Schematic view of an amperometric sensor for a reducing gas using a double oxygen-pump system.

where z is the charge number to reduce (or to oxidize) one molecule of the electroactive gas, V° the volume of the chamber, and I(t) the dc current during Δt (time delay to obtain $P_{in} \rightarrow 0$ or reverse). If the current is kept constant during Δt, Equation (10.70) simplifies to

$$P_{in} = \frac{RT\ I\ \Delta t}{zF\ V^{o}} \tag{10.71}$$

2. Different Kinds of Measurement

There are two kinds of coulometric sensors. Both work with two steps. In the first kind, the measurement is made during an electrochemical purge of the chamber (from P_{out} to zero) and in the second kind, the measurement is made during an electrochemical enrichment (from zero to P_{out}).

a. Electrochemical Purge Devices

The gas in the chamber and the external atmosphere are first brought into equilibrium, for example, by diffusion through a hole as in amperometric devices, (see Figure 10.11a). The time to reach equilibrium can be evaluated approximately. Equilibrium is either established after a fixed delay or is controlled galvanostatically until U = 0.[142] Next, electrochemical pumping is applied until the chamber is empty. If the current is constant, P_{out} follows directly from Δt. This step must be fast relative to diffusion of the electroactive species from the external atmosphere into the chamber. For obvious reasons, it is also necessary to avoid electrochemical reduction of the SIC. The empty state can be estimated from the measurement of the voltage U, but is masked in part by the ohmic drop. The use of a potentiometric integrated sensor, to control the value of P_{in}, greatly improves accuracy, but the construction of such a device would of course be more expensive.

b. Electrochemical Enrichment Devices

In this case, the chamber is closed and gas tight (for example, see Figure 10.14).[143] A current is first applied to pump the electroactive gas until a sufficient vacuum (P°) is obtained. A potentiometric sensor can be used, with the external gas as reference point. The relative vacuum (P^{o}/P_{in}) is evaluated from the gauge voltage E. Time is not measured during this first delay. Next, the direction of the current is reversed and Δt is measured until E = 0.

In this sensor, the second step can be slow, i.e., by choosing a small current, to improve the accuracy in the measurement of Δt. For a long delay, the seal materials must be insulators

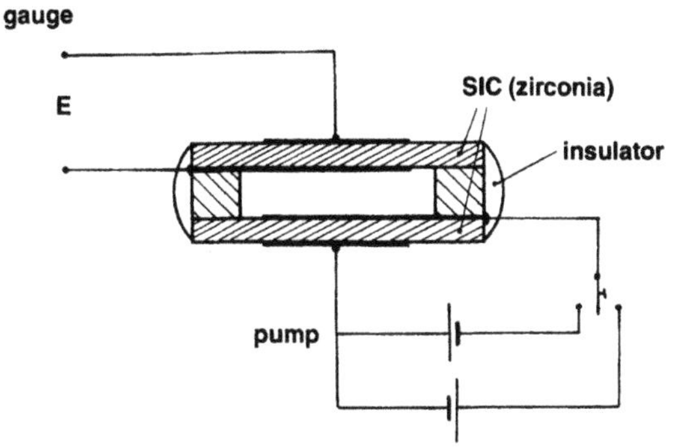

FIGURE 10.14. Closed-chamber device of a coulometric sensor (oxygen sensor).

because the use of metals would short circuit the cell, thereby generating a parasitic flux into the chamber electrochemically (a covering glass may be employed to improve the device which uses metal o-rings).

D. PUMP-GAUGE DEVICES

Pump-gauge devices are derived from amperometric and coulometric devices. They consist of a pumping part, to which a variable current can be applied, and a gauge part, which is used for measuring the resulting voltage. Only oxygen sensors are based on this principle.

1. Direct Current Mode

In the model proposed by Hetrick et al.,[144] a pumping-gauge system works in steady-state condition. The current is related to the pressure difference between the analyzed gas and the internal chamber,

$$I = 4FA\left(P_{out} - P_{in}\right) \tag{10.72}$$

where A is a coefficient (for instance, including the Knudsen coefficient and the area of the hole). The internal pressure P_{in} can be calculated from the gauge voltage E by using the Nernst equation,

$$P_{in} = P_{out}\exp\left(\frac{4FE}{RT}\right) \tag{10.73}$$

Equations (10.72) and (10.73) lead to

$$I = 4FA\,P_{out}\left(1 - \exp\left(\frac{4FE}{RT}\right)\right) \tag{10.74}$$

The current can be modified by an external source and a plot of I as a function of $\left(1-\exp\left(\frac{4FE}{RT}\right)\right)$ allows P_{out} to be measured.

2. Alternating Current Mode

Maskell et al.[145,146] have proposed an alternating current mode for amperometric devices. A sinusoidal current $I = I^{\circ}\sin\omega t$ is applied to the pumping part and the voltage is measured on the potentiometric part. A low frequency (lower than 10 Hz) is required to obtain equilibrium in the chamber. The corresponding voltage may be written as[147]

$$E_S = -\frac{RT}{4F} \ln\left(1 - \frac{RT\,I^o}{4F\,V^o\omega P_{out}} \cos^2\Psi\right) \tag{10.75}$$

where P_{out} is the oxygen partial pressure in the analyzed gas and ψ is a parameter depending on the geometrical factor of the pore, the diffusion coefficient and V°, the volume of the chamber. Equation (10.75) can be simplified if $RTI^o/4FV^o\omega P_{out}$ is small. The amplitude of the ac voltage is then proportional to P_{out}^{-1}.

Benammar and Maskell[148] have recently proposed a more sophisticated sensor to limit the parasitic phenomena. Their device is similar to the coulometric one, but it works in a pump-gauge tracking mode. As in the case of the amperometric sensor proposed by Maskell et al.,[145,146] an alternating current $I = I^o \sin \omega t$ is applied to the pumping part, with a low frequency (lower than 10 Hz) to obtain equilibrium in the chamber. The potentiometric voltage is equal to

$$E = \frac{RT}{4F}\left[\ln\frac{P_{out}}{P_{in}} - \ln\left(1 + \frac{RT\,I^o}{4F\,V^o\omega P_{in}} \cos\omega t\right)\right] \tag{10.76}$$

where V° is the internal volume, and P_{in} and P_{out} are the internal and external oxygen partial pressures, respectively. Using an electronic system to maintain the mean value of E equal to zero and to transform the emf into a dc voltage $E_{dc,}$ an approximate relation is obtained:[148]

$$E_{dc} = \left(\frac{R^2T^2\,I^o}{8\pi F^2\,V^o\omega}\right)\frac{1}{P_{out}} \tag{10.77}$$

The measured dc voltage is then proportional to P_{out}^{-1}.

E. CONDUCTOMETRIC SENSORS

In this section only conductometric sensors are discussed in which the ionic conductivity of the material plays a role. An example includes TiO_x, which exhibits a variable stoichiometry as a function of the oxygen pressure. Semiconductor materials, like SnO_2 or similar ones used in Figaro sensors for analysis of reducing species, work on principles of heterogeneous catalysis, without exhibiting ionic conduction in the bulk of the material, and therefore are not considered here.[149,150]

1. Principle

The working principle will be described on the TiO_x example because the stoichiometry of this material may change continuously in a large range, from x = 1.5 to x = 2, as a function of oxygen pressure. In a simple model, the defect chemical reaction can be written as

$$2Ti_{Ti}^{\times} + O_O^{\times} \Leftrightarrow 2Ti'_{Ti} + V_O^{\bullet\bullet} + 1/2\,O_2 \tag{10.78a}$$

Assuming that concentrations $Ti_{Ti}^{\times}$ and $O_O^{\times}$ remain constant, the equilibrium constant is given by

$$K_i^o \exp\left(\frac{-W_i}{kT}\right) = [Ti'_{Ti}]^2[V_O^{\bullet\bullet}]P_{O_2}^{1/2} \tag{10.78b}$$

where W_i is the enthalpy change of the reaction. Similarly, band electrons may be trapped, which equilibrium can be represented by

$$Ti'_{Ti} \Leftrightarrow Ti^{\times}_{Ti} + e' \tag{10.79a}$$

with the corresponding equilibrium constant

$$K_e^{\,o} \exp\left(\frac{-W_e}{kT}\right) = \frac{[e']}{[Ti'_{Ti}]} \tag{10.79b}$$

The electroneutrality condition reads

$$[Ti'_{Ti}] + [e'] = 2[V_O^{\bullet\bullet}] \tag{10.80a}$$

which, in a first approximation, becomes

$$[Ti'_{Ti}] = 2[V_O^{\bullet\bullet}] \tag{10.80b}$$

The concentration of free electrons is obtained from Equations (10.78b), (10.79b), and (10.80b):

$$[e'] = 2^{1/3}\, K_i^{o1/3} \exp\left(\frac{-W_i}{3kT}\right) P_{O_2}^{-1/6}\, K_e^{o} \exp\left(\frac{-W_e}{kT}\right) \tag{10.81}$$

The electronic and ionic conductivities are, respectively,

$$\sigma_e = K_e \exp\left(\frac{-W_{ae}}{kT}\right) P_{O_2}^{-1/6} \tag{10.82}$$

$$\sigma_O = K_O \exp\left(\frac{-W_{aO}}{kT}\right) P_{O_2}^{-1/6} \tag{10.83}$$

where W_{ae} and W_{aO} are the activation energies which take into account the reaction enthalpies W_i and W_e (Equations [10.78b] and [10.79b]).

The total conductivity σ is equal to $\sigma_e + \sigma_O$, but, in fact, the material is essentially an electronic conductor because the electron mobility is much higher than that of the oxygen ion. The ionic conduction does not play a determining role in the transducing law, but in the response time only, due to the mixed conduction. For another defect chemical reaction (e.g., formation of an F center or valency change of doping species), the law should be different. Generally, the conductivity is proportional to P^n, with n either positive or negative, being a function of the oxygen pressure. Good examples are provided in the literature and ranges in which the conductivity is proportional to $P^{-1/6}$, $P^{-1/4}$, or $P^{1/4}$ are observed.[151] In the intermediate range of pressure (P^* in Figure 10.15) the sensitivity is low. The oxygen partial pressure cannot be determined around this point. P^* is a function of temperature (see Figure 10.15).

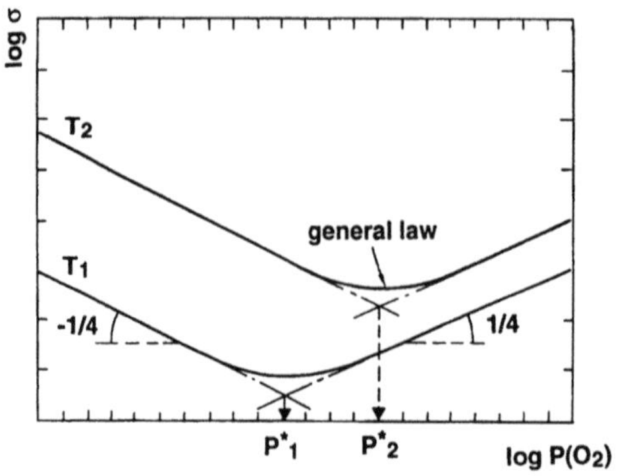

FIGURE 10.15. Schematic of the conductivity variation of TiO_x compound as a function of the oxygen pressure. For the case of the variation law:

$$\sigma = \sigma_e^o P^{-1/4} + \sigma_h^o P^{1/4};\ P^* = \left(\sigma_e^o / \sigma_h^o\right)^2.$$

2. Achievements

Materials such TiO_2, doped with Fe, Mn, Cr, or Ba oxides, or alkaline-earth ferrates, are used essentially for oxygen sensors.[152,153] The sensitive part is a sintered pellet which is porous to increase the exchange area with the gas and consequently to decrease the response time. In these devices, the area/volume ratio plays an important role. The grain size and the grain boundaries are two parameters which should remain constant as a function of time. The higher the temperature, the faster morphological changes of the ceramics will occur. The working temperature is generally higher than 600°C (at low temperature it becomes a semiconductor sensor). It was also proposed to use two sensors in a differential mode at the same temperature, one in contact with the analyzed atmosphere, the other insulated from it.

Thin-layer technologies for fabrication of conductometric sensors are being developed.[151,154] Such devices have two advantages: the response time is improved and mass production is less expensive.

It is also possible to use such materials for detection of a reducing gas. The local equilibrium at the interface may be written as

$$RG + 2Ti_{Ti}^{\times} + O_O^{\times} \Leftrightarrow RGO + 2Ti'_{Ti} + V_O^{\bullet\bullet} \tag{10.84}$$

where RG and RGO are the reduced and the oxidized forms of the gas-phase species, respectively. Combination with Equation (10.79) gives a similar conductivity law. Of course, such sensors are not selective, but are sensitive to RG/RGO couples, i.e., to the oxygen pressure corresponding to the equilibrium

$$RG + 1/2\ O_2 \Leftrightarrow RGO$$

III. MEASUREMENT CHARACTERISTICS

A. IMPEDANCE

The resistance of potentiometric sensors does not seem to be a crucial point a priori. Membranes with a very low conductivity (lower than 10^{-10} S cm^{-1}) are sometimes used in ISE devices. Nevertheless, a too-high resistance leads to a capacitive behavior of the sensor. The parasitic charges which appear in the circuit or electrical noise (intrinsic or extrinsic) may lead to erroneous results. To avoid this disadvantage, thin sensitive membranes may be used (pH or pNa glass membranes), but they are very fragile. Generally, a conductivity of about 10^{-5} to 10^{-7} S cm^{-1} is required. Miniaturization of devices allows a lower resistance of the SIC to be obtained, but the interface impedance is not fundamentally modified because it depends on the electrochemical kinetics, which is a function of the interfacial area.

Another delicate point is the input resistance of the voltmeter or analog-to-digital converter used to measure the potentiometric signal. The sensor gives an emf and no dc current must pass through it, because polarization effects at the electrodes or changes in the reference systems (internal or external, see for instance Equations [10.21], [10.23], or [10.24]) can occur and modify the voltage. It is generally advisable to use a minimum input resistance of about 10^{12} Ω. A high-impedance adapter (amplifier mounted as follower) is then necessary for usual voltmeters or analog-to-digital converters.

For amperometric sensors, the input resistance of the voltmeter is less crucial because the sensor works under current conditions. For instance, an input resistance of about 1000 times that of the sensor gives an error of about 1/1000. On the other hand, in two-electrode devices, the conductivity of the SIC must be as low as possible because an excessively high ohmic drop masks the variation of the voltage, and it becomes difficult to adjust the working voltage to the diffusion plateau. This last point is less crucial for coulometric sensors or for amperometric sensors using a reference electrode or having a potentiometric part integrated in the device (pump gauge).

B. RESPONSE TIME

Potentiometric sensors are very well suited to perform measurements in real time. For servo control, low response times avoid oscillation phenomena. The intrinsic response time is a function of the kinetics of different interfaces and of the electrical properties of the ionic materials (SIC and internal reference compounds). To simplify, one can model a potentiometric sensor by the equivalent electric circuit drawn in Figure 10.16 (this description is very simplified; for more details we refer to the specialized literature).[155] Each interface and each material are equivalent to a parallel RC circuit with its time constant τ_n. The transient response is given by

$$E = E_{t=0}\left(1 - \exp\frac{-t}{\tau_{eq}}\right) \tag{10.85}$$

where τ_{eq} is the equivalent time constant of the whole circuit. Of course, τ_{eq} is close to the time constant of the slowest mechanism. The fastest response times are a few milliseconds, but generally they are of the order of a few seconds or minutes. In a real situation, the response time is generally imposed by the kinetics to return to a new equilibrium state in the cell. Extrinsic phenomena occur, such as diffusion in liquid or gaseous phases, convection, and interaction with all compounds of the sensor and the cell (adsorption and desorption phenomena), and all these phenomena slow down the response. Special setups are necessary to diminish these phenomena and so to obtain short intrinsic response times.[155-157]

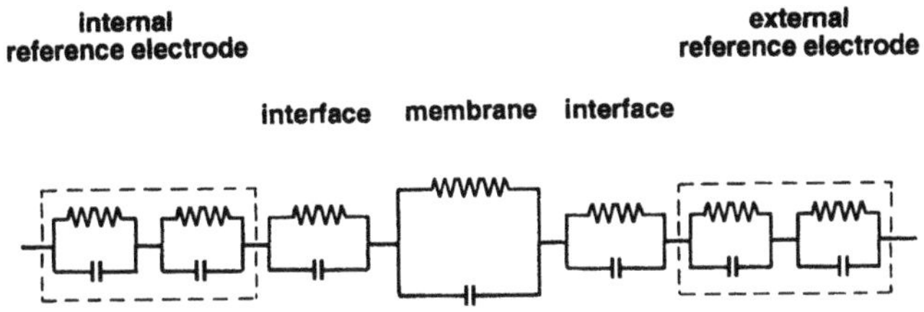

FIGURE 10.16. Simplified equivalent circuit for potentiometric measurement (ISE).

TABLE 10.3
Comparison of Signal Ratios for a Change of Two Orders of Magnitude of Activity (Concentration or Pressure)

Proportionality laws	log a	a	$a^{1/2}$	$a^{1/4}$	$a^{-1/4}$	$a^{-1/2}$	a^{-1}
Signal ratios	2	100	10	3.16	0.316	0.1	0.01

If amperometric sensors work with constant U, the response time can be short and they can be used in real-time measurements. On the other hand, for a multispecies analysis system, it is necessary to draw the entire I(U) curve (or to use a data acquisition system) to locate the different diffusion plateaus. Such sensors, like coulometric sensors, are more adapted for discontinuous measurements.

The conductometric sensors using substoichiometric oxides can work continuously. The intrinsic response time is a function of adsorption phenomena on the surface of the material and also of the equilibrium time in the bulk. It is related to the mixed conductivity of the material, just as semipermeability of the oxide.

C. SENSITIVITY

The widths of the measuring ranges depend on the kind of sensor. To cover an extended variation of activity or pressure, a law of logarithmic form or of $P^{-1/n}$ with $n > 1$ is interesting because the electrical signal changes in the range of 1 order of magnitude only, whatever the change of pressure or activity. On the other hand, it is necessary to change the calibration of the electrical setup if the sensor obeys a law in $P^{1/n}$ ($n > 0$), which is more difficult to manage in an automatic system. Table 10.3 compares the signal ratios for a change of 2 orders of magnitude of activity.

The sensitivity can be defined as the ratio $\delta P/\delta S$ (or $\delta C/\delta S$), where δS is the smallest variation of the measurable magnitude (e.g., electrical signal). For the different types of oxygen sensors, we have

- potentiometric sensors:

$$\frac{\delta P}{\delta E} = \frac{4F\,P}{RT} \tag{10.86}$$

  and the relative error in P will be a function of dE ($dP/P = (4F/RT)\,dE$), e.g.,
  if T = 1200 K: $dP/P = 39\,dE$ (if dE = 1 mV, $dP/P = 4\ 10^{-2}$)
  if T = 600 K: $dP/P = 77\,dE$ (if dE = 1 mV, $dP/P = 8\ 10^{-2}$)
- amperometric sensors, operating in Knudsen mode:

$$\frac{\delta P}{\delta I} = \frac{1}{4\,K'_d\,F\pi d^2} \tag{10.87}$$

  which is constant for given hole dimensions. The relative error is equal to that of I ($dP/P = dI/I$)
- coulometric sensors:

$$\frac{\delta P}{\delta t} = \frac{RT\,I}{4FV^o} \tag{10.88}$$

  which is proportional to I: the higher the current, the higher δP for a given δt. The relative error is equal to that of t ($dP/P = dt/t$).

- conductometric sensors, assuming $\sigma = \sigma^{\circ} P^{-\alpha}$:

$$\frac{\delta P}{\delta R} = \frac{1}{\alpha}\frac{P}{R} \tag{10.89}$$

and in this case, the relative error in P is related to that of R ($dP/P = \alpha^{-1}\, dR/R$), e.g., if $|\alpha|$ is equal to 1/4 or 1/6, $dP/P = 4$ to $6\, dR/R$.

IV. CONCLUSION AND PROSPECTS

Most solid state electrochemical sensors are interesting from a theoretical point of view because, generally, the responses obey thermodynamic laws. Potentiometric sensors of the first kind are good examples. Their main advantages are the reproducibility and the reliability of the measurements. The drift is very small and calibration is not necessary. The essential precaution is to use a very high-input impedance of the measuring setup to avoid any polarization effect and any change in the internal reference system. Amperometric sensors obey kinetic laws, depending on experimental parameters; for instance, porosity of the diffusion barrier (or geometric factors of the hole). If these parameters change as a function of time, the measurements are not reproducible and a periodic calibration becomes necessary.

The choice of a sensor is essentially a function of its use. For instance, for gas sensors, amperometric and coulometric devices are interesting for narrow ranges and potentiometric devices for large ranges of pressure.

Robustness and a good resistance to temperature treatments are the advantages of solid state membranes. They allow new devices to be conceived, for instance, in FIA tubular cells or for specific applications, in particular if the materials are very good ionic conductors.[20,158] Measurements in FIA cells can be made in differential mode by a change of solution during a short delay. In this manner, the voltage drift is a minor disadvantage.

Concerning future prospects, new fast ionic conductors could be used as membranes for ISE devices or in gauges of the second kind. The miniaturization of devices is a promising way for mass production, using microtechnologies. Solid materials are very well adapted for such devices. The major difficulty is the achievement of a reproducible composition of the SIC. Another advantage of such microdevices is a lowering of their impedance, but, as was mentioned previously, only the resistance of the SIC is decreased. The intrinsic response time is not modified any more.

The development of electronics becomes a very promising tool for sophisticated devices, for instance, for amperometric and coulometric sensors. Analytical methods such as voltamperometry, square wave polarography, etc., could lead to development of new types of sensors and to new concepts in the near future.

REFERENCES

1. Cremer, M., *Z. Biol. (Munich)*, 1906, *47*, 562.
2. Nikolskii, B.P., *Zh. Fiz. Khim.*, 1937, *10*, 495–503.
3. Eisenman, G., *Anal. Chem.*, 1968, *40*, 310–320.
4. Severinghaus, W. and Bradley, A.F., *J. Appl. Physiol.*, 1958, *13*, 515–520.
5. Kolthoff, I.M. and Sanders, H.L., *J. Am. Ceram. Soc.*, 1937, *59*, 416–420.
6. Pungor, E., Toth, K., and Havas, J., *Acta Chim. Acad. Sci. Hung.*, 1966, *48*, 17–22.
7. Frant, M.S. and Ross, J.W., *Science*, 1966, *154*, 1553–1555.
8. Déportes, C. and Darcy, M., *Silic. Ind.*, 1961, *26*, 499–504.
9. Kiukkola, K. and Wagner, C., *J. Electrochem. Soc.*, 1957, *104*, 379–387.

10. Weissbart, J. and Ruka, R., *Rev. Sci. Instrum.*, 1961, *32*, 593–595.
11. Peters, H. and Möbius, H.H., *Ger (East) Pat. 21 673*, 1961.
12. Rolland, P., Ph.D. Thesis, Paris 6, 1974.
13. Pelloux, A., Fabry, P., and Durante, P., *Sensors Actuators*, 1985, *7*, 245–252.
14. Gauthier, M., Chamberland, A., and Bélanger, A., *J. Electrochem. Soc.*, 1977, *124*, 1579–1583.
15. Velasco, G., Schnell, J.P., Siejka, J., Croset, M., *Sensors Actuators*, 1982, *2*, 371–384.
16. Ruka, R.J. and Panson, A.J., U.S. Patent 3,691,023, 1972.
17. Heyne, L., in *Measurement of Oxygen*, Degn, H., Baslev, I., and Brook, R., Eds., Elsevier, Amsterdam, 1976, 65–88.
18. Tien,T.Y., Stadler, N.L., Gibbons, E.F., and Zacmanidis, P.J., *Am. Ceram. Soc. Bull.*, 1975, *54*, 280–285.
19. Vlasov, Y.G., Bychkov, E.A., and Seleznev, B.L., *Sensors Actuators*, 1990, *2*, 23–31.
20. Fabry, P. and Siebert, E., in *Chemical Sensor Technology*, Vol. 4, Yamauchi, S., Ed., Elsevier, Amsterdam, 1992, 111–124.
21. Beekmans, N.M. and Heyne, L., U.S. Patent Specification 3,654,112, 1972.
22. Prudenziati, M. and Morten, B., *Sensors Actuators*, 1986, *10*, 65–82.
23. Van der Spiegel, J., Lauks, I., Chan, P., and Babic, D., *Sensors Actuators*, 1983, *4*, 291–298.
24. Fabry, P., Huang, Y.L., Caneiro, A., and Patrat, G., *Sensors Actuators B*, 1992, *6*, 299–303.
25. Ivanov, D., Currie, J., Bouchard, H., Lecours, A., Ardrian, J., Yelon, A., and Poulin, S., *Solid State Ionics*, 1994, *67*, 295–299.
26. Velasco, G., Schnell, J.P., and Croset, M., *Sensors Actuators*, 1982, *2*, 371–384.
27. Bergveld, P., *IEEE Trans. Biomed. Eng.*, 1970, *17*, 70–71.
28. Buck, R.P. and Hackerman, E., *Anal. Chem.*, 1977, *49*, 2315–2321.
29. Weppner, W., *Sensors Actuators*, 1987, *12*, 107–121.
30. *Solid State Electrochemistry and Its Applications to Sensors and Electronic Devices*, Goto, K.S., Ed., Elsevier, Amsterdam, 1988.
31. Kleitz, M., Siebert, E., Fabry, P., and Fouletier, J., in *Sensors: a Comprehensive Survey*, Vol. 2, Göpel, W., Hesse, H., Zemel, J.N., Eds., VCH Publishers, Weinheim, Germany, 1991, 341–428.
32. Wiemhöfer, H.D. and Cammann, K., in *Sensors: a Comprehensive Survey*, Vol. 2, Göpel, W., Hesse, H., Zemel, J.N., Eds., VCH Publishers, Weinheim, Germany, 1991, 159–189.
33. Möbius, H.H., in *Sensors: a Comprehensive Survey*, Vol. 3, Göpel, W., Hesse, H., Zemel, J.N., Eds., VCH Publishers, Weinheim, Germany, 1992, 1105–1154.
34. Azad, A.M., Akbar, S.A., Mhaisalkar, S.G., Birkefeld, L.D., and Goto, K.S., *J. Electrochem. Soc.*, 1992, *139*, 3690–3701.
35. Mari, C. and Barbi, G.B., in *Gas Sensors*, Sberveglieri, G., Ed., Kluwer Academic Publishers, Dordrecht, The Netherlands, 1992, 329–364.
36. *Ion-Selective Electrodes*, 2nd ed., Koryta, J. and Stulik, K., Eds., Cambridge University Press, Cambridge, U.K., 1983.
37. *Ion-Selective Electrode Methodology*, Vols. 1–2, Covington, A.K., Ed., CRC Press, Boca Raton, FL, 1979.
38. Koryta, J., *Anal. Chim. Acta*, 1990, *233*, 1–30.
39. Vlasov, Y.G., *J. Anal. Chem. of the USSR*, 1990, *45*, 923–934.
40. Solsky, R.L., *Anal. Chem.*, 1990, *62*, 21R–33R.
41. Janata, J., *Anal. Chem.*, 1990, *62*, 33R–44R.
42. Lundström, I., Van den Berg, A., Van der Schoot, B., Van den Vlekkert, H., Armgarth, M., and Nylander, C.I., in *Sensors: a Comprehensive Survey*, Vol. 2, Göpel, W., Hesse, H., and Zemel, J.N., Eds., VCH Publishers, Weinheim, Germany, 1991, 467–528.
43. Gauthier, M., Belanger, A., Meas, Y., and Kleitz, M., in *Solid Electrolytes*, Hagenmuller, P. and Van Gool, W., Eds., Academic Press: London, 1978, 497–517.
44. Kleitz, M., Pelloux, A., and Gauthier, M., in *Fast Ion Transport in Solids*, Vashista, P., Mundy, J.N., Shenoy, G.K., Eds., Elsevier, Amsterdam, 1979, 69–73.
45. Kleitz, M. and Siebert, E., in *Chemical Sensor Technology*, Vol. 2, Seiyama, T., Ed., Elsevier, Amsterdam, 1989, 151–170.
46. Göpel, W. and Wiemhofer, H.D., *Sensors Actuators B*, 1991, *4*, 365–372.
47. Weppner, W., *Mater. Sci. Eng. B., Solid State M.*, 1992, *15*, 48–45.
48. Rechnitz, G.A., *Pure Appl. Chem.*, 1973, *36*, 457–471.
49. *The Principle of Ion-Selective Electrodes and of Membrane Transport*, Morf., W.E., Ed., Elsevier, Amsterdam, 1981.
50. Buck, R.P., *Sensors Actuators*, 1981, *1*, 197–260.
51. *Modern Electrochemistry*, Vol. 1 & 2, Bockris, J. O'M. and Reddy, A.K.N., Eds., Plenum Press, New York, 1970.
52. *Principles of Electrochemistry*, 2nd Ed., Koryta, J., Dvorak, J., and Kavan, L., Eds., John Wiley & Sons, Chichester, 1993.

53. Oehme, F., in *Sensors: a Comprehensive Survey*, Vol. 2, Göpel, W., Hesse, H., and Zemel, J.N., Eds., VCH Publishers, Weinheim, Germany, 1991, 239–339.
54. Maienthal, E.J. and Taylor, J.K., in *Water and Water Pollution Handbook*, Vol. 4, Ciaccio, L.L., Ed., Marcel Dekker, New York, 1973, 1751–1800.
55. Buck, R.P., *Ion-Selective Electrode Methodology*, Vol. 1, Covington, A.K., Ed., CRC Press, Boca Raton, FL, 1979, 175–250.
56. *Principles of Chemical Sensors*, Janata, J., Ed., Plenum Press, New York, 1950.
57. Caneiro, A., Fabry, P., Khireddine, H., and Siebert, E., *Anal. Chem.*, 1991, *63*, 2550–2557.
58. *Chemical Sensors*, Edmonds, T.E., Ed., Chapman and Hall, New York, 1987.
59. *Chemical Sensors*, Edmonds, T.E., Ed., Blackie & Son Ltd., Glasgow, 1988.
60. Vlasov, Y.G., *Sensors Actuators B*, 1992, *7*, 501–504.
61. Vlasov, Y.G., Bychkov, E.A., and Legin, A.V., *Sensors Actuators B*, 1992, *10*, 55–60.
62. Neshkova, M.T., *Anal. Chim. Acta*, 1993, *273*, 255–265.
63. Legin, A.V., Bychkov, E.A., and Vlasov, Y.G., *Sensors Actuators B*, 1993, *15–16*, 184–187.
64. Nikolskii, B.P. and Materova, E.A., *Ion Selective Electrode Rev.*, 1985, *7*, 3–39.
65. Lyalin, O.O. and Turayeva, M.S., *Zh. Anal. Khim.*, 1976, *31*, 1879–1885.
66. Fjeldly, T.A. and Nagy, K., *J. Electrochem. Soc.*, 1980, *127*, 1299–1303.
67. Ravaine, D., Perera, G., and Hanane, Z., in *Chemical Sensors*, Seiyama, T., Fueki, K., Shiokava, J., and Suzuki, S., Eds., Kodansha/Elsevier, Tokyo, 1983, 521–526.
68. Fog. A. and Atlung, S., Intern. Patent WO 83 03304, 1983.
69. Déportes, C., Forestier, M., and Kahil, H., French Patent 84 19203, 1984.
70. Chu, W.F., Leonhard, V., Erdmann, H., and Ilgenstein, M., *Sensors Actuators B*, 1991, *4*, 321–324.
71. Leonhard, V., Erdmann, H., Ilgenstein, M., Cammann, K., and Krause, J., *Sensors Actuators B*, 1994, *18–19*, 329–332.
72. Fabry, P., Montero-Ocampo, C., and Armand, M. *Sensors Actuators*, 1988, *15*, 1–9.
73. Hammou, A., Fabry, P., Bonnat, M., Armand, M., and Montero-Ocampo, C., French Patent 86 09765, 1986.
74. Fabry, P., Gros, J.P., Million-Brodaz, J.F., and Kleitz, M., *Sensors Actuators*, 1988, *15*, 33–49.
75. *Glass Electrodes for Hydrogen and Other Cations*, Eisenman, G., Ed., Marcel Dekker, New York, 1967.
76. Macca, C. and Cakrt, M., *Anal. Chim. Acta*, 1983, *154*, 51–60.
77. Gadzekpo, V.P.Y. and Christian, G.D., *Anal. Chim. Acta*, 1984, *164*, 279–282.
78. Hiiro, K., Wakida, S., and Yamane, M., *Anal. Sci.*, 1988, *4*, 149–151.
79. Fouletier, J., Mantel, E., and Kleitz, M. *Solid State Ionics*, 1982, *6*, 1–13.
80. Maskell, W.C. and Steele, B.C.H., *J. Appl. Electrochem.*, 1986, *16*, 475–489.
81. Fouletier, J. and Siebert, E., *Ion Selective Electrode Rev.*, 1986, *8*, 133–151.
82. Maskell, W.C., in *Techniques and Mechanisms in Gas Sensing*, Moseley, P., Morris, J.O.W., and Williams, D.E., Eds., Adam Hilger, Bristol, 1991, 1–45.
83. Alcock, C.B., *Rev. Int. Hautes Temp. Réfract.*, 1992, *28*, 1–8.
84. Mari, C.M. and Barbi, G.B., in *Chemical Sensor Technology*, Vol. 4, Yamauchi, S., Ed., Elsevier, Amsterdam, 1992, 99–110.
85. Tieman, R.S. and Heireeman, W.R., *J. Anal. Lett.*, 1992, *25*, 807–819.
86. Aono, H., Sugimoto, E., Mori, Y., and Okajima, Y., *J. Electrochem. Soc.*, 1993, *140*, 3199–3203.
87. Miura, N. and Yamazoe, N., in *Chemical Sensor Technology*, Vol. 1, Seiyama, T., Ed., Elsevier, Amsterdam, 1988, 123–139.
88. Iwahara, H., in *Chemical Sensor Technology*, Vol.3, Yamazoe, N., Ed., Elsevier, Amsterdam, 1991, 117–129.
89. Chehab, S.F., Canaday, J.D., Kuriakose, A.K., Wheat, T.A., and Ahmad, A., *Solid State Ionics*, 1991, *45*, 299–310.
90. Iwahara, H., Uchida, H., Ogaki, K., and Nagato, H., *J. Electrochem. Soc.*, 1991, *138*, 295–299.
91. Kuo, CK., Tan, A.C., and Nicholson, P.S., *Solid State Ionics*, 1992, *53*, 58, 62.
92. Siebert, E., Fouletier, J., and Le Moigne, J., in *Proc. 2nd Int. Meeting on Chemical Sensors*, Aucouturier, J.L., et al., Eds., University of Bordeaux, 1986, 281–284.
93. Asano, M., Kuwano, J., and Kato, M., *J. Ceram. Soc. Jpn. Int. Ed.*, 1989, *97*, 1253–1258.
94. Yamazoe, N., Hisamoto, J., Miura, N., and Kuwata, S., *Sensors Actuators*, 1987, *12*, 415–423.
95. Sadaoka, Y., Sakai, Y., and Manabe, T., *Sensors Actuators B*, 1993, *13–14*, 532–535.
96. Maruyama, T., Sasaki, S., and Saito, Y., *Solid State Ionics*, 1987, *23*, 107–112.
97. Imanaka, N., Murata, T., Kawasato, T., and Adachi, G., *Sensors Actuators B*, 1993, *13–14*, 476–479.
98. Yao, S., Hosohara, S., Shinizu, Y., Miura, N., Futata, H., and Yamazoe, N., *Chem. Lett.*, 1991, *11*, 2069–2072.
99. Skeaff, J.M. and Dubreuil, A.A., *Sensors Actuators B*, 1993, *10*, 161–168.
100. Miura, N., Yao, S., Shimizu, Y., and Yamazoe, N., *Sensors Actuators B*, 1993, *13–14*, 387–390.
101. Kircherova, J. and Bale, C.W., *Solid State Ionics*, 1993, *59*, 109–115.
102. Kuwano, J., *Rep. Prog. Polym. Phys. Jpn.*, 1990, *33*, 411–414.
103. Nagata, K., Nishino, M., and Goto, K.S., *J. Electrochem. Soc.*, 1987, *134*, 1850–1854.

104. Inoue, T., Seki, N., Eguchi, K., and Arai, H., *J. Electrochem. Soc.*, 1990, *137*, 2523–2527.
105. Burkhard, D.J.M., Hanson, B., and Ulmer, G.C., *Solid State Ionics*, 1991, *48*, 333–339.
106. Gilderman, V.K., Andreeva, A.N., and Palguev, S.F., *Sensors Actuators B*, 1992, *7*, 738–742.
107. Eguchi, K, Inoue, T., Verda, M., Kaminae, J., and Arai, H., *Sensors Actuators B*, 1993, *13–14*, 38–40.
108. Kleitz, M., Iharada, T., Abraham, F., Mairesse, G., and Fouletier, J., *Sensors Actuators B*, 1993, *13–14*, 27–30.
109. Obayashi, H. and Okamoto, H. *Solid State Ionics*, 1981, *3–4*, 632–634.
110. Takahashi, T. and Iwahara, H., *Mater. Res. Bull.*, 1978, *13*, 1447–1453.
111. Nagashima, K., Ishimatsu, T., Hobo, T., and Asano, Y., *Bunseki Kagaku*, 1990, *39*, 229–232.
112. Pizzini, S. and Bianchi, G., *Chim. Ind. (Milan)*, 1973, *55*, 966–985.
113. Briot, F. and Vitter, G., French Patent 86 09778, 1986.
114. Goto, K., Ito, T., and Someno, M., *Trans. Metall. Soc. AIME*, 1969, *245*, 1662–1663.
115. Jacob, K. and Ramsesha, S.K., *Solid State Ionics*, 1989, *34*, 161–166.
116. Fouletier, J., Seinera, H., and Kleitz, M., *J. Appl. Electrochem.*, 1975, *5*, 177–185.
117. Baucke, F.G.K., *Glastech. Ber.*, 1983, *56K*, 307–312.
118. Pizzini, S., in *Fast Ion Transports in Solids*, Van Gool, W., Ed., North-Holland, Amsterdam, 1973, 461–475.
119. Mizusaki, J., Amano, K., Yamauchi, S., and Fueki, K., *Solid State Ionics*, 1987, *22*, 313–322.
120. Kleitz, M., Kloidt, T., and Dessemond, L., in *High Temperature Electrochemical Behavior of Fast Ion and Mixed Conductors*, Poulsen, F.W., Rentzen, J.J., Jacobsen, T., Skon, E., Ostergard, M.J.L., Eds., Riso Nat. Lab., Roskilde, Denmark, 1993, 89–118.
121. Worrell, W.L. and Iskoe, J.L., in *Fast Ion Transports in Solids*, Van Gool, W., Ed., North-Holland, Amsterdam, 1973, 512–521.
122. Wagner, C., in *Proc. Int. Comm. Electrochem. Thermo. and Kinetics (CITCE)*, Butterworth Scientific, London, 1957, 361–377.
123. Fouletier, J., Fabry, P., and Kleitz, M., *J. Electrochem. Soc.*, 1976, *123*, 204–213.
124. Pelloux, A., Quessada, J.P., Fouletier, J., Fabry, P., and Kleitz, M., *Solid State Ionics*, 1980, *1*, 343–354.
125. Siebert, E., Fouletier, J., and Vilminot, S., *Solid State Ionics*, 1983, *9–10*, 1291–1294.
126. Siebert, E., Fouletier, J., and Bonnat, M., *Solid State Ionics*, 1988, *28–30*, 1693–1696.
127. Kuwata, S., Miura, N., Yamazoe, N., and Seiyama, T., *Chem. Lett.*, 1984, 981–982.
128. Harke, S., Wiemhöfer, H.D., Göpel, W., *Sensors Actuators B*, 1990, *1*, 188–194.
129. Dietz, H., *Solid State Ionics*, 1982, *6*, 175–183.
130. Kaneyasu, K., Nakahara, T., and Takeuchi, T., *Sensors Actuators B*, 1993, *13–14*, 34–37.
131. Takeuchi, T. and Igarashi, I., in *Chemical Sensor Technology*, Vol. 1, Seiyama, T., Ed., Elsevier, Tokyo, 1988, 79–95.
132. Usui, T., Asad, A., Nakazawa, M., and Osanai, H., *J. Electrochem. Soc.*, 1989, *136*, 534–542.
133. Kondo, H., Takahashi, H., Saji, K., Takeuchi, T., and Igarashi, I., in *Proc. 6th Sensor Symposium: IEEE of Japan*, Tsukuba, 1986, 251–256.
134. Ishibashi, K., Kashina, I., Asada, A., *Sensors Actuators B*, 1993, *13–14*, 41–44.
135. Liaw, B. and Weppner, W., *Solid State Ionics*, 1990, *40–41*, 428–432.
136. Vinke, I.C., Seshan, K., Boukamp, B.A., de Vries,K.J., and Burggraaf, A.J. *Solid State Ionics*, 1989, *34*, 235–242.
137. Kuwano, J., Wakagi, A., and Kato, M., *Sensors Actuators B*, 1993, *13–14*, 608–609.
138. Osanai, H., Nakazawa, M., Isono, Y., Asada, A., Usui, T., and Kurumiya, Y., *Fujikura Tech. Rev.*, 1988, *17*, 34–42.
139. Miura, N., Haada, T., Shimizu, Y., and Yamazoe, N., *Sensors Actuators B*, 1990, *1*, 125–129.
140. Liu, J. and Weppner, W., *Sensors Actuators B*, 1992, *6*, 270–273.
141. Ohsuga, M. and Ohyama, Y., *Sensors Actuators*, 1986, *9*, 287–300.
142. Heyne, L., in *Measurement of Oxygen*, Degn, H., Baslev, I., Brook, R., Eds., Elsevier, Amsterdam, 1976, 65–88.
143. Franx, C., *Sensors Actuators*, 1985, *7*, 263–270.
144. Hetrick, R.E., Fate, W.A., and Wassel, W.C., *Appl. Phys. Lett.*, 1981, *38*, 390–392.
145. Maskell, W.C., Kaneko, H., and Steele, B.C.H., *J. Appl. Electrochem.*, 1987, *17*, 489–494.
146. Benammar, M. and Maskell, W.C., *Sensors Actuators B*, 1993, *12*, 195–198.
147. Benammar, M. and Maskell, W.C., *Sensors Actuators B*, 1993, *12*, 199–203.
148. Benammar, M. and Maskell, W.C., *Sensors Actuators B*, 1993, *15–16*, 162–165.
149. *Solid State Gas Sensors*, Moseley, P.T. and Tofield, B.C., Eds., Adam Hilger, Bristol, 1987.
150. Yamazoe, N. and Miura, N., in *Chemical Sensor Technology*, Vol. 4, Yamauchi, S., Ed., Elsevier, Amsterdam, 1992, 19–42.
151. Schönauer, U., *Techn. Messen.*, 1989, *56*, 260. (Reference from Göpel, W., Schierbaum, K.D., in *Sensors: a Comprehensive Survey*, Vol. 2, Göpel, W., Hesse, H., and Zemel, J.N., Eds., VCH Publishers, Weinheim, Germany, 1992, 429–466.)
152. Williams, D.E., in *Solid State Gas Sensors*, Mosely, P.T. and Tofield, B.C., Eds., Adam Hilger, Bristol, 1987, 71–123.

153. Moseley, P.T., *Sensors Actuators B,* 1992, *6,* 149–156.
154. Hunsko, J., Lantto, V., and Torvela, H., *Sensors Actuators B,* 1993, *15–16,* 245–248.
155. *Dynamic Characteristics of Ion-Selective Electrodes,* Lindner, E., Toth, K., and Pungor, E., Eds., CRC Press, Boca Raton, FL, 1988.
156. Attari, M., Fabry, P., Mallié, H., and Quezel, G., *Sensors Actuators B,* 1993, *15–16,* 173–178.
157. Sharna, A. and Pacey, P.D., *J. Electrochem. Soc.,* 1993, *140,* 2302–2308.
158. Alegret, S., Alonso, J., Bartroli, J., Machado, A.A.S.C., Lima, J.L.F.C., and Paulis, J.M., *Quim. Anal.,* 1987, *6,* 278–292.

Chapter 11

SOLID STATE BATTERIES

Christian Julien

CONTENTS

0-8493-8956-9/97/$0.00+$.50

LIST OF SYMBOLS AND ABBREVIATIONS

EV	electric vehicle
LIBES	Lithium Battery Energy Storage Technology Research Association
Mboe	million barrels oil equivalent
SRPE	solid redox polymerization electrode
SSE	solid-solution electrode
USABC	United States Advanced Battery Consortium
ZEV	zero-emission vehicle
C	capacity (A h)
E_W	specific energy density (W h kg^{-1})

I. INTRODUCTION

Solid state ionic materials have been extensively developed, and applications of solid electrolytes as well as insertion compounds have begun to converge into a coherent field during the last 10 years. Various designs of working devices are outlined in this chapter with the emphasis on either all solid state configurations or partial solid state ionic components. The large number of references reflects the great interest of researchers in energy storage devices, sensors, and optical and other electrochemical applications of solid state ionic conductors. For most types it is mentioned whether they are commercially available, items of intense current development, or one of the hot new research items.

Solid state ionic materials, which are solids which possess unusually high diffusion coefficients and conductances for specific ions, have assumed considerable importance in recent years for battery research. Several reviews on the development of solid electrolyte batteries have been written,[1-5] and some of the milestones are listed in Tables 11.1 and 11.2. There are two quite distinct applications for solid state ionic compounds in batteries. In addition to the electrolyte application, for which the electronic conductivity must be extremely low to avoid short circuiting the cell internally, there is also the possibility of using a solid state ionic compound as a cathode in which the diffusing cation dissolves to form an intercalation compound in the form of a solid solution electrode (SSE). For this type of application the solid should ideally have high values of both the ionic and electronic conductivities, thus it is a mixed conductor. These ionic cathodes may be distinguished from the more conventional cathodes which undergo reaction with a change of phase, e.g., PbO_2, NiOOH, etc., although the distinction is not entirely clear cut and intermediate situations may exist.

Batteries which utilize solid state ionics are of three general types: (1) All–solid state batteries in which the anode, electrolyte, and cathode are solids. A typical example of this battery type is the $Ag/RbAg_4I_5/RbI_3$ cell. Generally, solid state batteries are small primary or reserve batteries which operate at ambient temperature. (2) Solid electrolyte batteries with a liquid metal anode and/or a liquid cathode. These batteries can be ambient temperature systems, e.g., using a Na/Hg amalgam anode, although more generally the interest is in

TABLE 11.1
Primary Battery Developments

Date	Type	Chemistry
1800	Volta pile	Zn/Salt/Ag
1836	Daniell cell	$Zn/ZnSO_4$
		$CuSO_4/Cu$
1866	Leclanché cell	$Zn/NH_4Cl/MnO_2$
1878	Zinc air cell	$Zn/NaOH/O_2$
1945	Ruben cell	Zn/KOH/HgO
1955	Alkaline cell	$Zr/KOH/MnO_2$
1961	Silver–zinc cell	$Zn/KOH/Ag_2O$
1970–80	Lithium–iodine	$Li/LiI/I_2$
	Li/SSE	$Li/LiClO_4/MnO_2$
	Li/Soluble cathode	$Li/SOCl_2$

From Julien, C. and Nazri, G.A., *Solid State Batteries: Materials Design and Optimization*, Kluwer, Boston, 1994. With permission.

TABLE 11.2
Secondary Battery Developments

Date	Type	Chemistry
1860	Lead-acid	$PbO_2/H_2SO_4/Pb$
1900	Edison cell	Ni/2NiOOH/Fe
	Ni-Cd cell	Ni/2NiOOH/Cd
1965	Beta cell	$Na/\beta\text{-}Al_2O_3/S$
1970	Zinc-chlorine	$Zn/ZnCl_2/Cl_2$
1980–90	Li/SSE	$Li/PC\text{-}Li_2ClO_4/MX_2$
	Polymeric cells	$Li/PEO\text{-}LiClO_4/TiS_2$
	Glassy cells	$Li/Li^+\text{-glass}/TiS_2$
1991	Li microbatteries	$Li/Li^+\text{-glass}/TiS_2$
1992	Rocking-chair cells	$LiMn_2O_4$/elect./carbon
		$LiCoO_2$/elect./carbon
		$LiNiO_2$/elect./carbon

From Julien, C. and Nazri, G.A., *Solid State Batteries: Materials Design and Optimization*, Kluwer, Boston, 1994. With permission.

high-temperature systems for large-scale applications such as the $Na/\beta''\text{-}Al_2O_3/S$ battery. (3) Ionic cathode batteries in which the cathodes are being developed mostly for use with liquid electrolytes, in particular for lithium batteries.

This chapter surveys broadly the applications of solid state ionic materials as solid state battery components within the framework described above. It also highlights some recent solid state battery and microbattery developments. References to the literature are selective and it is by no means an exhaustive review of the subject. In the first part, we introduce and discuss the general requirements for solid state batteries. The second purpose is to investigate all the varieties of batteries which are commercially available or under high-level R&D. We have tried to maintain a balance between describing well established advanced systems and state-of-the-art developments which may or may not become of commercial importance. The main developments of primary and secondary electrochemical systems are listed in Tables 11.1 and 11.2, respectively. More details concerning solid state cells and thin-film batteries are given below.

A. ADVANTAGES OF SOLID STATE BATTERY TECHNOLOGY

Much of the research effort is aimed at developing rechargeable batteries that have high energy and power density. These batteries could reduce oil consumption by powering electric vehicles and storing electricity from generating plants for use during periods of peak demand.[6] In particular, specific energy densities of $E_W > 70$ W h kg^{-1} are desirable for urban automobile transport. This is to be compared with the value of 42 W h kg^{-1} characteristic of the Pb/PbO_2 cell and the value of 60 W h kg^{-1} of the Ni-Cd cell.[7] Rocking with a factor of three to six for the ratio of theoretical specific energy calculated (considering only the equivalent weights of cathode and anode batteries) to specific energy of a practical cell,[8] figures of merit for such applications are of the order 400 to 450 W h kg^{-1}.

In the case of conventional batteries, for instance, these systems contain a liquid electrolyte, generally a concentrated aqueous solution of potassium hydroxide or sulfuric acid. The liquid state offers very good contacts with the electrodes and high ionic conductivities, but anion and cation mobilities are of the same order of magnitude and their simultaneous flow gives rise to two major problems: (1) corrosion of the electrodes, (2) consumption of the solvent (water) by electrolysis during recharging and by corrosion during storage, making necessary periodic refilling. In addition, these two processes give off gases, thereby prohibiting the design of totally sealed systems. The resulting problems include leakage of the corrosive electrolyte and air entries which, even when kept to a minimum, deteriorate the electrolyte and the electrodes. A further drawback is the risk of electrode passivation — the formation of insulating layers of $PbSO_4$, $Zn(OH)_2$, etc., on the electrodes, also a consequence of the anion mobility.

In many solid electrolytes, the only mobile charge carrier is the cation A^+ while the counter ion is present on an immobile sublattice. Examples include β-alumina (Na_2O-11Al_2O_3), β″-alumina (Na_2O-5.33Al_2O_3), NASICON, lithium nitride, and inorganic glasses like lithioborate glasses. If blocking of the anion prevents passivation, corrosion, and solvent electrolysis reactions, it is possible to design totally sealed batteries, eliminating the deterioration of the electrolyte and the electrodes by the outside environment. The anion immobility makes it possible to increase the practical redox stability domain of the electrolyte far above its range fixed by thermodynamics. Under these conditions, the electrolyte can coexist with couples which are highly reducing at the negative electrode and highly oxidizing at the positive electrode.

It is possible to compare battery systems from the state of the three main components: the electrode A, the electrolyte, and the electrode B (Figure 11.1). All these media can be liquid, plastic, or solid. This is crucial because certain interfaces are difficult to handle. Common batteries have a solid–liquid–solid interface; the liquid–solid–liquid system corresponds to the Na-S battery using β-alumina as the electrolyte, which permits relatively easy manufacture. On the other side, the all-solid system involves difficult interface problems with crucial dimensional stability at each interface. These difficulties can be solved by using a polymer film as a plastic electrolyte; also, polyethylene oxide (PEO) film electrolytes are the key to a new battery design.

Microsolid state batteries in the form of thin films partly avoid the interface contact difficulty and can be used as devices in microelectronics. Recently, considerable attention has been focused on the preparation of solid state lithium batteries using solid polymer electrolytes which are made from polymer complexes formed by lithium salts and polymer ethers.[9] Ultrathin-film solid state lithium batteries have been fabricated using a thin solid polymer electrolyte film prepared by complexation of a plasma polymer and lithium perchlorate.[10]

B. POTENTIAL EFFECTS ON ENERGY CONSERVATION

Electric vehicles (EV) are good examples of the potential national benefits of widespread application of energy storage technologies.[11] The present European domestic fleet is about 200 million of vehicles which consume 3200 million barrels of oil equivalent (Mboe) per year. Anticipated improvements in the fleet-average vehicle fuel economy will be roughly offset by the projected growth in the size of the fleet, so transportation energy demand is not likely to change dramatically in the near future. If 10% of the vehicles were electrically

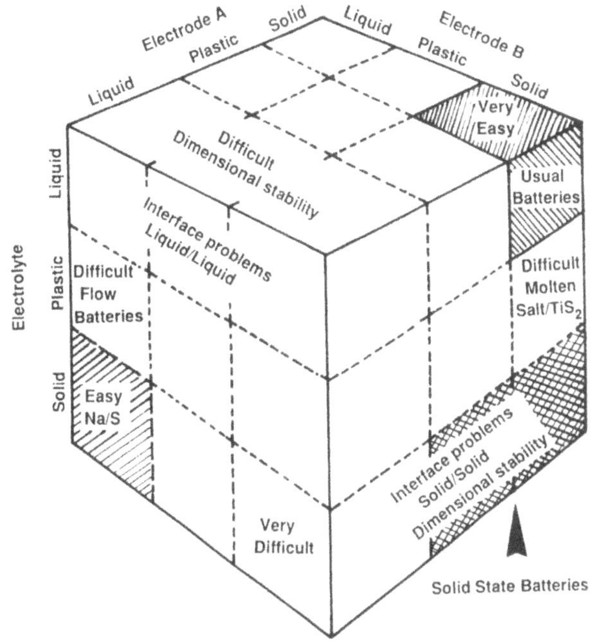

FIGURE 11.1. The three main components of electrochemical cells: the electrode A, the electrolyte, and the electrode B. All these media should be liquid, plastic, or solid. (From Julien, C. and Nazri, G.A., *Solid State Batteries: Materials Design and Optimization*, Kluwer, Boston, 1994. With permission.)

powered, however, petroleum consumption would be cut by 320 Mboe per year. This reduction corresponds to ECU's 9 billion per year at $20 per barrel.

EVs emit no pollutants, so emissions are shifted away from the vehicle tailpipe, where they are very difficult to control and regulate, to central utility generating stations, where they are much easier to control and regulate. At present, vehicles emit several million tons of sulfur oxides, nitrogen oxides, and hydrocarbons per year. Electrification of vehicles could eliminate all these pollutants if nuclear, solar, and/or hydroelectric power were used as the primary energy sources, and a significant portion of these pollutants would be eliminated if coal were used as the primary source.

The 640 Mboe associated with a 20% electric domestic vehicle fleet corresponds to about 2.2×10^{11} kg of domestic CO_2 production that would be avoided each year, provided that noncarbonaceous primary fuels, e.g., nuclear, solar, or hydroelectric, generate the electricity used to charge the batteries. The corresponding worldwide value is 6×10^{11} kg of CO_2 avoided each year, assuming a comparable replacement of the global vehicle fleet by electrics.

To spend electric energy efficiently and maintain batteries, EVs should be charged at night time so that minimum urgent quick charge may be necessary. One-day driving ranges of most personal cars are usually within 100 miles, for commuting, shopping, leisure, etc. Minimum performances of particle EVs should be (i) a cruising range of one-time charge of 200 miles at actual driving mode, (ii) a power-to-weight ratio of 50 kW/ton, (iii) a cruising range lifetime of a car of 120,000 miles, and (iv) a cost comparable with that of internal-combustion–engine vehicle.

C. REQUIREMENTS OF SOLID STATE BATTERY TECHNOLOGY

A detailed knowledge of the properties of the different components of solid state batteries is essential to the development of a physically based model capable of accurately predicting delivered capacity and end-of-life behavior for various battery designs.

The applications of modern energy storage systems reach into numerous aspects of daily life. Power sources can be classified into three classes: (1) high-power batteries (with a

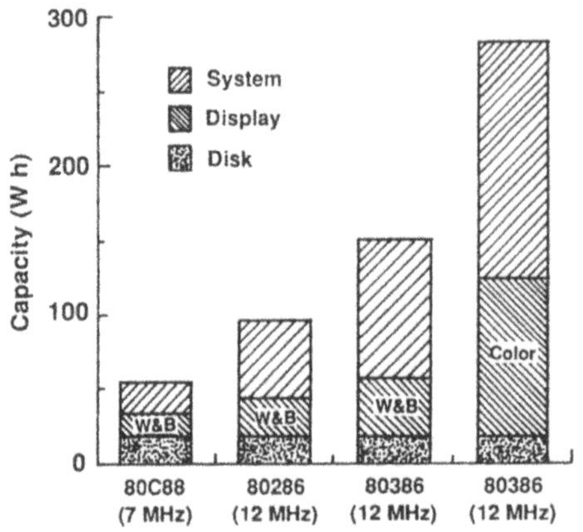

FIGURE 11.2. Capacity requirements vs. function for all–solid state batteries as laptop computer power sources. (From Julien, C. and Nazri, G.A., *Solid State Batteries: Materials Design and Optimization*, Kluwer, Boston, 1994. With permission.)

capacity C > 50 A h) are used in traction (locomotives, submarines, etc.), EVs, or load leveling; (2) miniature batteries (with a capacity range 0.2 to 2 A h) for powering hand-held devices, implantable medical equipment, telephones, computers, and other widely used products; and (3) microbatteries (with a typical capacity of 200 μA h) which can be associated with microelectronics.

Examination of the way that laptop computers are developed allows prediction of the future requirements for both battery power and energy of miniature systems (Figure 11.2). Typically for 4 h operation, with a 7-MHz system using a backlit, polarized black-and-white display, a 32-W h battery would be required. For 4 h operation of a 12-MHz system with the same black-and-white display, 76 W h would be needed, while for 8 h operation with a 12-MHz system and a VGA-color display, 280 W h is needed. With the Ni-Cd battery technology, the battery pack weighs 8 kg. A lithium–polymer technology can reduce this weight more than four times. Yet this is already perceived as precisely the direction in which laptop computer development must go, although the industry expectation is that this will required a decade to achieve.

Using thin-film technology, solid state microbatteries are compatible with microelectronics. One can in principle prepare an on-chip microbattery capable of maintaining its memory during a power outage. The combination of the thin-film configuration and the low current drain requirements (1 to 10 nA cm^{-2} for a C-MOS memory) allows for the use of electrolytes with much lower conductivities than are necessary in the Na/S cell applications, for example.[12]

One of the very few examples of a commercial solid state battery is the lithium heart pacemaker power source;[13] but many systems of potential applicability have been proposed during the last 15 years,[14] and are now in very advanced development for appearing on the market. The main problem areas in primary solid state batteries have been identified as (i) volume changes and modification of the geometrical arrangement, (ii) electrolyte resistance, (iii) discharge product resistance, (iv) material compatibility, and (v) manufacturability.[13]

For secondary batteries, highlights for some additional difficulties arising from the need to recycle the systems have been reviewed.[3] These include low diffusion coefficients for ionic transport within intercalation cathode materials, failure modes associated with high current densities, and exacerbations of interfacial contact problems on recycling.

Commercial viability clearly requires these sets of problems to be overcome, and for some proposed cell systems it can be seen from work reported herewith and elsewhere that the time for commercial application appears to be very close. In practice, certain solid state batteries present practical energy densities, i.e., including the weight of the casing and electrical connections in the calculation, of the order of 200 to 300 W h kg^{-1}. This value is eight times higher than that of lead batteries under the best conditions (Figure 11.3).

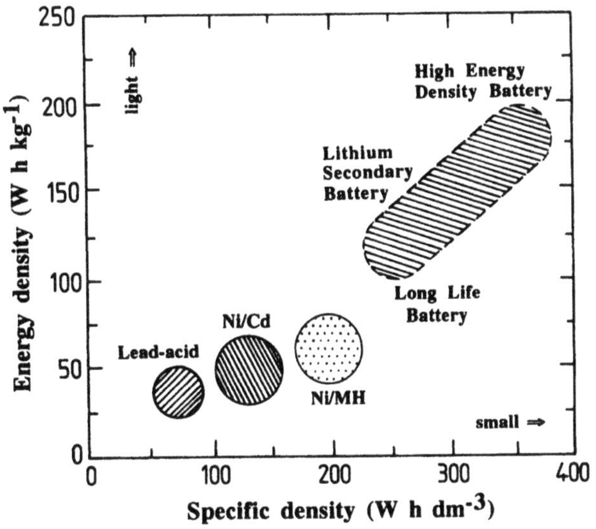

FIGURE 11.3. Energy density vs. specific density for secondary batteries. (From Julien, C. and Nazri, G.A., *Solid State Batteries: Materials Design and Optimization,* Kluwer, Boston, 1994. With permission.)

Solid state primary batteries can provide very long-life operation at low currents. The first example of such an application is the lithium-iodide solid state battery for cardiac pacemakers which is manufactured in the US by Catalyst Research Co., by Wilson Greatbatch, and by Medtronic Inc. The second example is lithium–glass battery, whose application envisaged is mainly as a power source for electronic computers, such as C-MOS memory backup. Cells commercially available are design XR2025HT by the Union Carbide group.

Solid state electrolyte cells have been developed and tested also for high-power purposes, for instance, in all-electric vehicle applications. The invention of the sodium/β-alumina/sulfur battery by Ford Motor Co. intensified interest in the commercial applications of solid electrolytes.

It is convenient to classify solid state batteries into four classes: high temperature, polymeric, lithium, and silver. Each of these will now be separately considered. Examples can be given for recent achievements in solid state batteries, and the characteristics of the major types of marketed solid state batteries are summarized in Table 11.3.

D. ADVANCED PROJECTS

New battery systems possessing improved characteristics regarding weight, size, life, recharge cycles, and meeting environmental and safety requirements are the goals of ongoing researches conducted in the U.S., in Europe, and in Japan, including universities, national laboratories, and industrial companies.

In 1989, the Join Opportunities for Unconventional of Long-term Energy supply (JOULE) program proposed by the Commission of the European Communities covered the projects provided in the field of research and technological development of nonnuclear energies and rational use of energy. The 122 million ECU available are used to finance cost-sharing research contracts for projects such as solar energy applications, fuel cells, and lithium batteries. R&D on solid lithium batteries with the aim to achieve a power density of 150 W h kg^{-1} and 1000 cycles with a capacity loss of not more than 20% are implemented by academic and industrial European partners.[15]

In 1991, the U.S. government entered into a partnership with the United States Advanced Battery Consortium (USABC) in a research and development effort to device new automobile battery systems and thereby accelerate the development and production of EVs for the mass market. The USABC was formed as a result of conclusions arrived at by the automobile manufacturers when confronted with the requirement to provide zero emission vehicles (ZEVs)

TABLE 11.3
Characteristics of Some Advanced Solid-Electrolyte Cells

System	Cell voltage (V)	Energy density W h cm^{-3}	Energy density W h kg^{-1}	Producer
$Ag/RbAg_4I_5/RbI_3$+ C	0.66	0.08	13	WGL[a]
$Li/LiI\text{-}Al_2O_3/PbI_2$	1.9	0.3	150	Duracell
$Li/LiI/I_2 + P_2V$ P	2.8	0.5	200	CRC[b],WGL,ETMI[c]
$Li/LiI\text{-}Al_2O_3/TiS_2$	2.3	0.9	520	Duracell
Na/β″-alumina/S	2.0	/	180	Ford, GE, Silent
Na/β″-alumina/$NiCl_2$	2.58	0.16	100	AEG[d]
$Li/LiI + SiO_2/Me_4NI_5$	2.75	0.4	125	WGL
$Li/P_2S_5\text{-}Li_2S\text{-}LiI/TiS_2$	2.5	0.21	/	UCAR[e], SAFT
$Li/PEO\text{-}LiClO_4/V_6O_{13}$	2.5	/	200	HQ[f]
$Li/MEEP\text{-}PEO/V_6O_{13}$	2.8	/	>150	Valence Tech. Inc.
Li-Al/LiCl-KCl/FeS	1.6	/	70	Varta
Li-Al/LiCl-KCl/FeS_2	1.6	/	95	SAFT, Gould

[a] Wilson Greatbatch Ltd.
[b] Catalyst Research Corp.
[c] Energy Technology, Medtronic Inc.
[d] This cell is named "ZEBRA-battery".
[e] Union Carbide Corp.
[f] Hydro-Quebec.

From Julien, C. and Nazri, G.A., *Solid State Batteries: Materials Design and Optimization,* Kluwer, Boston, 1994. With permission.

TABLE 11.4
The USABC Advanced Battery Criteria

Characteristic	Midterm	Long-term
Specific energy (Wh/kg)	80–100	200
Specific power (W/kg)	150–200	400
Power density (W/dm^3)	250	600
Lifetime (years)	5	10
Cycle life[a]	600	1000
Recharge time (hours)	<6	3–6
Ultimate price ($/kW h)	<150	<100

[a] At 80% depth of discharge.

From Julien, C. and Nazri, G.A., *Solid State Batteries: Materials Design and Optimization,* Kluwer, Boston, 1994. With permission.

by the late 1990s in response to new California automotive emission regulations. The California law requires that 2% of all new vehicles sold in the state must be ZEVs, and that by 2003 it must be 10%. This type of law affects 30 to 40% of the new-vehicle market in the U.S. The major objectives established by the USABC were the establishment of an advanced battery industry, the acceleration of the ZEVs market, and the development of batteries competitive with petroleum systems.[16] The USABC advanced battery technology midterm and long-term criteria are presented in Table 11.4. This consortium intends to work with four technologies: (i) nickel metal hydride cells developed by Ovonic and SAFT America, (ii) sodium sulfur batteries developed by Silent Power Inc., (iii) lithium iron disulfide cells developed by SAFT America, and (iv) lithium–polymer batteries developed by W.R. Grace and 3M.

In 1992 the Lithium Battery Energy Storage Technology Research Association (LIBES) was created in Japan. This program conducts the R&D on "dispersed battery energy storage

TABLE 11.5
The Target Performance of the LIBES Program in Japan

Characteristic	Long-life Li battery	High energy density Li battery
Energy density (Wh/kg)	120	180
Specific energy (W/dm^3)	240	360
Cycle life	3500	500
Energy efficiency	≥90	≥85
Usage	Load leveling	EV

From Julien, C. and Nazri, G.A., *Solid State Batteries: Materials Design and Optimization*, Kluwer, Boston, 1994. With permission.

technology" of the New Sun Shine Project of the Agency of Industrial Science and Technology, the MITI. A new 10-year R&D project with a budget of approximately Y14 billion was started. Targets of this project are (i) R&D of 20 to 30 kW h high-performance ambient-temperature lithium batteries with long life and high energy density and (ii) research on the total system of battery application technology (Table 11.5). In order to extend the life cycle, to enhance the energy density, to enlarge the power output, to improve the reliability, and to scale up lithium secondary battery, LIBES is going to challenge to such difficult subjects as research on materials, design, and manufacturing technologies for large-capacity cells, and research and development of module batteries based on these technological elements.

II. APPLICATIONS OF SOLID STATE IONICS TO BATTERIES

A. HIGH-TEMPERATURE CELLS

The alkali metals lithium and sodium are attractive as battery anodes on account of their high electrode potentials and their low atomic masses, which together result in excellent values for the battery specific energy. In this chapter we consider batteries which utilize solids (fast ion conductors) or fused salts as electrolytes and which operate at temperatures of 200 to 500°C.

Batteries that operate at elevated temperatures exhibit improved performance compared to ambient-temperature batteries. Figure 11.4 shows the cell voltage of two types of high-temperature batteries discharged at a high current. However, it is necessary to insulate them to prevent rapid heat loss to the environment.

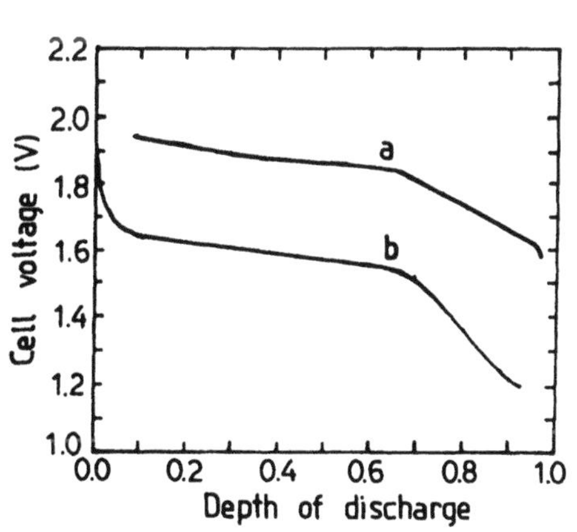

FIGURE 11.4. Cell voltage as a function of the depth of discharge for high-temperature batteries: (a) a Na/S battery discharged at 19-A rate and (b) a LiAl-Al/FeS_2 battery discharged at 15-A rate. (From Julien, C. and Nazri, G.A., *Solid State Batteries: Materials Design and Optimization*, Kluwer, Boston, 1994. With permission.)

1. Sodium Sulfur Batteries

The sodium sulfur battery, also known as the beta battery, owes its existence to the remarkable properties of β-alumina as a fast ion conductor for Na^+ ions. Discharge of the cell involves ionization of sodium atoms at the anode, followed by diffusion of Na^+ ions through the wall of the β-alumina to the cathode where reduction of sulfur occurs. At temperatures of around 300°C the ionic resistivity of β-alumina (Na_2O-$11Al_2O_3$) can be as low as 2 to 5 Ω cm, comparable to that of an aqueous sodium chloride solution, making it an ideal solid electrolyte for use in high-temperature Na batteries. The beta battery uses a molten Na negative electrode, the solid Na^+ ion-conducting electrolyte, and a molten sulfur–sodium polysulfide mixture as the positive electrode. At 300 to 350°C, which is the required temperature, the Na/S cells must be sealed from the atmosphere to prevent reaction with water and air.

In reality, Na^+-β-Al_2O_3 is not a single compound, but may consist of two or more nonstoichiometric phases, each with its own characteristic crystal structure. The most important phase, which has the highest ionic conductivity, is β″-alumina of typical composition $Na_{1.67}Mg_{0.67}Al_{10.33}O_{17}$ and is the preferred phase for battery operation.

The cell voltage of a beta cell, with a central liquid sodium anode, discharged at a 19-A rate is shown in Figure 11.4a. Discharge of the cell involves ionization of sodium atoms at the anode

$$2Na \Leftrightarrow 2Na^+ + 2e^-, \tag{11.1}$$

followed by diffusion of Na^+ ions through the β″-alumina ceramic to the cathode where reduction of sulfur occurs in two steps:

$$2Na^+ + 5S + 2e^- \Leftrightarrow Na_2S_5 \tag{11.2a}$$

$$\frac{2x}{5}Na^+ + \left(1-\frac{x}{5}\right)Na_2S_5 + \frac{2x}{5}e^- \Leftrightarrow Na_2S_{5-x}. \tag{11.2b}$$

The first reaction gives an open-circuit voltage of 2.08 V, whereas the second reaction gives 2.08 to 1.76 V.[8] In the first step, where the voltage is invariant, a mixture of Na_2S_5 and unreacted sulfur coexists. After complete conversion to Na_2S_5, further reaction is a single-phase process in which Na_2S_5 is converted progressively to Na_2S_{5-x} by addition of Na^+. The cell voltage falls linearly with the Na composition and the discharge is stopped at 1.76 V, corresponding to the composition of Na_2S_3. Current characteristics are with an energy density of 200 W h kg^{-1} and a capacity of 15 A h. The original interest has led to massive R&D programs in Europe (Chloride Silent Power, British Rail, UK AEA, CGE, Brown Boveri Co.), in the U.S. (General Electric Co., Ford Motor Co., Dow Chemical Co., Silent Power), and in Japan (Toshiba Electric, Yuasa Battery Co.). These have been working on the development of beta batteries for EV traction and load-leveling applications. Currently, the full-size cell is designed for a cycle of 5 h discharge and 7 h charge with a round-tip efficiency of 76%. Individual cells have regularly shown lives of 1000 to 2000 cycles. Despite their very intensive development, sodium sulfur batteries have been abandoned for some time because of the inability to maintain a consistent quality. Discharge of the cell involves ionization of sodium atoms at the anode, followed by diffusion of Na^+ ions through the wall of the beta alumina to the cathode where reduction of sulfur occurs.

Today a new interest is being generated because high-quality, fine-grained materials are now routinely produced, and it has been found that the addition of ZrO_2 helps refine the grain size and toughens the ceramic. Also, special sulfur electrode designs, including the use of graded resistivity shaped graphite felt current collectors, have improved the rechargeability of the sulfur electrode. A number of cell sizes and configurations have been considered, but 10 to 50 A h, inside-Na cell designs have become common in recent years. Efforts are under way in only three countries (Germany, U.S., and Japan) to develop this very promising battery for EV applications.[17]

A new beta cell has been developed by Beta-Power Inc. in the U.S. This cell is built with the concept of flat solid electrolyte disc instead of tubular design. This flat separator, typically 0.3 mm thick, is manufactured by a tape casting technique and allows high specific power of 600 W kg^{-1} and 1.65 kW cm^{-3} for a bipolar cell. This performance is a factor 5 to 6 higher than with the Na-S batteries of cylindrical cell design and is directly attributable to the thin flat plate design, which offers uniform discharge and excellent thermal conduction for sustained operation.[18]

Battery performance is more modest than that of an individual cell, owing in part to the need to provide a thermally insulating enclosure, and values around 100 and 130 W kg^{-1} are typical for a 50-kW h battery.[19] Such a battery should result in a range of about 125 miles for an EV under an urban driving schedule.

At Yuasa Battery Co. in Japan, a 50-kW, 400-kW h Na-S load-leveling battery has been constructed,[20] and it has demonstrated an energy efficiency of 85% for cycle of 8 h charge and 8 h discharge, and a 48.5-kW h m^{-2} energy-area ratio. Test of a 10-kW Na-S battery has started in 1992 at Hitachi Works. The battery is consists of 192 cells of 252 Ah capacity each. This battery can operate at a current of 345 A and delivers an energy density of 60 W h kg^{-1}.

2. Lithium Iron Sulfide Batteries

The lithium iron sulfide battery operates at about 400 to 500°C using a fused halide eutectic electrolyte immobilized in the pores of a suitable separator. This battery displays a number of attractive features compared to the Na-S battery, including prismatic flat-plate construction, ability to withstand numerous freeze–thaw cycles, cell failures in short-circuit conditions, ability to withstand overcharge, and low-cost materials and construction techniques. The major disadvantage is a somewhat lower performance. Although this battery is suitable for both EV and load-leveling applications, recent attention has focused on battery designs suitable for EV propulsion.[19]

The most commonly used electrolytes are the LiCl-KCl binary eutectic and the LiF-LiCl-LiI ternary lithium halides. With Li-Al alloy anodes and FeS_2 positive electrodes the discharge occurs in several discrete steps:

$$4Li + 3FeS_2 \rightarrow Li_4Fe_2S_5 + FeS, \tag{11.3}$$

$$2Li + Li_4Fe_2S_5 + FeS \rightarrow 3Li_2FeS_2, \tag{11.4}$$

$$6Li + 3Li_2FeS_2 \rightarrow 6Li_2S + 3Fe, \tag{11.5}$$

giving an open-circuit voltage of 2.1, 1.9, and 1.6 V, respectively (see Figure 11.4b). The use of Li-Al results in almost 50% decrease in the theoretical specific energy, but much better stability is achieved. Most development work has concentrated on the LiAl/FeS couple, and the Varta Battery Company (Germany) has produced a series of 140-A h cells with a specific energy of 100 W h kg^{-1} at a low discharge rate of 80 mA cm^{-2}, falling to density 50 W h kg^{-1} at high rates of current density 250 mA cm^{-2}.

The present version of LiAl//LiCl-LiBr-KBr/FeS_2 utilizes a dense FeS_2 electrode that only discharges to a stoichiometry corresponding to FeS (rather than to elemental Fe, which was the case of prior versions of this technology).[17] The melting point of LiCl-LiBr-KBr at 310°C permits cell operation at about 400°C. These innovations have resulted in greatly improved capacity retention, and cells are cycled for more than 1000 times. These cells are also developed in the U.S. and manufactured by Eagle-Picher, by Gould, and by SAFT America. Cells of 150- to 350-A h capacity yield specific energies of 70 to 95 W h kg^{-1} at a 4-h discharge rate.

There are still a number of unresolved scientific questions about the chemistry of LiAl/FeS cells and the mechanism of degradation and failure. In this system the separator is clearly a crucial component which must not only keep the electrode materials apart, but also allow good permeation of the electrolyte, and the most suitable materials are found to be boron nitride and zirconia in the form of woven cloths, but these are obviously very expensive options.

3. Sodium Chloride Batteries

The sodium chloride battery, which resembles the Na-S battery, is a recently developed high-temperature battery.[22] The major differences between the two systems are the positive electrode, which is an insoluble metal chloride in molten $NaAlCl_4$, and the temperature of operation, which is about 250°C. The cell configuration is $Na/\beta''\text{-}Al_2O_3/NaAlCl_4/MCl_2$, where M can be Fe, Ni, or possibly other transition metals. Both the $Na/FeCl_2$ and $Na/NiCl_2$ cells have been shown to be electrochemically reversible, and they exhibit a number of features that may be considered as improvements over those of Na-S cells.

Individual $Na/FeCl_2$ cells have demonstrated above 130 W h kg^{-1} and more than 1000 cycles, although their specific power is typically below 100 W kg^{-1}, particularly when at a low state of charge. Large EV batteries of about 25-kW h capacity have been constructed and tested.[22]

The features of this system include a lower operating temperature, lower than for the beta battery, ability to withstand limited overcharge and overdischarge, cell failures in short-circuit conditions, better safety characteristics, and higher cell voltage. Disadvantages include a slightly lower specific energy, lower specific power due to the reduced ionic conductivity of $\beta''\text{-}Al_2O_3$ at the lower temperature of operation. There also have a high resistance of the positive electrode and poor wetting of the $\beta''\text{-}Al_2O_3$ by sodium at the lower temperature. On balance, Na/MCl_2 batteries appear to offer some attractive characteristics, and their development is being pursued.

The ZEBRA-Battery developed by AEG Co. in Germany is an advanced high-energy battery based on nickel chloride as the positive electrode and sodium as the negative electrode. The cell reaction

$$2Na + NiCl_2 \Leftrightarrow Ni + 2NaCl \tag{11.6}$$

is fully reversible. No side reactions occur in the normal operations, a necessary condition for a long cycle life with stable capacity. Both electrodes are separated by a sodium ion-conducting electrolyte, the ceramic β''-alumina, which develops a sufficient low resistance at temperature of about 300°C. At this temperature the cell shows a voltage of 2.58 V. The energy density of ZEBRA-Batteries amounts to 100 W h kg^{-1} and 160 W h dm^{-3}, a fairly high value to meet the required daily range of a city car. In bench life testing, seven batteries with about 2500 cells were fully charged and discharged. During this test less than 2% of the cells failed up to 1000 cycles.

4. Lithium Chloride Batteries

Similar designs have been used in lithium chloride batteries, which operate at a temperature of 650°C. The cell has the form Li (liq.)/LiCl (liq.)/Cl_2 (g), carbon.The two electrodes, the liquid lithium anode and the porous carbon anode in which the chlorine gas is fed under pressure, are separated by a molten lithium chloride electrolyte. The overall cell reaction is

$$Li(liq.) + 1/2\ Cl_2\ (g) \rightarrow LiCl\ (liq.), \tag{11.7}$$

with an EMF of 3.46 V. This system delivers a theoretical energy density of 2.18 kW h kg^{-1} at the working temperature. The most serious problems with this system are, however, concerned with corrosion of cell components and the development of satisfactory seals.

5. Sodium–Sulfur–Glass Electrolyte

It is possible to replace β-alumina by a Na^+-conducting borate glass,[23] and other materials such as NASICON of composition $Na_{1-x}Si_xZr_2P_{3-x}O_{12}$ (with $1.8 < x < 2.4$) have been crystal engineered to maximize conductivity and ease of fabrication.[24]

Dow Chemical Co. has manufactured a Na-S-glass battery in which the electrolyte is a Na^+-borate glass formed by fibers. Cycle lives exceeding 500 cycles at 8% discharge have been obtained for individual cells, and a load-leveling battery has been designed which will use 12,500 gas-cooled cells of 0.8-kW h capacity.

B. POLYMERIC AND GLASS BATTERIES

Since Armand[9] suggested that solid polymer electrolytes derived from PEO might be used as electrolytes in thin-film, solid electrolyte batteries, major efforts have been made to produce polymer electrolytes with high conductivities at room temperature. Polymeric electrolytes are formed by complexes between salts of alkali metals and polymers containing solvating heteroatoms. The most common examples concern complexes between PEO which forms the (CH_2-CH_2-O-CH_2-CH_2) chains. The solvating heteroatom, here oxygen, acts as a donor for the cation M^+. The anion X^-, generally of large dimension, stabilizes the PEO-M^+ complex. The first polymer electrolytes had conductivities of less than 10^{-5} S cm^{-1} at 25°C, far too low for use in normal battery applications. Recently, however, electrolyte compositions have been produced which exhibit conductivity characteristics competitive with the properties of non-aqueous liquid electrolytes.[25] These new materials have stimulated major development efforts in polymer electrolyte battery technology. The PEO-MX complexes may be considered as plastic electrolytes, i.e., a compromise between liquid and solid crystalline electrolytes. Most of the polymeric batteries are of the form Li/PEO-Li salt/IC, where IC can be an intercalation compound, but also a composite electrode including high amount of active intercalation material. From an examination of the temperature dependence of ionic conductivity for polymer complexes, one can remark the relatively high conductivity at room temperature of the complexes based on polyphosphazene, due to the low value of the vitreous transition temperature, T_g, and thus, to the flexibility of the chains of the polymer, in turn related to the character of the P-N bonds.[26] Figure 11.5 shows the cell voltage of some polymer batteries which are intensively studied.

1. Lithium–Polymer Intercalation Compound Batteries

An all–solid state construction uses a Li salt-doped PEO film to separate the lithium electrode from an ion insertion-type positive electrode such as V_6O_{13},[25] TiS_2,[27] MnO_2, Cr_3O_8, or LiV_3O_8.[28] In many development cells, the positive electrode is a composite and consists of small particles of insertion compound bound together with polymer electrolyte and carbon, which improves its ionic and electronic conductivities, respectively. The composite electrode (50 to 75 μm thick) is deposited on a thin copper or nickel current collector less than 25 μm and a film (25 to 50 μm) of $[(C_2H_4O)_9\ LiCF_3SO_3]_n$ polymer electrolyte completes the lithium cell. Another possibility is to roll the cell assembly.

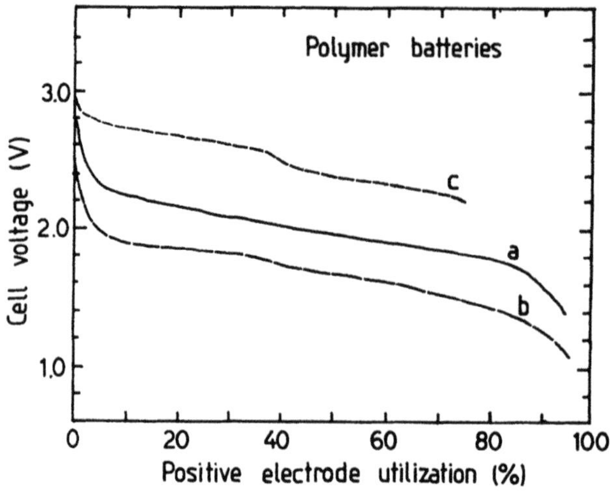

FIGURE 11.5. Discharge curves as a function of the cathode utilization of polymer batteries: (a) Li/PEO/TiS_2, (b) Li-Al/PEO/TiS_2, and (c) Li/$(PEO)_8(LiCF_3)_2$/$[-S(R)_nO(R)_nS-]_n$ cells having a theoretical capacity of about 3 mA h. (From Julien, C. and Nazri, G.A., *Solid State Batteries: Materials Design and Optimization*, Kluwer, Boston, 1994. With permission.)

A cell of this type using V_6O_{13} as the active material of the composite positive electrode Li/$(PEO)_9LiCF_3SO_3$/V_6O_{13}+PEO+C would have a potential of about 2.8 V. During discharge, at current densities of around 1 mA cm^{-2}, the voltage stabilizes at around 2 V. The practical energy density is in the order of 200 W h kg^{-1}, the power density reaching 0.1 W g^{-1}. The insertion process implies the movement of the sole Li^+ ion alone, and this underlines the importance of the availability of an electrolyte with a purely Li^+ transport. This condition is not fulfilled in the case of $(PEO)_9LiCF_3SO_3$, where the transport number of Li^+, t_{Li^+}, is approximately equal to 0.5.[29]

The system examined in the French–Canadian Project considers the following sequence Li/$(PEO)_8$ $LiClO_4$/TiS_2+PEO, C, which is based on the intercalation of lithium into titanium disulfide lattice.[25,30] The process implies the movement of Li^+ ions, this again making advisable the use of an electrolyte with an essentially pure Li^+ transport. This condition is not realized in $(PEO)_8$ $LiClO_4$, where t_{Li^+} is about 0.3.[31,32]

Figure 11.5 shows the discharge curve of Li/PEO-$LiClO_4$/TiO_2 electrochemical cell as a function of the cycling number. Initial cell cycling was carried out between the voltage limits 3.0 and 1.2 V at a C/8 rate.* The primary discharge of anatase results in two distinct plateaus; the final composition is $Li_{1.0}TiO_2$, corresponding to a theoretical energy density of 565 W h kg^{-1}. The comparison of discharge curve of a Li/PEO/TiS_2, with that of similar cells using an Li-Al negative electrode, is shown in Figure 11.5. These cells have area of 3.8 cm^2 and theoretical capacity of 3.3 mA h.[25]

Polymer electrolytes which are based on simple polyethers are likely to have a wide range of electrochemical stability known as the stability window. Stability windows up to 4 V have been found for lithium–polymer electrolytes such as the $(PEO)_8$-$LiClO_4$, the $(PEO)_7$-$LiBF_4$, and the $(PEO)_9$-$LiCF_3SO_3$ complexes. The main application envisaged is storage batteries for EV traction. Prototypes implemented to date have operated over several hundred cycles with energy efficiencies of 80 to 90%.[25] An alternative type of a polymeric battery is one that contains an electronically conducting polymer such as polyacethylene or polypyrrole as a cathode material, and an all-polymeric battery based on the material $[Na(CH)]_x$/PEO:NaI/$[(CH)I_3]_x$ has been described by Chiang.[33] The polymer salt phases transform to resistive phases at lower

* The C-rate is a method for expressing the rate of charge or discharge of a cell or battery. A cell discharging at a C-rate of τ will deliver its nominal rated capacity in $1/\tau$ h; e.g., if the rated capacity is 2 A h, a discharge rate of C/1 corresponds to a discharge current of 2 A, a rate of C/10 to 0.2 A, etc.

temperatures, and consequently little development has been reported at normal ambient temperatures. Cells have been reported operated at 26°C with MoO_2 cathodes, from which it is projected that energy densities in the range 0.1 to 0.2 W h cm^{-3} may be achieved.[25]

Recently, Gould Co. introduced a primary lithium battery utilizing a solid electrolyte. This consists of a highly propylene carbonate (PC) plasticized film of PEO and $LiCF_3SO_3$. The cathode is based on MnO_2, while metallic lithium is used as the anode. To date the product appears to be limited in the marketplace to special applications. The first commercial available polymer electrolyte lithium rechargeable battery is being introduced by Valence Technology in San Jose (U.S.). The electrolyte consists of a radiation crosslinked polymer formed from a mixture of a liquid prepolymeric polyethylenically unsaturated compound, a radiation-inert ionically conducting liquid, and a lithium salt. The cathode is based on V_6O_{13}.

A solid-polymer "credit-card" battery has been developed by Yuasa Battery Co. in Japan.[34] This 0.1-mm-thick battery is composed of MnO_2, a solid-polymer electrolyte, and lithium, and has a volumetric energy density of 400 W h dm^{-3}. It retains 70% of its capacity over 35 cycles of charge/discharge.

2. Solid Redox Polymerized Electrode Batteries

The superior mechanical properties of elastomeric electrolytes, combined with the simplicity of fabrication of thin separator films from these materials, has led to very active pursuit of solid state cells based on solid polymeric materials. However, the development of a novel class of solid state organic positive electrodes which not only operate on an entirely new principle for energy storage, but have demonstrated performances in a number of instances exceeding those for intercalation electrodes in solid state lithium and sodium batteries. These new electrodes have been called solid redox polymerization electrodes (SRPE) and are based on polysulfide polymers $[-S(R)_nO\ (R)_nS-]_n$ where R is an organic radical such as CH_3, C_6H_5, CF_3, etc.[35-37] The redox mechanism for the positive electrode is in essence a redox dimerization/scission reaction which occurs in two steps as

$$RS^- = RS^* + e^- \tag{11.8}$$

$$RS^* + RS^- = RSSR + e^- . \tag{11.9}$$

The rate-limiting step in this mechanism is electron transfer. The standard rate constant is a strong function of R with higher rates of electron transfer being observed with the increasing electron-withdrawing nature of the organic moiety.[36] The ability of thio-groups to undergo reversible dimerization/scission can be extended to polymers where oxidation leads either to reversible inter-intramolecular crosslinking or to reversible electro-polymerization.

All–solid state alkali metal/SRPE cells were constructed by sandwiching a polymeric electrolyte between the thin-film composite SRPE cathode and a thin foil of alkali metal, i.e., Li or Na.[37] The theoretical energy densities for lithium- and sodium-based batteries using $(SRS)_n$ of the SRPEs are of 990 W h kg^{-1} (an open-circuit voltage of 3.0 V) and 750 W h kg^{-1} (an open-circuit voltage of 2.7 V), respectively. The equivalent weight of this $(SRS)_n$ electrode is 74 g Li/PEO/$(SRS)_n$. Batteries are operating at 100°C. Such a cell, subjected to extended cycling tests at a discharge current density of 0.125 mA cm^{-2}, corresponding to a rate of C/4.8, was operating at a steady-state energy density of approximately 260 W h kg^{-1} and a steady-state power output of 160 W kg^{-1}. A prototype battery has been discharged at a high current density rate as shown in Figure 11.5c.

3. Alkali Glass Batteries

On the application level, two types of batteries using vitreous electrolytes have recently reached the predevelopment stage. The first type is the sodium sulfur battery, operating at

300°C in which the sodium and sulfur are in liquid form separated by 8-μm-thick hollow glass fibers. Each cell has a voltage of around 2 V and the vitreous electrolyte allows current densities of up to 2 to 4 mA cm^{-2}. Cells having capacity of 5 A h incorporating 8000 fibers have been built and tested over several operating years. The vitreous electrolyte in this case is sodium silicoborate containing a small amount of dissolved sodium chloride. The prototypes are developed in the U.S. by Dow Chemical for use in fixed electricity storage facilities.[23]

The second type of a glassy battery operating at ambient temperature is the lithium battery, involving a lithium metal anode and a titanium disulfide (TiS_2) cathode, with a vitreous solid state electrolyte. Such batteries have been developed first by SAFT in France,[38] and then by Union Carbide in the U.S.[39] They are commercialized by the two companies.

Lithium glasses appear to be promising alternatives to polymer electrolytes in lithium intercalation cathode systems,[40] although the lack of deformability caused by their hardness can be a major problem.[41] A battery based on a mixture with the composition $2.5LiI+Li_4P_2S_7$ glass solid electrolyte having an ionic conductivity of 2×10^{-3} S cm^{-1} at 25°C has been developed by Eveready Battery Co.[39] Such a battery has been designed for operating in a wide range of temperature from –55 to 125°C, which is a condition that traditional aqueous or nonaqueous systems, even polymeric batteries, cannot satisfy. The purpose of this battery is for CMOS memory backup for which energy density is not important. Questions of cell stability over years of in use service must be answered. A solid state cell having a Li/5LiI-$Li_4P_2S_7/TiS_2$-SE configuration has an open circuit voltage of 2.5 V and furnishes a specific energy density of 150 to 200 W h kg^{-1}. Such a battery, which is designed with 50 mA h capacity delivers continuous current densities up to 0.1 mA cm^{-2} and allows pulsed current densities up to 10 mA cm^{-2}.When the temperature operating range is extended to 200°C, LiI is replaced by LiBr in the solid electrolytes. These cells are commercially available and are termed XR2025HT by the Eveready Battery Co. in the U.S. Performance of 40 mA h cells is highly satisfactory. Voltages of around 2 V are obtained.

C. SOLID STATE PRIMARY LITHIUM BATTERIES

Commercially available solid electrolyte batteries use a lithium anode which is attractive because this metal is strongly electropositive. They generally exhibit high thermal stability, low rates of self-discharge (shelf life of 5 to 10 years or better), the ability to operate over a wide range of environmental conditions (temperature, pressure, and acceleration), and high energy densities of 0.3 to 0.7 W h cm^{-3}. However, limitations associated with a complete solid state battery include relatively low power capacity due to the high impedance of most lithium-conducting solid electrolytes. The three commercial solid electrolyte battery systems are based on the solid electrolyte LiI, either formed *in situ* during cell manufacture or dispersed with alumina. They include the solid state chalcogenide glass battery which was described above.

1. Lithium–Iodine Cells

The lithium–iodine battery has been used to power more than 3.5 million cardiac pacemakers since its introduction in 1972. During this time the lithium–iodine system has established a record of reliability and performance unsurpassed by any other electrochemical power source.

The lithium–iodine battery has a solid anode of lithium and a polyphase cathode of poly-2-vinyl-pyridine (P2VP), which is largely iodine (at 90% by weight). The solid electrolyte is constituted by a thin film of LiI. The discharge reaction is given by

$$2Li + P2VP \cdot nI_2 \rightarrow P2VP \cdot (n-1)I_2 + 2LiI. \tag{11.10}$$

During the electrochemical reaction a thin membrane of LiI is formed and grown as the discharge proceeds. This cell has an open-circuit voltage of 2.8 V. The theoretical specific energy for the $Li/LiI/I_2$ (P2VP) cell is 1.9 W h cm^{-3}. The electrolyte ionic conductivity is 6.5×10^{-7} S cm^{-1} at 25°C, and the energy density is 100 to 200 W h kg^{-1}.[42] As the battery discharges, more lithium iodide forms and the ohmic resistance of the cell rises exponentially with discharge capacity. The behavior of the cathode is somewhat complex. As a cell discharges, the net reaction removes crystalline iodine from the cathode, causing a decrease in the resistance. This cathode resistance continues to decrease until all the crystalline iodine is consumed. If discharge proceeds beyond this composition into the single-phase region of the phase diagram, the cathode resistance increases rapidly and soon dominates the cell resistance occurs at a residual cathode weight ratio of about 8.[43]

The volume change accompanying the cell discharge is –12% if the cathode is 91% iodide by weight. This volume change may be accommodated by the formation of a porous discharge product or by the formation of macroscopic voids in the cell. Such batteries are used as power sources for implantable cardiac pacemakers, operating at 37°C. They are commercialized by Catalyst Research Co., by Wilson Greatbatch Inc., and Medtronic Inc. in the U.S. The lithium-iodine batteries have extended system lives up to 10 years for 120- to 250-mA h capacities. Power sources for portable monitoring or recording instruments have a nominal capacity of 15 A h or less, and most have deliverable capacities under 5 A h.

Batteries of medium capacities, i.e., up to around 1 A h, can be used for random-access memory power supplies in electronics. Similar batteries using Li/Br have also been built. The greater electronegativity of bromine gives rise to voltages of the order of 3.5 V and energy densities as high as 1.25 W h cm^{-3}. Their practical application is, however, limited by the low conductivity of the LiBr films formed.

2. $Li/LiI\text{-}Al_2O_3/PbI_2$ Cells

These batteries are recommended for low-rate operations and they are particularly suited for applications requiring long life under low drain or open-circuit conditions. Different cathodes have been used in these commercial solid state cells. A mixture of PbI_2+Pb or PbI_2+PbS+Pb has been used, and a new system under development utilizes a mixture of TiS_2 + S or As_2S_3, which increases the energy density. The solid electrolyte is a dispersion of LiI and LiOH with alumina. Lithium ionic conductivities as high as 10^{-4} S cm^{-1} have been reported in such a dispersion at temperature of 25°C.[44]

The discharge properties of these solid state batteries are characterized by an open-circuit voltage of 1.9 V and an energy density of 75 to 150 W h kg^{-1}. A three-cell battery design delivers 6 V and offers a capacity of 140 mA h for pacemaker power sources. This system is manufactured by Duracell International. In typical CMOS memory applications the 350-mA h cell can be used to 1-V cutoff.

3. Li/LiI (SiO_2, H_2O)/Me_4NI_5 + C Cells

This cell was evaluated for use in cardiac pulse generators and exhibited a voltage of 2.75 V and was projected to have an energy density of 0.4 W h cm^{-3}.[45] The cathode is a mixture of carbon and tetramethyl-ammonium penta-iodide (Me_4NI_5).

4. Lithium Bromine Trifluoride Battery

Great advances have been made over the past decade based on the unconventional approach of combining alkali metals with strongly oxidizing liquid, e.g., SO_2, $SOCl_2$, or BrF_3, which acts simultaneously as electrolyte solvent and cathode depolarizer. BrF_3 is a very reactive liquid at room temperature, and the concept of a Li/BrF_3 cell has appeared.[45] The cell reaction has been represented by

$$3Li + BrF_3 \rightarrow 3LiF + 1/2Br_2 . \quad (11.11)$$

Thus no electrolyte salt is necessary because lithium was found to be stable in BrF_3 due to the formation of the protective surface layer.

Upon cell activation a potential of 5 V is established which increases them to over 5.1 V, giving theoretical energy densities of 2680 W h kg^{-1} and 4480 W h dm^{-3}. A typical discharge of a Li/BrF_3/C cell is achieved at 5 mA cm^{-2} current density and capacity of 5 mA h cm^{-2} has been measured. The cell performance is controlled by the buildup of the reaction product layer at the anode–electrolyte interface, leading to an increase in impedance. The BrF_3 electrolyte can also be modified by dissolution of various fluorides, e.g., $LiAsF_6$, $LiPF_6$, $LiSbF_6$ or $LiBF_4$. Such a cell is actually strongly developed by EIC Laboratories Inc. in the U.S.

D. SOLID STATE SECONDARY LITHIUM BATTERIES

A question can be raised at this point as to the usefulness of a rechargeable solid state secondary cell with such a low mA h output. Matsushita Electric Industrial Co. (Japan) has developed a 20-mm-diameter secondary cell of the type $LiSn_xBi_y$/$LiClO_4$ in PC/C (activated carbon) which has a 1-mA h rating.[46] More recently, Bridgestone and Seiko have jointly developed a Li/polyaniline (AL2016) secondary cell with a capacity of 3 mA h.

At Eveready Battery Co. (U.S.) a Li/chalcogenide glass/TiS_2 carbon is under active development.[47] The target of the R&D is the CMOS memory backup market. This cell is based on phosphorous chalcogenide glasses $Li_8P_4O_{0.25}S_{13.75}$ mixed with LiI and on a solid-solution composite electrode TiS_2–solid electrolyte-black carbon with percentages by weight of 51:42:7. The cathode capacity ranges from 1.0 to 9.5 mA h. The cell packaging is a standard-sized XR2016 coin cell. The resistance of the cell at 21°C is between 25 and 100 Ω, depending upon where the cell is in the charge/discharge cycle. More than 200 cycles have been obtained in experimental cells. It seems that the phosphorous chalcogenide-based electrolytes have additional advantages beyond their plasticity and ease of fabrication and use. The use of vitreous electrolyte network formers SiS_2, B_2S_3, GeS_2, etc., is suspect from the viewpoint of interface stability.[47]

E. SECONDARY INSERTION CATHODE LITHIUM BATTERIES

Of particular interest for secondary-battery electrodes is the case of an electron donor system. The first experimental investigation of this concept includes the electrochemical reaction which takes place in a Li/Li_xTiS_2 cell in the composition range $0 \le x \le 1$ (see Figure 11.6).

The redox reactions may occur on the host cation, the host anion, or the guest species. For example, insertion of lithium into the oxospinel $Li[Mn_2]\ O_4$ reduces the host cation to give $Li_{1+x}[Mn^{4+}_{1-x}\ Mn^{3+}_{1+x}]\ O_4$, whereas insertion into the thiospinel $Cu\ [Cr_2]\ S_4$ reduces the host anion to give Li_x+Cu^+[$Cr_2$3+] S_4(7–x)–. In the case of the chevrel phase $Li^+(Mo_6S_8)^-$, further insertion of lithium results in the formation of lithium trimers, thus reducing the guest species:[48]

$$2xLi^+ + 2xe^- + Li^+(Mo_6S_8)^- \Leftrightarrow x(Li_3)^+(Mo_6S_8)^- + (1-x)Li^+(Mo_6S_8)^- . \quad (11.12)$$

Furthermore, the topotactic insertion/extraction reactions may occur by diffusion in one-dimensional channels as in the hexagonal tungsten bronze, in two-dimensional as in layered transition metal dichalcogenides, or in three-dimensional as in the close-packed spinel structure. There are a large number of candidate materials which can be used in lithium insertion batteries. As shown in Figure 11.7, one can classified these materials as a function of the delivered cell voltage vs. Li. The most attractive family comes from the lithiated transition metal oxide which are now used in lithium metal-free batteries.

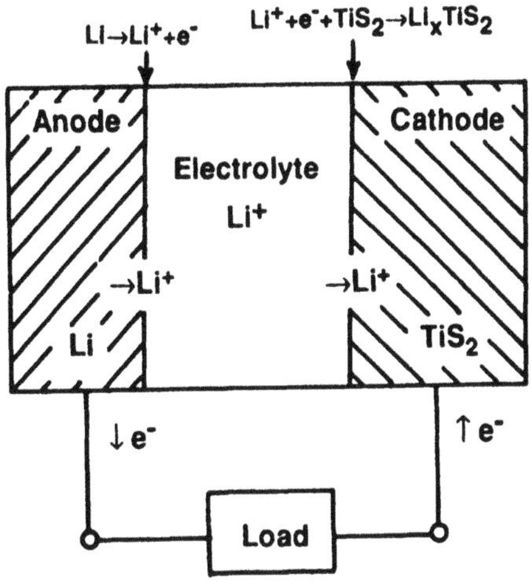

FIGURE 11.6. Schematic diagram of a solid state electrolyte battery providing power to an external circuit. The electrochemical redox process is that for an intercalation material, Li_xTiS_2. (From Julien, C. and Nazri, G.A., *Solid State Batteries: Materials Design and Optimization*, Kluwer, Boston, 1994. With permission.)

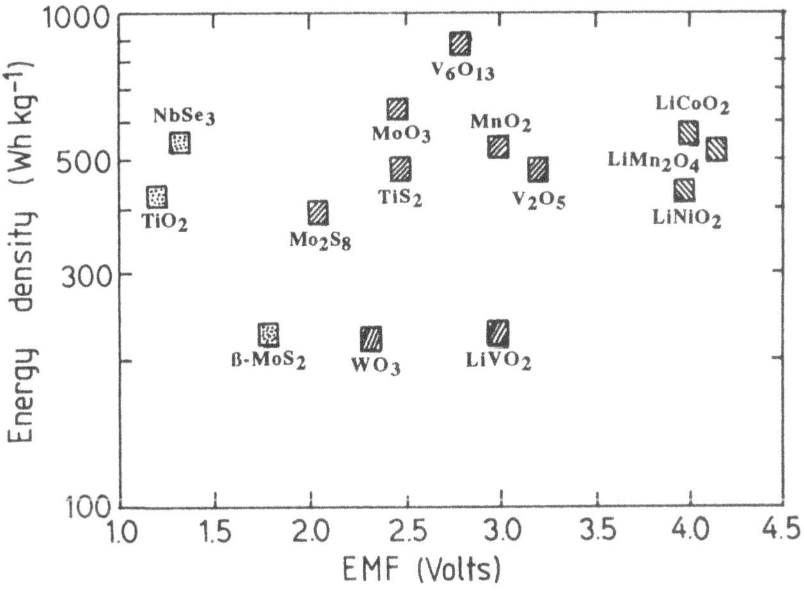

FIGURE 11.7. Energy density vs. cell voltage of various Li/intercalation compounds. (From Julien, C. and Nazri, G.A., *Solid State Batteries: Materials Design and Optimization*, Kluwer, Boston, 1994. With permission.)

The use of the light, highly electropositive alkali metals, such as lithium, for the negative electrode and an ion insertion compound for the positive electrode gives attractive possibilities for high-energy density cells. The electrolyte generally consists of a solution of a lithium salt, e.g., $LiClO_4$, LiCl, or $LiBF_4$ in an organic solvent such as dimethyoxyethane (DME), PC, or dimethyl-tetrahydrofuran (THF).

Lithium organic liquid electrolyte batteries have recently gained considerable importance as high-energy density power sources for a variety of terrestrial and space applications. Commercially available organic liquid electrolyte primary batteries include Li/SO_2, Li/V_2O_5, $Li/(CF)_x$, Li/MnO_2, Li/FeS_2, Li/CuS, and Li/CuO. Prototypes of organic liquid electrolyte secondary batteries such as the Li/TiS_2 and Li/MoS_2 systems, depending upon cell sizes, can

TABLE 11.6
Characteristics of the Major Types of Ambient-Temperature Lithium-Insertion Cathode Batteries

System	Cell voltage (V)	Energy density (W h kg^{-1})
$Li/LiBF_4$-PC/CF_x	2.8	320
$Li/LiClO_4$-PC/Ag_2CrO_4	3.5	200
$Li/LiClO_4$-PC/TiS_2	2.5	140
Li/organic-elect./MoS_2	2.3	65
Li/PC-based/$NbSe_3$	1.9	245

From Julien, C. and Nazri, G.A., *Solid State Batteries: Materials Design and Optimization*, Kluwer, Boston, 1994. With permission.

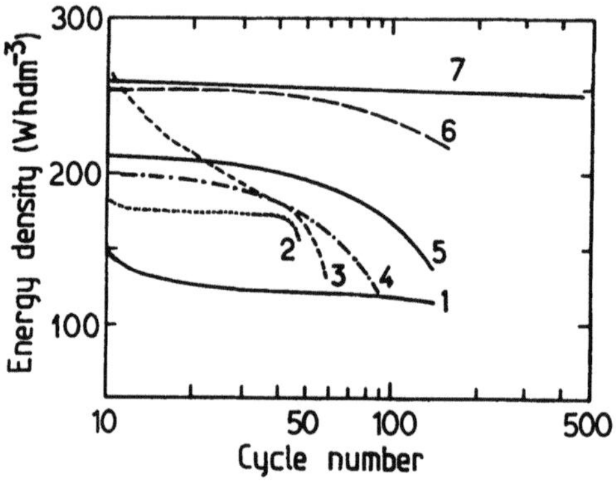

FIGURE 11.8. Cycle life vs. energy density of some secondary lithium batteries: (1) Li/MoS_2 (C-size Molicel), (2) Li/V_2O_5 (SAFT), (3) Li/MoS_2 (AA-size Molicel), (4) Li/TiS_2 (Eveready), Li/Mo_6S_8 (Molicel), (6) $Li/NbSe_3$ (AT&T), and (7) $LiCoO_2/Li_xC_6$ (Sony Energytec). (From Julien, C. and Nazri, G.A., *Solid State Batteries: Materials Design and Optimization*, Kluwer, Boston, 1994. With permission.)

deliver specific energies of 90 to 140 W h kg^{-1} and volumetric energy densities of 0.16 to 0.2 W cm^{-3}. A cell utilizing the THF-$LiAsF_6$ (1.5 M) solution ran more than 225 cycles when cycle at a depth of discharge of 60%. This cell, having a capacity of 5 A h, had a current density of 0.5 mA cm^{-2}.[49] The characteristics of the major types of ambient-temperature lithium insertion cathode batteries are summarized in Table 11.6.

Rechargeable batteries of various types have now been developed around the world. They are considered in the following sections. The volumetric energy density in W h dm^{-3} vs. cycle life of some advanced commercial battery systems is shown in Figure 11.8.

1. Li/TiS_2 Battery

Most recent effort have been devoted to the development of small cells or batteries using Li/Li^+-LOE/TiS_2 geometry where Li^+-LOE is a lithium liquid organic electrolyte.

A spirally wound AA-size Li/TiS_2 cell has been constructed by Grace Co. (U.S.). At 200-mA discharge rate, 1 A h is delivered to 1.7 V.[49] In C-size, a Li/TiS_2 cell built by EIC Laboratories Inc. (U.S.), a capacity of 1.6 A h is obtained. This cell operates in the temperature range from –20 to +20°C.[50] Other systems are constructed using solid vitreous or polymeric electrolytes, as it has been shown above. A miniature cell was manufactured by Exxon in two sizes with capacities of 25 and 90 mA h.

2. Li/MoS_2 Battery

As the first commercially available rechargeable lithium power source system, the Li/MoS_2-type battery (MOLICEL™) is manufactured by Moli Energy Ltd. in Canada. This

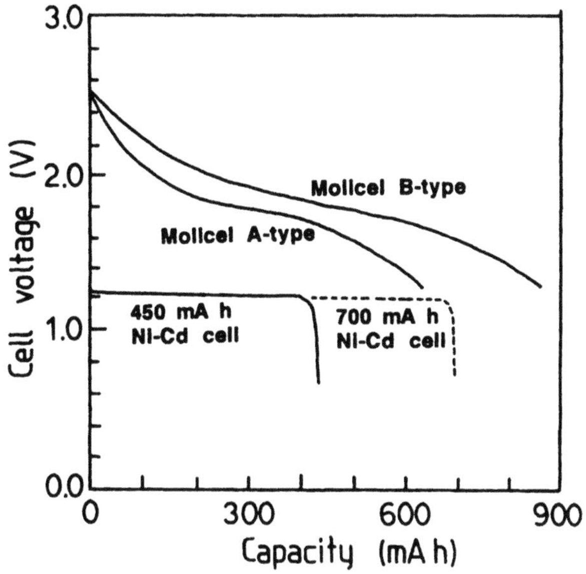

FIGURE 11.9. Electrochemical characteristics of Molicel employing MoS_2 electrode (A-type), MnO_2 electrode (B-type) compared to Ni-Cd cells. (From Julien, C. and Nazri, G.A., *Solid State Batteries: Materials Design and Optimization*, Kluwer, Boston, 1994. With permission.)

cell delivers performance and operating features far superior to conventional secondary batteries (Figure 11.9). This battery uses a lithium anode, a PC-based electrolyte solution, and a processed MoS_2 cathode.[47] The discharge reaction is

$$xLi^+ + xe^- + MoS_2 \Leftrightarrow Li_xMoS_2. \tag{11.13}$$

The electrolyte is a nonaqueous solution which permits high electrolyte-to-electrode ion transfer. Sustained drain rates of several amperes at a cell voltage between 2.3 and 1.3 V can be obtained. The energy density is in the 60- to 65-W h kg^{-1} range at a discharge rate of C/3 (approximately 800 mA). The total capacity of a C-size cell is 3.7 A h and the number of realizable cycles is dependent upon the charge and discharge conditions. Within a wide range of conditions a cell can be expected to deliver in excess of 150 complete cycles.[52] Li/MoS_2 batteries are especially manufactured as the power source for pocket telephones in Japan.

3. Li-$NbSe_3$ Battery

A more energetic lithium rechargeable system under study and development at AT&T is the Li-$NbSe_3$ system, termed FARADAY cell.[53] The AA-size cylindrical cell is designed for operating over 200 cycles at a typical current of 400 mA to a cutoff capacity of 0.7 A h. The cell reaction may be written

$$3Li^+ + 3e^- + NbSe_3 \Leftrightarrow Li_3NbSe_3. \tag{11.14}$$

The ability to incorporate three lithium gives a relatively high theoretical energy density of 1600 W h dm^{-3} for the Li-$NbSe_3$ couple. A practical energy density of 200 W h dm^{-3} is achieved with the possibility of 350 cycles of charge/discharge.

4. Li-V_2O_5 Battery

A coin-type lithium-V_2O_5 battery to provide all the power necessary for single-cell memory backup of microprocessor-based electronic equipment has been developed by

TABLE 11.7
State-of-the-Art for Lithium Rechargeable Batteries (Cylindrical Cells) Developed or Commercialized in the World

Country	Company	Performances (W h dm^{-3})	Cycle	Positive electrode	Market[a]
France	SAFT	175	50	V_2O_5	M
Japan	Sony	220	200	MnO_2	C
	Sanyo	—	—	MnO_2	C
	Matsushita	220	40	V_2O_5	C
				MnO_2	R&D
Canada	Moli Energy	60	100	MoS_2	C
		100	200	MnO_2	C
		100	150	Mo_6S_8	C
U.S.	Eveready	180	100	TiS_2	C+M
	Honeywell	175	50	V_2O_5	M
	AT&T	200	350	$NbSe_3$	S
	Grace	230	100	TiS_2	—
	Valence Tech.	—	200	V_6O_{13}	C+M

[a] M = military application, C = consumer, S = space.

From Julien, C. and Nazri, G.A., *Solid State Batteries: Materials Design and Optimization,* Kluwer, Boston, 1994. With permission.

Matsushita Micro Battery Co. in Japan. The cathode is constructed by a mixture of vanadium pentoxide and 5% carbon black. During the discharge a depth of x = 3 is obtained with an average voltage of about 2 V. The battery has a relatively flat discharge curve at an output voltage of 3 V and an electrical capacity of approximately 36 mA h at a discharge rate of 1 mA.[54] It also proved able to withstand up to 50 deep charge/discharge cycles and up to 1000 shallow charge/discharge cycles for which the depth of discharge set around 10%. A C-size Li/V_2O_5 cell is also constructed by SAFT in France (see Table 11.7). This cell delivers a capacity of 1.4 A h.[55]

5. Li/MnO_2 Battery

Many organizations (Moli Energy Ltd., Sony Energytec, Sanyo, Fuji, NTT, Matsushita, Varta, SAFT, etc.) embarked on ambitious programs to develop small rechargeable Li cells including a manganese dioxide cathode as a stepping stone to the ultimate goal of a technology scalable to EV-sized batteries.

MnO_2 is currently being investigated as a positive material for high-energy density, low-cost lithium secondary batteries. Although initial stages of the investigation showed that MnO_2 had limited rechargeability, various improvements have been recently made.[56-59] It was found that a composition of lithium containing manganese dioxide (CDMO) exhibits superior rechargeability, and this led to the development of a high-energy density flat-type lithium secondary battery. This type of cell is manufactured by Sanyo in Japan. CDMO is prepared from Li salts and MnO_2 by a heat-treatment method, and the material structure consists of Li_2MnO_3 and λ-β-MnO_2 "composite dimensional manganese oxide". A recent report has shown that Li_2MnO_3 is an electrochemically inactive material and the $Li_xMn_{2-y}O_4$ ($0 \leq x \leq 1.33$) is a electrochemically spinel phase. The end members of this system are the stoichiometric spinel compound $Li_4Mn_5O_{12}$ and the defect spinel λ-MnO_2.

Rechargeable Li-MnO_2 AA-size cells are manufactured by Moli Energy Ltd. under the name MOLICEL™ (see Figure 11.9). This cell uses an electrolyte consisting of 1M-$LiAsF_6$ in PC/ethylene carbonate (1:1). A typical cycle life between 3.4 and 2.4 V has been achieved with 180 and 60 mA discharge and recharge currents, respectively. Minor capacity fluctuations throughout the cycle life have been detected at an operating temperature of 24°C.

AA-size cells constructed by Battery Technologies Inc. in the U.S. show very good endurance tests. These cells accumulate up to 400 cycles on the 24-Ω test and up to 250 cycles on the 10-Ω test.[60] The Li-MnO_2 cells developed by Sanyo Electric Co. in Japan are termed ML 2430 and ML 2016 with nominal capacity of 70 and 20 mA h, respectively. About 200 cycles are obtained at a capacity of 45 mA h. The flat-type battery ML 2430 with a nominal voltage of 3 V can operate over 3000 cycles at a low discharge capacity of 1 mA h. European efforts are also directed toward Li-MnO_2 secondary batteries.[61] Researches are developed by SAFT in France and Varta in Germany (see Table 11.7).

6. Other Items

A new AA-size Li/Mo_6S_8 rechargeable cell has been developed to the prototype stage by Moli Energy Ltd.[51] This cell can deliver about 2 W h of energy and is capable of a high-rate discharge up to about the 2C rate. This AA-size cell delivers 100 W h kg^{-1} with a typical life of 200 cycles.

The material LiI-Al_2O_3 is an example of a polyphase solid electrolyte.[60] The intimate mixing of small particles of an ion-conducting phase with a nonconducting phase has been found to enhance the total conductivity. Although explanations for the mechanisms of lithium ion transport in these polyphase electrolytes remain controversial, there is general agreement that the mechanism is associated with the high specific surface area of the interparticle contact region. It should be noted that the LiI-Al_2O_3 polyphase electrolyte attains a specific conductivity of 10^{-3} S cm^{-1} around 100°C, and at this temperature is sufficiently plastic to ensure good interfacial contact with solid insertion electrode materials. This material has been incorporated in the solid state cell Li-Si/LiI (Al_2O_3)/TiS_2, Sb_2S_3, Bi with operated at 300°C. Preliminary design studies indicated that practical energy densities of 200 and 500 W h kg^{-1} could be realized with this cell, but only limited cycling data have been reported, and the high temperature of operation appears to be a disadvantage.

From a specific energy point of view, all–solid state batteries with insertion cathodes have the advantage that overall process only includes a minimum of components, but severe limitations were met in the design of batteries. In order to optimize the interfacial area and hence reduce its resistance, a composite electrode is commonly used in which the electronic and ionic conductors are mixed as powders and subjected to heat or pressure treatment in order to form the electrode.[61]

F. LIQUID ELECTROLYTE PRIMARY LITHIUM BATTERIES

Four main groups of systems may be distinguished: polycarbon fluorides, sulfides, oxosalts, and oxides. Oxosalt batteries are in a range of commercial power sources based on silver chromate, Ag_2CrO_4, with the discharge process

$$2Li + Ag_2CrO_4 \rightarrow 2Ag + Li_2CrO_4, \tag{11.15}$$

The nominal voltage is 3.5 V and specific energy of 200 W h kg^{-1} or 575 W h dm^{-3} to a 2.5-V cutoff is estimated. Button cells are produced by SAFT. Sulfide electrodes have the advantage over the corresponding oxides that most of them are good electronic conductors, and hence sulfide-based batteries do not usually require the addition of carbon into the cathode. Batteries based on cupric sulfide cells (three in series) have been developed for use in a cardiac pacemaker. Reduction of CuS takes place in two steps,

$$2CuS + 2Li \rightarrow Cu_2S + Li_2S, \tag{11.16}$$

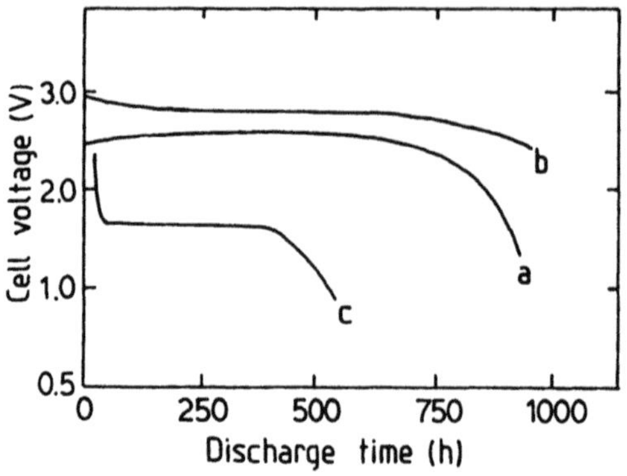

FIGURE 11.10. Discharge curves for lithium-organic batteries: (a) $Li/(CF_x)_n$ button cell under load of 13 kΩ, (b) Li/MnO_2 button cell (Varta CR 2025) under load of 15 kΩ, and (c) Li/CuO button cell (SAFT LC 01) at 75-kΩ load. (From Julien, C. and Nazri, G.A., *Solid State Batteries: Materials Design and Optimization,* Kluwer, Boston, 1994. With permission.)

$$Cu_2S + 2Li \rightarrow 2Cu + Li_2S, \quad (11.17)$$

so that the discharge curve exhibits two plateaus at 2.12 and 1.75 V.

Figure 11.10 shows discharge curve of various lithium-liquid electrolyte primary batteries: $Li/(CF_x)_n$, Li/MnO_2, and Li/CuO. These discharge curves have been obtained under load of 13-75 kΩ.

1. Lithium–Polycarbon Fluorides Cell

Polycarbon fluorides of the general formula $(CF_x)_n$ can be obtained by direct fluorination of carbon black or other varieties of carbon at high temperatures. A specific energy density of 2600 W h kg^{-1} can be achieved with these materials. Lithium cells with polycarbon fluoride cathodes have an open-circuit voltage in the range of 2.8 to 3.3 V, depending on the composition of the cathode material. A typical cell reaction may be written

$$nxLi + (CF_x)_n \rightarrow nC + nxLiF, \quad (11.18)$$

Cells based on polycarbon fluorides are manufactured commercially in various forms. This system is developed by Matsushita Electric Industrial Co. and is designed as a BR 435 cylindrical cell. New advanced cells constructed by Nippon Steel Co. use carbon fibers as electrodes and are found rechargeable batteries. Cells for military applications have been produced in the U.S. by Eagle Picher and by Yardney Electric. There are spiral-wound cylindrical cells with a largest cell capacity of 5 A h.

2. Lithium Oxide-Compounds Cell

In many cases the discharge mechanisms involved in lithium oxide cells are still not fully understood. The discharge reaction can be described as a formal displacement process,

$$2Li + MO \rightarrow Li_2O + M, \quad (11.19)$$

where MO is an oxide material such as CuO, MnO_2, Bi_2O_3 or Pb_3O_4. The theoretical capacities of various oxides are listed in Table 11.8. Li/MnO_2 primary cells are manufactured by Sanyo as button cells. The open-circuit voltage is in the range 3.0 to 3.5 V and a practical energy density of 500 W h dm^{-3} is obtained. Varta has developed a CR 2025-type cell and SAFT has manufactured a LM 2020-type cell.

TABLE 11.8
Product and Theoretical Capacity of Oxide Materials where Displacement Reactions are Involved

Active material	Product	Capacity (Ah kg^{-1})
CuO	Cu	670
MnO_2	Mn_2O_3	310
Bi_2O_3	Bi	350
Pb_3O_4	Pb	310

From Julien, C. and Nazri, G.A., *Solid State Batteries: Materials Design and Optimization*, Kluwer, Boston, 1994. With permission.

The Li/CuO button cell (LC01) constructed by SAFT exhibits a single step which may be attributed to the simple displacement reaction:

$$2Li + CuO \rightarrow Li_2O + Cu. \tag{11.20}$$

This Li/CuO cell has an open-circuit voltage of 1.5 V and has the highest specific energy of all solid cathode lithium-based cells. Practical value of 750 W h dm^{-3} is obtained. The liquid electrolyte varies from manufacturer to manufacturer, but $LiClO_4$ in dioxolane is very often used. The cylindrical cells manufactured by SAFT have practical capacities in the range 0.5 to 3.9 A h.

G. SILVER AND COPPER BATTERIES

The development of silver or copper anode batteries as early as the 1960s was due to the existence of good silver or copper cation-conducting electrolytes and the relatively simple manufacturing processes involved.

1. Silver Cells

Crystalline solid electrolyte cells of Ag^+ ions were early reported by Takahashi and Yamamoto,[62] Foley,[63] and Owens[64] in complete detail. Following the initial discovery of Ag_4RbI_5 solid electrolyte by Owens in 1970, a range of other related materials were developed.[65] This material exhibits an unusually high ionic conductivity of 0.26 S cm^{-1} at 25°C. This permits cell discharge at much higher current drains than those available with LiI-based cells. The best-known example is the battery operating according to the electrochemical chain $Ag/RbAg_4I_5/RbI_3$, giving an open-circuit voltage of 0.66 V at 25°C and a practical energy density of 4.4 W h kg^{-1}. The selection of RbI_3 as the cathode material rather than iodine is due to the parasitic reaction possible between the iodine and the electrolyte, leading to the formation of a poorly conducting film.

Silver cell using fast ion-conducting glasses in the systems $AgI\text{-}Ag_2Ch\text{-}P_2Ch$, where Ch is O, S, or Se, have been characterized.[66] The glass with composition $7AgI\text{-}1Ag_2Se\text{-}2P_2Se_5$ exhibits an ionic conductivity of 10^{-2} S cm^{-1} at room temperature. Value of open-circuit voltage of the cell $Ag/7AgI\text{-}1Ag_2Se\text{-}2P_2Se_5/I_2+C$ implies that the cell reaction is essentially the same as that of $Ag/AgI/I_2$.[62] Both anode and cathode materials were composite, i.e., a mixture of active material and solid electrolyte in powder form. A current density of 1 mA cm^{-2} can be used without serious polarization. The reason for the voltage decrease at higher current densities is attributed to cathodic and anodic polarization; improving the electrode assembly could serve the higher current densities in such a cell.[67]

Another type of solid state cell utilizing Ag^+ ion-conducting glasses has been reported by Minami.[68] This cell is constructed with a NbS_2 intercalation cathode in the pellet form, and a flash-evaporated film of the glass AgI-Ag_2O-B_2O_3 which has a ionic conductivity of 10^{-2} S cm^{-1} at room temperature. The open-circuit voltage ranged from 0.35 to 0.39 V. The discharge–charge cycles at current density of 3.3 μA cm^{-2} were repeated several times, and no degradation was observed under such conditions. This type of cell is a secondary battery.[68]

Organo-mineral compounds such as $(CH_3)_4NI_9$ or $(CH_3)_4NI_5$ can be used to increase the energy density slightly. Furthermore, the ionic conductivity of the reaction products $((CH_3)_4N)_2Ag_{13}I_{15}$ is high, offering good charge transfer continuity.[69] The Ag/I_2 batteries were inherently deficient in two significant areas, voltage and size; consequently no commercial application resulted.

2. Copper Cells

It is possible to replace silver by copper in many of the silver electrolytes. Cells based on a particular CuI/organic iodide electrolyte have been described.[70] Those involving $Cu_{16}Rb_4I_7Cl_{13}$ have been recently reviewed.[71]

The first report for the cells with copper ion conductors was presented by Takahashi and Yamamoto.[72] They examined four types of solid state cells with the following electrochemical chains: Cu+X/X/$CuBr_2$+X+Graphite, and Cu+X/X/Ch, Cu_2S+X+Graphite, where X is 7CuBr-$C_6H_{12}N_4CH_3Br$ and Ch is a chalcogen (S, Se. or Te). The open circuit voltages of these cells were 0.8, 0.448, 0.373, and 0.258 V, respectively, at 25°C. The value in the initial open-circuit voltage of the second cell is in good agreement from the free-energy variation of the reaction

$$Cu + CuBr_2 \rightarrow 2CuBr. \tag{11.21}$$

However, the cell performance is time dependent, the open-circuit voltage decreased gradually, and the cell resistance increased. The anode capacity of these cells was 17 mA h, and the cathode capacities 340, 130, and 84 mA h, respectively. The rapid decrease in the potential of cell including sulfur discharged at a rate of 54 μA cm^{-2} is due to the high resistance of sulfur. The best performance was obtained with cell including selenium; a capacity of 26 mA h was found to the cut-off voltage of 0.26 V, corresponding to a cathode efficiency of 18%.

Recently, copper solid electrolyte cells have been designed using intercalation cathodes. The first approach for the rechargeable cells with a copper ion conductor has been done by Lazzari and Razzini[73] and by Scrosati and Voso.[74] The cells consisted of a high copper ion-conducting solid, 0.94CuBr-0.06$C_6H_{12}N_4CH_3Br$, and an intercalation cathode $Cu_{0.4}TiS_2$, and an intercalation anode $Cu_{0.8}TiS_2$. In this cell, the kinetic limitations noticed with copper electrodes should be avoided by the use of intercalated electrodes on both sides with resulting improved cycling capabilities. Discharges at a rate of 20 μA cm^{-2} have been carried out over 135 cycles without any appreciable deterioration.

Kanno et al.[75] have studied a series of cells using intercalation metal chalcogenide material, MX_2, as cathode and Chevrel phase as anode, of the type $Cu_4Mo_6S_8$/$Rb_4Cu_{16}I_{6.8}Cl_{13.2}$/$MX_2$. Cathodes were composite, the anode was a mixture of equal amounts of $Cu_4Mo_6S_8$ and electrolyte, and the cathode was fabricated in the proportion 2:3 of MX_2 compound and electrolyte. The discharge capacity of NbS_2 is remarkable. The cathode utilization is over 30% in e/NbS_2 to the cut-off voltage of 0.3 V. According to the electron diffraction study, Cu_xNbS_2 ($0 < x < 0.9$) forms a continuous nonstoichiometric compound over the whole composition range, showing ordered and disordered arrangements in copper metal, depending on composition and temperature.[76] So, the formation of a wide-range solid solution is advantageous to the efficiency of a secondary battery.

The electrochemical chain Cu/$Rb_4Cu_{16}I_{6.8}Cl_{13.2}$/$TiS_2$ constitutes a cell with a capacity of 7.8 mA h using 0.2 g TiS_2 an was discharged at a rate of 75 μA cm^{-2}. During the discharge at 25°C the cell voltage is 0.59 V and a current of 10 μA allows 100 cycles.[77] The copper cell manufactured by Matsushita in Japan has a life of up to 300 cycles. Japan Synthetic Rubber Co. received attention for a solid state secondary cell based on $RbCu_4Cl_3I_2$.[78]

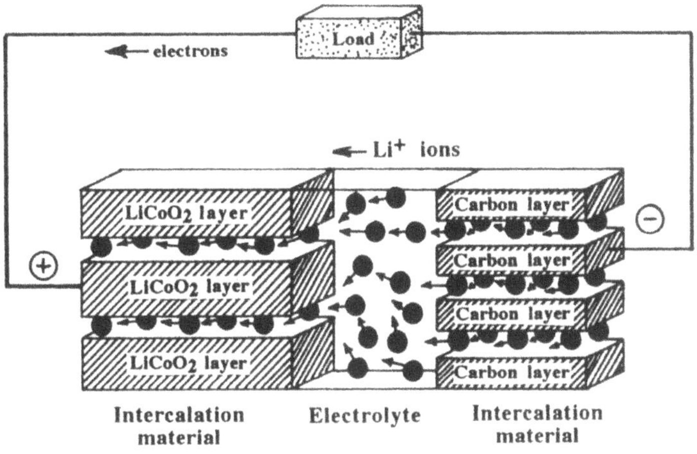

FIGURE 11.11. Schematic representation of a metal-free lithium battery (or rocking-chair battery) during the discharge. The Li^+ ions rock between the two intercalation materials: $LiCoO_2$ and Li_xC_6 electrodes. The immediate advantages expected for such a battery are the high energy density and the safety behavior. (From Julien, C. and Nazri, G.A., *Solid State Batteries: Materials Design and Optimization,* Kluwer, Boston, 1994. With permission.)

III. LITHIUM METAL-FREE RECHARGEABLE BATTERIES

A. PRINCIPLE

During the past 15 years nonaqueous lithium cells have been developed especially in the 3-V primary systems. The practical lithium cells, such as Li/MnO_2, $Li/(CF)_n$, and so forth, have been expanding their use in electronic devices.

It is now recognized that ambient-temperature nonaqueous rechargeable lithium cells using lithium metal as the negative electrode exhibit problems which hinder their wide utilization in the consumer market. These problems arise mainly from three factors:

(i) The short cycle life for deeply discharged cells; the dendritic formation during charging creates dead-lithium which reduces considerably the battery cycle life.[79] The dead-Li, which is electrochemically inactive, is reported to be chemically active;[80] the freshly deposited lithium dendrites react with the electrolyte and become isolated from the bulk lithium metal, resulting in lower coulombic efficiencies of lithium electrodes.[79] The main result is that Li batteries generally have a lower cycle life than Ni-Cd cells. Even in commercial Li/MoS_2 there is a three- to fourfold stoichiometric excess of Li compared to that needed to fully intercalate the cathode.[81] If the Li cycling efficiency is 99%, then 1% of the cycled Li is lost on each cycle. Therefore, to obtain 200 deep discharge cycles, at least a threefold excess of Li is required.

(ii) The high cost of the technology due to the need for assembly equipment to be located in dry rooms, the relatively expensive separators, and the need for excess Li metal.

(iii) The unsafe operating characteristics due to the high reactivity of lithium metal.[82]

To alleviate this problem, an innovative secondary lithium cell system (Figure 11.11) has been proposed that uses another intercalation compound as the negative electrode.[83-94] Among the alternative materials that could replace lithium metal, carbon provides the best compromise between large specific capacity and reversible cycling behavior,[90,91] but other materials have also been proposed (see Table 11.9).

B. ELECTRODES FOR ROCKING-CHAIR BATTERIES

This new generation of rechargeable lithium cells is called in different ways, i.e., *rocking-chair*, shuttlecock, seesaw, or lithium-ion technology. The rocking-chair batteries do not

TABLE 11.9
Characteristics of Some Lithium Metal-Free Rechargeable Batteries

Anode	Electrolyte[a]	Cathode	Current density ($mA\ cm^{-2}$)	Cycles	Producer
WO_2	$LiClO_4$-PC	TiS_2	0.1	68	Rome Univ.
MoO_2	$LiAsF_6$-PC	$LiCoO_2$	0.1	—	AT&T
WO_2	$LiAsF_6$-PC	$LiCoO_2$	0.1	—	AT&T
Li_xC_6	Li-HPE[b]	$LiMn_2O_4$	0.8	800	AT&T
$Li_9Mo_6Se_6$	$LiAsF_6$-THF	Mo_6Se_6	1.2	10	Bell Comm.
TiS_2	$LiAsF_6$-AN	$LiCoO_2$	0.5	500	U.S. Army[c]
Coke	$LiN(CF_3SO_2)_2$	$LiNiO_2$	1.0	1000	Moli Energy
Carbon	—	$LiNiO_2$		1000	SAFT
Li_xC_6	SPE[d]	$LiNiO_2$	0.12	50	EIC Lab.
Li_xC_6	$LiPF_6$-PC-DEC	$LiCoO_2$	0.5	100	Sony[e]

[a] PC: propylene carbonate, THF: tetrahydrofuran, AN: acetonitrile, DEC: diethyl carbonate.
[b] Li-HPE is a hybrid plastic electrolyte which is a two-component mixture of a Li salt-doped polymer matrix swollen with a liquid electrolyte.
[c] Thin-film battery.
[d] SPE is a solid polymer electrolyte with PAN-EC-PC-$LiN(SO_2CF_3)_2$.
[e] Coin cell in the market in commercial quantities (Li ion by Sony).

From Julien, C. and Nazri, G.A., *Solid State Batteries: Materials Design and Optimization*, Kluwer, Boston, 1994. With permission.

require a stringent manufacturing environment, because the starting electrode materials, i.e., the lithiated manganese oxide and the carbon, for example, are stable in ambient atmosphere. The cell is assembled in its discharged state, where the output voltage is close to zero volt, and, consequently, can be handled before use without any fear of irreversible degradation due to a short circuit. The battery is activated during the first charge. This is similar to the well-known and widely used Ni-Cd batteries that need to be charged prior to their first use. During charging, the Li^+ ions are deintercalated from the Li-bearing cathode, and intercalated into the carbon anode. Thus the amount of cathode and anode materials are directly correlated. It can be seen that the output voltage of such a cell is determined by the difference between the electrochemical potential of Li within the two insertion compounds. Thus one needs to have, as the cathode and anode, intercalation compounds that can intercalate Li at high and low voltage, respectively. In the case of a $LiMn_2O_4$/C cell, the voltage can be cycled between 4.35 and 2.3 V.

Rocking-chair batteries can be made using two different intercalation compounds as the electrodes in which the chemical potential of the intercalant differ by several electron volts. For the anode, the chemical potential of the intercalated Li should be close to that of Li metal, like it is in Li_xC_6.[90,92-96] Other anode materials such as MoO_2, WO_2, TiS_2, and $Li_9Mo_6Se_6$ have been reported in the literature.[86,88-89,93]

An inherent problem with the rocking-chair batteries using transition-metal oxide cathodes is the decomposition of common nonaqueous electrolytes for voltages greater than 4.5 and 4.3 V at room temperature and 55°C, respectively. To overcome this problem, a novel $LiClO_4$-free electrolyte has been used that allows safe operation up to 5 V, both at room temperature and at 55°C.[96]

C. ROCKING-CHAIR BATTERIES

The development of ambient-temperature lithium metal-free batteries has remained a formidable challenge. However, at the time of this writing, there is only one rechargeable cell type which is not a coin cell on the market in commercial quantities. It is the so-called "Lithium-Ion Cell" by Sony.[97] This system appears to be on the way to becoming a standard. It is being used at present to power cellular telephones, video tape recorders, and notebook

TABLE 11.10
Performance of Different Types of Lithium Metal-Free Rechargeable Batteries

Type	Capacity (A h kg^{-1})	Voltage[a] (V)	Energy (W h kg^{-1})
$LiCoO_2$-C	22	3.6	78
$LiMn_2O_4$-C	25	3.7	91
$LiNiO_2$-C	25	3.2	80
Ni-Cd	25	1.2	30
Ni-MH	40	1.26	51

[a] At mid-discharge.

From Julien, C. and Nazri, G.A., *Solid State Batteries: Materials Design and Optimization*, Kluwer, Boston, 1994. With permission.

computers. In its batteries Sony has incorporated an additional safety feature in the form of a sophisticated electronic circuit. This circuit monitors the individual voltage of the series-connected cells in the battery and prevents overcharge and overdischarge of individual cells. Furthermore, it limits the discharge current to a maximum value.

Performances of the three main types of rocking-chair batteries are presented in Table 11.10 and compared with the Ni-Cd and Ni-MH cells. In 1992, AT&T Bellcore demonstrates the feasibility of the $Li_xC_6/LiMn_2O_4$ couple for the development of a 0.4-A h lithium-ion rechargeable battery. This AA cell has been cycled 400 times at a C/1.5 rate between 2 and 4.6 V. Asahi/Toshiba jointly announced that they are going into production of the Li-ion battery soon for commercial and consumer applications.

Solid state C/$LiNiO_2$ batteries employing a composite polymer electrolyte have been fabricated by EIC Laboratories Inc. The cycling behavior of a C/$LiNiO_2$ cell with PAN-EC-PC-$LiN(SO_4CF_3)_2$ has been tested at a C/2 rate corresponding to a current density of 0.25 mA cm^{-2}. The cell has a middischarge voltage of about 3.0 V and a specific energy of 337 W h kg^{-1}. The coulombic efficiency was close to 100% over 120 cycles.

A new rechargeable plastic Li-ion battery has been recently introduced by Bellcore. The C/$LiMn_2O_4$ plastic cell uses a fluorinated polymer matrix in which the liquid electrolyte is so tightly absorbed that the working battery looks and feels like ordinary plastic. This battery is built from single plastic laminate electrodes and electrolyte. At the present time more than 800 cycles have been achieved. The plastic C/$LiMn_2O_4$ of 0.58-A h capacity is able to deliver 90% of its energy at 1C rate. Yardney Technical Products Inc. has developed a prismatic design, 3-A h Li-ion cells which provide a specific energy of 87 W h kg^{-1}. The average operating voltage of the developed cell is 3.7 V at C/3 rate.

Rocking-chair batteries developed at Moli Energy Ltd.[91] show high energy density, long cycle life, excellent high-temperature performance, and low self-discharge rates. The cell arrangement, i.e., coke/PAN-$LiN(CF_3SO_2)_2$/$LiNiO_2$, is easily assembled.

Figure 11.12 shows the discharge behavior of 4-V rocking-chair cells, i.e., a $LiMn_2O_4$/coke cell and a $LiNiO_2$/coke cell, as compared with the discharge behavior of commercialized D-size Ni-Cd batteries and of the D-size Li-ion rechargeable batteries $LiCoO_2$/carbon prototype of Sony Energytec Co.[97] It can be remarked that $LiMn_2O_4$/carbon cells have almost the same specific capacity as Ni-Cd.[90] Furthermore, because of their larger output voltage (3.7 V compared to 1.2 V for Ni-Cd), their specific energy is about three times higher (85 mW h g^{-1} compared to 30 mW h g^{-1} for Ni-Cd).

IV. MICROBATTERIES

Thin-film cell-based microbatteries having form factors and voltages compatible with microelectronics applications are evolving as viable power sources. This section provides an

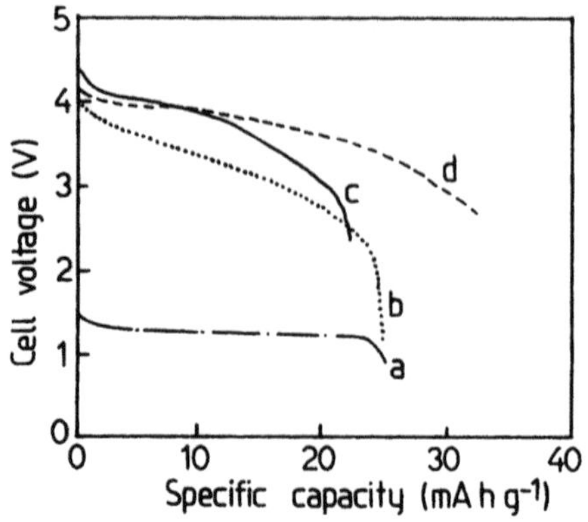

FIGURE 11.12. Discharge curves vs. specific energy of (a) commercialized D-size Ni-Cd battery, (b) D-size Li-ion rechargeable batteries $LiCoO_2$/carbon announced by Sony Energytec, (c) $LiMn_2O_4$/coke cell, and (d) $LiNiO_2$/coke cell. (From Julien, C. and Nazri, G.A., *Solid State Batteries: Materials Design and Optimization*, Kluwer, Boston, 1994. With permission.)

overview of the current state of development of these new microelectronic components. It also aims to introduce thin-film rechargeable batteries to the semiconductor device community and to stimulate new ideas for their application.

A. SILVER AND COPPER MICROBATTERIES

Due to the high ionic conductivity of silver and copper solid electrolytes, several silver and copper microbatteries have been fabricated early, and it was found that they are impractical, with a low energy density and high cost. Thin films of silver compounds as the microbattery electrolyte were reviewed by Kennedy.[98]

Vouros and Masters[99] have reported a thin-film silver battery. They have patented a thin-film electrochemical cell of the type Ag/AgI/Pt which gives an open circuit voltage of 550 mV. Takahashi and Yamamoto[100] have fabricated a cell of the type $Ag/Ag_3SI/I_2$, carbon, giving interesting results. Six of these cells were stacked entirely by vacuum evaporation and deliver the open-circuit voltage of 1.2 V at 25°C. Although a high current density of 100 $\mu A\ mm^{-2}$ could be drawn from these cells, the main drawback was the problem due to iodine oxidizing the solid electrolyte. Recently, Minami[101] has reported other silver conductors used in batteries and electrochromic devices, but without a better performances. Another type of silver battery has been proposed by Chandra et al.[102] The electrochemical chain $Ag/NH_4Ag_4I_5/(C+RbAg_4I_5+KI_3)$ was used. The open-circuit voltage of 0.6 V is lower than that obtained with the classical $Ag/Ag^+/I_2$ electrochemical chain (0.69 V). The electrolyte in this case is not a pure ionic conductor.

Double-layer capacitors which utilize either Cu^+ or Ag^+ ion conductor have been fabricated and used as rechargeable power sources.[103] The double-layer capacitors have a positive electrode of either Cu or Ag and the electrodes are separated by solid electrolyte layer of $Rb_2Cu_8I_3Cl_7$ or $RbAg_4I_5$. These devices have a high capacity and low leakage current. These have been proposed as standby power sources for RAM devices.

B. LITHIUM MICROBATTERIES

1. Lithium Electrode Thin Films

Lithium is one of the highly reactive metals, and there are few works on the deposition process of lithium films in the literature. Li films are deposited as vapor rather than sputtered due to the low melting point of the Li metal, e.g., 178°C under normal conditions. In their prior works, Liang et al.[104,105] have shown that depositions of lithium anodes have to be controlled carefully to obtain satisfactory films. The lithium films were deposited on LiI solid electrolyte film, and the substrate was cooled to –20°C to avoid diffusion of Li atoms in LiI.

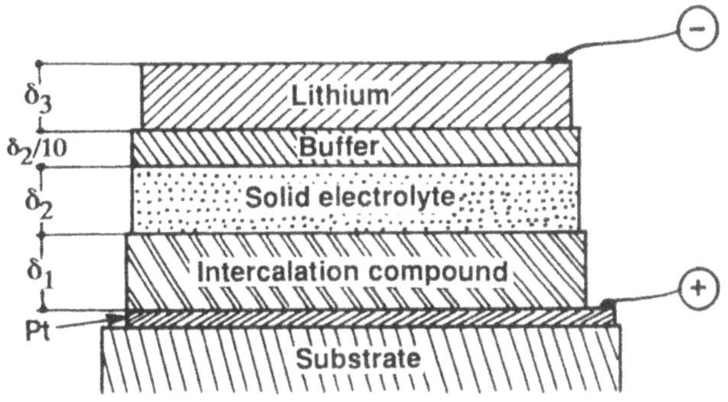

FIGURE 11.13. Schematic representation of a lithium microbattery. It is worth noting that the volume of the intercalation compound determines the cell capacity and that a buffer layer can be inserted between the lithium film and the solid electrolyte film to prevent chemical reaction. (From Julien, C. and Nazri, G.A., *Solid State Batteries: Materials Design and Optimization,* Kluwer, Boston, 1994. With permission.)

In their thin-film battery, Kanehori et al.[106] have deposited lithium films under moderate vacuum pressure conditions at 10^{-4} Pa, but did not give information about lithium evaporation temperature. Rabardel et al.[107] have grown lithium films using a special evaporation apparatus which allowed cooling of the substrate to liquid nitrogen temperature. To get homogeneous films, the lithium source was heated up to 430°C and the deposition rate was kept at 0.3 mm h^{-1}. In their high-performance microbattery, Jones and Akridge[108] have deposited lithium films at a fast rate of 0.1 μm s^{-1} on the top of the sandwich.

2. Lithium Microbatteries with Chalcogenide Cathode

Lithium solid state microbatteries, which consist of a thin-film electrolyte sandwiched between two thin-film electrodes (Figure 11.13), have been realized and give high performances.[106,107-120] Among many attempts presented, one finds consistent effort toward the development of complete thin-film solid state battery, and the most promising device seems to be that developed by Jones and Akridge.[108]

The prior lithium battery was proposed by Liang and Epstein.[104] This battery was based on the Li/LiI/AgI electrochemical chain which exhibited an open circuit voltage of 2.1 V and short-circuit currents of greater than 100 μA cm^{-2} at 25°C.

Through the different patents, one can observe the technology evolution in the solid state microbattery field. The solid cell constructed at Duracell and patented by Rea and Davis[112] showed an available capacity of 88% with the use of bismuth cathode, but the reversibility seems to be limited to ten cycles only.

Recently, the microbattery designs get attention on the use of more porous cathodic materials, which seem to allow better reversibility and high current densities. In these systems, quite low-conducting thin-film glasses are utilized as the electrolyte. Because of the layer thickness in a microbattery (1 to 3 μm), a solid electrolyte with a conductivity of 10^{-6} S cm^{-1} would be acceptable. For example, a battery of 1 cm^2 area using this electrolyte would only add 50 Ω to the cell resistance for every micron of electrolyte thickness.

Microbatteries developed by Kanehori et al.[106] have shown interesting features. The secondary thin-film lithium cells were fabricated by deposition of successive layers of amorphous $Li_{3.6}Si_{0.6}P_{0.4}O_4$ as the solid electrolyte and TiS_2 films deposited by low pressure CVD (condensed vapor deposition) as the cathode. Cells showed a discharge capacity of about 150 μA h cm^{-2} which is 80% of the theoretical value for TiS_2 powder. It was also demonstrated by the authors that these cells could endure 2000 charge/discharge cycles. But this system faces two major difficulties: (i) the mechanical stability of the intercalation cathode TiS_2 deteriorates after a large number of cycles and (ii) the stiffness of the electrolyte material reduces its capacity to accommodate the volume expansion of the cathode after intercalation. The major problem still remains the necessity of perfect physical contact between the solid electrolyte and the electrode.

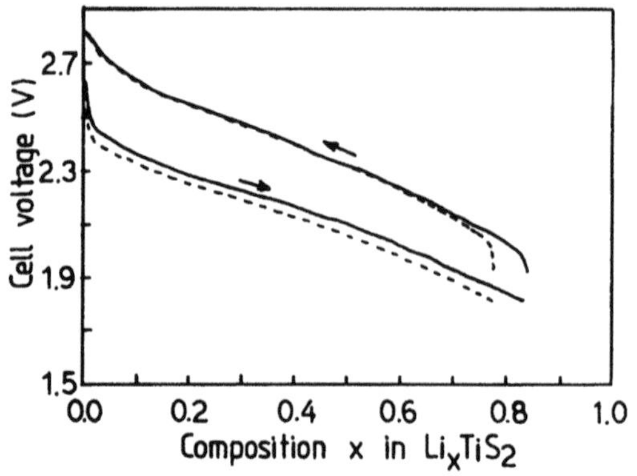

FIGURE 11.14. Secondary performance of a solid state Li/TiS_2 microbattery. The discharge/charge curve after 1000 cycles (full line) and after 4000 cycles (dashed line). (Adapted from Jones, S.D. and Akridge, J.R., *J. Power Sources*, 1993, 43/44, 505.)

Jourdaine et al.[113] proposed a microbattery using a mixed-conducting glass. The system V_2O_5-P_2O_5 or V_2O_5-TeO_2 had the positive electrode. Thermal evaporation of this material, of lithium borophosphate (LiOP) glass as electrolyte and lithium as the negative electrode formed a microbattery which was able to deliver a few μA cm^{-2} current density at an average cell voltage of 1.3 V.

In an advanced microbattery, Meunier et al.[109] have used RF-sputtering technique to grow a new titanium oxysulfide thin-film on which a sputtered glass of composition B_2O_3-$0.8Li_2O$-$0.8Li_2SO_4$ was deposited. A thermal evaporation of lithium made a microbattery complete. A current density of up to 60 μA cm^{-2} at an average cell voltage of 2 V could be easily obtained. This system is perfectly reversible; more than 100 cycles have been obtained without any trouble.

More recently, Jones and Akridge[108] have given the most promising example of lithium microbattery. The microbattery is constructed using cathode sputtering for deposition of the contacts, TiS_2 cathode, and oxide–sulfide solid electrolyte, while high-vacuum vapor deposition is used for a LiI layer and Li anode. The complete battery has an overall thickness of approximately 10 μm and an open-circuit voltage near 2.5 V. The sputtered cathode has a stoichiometry $TiS_{2.09}$ with only a slight excess of sulfur. The high degree of porosity is favorable to high lithium diffusivity and high current densities ranging from 10 to 135 μA cm^{-2}. The energy density is 230 W h dm^{-3} for a cathode thickness of 4 μm. The reduced sulfide content solid electrolyte $6LiI$-$4Li_3PO_4$-P_2S_5 has an ionic conductivity of 2×10^{-5} S cm^{-1}. The microbattery is capable of supplying current pulses of 2 s duration at current densities greater than 2 mA cm^{-2}. The microbattery also shows excellent secondary performances; cells routinely give over 4000 cycles (Figure 11.14).

Thin-film solid state microbatteries with lithium borate glass, B_2O_3-$0.8Li_2O$, as a solid electrolyte and InSe as a cathodic material have been built using the flash-evaporation system,[114] with an area of 1 cm^2 and a cathode mass of 1.5 to 8 mg. The microbattery has been formed on silica substrate by the deposition of the successive layers: platinum current collectors (0.2 μm thick), InSe cathode (5 μm thick), lithio-borate glass, and lithium film (4 μm thick). The electrolyte film of thickness of 0.5 μm was made from $BLiO_2$+Li_2O, evaporated at 1100°C, while the metallic lithium anode film was fabricated at the rate of 5 μm h^{-1} from a boat heated at 500°C. An amorphous $B_2O_{3\text{-}3}Li_2O$ film was grown which exhibits an ionic conductivity of 4.5 × 10-8 S cm^{-1} at room temperature and an activation energy of 0.62 eV.

Lithium microbatteries grown with the successive layers Pt/Li/B_2O_3-$3Li_2O$/InSe/Pt, which have a theoretical electrochemical capacity of 0.28 mA h, have been discharged at a small constant current density of 0.2 μA cm^{-2}. These Li microbatteries were encapsulated using a silica slide sealed with the battery substrate by oxygen-free epoxy. The initial voltage was about 2.2 V, and the prompt drop in voltage in the initial stage has been attributed to ohmic polarization. After that, the discharge voltage became stable and a plateau appeared at 1.45 V. These cells give a practical capacity of 30 μA h cm^{-2}. Analysis of the experimental data using the Butler–Volmer formalism gives an exchange current $i_o = i/0.1$ and a chemical lithium diffusion coefficient for the lithium ions of 5×10^{-15} cm^2 s^{-1}.[115]

3. Lithium Microbatteries with Oxide Cathode

Ohtsuka et al.[116] have fabricated Li/MoO_{3-x} microbatteries by RF-sputtering technique using Li_2O-V_2O_5-SiO_2 as the electrolyte. Li/MoO_{3-x} cells showed good cathode utilization and rechargeability. The electrolyte film of conductivity 10^{-8} S cm^{-1} was found to be stable with the electrodes. However, the cells suffered from self-discharge due to high electronic conductivity of the electrolyte film. The discharge capacity was small, decreasing from 18 μA h cm^{-2} at the first cycle to 14 μA h cm^{-2} at the 98th cycle.

An all–solid state cell employing $LiAlCl_4$ as the lithium solid electrolyte and using lithium intercalating compounds, i.e., TiS_2 and $LiCoO_2$ thin films as the electroactive materials in the rocking-chair configuration, has been described by Plichta et al.[117] Thin films of TiS_2 were prepared by chemical vapor deposition onto aluminum substrates while thin films of $LiCoO_2$ on aluminum were prepared by organometallic decomposition. The cell was operated at 100°C and exhibited an open circuit voltage of 2.1 V in the charged state. The cell showed excellent discharge characteristics at current densities up to 0.1 mA cm^{-2} and only a slight loss of capacity over 100 charge/discharge cycles.

Rechargeable thin-film lithium batteries consisting of lithium metal anodes, an amorphous inorganic electrolyte, and cathodes of lithium intercalation compounds have been fabricated by Bates et al.[118,119] These include Li-TiS_2, Li-V_2O_5, and Li-$Li_xMn_2O_4$ cells. The solid-film electrolyte of composition $Li_{2.9}PO_{3.3}N_{0.46}$ (Lipon) is obtained by RF-magnetron sputtering of Li_3PO_4 in N_2 atmosphere. This material, which shows a good stability, has a conductivity of 2 μS cm^{-2} at room temperature. A Li-TiS_2 cell with Lipon electrolyte was cycled over 4000 times before failing due to a short circuit.

Thin-film batteries based on $LiMn_2O_4$ and glass electrolyte are fabricated and tested by Bellcore.[120] Electrolyte films of LiBP and lithium phosphorus oxynitride (Lipon) are grown using electron beam evaporation. The oxide-based glassy electrolyte films are compatible with the spinel $LiMn_2O_4$ cathode, have higher chemical and mechanical stability above 4 V than the sulfide-based glasses, and have an ionic conductivity of 1 to 5×10^{-6} S cm^{-1}. Thin film batteries have been cycled at current densities of up to 70 μA cm^{-2} over 150 cycles.

Thin-film V_2O_5 electrodes have been studied for lithium microbatteries by West et al.[121] Thin-film cathode were prepared by three techniques: RF-sputtering, physical vapor deposition, and solvent casting. Microbatteries have been tested using V_2O_5 films, a PEO-based polymer electrolyte, and a large excess of Li metal. The discharge behavior has been studied as a function of the oxygen content in the chamber during the sputtering growth. Good cycling performance has been obtained at 50 μA cm^{-2} current density.

Julien and Gorenstein[122] have investigated thin-film microbatteries with various cathode films such as MoO_3, V_2O_5, and V_6O_{13} grown by the flash-evaporation method. Particular attention was paid to the structural and transport properties of these materials, and their behavior in electrochemical cells have been studied in relation with their physical properties. Thermodynamics and kinetics have been investigated as functions of the growth conditions of thin-film components. Representative voltage curves as a function of the charge per volume for Li/InSe, Li/TiS_2, Li/MoO_3, Li/α-V_2O_5, and Li/V_6O_{13} microbatteries are shown in Figure 11.15.[122] For TiS_2, α-V_2O_5, MoO_3, and V_6O_{13}, the open-circuit voltages at full charge are about 2.5, 3.0, 3.3 and 3.7 V, respectively. Discharge curves shown in Figure 11.15 are characteristics of intercalation cathodes, and the magnitude of the voltage is a function of the material and of the amount of lithium intercalated in the host structure.

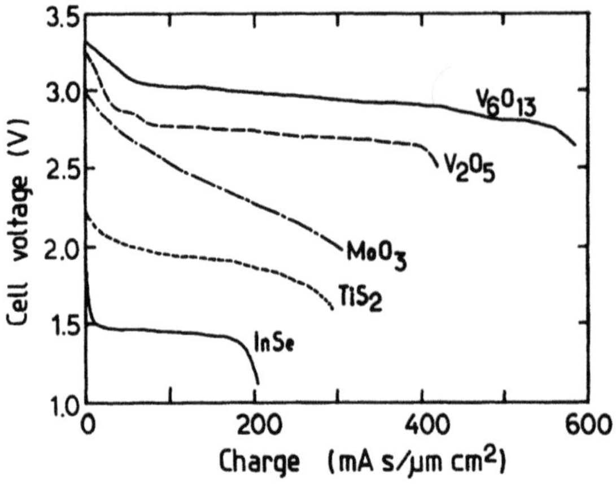

FIGURE 11.15. Cell voltage vs. capacity for thin-film lithium batteries having cathode materials: InSe, TiS_2, α-V_2O_5, $Li_xMn_2O_4$, and V_6O_{13}. (From Julien, C. and Nazri, G.A., *Solid State Batteries: Materials Design and Optimization,* Kluwer, Boston, 1994. With permission.)

REFERENCES

1. Owens, B.B., *Adv. Electrochem. Electrochem. Eng.*, 1991, *8*, 1.
2. Liang, C.C., in *Fast Ion Transport in Solids,* van Gool, W., Ed., North-Holland, Amsterdam, 1973, 19.
3. *Solid State Batteries,* NATO-ASI, Series, Ser. *E101,* Sequeira, C.A.C. and Hooper, A., Eds., Martinus Nijhoff, Dordrecht, 1985.
4. Julien, C., *Mater. Sci. Eng. B,* 1990, *6,* 9.
5. Julien, C. and Nazri, G.A., *Solid State Batteries: Materials Design and Optimization,* Kluwer, Boston, 1994.
6. Cairns, E.J., in *Materials for Advanced Batteries,* Murphy, D.W., Broadhead, J., and Steele, B.C.H., Eds., Plenum Press, New York, 1980, 3.
7. Caulder, S.M. and Simon, A.C., in *Materials for Advanced Batteries,* Murphy, D.W., Broadhead, J., and Steele, B.C.H., Eds., Plenum Press, New York, 1980, 199.
8. Cairns, E.J. and Shimotake, P., *Science,* 1969, *164,* 1347.
9. Armand, M.B., *Solid State Ionics,* 1983, *9–10,* 745.
10. Ogumi, O., Uchimoto, Y., Takehara, Z., and Kamamori, Y., *J. Electrochem. Soc.*, 1988, *135,* 2649.
11. McLarnon, F.R. and Cairns, E.J., in *Annual Review of Energy,* Vol. 14, Hollander, J.M., Ed., Annual Rev. Inc., Palo Alto, 1989.
12. Julien, C., Massot, M., Balkanski, M., and Tuller, H.L., in *Glasses for Electronic Applications, Ceramic Transactions,* Nair, K.M., Ed., The American Ceramic Society, Westerville, Ohio, 1991, 51.
13. Owens. B.B., Skarstad, P.M., and Unterecker, D.F., in *Handbook of Batteries and Fuel Cells,* Linden, D., Ed., McGraw-Hill, New York, 1984, 12–1.
14. Vincent, C.A., Bonino, F., Lazzari, M., and Scrosati, B., *Modern Batteries,* Edward Arnold, London, 1984.
15. Official Journal of the European Communities, 1988, No. S 228/93.
16. Williams, J.F., USABC Presentation to the California Air Resources Board Public Meeting, Carb, 1994.
17. Jensen, J., *Energy Storage,* Butterworths, London, 1980.
18. Koenig, A.A., *J. Electrochem. Soc.*, 1990, *137,* 376C.
19. Brown, P.J., Kirk, R.S., and Patil, P.G., *Proc. 23rd Intersoc. Energy Conv. Eng. Conf. Denver,* American Society of Mechanical Engineering, New York, 1988, 271.
20. Kita, A., Nomura, E., Matsui, K., Kagawa, H., and Takashima, K., *Proc. 23rd Intersoc. Energy Conv. Eng. Conf. Denver,* American Society of Mechanical Engineering, New York, 1988, 317.
21. Kaun, T.D., Holifield, T.F., Nogohosian, M., and Nelson, P.A., *J. Electrochem. Soc.*, 1988, *135,* 344C.
22. Bones, R.J., Coetzer, J., Galloway, R.G., and Teagle, D.A., *J. Electrochem. Soc.*, 1987, *134,* 2379.
23. Levine, C.A., in *Applications of Solid Electrolyte,* Takahashi, T. and Kozawa, A., Eds., JEC Press, Cleveland, 1980, 99.
24. Goodenough, J.B., Hong, H.Y.P., and Kafalos, J.A., *Mater. Res. Bull.*, 1976, *11,* 204.
25. Gauthier, M., Fauteux, D., Vassort, G., Belanger, A., Duval, M., Ricoux, P., Gabano, J.M., Muller, D., Rigaud, P., Armand, M.B., and Deroo, D., *J. Electrochem. Soc.*, 1985, *132,* 1333.

26. Armand, M., in *Solid State Batteries,* NATO ASI Series, Ser. *E101,* Sequeira, C.A.C. and Hooper, A., Eds., Martinus Nijhoff, Dordrecht, 1985, 63.
27. Hooper, A., Powell, R.J., Marshall, T.J., and Neat, R.J., *J. Power Sources,* 1989, *27,* 3.
28. Bonino, F., Ottaviani, M., Scrosati, B., and Pistoia, G., *J. Electrochem. Soc.,* 1988, *135,* 12.
29. Douridah, A., Dalard, F., Deroo, D., Armand, M., *Solid State Ionics,* 1986, *18–19,* 287.
30. Gauthier, M., Fauteux, D., Vassort, G., Belanger, A., Duval, M., Ricoux, P., Gabano, J.M., Muller, D., Rigaud, P., Armand, M.B., and Deroo, D., *J. Power Sources,* 1985, *14,* 23.
31. Macklin, W.J. and Neat, R.J., *Solid State Ionics,* 1992, *53–56,* 694.
32. Ferloni, P., Chiodelli, G., Magistris, A., and Sanesi, M., *Solid State Ionics,* 1986, *18–19,* 258.
33. Chiang, C.K., *Polymer,* 1981, *22,* 1454.
34. Noda, T., Kato, S., Yoshihisa, Y., Takeuchi, K., and Murata, K., *J. Power Sources,* 1993, *43–44,* 89.
35. Liu, M., Visco, S.J., and DeJonghe, L.C., *J. Electrochem. Soc.,* 1989, *136,* 2570.
36. Liu, M., Visco, S.J., and DeJonghe, L.C., *J. Electrochem. Soc.,* 1990, *137,* 750.
37. Liu, M., Visco, S.J., and DeJonghe, L.C., *J. Electrochem. Soc.,* 1990, *137,* 367C.
38. Malugani, J.P., Fahys, B., Mercier, R., Robert, G., Duchange, J.P., Baudry, S., Broussely, M., and Gabaus, J.P., *Solid State Ionics,* 1983, *9–10,* 659.
39. Akridge, J.R. and Vourlis, H., *Solid State Ionics,* 1986, *18–19,* 1082.
40. Levasseur, A., in Materials for *Solid State Batteries,* Chowdari, B.V.R. and Radhakrishna, S., Eds., World Scientific, Singapore, 1986, 97.
41. Owen, J.R., in *Solid State Batteries,* NATO ASI Series, Ser. *E101,* Sequeira, C.A.C. and Hooper, A., Eds., Martinus Nighoff, Dordrecht, 1985, 413.
42. Schlaikjer, C.R. and Liang, C.C., *J. Electrochem. Soc.,* 1971, *118,* 1447.
43. Skarstad, P.M. and Schmidt, C.L., *J. Electrochem. Soc.,* 1990, *137,* 367C.
44. Liang, C.C., Joshi, A.V., and Hamilton, W.E., *J. Appl. Electrochem.,* 1978, *8,* 445.
45. Park, K.H., Miles, M.H., Bliss, D.E., Stilwell, D., Hollins, R.A., and Rhein, R.A., *J. Electrochem. Soc.,* 1988, *135,* 2901.
46. Matsuda, M., *J. Electrochem. Soc.,* 1984, *131,* 104.
47. Akridge, J.R. and Vourlis, H., in *Micro-Solid State Batteries,* NATO ASI Series, Ser. *B217,* Akridge, J.R. and Balkanski, M., Eds., Plenum Press, New York, 1990, 353.
48. Goodenough, J.R., in *Micro-Solid State Batteries,* NATO ASI Series, Ser. *B217,* Akridge, J.R. and Balkanski, M., Eds., Plenum Press, New York, 1990, 213.
49. Anderman, M., Lundquist, J.T., Johnson, S.L., and Giovannoi, T.R., *J. Power Sources,* 1989, *26,* 309.
50. Abraham, K.M., Pasquariello, D.M., and Schwartz, D.A., *J. Power Sources,* 1989, *26,* 247.
51. Stiles, J.A.R., *J. Power Sources,* 1989, *26,* 65.
52. Py, M.A. and Haering, R.R., *Can. J. Phys.,* 1983, *61,* 76.
53. Trumbore, F.A., *J. Power Sources,* 1989, *26,* 65.
54. Koshita, N., Ikehata, T., and Takata, K., *J. Electrochem. Soc.,* 1990, *137,* 369C.
55. Labat, J., Dechenaux, V., Jumel, Y. and Gabano, J.P., *J. Electrochem. Soc.,* 1987, *134,* 406C.
56. Pistoia, G., *J. Electrochem. Soc.,* 1982, *129,* 1861.
57. Nohma, T., Saito, T., Furukawa, N., and Ikeda, H., *J. Power Sources,* 1989, *26,* 389.
58. Roberge, P.R., Beaudoin, R., Verville, G., and Smit, J., *J. Power Sources,* 1989, *27,* 177.
59. Nohma, T. and Furukawa, N., *J. Electrochem. Soc.,* 1990, *137,* 369C.
60. Wagner, J.B., Jr., *Mater. Res. Bull.,* 1981, *15,* 1691.
61. Julien, C., Saikh, S.I., and Balkanski, M., *Mater. Sci. Eng. B,* 1992, *14,* 121.
62. Takahashi, T. and Yamamoto, O., *Electrochim. Acta,* 1966, *11,* 779.
63. Foley, R.T., *J. Electrochem. Soc.,* 1969, *116,* 13.
64. Owens, B.B., in *Advances in Electrochemistry and Electrochemical Engineering,* Vol. 8, Tobias, C.W., Ed., John Wiley & Sons, New York, 1971, 1.
65. Owens, B.B., *J. Electrochem. Soc.,* 1970, *117,* 1536.
66. Minami, T., in *Materials for Solid State Batteries,* Chowdari, B.V.R. and Radhakrishna, S., Eds., World Scientific, Singapore, 1986, 169.
67. Minami, T., Katsuda, T., and Tanaka, M., *J. Electrochem. Soc.,* 1980, *127,* 1308.
68. Minami, T., in *Materials for Solid State Batteries,* Chowdari, B.V.R. and Radhakrishna, S., Eds., World Scientific, Singapore, 1986, 181.
69. Geller, S. and Lind, M.D., *J. Chem. Phys.,* 1970, *52,* 5854.
70. Hakwood, P. and Linford, R.G., *Chem. Ind.,* 1981, 523.
71. Yamamoto, O., Takede, Y., and Kanno, R., in *Materials for Solid State Batteries,* Chowdari, B.V.R. and Radhakrishna, S., Eds., World Scientific, Singapore, 1986, 275.
72. Takahashi, T. and Yamamoto, O., *J. Appl. Electrochem.,* 1977, *7,* 37.
73. Lazzari, M. and Razzini, G., *J. Power Sources,* 1976–77, *1,* 57.
74. Scrosati, B. and Voso, M.A., *J. Electrochem. Soc.,* 1979, *126,* 699.
75. Kanno, R., Takeda, Y., Oda, Y., Ikeda, H., and Yamamoto, O., *Solid State Ionics,* 1986, *18–19,* 1068.

76. Boswell, F.W., Prodan, A., Corbett, J.M., *Phys. Status Solidi A,* 1976, *35,* 591.
77. Kanno, R., Takeda, Y., Imura, M., and Yamamoto, O., *J. Appl. Electrochem.,* 1982, *12,* 681.
78. Sotomura, T., *Prog. Batteries Solar Cells,* 1987, *6,* 26.
79. Yoshimatsu, I., Hirai, T., and Yamaki, J., *J. Electrochem. Soc.,* 1988, *135,* 2422.
80. Selim, R. and Bro, P., *J. Electrochem. Soc.,* 1974, *121,* 1457.
81. Laman, F.C. and Brandt, K., *J. Power Sources,* 1988, *24,* 195.
82. Wilkinson, D.P., Dahn, J.R., Von Sacken, U., and Fouchard, D.T., *J. Electrochem. Soc.,* 1990, *137,* 370C.
83. Murphy, D.W., DiSalvo, F.J., Carides, J.N., and Waszszak, J.V., *Mater. Res. Bull.,* 1978, *13,* 1395.
84. Murphy, D.W. and Carides, J.N., *J. Electrochem. Soc.,* 1979, *126,* 349.
85. Lazzari, M. and Scrosati, B., *J. Electrochem. Soc.,* 1980, *127,* 773.
86. Armand, M., in *Materials for Advanced Batteries,* Murphy, D.W., Broadhead, J., and Steele, B.C.H., Eds., Plenum Press, New York, 1980, 145.
87. Mizushima, K., Jones, P.C., Wiseman, P.J., and Goodenough, J.B., *Mater. Res. Bull.,* 1980, *15,* 783.
88. Auborn, J.J. and Barbario, Y.L., *J. Electrochem. Soc.,* 1987, *134,* 638.
89. Tarascon, J.M., *J. Electrochem. Soc.,* 1985, *132,* 2089.
90. Guyomard, D. and Tarascon, J.M., *J. Electrochem. Soc.,* 1992, *139,* 937.
91. Dahn, J.R., Von Sacken, U., Juzhow, M.W., and Al-Janaby, H., *J. Electrochem. Soc.,* 1991, *138,* 2207.
92. Tarascon, J.M. and Guyomard, D., *J. Electrochem. Soc.,* 1991, *138,* 2864.
93. Plichta, E.J. and Behl, W.K., *J. Electrochem. Soc.,* 1993, *140,* 46.
94. Ohzuku, T., in *Lithium Batteries,* Pistoia, G., Ed., Elsevier: Oxford, 1993, 239.
95. Sekai, K., Azuma, H., Omaru, A., Fujita, S., Imoto, H., Endo, T., Yamaura, K., Nishi, Y., Mashiko, S., and Yokogawa, M., *J. Power Sources,* 1993, *43–44,* 241.
96. Tarascon, J.M., Guyomard, D., and Baker, G.L., *J. Power Sources,* 1993, *43–44,* 689.
97. Nagaura, T., *4th Int. Rechargeable Battery Seminar,* Deerfield Beach, FL, 1990.
98. Kennedy, J.H., *Thin Solid Films,* 1977, *43,* 41
99. Vouros, P. and Masters, J., *J. Electrochem. Soc.,* 1969, *116,* 880.
100. Takahashi, T. and Yamamoto, O., U.S. Patent 3,558,357, 1971.
101. Minami, T., in Materials for *Solid State Batteries,* Chowdari, B.V.R. and Radhakrishna, S., Eds., World Scientific, Singapore, 1986, 181.
102. Chandra, S., Agrawal, R.C., and Pandey, R.K., *Natl. Acad. Sci. Lett.,* 1978, *1,* 112.
103. Sekido, S. and Ninomiya, Y., *Solid State Ionics,* 1981, *3–4,* 153.
104. Liang, C.C., Epstein, J., and Boyle, G.H., *J. Electrochem. Soc.,* 1969, *116,* 1452.
105. Liang, C.C. and Bro, P., *J. Electrochem. Soc.,* 1969, *116,* 880.
106. Kanehori, K., Matsumoto, K., Miyauchi, K., and Kudo, T., *Solid State Ionics,* 1983, *9–10,* 1445.
107. Rabardel, L., Levasseur, A., Delmas, C., and Hagenmuller, P., *Le Vide-Les Couches Minces,* 1983, *217,* 289.
108. Jones, S.D. and Akridge, J.R., *Solid State Ionics,* 1992, *53–56,* 628.
109. Meunier, G., Dormoy, R., and Levasseur, A., *Mater. Sci. Eng. B,* 1989, *3,* 13.
110. Levasseur, A., Kbala, M., Hagenmuller, P., Couturier, G., and Danto, Y., *Solid State Ionics,* 1985, *9–10,* 1439.
111. Balkanski, M., Julien, C., and Emery, J.Y., *J. Power Sources,* 1989, *26,* 645.
112. Rea, J.R. and Davis, A., U.S. Patent 4,299,890, 1981.
113. Jourdaine, L., Souquet, J.L., Delord, V., and Ribes, M., *Solid State Ionics,* 1988, *28–30,* 1490.
114. Julien, C., Khelfa, A., Benramdane, N., Guesdon, J.P., Dzwonkowski, P., Samaras, I., and Balkanski, M., *Mater. Sci. Eng. B,* 1994, *23,* 105.
115. Julien, C., in *Lithium Batteries,* Pistoia, G., Ed., Elsevier, Oxford, 1993, 167.
116. Ohtsuka, H., Okada, S., and Yamaki, J., *Solid State Ionics,* 1990, *40–41,* 964.
117. Plichta, E.J., Behl, W.K., Vujic, D., Chang, W.H.S., and Schleich, D.M., *J. Electrochem. Soc.,* 1992, *139,* 1509.
118. Bates, J.B., Dudney, N.J., Gruzalski, G.R., Zuhr, R.A., Choudhury, A., Luck, C.F., and Robertson, J.D., *J. Power Sources,* 1993, *43–44,* 103.
119. Bates, J.B., Gruzalski, G.R., Dudney, N.J., Luck, C.F., Yu, X.H., and Jones, S.D., *Solid State Technol.,* July 1993, 59.
120. Shokoohi, F.K. and Tarascon, J.M., U.S. Patent 5,110,696, 1992.
121. West, K., Zachau-Christiansen, B., Jacobsen, T., and Skaarup, S., *J. Power Sources,* 1993, *43–44,* 127.
122. Julien, C. and Gorenstein, A., *1st Euroconference on Solid State Ionics,* Zakynthos, 1994.

Chapter 12

SOLID OXIDE FUEL CELLS

Abdelkader Hammou and Jacques Guindet

CONTENTS

I. FOREWORD

The main objective of this chapter is to describe the state of the art in solid oxide fuel cells (SOFC), with an emphasis on materials (electrolytes, electrodes, interconnects), interfacial reactions, and cell configurations. Several overviews have already been published on the subject and provide relevant background information.[1-5]

II. INTRODUCTION

The SOFC was derived from the Nernst glower, originally intended as a commercial light source to replace carbon filament lamps at the end of the last century.[6] This device made use of the Nernst mass composed of yttria-stabilized zirconia (YSZ), a predominantly ionic

0-8493-8956-9/97/$0.00+$.50

conductor in air. The solid state and the invariance of YSZ as an electrolyte led Baur and Preis,[7] in 1937, to develop the first SOFC using coke and magnetite as a fuel and oxidant, respectively.

A first period of intense activity in SOFC research began in the early 1960s with extensive research programs driven by new energy needs, mainly for space, submarine, and military applications. At that time, basic research dealt with the improvement of electrolyte conductivity and the first steps in SOFC technology. A second period of high activity began in the mid-1980s and goes on today, focusing on electrode materials and technology. A large number of industrial countries are carrying out national or joint international SOFC programs covering both basic research and development and involving public or private institutions and companies. In the U.S., the Department of Energy (DOE), the Electric Power Research Institute (EPRI) and the Gas Research Institute (GRI) are actively involved in SOFC research and development, funding several contractors such as Westinghouse Electrical Corporation, Allied Signal, Argonne National Laboratory, Pacific Northwest Laboratories, and the University of Missouri–Rolla.[8] In Japan, SOFC research and development is carried out by a large number of universities, national and private laboratories, and companies in many industries such as ceramic, electrical and plant engineering, and oil refining. Electrical power and gas utilities, in particular, are actively involved. They are supported financially by the New Energy Development Organisation (NEDO) and are in some cases organized in consortiums.[9] In Europe, the European Community started a 2-year exploratory R&D program in 1987 to investigate new structures taking into account the technical feasibility, the possibility for cost reduction, and the potential for cheap mass production. The target is the production of a 200-kW unit by 1997. Siemens and Dornier (Germany), ECN (Netherlands), GEC (U.K.) and several institutes and universities take part in this program.[10] Moreover, national SOFC efforts are being carried out simultaneously in Germany, The Netherlands, Denmark, Switzerland, and Norway. In the ex-USSR, research programs dealing with SOFCs are being conducted in the institute of electrochemistry in Sverdlovsk.[11] finally, in Australia, the Ceramic Fuel Cell Company has recently been formed by a consortium to provide a focus for research, development, and commercialization of SOFC technology.[12]

III. ADVANTAGES AND DRAWBACKS OF SOLID OXIDE FUEL CELLS

An SOFC may be defined as a ceramic multilayer system working at high temperature using gaseous fuel and oxidant. Such characteristics offer a number of advantages over traditional generators and other types of fuel cells:

- The use of expensive catalysts such as platinum or ruthenium is not necessary.
- The high quality of exhaust heat (800 to 900°C) is very useful for cogeneration applications in industry.
- High efficiency for electricity production (50%) can be achieved in combined cycles.
- Internal reforming of natural gas may considerably reduce costs.
- Electrolyte loss maintenance as well as electrode corrosion are eliminated, which is not the case for phosphoric acid or molten carbonate fuel cells (PAFC, MCFC).
- The cell life is increased due to higher tolerance to impurities such as sulfur in fuel.
- CO_2 emission is considerably reduced.
- Chemical cogeneration is possible in producing electricity and chemical compounds when appropriate electrocatalytic anodes are used.
- SOFCs can be used as high-temperature water electrolyzers without major modifications.
- Finally, SOFCs offer flexibility in the planning and siting of power generation capacity as a result of their modular nature.

Unfortunately, SOFCs also have a number of serious drawbacks which are to a large extent responsible for their relatively slow development:

- Electrolyte resistivity and electrode polarization are still too high.
- The formation of low conducting phases by solid state reactions at the cathode/electrolyte interface must be reduced or even avoided.
- The brittleness of the ceramic components makes it difficult to use cell sizes larger than 0.2 m^2. For PAFCs and MCFCs, the standard size lies around 0.5 to 1 m^2. This limitation is a major problem for scaling up SOFC power plants to megawatt size.
- Introduction on the energy market would presently involve a high capital cost-to-performance ratio.

IV. ELECTROLYTES

The required properties of electrolytes used in SOFCs are mainly fixed by the high operating temperature which dictates constraints of different types:

- electrochemical: high ionic conductivity (≥0.1 S cm^{-1} at 1000°C), low electronic transference number (<10^{-3}) and large electrolytic domain
- chemical: electrolyte must be stable with respect to electrode materials, oxygen, and fuel gas
- thermal: phase stability and a good match of thermal expansion coefficient (TEC) with other cell components
- mechanical: fracture toughness (>400 MPa at room temperature) and gas tightness.

The above mentioned constraints have restricted the choice to oxide-based ceramics where the charge carrier is an ion associated with the oxidant (O_2) or the fuel (H_2, hydrocarbons).

The most investigated oxide-conducting solid electrolytes for potential use in SOFCs belong to the fluorite-type solid solutions with the general formula MO^{2}-M′O or MO^{2}-M''_2O_3, where MO_2 is the basic oxide and M′O or M''_2O_3 the dopant with

M = Zr, Hf, Ce
M′ = Ca
M″ = Sc, Y, Ln (Rare Earth)

Oxygen vacancies, created by doping, make the migration of O^{2-} species possible in these materials. After a long period of research aimed at finding suitable electrolytes for SOFCs, doped zirconia has been selected. The salient features of its electrical conductivity are the following:

- Isothermal variation of conductivity shows a maximum located at 9 mol% of dopant for ZrO_2-M_2O_3 systems[13] see Figure 12.1 (see also Chapter 6 of this handbook).
- The activation energy is close to 0.8 eV for the composition of maximum conductivity.
- For a given composition, the conductivity increases as the radius of the dopant cation approaches that of Zr^{4+}. Thus, the best conductivity in zirconia-based systems is obtained with the ZrO_2-Sc_2O_3 system ($R_{Zr}4+$ = 0.80 Å and $R_{Sc}3+$ = 0.81 Å).
- The electrolytic domain extends over several orders of magnitude of oxygen partial pressure with a significant influence of temperature. In highly reducing media, n-type electronic conductivity occurs.
- Finally, a decrease of ionic conductivity is observed with time in the working temperature range of SOFCs.[14]

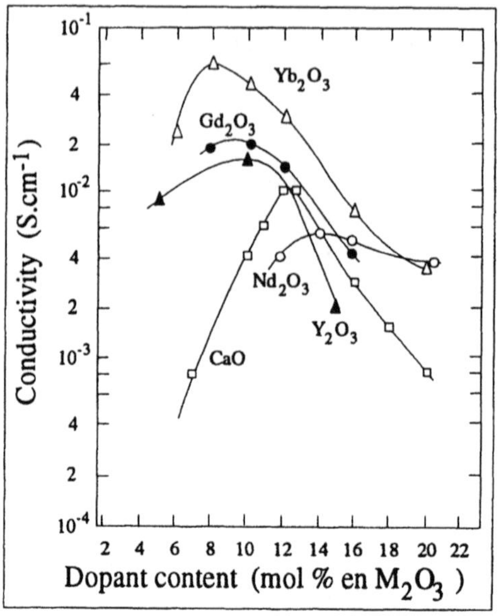

FIGURE 12.1. Dependence of electrical conductivity on composition for ZrO_2-based solid solutions at 800°C. (From Takahashi, T., *Physics of Electrolytes*, Vol. 2, Hladic, J., Ed., Academic Press, London, 1972, 980–1049. With permission.)

Taking into account the constraints mentioned above, there is general agreement on the use of YSZ with the $(ZrO_2)_{0.92}(Y_2O_3)_{0.08}$ composition as a solid electrolyte. Nevertheless, it is commonly considered that the current operating temperature of SOFCs must be lowered in order to minimize component degradation with time and to reduce the cost of the auxiliary parts such as heat exchangers, piping, etc. The following options have been retained:

- reduction of the ohmic losses due to electrolyte resistance. Many efforts have been made to develop processes to produce gas-tight layers of YSZ, with appropriate thickness and microstructure showing a minimum resistance
- to find more conductive electrolytes with lower activation energy. It is with this aim in view that several oxide systems were investigated. In the zirconia family, pyrochlore solid solutions of the general formula $M_2Zr_2O_7$ (M = rare earth) were studied with the possibility of substituting zirconium by a transition metal.[15-19] In a recent study,[18] it was demonstrated that one can vary both the magnitude and ratio of ionic and electronic conductivities in $Gd_2(Zr_xTi_{1-x})_2O_7$ and $Y_2(Zr_xTi_{1-x})_2O_7$ through systematic change in x. Ionic conductivity predominates for high x values, but it is still lower than that of YSZ. A factor of approximately 1/5 is observed at 1000°C. Although CeO_2- and Bi_2O_3-based solid solutions show higher ionic conductivities than YSZ, their tendency to be reduced in the presence of fuel gas excludes them for use in SOFCs.[20-32] Finally, the oxides with a perovskite structure investigated in the framework of SOFCs includes $CaTi_{1-x}Al_xO_{3-d}$, $CaTi_{1-x}Mg_xO_{3-d}$, and $Ln_{1-x}Ca_xAlO_{3-d}$, with Ln = Pr, Nd, Gd, Sm, La, Y, and Yb. Interesting results were achieved for the $Nd_{0.9}Ca_{0.1}Al_{0.5}Ga_{0.5}O_3$ composition.[33-37] Its conductivity is 4×10^{-2} Scm^{-1} at 950°C and the ionic transference number is estimated to be 1.0 over a wide range of oxygen partial pressures (1 to 10^{-20} atm).

The brittleness of YSZ is a major barrier to scaling up SOFC power plants to megawatt size. Increasing the cell surface area is currently a real challenge for the promotion of SOFCs. This can be achieved by improving both the fracture strength and the toughness of the electrolyte. Most of the studies concerning this aspect were carried out on tetragonal (TZP), partially (PSZ), and fully (YSZ) stabilized zirconia mixed with an appropriate amount of Al_2O_3 to form composite materials.[37-40] Fracture strength and fracture toughness of both TZP and YSZ are greatly improved by addition of alumina. Small additions (<0.6 mol% Al_2O_3) are effective for densification of YSZ with a slight increase of electrical conductivity and grain size. For large additions, higher than the solubility limit, the average grain size decreases with increasing Al_2O_3 content.[39,40] This behavior is shown in Figure 12.2. The composition dependence of the three-point bending strength of YSZ-Al_2O_3 composites at various temperatures is shown in Figure 12.3. The enhancement of bending strength is explained simply by the replacement of YSZ with Al_2O_3 because the bending strength of alumina is higher than that of YSZ. Moreover, alumina additions slightly increase both toughness and hardness, a crack deflection process being observed[63] (see Figure 12.4). Toughness variations may be theoretically explained if spherical particles of alumina are considered. The effect of the particle size of Al_2O_3 seems to be unimportant. When 20 wt% Al_2O_3 is added, a gain of 50% in bending strength is achieved. For this composition, the measured conductivity is 0.1 S cm^{-1} at 1000°C and the bending strength is equal to 33 kg/mm^2 at room temperature. For comparison, the values of these two properties, measured under the same conditions, are 1.5 S cm^{-1} and 24 kgf/mm^2 without Al_2O_3, respectively. Finally, to improve the mechanical properties of zirconia also, MgO was added. Abnormally high values were obtained in the case of the $(ZrO_2)_{0.8}(Y_2O_3)_{0.04}(MgO)_{0.04}$. Since this composition is close to the two-phase (tetragonal + fluorite) region, it is possible that undetected tetragonal zirconia has precipitated and is at the origin of the better mechanical performance. A conductivity value of 9×10^{-2} S cm^{-1} was achieved at 1000°C. These results seem very promising for the use of such electrolytes in SOFCs, mainly in the planar configuration.

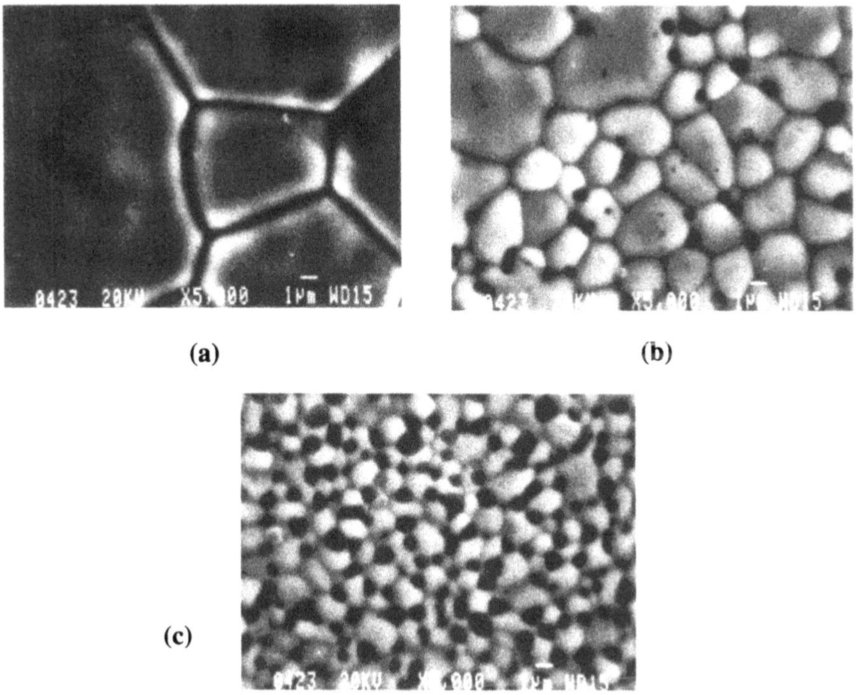

FIGURE 12.2. Scanning electron micrographs of (a) 8YSZ and the composites of (b) 8YSZ and 3 wt% Al_2O_3 (0.23 m), (c) 8YSZ and 20 wt% Al_2O_3. (From Ishizaki, F., Yoshida, T., and Sakurada, S., *Proceedings of the First International Symposium on Solid Oxide Fuel Cells,* Singhal, S.C., Ed., The Electrochemical Society, Pennington, NJ, 1989, 3–14. With permission.)

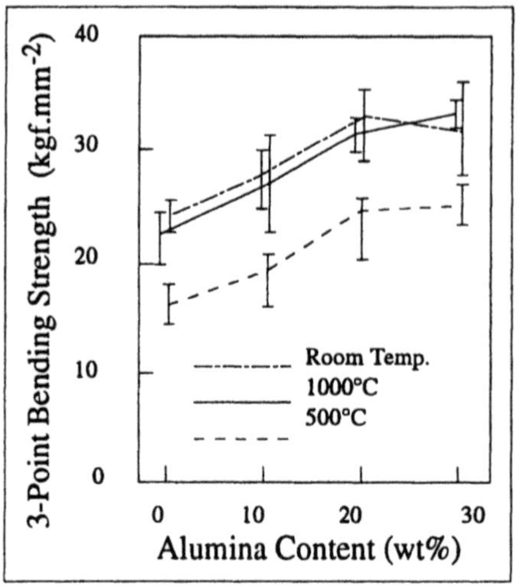

FIGURE 12.3. Dependence of three-point bending strength on composition for YSZ-Al_2O_3 composites. (From Yamamoto, O., Takeda, Y., Imanishi, Kawahara, T., Shen, G.Q., Mori, M., and Abe, T., *Proceedings of the Second International Symposium on Solid Oxide Fuel Cells,* Gross, F., Zeghers, P., Singhal, S.C., and Iwahara, H., Eds., Office for Official Publications of the European Communities, Luxembourg, 1991, 437–444. With permission.)

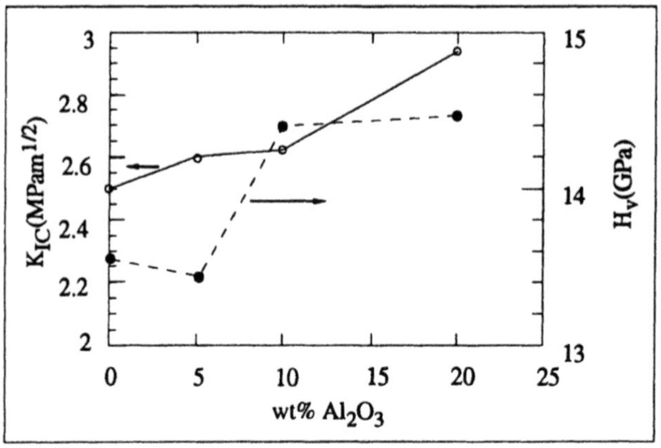

FIGURE 12.4. Dependence of toughness and Vickers hardness of YSZ-Al_2O_3 composites on Al_2O_3 content. (From Minh, N.Q., Armstrong, T.R., Esopa, J.R., Guiheen, J.V., Horne, C.R., Liu, F.S., Stillwagon, T.L., and Van Ackeren, J.J., *Proceedings of the Second International Symposium on Solid Oxide Fuel Cells,* Gross, F., Zeghers, P., Singhal, S.C., and Iwahara, H., Eds., Office for Official Publications of the European Communities, Luxembourg, 1991, 93–98. With permission.)

Another way to improve mechanical properties is based on the fabrication processes. It has been reported that plastic deformation processes significantly improve the bending strength of cubic zirconia foils.[41] A toughness of 2.4 $MPm^{1/2}$ has been achieved. In this technique, the most important parameter is the agglomerates breakdown during mixing because the survival of large agglomerates may cause defects in the final products. Powders with small, weak agglomerates must be selected. The influence of agglomerates on bending strength is shown in Figure 12.5.[42]

Very few data are available describing electrical properties of the $A_2B_2O_5$ systems with brownmillerite structure, which can be considered as being derived from the perovskite structure ABO_3 by removing one sixth of the oxygen atoms.[43] In the brownmillerite structure, the anion vacancies are usually ordered, but may become disordered at high temperature. It

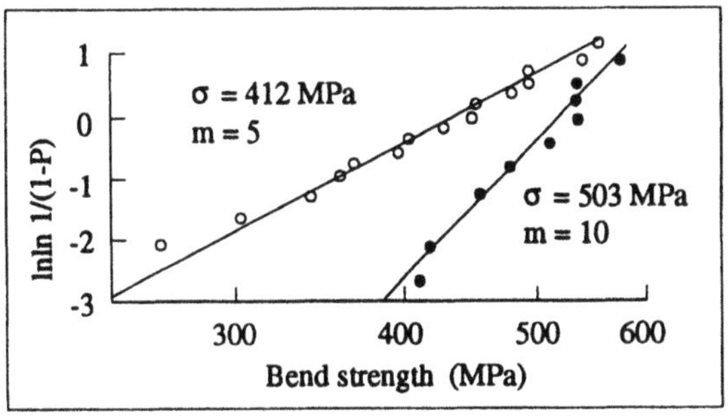

FIGURE 12.5. Weibull plots for three-point bending strength of zirconia foils (200 mm). The left curve shows results for an agglomerated 8YSZ powder. The right curve demonstrates the improvement gained by using disagglomerated powder. (From Kendall, K., *Proceedings of the Fuel Cell Seminar 1992,* Courtesy Associates, Washington, D.C., 1992, 265–268. With permission.)

is of interest to investigate oxide solid solutions showing this structure, focusing work on the decrease of the temperature at which the anionic vacancies disorder.

V. CATHODE MATERIALS AND INTERFACIAL REACTIONS

It is currently considered that the improvement of SOFC performance requires solutions to the problems of ohmic and polarization losses at the cathodic side of the cell, arising essentially from the high operating temperature. The requirements to be met mostly by the cathode material are as follows:

- high electronic conductivity (>100 S cm^{-1})
- nonnegligible anionic (oxide) conductivity (~10^{-1} S cm^{-1})
- high catalytic activity for oxygen molecule dissociation and oxygen reduction
- chemical stability with regard to electrolyte and interconnection material
- match of the TEC with the other cell components
- ability to be formed into films with the desired microstructure and a good adherence to the adjacent components

A number of the above requirements are more or less fulfilled by the ABO_3 oxide systems of perovskite structure, where A is a rare earth element and B a transition metal (Fe, Ni, Co, Mn).[44-50] $LaMnO_3$-based systems are considered to be the most promising cathode materials and so were extensively studied. In most cases, alkaline earth (Sr^{2+},Ca^{2+}) cations are partially substituted for the rare earth element. The charge compensation operates by a valence change of the transition metal cations and, under certain operating conditions, by the formation of oxygen vacancies. Thus, the cathode material behaves as a mixed (ionic + electronic) conductor. The transition metal Mn may also be substituted by other transition metals, mainly Co.[51-55] Substitution operations are intended to "tailor" the most appropriate cathode with regard to the above mentioned requirements. As the $La_{1-x}Sr_xMnO_3$ is the most widely studied system, it will be described in detail. Introduction of Sr-dopant significantly modifies several properties:

The electrical conductivity increases with increasing x. A conductivity maximum is observed for $x \approx 0.5$ as shown in Figure 12.6.[56-58] A semiconducting type of behavior is observed for low Sr contents ($x < 0.4$) and a metallic behavior for the most conductive compositions. Substituting cobalt for manganese significantly improves conductivity.[58,59] Figure 12.7 shows the temperature dependence of electrical conductivity of the $La_{1-x}Sr_xMO_3$ systems with M = Mn, Co, Fe, and x = 0.3, 0.5, 0.7. Co doped lanthanum manganites show the highest conductivities.[60] Note also that at x = 0.5 the conductivity is highest for a given transition metal.

The TEC is an increasing function of x.[61,62] That is a serious drawback which can be partially overcome by substituting smaller cations (Y^{3+},Ca^{2+}) for La^{3+}.[63]

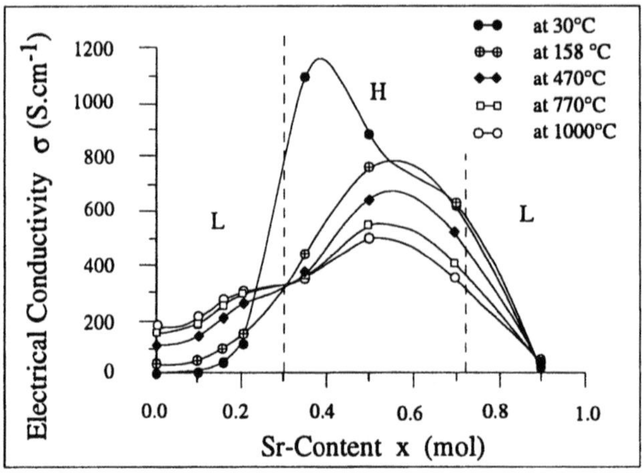

FIGURE 12.6. Electrical conductivity as a function of composition in the $La_{1-x}Sr_xMnO_3$ system. (From Li, Z., Behruzi, M., Fuerst, L., and Stover, D. *Proceedings of the Third International Symposium on Solid Oxide Fuel Cells,* Singhal, S.C. and Iwahara, H., Eds., The Electrochemical Society, Pennington, NJ, 1993, 171–179. With permission.)

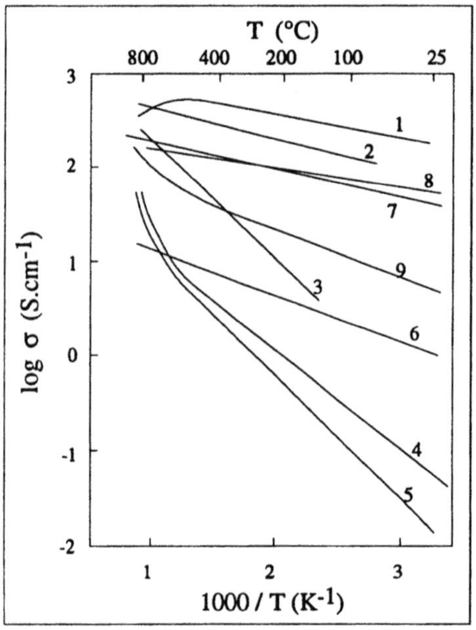

FIGURE 12.7. Temperature dependence of the electrical conductivity of sputtered films of $La_{1-x}Sr_xMO_3$:

1 — $La_{0.3}Sr_{0.7}CoO_3$	2 — $La_{0.5}Sr_{0.5}CoO_3$
3 — $La_{0.7}Sr_{0.3}CoO_3$	4 — $La_{0.3}Sr_{0.7}FeO_3$
5 — $La_{0.5}Sr_{0.5}FeO_3$	6 — $La_{0.7}Sr_{0.3}FeO_3$
7 — $La_{0.3}Sr_{0.7}MnO_3$	8 — $La_{0.5}Sr_{0.5}MnO_3$
9 — $La_{0.7}Sr_{0.3}MnO_3$	

(From Takeda, Y., Kanno, R., Noda, M., and Yamamoto, O., *J. Electrochem. Soc.*, 1987, *11*, 2656–2661. With permission.)

Thermogravimetric measurements indicate little dependence of oxygen weight loss of oxygen partial pressure. On the contrary, the oxygen weight loss increases with increasing Sr content at low P_{O2}[64,65] (see Figure 12.8). The oxygen loss reaction is reversible if the decomposition pressure of the material is not reached. In this domain, the manganite becomes a mixed conductor due to the formation of oxygen vacancies.

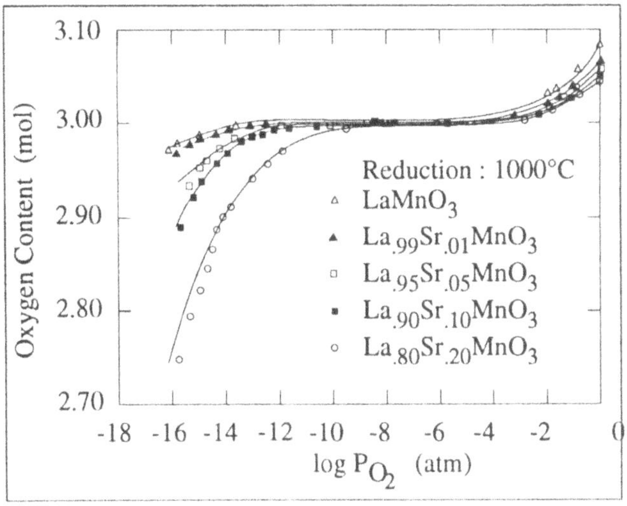

FIGURE 12.8. Oxygen content in $La_{1-x}Sr_xMnO_3$ vs. oxygen partial pressure at 1000°C. (From Kuo, J.H., Anderson, H.U., and Sparlin, D.M., *J. Solid State Chem.*, 1989, *83*, 52–60. With permission.)

With increasing x, the chemical reactivity at the cathode/zirconia interface increases, giving rise to the formation of $La_2Zr_2O_7$ and/or $SrZrO_3$. Also reported is the formation of manganese oxide.[62] The amount of these poorly conductive phases depends strongly on firing temperature and time. There is no agreement in the published literature as to the respective effects of temperature, cathode composition, and microstructure. Substitution of cobalt for manganese results in an appreciable increase of the interfacial reactivity. The formation of the secondary phases $La_2Zr_2O_7$ and/or $SrZrO_3$ does not only depend on the La to Sr ratio, but even more on the Mn:Co ratio.[55] As an example, we report here the results obtained for six $La_{1-x}Sr_xMn_{1-y}Co_yO_3$ compositions with x = 0.2, 0.5 and y = 0.05, 1. Interfacial reactions were observed at 1200, 1300, and 1350°C. The following conclusions have been drawn:

- Co-free compositions are less reactive to the formation of $SrZrO_3$ and traces of $La_2Zr_2O_7$; $SrZrO_3$ content increases with x.
- Mn-free compositions show cobalt oxide in addition to $La_2Zr_2O_7$ and $SrZrO_3$.
- Sr-rich (x = 0.5) stoichiometries tend to have higher amounts of secondary phases.
- Co-containing cathodes with x = 0.2 (La-rich) favor the formation of $La_2Zr_2O_7$ in addition to $SrZrO_3$, while cathodes with x = 0.5 (Sr-rich) show large amounts of $SrZrO_3$ in addition to $La_2Zr_2O_7$. Figures 12.9 and 12.10 illustrate all these results.

With increasing x in $La_{1-x}Sr_xMnO_3$, the overpotential decreases with the onset of a minimum for the composition x ≈ 0.5 as shown in Figure 12.11. As for electrical conductivity, the best performance is achieved with Co-containing lanthanum manganite. Hammouche[57] has shown that the cathodic polarization curve, schematically drawn in Figure 12.12, can be divided into two domains separated by a specific transition potential E_t. At low cathodic polarization ($E > E_t$), the oxide cathode behaves as a classic metallic electrode, i.e., platinum. In this region, the reactions involved are

$$1/2\ O_{2(g)} + s_{(ed)} \rightarrow O_{ad(ed)} \qquad \text{ed : electrode}$$

$$\text{el : electrolyte}$$

$$O_{ad(ed)} + V_{O(el)}^{\bullet\bullet} \rightarrow 2\ h^{\bullet} + O_{O(el)}^{x} + s_{(ed)}$$

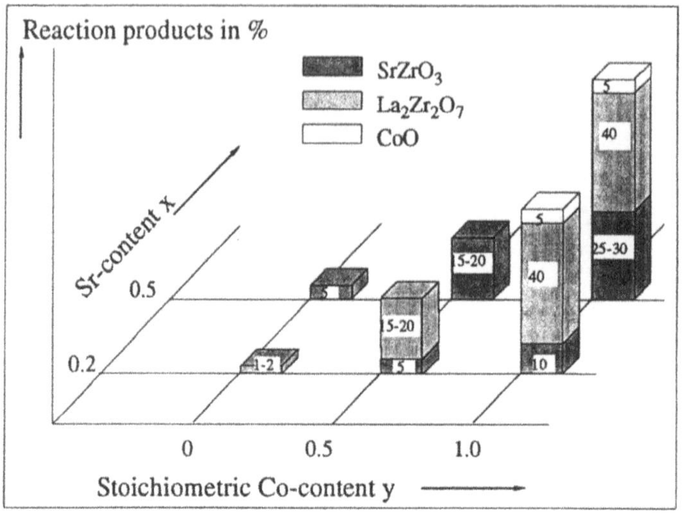

FIGURE 12.9. Secondary phases in $La_{1-x}Sr_xMn_{1-y}Co_yO_3$/YSZ powder mixtures (reaction at 1200°C for 5 h). (From Ivers-Tiffée, E., Schiessl, M., Oel, H.J., and Wersing, W. *Proceedings of the Third International Symposium on Solid Oxide Fuel Cells,* Singhal, S.C. and Iwahara, H., Eds., The Electrochemical Society, Pennington, NJ, 1993, 613–622. With permission.)

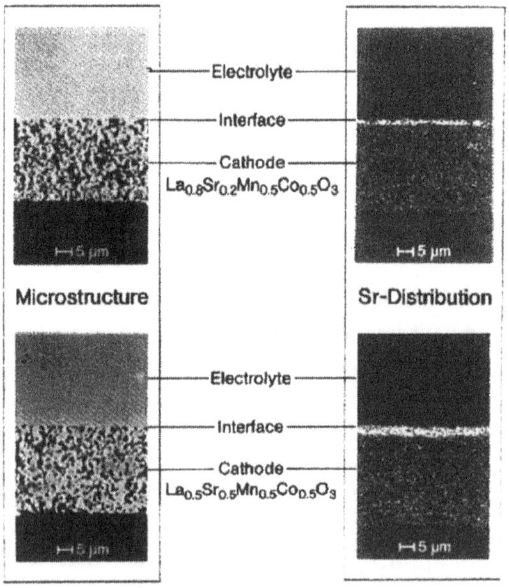

FIGURE 12.10. Microprobe analysis of cathode/electrolyte interfaces (reaction at 1300°C for 5 h). (From Ivers-Tiffée, E., Schiessl, M., Oel, H.J., and Wersing, W. *Proceedings of the Third International Symposium on Solid Oxide Fuel Cells,* Singhal, S.C. and Iwahara, H., Eds., The Electrochemical Society, Pennington, NJ, 1993, 613–622. With permission.)

where s refers to an active site on the electrode surface. These reactions take place at the gas/cathode/electrolyte triple-point contact. A SOFC using Sr-doped lanthanum manganites functions in this domain. To widen the electrode/electrolyte interface, it is suggested to use cathode-zirconia composites.[66] At the potential E_t, a sharp transition is observed. It is proposed that the sudden increase of current results from a major change of mechanism leading to the onset of an important electrocatalytic effect. The formation of oxygen vacancies in the

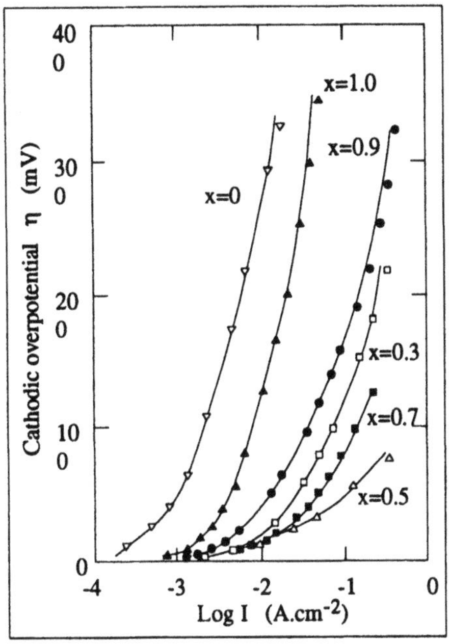

FIGURE 12.11. Cathodic polarization curves in air at 800°C for $La_{1-x}Sr_xMnO_3$ electrodes. (From Takeda, Y., Kanno, R., Noda, M., and Yamamoto, O., *J. Electrochem. Soc.*, 1987, *11*, 2656–2661. With permission.)

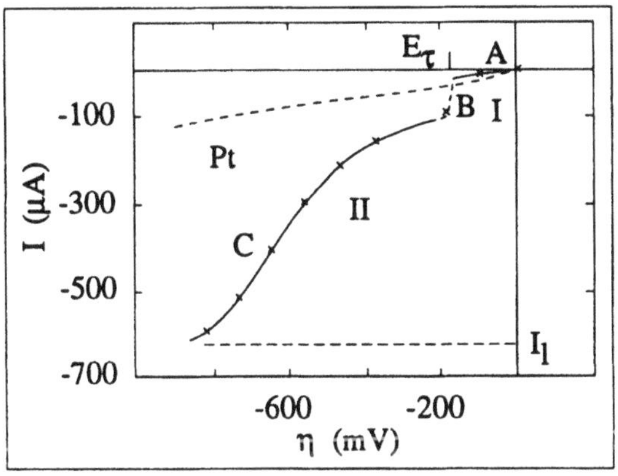

FIGURE 12.12. Steady-state current–voltage characteristics of Pt and $La_{0.7}Sr_{0.3}MnO_3$ cathodes in air at 960°C. (From Hammouche, A., thesis, Grenoble University, 1989. With permission.)

manganite is at the origin of such a behavior. This assertion has been supported by the analysis of the results obtained independently in cyclic voltammetry, transient electrode response, isothermal thermogravimetry, and impedance spectroscopy.[57] In this potential domain, the electrochemical reaction can be described by the following sequence,

$$1/2\ O_{2(g)} + s_{(ed)} \rightarrow O_{ad(ed)}$$

$$O_{ad(ed)} + V_{O(el)}^{\bullet\bullet} \rightarrow 2\ h^{\bullet} + O_{O(el)}^{x} + s_{(ed)}$$

$$O_{O(ed)}^{x} \rightarrow O_{int}^{x}$$

$$O_{int}^{x} + V_{O(el)}^{\bullet\bullet} \rightarrow O_{int}^{x} + V_{O(ed)}^{\bullet\bullet}$$

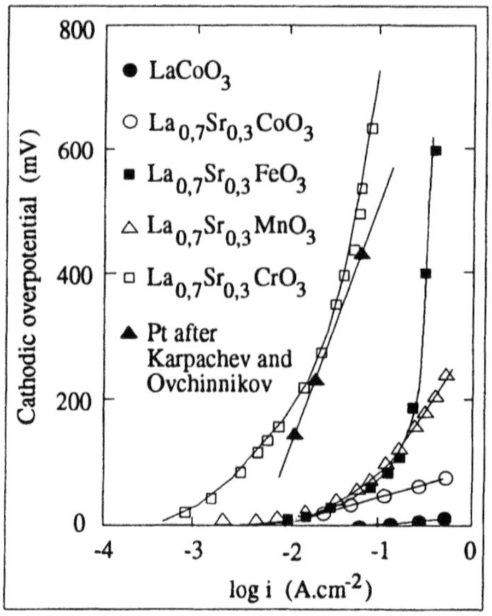

FIGURE 12.13. Cathodic polarization for sputtered $La_{1-x}Sr_xMO_3$ films in air at 800°C. (From Yamamoto, O., Takeda, Y., Kanno, R., and Noda, M., *Solid State Ionics,* 1987, *22,* 241–246. With permission.)

where int refers to the manganite/zirconia interface. The electrode reaction takes place on both the triple-point contact zone and the whole surface of the cathode, which behaves as a mixed conductor. E_t is function of dopant content in the A and, more sensitively, the B sites of the perovskite structure. This parameter could be considered as a measure of the electrocatalytic effect of the cathode under polarization. The higher its algebraic value, the better the electrocatalytic effect. The same tendency is observed for the oxygen self-diffusion coefficient, which could be more or less attributed to the electrocatalytic effect. Figure 12.13 illustrates this tendency showing that cobalt has a particularly pronounced effect.[67] Another approach to study oxygen reduction kinetics utilizes the results of electrode resistance and electrode capacitance deduced from impedance spectroscopy measurements. The dependence of these two factors on temperature and oxygen partial pressure is used for the determination of oxygen reduction mechanism(s). Although large efforts are made in this basic aspect of research in SOFCs, there is no general agreement in the published literature on the mechanism(s) of oxygen reduction at the manganite/zirconia interface.[68-71] This situation may be partially explained by the differences in characteristics of the investigated cathodes such as composition, microstructure, thickness, sintering conditions, oxygen partial pressure, cathodic polarization domain, etc.

With a view to minimize or even to avoid the formation of poorly conducting phases at the cathode/electrolyte interface, several solutions have been attempted, including:

- lowering the stoichiometry of the A site in the doped ABO_3 perovskites.[72-73]
- using some specific compositions of the $La_{1-x}Ca_xMnO_3$. It is reported that $La_{0.6}Ca_{0.4}MnO_3$ does not react with YSZ below 1200°C.[74-77] Nevertheless, lower Ca-containing solid solutions react at 1400°C with the formation of $La_2Zr_2O_7$.[77]
- inserting a chemically stable and good oxygen ion conducting buffer between the cathode and the electrolyte. Obviously, the selected cathode must have a high electrical conductivity and corresponding catalytic activity. Some experiments were made using the $La_{1-x}Sr_xCo_{1-y}Fe_yO_3$ (LCSF) family which exhibits high oxide ion conductivity (10^{-2} to 1 S cm^{-1} at 1073 K) in addition to high electronic conductivity (10^2 to 10^3 S cm^{-1}). The introduction of a $(CeO_2)_{0.8}(SmO_{1.5})_{0.2}$ buffer layer between the $La_{0.6}Sr_{0.4}Co_{0.2}Fe_{0.8}O_3$ cathode and YSZ considerably reduces the formation of resistive

YSZ / M, H_2

M = △ Mn, ▽ Fe, △ Co, ⊕ Ni, ○ Ni-YSZ,
    ▽ Ru, ⊗ Rh, ⊡ Pd, □ Pt, ⊠ Au

SDC / M', H_2

M' = ■ Pt, ● Ni-YSZ

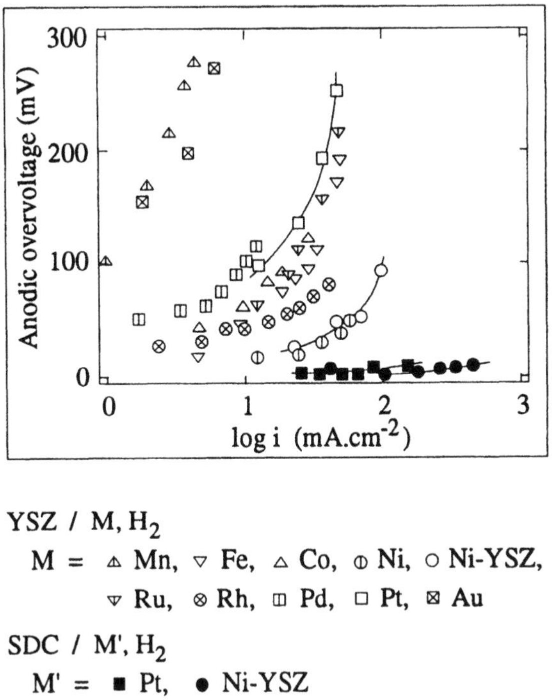

FIGURE 12.14. Anodic polarization curve at 800°C for the YSZ/M, H_2 electrode. (From Eguchi, K., Setoguchi, M., Sawano, M., Tamura, S., and Arai, H., *Proceedings of the Second International Symposium on Solid Oxide Fuel Cells,* Gross, F., Zeghers, P., Singhal, S.C., and Iwahara, H., Eds., Office for Official Publications of the European Communities, Luxembourg, 1991, 603–610. With permission.)

phases at the interfaces.[54] These promising results led the authors to suggest the use of a two-layered thin film of YSZ and Sm-doped ceria on the LCSF cathode substrate for SOFCs working at intermediate temperature.

VI. ANODE MATERIALS

Anode materials require high stability in reducing atmospheres. Transition metals are the best candidates because of their high catalytic activity. However, the TEC difference between the metals and the YSZ electrolyte prevents their use as porous layers because they tend to detach from the electrolyte during thermal cycling, causing a large increase of electrode resistance. This can be partially overcome by sintering fine-grained metal-YSZ cermets on the electrolyte surface. Taking into account volatility, chemical stability, catalytic activity, and cost, nickel appears to be the best candidate as a metallic anode. Figure 12.14 shows the steady state characteristics obtained by Eguchi et al.,[78] for different metals showing the high performance of nickel. With respect to the H_2 oxidation rate, metallic anodes may be classified as follows:

$$Ni > Fe > Ru > Co > Pt = Pd > Au > Mn$$

Suzuki et al.[79] have developed Ru-YSZ anodes which show significantly lower polarization compared with the conventional anodes using Ni-YSZ cermets. Moreover, tests using Ru/Al_2O_3 have revealed that under SOFC anode operational conditions, Ru metal has a high-steam reforming reaction activity, carbon deposition resistance, and sintering resistance. Dees et al.[80] have prepared Ni-YSZ cermets by mixing NiO and zirconia. Then NiO is reduced to

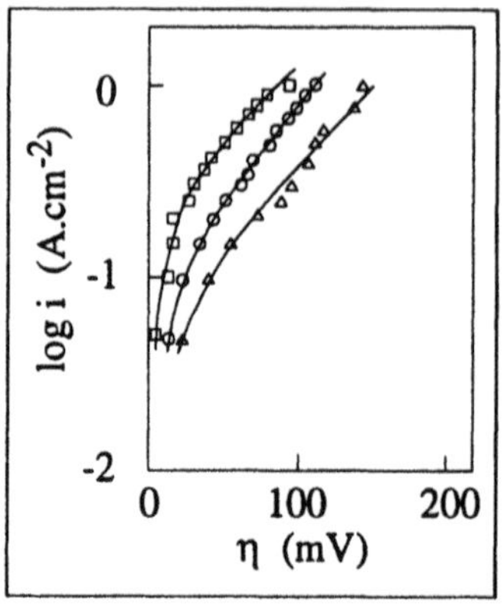

△ Ni-YSZ prepared by the powder mixing process
○ Ni-YSZ prepared by the drip-pyrolysis process
□ $Ni_{0.8}Mg_{0.2}O$-YSZ prepared by the drip-pyrolysis process

FIGURE 12.15. Tafel plots of the anodic overvoltage for three kinds of electrode: Ni-YSZ prepared by powder mixing process; Ni-YSZ prepared by drip- pyrolysis process; and $(Ni_{0.8}Mg_{\cdot 0.2})$ O-YSZ prepared by the drip-pyrolysis process. (From Okumura, K., Yamamoto, Y., Fukui, T., Hanyu, S., Kubo, Y., Esaki, Y., Hattori, M., Kusunoki, A., and Takeuchi, S., *Proceedings of the Third International Symposium on Solid Oxide Fuel Cells,* Singhal, S.C. and Iwahara, H., Eds., The Electrochemical Society, Pennington, NJ, 1993, 444–453. With permission.)

the metallic form by heating the samples in a reducing atmosphere of H_2, H_2O, and He at 1000°C. A large increase in porosity is observed after reduction. Middleton et al.[81] found that Ni particles larger than 3 µm retain a core of NiO which increases the cermet resistance. Recently, Okumura et al.[82] have studied the influence of microstructure on the anodic overvoltage in preparing Ni-YSZ and $(Ni_{0.8}Mg_{0.2}O)$/YSZ by a drip pyrolysis process. Figure 12.15 shows the steady state characteristics. The Mg-containing anode exhibits 20 mV overpotential at 300 mA cm^{-2} against 40 mV for the Ni-YSZ system. This is mainly attributed to the finely porous structure of the (Ni-Mg)O solid solution which induces a greater specific surface area, about 30 times higher than that of NiO.

The kinetics of H_2 oxidation has been investigated on a Ni-YSZ cermet using impedance spectroscopy at zero dc polarization.[83] The electrode response appears as two semicircles. The one in the high-frequency range is assumed to arise partly from the transfer of ions across the three-phase line and partly from the resistance inside the electrode particles. The semicircle observed at low frequency is attributed to a chemical reaction resistance. The following reaction mechanism is suggested:

$$H_2 \rightleftarrows 2\ H_{ad,Ni}$$

$$2 \times \left(H_{ad,Ni} \rightleftarrows H^+_{ad,Ni} + e^-\right)$$

Diffusion of $H_{ad,Ni}$ to the Ni-YSZ boundary followed by proton transfer:

$$2 \times \left(H^+_{ad,Ni} + O^{2-}_{YSZ} \rightleftarrows OH^-_{YSZ}\right)$$

$$2\ OH^-_{YSZ} \rightleftarrows H_2O + O^{2-}_{YSZ}$$

Using a ball-shaped nickel working electrode, Guindet et al.[84] plotted current–voltage curves, each point being characterized by impedance spectroscopy. All the measurements

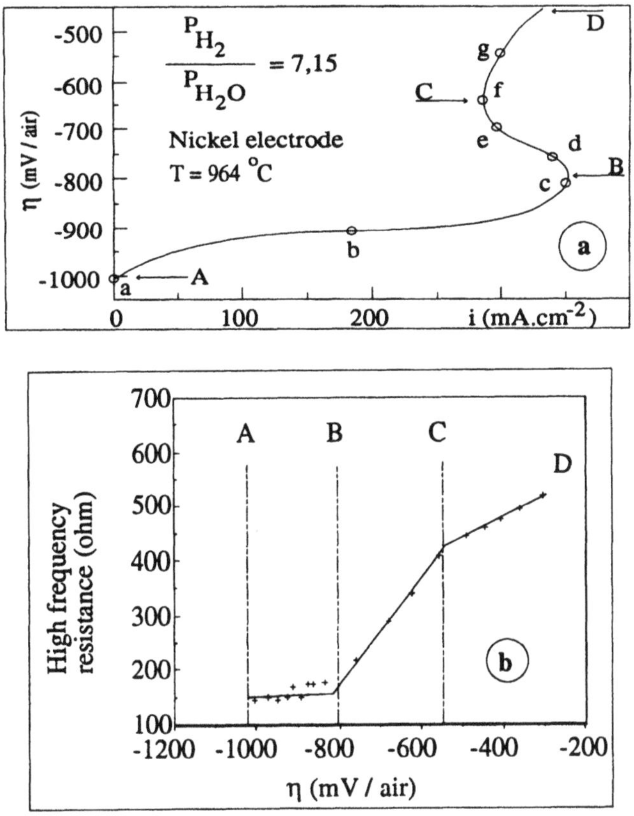

FIGURE 12.16. Typical anodic polarization curve (a) and ohmic resistance curve (b) for the YSZ/Ni, H_2/H_2O electrode. (From Guindet, J., thesis, Grenoble University, 1991. With permission.)

were performed using a three-electrode cell. Figure 12.16 shows a typical polarization curve which can be divided into three polarization domains, AB, BC, and CD. The current peak observed at point B is attributed to the metallic electrode passivation. The AB domain can be attributed unambiguously to H_2 oxidation on nickel. The variation of the current density i vs. overvoltage η is described by the following Tafel-type law:

$$i = i_0 \exp\left(\frac{2F}{RT}\eta\right)$$

with i_o = 17.8 mA cm^{-2}. A similar behavior has been observed by Kawada et al.[85] at 1000°C with an exchange current density equal to 32 mA cm^{-2}. In this polarization domain the ohmic drop, arising from the electrolyte resistance, is constant. Impedance diagrams show two semicircles which can be attributed to charge transfer and adsorption processes. In the BC domain, the current density is a decreasing function of the overvoltage. This behavior is similar to that observed for passivation phenomena and may be explained by the oxidation of Ni metal, leading to a large increase of the ohmic drop. Electrochemical oscillations were observed in this domain as well. The impedance diagram recorded at point C is characteristic for passivation.[86-91] Finally, in the CD domain, it is suggested that H_2 oxidation takes place on the NiO electrode, whose thickness continues to increase.

To decrease the anodic overpotential, it was suggested to insert a mixed conductor between YSZ and metallic conductors. Tedmon et al.[92] reported a significant decrease of polarization when ceria-based solid solutions like $(CeO_2)_{0.6}(LaO_{1.5})_{0.4}$ are used. This effect was attributed to mixed conduction resulting from the partial reduction of Ce^{4+} to Ce^{3+} in the reducing

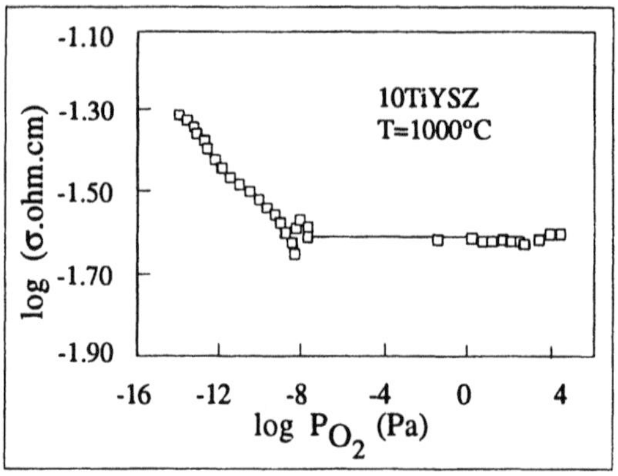

FIGURE 12.17. Dependence of electrical conductivity on oxygen partial pressure for $(YSZ)_{0.9}$-$(TiO_2)_{0.1}$ at 1000°C. (From Marques, R.M.C., Frade, J.R., and Marques, F.M.B., *Proceedings of the Third International Symposium on Solid Oxide Fuel Cells,* Singhal, S.C. and Iwahara, H., Eds., The Electrochemical Society, Pennington, NJ, 1993, 513–522. With permission.)

operating conditions. Miyamoto et al.[93] have observed a significant decrease of the ohmic drop when YSZ is used with a thin modified layer of mixed conductors, but polarization characteristics did not change in comparison with the Ni-YSZ anode. A comparative study of ceria and titania-doped YSZ has shown that titania additions in the order of 10 mol% are effective in increasing electronic conductivity within the solid solution as shown in Figure 12.17.[94-95] This suggests that such solid solutions may be good candidates as anode cermet components for SOFCs.

One of the main objectives of SOFCs in the future is the use of gaseous fuel mixtures of CO-H_2-H_2O produced from coal gasification plants or by steam reforming of hydrocarbons, especially methane. Very few data are available on the direct oxidation of methane in SOFCs.[96-100] Although nickel fulfills major requirements for anode materials when H_2 and CO are employed as fuels, its use for direct oxidation of methane encourages carbon deposition. Steele et al.[97] have shown that SOFCs with ceramic anodes can be operated with 100% CH_4 without any carbon deposition. This indicates that CH_4 can be oxidized in SOFCs without the incorporation of *ex situ* or *in situ* reforming stages. Water produced at the anode participates in the reforming reaction:

$$CH_4 + H_2O \rightleftarrows CO + 3H_2$$

This reaction is endothermic and therefore could introduce local thermal disturbances which can affect the operating conditions of the cell. On the other hand, it was found that anodic polarization is mildly affected by varying the H_2O/CH_4 ratio between 0.4 and 2. A shallow minimum is observed for the H_2O/CH_4 ratio near 1.6.[101] Finally, it has to be recalled that H_2 is more easily oxidized than CO, both being derived from the reformed CH_4.

VII. INTERCONNECTION MATERIALS (INTERCONNECTS)

Interconnection materials are necessary to combine single cells to form stacks by connecting the cathode material of one cell to the anode material of the adjacent one. By this function, they are in contact with an oxidizing and a reducing medium at the cathode and the anode respectively. The requirements to be met, especially for the planar SOFC configuration, are

TABLE 12.1
Thermal Expansion Coefficients (TEC) of $LaCrO_3$ as a Function of Dopants Between 25 to 1000°C.

Dopant	TEC ($\times 10^{-6}$/°C)
$LaCrO_3$	9.5
2% Sr	10.2
10% Co, 20% Ca	11.1
10% Co, 30% Ca	10.4
YSZ	10.3

From Anderson, H.U., *Solid State Ionics,* 1992, *52,* 33–41. With permission.

- high electronic conductivity (>1 S cm^{-1}) with a small variation within the oxygen partial pressure range from air to fuel gas
- chemical and phase stability in both air and fuel atmospheres during fabrication and operation
- TEC close to that of zirconia and other cell components
- resistance to thermal shock
- physical and electrochemical gas tightness

Despite large research efforts devoted to interconnects during the last decade, there is still no component that satisfies all the above requirements. At present, the most significant interconnection material is doped $LaCrO_3$ which belongs to the perovskite family ABO_3. Two major problems are still to be solved: (i) thermal expansion mismatch with zirconia; (ii) poor sinterability in oxidizing conditions.

A better TEC match can be successfully achieved by appropriate doping as shown in Table 12.1.[102]

Sintering and densification problems are much more serious. Generally, the near full density of chromites is achieved in sintering oxide powders in reducing atmosphere (10^{-12} to 10^{-9} atm) at relatively high temperature (~1775°C).[103] These conditions are not suitable in terms of cost or for cosintering interconnection materials and other SOFC components, in particular the cathode material. Low density and substantial open porosity are obtained for lanthanum chromites when they are sintered in air, even at temperatures higher than 1990 K. Ideally, it is desirable to sinter chromites at temperatures not exceeding 1400°C in air with additives that will improve the properties and not cause degradation or interaction with other cell components. Over the years, the problem of sinterability and densification has been addressed by a number of investigators. To solve it, sintering aids were first introduced. Meadowcroft[104] showed that the sinterability of Sr-doped $LaCrO_3$ can be significantly improved by the addition of $SrCO_3$ before sintering. The maximum beneficial effect is observed when 4 to 6 mol% is added. The improvement is probably due to the formation of $SrCrO_4$ at an intermediate temperature followed by melting and liquid phase sintering. Flandermeyer et al.[105] used low melting eutectics as well as metal (La,Y,Mg) fluorides up to 8 to 10 wt% to increase the density of sintered compacts. Co and/or Ni substitutions for chromium have been shown to promote a liquid-phase sintering mechanism resulting in remarkable improvement of density at low temperature.[106-109] For example, powder from the $La_{1-x}Ca_xCr_{1-y}Co_yO_3$ system with $x > 0.1$ and $y > 0.1$ can be densified in air at temperatures below 1400°C. Several experiments were performed on Ca-substituted $LaCrO_3$ or $YCrO_3$ with either slight A-site enrichment or depletion in the perovskite structure.[110-111] They have shown that a 2-mol% chromium deficiency drastically improves densification without any modification of the chemical properties.

In doped lanthanum chromite, the electrical conductivity increases with increasing Co and Ca content. The doped materials show a positive Seebeck coefficient, indicating p-type conductivity.[112,113] Conductivity values as high as 60 S cm^{-1} have been achieved in air at 1000°C. A decrease of conductivity was observed at low oxygen partial pressure and was attributed to the formation of oxygen vacancies.[108,114] Under these conditions, the interconnect behaves as a mixed conductor and is subject to oxygen semipermeability. This aspect has been approached by some investigators.[115,116] Yokokawa et al.[116] report that the oxygen permeation current density through lanthanum calcium chromite is of the order of 1 mA cm^{-2} in an oxygen potential gradient of 0.2 to 10^{-15} atm. This result was interpreted in terms of a point defect model and a vacancy diffusion coefficient of 1×10^{-7} cm^2 s^{-1}. SOFC evaluation has to take into account this oxygen leak, which reduces cell performance. Finally, on the basis of the results obtained on calcium- and nickel-doped lanthanum chromite, Christie et al.[109] claim that Ca-doped $LaCrO_3$-based materials which use a liquid-phase sintering mechanism to facilitate densification are probably not stable in the presence of CO_2. The authors emphasize the need to evaluate interconnect materials under realistic conditions, particularly in SOFC systems running on fuels other than hydrogen.

Particularly in the SOFC planar configuration, metallic interconnects have been developed in the form of metal alloys which fulfill the more severe requirements, mainly electrical conductivity, gas tightness, chemical stability at the oxygen as well as the fuel gas side, and low-cost fabrication. The use of metallic compounds is also advantageous with regard to their high thermal conductivity, which could help to even out the cell temperature. However, bonding to ceramics and matching of thermal expansion coefficient are still to be improved. An oxide dispersion-strengthened (ODS), Cr-based alloy has been successfully tested.[117] After 5000 h of testing in operational gases (steam-reformed natural gas as well as coal gas), a thin, stable surface layer protects the alloy. Metallic interconnect materials may also be used in a cermet form. Seto et al.[118] report that a cermet of 60 vol% alumina and 40% alloy (Inconel 600) was best suited to the requirements of a planar SOFC, the cells using the cermet being equal in performance to those using metallic interconnectors.

VIII. SOLID OXIDE FUEL CELL CONFIGURATIONS AND PERFORMANCE

As far as the technological aspect of SOFCs is concerned, three main designs are presently seeing rapid development: tubular, planar and monolithic configurations. The cell components in these different designs are the same except for the interconnection material, which can be ceramic or metallic:

- electrolyte: $(ZrO_2)_{1-x}$ $(Y_2O_3)_x$ with x = 0.08
- anode: Ni/YSZ cermet with a nickel volume close to 50% and a porosity of 40%
- cathode: $La_{1-x}Sr_xMnO_3$. Generally, x = 0.16 is selected, but some tests are performed with higher Sr contents characterized by a higher bulk conductivity and lower cathodic polarization. A porosity of 30 to 40% is commonly used.
- interconnection material: M-doped $LaCrO_3$ (M = Mg, Sr) or metallic superalloys.

A. THE TUBULAR CONFIGURATION

Initially, the development programs for the fabrication of SOFC modules started using a relatively thick electrolyte, the cell being arranged in a "bell and spigot" tubular design.[119] To achieve mechanical strength, the self-supporting criteria dictate a minimum thickness around 0.4 mm, leading to an important ohmic drop under SOFC operating conditions. Over the last 20 years, the tubular design has been completely revised by Westinghouse,[120] allowing

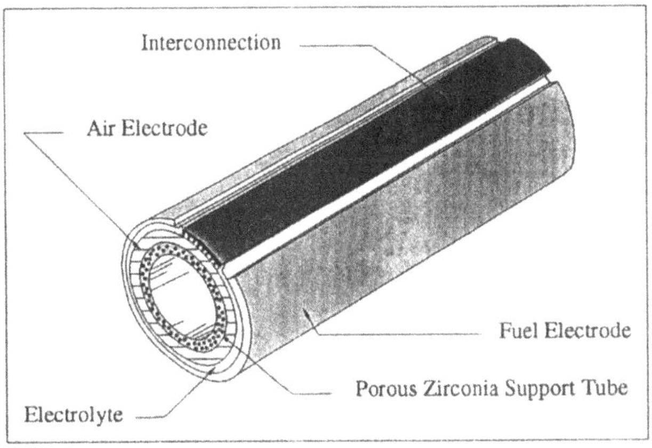

FIGURE 12.18. Schematic representation of the solid oxide fuel cell tubular configuration (Westinghouse). (From Singhal, S.C., *Proceedings of the Second International Symposium on Solid Oxide Fuel Cells,* Gross, F., Zeghers, P., Singhal, S.C., and Iwahara, H., Eds., Office for Official Publications of the European Communities, Luxembourg, 1991, 25–33. With permission.)

the fabrication of a simplified assembly suitable for mass-production processes. Figure 12.18 illustrates the configuration of the single tubular cell.[121] A porous calcia-stabilized zirconia tube closed at one end forms the porous support tube (PST). Its sintering is conducted at 1650°C in air. The first PSTs were 30 cm long with a thickness of 1.5 cm, acceptable according to engineering studies. The PST is overlaid with a 0.5- to 1.4-mm-thick porous cathode of Sr-doped lanthanum manganite. Then a gas-tight layer of YSZ electrolyte 40 μm thick covers the cathode, except for a strip 9 mm wide along the active cell length. The interconnection material based on Mg-doped $LaCrO_3$ is deposited in this strip. Finally, the whole electrolyte area is covered by the anode material formed by a Ni/YSZ cermet. To avoid internal short circuits, a narrow zone is laid out near the interconnection material. The cell components, materials and fabrication process related to the Westinghouse tubular geometry are summarized in Table 12.2.[122] To increase the power output of the tubular single cell, development efforts are currently being made in two directions:

- increasing the active length from 30 to 100 cm (currently in production)
- reducing the thickness of the PST because the inherent resistance caused by the oxygen diffusion toward the cathode is still high. Development efforts have allowed a reduction in the thickness to 1.2 mm.

TABLE 12.2
Cell Components, Materials, and Fabrication Processes for the Tubular Configuration of SOFC (Westinghouse)

Component	Material	Thickness	Fabrication process
Support tube	$ZrO_2(CaO)$	1.2 mm	Extrusion sintering
Air electrode	$La(Sr)MnO_3$	1.4 mm	Slurry coat-sintering
Electrolyte	$ZrO_2(Y_2O_3)$	40 μm	Electrochemical vapor deposition
Interconnection	$LaCr(Mg)O_3$	40 μm	Electrochemical vapor deposition
Fuel electrode	$Ni\text{-}ZrO_2(Y_2O_3)$	100 μm	Slurry coat-electrochemical vapor deposition

From Singhal, S.C., *Proceedings of the Second International Symposium on Solid Oxide Fuel Cells,* Gross, F., Zeghers, P., Singhal, S.C., and Iwahara, H., Eds., Office for Official Publications of the European Communities, Luxembourg, 1991, 25–33. With permission.

Recently, a major improvement in the tubular design has been introduced, definitively eliminating the PST: the present SOFC technology makes use of an air electrode sufficiently thick to support the whole cell. The resistance associated with oxygen diffusion toward the cathode is eliminated, which significantly improves the cell efficiency. A sixfold enhancement in cell power output may result from combining the cell length increase and the utilization of an air electrode-supported cell. Figure 12.19 shows the current–voltage curve of a single tubular state-of-the-art SOFC over the temperature range 875 to 1000°C at constant oxidant and fuel utilizations. The cell in question has a thin-wall PST (1.2 mm thick) with 50 cm of active length.[122] As dictated by thermodynamics, lower temperatures extrapolate to higher open-circuit voltages. Also, as expected, at high temperature, higher voltages are obtained at higher current densities due to a better electrolyte conductivity and lower electrode polarization. The power output dependence of the current density is shown in Figure 12.20, which shows a peak power density of approximately 0.252 W/cm^2 at 1000°C.

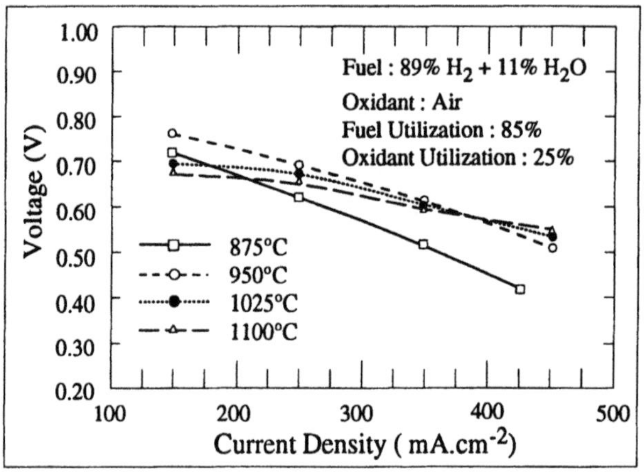

FIGURE 12.19. Current-voltage curve of a tubular solid oxide fuel cell at different temperatures. (From Singhal, S.C., *Proceedings of the Second International Symposium on Solid Oxide Fuel Cells,* Gross, F., Zeghers, P., Singhal, S.C., and Iwahara, H., Eds., Office for Official Publications of the European Communities, Luxembourg, 1991, 25–33. With permission.)

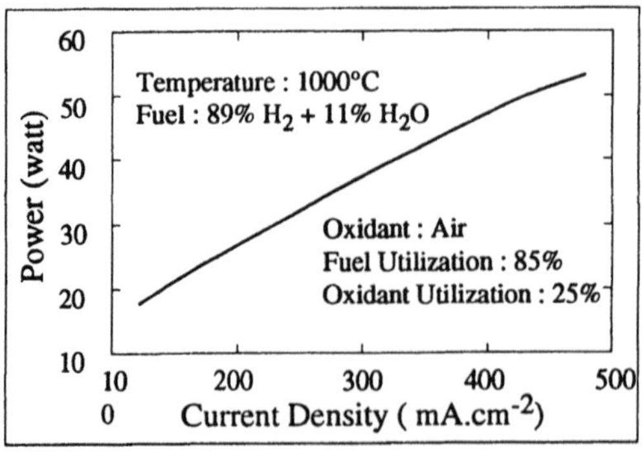

FIGURE 12.20. Power output dependence on current for a 50-cm active cell length. (From Singhal, S.C., *Proceedings of the Second International Symposium on Solid Oxide Fuel Cells,* Gross, F., Zeghers, P., Singhal, S.C., and Iwahara, H., Eds., Office for Official Publications of the European Communities, Luxembourg, 1991, 25–33. With permission.)

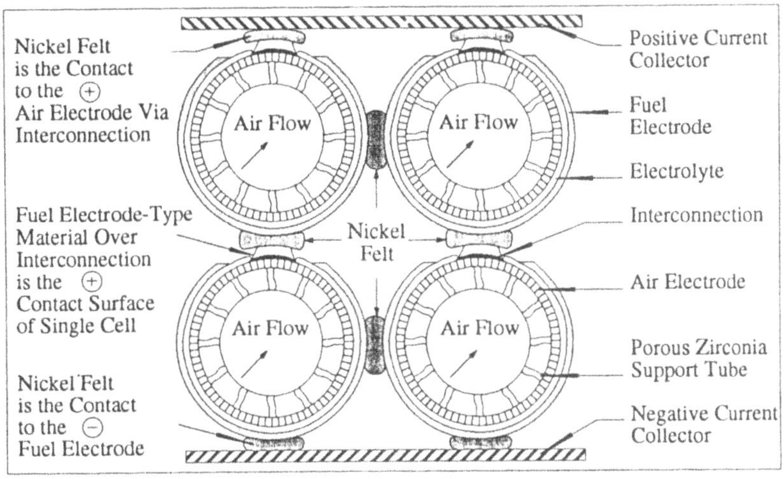

FIGURE 12.21. Cross section perpendicular to cell axis for a multicell assembly with tubular configuration (Westinghouse). (From Brown, J.T., *IEEE Trans. on Energy Conv.*, 1988, 2, 193–197. With permission.)

In the Westinghouse tubular technology, the arrangement of single cells in stacks is made by connecting cells with ductile nickel felt pads which are in permanent contact with the reducing fuel atmosphere. These pads, made of nickel fibers sinter-bonded to each other and to the nickel metal of the cermet anode, provide a mechanically compliant and low-electrical resistance connection between single cells. This arrangement is schematically illustrated in Figure 12.21.[121] As the PST is closed at one end, oxidant (air) is injected through a ceramic tube inserted in the tubular fuel cell and flows through the annular space lying between the cell and the injector tube. Fuel flows on the outside of the cell from the closed end and is electrochemically oxidized while flowing to the open side of the cell. This arrangement makes it easy to burn the depleted fuel gas by the oxygen-depleted air. Typically, 50 to 90% of the fuel is converted electrochemically. Part of the depleted fuel is recirculated in the fuel stream and the rest combusted to preheat air and/or fuel coming into the fuel cell. Figure 12.22 shows the basic design concept of the Westinghouse SOFC generator.[123] It is worth noting that this arrangement requires no sealing.

Evidence that the tubular configuration is currently coming to maturity is the construction of a Pre-Pilot Manufacturing Facility (Westinghouse) dedicated to SOFC manufacturing development, moving SOFC technology from a laboratory environment to a manufacturing environment.[124]

Table 12.3 summarizes the different size generators that have been developed up to now, showing that a number of tests have been performed by prospective customers. The Osaka Gas and Tokyo Gas companies have developed a 25 kW-class SOFC cogeneration system jointly with Westinghouse.[125] The specifications and operating conditions of this system are shown in Table 12.4. The system consists of two independently operable modules. Each module includes 576 SOFC single cells of 50 cm active length. Natural gas is used as a fuel. The system is able to provide 33 kW net power and 27 kW steam at 8 kg/cm^2 gauge pressure at peak operation. The estimated performance of this 25-kW SOFC system is shown in Figure 12.23.[125] A maximum net efficiency of 35% is obtained at the thermal balance point where the system balances thermally without external heat input. Note that it is planned to begin testing a 100-kW generator in early 1994.

Although the Westinghouse process for manufacturing tubular cells is highly developed, as already discussed, it is generally considered too expensive to be competitive commercially.

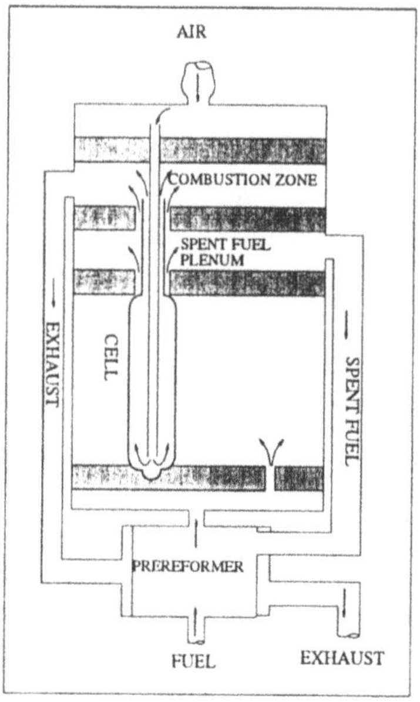

FIGURE 12.22. Westinghouse seal-less generator concept with integrated prereformer. (From Takeuchi, S., Kusunoki, A., Matsubara, H., Kikuoka, J., Ohtsuki, J., Satomi, T., and Shinosaki, K., *Proceedings of the Third International Symposium on Solid Oxide Fuel Cells,* Singhal, S.C. and Iwahara, H., Eds., The Electrochemical Society, Pennington, NJ, 1993, 678–683. With permission.)

TABLE 12.3
Summary of Westinghouse SOFC Generator Systems

Customer	Size (kW)	No. of cells	Cell length (cm)	Test time (h)	Year
U.S. DOE	0.4	24	30	2,000	1984
U.S. DOE	5.0	324	30	500	1986
TVA	0.4	24	30	1,760	1987
Tokyo Gas (TG)	3.0	144	36	5,000	1987–88
Osaka Gas (OG)	3.0	144	36	3,700	1987–88
GRI	3.0	144	36	5.400	1989–90
U.S. DOE	20.0	576	50	3.355	1990–91
KEPCO/TG/OG	25.0*	1.152	50	—	1992
TG/OG	25.0*	1.152	50	—	1992

* Nominal rating. Capable of producing 40 kW at peak operation.

From Singhal, S.C., *Proceedings of the Second International Symposium on Solid Oxide Fuel Cells,* Gross, F., Zeghers, P., Singhal, S.C., and Iwahara, H., Eds., Office for Official Publications of the European Communities, Luxembourg, 1991, 25–33. With permission.

The main reason resides in the number of costly high-temperature unit operations and low-pressure electrochemical vapor deposition (EVD) operations. In addition, the active surface-to-volume ratio is approximately 1 cm^2/cm^3, which is considered to be a modest performance. This parameter could be increased with corresponding increases of both volume power density and area power density. So, in the last decade, many efforts have been made to find new concepts based more or less on planar cell structures.

TABLE 12.4
Specifications of 25 kW-Class SOFC Cogeneration System

Generator	20 kW-class module × 2	
	Reforming: internal (prereformer)	
	Operating temperature: 1000°C	
	Pressure: atmospheric	
Electric output	Net ac power	25 kW at thermal balance point
		33 kW at maximum
Heat recovery	Steam 8 kg/cm^2 27 kW	
Fuel	Town Gas 13A (natural gas)	
Control	Automatic	
Configuration	Package type	

From Shinosaki, K., Washio, S., Satomi, T., and Koike, S., *Proceedings of the Third International Symposium on Solid Oxide Fuel Cells,* Singhal, S.C. and Iwahara, H., Eds., The Electrochemical Society, Pennington, NJ, 1993, 684–689. With permission.

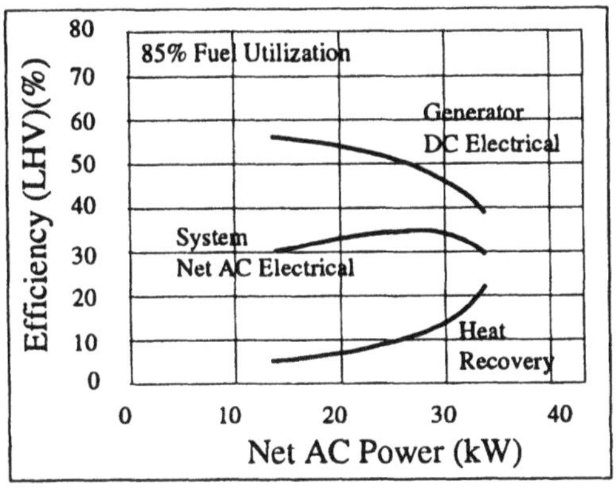

FIGURE 12.23. Estimated performance of the Westinghouse 25 kW-class solid oxide fuel cell cogeneration system developed by Westinghouse jointly with Tokyo Gas and Osaka Gas. (From Shinosaki, K., Washio, S., Satomi, T., and Koike, S., *Proceedings of the Third International Symposium on Solid Oxide Fuel Cells,* Singhal, S.C. and Iwahara, H., Eds., The Electrochemical Society, Pennington, NJ, 1993, 684–689. With permission.)

B. MONOLITHIC SOLID OXIDE FUEL CELLS (MSOFCs)

The MSOFC concept was initiated at Argonne National Laboratory.[126,127] The cell consists of a honeycomb-like array of fuel and oxidant channels that looks like corrugated paperboard. Co-flow and cross-flow versions of this geometry have been investigated. Figure 12.24 illustrates the cross-flow version.[128,129] Oxidant and fuel channels are formed from corrugated anode and cathode layers. These layers are separated alternatingly by flat multilayer laminates of the active cell cathode/electrolyte/anode components and multilayer laminates in the following sequence: anode/interconnection material/cathode. To sinter the different materials and create permanent and tight bonds, green materials are co-fired at temperatures of about 1300 to 1400°C. It is expected that an active surface area per unit volume of 10 cm^2/cm^3 can be achieved. Potentially, MSOFCs may be able to operate at an average current density four times higher than that of the tubular design and at around 50 mV higher cell voltages. With further development, a power level greater than 2.5 W/cm^2 is expected. Figure 12.25 shows the performance of a MSOFC single cell operating at 1000°C using H_2 as fuel and air as

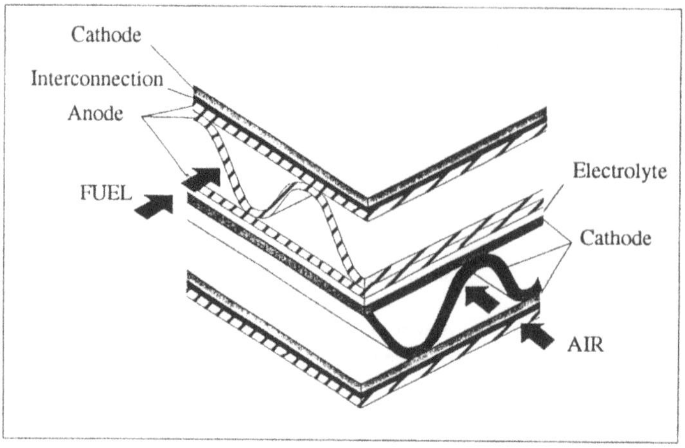

FIGURE 12.24. Cross-flow monolithic solid oxide fuel cell. (From Minh, N.Q., Horne, C.R., Lin, F., and Staszak, P.R., *Proceedings of the First International Symposium on Solid Oxide Fuel Cells,* Singhal, S.C., Ed., The Electrochemical Society, Pennington, NJ, 1989, 307–316. With permission.)

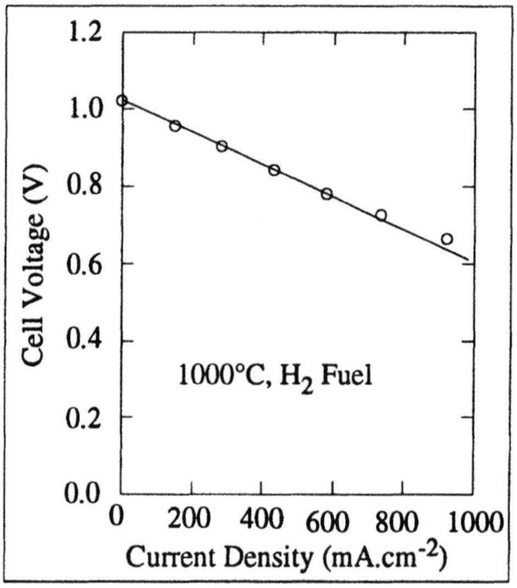

FIGURE 12.25. Current-voltage characteristics of monolithic single solid oxide fuel cell at 1000°C using pure H_2. (From Brown, J.T., *High Conductivity Solid Ionic Conductors Recent Trends and Applications,* Takahashi, T., Ed., World Scientific, Singapore, 1989, 630–63. With permission.)

oxidant.[3] A current density of 1 A/cm^2 at 0.6 V can be achieved, which is a remarkable result. Although important improvements have been observed for MSOFC multicell stacks, their performance remains significantly lower than that of the single cells.[130] This is mainly due to interactions among the materials during the co-sintering process and poor densification of the lanthanum chromite-based materials in laminated form. The principal challenge is now to develop the technology required to fabricate high-performance stacks. Recent experimental results show that the interconnection material, laminated to thin anode and cathode layers (<15 mm), can be sintered to high density at 1400°C in air.[131] The electrode pore volume is believed to fill with liquid phases from the interconnection material and densify, thus allowing the latter also to densify. Nevertheless, the confirmation of the performance has not yet been established.

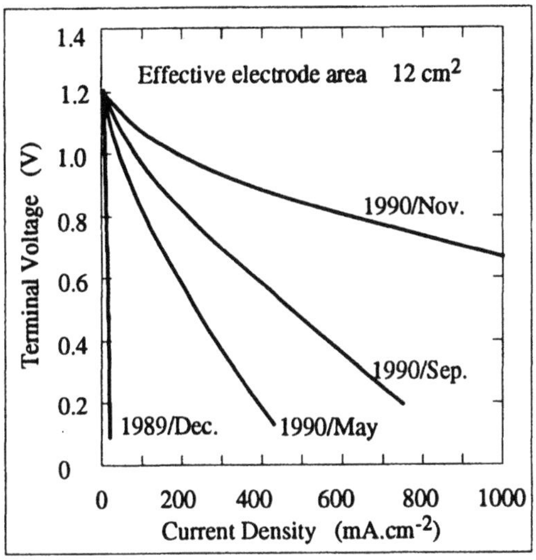

FIGURE 12.26. Improvement of the single monolithic cell performance with time at 1000°C. (From Takagi, H., Kobayashi, S., Shiratori, A, Nishida, K., and Sakabe, Y., *Proceedings of the Second International Symposium on Solid Oxide Fuel Cells,* Gross, F., Zeghers, P., Singhal, S.C., and Iwahara, H., Eds., Office for Official Publications of the European Communities, Luxembourg, 1991, 99–103. With permission.)

An as-monolithic SOFC has recently been investigated where the single cell is composed of an electrolytic membrane with anode and cathode, two porous distributors surrounded by gas-tight materials (spacer), and two interconnects.[132] The distributors are composed of a ceramic foam with about 90% porosity. The fuel gas or oxidant gas flows through the open pores in the distributors. The single cell performance is illustrated in Figure 12.26,[132] showing remarkable improvement with time. No data for multicell stacks have been reported for this technology.

C. THE PLANAR (BIPOLAR PLATE) DESIGN

A sketch of the planar design is shown in Figure 12.27.[133] The cell consists of a cross-flow arrangement where the gas-tight separation is achieved by dense ceramic or metallic plates with grooves for oxidant and fuel supply to the appropriate electrode. The active electrochemical cell is formed by a porous cathode (P: positive), a dense and thin electrolyte (E), and a porous anode (N: negative) forming a composite flat multilayer, commonly called PEN. For the PEN fabrication, several methodologies have been investigated. Some of them involve forming the thin green layers of the three components, cutting, laminating, and co-firing.[134] In such a procedure, the mismatch of sintering shrinkage profiles and the thermal expansion parameters of the PEN-forming components have to be minimized to achieve a crack-free electrolyte and a good flatness of the multilayer. In another PEN fabrication procedure, optimized cathode and anode structures are tape casted[135] or screen printed[136] on the dense sintered electrolyte and are co-fired at around 1200°C. Figure 12.28 shows a typical SEM micrograph of a PEN cross section. A current–voltage plot of a planar single cell of 3×5 cm^2 is shown in Figure 12.29. A power density of 0.4 W/cm^2 at a cell voltage of 0.7 V can be achieved with pure H_2 as a fuel at 1000°C.[136]

To transfer the performance characteristics exhibited by single cells, numerous organizations are investigating a wide range of planar SOFC configurations focusing on different interconnect (separator) designs and fabrication processes. As far as the interconnect is concerned, two approaches are currently investigated: ceramic or metallic.

The ceramic-type interconnect commonly used for planar SOFCs is based on lanthanum chromite. Recently, a $La_xCa_{1-x}Cr_yCo_{1-y}O_3$ solid solution has been developed. It is readily

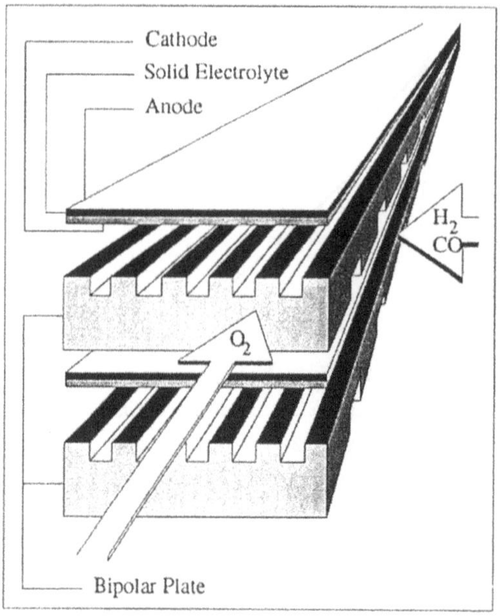

FIGURE 12.27. Bipolar plate design of solid oxide fuel cell.

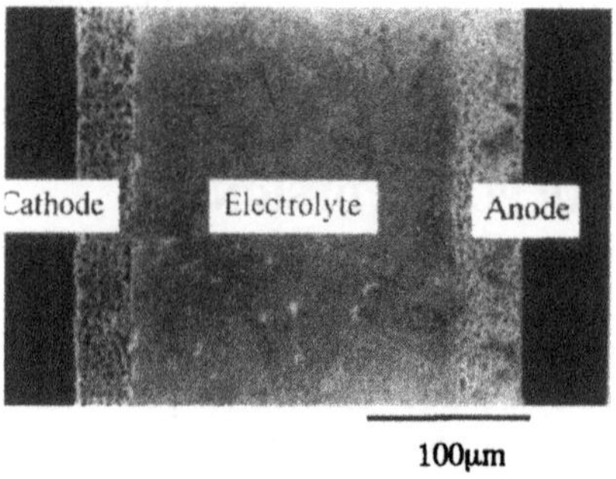

FIGURE 12.28. Cross section of anode/electrolyte/cathode multilayer. (From Tagaki, H, Taira, H., Shiratori, A., Koyabashi, S., Sugimoto, Y., Sakamoto, S., and Tomono, K., *Proceedings of the Third International Symposium on Solid Oxide Fuel Cells,* Singhal, S.C. and Iwahara, H., Eds., The Electrochemical Society, Pennington, NJ, 1993, 738–743. With permission.)

sintered in air at 1350 to 1500°C and appears to have superior stability and electrical properties compared to previous lanthanum chromite materials.[137] Densities greater than 95% of theoretical are routinely achieved. The interconnect is machined into the desired shape to form the channels for gas flow. Figure 12.30 shows a sectional view of a 25-cell stack of 10 × 10 cm^2 and its performance. A maximum output of 421 W is obtained.

The metallic-type interconnect was invented at Siemens laboratories[137] and is currently being developed by some organizations.[137-138] The Siemens multiple-cell array concept is shown in Figure 12.31. It can be seen that multiple cells are made up of individual PEN cells. The metallic bipolar plate combines the separation of the fuel gas from the air with a high resistive contact between adjacent cells. In addition, the relatively high thermal conductivity

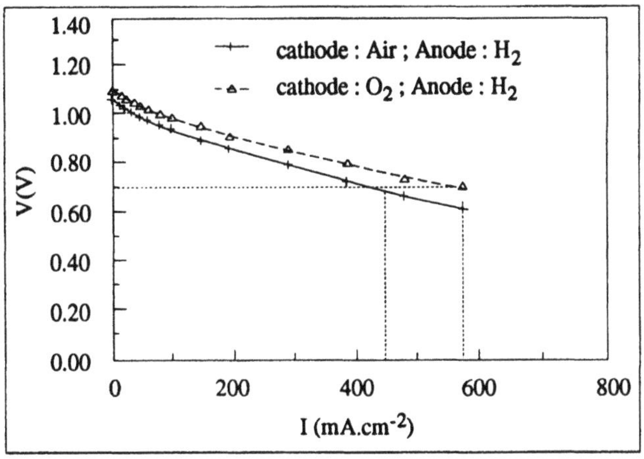

FIGURE 12.29. Current–voltage characteristics of a single planar cell (5 × 5 cm²). (From Van Berkel, F.P.F., Van Heuveln, F.H., and Huijmans, J.P.P., *Proceedings of the Third International Symposium on Solid Oxide Fuel Cells*, Singhal, S.C. and Iwahara, H., Eds., The Electrochemical Society, Pennington, NJ, 1993, 744–751. With permission.)

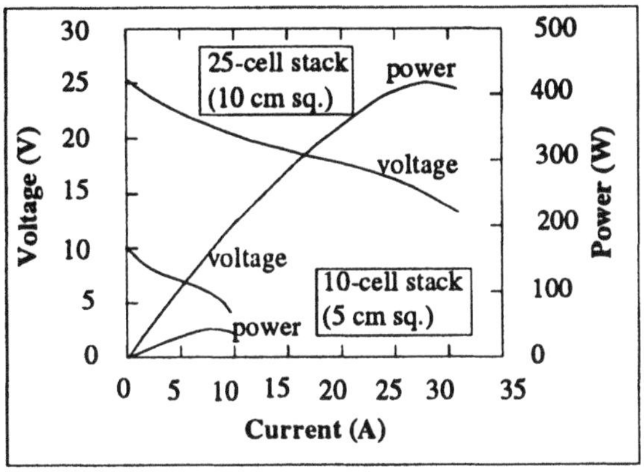

FIGURE 12.30. Performance curves of a ten-cell stack of 5 cm² cell and a 25-cell stack of 10 cm² cell. (From Hikita, T., Hishimuma, M., Kawashima, T., Yasuda, I., Koyama, T., and Matsuzaki, Y., *Proceedings of the Third International Symposium on Solid Oxide Fuel Cells*, Singhal, S.C. and Iwahara, H., Eds., The Electrochemical Society, Pennington, NJ, 1993, 714–723. With permission.)

of metal leads to a more uniform heat distribution in the SOFC stack. The metal alloys used now fulfill the main requirements, i.e., good match of thermal expansion coefficient and good resistance to corrosion. Figure 12.32 shows the performance of three stacks with one, four, and ten layers with four cells per layer. The current densities of stacks are in the same range as those of single cells. The corresponding power outputs are reported in Figure 12.33. Finally, in the Mitsui version of the planar configuration,[139] current collectors for the air side and fuel side are different. The materials of both current collectors are very similar in their composition to those of the electrode attached. The gas separator is made up of some electrically nonconductive ceramic component. They mainly consist of the $MgAl_2O_4$ spinel material.

Most of the planar SOFC designs require high-temperature seals, chemically stable in both fuel and oxidant atmospheres. Very few data concerning this aspect are available in the published literature. There is no doubt that the success of the planar configuration will greatly depend on how the problem of the high-temperature, leak-tight seal can be satisfactorily solved.

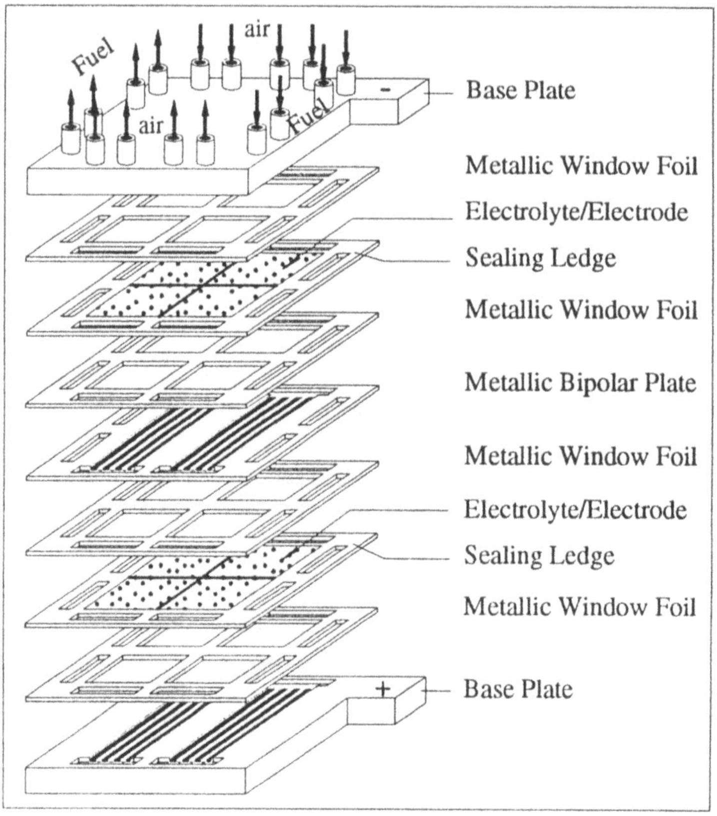

FIGURE 12.31. Multiple-cell array concept for solid oxide fuel cell reactor (Siemens). (From Drenckhahn, W. and Wollmar, H.E., *Fuel Cell Seminar 1992,* Courtesy Associates, Washington, D.C., 1992, 419–422. With permission.)

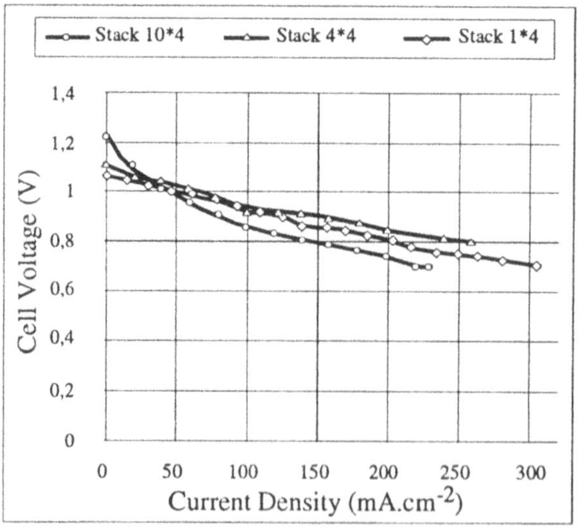

FIGURE 12.32. Average cell performances of three different stacks (Siemens). (From Drenckhahn, W. and Wollmar, H.E., *Fuel Cell Seminar 1992,* Courtesy Associates, Washington, D.C., 1992, 419–422. With permission.)

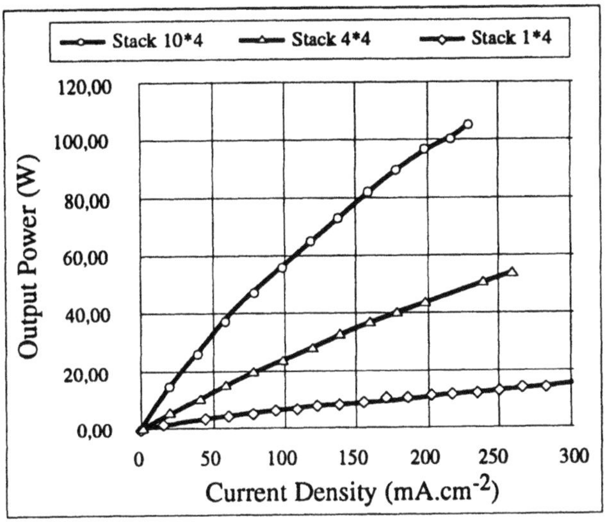

FIGURE 12.33. Dependence of output power on current density for three different stacks (Siemens). (From Drenckhahn, W. and Wollmar, H.E., *Fuel Cell Seminar 1992*, Courtesy Associates, Washington, D.C., 1992, 419–422. With permission.)

IX. COMPARATIVE PERFORMANCE EVALUATION

A comparative evaluation of the performance potentials of a number of SOFC configurations has been recently published.[140] The analyzed configurations are different in their design, contact arrangement, and the interconnect used. Identical PEN materials and operating conditions were taken into account to determine the power losses of the different designs. Performance was analyzed by identifying the cross-plane and in-plane resistances of the SOFC elements. Table 12.5 summarizes the results of the three main configurations described previously: tubular (Westinghouse), monolithic (Argonne), and planar (Siemens).

TABLE 12.5
Potential Cell Performance of 400 mA cm^{-2} at 1000°C

Configuration	Cell voltage (V)	Power density (W/cm^2)	Cell efficiency (%)
Tubular	0.65	0.261	50
Monolithic	0.84	0.335	65
Planar	0.8	0.318	61

X. MANUFACTURING PROCESSES

The commercialization of SOFC generators requires fabrication techniques to be as simple as possible, fast, and moderate in terms of cost. On the other hand, the technique chosen may influence the SOFC configuration (tubular, monolithic, planar). A wide range of manufacturing processes have been investigated to fabricate the multilayer cell.[141] Roughly, these processes can be divided into two categories: gas phase and wet-type methods.

The gas-phase route includes:

- EVD combined with classical vapor deposition (CVD)[142-148]
- chemical aerosol deposition technology (CADT)[141,149]

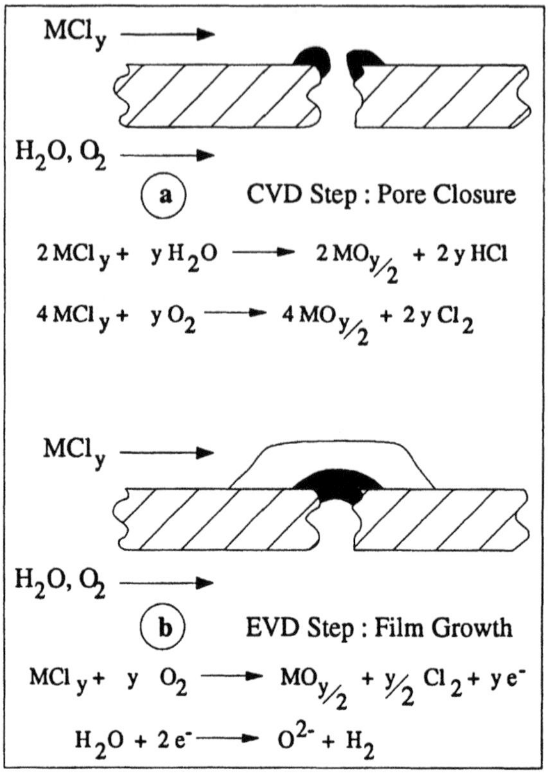

FIGURE 12.34. Principle of electrochemical vapor deposition. (From Pal, U.B. and Singhal S.C., *Proceedings of the First International Symposium on Solid Oxide Fuel Cells,* Singhal, S.C., Ed., The Electrochemical Society, Pennington, NJ, 1989, 41–56. With permission.)

- spray pyrolysis[150]
- plasma spraying[151,152]
- laser spraying[153]
- laser evaporation[154]
- radio frequency sputtering[155-157]
- electron beam evaporation[158,159]

Among these methods, only the CVD–EVD technique can be considered as a "state-of-the-art" gas-phase method, and so only this will be described, with details for the case of the deposit of YSZ. It was developed first by Westinghouse[160] and it has demonstrated the ability to reproducibly deposit relatively thin films of numerous ionic (electrolytes) or electronic conducting (interconnects) layers on porous supports. The principle of the CVD–EVD technique is shown in Figure 12.34.[145] The driving force is the oxygen chemical potential gradient, and the growth of the YSZ layer proceeds in two steps. The first is a conventional CVD reaction. The metal chlorides $ZrCl_4$ and YCl_3 are volatilized in a predetermined ratio and passed over the outer surface of the porous air electrode (Sr-doped lanthanum manganite) which covers the porous support tube (calcia-stabilized zirconia). The oxygen partial pressure in the chloride gaseous mixture is estimated to be 10^{-18} atm. Water vapor, containing in some cases oxygen at a partial pressure of about 10^{-4} atm, flows inside the support tube. The operating temperature is about 1400°C. By molecular diffusion of metal chlorides and water vapor (oxygen) through the porous support tube and the cathode material, the following reactions take place:

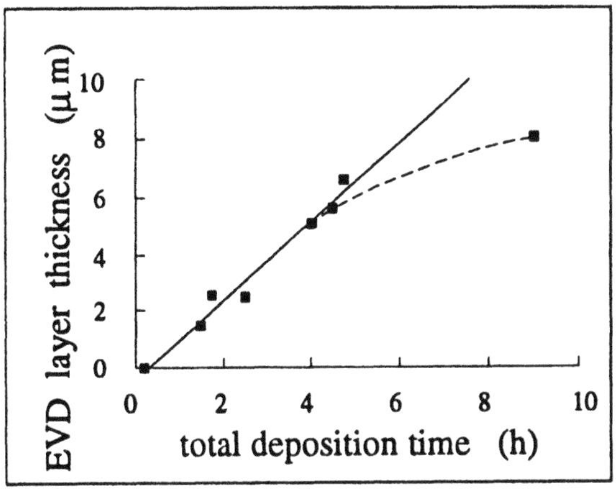

FIGURE 12.35. Growth of YSZ film thickness by EVD as function of time. (From de Vries, K.J., Kuipers, R.A., and de Haart, L.G.J., *Proceedings of the Second International Symposium on Solid Oxide Fuel Cells,* Gross, F., Zeghers, P., Singhal, S.C., and Iwahara, H., Eds., Office for Official Publications of the European Communities, Luxembourg, 1991, 135–143. With permission.)

$$2\ MCl_y + y\ H_2O \rightarrow 2\ MO_y/2 + 2y\ HCl$$

$$4\ MCl_y + y\ O_2 \rightarrow 4\ MO_y/2 + 2y\ Cl_2$$

where M is a metallic cation (M = Zr, Y) and y the valency of M. YSZ formed by this reaction fills the cathode pores. It was shown that the films produced during this CVD stage penetrate no more than two to three pore diameters from the chloride face of the substrate.[144] Hence pore closure is achieved and the actual EVD starts, involving the following electrochemical reactions:

$$MCl_y + \frac{y}{2}O^{2-} \rightarrow MO_{y/2} + \frac{y}{2}Cl_2 + y\,e^-$$

on the chloride side and

$$H_2O + 2e^- \rightarrow O^{2-} + H_2$$

$$\tfrac{1}{2}O_2 + 2e^- \rightarrow O^{2-}$$

on the steam side. O^{2-} is transported from the steam side to the chloride side where it reacts, electrons being transported in the opposite direction. From kinetic studies, it was concluded that the CVD–EVD growth process is governed by:

- a linear time dependency in the pore closure step. The rate-limiting step is considered to be the gas diffusion in the substrate pores.
- a parabolic time dependence in the film growth step. The limiting step is the electrochemical bulk transport of oxygen through the growing YSZ layer. This behavior is illustrated in Figure 12.35. In addition, it was noticed that the moment of change in the rate-limiting step depends strongly on the substrate pore dimensions.[148]

In conclusion, the CVD–EVD technique is at present a key technology for depositing uniform gas-tight fine-grained films of electrolytes and interconnects mainly in the tubular configuration. However, it is commonly considered as too costly due to the high temperature reaction, the presence of corrosive gases, the relatively low deposition rate, and the number of firing operations as far as the tubular design is concerned.

The wet process consists of the powder preparation of the four active components, the formation of green sheets, and finally the sintering operation.[161,162] The component powders are first prepared by conventional methods such as coprecipitation (hydroxides, oxalates, etc.), sol-gel, spray drying, and freeze drying. They must be tailored to obtain the desired surface area, particle size, size distribution, and morphology. Then, the ceramic powders are mixed with binders, plasticizers, and organic solvents to form the green sheets. At this step, single or multilayer green tapes are manufactured using different processes such as extrusion moulding, roll-calandering, or doctor blade techniques. Although the wet-type method applies to both planar and monolithic configurations, it is best suited to the latter. Today, in MSOFC, two multilayer ceramics, anode/electrolyte/cathode (A/E/C) and anode/interconnect/cathode (A/I/C) are manufactured separately.[161] Corrugations are made by compression moulding and stacking by green-state bonding. Finally, the multilayer structures are co-fired at around 1400°C to burn out the organic additives and sinter. For the A/E/C trilayer, it is reported that the desired characteristics have been achieved, the typical interfacial resistance being reduced to less than 0.5 ohm cm^{-2}. For the A/I/C trilayer, suitable processes for co-firing still need to be developed to solve the problem of poor densification for the interconnects. It is commonly accepted that a liquid phase appears and acts as an aid for interconnect densification. When the latter is incorporated in the A/I/C trilayer, the liquid phase tends to migrate into the porous electrode material, rendering the interconnect porous itself.

XI. CONCLUSION

The state-of-the-art SOFC operates at 1000°C, providing high efficiency, low emissions, and the opportunity for high-temperature heat recovery. Today, the tubular design fulfills the life time requirement satisfactorily. However, the high operating temperature is considered as the most serious problem to be solved in order to reduce the capital cost of SOFCs and make them competitive with other power generators. The operating temperature must be decreased to around 800°C, which will in particular allow the use of planar configurations with metallic interconnects, more efficient gas-tight seals, and less costly ceramic processing techniques.

ACKNOWLEDGMENT

The authors gratefully acknowledge all the scientific edition institutions (Academic Press, Electrochemical Society, Inc., Office for Official Publications of the European Communities, Pergamon Press, World Scientific Publishing Co.), and all the authors who granted them the permission to reproduce in this article some of their published figures.

REFERENCES

1. Takahashi, T., Iwahara, H., and Suzuki, Y., *Proceedings of the Third International Symposium on Solid Oxide Fuel Cells,* Presses Académiques Européennes, Brussels, 1969, 113–119.
2. Appleby, A.J. and Foulkes, F.R., *Fuel Cell Handbook,* Van Nostrand Reinhold, New York, 1989.
3. Brown, J.T., *High Conductivity Solid Ionic Conductors: Recent Trends and Applications,* Takahashi, T., Ed., World Scientific, Singapore, 1989, 630–63.
4. Hammou, A., *Advances in Electrochemical Science and Engineering,* Vol. 2, Gerischer, H. and Tobias, C.W., Eds., VCH Publishers, Weinheim, 1992, 87–139.

5. Minh, N.Q., *J. Am. Ceram. Soc.*, 1993, *76*, 563–588.
6. Nernst, W., *Z. Elektrochem.*, 1899, *6*, 41.
7. Baur, E. and Preis, H.Z., *Z. Elektrochem.*, 1937, *43*, 727–732.
8. Traub Hoore, D., *Proceedings of the Third International Symposium on Solid Oxide Fuel Cells*, Singhal, S.C. and Iwahara, H., Eds., The Electrochemical Society, Pennington, NJ, 1993, 3–5.
9. Tagawa, H.D., *Proceedings of the Third International Symposium on Solid Oxide Fuel Cells*, Singhal, S.C. and Iwahara, H., Eds., The Electrochemical Society, Pennington, NJ, 1993, 6–15.
10. Gross, F. *Proceedings of the Second International Symposium on Solid Oxide Fuel Cells*, Gross, F., Zeghers, P., Singhal, S.C., and Iwahara, H., Eds., Office for Official Publications of the European Communities, Luxembourg, 1991, 7–23.
11. Demin, A.K., Kuzin, B.L., Lipilin, A.S., Neuimin, A.D., Perfiliev, M.V., and Somov, S.I., *Proceedings of the Second International Symposium on Solid Oxide Fuel Cells*, Gross, F., Zeghers, P., Singhal, S.C., and Iwahara, H., Eds., Office for Official Publications of the European Communities, Luxembourg, 1991, 67–73.
12. Badwal, S.P.S. and Foger, K., *Proceedings of the Third International Symposium on Solid Oxide Fuel Cells*, Singhal, S.C. and Iwahara, H., Eds., The Electrochemical Society, Pennington, NJ, 1993, 21–26.
13. Takahashi, T., *Physics of Electrolytes*, Vol. 2, Hladik, J., Ed., Academic Press, London, 1972, 989–1049.
14. Baukal, W., *Electrochim. Acta*, 1969, *14*, 1071–1080.
15. Van Dijk, T., De Vries, K.J., and Burggraaf, A.J., *Phys. Status Solidi*, 1980, *b 101*, 765–774.
16. Van Dijk, T., De Vries K.J., and Burggraaf, A.J., *Phys. Status Solidi*, 1980, *a 58*, 115–125.
17. Moon, P.K. and Tuller, H.L., *Solid State Ionics*, 1988, *28–30*, 470–474.
18. Moon, P.K. and Tuller, H.L., *Proceedings of the First International Symposium on Solid Oxide Fuel Cells*, Singhal, S.C., Ed., The Electrochemical Society, Pennington, NJ, 1989, 30–40.
19. Spears, M., Kramer, S., Tuller, H.L., and Moon, P.K., *Ionic and Mixed Conductors*, Ramanarayanan, T.A. and Tuller, H.L., Ed., *The Electrochemical Society Softbound Proceedings Series*, Pennington, NJ, 1991, 32–38.
20. Tuller, H.L. and Nowick, A.S., *J. Electrochem. Soc.*, 1975, *122*, 255–259.
21. El Adham, K. and Hammou, A., *J. Chim. Phys.*, 1982, *79*, 633–639.
22. Takahashi, T. and Iwahara, H., *J. Appl. Electrochem.*, 1973, *3*, 65–72.
23. Verkerk, M.J. and Burggraaf, A.J., *J. Electrochem. Soc.*, 1980, *10*, 81–90.
24. Duran, P., Jurado, J.R., Moure, C., Valverde, N., and Steele, B.C.H., *Mat. Chem. Phys.*, 1987, *18*, 287–294.
25. Dordor, P.J., Tanaka, J., and Watanabe, A., *Solid State Ionics*, 1981, *3/4*, 463–467.
26. Verkerk, M.J., Keizer, K., and Burggraaf, A.J., *J. Appl. Electrochem.*, 1980, *10*, 81–90.
27. Cahen, H.T., Van den Belt, T.G.M., De Witt, J.W.H., and Broers, J.H.J., *Solid State Ionics*, 1980, *1*, 411–423.
28. Washman, E.D., Jiang, N., Mason, D.M., and Stevenson, D.A., *Proceedings of the First International Symposium on Solid Oxide Fuel Cells*, Singhal, S.C., Ed., The Electrochemical Society, Pennington, NJ, 1989, 15–21.
29. Honnart, F., Boivin, J.C., Thomas, D., and De Vries, K.J., *Solid State Ionics*, 1983, *9/10*, 921–928.
30. Dumelié, M., Nowogrocki, G., and Boivin, J.C., *Solid State Ionics*, 1988, *28/30*, 524–528.
31. Antonucci, V., Ielo, I., Giordano, N., Chiodelli, G., and Magistris, A., *Proceedings of the First International Symposium on Solid Oxide Fuel Cells*, Singhal, S.C., Ed., The Electrochemical Society, Pennington, NJ, 1989, 769–776.
32. Coates, R.V., *J. Appl. Electrochem.*, 1964, *14*, 346–350.
33. Forrat, F., Dauge, G. Trécoux, P., Danner, G., and Christen, M., *C.R. Acad. Sci. Paris*, 1964, *259*, 2813–2816.
34. Patterson, F.K., Moeller, C.W., and Ward, R., *Inorg. Chem.*, 1963, *2*, 196–198.
35. Takahashi, T. and Iwahara, H., *Research in Effective Use of Energy*, Sp. Publ. Ministry of Education Science and Culture of Japan 1982, *3*, 727–734.
36. Matsuda, H., Ishihara, T., Mizuhara, Y., and Takita, Y., *Proceedings of the Third International Symposium on Solid Oxide Fuel Cells*, Singhal, S.C. and Iwahara, H., Eds., The Electrochemical Society, Pennington, NJ, 1993, 129–136.
37. Sato, T. and Shimada, M., *J. Mater. Sci.*, 1985, *20*, 3988–3991.
38. Rajendran, S., Rossel, H.J., and Sanders, J.V., *J. Mater. Sci.*, 1989, *24*, 1195–1202.
39. Ishizaki, F., Yoshida T., and Sakurada, S., *Proceedings of the First International Symposium on Solid Oxide Fuel Cells*, Singhal, S.C., Ed., The Electrochemical Society, Pennington, NJ, 1989, 3–14.
40. Yamamoto, O., Takeda, Y., Imanishi, N., Kawahara, T., Shen, G.Q., Mori, M., and Abe, T., *Proceedings of the Second International Symposium on Solid Oxide Fuel Cells*, Gross, F., Zeghers, P., Singhal, S.C., and Iwahara, H., Eds., Office for Official Publications of the European Communities, Luxembourg, 1991, 437–444.
41. Kendall, K., *Bull. Am. Cer. Soc.*, 1991, *70*, 1159–60.
42. Kendall, K., *Proceedings of the Fuel Cell Seminar 1992*, Courtesy Associates, Washington, D.C., 1992, 265–268.
43. Steele, B.C.H., *Mater. Sci. Eng.*, 1992, *B13*, 79–87
44. White, D.W., *General Electric R&D*, Rep. 1978, 254.

45. Fischer, W., Rohr, F.J., and Steiner, R., *Brown Boveri Co Nachr.,* 1972, *54,* 4–12.
46. Tedmon, C.S., Spacil, H.S., and Mittof, S.P., *J. Electrochem. Soc.,* 1969, *116,* 1170–1175.
47. Ohno, Y., Nagata, S., and Sato, H., *Solid State Ionics,* 1981, *3/4,* 439–442.
48. Isaacs, H.S. and Olmer, L.J., *J. Electrochem. Soc.,* 1982, *129,* 436–443.
49. Kuzin, B.L. and Vlasov, A.N., *Elecktrokhimiya,* 1984, *20,* 1636–1642.
50. Priestnall, M.A. and Steele, B.C.H., *Proceedings of the First International Symposium on Solid Oxide Fuel Cells,* Singhal, S.C., Ed., The Electrochemical Society, Pennington, NJ, 1989, 157–169.
51. Steele, B.C.H., Carter, S., Kajda, J., Kontoulis, I., and Kilner, J.A. *Proceedings of the Second International Symposium on Solid Oxide Fuel Cells,* Gross, F., Zeghers, P., Singhal, S.C., and Iwahara, H., Eds., Office for Official Publications of the European Communities, Luxembourg, 1991, 517–525.
52. Eguchi, K., Inoue, T., Ueda, M., Kamimae, J., and Arai, H., *Proceedings of the Second International Symposium on Solid Oxide Fuel Cells,* Gross, F., Zeghers, P., Singhal, S.C., and Iwahara, H., Eds., Office for Official Publications of the European Communities, Luxembourg, 1991, 697–704.
53. Tai, L.W., Nasrallah, M.M., and Anderson, H.U., *Proceedings of the Third International Symposium on Solid Oxide Fuel Cells,* Singhal, S.C. and Iwahara, H., Eds., The Electrochemical Society, Pennington, NJ, 1993, 241–251.
54. Chen, C.C., Nasrallah, M.M., and Anderson, H.U. *Proceedings of the Third International Symposium on Solid Oxide Fuel Cells,* Singhal, S.C. and Iwahara, H., Eds., The Electrochemical Society, Pennington, NJ, 1993, 598–612.
55. Ivers-Tiffée, E., Schiessl, M., Oel, H.J., and Wersing, W. *Proceedings of the Third International Symposium on Solid Oxide Fuel Cells,* Singhal, S.C. and Iwahara, H., Eds., The Electrochemical Society, Pennington, NJ, 1993, 613–622.
56. Li, Z., Behruzi, M., Fuerst, L., and Stover, D. *Proceedings of the Third International Symposium on Solid Oxide Fuel Cells,* Singhal, S.C. and Iwahara, H., Eds., The Electrochemical Society, Pennington, NJ, 1993, 171–179.
57. Hammouche, A., thesis, Grenoble University, 1989.
58. Mackor, A., Spee, C.I.M.A., Van der Zouwen-Assink, E.A., Batista, J.L., and Schoonman, J., *Proceedings of 25th Intersoc. Energy Conv. Eng. Conf.,* 1990, *3,* 251.
59. Wersing, W., Ivers-Tiffee, E., Schiessl, M., and Greiner, H., *Proceed. SOFC Nagoya,* 1989, 21–26.
60. Takeda, Y., Kanno, R., Noda, M., and Yamamoto, O., *J. Electrochem. Soc.,* 1987, *11,* 2656–2661.
61. Hammouche, A., Siebert, E., and Hammou, A., *Mater. Res. Bull.,* 1989, *24,* 367–380.
62. Mackor, A., Koster, I.P.M., Kraaijkamp, J.G., Gerretsen, J., and Van Eijk, J.P.G.M., *Proceedings of the Second International Symposium on Solid Oxide Fuel Cells,* Gross, F., Zeghers, P., Singhal, S.C., and Iwahara, H., Eds., Office for Official Publications of the European Communities, Luxembourg, 1991, 463–471.
63. Scotti, C. Gharbage, B., Lauret, H., Lévy, M., and Hammou, A., *Mater. Res. Bull.,* 1993, 28, 1215–1220.
64. Kuo, J.H., Anderson, H.U., and Sparlin, D.M., *J. Solid State Chem.,* 1989, *83,* 52–60.
65. Hammouche A., Siebert, E., Hammou, A., Kleitz, M., and Caneiro, A., *J. Electrochem. Soc.,* 1991, *138/5,* 1212–1216.
66. Kenjo, T. and Nishiya, M., *Solid State Ionics,* 1992, *57,* 295–302.
67. Yamamoto, O., Takeda, Y., Kanno, R., and Noda, M., *Solid State Ionics,* 1987, *22,* 241–246.
68. Kleitz, M., Kloidt, T., and Dessemond, L., *14th Int. Symp. on Materials Science,* Poulsen, F.W., Bentsen, J.J., Jacobsen, T., Skon, E., and Ostergard, M.J.L., Eds., Riso Roskilde, Denmark, 1993, 89–116.
69. Misuzaki, J., Tagawa, H. Sato, T., and Narita, H., *14th Int. Symp. on Materials Science,* Poulsen, F.W., Bentsen, J.J., Jacobsen, T., Skon, E., and Ostergard, M.J.L., Eds., Riso Roskilde, Denmark, 1993, 343–350.
70. Mogensen, M., *14th Int. Symp. on Materials Science,* Poulsen, F.W., Bentsen, J.J., Jacobsen, T., Skon, E., and Ostergard, M.J.L., Eds., Riso Roskilde, Denmark, 1993, 117–135.
71. Perfiliev, M.V., *Solid State Ionics,* 1983, *9/10,* 756–776.
72. Marques, F., *Fourth Periodic Report JOUE-0044C,* Office for Official Publications of the European Communities, Luxembourg, 1992, 4–5.
73. Gharbage, B., Pagnier, T., and Hammou, A., *14th Int. Symp. on Materials Science,* Poulsen, F.W., Bentsen, J.J., Jacobsen, T., Skon, E., and Ostergard, M.J.L., Eds., Riso Roskilde, Denmark, 1993, 249–256.
74. Yamamoto, O., Takeda, Y., Kanno, R., and Kojima, T., *Proceedings of the First International Symposium on Solid Oxide Fuel Cells,* Singhal, S.C., Ed., The Electrochemical Society, Pennington, NJ, 1989, 242–253.
75. Tagawa, H., Misuzaki, J., Katou, M., Hirano, K. Sawata, A., and Tsuneyoshi, K., *Proceedings of the Second International Symposium on Solid Oxide Fuel Cells,* Gross, F., Zeghers, P., Singhal, S.C., and Iwahara, H., Eds., Office for Official Publications of the European Communities, Luxembourg, 1991, 681–688.
76. Yamamoto, O., Takeda, Y. Imanishi, N., and Sakaki, Y., *Proceedings of the Third International Symposium on Solid Oxide Fuel Cells,* Singhal, S.C. and Iwahara, H., Eds., The Electrochemical Society, Pennington, NJ, 1993, 205–212.
77. Kaneko, H., Tainatsu, H., Wada, K., and Iwamoto, E. *Proceedings of the Second International Symposium on Solid Oxide Fuel Cells,* Gross, F., Zeghers, P., Singhal, S.C., and Iwahara, H., Eds., Office for Official Publications of the European Communities, Luxembourg, 1991, 673–680.

78. Eguchi, K., Setoguchi, M., Sawano, M., Tamura, S., and Arai, H., *Proceedings of the Second International Symposium on Solid Oxide Fuel Cells,* Gross, F., Zeghers, P., Singhal, S.C., and Iwahara, H., Eds., Office for Official Publications of the European Communities, Luxembourg, 1991, 603–610.
79. Suzuki, M., Sasaki, H., Otoshi, S., and Ippomatsu, M., *Proceedings of the Second International Symposium on Solid Oxide Fuel Cells,* Gross, F., Zeghers, P., Singhal, S.C., and Iwahara, H., Eds., Office for Official Publications of the European Communities, Luxembourg, 1991, 585–591.
80. Dees, D.W., Claar, T.D., Easler, T.E., Fee, D.C., and Mrazek, F.C., *J. Electochem. Soc.,* 1987, *134,* 2141–2146.
81. Middleton, P.H., Stirsen, M.E., and Steele, B.C.H., *Proceedings of the First International Symposium on Solid Oxide Fuel Cells,* Singhal, S.C., Ed., The Electrochemical Society, Pennington, NJ, 1989, 90–98.
82. Okumura, K., Yamamoto, Y., Fukui, T., Hanyu, S., Kubo, Y., Esaki, Y., Hattori, M., Kusunoki, A., and Takeuchi, S., *Proceedings of the Third International Symposium on Solid Oxide Fuel Cells,* Singhal, S.C. and Iwahara, H., Eds., The Electrochemical Society, Pennington, NJ, 1993, 444–453.
83. Mogensen, M. and Lindegaard, T., *Proceedings of the Third International Symposium on Solid Oxide Fuel Cells,* Singhal, S.C. and Iwahara, H., Eds., The Electrochemical Society, Pennington, NJ, 1993, 484–493.
84. Guindet, J., thesis, Grenoble University, 1991.
85. Kawada, T., Sakai, N., Yokokawa, H., and Dokiya, M., *J. Electrochem. Soc.,* 1990, *137,* 3042–3047.
86. Epelboin, I., Gabrielli, C., Keddam, M., and Takenouti, H., in *Comprehensive Treatise of Electrochemistry,* Plenum Press, New York, 1981, 151–192.
87. Schouler, E.J.L., thesis, Grenoble University, 1979.
88. Epelboin, I. and Wiart, R., *J. Electrochem. Soc.,* 1971, *118,* 1577–1582.
89. Diard, J.P., Le Gorrec, B., Montella, C., and Saint-Aman, E., *Electrochem. Acta,* 1982, *27,* 1055–1059.
90. Kado, T. and Kunitomi, N., *J. Electrochem. Soc.,* 1991, *138,* 3312–3321.
91. Chabli, A., Diard, J.P., and Le Gorrec, B., *Surf. Tech.,* 1982, *15,* 357–361.
92. Tedmon, C.S., Spacil, H.S., and Mittof, S.P., *General Electric,* Rep. 69-C-056, 1969.
93. Miyamoto, H., Sumi, M., Mori, K., Koshiro, I., Nanjo, F., and Funatsu, M., *Proceedings of the Third International Symposium on Solid Oxide Fuel Cells,* Singhal, S.C. and Iwahara, H., Eds., The Electrochemical Society, Pennington, NJ, 1993, 504–512.
94. Marques, R.M.C., Frade, J.R., and Marques, F.M.B., *Proceedings of the Third International Symposium on Solid Oxide Fuel Cells,* Singhal, S.C. and Iwahara, H., Eds., The Electrochemical Society, Pennington, NJ, 1993, 513–522.
95. Colomer, M.T., Jurado, J.R., Marques, R.M.C., and Marques, F.M.B., *Proceedings of the Third International Symposium on Solid Oxide Fuel Cells,* Singhal, S.C. and Iwahara, H., Eds., The Electrochemical Society, Pennington, NJ, 1993, 523–532.
96. Nguyen, B.C., Lin, T.A., and Mason, D.M., *J. Electrochem. Soc.,* 1986, *133,* 1807–1815.
97. Steele, B.C.H., Kelly, I., Middleton, H., and Rudkin, R., *Solid State Ionics,* 1988, *28/30,* 1547–1552.
98. Mogensen, M. and Bentzen, J.J., *Proceedings of the First International Symposium on Solid Oxide Fuel Cells,* Singhal, S.C., Ed., The Electrochemical Society, Pennington, NJ, 1989, 99–105.
99. Hildrum, R., Middleton, H., Brustad, M., and Tunold, R., *14th Int. Symp. on Materials Science,* Poulsen, F.W., Bentsen, J.J., Jacobsen, T., Skon, E., Ostergard, M.J.L., Eds., Riso Roskilde, Denmark, 1993, 99–105.
100. Swider, K.E. and Worrel, W.L., *Proceedings of the Second International Symposium on Solid Oxide Fuel Cells,* Gross, F., Zeghers, P., Singhal, S.C., and Iwahara, H., Eds., Office for Official Publications of the European Communities, Luxembourg, 1991, 593–601.
101. Yentekakis, I.V., Neophytides, N.G., Koloyiannis, A.C., and Vayenas, C.G., *Proceedings of the Third International Symposium on Solid Oxide Fuel Cells,* Singhal, S.C. and Iwahara, H., Eds., The Electrochemical Society, Pennington, NJ, 1993, 904–912.
102. Anderson, H.U., *Solid State Ionics,* 1992, *52,* 33–41.
103. Groupp, L. and Anderson, H.U., *J. Am. Ceram. Soc.,* 1976, *59,* 449–453.
104. Meadowcroft, D.B., *Br. J. Appl. Phys.,* 1969, *2,* 1225–1233.
105. Flandermeyer, B.K., Nasrallah, M.M., Sparlin, D.M., and Anderson, H.U., *Transport in Nonstoichrometric Compounds,* Stubican, V. and Sinkovich, G., Eds., Plenum Press, New York, 1985, 17–25.
106. Koc, R. and Anderson, H.U., *J. Eur. Ceram. Soc.,* 1992, *9,* 285–292.
107. Nasrallah, M.M., Carter, J.D., Anderson, H.U., and Koc, R., *Proceedings of the Second International Symposium on Solid Oxide Fuel Cells,* Gross, F., Zeghers, P., Singhal, S.C., and Iwahara, H., Eds., Office for Official Publications of the European Communities, Luxembourg, 1991, 637–644.
108. Christiansen, N., Gordes, P., Alstrup, N.C., and Mogensen, G. *Proceedings of the Third International Symposium on Solid Oxide Fuel Cells,* Singhal, S.C. and Iwahara, H., Eds., The Electrochemical Society, Pennington, NJ, 1993, 401–413.
109. Christie, G.M., Middleton, P.H., and Steele, B.C.H. *Proceedings of the Third International Symposium on Solid Oxide Fuel Cells,* Singhal, S.C. and Iwahara, H., Eds., The Electrochemical Society, Pennington, NJ, 1993, 315–324.

110. Chick, L.A., Armstrong, T.R., Mc Cready, D.E., Coffey, G.W., Maupin, G.D., and Bates, J.L. *Proceedings of the Third International Symposium on Solid Oxide Fuel Cells,* Singhal, S.C. and Iwahara, H., Eds., The Electrochemical Society, Pennington, NJ, 1993, 374–384.
111. Sakai, N., Kawada, T., Yokohawa, H., Dokiya, M., and Iwata, T., *Solid State Ionics,* 1990, *40/41,* 394–397.
112. Karim D.P. and Aldred, A.T., *Phys. Rev.,* 1979, *B20,* 2255–2263.
113. Koc, R. and Anderson, H.U., *J. Mater. Sci.,* 1992, *27,* 5477–5482.
114. Mori, M., Itoh, H., Mori, N., and Abe, T., *Proceedings of the Third International Symposium on Solid Oxide Fuel Cells,* Singhal, S.C. and Iwahara, H., Eds., The Electrochemical Society, Pennington, NJ, 1993, 325–334.
115. Yasuda, I. and Hikita, T., *Proceedings of the Second International Symposium on Solid Oxide Fuel Cells,* Gross, F., Zeghers, P., Singhal, S.C., and Iwahara, H., Eds., Office for Official Publications of the European Communities, Luxembourg, 1991, 645–652.
116. Yokokawa, H., Horita, T., Sakaï, N., van Hassel, B.A., Kawada, T., and Dokiya, M., *Proceedings of the Third International Symposium on Solid Oxide Fuel Cells,* Singhal, S.C. and Iwahara, H., Eds., The Electrochemical Society, Pennington, NJ, 1993, 364–369.
117. Drenckhahn, W. and Wollmar, H.E., *Fuel Cell Seminar 1992,* Courtesy Associates, Washington, D.C., 1992, 419–422.
118. Seto, H., Miyata, T., Tsunoda, A., Yoshida, T., and Sakurada, S. *Proceedings of the Third International Symposium on Solid Oxide Fuel Cells,* Singhal, S.C. and Iwahara, H., Eds., The Electrochemical Society, Pennington, NJ, 1993, 421–427.
119. Rohr, F.J., *Solid Electrolytes,* Hagenmuller, P. and Van Gool, W., Eds., Academic Press: New York, 1978, 431–450.
120. Isenberg, A.O., *Proceedings of the High Temperature Solid Oxide-Electrolytes Conference,* BNL 51728, Brookhaven National Laboratory, Associated Universities, 1983, 5–15.
121. Brown, J.T., *IEEE Trans. on Energy Conv.,* 1988, *2,* 193–197.
122. Singhal, S.C., *Proceedings of the Second International Symposium on Solid Oxide Fuel Cells,* Gross, F., Zeghers, P., Singhal, S.C., and Iwahara, H., Eds., Office for Official Publications of the European Communities, Luxembourg, 1991, 25–33.
123. Takeuchi, S., Kusunoki, A., Matsubara, H., Kikuoka, J., Ohtsuki, J., Satomi, T., and Shinosaki, K., *Proceedings of the Third International Symposium on Solid Oxide Fuel Cells,* Singhal, S.C. and Iwahara, H., Eds., The Electrochemical Society, Pennington, NJ, 1993, 678–683.
124. Ray, E.R., *Fuel Cell Seminar 1992,* Courtesy Associates: Washington, D.C., 1992, 415–418.
125. Shinosaki, K., Washio, S., Satomi, T., and Koike, S., *Proceedings of the Third International Symposium on Solid Oxide Fuel Cells,* Singhal, S.C. and Iwahara, H., Eds., The Electrochemical Society, Pennington, NJ, 1993, 684–689.
126. Fee, D.C., Steunenberg, R.K., Claar, T.D., Poeppel, R.B., and Ackerman, J.P., *Fuel Cell Seminar,* Courtesy Associates, Washington, D.C., 1983, 74–78.
127. Mc Pheeters, C.C., Fee, D.C., Poeppel, R.B., Claar, T.D., Busch, D.E., Flandermeyer, B.K., Easler, T.E., Durek, J.T., and Picciolo, J.J., *Fuel Cell Seminar,* Courtesy Associates, Washington D.C., 1986, 44–47.
128. Minh, N.Q., Horne, C.R., Lin, F., and Staszak, P.R., *Proceedings of the First International Symposium on Solid Oxide Fuel Cells,* Singhal, S.C., Ed., The Electrochemical Society, Pennington, NJ, 1989, 307–316.
129. Myles, K.M., *Proceedings of the Second International Symposium on Solid Oxide Fuel Cells,* Gross, F., Zeghers, P., Singhal, S.C., and Iwahara, H., Eds., Office for Official Publications of the European Communities, Luxembourg, 1991, 85–92.
130. Myles, K.M. and Mc Pheeters, C.C., *J. Power Sources,* 1990, *29,* 311–319.
131. Minh, N.Q., Amiro, A., Armstrong, T.R., Esopa, J.R., Guiheen, J.V., Horne, C.R., and Van Ackeren, J.J., *Fuel Cell Seminar 1992,* Courtesy Associates, Washington, D.C., 1992, 607–610.
132. Takagi, H., Kobayashi, S., Shiratori, A, Nishida, K., and Sakabe, Y., *Proceedings of the Second International Symposium on Solid Oxide Fuel Cells,* Gross, F., Zeghers, P., Singhal, S.C., and Iwahara, H., Eds., Office for Official Publications of the European Communities, Luxembourg, 1991, 99–103.
133. Barringer E.A., Heitzenrater, D.R., and Tharp, M.R., *Proceedings of the Third International Symposium on Solid Oxide Fuel Cells,* Singhal, S.C. and Iwahara, H., Eds., The Electrochemical Society, Pennington, NJ, 1993, 771–781.
134. Tagaki, H, Taira, H., Shiratori, A., Koyabashi, S., Sugimoto, Y., Sakamoto, S., and Tomono, K., *Proceedings of the Third International Symposium on Solid Oxide Fuel Cells,* Singhal, S.C. and Iwahara, H., Eds., The Electrochemical Society, Pennington, NJ, 1993, 738–743.
135. Van Berkel, F.P.F., Van Heuveln, F.H., and Huijmans, J.P.P., *Proceedings of the Third International Symposium on Solid Oxide Fuel Cells,* Singhal, S.C. and Iwahara, H., Eds., The Electrochemical Society, Pennington, NJ, 1993, 744–751.
136. Hikita, T., Hishimuma, M., Kawashima, T., Yasuda, I., Koyama, T., and Matsuzaki, Y., *Proceedings of the Third International Symposium on Solid Oxide Fuel Cells,* Singhal, S.C. and Iwahara, H., Eds., The Electrochemical Society, Pennington, NJ, 1993, 714–723.

137. Ivers-Tiffee, E., Wersing, W., and Reichelt, B., *Fuel Cell Seminar,* Courtesy Associates, Washington, D.C., 1990, 137–140.
138. Akiyama, Y., Yasuo, T., Ishida, N., Taniguchi, S., and Saito, T., *Proceedings of the Third International Symposium on Solid Oxide Fuel Cells,* Singhal, S.C. and Iwahara, H., Eds., The Electrochemical Society, Pennington, NJ, 1993, 724–731.
139. Shimotsu, M., Isumi, M., and Murata, K., *Proceedings of the Third International Symposium on Solid Oxide Fuel Cells,* Singhal, S.C. and Iwahara, H., Eds., The Electrochemical Society, Pennington, NJ, 1993, 732–737.
140. Bossel, U.G., *Proceedings of the Third International Symposium on Solid Oxide Fuel Cells,* Singhal, S.C. and Iwahara, H., Eds., The Electrochemical Society, Pennington, NJ, 1993, 833–840.
141. Van Dieten, V.E.J., Walterbos, P.H.M., and Schoonman, J., *Proceedings of the Second International Symposium on Solid Oxide Fuel Cells,* Gross, F., Zeghers, P., Singhal, S.C., and Iwahara, H., Eds., Office for Official Publications of the European Communities, Luxembourg, 1991, 183–191.
142. Isenberg, A.O., *Solid State Ionics,* 1981, *3/4,* 431–437.
143. Feduska, W. and Isenberg, A.O., *J. Power Sources,* 1983, *10,* 89–102.
144. Carolan, M.F. and Michaels, J.N., *Solid State Ionics,* 1987, *25,* 205–216.
145. Pal, U.B. and Singhal S.C., *Proceedings of the First International Symposium on Solid Oxide Fuel Cells,* Singhal, S.C., Ed., The Electrochemical Society, Pennington, NJ, 1989, 41–56.
146. Dekker, J.P., Kiwiet, N.J., and Schoonman, J., *Proceedings of the First International Symposium on Solid Oxide Fuel Cells,* Singhal, S.C., Ed., The Electrochemical Society, Pennington, NJ, 1989, 57–66.
147. Lin, Y.S., de Haart, L.G.J., de Vries, K.J., and Burggraaf, A.J., *J. Electrochem. Soc.,* 1990, *37,* 3960–3966.
148. de Vries, K.J., Kuipers, R.A., and de Haart, L.G.J., *Proceedings of the Second International Symposium on Solid Oxide Fuel Cells,* Gross, F., Zeghers, P., Singhal, S.C., and Iwahara, H., Eds., Office for Official Publications of the European Communities, Luxembourg, 1991, 135–143.
149. Gharbage, H., Henault, M., Pagnier, T., and Hammou, A., *Mater. Res. Bull.,* 1991, *26,* 1001–1007.
150. Arai, H., *Proceedings of International Symposium on SOFC,* Yamamoto, D., Dokiya, M., and Tagawa, H., Eds., Science House, Tokyo 1989, 9–14.
151. Rohr, F.J., *Proceedings of the Workshop on High Temperature SOFC,* Isaacs, H.S., Sunivasan, S. and Harry, I.L., Eds., BNL, Associated Universities, Upton, New York, 1977, 122–138.
152. Arai, H., Eguchi, K., Setoguchi, T., Yamaguchi, R., Hashimoto, K., and Yoshimura, H., *Proceedings of the Second International Symposium on Solid Oxide Fuel Cells,* Gross, F., Zeghers, P., Singhal, S.C., and Iwahara, H., Eds., Office for Official Publications of the European Communities, Luxembourg, 1991, 167–174.
153. Tsukamoto, K., Uchiyama, F., Kaga, Y., Ohno, Y., Yanagisawa, T., Monma, A., Takahagi, Y., Lain, M.J., and Nakajima, T., *Solid State Ionics,* 1990, *40/41,* 1003.
154. Nakagawa, H., Kosuge, S., Tsuneisumi, H., Matsuda, E., Mikara, H., and Sato, Y. *Proceedings of the First International Symposium on Solid Oxide Fuel Cells,* Singhal, S.C., Ed., The Electrochemical Society, Pennington, NJ, 1989, 71–78.
155. Negishi, A., Nozaki, K., and Ozawa, T., *Solid State Ionics,* 1981, *3/4,* 443–446.
156. Thiele, E.S., Wang, L.S., Mason, T.O., and Barnett, S.A., *J. Vac. Sci. Technol.,* 1991, *9/6,* 3054–3060.
157. Kandhar, A. and Elangovan, S. *Denki Kagaku,* 1990, *6,* 551–556.
158. Markin, T.L., Bones, N.J., and Dell, R.M., *Superionic Conductors,* Mahan, G. and Roth, W., Eds., Plenum Press, New York, 1976, 15–35.
159. Yamazaki, Y., Namikawa, T., and Michibata, H., *Proceedings of the Second International Symposium on Solid Oxide Fuel Cells,* Gross, F., Zeghers, P., Singhal, S.C., and Iwahara, H., Eds., Office for Official Publications of the European Communities, Luxembourg, 1991, 175–181.
160. Isenberg, A.O., *Electrochemical Society Symposium Electrode Materials, Processes for Energy Conversion and Storage,* 1977, 77(6), 572–581.
161. Dokiya, M., Sakai, N., Kawada, T., Yokokama, H., and Anzai, I., *Proceedings of the Second International Symposium on Solid Oxide Fuel Cells,* Gross, F., Zeghers, P., Singhal, S.C., and Iwahara, H., Eds., Office for Official Publications of the European Communities, Luxembourg, 1991, 127–134.
162. Minh, N.Q., Armstrong, T.R., Esopa, J.R., Guiheen, J.V., Horne, C.R., Liu, F.S., Stillwagon, T.L., and Van Ackeren, J.J., *Proceedings of the Second International Symposium on Solid Oxide Fuel Cells,* Gross, F., Zeghers, P., Singhal, S.C., and Iwahara, H., Eds., Office for Official Publications of the European Communities, Luxembourg, 1991, 93–98.
163. Jurado, J.R., *Fourth Periodic Report, Contract JOULE — 0044C,* Joule Programme CEC, 1992, 12.

Chapter 13

ELECTROCATALYSIS AND ELECTROCHEMICAL REACTORS

Constantinos G. Vayenas, Symeon I. Bebelis, I.V. Yentekakis, and S.N. Neophytides

CONTENTS

0-8493-8956-9/97/$0.00+$.50

LIST OF ABBREVIATIONS AND SYMBOLS

Abbreviations

CSTR	continuously stirred tank reactor
emf	electromotive force
NEMCA	nonfaradaic electrochemical modification of catalytic activity
OCM	oxidative coupling of methane
rds	rate-determining step
SEP	solid electrolyte potentiometry
SOFC	solid oxide fuel cell
TOF	turnover frequency
tpb	three-phase-boundary
UHV	ultra-high vacuum
XPS	x-ray photoelectron spectroscopy
YSZ	yttria-stabilized zirconium oxide

Symbols

a_O	oxygen activity
ΔG	change in free enthalpy
ΔH	change in enthalpy
ΔS	change in entropy
E	cell voltage
E_{act}	activation energy
$e\Phi$	work function
I	current
I_0	exchange current
i	current density
i_0	exchange current density
i_L	limiting current density
K_O	oxygen adsorption equilibrium constant
l	length of tpb
n	number of electrons transferred in overall cell reaction
P	power density
P_i	promotion index
P_{O_2}	oxygen partial pressure
R, R′	resistance, area-specific resistance
r	reaction rate
s	number of electrons transferred in rds step
V	electrode potential

Greek symbols

θ	degree of coverage
η	overvoltage, polarization
Λ	faradaic efficiency
φ	inner (or Galvani) potential
ρ	rate enhancement ratio
ε	fuel cell efficiency

Ψ	Volta electrode potential
υ	sweep rate in voltammetry
ν	stoichiometric number = number of times rds occurs
γ	number of electrons transferred *before* the rds
τ	time constant
α_a, α_c	(apparent) charge transfer coefficient for anodic, cathodic reaction
θ_O	degree of coverage by oxygen
β	symmetry factor
μ	chemical potential

Sub- and superscripts

a, c	anode, cathode
act	activation
c	electrode–electrolyte contact resistance
conc	concentration
el	electrode
i	solid electrolyte
ohm	ohmic
p	peak
R	reference electrode
rev	reversible
TH	thermoneutral
W	working electrode

I. CHARGE TRANSFER AND THE NATURE OF OVERPOTENTIALS IN SOLID STATE ELECTROCHEMISTRY

During the last 3 decades solid electrolyte galvanic cells have been studied intensively both for sensor applications[1-5] and for their potential use as power-producing devices.[6] Other emerging applications include their use for chemical cogeneration, i.e., the simultaneous production of power and chemicals[7-12] and for enhancing and controlling the catalytic properties of metal and metal oxide catalysts via the effect of non-faradaic electrochemical modification of catalytic activity (NEMCA) or *in situ* controlled promotion of catalysis.[13-27] All these applications depend crucially on electrocatalytic (net charge transfer) reactions occurring primarily at the solid electrolyte/electrode/gas three-phase boundaries (tpb) and on catalytic (no net charge transfer) reactions occurring primarily on the gas-exposed electrode surface. Improving our current understanding of electrocatalysis and catalysis in solid electrolyte galvanic cells is of considerable theoretical and practical importance. Some fundamental electrochemical definitions and relations of importance for this subject are introduced and discussed in this chapter. For a more complete treatment see Chapters 2 and 5 of this handbook.

A. GENERAL CONSIDERATIONS

The operating principle of a current state-of-the-art power-producing solid electrolyte fuel cell is shown schematically in Figure 13.1. The positive electrode (cathode) acts as an electrocatalyst, i.e., it promotes the electrocatalytic (net charge transfer) reduction of $O_2(g)$ to O^{2-}:

$$O_2(g) + 4e^- \rightarrow 2O^{2-} \tag{13.1}$$

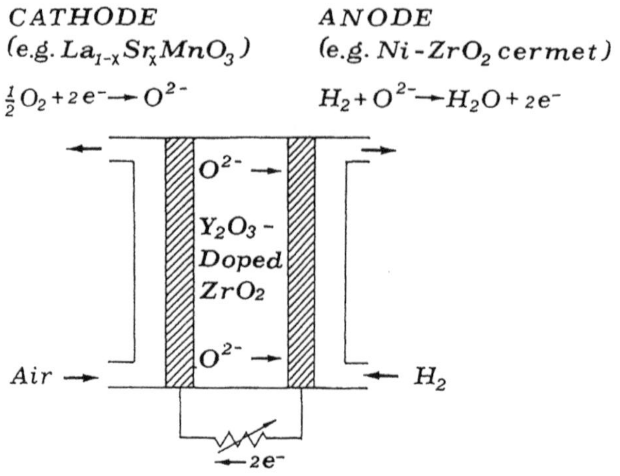

FIGURE 13.1. Operating principle of a solid oxide fuel cell. (Reprinted with permission from Vayenas, C.G., Bebelis, S., and Kyriazis, C.C., *CHEMTECH,* 1991, *21,* 422. Copyright 1991 The American Chemical Society.)

or, in Kröger–Vink notation:

$$O_2(g) + 2V_O^{\bullet\bullet} + 4e' \rightarrow 2O_O \tag{13.2}$$

where $V_O^{\cdot\cdot}$ stands for an O^{2-} vacancy in the yttria-stabilized zirconia (YSZ) lattice and O_O denotes an oxygen anion O^{2-} in the lattice.

Although several metals, such as Pt and Ag, can also act as electrocatalysts for Reaction (13.1), the most commonly used cathodic electrocatalysts in solid oxide fuel cells (SOFC) are perovskites such as $La_{1-x}Sr_xMnO_{3-\delta}$ and $La_{1-x}Sr_xCoO_{3-\delta}$. These materials are mixed ionic–electronic conductors (for a more extensive discussion see Chapter 7 of this handbook) and thus the electrocatalytic sites are not restricted to the geometric tpb, but can also exist on the gas-exposed electrode surface as well, forming an electrocatalytically active zone which can, in principle, extend over the entire gas-exposed electrode surface.

The negative electrode (anode) acts as an electrocatalyst for the reaction of O^{2-} with the fuel, e.g., H_2:

$$2H_2 + 2O^{2-} \rightarrow 2H_2O + 4e^- \tag{13.3}$$

A Ni-YSZ cermet is used as the anodic electrocatalyst in state-of-the art SOFC for two reasons: first because Ni is an active electrocatalyst for Reaction (13.3) and the porous cermet structure consisting of Ni and YSZ particles, typically 1 μm in size, has a large tpb length, and the electrocatalytic Reaction (13.3) takes place at the tpb. Consequently the use of Ni in a Ni-YSZ cermet to carry out electrocatalytic reactions is analogous to the use of Ni or other metals in a highly dispersed form on SiO_2 or Al_2O_3 (supported catalysts) to carry out catalytic reactions.[28] The second reason for using a Ni-YSZ cermet as the anode is that Ni is also an active catalyst for the steam-reforming reaction:

$$CH_4 + H_2O \rightarrow CO + 3H_2 \tag{13.4}$$

and for the water–gas-shift reaction:

$$CO + H_2O \rightarrow CO_2 + H_2 \tag{13.5}$$

This permits efficient SOFC operation with CH_4, and also natural gas, as the fuel. Direct electrocatalytic CH_4 oxidation at the tpb is too slow for practical SOFC units with any known electrocatalyst, but, due to the catalytic properties of the Ni surface, CH_4 is converted to CO_2 and H_2 via Reactions (13.4) and (13.5) and the latter is then easily oxidized electrocatalytically at the tpb.

Fuel cells such as the one shown in Figure 13.1 convert H_2 to H_2O or CH_4 to CO_2 and H_2O and produce electrical power with no intermediate combustion cycle. Thus their thermodynamic efficiency compares favorably with thermal power generation which is limited by Carnot-type constraints. One important advantage of solid electrolyte fuel cells is that due to their high operating temperatures (typically 800 to 1000°C), they offer the possibility of "internal reforming" via Reactions (13.4) and (13.5) which allows for the use of fuels such as CH_4 or natural gas without a separate external reformer.

B. THERMODYNAMIC CONSIDERATIONS

The maximum work obtainable from an isothermal continuous process is the negative of the Gibbs free energy change between the product and reactant streams $-\Delta G$. When an oxidation reaction, e.g.,

$$H_2 + 1/2\ O_2 \rightarrow H_2O \tag{13.6}$$

$$CH_4 + 2O_2 \rightarrow CO_2 + 2H_2O \tag{13.7}$$

is carried out in a fuel cell, the useful electrical work produced tends to the upper limit of $-\Delta G$ when reversible operation is approached, i.e., when the operating cell voltage E approaches the open-circuit reversible voltage E_{rev}:

$$E_{rev} = -\Delta G/nF \tag{13.8}$$

where n is the number of electrons involved in the oxidation reaction, e.g., 2 for Reaction (13.6) and 4 for Reaction (13.7).

The fuel cell efficiency ε is defined by:

$$\varepsilon = \frac{W}{-\Delta H^{o}} = \frac{E}{E_{TH}} = \frac{\Delta G}{\Delta H^{o}} \cdot \frac{E}{E_{rev}} \tag{13.9}$$

where E_{TH} is the "thermoneutral" voltage defined as $E_{TH} = (-\Delta H^{o})/nF$. According to Equation (13.9), ε can exceed unity for reactions with a positive entropy change ΔS^{o} such as the ammonia oxidation to NO.[10] Under such operating conditions ($\varepsilon > 1$) a fuel cell absorbs heat from the environment instead of producing heat.

When the external resistive load of the fuel cell is finite and electrical power is produced, then in general the operating voltage E drops below the reversible voltage E_{rev} and the difference η:

$$\eta = E_{rev} - E \tag{13.10}$$

is called the cell overpotential (Figure 13.2).

One of the key problems in electrochemical power-producing devices, which hampered commercialization for years, is the thorough understanding and minimization of cell overpotential or "polarization."

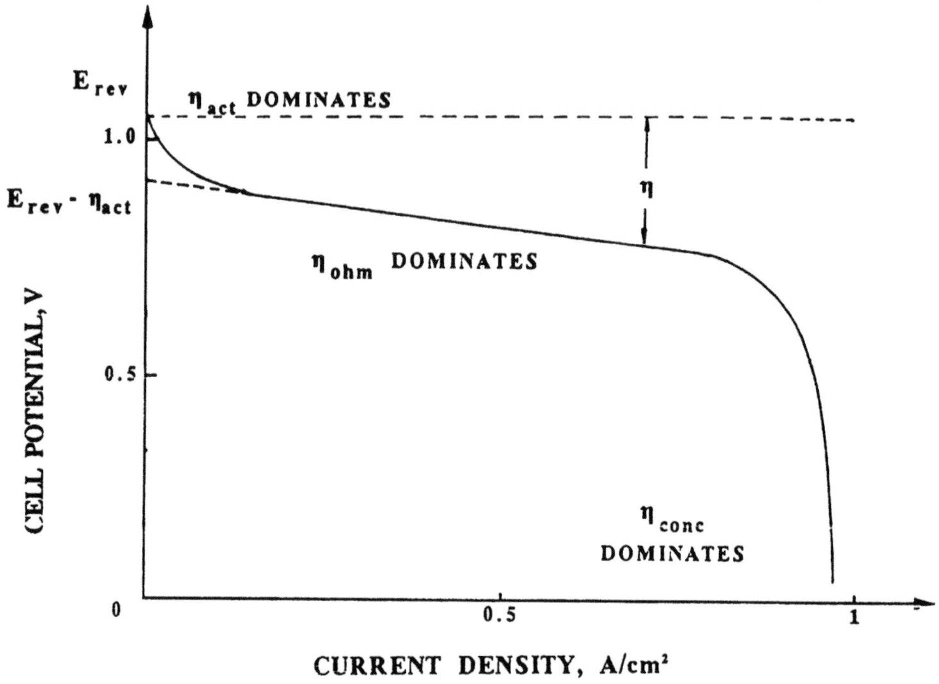

FIGURE 13.2. Typical current–voltage curve of a fuel cell obtained by varying the external load. (Reprinted with permission from Debenedetti, P.G. and Vayenas, C.E., *Chem. Eng. Sci.*, 1983, *38*, 1817. Copyright 1983 Elsevier Science Ltd.)

C. TYPES OF OVERPOTENTIAL

The cell overpotential η can be considered as the sum of three major components termed ohmic overpotential η_{ohm}, concentration overpotential η_{conc}, and activation overpotential η_{act}:

$$\eta = \eta_{ohm} + \eta_{conc} + \eta_{act} \tag{13.11}$$

The ohmic overpotential can be measured via ac impedance spectroscopy or via the current interruption technique in conjunction with a recording oscilloscope and is proportional to the cell current I. It is due to the ohmic resistance R_{el} of the electrodes, of the solid electrolyte R_i, and the electrode–electrolyte contact resistance R_c:

$$\eta_{ohm} = I\left(R_{el} + R_i + R_c\right) \tag{13.12}$$

By introducing the current density i (A/cm²) and the area-specific resistances, R'_{el}, R'_i, and R'_c ($\Omega\cdot cm^2$), one can rewrite Equation (13.12) as:

$$\eta_{ohm} = i\left(R'_{el} + R'_i + R'_c\right) \tag{13.12a}$$

To minimize the ohmic overpotential, one has to use thin, highly conductive solid electrolytes and highly conductive electrodes with a good adherence to the solid electrolyte to minimize the contact resistance.

The cell concentration overpotential η_{conc} is the sum of the concentration overpotentials at the anode and at the cathode and is caused by slow mass transfer between the gas phase and the tpb. For an SOFC operating on H_2 fuel it can be expressed as:

$$\eta_{conc} = \frac{-RT}{4F}\left[\ln\left(1-\frac{i}{i_{L,c}}\right)+2\ln\left(1-\frac{i}{i_{L,a}}\right)\right] \quad (13.13)$$

where $i_{L,a}$ and $i_{L,c}$ are the anodic and cathodic limiting current densities. When mass transfer in the gas phase is rate limiting, then $i_{L,a}$ and $i_{L,c}$ can be computed for many geometries of the anodic and cathodic compartments via standard analytical or empirical mass transfer correlations.[30,31] When surface diffusion on the electrode surface to the tpb is rate limiting then $i_{L,a}$ and $i_{L,c}$ must be measured experimentally. The concentration overpotential can be minimized by appropriate design of the anodic and cathodic compartments and by the use of porous electrodes. In this way, limiting current density values $i_{L,a}$ and $i_{L,c}$ of at least 2 to 3 A/cm^2 can be achieved so that in view of Equation (13.13), η_{conc} can be rather small (<100 mV) even for current densities up to 0.7 A/cm^2, which is close to the maximum current densities obtained with state-of-the-art SOFC units.[32] In general, due to the high values of gaseous and surface diffusivities at the operating temperature of SOFCs relative to liquid-phase diffusivities, the concentration overpotential is a less serious problem in SOFCs than in low-temperature aqueous electrolyte fuel cells.

The cell activation overpotential η_{act} is the sum of the activation overpotentials at the anode and cathode and is caused by slow electrocatalytic (charge transfer) reactions at both electrodes. The anodic and cathodic activation overpotentials, $\eta_{act,a}$ and $\eta_{act,c}$, respectively, can be related to the current density i via the Butler–Volmer equation:

$$i = i_{0,a}\left[\exp\left(\frac{\alpha_{a,a}F\eta_{act,a}}{RT}\right)-\exp\left(\frac{-\alpha_{c,a}F\eta_{act,a}}{RT}\right)\right] \quad (13.14a)$$

$$i = i_{0,c}\left[\exp\left(\frac{\alpha_{a,c}F\eta_{act,c}}{RT}\right)-\exp\left(\frac{-\alpha_{c,c}F\eta_{act,c}}{RT}\right)\right] \quad (13.14b)$$

where $i_{0,a}$, $i_{0,c}$ are the exchange current densities of the anode and cathode, respectively, $\alpha_{a,a}$ and $\alpha_{c,a}$ are the anodic ($i > 0$) and cathodic ($i < 0$) charge transfer coefficients of the anode, and $\alpha_{a,c}$ and $\alpha_{c,c}$ are the anodic and cathodic charge transfer coefficients of the cathode. The anodic and cathodic charge transfer coefficients a_a and a_c depend on the electrocatalytic reaction mechanism and typically take values between zero and two.[33,34] A zero value (a rather rare case) implies that the rate-limiting step is catalytic or in general involves no net charge transfer, in which case the term chemical overpotential can also be used.

The exchange current densities $i_{0,a}$ and $i_{0,c}$ of the anode and cathode, respectively, are of crucial importance as they determine the magnitude of $\eta_{act,a}$ and $\eta_{act,c}$ via Equation (13.14). A good electrocatalyst is characterized by a high value of the exchange current density i_0.

Two points need to be emphasized about the well-known Butler–Volmer equation:

$$i = i_0\left[\exp\left(\frac{\alpha_a F\eta}{RT}\right)-\exp\left(\frac{-\alpha_c F\eta}{RT}\right)\right] \quad (13.14c)$$

and for its limiting high-field or Tafel approximation[30] to which it reduces for $|\eta|$>200mV:

$$\eta = \frac{RT}{\alpha_a F}\ln(i/i_0) \quad ; \quad i>0 \quad (13.15a)$$

$$\eta = -\frac{RT}{\alpha_c F}\ln(-i/i_0) \quad ; \quad i<0 \tag{13.15b}$$

Firstly, is a kinetic expression, a rate law, such as, e.g., the Langmuir–Hinshelwood–Hougen–Watson rate expressions in heterogeneous catalysis,[35] and as such has no universal applicability. It is derived on the basis of mass action kinetics[30,34] and does reduce to the fundamental thermodynamic Nernst equation for $i = 0$, thus $\eta = 0$.[30] Nevertheless, experimental deviations can be expected as with any other, even most successful, rate expression.

Secondly, the parameters i_0, α_a, and α_c, which are usually treated as constants, are in reality dependent on the coverage of species adsorbed at the tpb. Consequently if the coverage of these species changes significantly with η, then i_0, but also α_a and α_c, can vary significantly.

Figure 13.2 shows a typical current–voltage curve of a fuel cell obtained by varying the external resistive load which shows that, as expected from Equations (13.12) to (13.14), the concentration overpotential becomes important at very high current densities, while the activation and ohmic overpotentials dominate at low and intermediate current densities where the cell power density $P = Ei$ is maximized. State-of-the-art SOFC units can produce power densities of the order of 0.3 W/cm^2 and, to a good approximation, E decreases linearly with i with a slope, denoted by R'_{cell}, as low as 0.5 Ω cm^2. This area-specific resistance provides a good measure of the practical usefulness of a SOFC unit. Obtaining such small R'_{cell} values (<0.5 Ω cm^2) which are necessary for technological SOFC power-producing applications, depends crucially on minimizing the cell ohmic and anodic and cathodic activation overpotentials.

D. EXCHANGE CURRENT DENSITY AND ELECTROCATALYTIC ACTIVITY

The magnitude of the activation overpotential for an electrocatalytic reaction occurring at the electrode/solid electrolyte tpb or interface depends crucially on the exchange current density i_0 (Equation 13.14). In order to measure the overpotential of a working electrode (W), anode, or cathode, and thus extract i_0, α_a, and α_c from the Butler–Volmer equation, it is necessary to utilize a reference (R) electrode.[13,30] When a current I flows between the working electrode and a counter electrode, then the potential of the working electrode V_{WR} with respect to the reference electrode deviates from its open-circuit ($I = 0$) value V^o_{WR} and the working electrode overpotential η_{WR} is defined from:

$$\eta_{WR} = V_{WR} - V^o_{WR} \tag{13.16}$$

The overpotentials η_W and η_R of the working and reference electrodes (the latter vanishes for an ideal reference electrode) are defined as the deviations of the inner (or Galvani) potentials[13,16] of these electrodes from their open-circuit values:

$$\eta_W = \varphi_W - \varphi^o_W \quad ; \quad \eta_R = \varphi_R - \varphi^o_R \tag{13.17}$$

One can write Equation (13.16) in the form:

$$\eta_{WR} = \eta_W + \eta_R + \eta_{ohmic,WR} \tag{13.18}$$

Ideally, no current flows through the reference electrode; therefore in principle $\eta_R = 0$ and $\eta_{ohmic,WR} = 0$. In practice, the first assumption is usually adequate for reasonably nonpolarizable reference electrodes, since the parasitic (uncompensated) current flowing via the reference electrode is usually very small. The ohmic drop, however, between the working and reference electrodes, i.e., $\eta_{ohmic,WR}$, may in general not be negligible[13,30] and must be

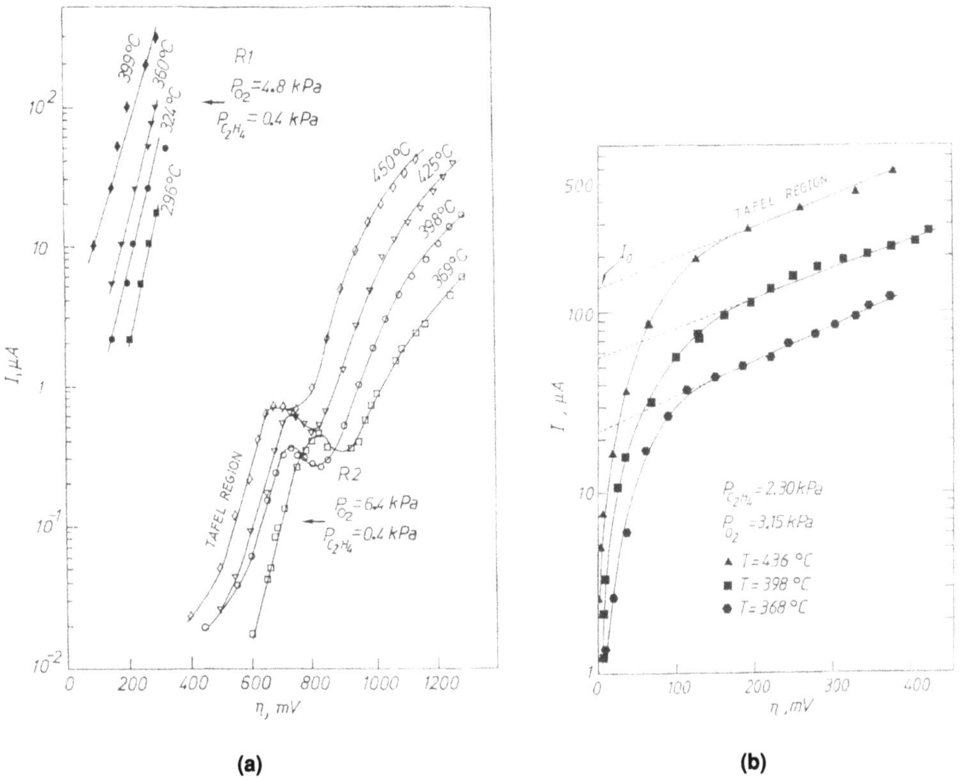

FIGURE 13.3. Typical Tafel plots for Pt catalyst/YSZ (a) and Ag catalyst/YSZ (b) interfaces. (From Bebelis, S. and Vayenas, C.G., *J. Catal.*, 1989, 118, 125; Vayenas, C.G. et al., *Catal. Today*, 1992, 11, 303. With permission.)

determined in each case using, e.g., the current interruption technique in conjunction with a recording oscilloscope.[13,36] The ohmic component decays to zero within typically less than 1 μs, and the remaining part of η_{WR} is η_W. As in aqueous electrochemistry, the reference electrode must be placed as near to the catalyst as possible to minimize $\eta_{ohm,WR}$.

It is worth emphasizing that although overpotentials are usually associated in electrochemistry with electrode/electrolyte interfaces, in reality they refer to, and are measured as, deviations of the potentials of the electrodes only. Thus the concept of overpotential must be associated with an electrode and not with an electrode/electrolyte interface, although the nature of the interface will, in general, dictate the magnitude of the overpotential. This becomes an important point in Section III (catalysis).

The usual procedure for measuring the exchange current density i_0 is then to measure η as a function of I and to plot ln I vs. η_W (Tafel plot). Such plots are shown in Figure 13.3 for Pt and Ag electrodes deposited on YSZ. From the slopes of the linear part of these plots ($|\eta| > 200$ mV, in which case Equation 13.15 are valid) one obtains the transfer coefficients α_a and α_c. By extrapolating the linear part of the plot to $\eta = 0$, one obtains i_0. One can then plot i vs. η and use the "low field" approximation of the Butler–Volmer equation which is valid for $|\eta| < 10$ mV, i.e.,

$$i/i_0 = (\alpha_a + \alpha_c)F\eta/RT \tag{13.19}$$

in order to check the accuracy of the extracted i_0, α_a, and α_c values.

The exchange current density i_0 is a measure of the electrocatalytic activity of the tpb or, more generally, the electrode/solid electrolyte interface for a given electrocatalytic reaction. It expresses the (equal and opposite) rates of the forward (anodic) and backward (cathodic)

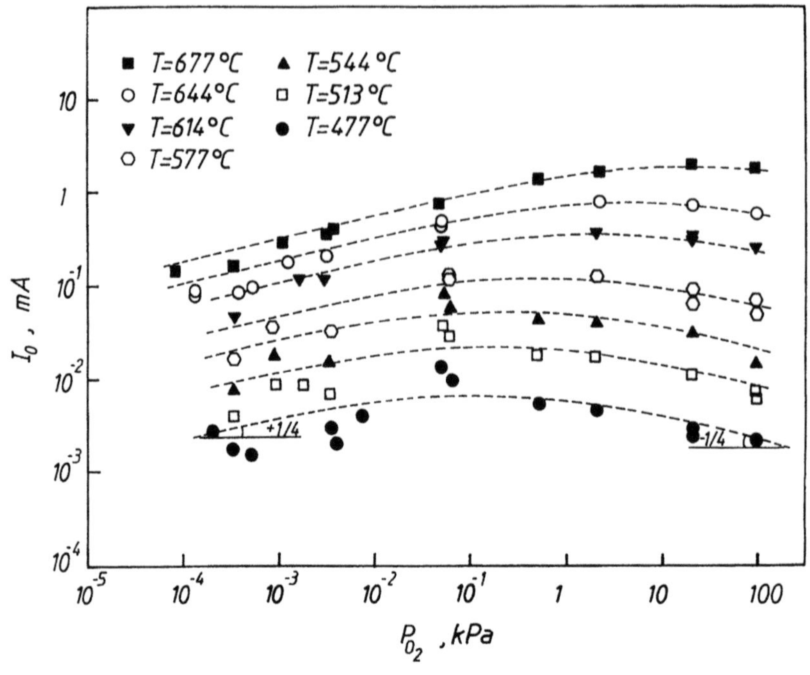

FIGURE 13.4. Effect of temperature and P_{O_2} on the I_0 of a Pt catalyst film (A = 2 cm^2) on YSZ. (From Manton, M., Ph.D. thesis, MIT, Cambridge, MA, 1984.)

electrocatalytic reaction rates when no net current crosses the electrode/solid electrolyte interface or, equivalently, the tpb. It has recently been shown that quite often, the exchange current I_0 is proportional to the length l of the tpb[37-39] as one would intuitively expect. The tpb length l can be estimated via cyclic voltammetry[37] and electron microscopy.[38] In principle, different electrocatalysts should be compared via the parameter I_0/l. In practice this is seldom done due to the difficulty in measuring l accurately and the comparison is usually made via the exchange current density $i_0 = I_0/A$, where A is the superficial surface area of the electrode/electrolyte interface.

The exchange current density i_0 is usually strongly temperature dependent. It increases with temperature with an activation energy which is typically 140 to 180 kJ/mol for Pt/YSZ and 80 to 100 kJ/mol for Ag/YSZ.[13,36] The i_0 dependence on gaseous composition is usually complex. Figure 13.4 shows the measured i_0 dependence on P_{O_2} and T for Pt/YSZ films.[40] These results can be described adequately on the basis of Langmuir-type dissociative oxygen chemisorption at the tpb, i.e:

$$\theta_O = K_O P_{O_2}^{1/2} \Big/ \left(1 + K_O P_{O_2}^{1/2}\right) \tag{13.20}$$

where θ_O is the coverage of atomic oxygen at the tpb and K_O is the adsorption equilibrium constant of oxygen on Pt. It can be shown[36,40] that:

$$i_0 = c\left[\theta_O\left(1 - \theta_O\right)\right]^{1/2} \tag{13.21a}$$

where c is a constant, or, equivalently:

$$i_0 = cK_O^{1/2} P_{O_2}^{1/4} \Big/ \left(1 + K_O P_{O_2}^{1/2}\right) \tag{13.21b}$$

which explains nicely the observed i_0 maximum and the fact that i_0 is proportional to $P_{O_2}^{1/4}$ for low P_{O_2} and $P_{O_2}^{-1/4}$ for high P_{O_2} (Figure 13.4). According to this successful model, the i_0 maximum corresponds to $\theta_O = 1/2$.

Extraction of I_0 and of the transfer coefficients α_a and α_c from steady-state current-overpotential data is frequently complicated by significant changes in surface coverage of oxygen at the tpb or by surface oxide formation.[40,41] It has been shown recently[41] that several of these problems can be overcome by the use of cyclic voltammetry in conjunction with the equations:

$$\frac{dE_{p,a}}{d\ln\upsilon} = \frac{RT}{\alpha_a F} \tag{13.22a}$$

$$\frac{dE_{p,c}}{d\ln\upsilon} = -\frac{RT}{\alpha_c F} \tag{13.22b}$$

where υ is the sweep rate (V/s) and $E_{p,a}$, $E_{p,c}$ are the potentials corresponding to the anodic and cathodic oxygen peaks, respectively.[42] In this way it has been shown that for porous Pt/YSZ electrodes exposed to oxygen $\alpha_a = \alpha_c = 1$.[41]

The transfer coefficients α_a and α_c convey useful mechanistic information and can be used for mechanism discrimination. For a one-electron transfer process (n = 1), e.g.,:

$$H^+ + e^- \rightarrow H(a) \tag{13.23}$$

the Butler–Volmer equation is

$$i/i_0 = \exp(\beta F\eta/RT) - \exp[-(1-\beta)F\eta/RT] \tag{13.24}$$

where β is the symmetry factor ($0 < \beta < 1$), which is frequently found to equal 0.5.[33,34]

For multielectron transfer processes ($n \geq 2$, e.g., oxygen reduction) the Butler–Volmer equation is

$$i/i_0 = \exp(\alpha_a F\eta/RT) - \exp[-\alpha_c F\eta/RT] \tag{13.25}$$

It can be shown[30,33] that α_a and α_c are related to the symmetry factor β of the rate-determining step (rds) via:

$$\alpha_a = \frac{n-\gamma}{\nu} - s\beta \tag{13.26a}$$

$$\alpha_c = \frac{\gamma}{\nu} + s\beta \tag{13.26b}$$

where:

- n = total number of electrons transferred in the overall electrocatalytic reaction
- ν = stoichiometric number, i.e., number of times the rds occurs for one act of the overall reaction
- γ = number of electrons transferred before the rds in the forward reaction
- s = number of electrons transferred in the rds (0 or 1)

It follows from Equation (13.26) that

$$\alpha_a + \alpha_c = n/\nu \tag{13.27}$$

The above equations can be quite useful for discriminating between various mechanisms for a given overall electrocatalytic reaction. They give rise to several well-known Tafel slopes, the most common of which (2RT/F, RT/F, 2RT/3F, RT/2F, and RT/4F) have been observed.[33,34] However, the derivation of Equation (13.26) is based on three important assumptions:

I. A single clearly defined rds
II. Low coverages ($\theta \ll 1$) which can be described by a linear isotherm
III. A single reaction pathway

II. ELECTROCATALYTIC OPERATION OF SOLID ELECTROLYTE CELLS

A. ELECTROCATALYSIS FOR THE PRODUCTION OF CHEMICALS

The preparation, structure, and performance of SOFC used for power generation is described in Chapter 12 of this handbook. Both the electrocatalysis and the catalysis of the Ni/YSZ anode are reasonably well understood and the same applies for the electrocatalysis of the $La_{1-x}Sr_xMnO_{3-\delta}$ and $La_{1-x}Sr_xCoO_{3-\delta}$ cathodes. State-of-the-art SOFC units with anodic and cathodic overpotentials of less than 150 mV each at current densities up to 0.5 A/cm^2 at T = 950°C are currently available.[6]

In recent years it has been shown that solid electrolyte fuel cells with appropriate electrocatalytic anodes can be used not only for power generation via oxidation of H_2 and CH_4, but also for "chemical cogeneration", i.e., for the simultaneous production of power and useful chemicals. This mode of operation, first demonstrated for the case of NH_3 oxidation to NO,[8-10] combines the concepts of a fuel cell and of a chemical reactor (Figure 13.5). The economics of chemical cogeneration have been modeled and discussed recently.[7] The economics appear promising for a few highly exothermic reactions which can be carried out at temperatures above 700°C, such as the oxidation of H_2S to sulfur and SO_2[11] and of NH_3 to NO.[8-10] It is likely that if SOFC units operating on H_2 or natural gas become commercially available, then they could be used with appropriate anodic electrocatalysts by some chemical industries.

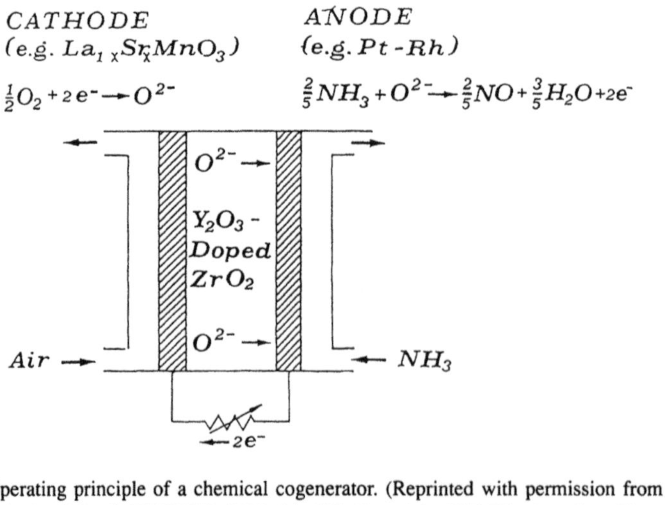

FIGURE 13.5. Operating principle of a chemical cogenerator. (Reprinted with permission from Vayenas, C.G., Bebelis, S., and Kyriazis, C.C., *CHEMTECH*, 1991, *21*, 422. Copyright 1991 The American Chemical Society.)

TABLE 13.1
Electrocatalytic Reactions Investigated in Doped ZrO_2 Solid Electrolyte Fuel Cells for Chemical Cogeneration

	Electrocatalyst	Ref.
$2NH_3 + 5O^{2-} \rightarrow 2NO + 3H_2O + 10e^-$	Pt, Pt-Rh	8–10
$CH_3OH + O^{2-} \rightarrow H_2CO + H_2O + 2e^-$	Ag	12
$C_6H_5\text{–}CH_2CH_3 + O^{2-} \rightarrow C_6H_5\text{-}CH{=}CH_2 + H_2O + 2e^-$	Pt, Fe_2O_3	43,44
$H_2S + 3O^{2-} \rightarrow SO_2 + H_2O + 6e^-$	Pt	11
$C_3H_6 + O^{2-} \rightarrow C_3$ dimers $+ 2e^-$	Bi_2O_3-La_2O_3	45
$CH_4 + NH_3 + 3O^{2-} \rightarrow HCN + 3H_2O + 6e^-$	Pt, Pt-Rh	46
$2CH_4 + 2O^{2-} \rightarrow C_2H_4 + 2H_2O + 4e^-$	Ag, Ag-Sm_2O_3	59

Table 13.1 lists the anodic reactions which have been studied so far in small cogenerative SOFC units. One simple and interesting rule which has emerged from these studies is that the selection of the anodic electrocatalyst for a selective electrocatalytic oxidation can be based on the heterogeneous catalytic literature for the corresponding selective catalytic oxidation. Thus the selectivity of Pt and Pt-Rh alloy electrocatalysts for the anodic NH_3 oxidation to NO:

$$2NH_3 + 5O^{2-} \rightarrow 2NO + 3H_2O + 10e^- \tag{13.28}$$

turns out to be comparable (>95%) with the selectivity of Pt and Pt-Rh alloy catalysts for the corresponding commercial catalytic oxidation where oxygen is co-fed with NH_3 in the gas phase.[8-10] The same applies for Ag which turns out to be equally selective as an electrocatalyst for the anodic partial oxidation of methanol to formaldehyde:[12]

$$CH_3OH + O^{2-} \rightarrow H_2CO + H_2O + 2e^- \tag{13.29}$$

as it is a catalyst for the corresponding heterogeneous catalytic reaction:

$$CH_3OH + 1/2\,O_2 \rightarrow H_2CO + H_2O \tag{13.30}$$

Another reaction which is of considerable interest from the view point of chemical cogeneration is the oxidative coupling of methane (OCM) to C_2 hydrocarbons ethane and ethylene:[47]

$$2CH_4 \xrightarrow{O^{2-}} C_2H_6 \xrightarrow{O^{2-}} C_2H_4 \xrightarrow{O^{2-}} 2CO_2 \tag{13.31}$$

There have been several OCM studies utilizing SOFC reactors[48-58] and a recent review.[48] In many of these studies electrical power was supplied to the cell to increase cell current. This, however, tends in general to decrease the selectivity and yield of C_2 hydrocarbons. Very recently it was found that it is possible to obtain C_2 yields up to 88% by means of a novel gas-recycle SOFC reactor–separator.[59] The C_2H_4 yield is up to 85%.[59] These systems are of considerable technological interest.

Aside from chemical cogeneration studies, where the electrocatalytic anodic and cathodic reactions are driven by the voltage spontaneously generated by the solid electrolyte cell, several other electrocatalytic reactions, including the OCM reaction, have been investigated in solid electrolyte cells[52-58,60-67] and are listed in Table 13.2. Earlier studies by Schouler and co-workers focused mainly on the investigation of electrocatalysts for H_2O electrolysis.[61] Gür and Huggins[64-66] and Mason and co-workers[63] were first to show that other electrocatalytic reactions, such as NO decomposition and CO hydrogenation can be carried out in zirconia

TABLE 13.2
Electrocatalytic Reactions Investigated in Doped ZrO_2 Solid Electrolyte Fuel Cells with External Potential Application

	Electrocatalyst	Ref.
$H_2O + 2e^- \rightarrow H_2 + O^{2-}$	Ni	60–62
$2NO + 4e^- \rightarrow N_2 + 2O^{2-}$	Pt, Au	63,64
$CO + 2H_2 + 2e^- \rightarrow CH_4 + O^{2-}$	Pt, Ni	65,66
$C_3H_6 + O^{2-} \rightarrow C_3H_6O + 2e^-$	Au	67
$CH_4 + O^{2-} \rightarrow C_2H_6, C_2H_4, CO, CO_2, H_2O + 2e^-$	Ag, Ag-MgO, Ag-Bi_2O_3, Ag-Sm_2O_3	48–59

cells. More recently Otsuka et al.,[52,54,55] Eng and Stoukides,[50] Seimanides and Stoukides,[53] Belyaev et al.,[56] and Vayenas et al.,[59] have concentrated on the investigation of the OCM reaction using a variety of metal and metal oxide electrodes deposited on YSZ. The use of protonic conductors to enhance the OCM reaction has also been explored,[68,69] which is in principle a very interesting idea.

B. ELECTROCHEMICAL REACTOR ANALYSIS AND DESIGN

SOFCs can be viewed as a special type of a chemical reactor which generates electrical power in addition to heat. The first rigorous engineering modeling of SOFC units was carried out in 1983 by Debenedetti and Vayenas,[29] who assumed well-mixed (CSTR-type) anodic and cathodic compartments, and isothermality within the SOFC structure for the overall reaction:

$$A + 1/2O_2 \rightarrow B \tag{13.32}$$

e.g., H_2 or CO oxidation, and showed that the following dimensionless mass, energy, and electron balances govern the SOFC behavior:

Fuel (A) and oxygen mass balances:

$$X_A = A_1 \cdot \xi \tag{13.33a}$$

$$X_{O_2} = A_1 \cdot A_2 \cdot \xi \tag{13.33b}$$

Energy balance:

$$\begin{aligned} A_1A_3(1 - A_4\xi)\xi = (1 + A_6 - A_1A_2A_7\xi)(\Theta - 1) - \\ -A_5(\Theta_1^o - 1) - A_6(\Theta_2^o - 1) + A_8(\Theta - \Theta_c) \end{aligned} \tag{13.33c}$$

Electron balance:

$$\left\{A_4(1 + A_9) + \exp\left[A_{12}\left(\frac{1}{\Theta} - 1\right)\right]\right\} \cdot \xi = A_{10^+} \tag{13.33d}$$

$$+ A_{11}\Theta\left\{\ln\frac{P_o y_{O_2}^o (1 - X_{O_2})(1 - X_A)^2}{(1 - y_{O_2}^o X_{O_2})X_A^2} + 2\ln(1 - A_{13}\xi) + \right.$$

$$\left. + \ln(1 - A_{14}\xi) - A_{15}\left[\ln(A_{16}\xi) + A_{17}\ln(A_{18}\xi)\right]\right\}$$

where the dimensionless current ξ and temperature Θ are defined by:

$$\xi = I\rho_o\delta / A_{el}E_{TH} \tag{13.34a}$$

$$\Theta = T/T_o \tag{13.34b}$$

All symbols appearing in Equations (13.33) and (13.34) are defined in Table 13.3. The dimensionless parameters A_1 to A_4 and A_6 to A_9 play a key role in SOFC performance. Their physical significance and importance for scale-up is discussed in Reference 29. The left- and right-hand sides of the dimensionless energy balance (13.33c) can be viewed as a heat generation and heat removal term, respectively. The dimensionless electron balance (13.33d) can be viewed as a dimensionless kinetic expression where the right-hand side represents the driving force for current flow, thus for reaction. Numerical solution of Equations (13.33) with realistic parameter values has shown that, similar to chemical reactors, SOFC units can exhibit steady-state multiplicity as well as ignition and extinction phenomena over a wide range of design and operating parameters.[29]

The above model is valid both for power-producing and chemical-producing SOFC units. It is a lumped-parameter CSTR-type model, i.e., it assumes uniform gas composition and temperature within the SOFC structure. When the first flat plate or cross-flow monolithic SOFC units were fabricated and tested,[70] a new two-dimensional model was developed to account for the variation in gas-phase composition and temperature within the SOFC structure.[71] The governing equations of this two-dimensional mixing cell model are not given here due to space limitations, but can be found in Reference[71] as well as in more recent papers where in addition to cross-flow SOFC units, counter-flow, and concurrent-flow SOFC units are also modeled.[72] Various improvements and modifications of the original model have been published recently,[73,74] also accounting for SOFC operation with internal reforming.[74]

III. CATALYSIS ON THE ELECTRODES OF SOLID ELECTROLYTE CELLS

A. POTENTIOMETRIC INVESTIGATIONS

The interesting role which solid electrolytes can play in the study of heterogeneous catalysis was first recognized by Wagner,[75] who proposed the use of solid electrolyte cells for the measurement of the activity of oxygen on metal and metal oxide catalysts. This technique, first used to study the mechanism of SO_2 oxidation on noble metals,[76] was subsequently called solid electrolyte potentiometry (SEP).[77] It has been used in conjunction with kinetic measurements to study the mechanism of several catalytic reactions on metals[77-91] and, more recently, metal oxides.[92-96] It is particularly suitable for the study of oscillatory catalytic reactions.[77,81,83,87-89] The SEP literature has recently been reviewed.[92,97-99]

Figure 13.6 shows a typical experimental arrangement for the use of SEP. The bottom porous metal electrode (e.g., Pt) of the oxygen ion-conducting solid electrolyte cell is exposed to ambient air and serves as a reference electrode (R). The top electrode, which we denote by W (for "working"), acts both as an electrode and as a catalyst for the catalytic reaction to be studied. The open-circuit potential V^o_{WR} of the catalyst electrode relative to the reference electrode can be written as:

$$V^o_{WR} = (1/4F)\left[\mu_{O_2,W} - \mu_{O_2,R}\right] \tag{13.35}$$

where $\mu_{O_2,W}$ and $\mu_{O_2,R}$ are the chemical potentials of oxygen at the catalyst and reference electrode surfaces, respectively. Equation (13.35) is derived on the following four assumptions:

TABLE 13.3
Dimensionless and Dimensional Parameters Appearing in the Governing Mass, Energy and Electron Balance Equations of Single-Cell SOFC Units (Equations 13.33 and 13.34)

A_1	$E_{TH} \cdot A_{el}/2F\rho_o\delta N_1 y_A^o$
A_2	$N_1 y_A^o/2N_2 y_{O_2}^o$
A_3	$(-\Delta H^o)y_A^o/\bar{C}_{pl} \cdot T_o$
A_4	$R_{ex} \cdot A_{el}/\rho_o\delta$
A_5	$\bar{C}_{p,1}^o/\bar{C}_{p,1}$
A_6	$N_2 \cdot \bar{C}_{p,2}/N_1 \cdot \bar{C}_{p,1}$
A_7	$N_2 \cdot y_{O_2}^o \cdot \bar{C}_{p,2}/N_1\bar{C}_{p,1}$
A_8	$U \cdot A_{ex}/N_1 \cdot \bar{C}_{p,1} \cdot T_o$
A_9	R_{el}/R_{ex}
A_{10}	E^o/E_{TH}
A_{11}	$RT_o/4FE_{TH}$
A_{12}	E^*/RT_o
A_{13}	$E_{TH} \cdot A_{el}/\rho_o \cdot \delta \cdot I_{L,a}$
A_{14}	$E_{TH} \cdot A_{el}/\rho_o \cdot \delta \cdot I_{L,c}$
A_{15}	$4/\alpha_a$
A_{16}	$E_{TH} \cdot A_{el}/\rho_o \cdot \delta \cdot I_{0,a}$
A_{17}	α_a/α_c
A_{18}	$E_{TH} \cdot A_{el}/\rho_o \cdot \delta \cdot I_{0,c}$
A_{el}	Superficial electrode surface area per unit cell, m^2
A_{ex}	Heat loss surface area per unit cell, m^2
$\bar{C}_p$	Temperature and composition averaged molar specific heat, J/mol K
E	Operating cell voltage, V
E^o	Reversible cell voltage corresponding to unit activity of reactants and products, V
E_{rev}	Reversible cell voltage, V
E_{TH}	Thermoneutral cell voltage ($=(-\Delta H^o)/nF$), V
E^*	Activation energy for O^{2-} conduction in solid electrolyte, J/mol
F	Faraday constant, 96,500 Cb/g-equivalent
I	Current per unit cell, A
$I_{0,a}$	Anodic exchange current, A
$I_{0,c}$	Cathodic exchange current, A
$I_{L,a}$	Limiting anodic current corresponding to completely mass transfer controlled anodic operation, A
$I_{L,c}$	Limiting cathodic current, A
ξ	Dimensionless current per unit cell ($=I\rho_o\delta/A_{el} \cdot E_{TH}$)
n	Electrons transferred per molecule A converted
N_1	Fuel stream molar flow rate, mol/s
N_2	Oxygen stream feed molar flow rate, mol/s

TABLE 13.3 (continued)
Dimensionless and Dimensional Parameters Appearing in the Governing Mass, Energy and Electron Balance Equations of Single-Cell SOFC Units (Equations 13.33 and 13.34)

P_o	Operating cell pressure, bar
R	Gas constant, 8.314 J/mol K
R_{el}	Electrode resistance, Ω
R_i	Electrolyte resistance, Ω
R_{ex}	External load, Ω
T	Absolute temperature, K
U	Overall heat transfer coefficient, W/m² K
W	Work produced per mol of A converted, J
X_A	Fuel conversion
X_{O2}	Oxygen conversion
y_A^o	Fuel mol fraction in feed
$y_{O_2}^o$	Oxygen mol fraction in feed
Greek Symbols	
α_a, α_c	Transfer coefficients for anodic and cathodic overpotential, dimensionless
ΔG	Gibbs free energy change for Reaction (13.32), J/mol A
ΔH^o	Standard enthalpy change for Reaction (13.32) J/mol A at T_o
ΔS^o	Standard entropy change for Reaction (13.32), J/K mol A at T_o
δ	Electrolyte thickness, m
Θ	Dimensionless temperature
ρ	Electrolyte resistivity, Ω
ρ_o	Electrolyte resistivity at reference temperature T_o, Ω
Subscripts	
1	Fuel stream
2	Air stream
c	Ambient conditions
o	Reference conditions
Superscript	
o	Feed conditions

1. The solid electrolyte is a pure O^{2-} conductor.
2. The catalyst and reference electrodes are made of the same bulk material.
3. The dominant electrocatalytic reaction taking place at the metal-solid electrolyte-gas three-phase-boundaries (tpb) is:

$$O(a) + 2e^- \rightleftharpoons O^{2-} \tag{13.36}$$

or

$$O_2(a) + 4e^- \rightleftharpoons 2O^{2-} \tag{13.37}$$

where O(a) and O_2(a) denote oxygen dissociatively or molecularly adsorbed at the tpb, respectively. It is also assumed that equilibrium is established for Reaction (13.36) or (13.37).

4. There are no concentration or chemical potential gradients within the porous catalyst film so that $\mu_{O_2,W}$ is the same at the tpb and over the entire catalyst surface. This implies that thin (e.g., 5 to 10 μm) and sufficiently porous metal or metal oxide catalyst films must be used in order to ensure the absence of internal diffusional limitations.

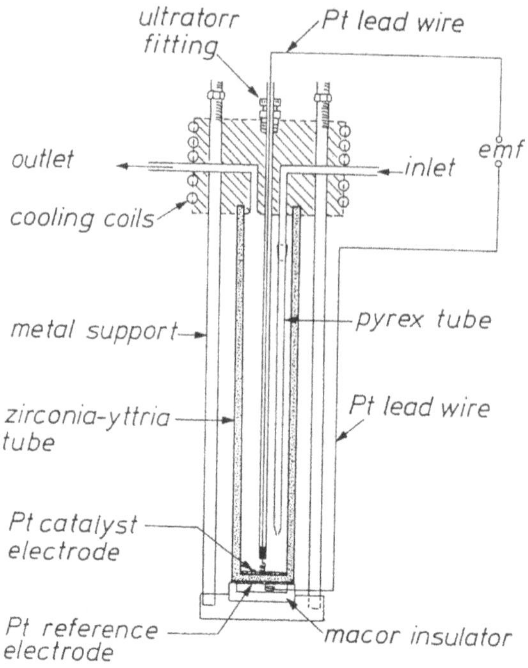

FIGURE 13.6. Schematic of a solid electrolyte reactor used for simultaneous kinetic and SEP studies. (From Yentekakis, I.V., Neophytides, S., and Vayenas, C.G., *J. Catal.*, 1988, *111*, 152. With permission.)

Assumption 3 is certainly valid for the reference electrode, but may not always be valid for the catalyst electrode: if one of the reactants adsorbs strongly on the catalyst surface and has a high affinity for reaction with O^{2-} (e.g., H_2 or CO under fuel-rich conditions), then other electrocatalytic reactions, such as

$$H_2 + O^{2-} \rightleftharpoons H_2O + 2e^- \tag{13.38}$$

$$CO + O^{2-} \rightleftharpoons CO_2 + 2e^- \tag{13.39}$$

may also take place at the tpb, leading to the establishment of mixed potentials. In this case V^o_{WR} provides only a qualitative measure of surface activities. This point has been discussed in the past.[84,85] In practice this means that one must be rather cautious when using Equation (13.35) to treat very negative V^o_{WR} values, e.g., less than –400 mV, unless one proves the validity of assumption 3 via electrokinetic measurements by utilizing a three-electrode system, which is not an easy experiment.[84,85] When assumption 3 is also satisfied, in addition to 1, 2, and 4, then the emf V^o_{WR} provides indeed an *in situ* quantitative measure of the chemical potential or activity of oxygen adsorbed on the catalyst under reaction conditions, and this information can indeed be quite useful.[77,87-89]

Before discussing how Equation (13.35) has been used in the past to analyze SEP data, it is important to discuss first some recent findings. It was recently found both theoretically[13,16] and experimentally by means of a Kelvin probe[17,100] that the emf V^o_{WR} of solid electrolyte cells provides a direct measure of the difference in the work function $e\Phi$ of the gas-exposed surfaces of the working and reference electrodes:

$$eV^o_{WR} = e\Phi_W - e\Phi_R \tag{13.40}$$

Equation (13.40) is always valid provided the catalyst and reference electrodes are made of the same material.[17,100] It is also valid when other types of solid electrolytes are used, e.g., β''-Al_2O_3, a Na^+ conductor.[17,100] Equation (13.40) is much more general and fundamental than Equation (13.35), as it does not require the establishment of any specific electrochemical equilibrium. It shows that *solid electrolyte cells are work function probes for their gas-exposed electrode surfaces*. It also shows that SEP is essentially a work function measuring technique and that several aspects of the SEP literature reviewed in References 92,97–99 must be reexamined in the light of these findings. One can still use SEP to extract information about surface activities, provided the nature of the electrocatalytic reaction at the tpb is well known, but, even when this is not the case, the cell emf still provides a direct measure of the work function difference between the two gas-exposed electrode surfaces. Equation (13.40) also holds under closed-circuit conditions, as is further discussed in Section III.B.

Equation (13.35) has been used in the following way in previous SEP studies.[78-82,87] First one notes that the chemical potential of oxygen adsorbed on the reference electrode is given by

$$\mu_{O_2,R} = \mu^{o}_{O_{2(g)}} + RT \ln (0.21) \tag{13.41}$$

where $\mu^{o}_{O_{2(g)}}$ is the standard chemical potential of gaseous oxygen at the temperature T, and 0.21 (bar) corresponds to the oxygen activity in the reference gas, i.e., air. Defining the activity of atomic oxygen adsorbed on the catalyst a_O from:

$$\mu_{O_2,W} = \mu^{o}_{O_{2(g)}} + RT \ln a_O^2 \tag{13.42}$$

one obtains

$$V^{o}_{WR} = (RT/2F) \ln \left[a_O/(0.21)^{1/2}\right] \tag{13.43}$$

Consequently, by measuring V^{o}_{WR} and T one computes a_O which expresses the square root of the partial pressure of gaseous oxygen that would be in thermodynamic equilibrium with oxygen adsorbed on the catalyst surface, if such an equilibrium were established. By comparing the potentiometrically obtained surface oxygen activity with the independently measured gas-phase oxygen activity P_{O_2}, it is possible to extract useful information about the rds of the catalytic reaction. Thus if $a_O^2 = P_{O_2}$, then the oxygen adsorption step is in equilibrium and cannot be rate limiting. If, however $a_O^2 < P_{O_2}$, as is often found to be the case, then oxygen adsorption is rate limiting.

Despite its limitations, SEP is one of the very few techniques which can be used to extract *in situ* information about adsorbed species on catalyst surfaces without UHV requirements. It is particularly useful for the study of oscillatory reactions. Figure 13.7 shows typical rate and V^{o}_{WR}, thus also $e\Phi$ oscillations during CO oxidation on Pt.[89,97] Figure 13.8 shows the dependence of the frequency and amplitude of the rate and V^{o}_{WR} oscillations on a_O. It can be seen that the high-frequency transition (bifurcation) between oscillatory and nonoscillatory states occurs near the decomposition pressure a_O^* of surface PtO_2 which is given by:

$$\ln \left(a_O^*/\text{kPa}^{1/2}\right) = 14.8 - 12000/T \tag{13.44}$$

This dissociation pressure expression has been obtained from similar experiments during C_2H_4 oxidation on Pt,[87,88,101] which is also an oscillatory reaction. This expression is in good agreement with independent high-precision resistance measurements[102] and with XPS data,

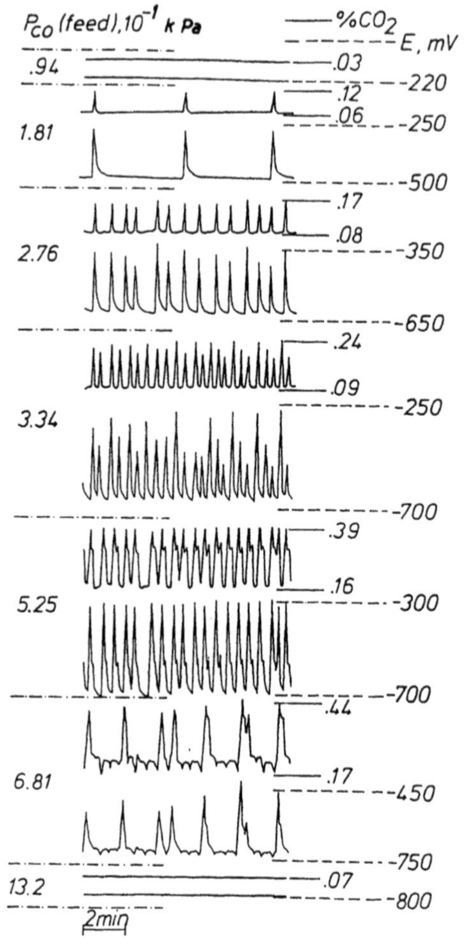

FIGURE 13.7. Effect of inlet CO partial pressure on reaction rate and on V^o_{WR} and eΦ oscillations during CO oxidation on Pt; P_{O_2} = 5.4 kPa, T = 337°C, total molar flow rate 2.7×10^{-4} mol/s. (From Yentekakis, I.V., Neophytides, S., and Vayenas, C.G., *J. Catal.*, 1988, *111*, 152. With permission.)

which have positively confirmed the existence of surface PtO_2.[103-105] The information derived by SEP, i.e., that the transition between oscillatory and nonoscillatory states takes place very near the dissociation pressure of PtO_2, provided the first evidence that the rate, V^o_{WR}, and eΦ oscillations during C_2H_4 and CO oxidation on Pt under atmospheric pressure conditions are due to the formation and decomposition of surface PtO_2.[88,101] This has led to mathematical models which describe the oscillatory phenomena in a semiquantitative manner.[87,106]

More recently, SEP measurements have been extended to metal oxide catalyst systems,[90,91,93-96] such as Cu oxide used for the partial oxidation of propene to acrolein (Figure 13.9).

B. ELECTROCHEMICAL ACTIVATION OF CATALYZED REACTIONS

During the last few years a new application of solid electrolytes has emerged. It was found that the catalytic activity and selectivity of the gas-exposed electrode surface of metal electrodes in solid electrolyte cells is altered dramatically and reversibly upon polarizing the metal/solid electrolyte interface. The induced steady-state change in *catalytic* rate can be up to 9000% higher than the normal (open-circuit) catalytic rate and up to 3×10^5 higher than the steady-state rate of ion supply.[14,107] This new effect of non-faradaic electrochemical modification of catalytic activity (NEMCA) has been already demonstrated for more than

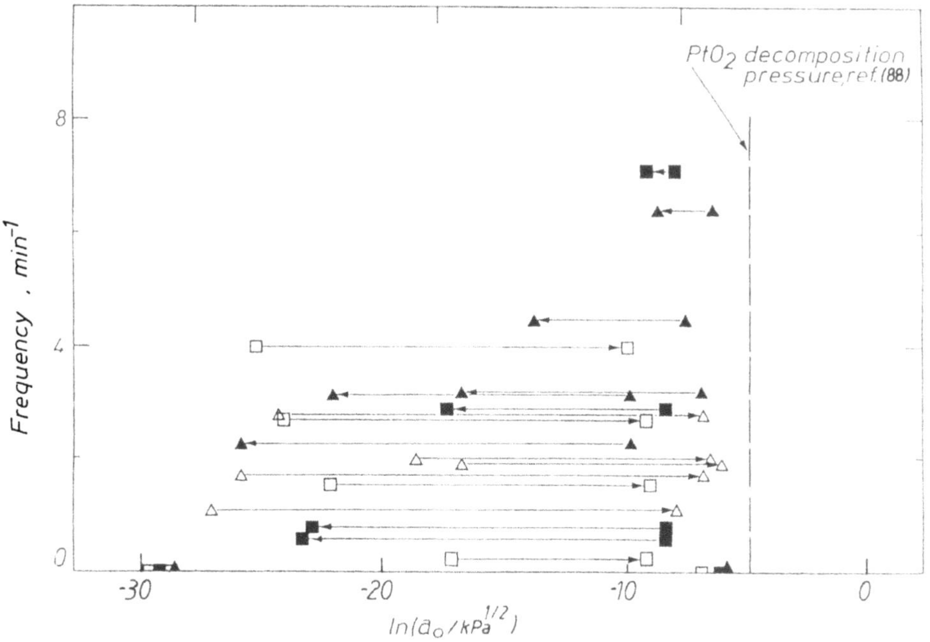

FIGURE 13.8. Effect of oxygen activity a_O on the frequency and amplitude of a_O oscillations during CO oxidation on Pt. Open and filled symbols correspond to oscillations obtained on the preoxidized and prereduced Pt catalyst film; T = 610 K. (From Yentekakis, I.V., Neophytides, S., and Vayenas, C.G., *J. Catal.*, 1988, *111*, 152. With permission.)

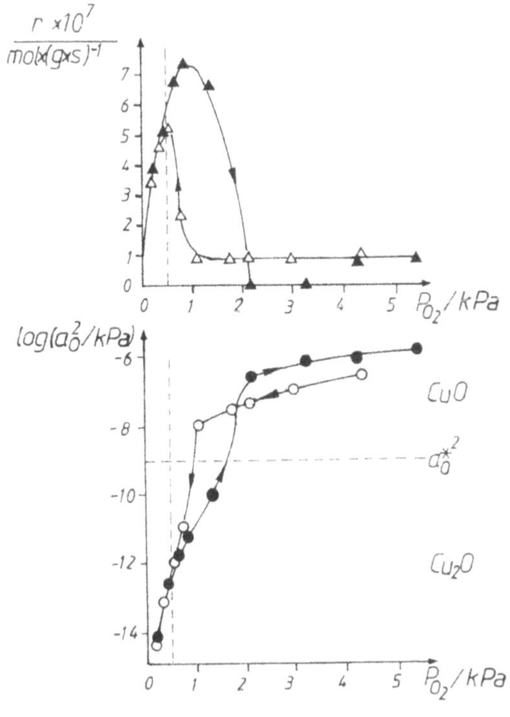

FIGURE 13.9. Effect of P_{O_2} on the rate of acrolein formation and on the oxygen activity on a Cu oxide catalyst during oxidation of propene, indicating rate and a_O hysteresis. The rate of by-product CO_2 formation increases monotonically with P_{O_2}. (From Vayenas, C.G., Bebelis, S., Yentekakis, I.V., and Lintz, H.-G., *Catal. Today*, 1992, *11*, 303. With permission.)

40 catalytic reactions on Pt, Pd, Rh, Ag, Au, and Ni surfaces by using O^{2-}, F^-, Na^+, and H^+ conducting solid electrolytes. There are also recent demonstrations for aqueous electrolyte systems.[20,26] In this section the common features of NEMCA studies are summarized and the origin of the effect is discussed in the light of recent *in situ* work function and XPS measurements which have shown that:

I. Solid electrolyte cells with metal electrodes are work function probes and work function controllers, via potential application, for their gas-exposed electrode surfaces.
II. NEMCA is due to an electrochemically driven and controlled back-spillover of ions from the solid electrolyte onto the gas-exposed electrode surface. These back-spillover ions establish an effective electrochemical double layer and act as promoters for catalytic reactions. This interfacing of electrochemistry and catalysis offers several exciting theoretical and technological possibilities. We note that in the catalytic literature, the term spillover usually denotes migration of a species from a metal to a support, while the term back-spillover denotes migration in the opposite direction and is thus more appropriate here.

1. Introduction

During the last few years it has become apparent that the "active" use of solid electrolyte cells offers some very interesting possibilities in heterogeneous catalysis: it was found that solid electrolyte cells can be used not only to study, but also to influence catalytic phenomena on metal surfaces in a very pronounced and reversible manner. Work in this area prior to 1988 had been reviewed.[97] Then in 1988 the first reports on NEMCA appeared in the literature.[14,15] Since then the NEMCA effect has been described for more than 40 catalytic reactions,[13-25] and work prior to 1992 has been reviewed in a monograph.[13] In addition to the group which first reported this novel effect,[13-20] the groups of Sobyanin,[21,22] Lambert et al.,[23] Stoukides et al.,[24] and Haller et al.[25] have also contributed recently to the NEMCA literature. Very recently, the NEMCA effect was also demonstrated in an aqueous electrolyte system by Anastasijevic et al.[26] and by Neophytides et al.[20] The term "Electrochemical Promotion in Catalysis" has also been proposed by Pritchard[27] to describe the NEMCA effect.

It has been found that the catalytic activity and selectivity of porous metal catalyst films deposited on solid electrolytes can be altered in a dramatic, reversible, and, to some extent, predictable[13] manner by carrying out the catalytic reaction in solid electrolyte cells of the type:

gaseous reactants, metal catalyst | solid electrolyte | metal, O_2

(e.g., C_2H_4+O_2) (e.g., Pt) (e.g., YSZ) (e.g., Ag)

where the metal catalyst also serves as an electrode and by applying currents or potentials to the cell with a concomitant supply or removal of ions, e.g., O^{2-}, F^-, Na^+, H^+, to or from the catalyst surface.

The NEMCA effect was first demonstrated on Pt and Ag electrodes using 8 mol% Y_2O_3-stabilized ZrO_2 (YSZ), an O^{2-} conductor, as the solid electrolyte.[14-16] Electrochemical O^{2-} pumping to the catalyst electrode was found to cause an up to 60-fold (6000%) steady-state reversible enhancement in the rate of C_2H_4 oxidation.[16] Furthermore, this steady-state rate increase was found to be up to 3×10^5 times higher than the steady-state rate of supply of O^{2-} to the catalyst, i.e., the enhancement factor Λ[13,16] or apparent faradaic efficiency for the process is 3×10^5.[16] More recently, the rate of C_2H_4 oxidation on Rh electrodes supported on YSZ was found to reversibly increase by a factor of 90, again with faradaic efficiency values of the order of 10^5.[107,108]

The effect of NEMCA or electrochemical promotion[27] does not appear to be limited to any particular metal, solid electrolyte, or group of catalytic reactions. Thus, in addition to

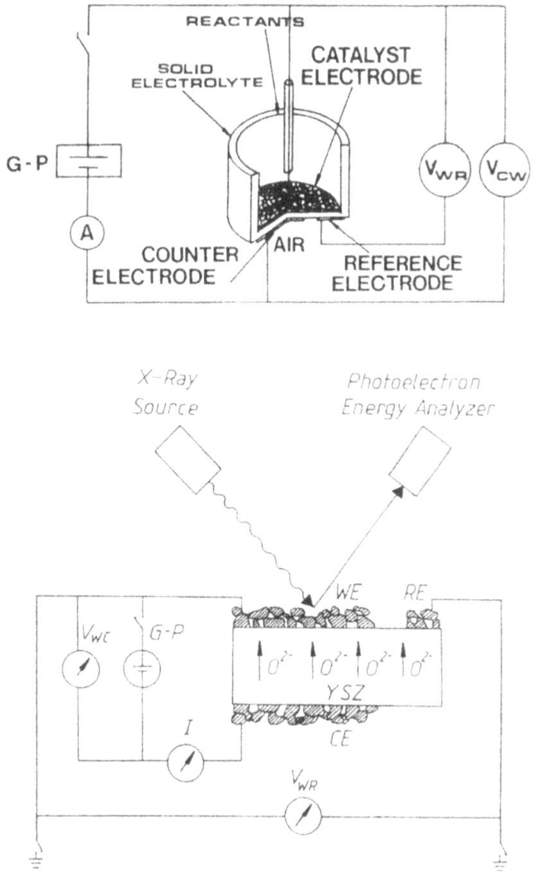

FIGURE 13.10. Schematic of the experimental setup for NEMCA studies (a), and for using XPS (b) G-P.: galvanostat-potentiostat. (From Vayenas, C.G., Ladas, S., Bebelis, S., Yentekakis, I.V., Neophytides, S., Jiang, Y., Karavasilis, Ch., and Pliangos, C., *Electrochim. Acta*, 1994, *39*, 1849. With permission.)

O^{2-}-conducting solid electrolytes,[13-16,18,19] the NEMCA effect has also been demonstrated using Na^+-conducting solid electrolytes such as β''-Al_2O_3[17,23], H^+-conducting solid electrolytes such as $CsHSO_4$,[21] and, very recently, F^--conducting solid electrolytes, such as CaF_2.[109] In addition to complete[15,16] and partial oxidation reactions[13] the NEMCA effect has been demonstrated for dehydrogenation, hydrogenation, and decomposition reactions.[13,21]

2. Experimental Setup

The basic experimental setup is shown schematically in Figure 13.10a. The metal working catalyst electrode, usually in the form of a porous metal film 3 to 20 μm in thickness, is deposited on the surface of a ceramic solid electrolyte (e.g., YSZ, an O^{2-} conductor, or β''-Al_2O_3, a Na^+ conductor). Catalyst, counter, and reference electrode preparation and characterization details have been presented in detail elsewhere,[13] together with the analytical system for on-line monitoring of the rates of catalytic reactions by means of gas chromatography, mass spectrometry and IR spectroscopy.

The superficial surface area of the metal working catalyst electrode is typically 2 cm^2 and its true gas-exposed surface area is typically 5 to 10^3 cm^2 as measured via surface titration of oxygen with CO or C_2H_4.[13-19] The catalyst electrode is exposed to the reactive gas mixture (e.g., C_2H_4 + O_2) in a continuous-flow, gradientless reactor (CSTR). Under open-circuit conditions ($I = 0$), it acts as a regular catalyst for the catalytic reaction under study, e.g., C_2H_4 oxidation. The counter and reference electrodes are usually exposed to ambient air.

A galvanostat or potentiostat is used to apply constant currents between the catalyst and the counter electrode or constant potentials between the catalyst and reference electrodes, respectively. In this way, ions (O^{2-} in the case of YSZ, Na^+ in the case of β''-Al_2O_3) are supplied from (or to) the solid electrolyte to (or from) the catalyst electrode surface. The current is defined positive when anions are supplied to or cations removed from the catalyst electrode. There is convincing evidence that these ions (together with their compensating [screening] charge in the metal, thus forming surface dipoles) migrate (back-spillover) onto the gas-exposed catalyst electrode surface.[13,17,110] Thus the solid electrolyte acts as an active catalyst support and establishes an effective electrochemical double layer on the gas-exposed, i.e., catalytically active, electrode surface.

The (average) work function of the gas-exposed catalyst electrode surface can be measured *in situ,* i.e., during reaction at atmospheric pressure and temperatures up to 300°C, by means of a Kelvin probe (vibrating condenser method) using, e.g., a Besocke Delta-Phi Kelvin probe with a Au vibrating disc as described in detail elsewhere.[17,100]

XPS is a useful tool for investigating metal electrode surfaces under conditions of electrochemical O^{2-} pumping. The experimental setup is shown schematically in Figure 13.10b. A 2-mm-thick YSZ slab (10 x 13 mm) with a Pt catalyst electrode film, a Pt reference electrode, and a Ag counter electrode were mounted on a resistively heated Mo holder in an UHV chamber (base pressure 5×10^{-10} Torr) and the catalyst-electrode film (9 × 9 mm) was examined at temperatures 25 to 525°C by XPS using a Leybold HS-12 analyzer operated at constant ΔE mode with 100-eV pass energy and a sampling area of 5 × 3 mm. Electron binding energies were referenced to the metallic Pt $4f_{7/2}$ peak of the grounded catalyst electrode at 71.1 eV, which always remained unchanged with no trace of an oxidic component. Further experimental details are given elsewhere.[110]

3. Catalytic Rate Modification

Figure 13.11 shows a typical NEMCA experiment carried out in the setup depicted in Figure 13.10a. The catalytic reaction under study is the complete oxidation of C_2H_4 on Pt:[16]

$$CH_2 = CH_2 + 3O_2 \rightarrow 2CO_2 + 2H_2O \tag{13.45}$$

The figure shows a typical galvanostatic transient, i.e., it depicts the transient effect of a constant applied current on the rate of C_2H_4 oxidation (expressed in g-atom O/s).

The Pt catalyst film with a surface area corresponding to $N = 4.2 \cdot 10^{-9}$ g-atom Pt, as measured by surface titration techniques,[13] is deposited on YSZ and is exposed to P_{O_2} = 4.6 kPa, $P_{C_2H_4}$ = 0.36 kPa in the CSTR-type flow reactor depicted schematically in Figure 13.10a. Initially ($t < 0$), the circuit is open ($I = 0$) and the open-circuit catalytic rate r_o is $1.5 \cdot 10^{-8}$ g-atom O/s. The corresponding turnover frequency (TOF), i.e., the number of oxygen atoms reacting per site per second, is 3.57 s^{-1}.

Then at $t = 0$ a galvanostat is used to apply a constant current of +1 μA between the catalyst and the counter electrode (Figure 13.10a). Now oxygen ions O^{2-} are supplied to the catalyst/gas/solid electrolyte tpb at a rate $G_O = I/2F = 5.2 \cdot 10^{-12}$ g-atom O/s. The catalytic rate starts increasing (Figure 13.11) and within 25 min gradually reaches a value $r = 40 \cdot 10^{-8}$ g-atom O/s, which is 26 times larger than r_o. The new TOF is 95.2 s^{-1}. The increase in catalytic rate $\Delta r = r - r_o = 38.5 \cdot 10^{-8}$ g-atom O/s is 74,000 times larger than I/2F. This means that each O^{2-} supplied to the Pt catalyst causes at steady-state 74,000 additional chemisorbed oxygen atoms to react with C_2H_4 to form CO_2 and H_2O. This is why this novel effect has been termed the non-faradaic Electrochemical Modification of Catalytic Activity (NEMCA).

There is an important observation to be made regarding the time required for the rate to approach its steady-state value. Since catalytic rate transients obtained during galvanostatic

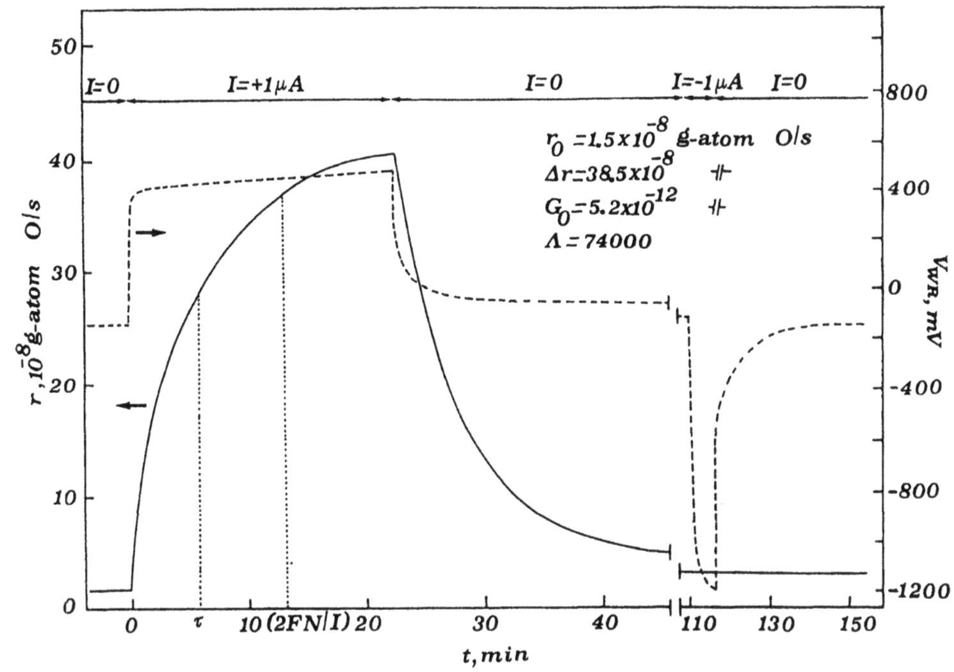

FIGURE 13.11. Rate and catalyst potential response to step changes in applied current during C_2H_4 oxidation on Pt; T = 370°C, P_{O_2} = 4.6 kPa, $P_{C_2H_4}$ = 0.36 kPa. The steady- state rate increase Δr is 74,000 times higher than the steady-state rate of supply of O^{2-} to the catalyst (Λ = 74,000). (From Bebelis, S. and Vayenas, C.G., *J. Catal.*, 1989, *118*, 125. With permission.)

(i.e., constant current) operation are found in NEMCA studies to be usually, but not always,[13] of the type:

$$\Delta r = \Delta r_{max}\left[1 - \exp(-t/\tau)\right] \tag{13.46}$$

i.e., similar to the response of a first-order system with a characteristic time constant τ, one can define the NEMCA time constant τ as the time required for Δr to reach 63% of its maximum, i.e., steady-state value. As shown in Figure 13.11, τ is of the order of 2FN/I, and this turns out to be a general observation in NEMCA studies utilizing doped ZrO_2, i.e.:

$$\tau \approx 2FN/I \tag{13.47}$$

This observation shows that NEMCA is a catalytic effect, i.e., it takes place over the entire gas-exposed catalyst surface, and is not an electrocatalytic effect localized at the tpb metal/solid electrolyte/gas. This is because 2FN/I is the time required to form a monolayer of an oxygen species on a surface with N sites when it is supplied at a rate I/2F. The fact that τ is found to be smaller than 2FN/I, but of the same order of magnitude, shows that only a fraction of the surface is occupied by oxygen back-spillover species, as discussed in detail elsewhere.[13] It is worth noting that if NEMCA were restricted to the tpb, i.e., if the observed rate increase were due to an electrocatalytic reaction, then τ would be practically zero during galvanostatic transients.

As shown in Figure 13.11, NEMCA is reversible, i.e., upon current interruption the catalytic rate returns to its initial value within roughly 100 min. The rate relaxation curve upon current interruption conveys valuable information about the kinetics of reaction and

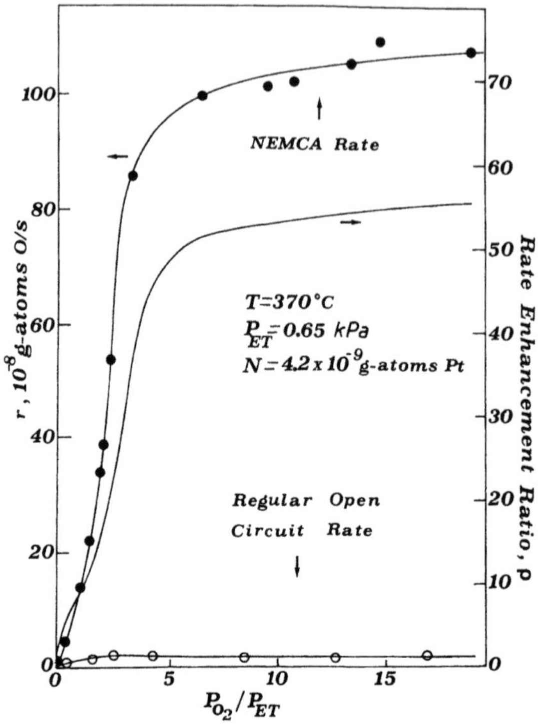

FIGURE 13.12. Effect of gaseous composition on the regular (open-circuit) steady-state rate of C_2H_4 oxidation on Pt and on NEMCA-induced catalytic rate when the catalyst film is maintained at V_{WR} = 1V. (From Bebelis, S. and Vayenas, C.G., *J. Catal.*, 1989, *118*, 125. With permission.)

desorption of the promoting oxygen species as discussed in detail elsewhere.[108,109] Negative current application has practically no effect on the rate of this particular reaction.

4. Effect of Gaseous Composition on Regular (Open-Circuit) and NEMCA-Induced Reaction Rate

Figure 13.12 shows the effect of O_2 to C_2H_4 ratio on the regular (open-circuit) steady-state rate of C_2H_4 oxidation on Pt and on the NEMCA-induced rate when the same catalyst film is maintained at a potential of +1 V (V_{WR} = +1 V) with respect to the reference Pt/air electrode (Figure 13.10a). It can be seen that the effect is much more pronounced at high $P_{O_2}/P_{C_2H_4}$ ratios, i.e., for high oxygen coverages, where the NEMCA-induced reaction rate or TOF values are a factor of 55 higher than the corresponding open-circuit rate. A quantitative description and explanation of the NEMCA behavior of C_2H_4 oxidation on Pt can be found elsewhere.[13,16]

Figure 13.13 shows the effect of C_2H_4 partial pressure at constant P_{O_2} on the rate of C_2H_4 oxidation on Rh at various imposed values of V_{WR}. Increasing V_{WR} causes up to 90-fold (9000%) rate enhancement relative to the open-circuit rate value. The dramatic rate enhancement with increasing V_{WR} depicted in Figures 13.12 and 13.13 is due to the weakening of the metal-covalently chemisorbed oxygen chemisorptive bond, cleavage of which is rate limiting, as discussed in detail elsewhere.[13,16,108]

5. Definitions and the Role of the Exchange Current I_0

Table 13.4 provides a list of the reactions which have been studied already and shown to exhibit NEMCA. In order to compare different catalytic reactions, it is useful to define two dimensionless parameters, i.e., the enhancement factor or faradaic efficiency Λ and the rate enhancement ratio ρ.[13] The former is defined as:

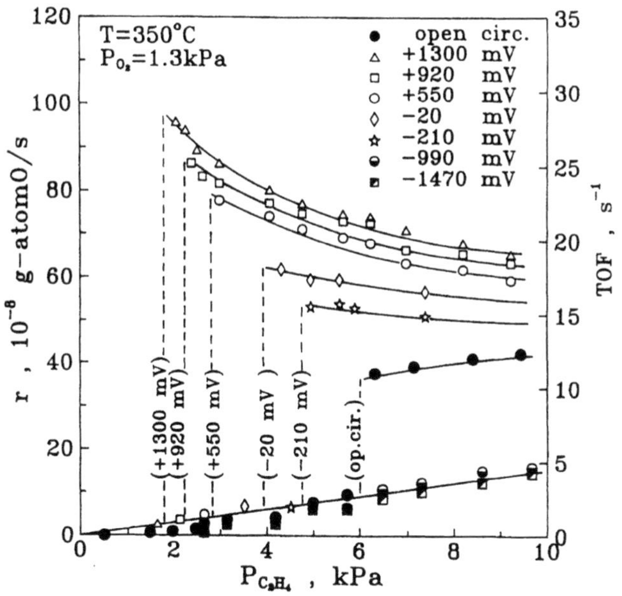

FIGURE 13.13. Effect of gaseous composition and catalyst potential V_{WR} on the steady-state rate of C_2H_4 oxidation on Rh. (From Vayenas, C.G., Ladas, S., Bebelis, S., Yentekakis, I.V., Neophytides, S., Jiang, Y., Karavasilis, Ch., and Pliangos, C., *Electrochim. Acta,* 1994, *39,* 1849. With permission.)

$$\Lambda = \Delta r/(I/2F) \tag{13.48}$$

where the change in catalytic rate Δr is expressed in terms of g-atom of O. More generally, Λ is computed by expressing Δr in g-equivalent and dividing by I/F.

A catalytic reaction is said to exhibit NEMCA when $|\Lambda| > 1$. When $\Lambda > 1$ as, e.g., in the case of C_2H_4 oxidation on Pt, the reaction is said to exhibit positive or electrophobic NEMCA behavior. When $\Lambda < -1$, then the reaction is said to exhibit electrophilic behavior.

The rate enhancement ratio ρ is defined from:

$$\rho = r/r_o \tag{13.49}$$

In the C_2H_4 oxidation example presented in Figure 13.11 and discussed above the Λ and ρ values at steady state are $\Lambda = 74{,}000$ and $\rho = 26$.

As it turns out experimentally (Figure 13.14) and can be explained theoretically,[13] one can estimate or predict the order of magnitude of the absolute value of the enhancement factor Λ for any given reaction, catalyst, and catalyst/solid electrolyte interface from:

$$|\Lambda| \approx 2Fr_o/I_0 \tag{13.50}$$

where I_0 is the exchange current of the metal/solid electrolyte interface.

As noted in Section I.D, the parameter I_0 can be easily determined from standard ln I vs. η (Tafel) plots.[13,16] The overpotential η of the catalyst electrode is defined, according to Section I.D, from:

$$\eta = V_{WR} - V_{WR}^{o} \tag{13.51}$$

TABLE 13.4
Catalytic Reactions Found to Exhibit the NEMCA Effect

I. Positive (electrophobic) NEMCA effect ($\Delta r > 0$ with $I > 0$*, $\Lambda > 0$, $\Delta e\Phi > 0$)

Reactants	Products	Catalyst	Electrolyte	T [°C]	Λ	$\rho = r/r_o$	P_i
C_2H_4,O_2	C_2H_4O, CO_2	Ag	ZrO_2-Y_2O_3	320–470	≤300	≤30**	30
C_2H_4,O_2	C_2H_4O, CO_2	Ag	β″-Al_2O_3	350–410	≤3·10³	≤4**	–20
C_3H_6,O_2	C_3H_6O, CO_2	Ag	ZrO_2-Y_2O_3	320–420	≤300	≤2**	1
C_2H_4,O_2	CO_2	Pt	ZrO_2-Y_2O_3	260–450	≤3·10⁵	≤55	55
C_2H_4,O_2	CO_2	Rh	ZrO_2-Y_2O_3	250–400	≤5·10⁴	≤90	90
C_2H_4,O_2	CO_2	Pt	β″-Al_2O_3	180–300	≤5·10⁴	≤4	–30
C_2H_4,O_2	CO_2	Pt	TiO_2	450–600	≤5·10³	≤20	20
CO,O_2	CO_2	Pt	ZrO_2-Y_2O_3	300–550	≤2·10³	≤3	3
CO,O_2	CO_2	Pt	CaF_2	500–700	≤200	≤2.5	2
CO,O_2	CO_2	Pt	β″-Al_2O_3	300–450	≤10⁴	≤8	–30 to 250
CO,O_2	CO_2	Ag	ZrO_2-Y_2O_3	350–450	≤20	≤5	4
CO,O_2	CO_2	Pd	ZrO_2-Y_2O_3	400–550	≤10³	≤2	1
CH_3OH,O_2	H_2CO,CO_2	Pt	ZrO_2-Y_2O_3	300–500	≤10⁴	≤4**	3
CH_4,O_2	CO_2,C_2H_4, C_2H_6	Ag	ZrO_2-Y_2O_3	650–850	≤5	≤30**	30
CH_4,O_2	CO_2	Pt	ZrO_2-Y_2O_3	600–750	≤5	≤70**	70
CO_2,H_2	CH_4,CO	Rh	ZrO_2-Y_2O_3	390–450	≤40	≤2**	1
CH_4,H_2O	CO,CO_2,H_2	Ni	ZrO_2-Y_2O_3	600–900	≤12	≤2**	1
H_2,O_2	H_2O	Pt	KOH-H_2O	25–50	≤20	≤6	5

II. Negative (electrophilic) NEMCA effect ($\Delta r > 0$ with $I < 0$*, $\Lambda < 0$, $\Delta e\Phi < 0$)

Reactants	Products	Catalyst	Electrolyte	T [°C]	Λ	$\rho = r/r_o$	P_i
CO,O_2	CO_2	Pt	ZrO_2-Y_2O_3	300–550	≤–500	≤6	–5
CH_3OH,O_2	H_2CO,CO_2	Pt	ZrO_2-Y_2O_3	300–550	≤–10⁴	≤15**	–15
CH_3OH	H_2CO,CO,CH_4	Ag	ZrO_2-Y_2O_3	550–750	≤–25	≤6**	–5
CH_3OH	H_2CO,CO,CH_4	Pt	ZrO_2-Y_2O_3	400–500	≤–10	≤3**	–3
CH_4,O_2	CO_2	Pt	ZrO_2-Y_2O_3	600–750	≤–5	≤30	–30
CO,O_2	CO_2	Ag	ZrO_2-Y_2O_3	350–450	≤–800	≤15	–15
C_2H_4,O_2	CO_2	Pt	TiO_2	450–600	≤–5·10³	≤20	–20

* Current is defined positive when O^{2-} is supplied to or Na^+ removed from the catalyst surface.
** Change in product selectivity observed.

where V_{WR} is the catalyst (working electrode, W) potential with respect to a reference (R) electrode. The overpotential η is related to current I via the classical Butler–Volmer equation:[30]

$$(I/I_0) = \exp(\alpha_a F\eta/RT) - \exp(-\alpha_c F\eta/RT) \qquad (13.52)$$

where α_a and α_c are the anodic and cathodic transfer coefficients, respectively. Thus by measuring η as a function of I one can extract I_0, α_a, and α_c. Physically, I_0 expresses the (equal under open-circuit conditions) rates of the electronation and deelectronation reaction at the tpb, e.g.,:

$$O^{2-} \rightleftharpoons O(a) + 2e^- \qquad (13.53)$$

where O(a) stands for oxygen adsorbed on the metal catalyst in the vicinity of the tpb. Thus, the exchange current I_0 is a measure of the nonpolarizability of the metal/solid electrolyte interface.

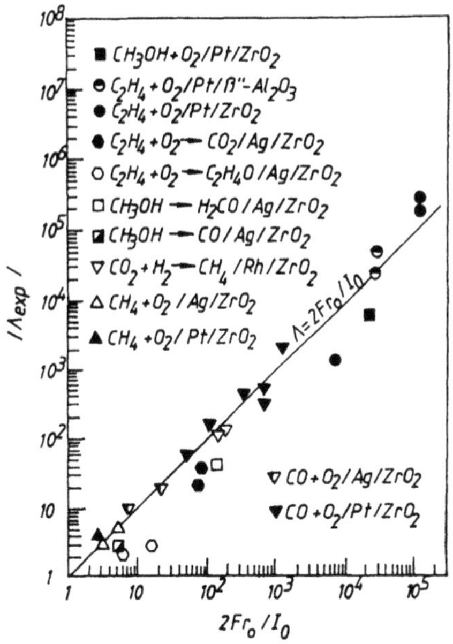

FIGURE 13.14. Comparison of predicted and measured enhancement factor Λ values for the first 12 catalytic reactions found to exhibit the NEMCA effect. (From Vayenas, C.G., Bebelis, S., Yentekakis, I.V., Tsiakaras, P., Karasali, H., and Karavasilis, Ch., in *Proceedings of the 10th Int. Congress on Catalysis,* Guczi, L., Solymosi, F., and Tetényi, P., Eds., Elsevier Science Publishers, B.V., Amsterdam, 1993, 2139. With permission.)

As shown in Figure 13.14, Equation (13.50) shows good agreement with experiment for all catalytic reactions studied so far. The agreement extends over more than 5 orders of magnitude. Thus, contrary to fuel cell applications where nonpolarizable, i.e., high I_0 electrode/electrolyte interfaces are desirable to minimize activation overpotential losses, the opposite holds for catalytic applications, i.e., I_0 must be low in order to obtain high Λ values, i.e., a strong non-faradaic rate enhancement.

Although Λ is an important parameter for determining whether or not a reaction exhibits NEMCA, it is not a fundamental one. The reason is that for the same catalytic reaction on the same catalyst material one can obtain significantly different |Λ| values by varying I_0 (Equation [13.50]). The parameter I_0 is proportional to the tpb length[37] and can thus be controlled during catalyst film preparation by varying the sintering temperature and in this way control the metal crystallite size and tpb length.[13]

From a catalytic point of view, a very important parameter which can be obtained via NEMCA is the promotion index P_i of the doping species defined from:

$$P_i = \frac{\Delta r / r_o}{\Delta \theta_i} \tag{13.54}$$

where θ_i is the coverage of the back-spillover promoting species introduced on the catalytic surface[110] (e.g., $O^{\delta-}$, $Na^{\delta+}$, F^-, etc.). When $P_i > 0$, then the back-spillover species has a promoting effect on the catalytic reaction under study. When $P_i < 0$, then the back-spillover species has a poisoning effect on the catalytic reaction. When the back-spillover species does not react appreciably with any of the reactants, as, e.g., in the case of $Na^{\delta+}$, then $\Delta\theta_{Na}$ can be measured accurately via coulometry.[23] Very recently, a method based on the current interruption technique has been developed to measure also $\Delta\theta_i$ for the case of $O^{\delta-}$ and F^- back-spillover species[109] which react with the reactants of the catalytic reaction at a rate Λ times slower than the NEMCA-induced catalytic rate.[109] The method has been applied to only

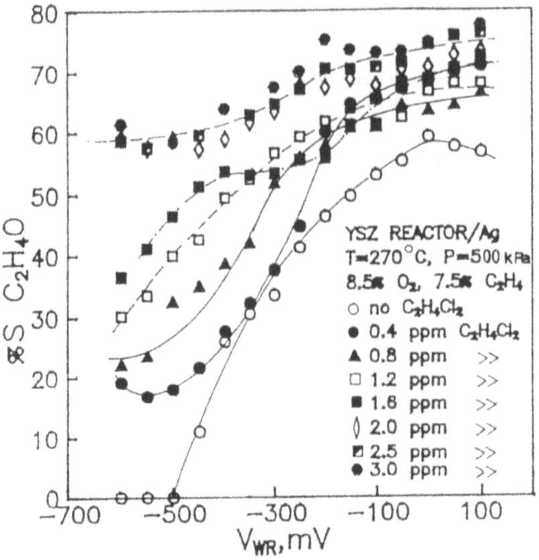

FIGURE 13.15. Effect of catalyst potential V_{WR} on the selectivity to ethylene oxide during C_2H_4 oxidation on Ag at various levels of gas-phase "moderator" $C_2H_4Cl_2$. (From Vayenas, C.G., Ladas, S., Bebelis, S., Yentekakis, I.V., Neophytides, S., Jiang, Y., Karavasilis, Ch., and Pliangos, C., *Electrochim. Acta,* 1994, *39,* 1849. With permission.)

very few reactions yet, and thus the P_i values listed in Table 13.4 for the case of ZrO_2-Y_2O_3 solid electrolytes as the ion donor are based on the approximation[110] $\Delta\theta_i = 1$ for the maximum measured ρ value for each reaction, in which case P_i equals (ρ – 1).

6. Selectivity Modification

One of the most promising applications of NEMCA is in product selectivity modification. An example is shown in Figure 13.15 for the case of C_2H_4 oxidation on Ag.[107] The figure shows the effect of varying the catalyst potential V_{WR} on the selectivity to ethylene oxide (the other products being CO_2 and for $V_{WR} < -0.4$ V some acetaldehyde) at various levels of addition of gas-phase chlorinated hydrocarbon "moderators". With no 1,2-$C_2H_4Cl_2$ present in the feed the selectivity to ethylene oxide is varied between 0 and 56% by varying V_{WR}. Combination of NEMCA and 1,2-$C_2H_4Cl_2$ addition gives selectivities well above 75% when YSZ is used as the solid electrolyte and up to 88% when using β″-Al_2O_3 as the ion donor.[107] The beneficial effect of increasing eΦ and Cl coverage is due to the weakening of the binding strength of chemisorbed atomic oxygen[13] which makes it more selective for epoxidation.[13,18]

7. Work Function Measurements: An Additional Meaning of the EMF of Solid Electrolyte Cells with Metal Electrodes

One of the key steps in understanding the origin of NEMCA was the realization that solid electrolyte cells can be used both to monitor and to control the work function of the gas-exposed surfaces of their electrodes.[13,17] It was shown both theoretically[13] and experimentally[17,100] that:

$$eV^{o}_{WR} = e\Phi_W - e\Phi_R \tag{13.40}$$

and

$$e\Delta V_{WR} = \Delta\left(e\Phi_W\right) \tag{13.55}$$

where $e\Phi_W$ is the catalyst surface work function and $e\Phi_R$ is the work function of the reference electrode surface. The reference electrode must be of the same material as the catalyst for Equation (13.40) to hold, but Equation (13.55) is not subject to this restriction. The derivation of Equations (13.40) and (13.55) is quite straightforward.[13] It is based on the standard definition of the work function[13,30,111,112] and on the fact that the average Volta electrode potential Ψ vanishes at the electrode/gas interface, since no net charge can be sustained there.[13]

The validity of Equations (13.40) and (13.55) was demonstrated by using a Kelvin probe to measure *in situ* $e\Phi$ on catalyst surfaces subject to electrochemical promotion.[17,100]

Therefore, by applying currents or potentials in NEMCA experiments and thus by varying V_{WR}, one also varies the average catalyst surface work function $e\Phi$ (Equation 13.55). Positive currents increase $e\Phi$ and negative currents decrease it. Physically, the variation in $e\Phi$ is primarily due to back-spillover of ions to or from the catalyst surface.

8. Dependence of Catalytic Rates and Activation Energies on eΦ

In view of Equation (13.55), it follows that NEMCA experiments permit one to directly examine the effect of catalyst work function $e\Phi$ on catalytic rates. From a fundamental viewpoint the most interesting finding of all previous NEMCA studies is that over wide ranges of catalyst, work function $e\Phi$ catalytic rates depend exponentially on $e\Phi$, and catalytic activation energies vary linearly with $e\Phi$.[13,17]

A typical example is shown in Figure 13.16 for the catalytic oxidation of C_2H_4 and of CH_4 on Pt. Both reactions exhibit electrophobic behavior which is due to the weakening of the Pt=O chemisorptive bond with increasing $e\Phi$.[13,16,19] Chemisorbed atomic oxygen is an electron acceptor, thus increasing $e\Phi$ causes a weakening in the Pt=O bond, cleavage of which is involved in the rate-limiting step of the catalytic oxidation, and thus a linear decrease in activation energy and an exponential increase in catalytic rate is observed. In general, increasing $e\Phi$ weakens the chemisorptive bond of electron acceptor adsorbates such as oxygen.[13] Figure 13.13 provides another example. The abrupt rate increases are due to reduction of surface Rh oxide.[108] Increasing $e\Phi$ weakens the Rh–O bond and destabilizes the oxide, thus causing the observed dramatic rate enhancement.

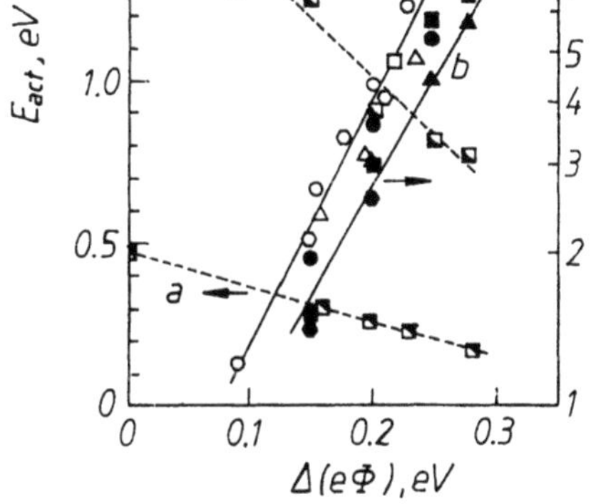

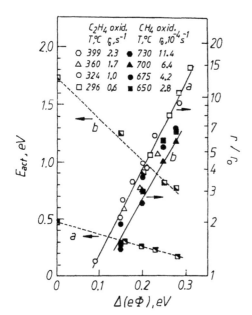

FIGURE 13.16. Effect of catalyst work function $e\Phi$ on the activation energy E_{act} and catalytic rate enhancement ratio r/r_o for C_2H_4 oxidation on Pt (a) and CH_4 oxidation on Pt (b). (From Vayenas, C.G., Electrochemical activation of catalytic reactions, in *Elementary reaction steps in Heterogeneous Catalysis*, Joyner, R.W. and van Santen, R.A., Eds., NATO ASI Series, Kluwer Academic Publishers, Dordrecht, 1993, 73. With permission.)

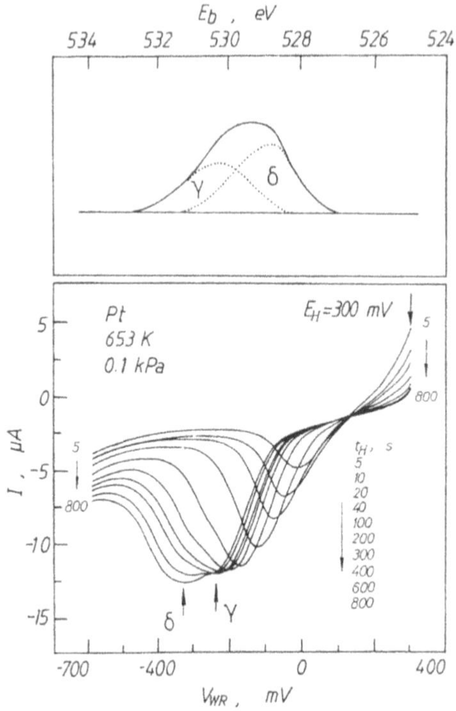

FIGURE 13.17. Top: O1s photoelectron spectrum of oxygen adsorbed on a Pt electrode supported on YSZ under UHV conditions after applying a constant overpotential ΔV_{WR} = 1.2 V corresponding to a steady state current I = 40 μA for 15 min at 673 K. The same O1s spectrum was maintained after turning off the potentiostat and rapidly cooling to 400 K. The γ state is normally chemisorbed atomic oxygen (E_b = 530.2 eV) and the δ state is spillover oxidic oxygen (E_b = 528.8 eV).[110] Bottom: linear potential sweep voltammogram obtained at T = 653 K and P_{O_2} = 0.1 kPa on a Pt electrode supported on YSZ showing the effect of holding time t_H at V_{WR} = 300 mV on the reduction of the γ and δ states of adsorbed oxygen; sweep rate: 30 mV/s.[41]

9. XPS Spectroscopic and Voltammetric Identification of Back-spillover Ions as the Cause of NEMCA

The first XPS investigation of Ag electrodes on YSZ under O^{2-} pumping conditions was published in 1983.[113] That study provided direct evidence for the creation of back-spillover oxide ions on Ag (O1s at 529.2 eV) upon applying positive currents. More recently, Göpel and co-workers have used XPS, UPS, and EELS to study Ag/YSZ catalyst surfaces under NEMCA conditions.[114] Their XPS spectra are similar to those in Reference 113.

Very recently, a similar detailed XPS study was performed on Pt films interfaced with YSZ.[110] This study showed that:

I. Back-spillover oxide ions (O1s at 528.8 eV) are generated on the gas-exposed electrode surface upon positive current application (peak δ in Figure 13.17, top).

II. Normally chemisorbed atomic oxygen (O1s at 530.2 eV) is also formed upon positive current application (peak γ in Figure 13.17, top). The coverages of the γ and δ states of oxygen are comparable and of the order of 0.5 each.[110]

III. Oxidic back-spillover oxygen (δ state) is *less* reactive than normally chemisorbed atomic oxygen (γ state) with the reducing (H_2 and CO) UHV background.[110]

These observations provide a straightforward explanation for the origin of NEMCA when using YSZ. Backspillover oxide ions (O^{2-} or O^-) generated at the tpb upon electrochemical O^{2-} pumping to the catalyst spread over the gas-exposed catalyst/electrode surface. They are accompanied by their compensating charge in the metal, thus forming back-spillover dipoles. They thus establish an effective electrochemical double layer which increases the catalyst surface work function and affects the strength of chemisorptive bonds such as that of normally chemisorbed oxygen via through-the-metal or through-the-vacuum interactions. This change in chemisorptive bond strength causes the observed dramatic changes in catalytic rates. It thus appears that the physicochemical origin of NEMCA is closely related to the very interesting electrical polarization- (0.3 V) and work function change-induced effects on chemisorption recently observed by Xu et al.[115] on well-characterized surfaces under UHV conditions.

The creation of two types of chemisorbed oxygen on Pt surfaces subject to NEMCA conditions has been recently confirmed by means of linear potential sweep voltammetry (Figure 13.17, bottom).[41] The first oxygen reduction peak corresponds to normally chemisorbed oxygen (γ state) and the second reduction peak, which appears only after prolonged application of a positive current,[41] must correspond to the δ state of oxygen, i.e., back-spillover oxidic oxygen. Recent *in situ* XPS investigation of Pt films deposited on β''-Al_2O_3 has similarly shown[118] that electrochemically controlled Na back-spillover is the origin of NEMCA when using β''-Al_2O_3 as the solid electrolyte.

IV. CONCLUDING REMARKS

Electrocatalysis plays an important role in the efficient operation of solid electrolyte fuel cells used for power generation. Catalysis at the anode also has a significant role. In addition to power generation, solid electrolyte electrochemical reactors can also be used for chemical cogeneration, i.e., for the simultaneous production of electrical power and chemicals. If solid electrolyte fuel cells become commercially available in a large scale, then chemical cogeneration could become an attractive option for the industrial production of some important chemicals.

Electrocatalysis in solid electrolyte cells can also be used to activate catalysis on the gas-exposed electrode surfaces. This novel application of solid electrolytes (NEMCA effect or electrochemical promotion) is of great importance both in electrochemistry[119] and in heterogeneous catalysis.[27,120] The solid electrolyte is used as a reversible promoter donor, via electrocatalysis, to precisely tune and control the catalytic activity and product selectivity of the electrodes. Aside from several potential technological applications, this new application of electrocatalysis allows for a systematic study of the action of promoters in heterogeneous catalysis.

ACKNOWLEDGMENT

Sincere thanks are expressed to the CEC Science, Joule, and Human Capital and Mobility programs for financial support.

REFERENCES

1. Alcock, C.B., *Solid State Ionics*, 1992, *53–56*, 3.
2. Worrell, W.L., in *Electrochemistry and Solid State Science Education*, Smyrl, W.R. and Mc Larnon, F., Eds., The Electrochemical Society, Pennington, NJ, 1986.
3. Liaw, B.Y., Liu, J., Menne, A., and Weppner, W., *Solid State Ionics*, 1992, *53–56*, 18.
4. Göpel,W., Hesse, J., and Zemel, J.N., Eds., *Sensors: a Comprehensive Survey*, Vol. 1,2,3, VCH Publishers, Mannheim 1989, 1991.

5. *Proc. of the Symposium on Chemical Sensors II*, Butler, M., Ricco, A., and Yamazoe, N., Eds., Proc. Vol. 93–7, The Electrochemical Society, Pennington, NJ, 1993.
6. *Proc. of the 3rd Int. Symposium on Solid Oxide Fuel Cells*, Singhal, S.C. and Iwahara, H., Eds., Vol. 93-4, The Electrochemical Society, Pennington, NJ, 1993.
7. Vayenas, C.G., Bebelis, S., and Kyriazis, C.C., *CHEMTECH*, 1991, *21*, 422.
8. Vayenas, C.G. and Farr, R.D., *Science*, 1980, *208*, 593.
9. Farr, R.D. and Vayenas, C.G., *J. Electrochem.*, 1980, *127*, 1478.
10. Sigal, C. and Vayenas, C.G., *Solid State Ionics*, 1981, *5*, 567.
11. Yentekakis, I.V. and Vayenas, C.G., *J. Electrochem. Soc.*, 1989, *136*, 996.
12. Neophytides, S. and Vayenas, C.G., *J. Electrochem. Soc.*, 1990, *137*, 839.
13. Vayenas, C.G., Bebelis, S., Yentekakis, I.V., and Lintz, H.-G., *Catal. Today*, 1992, *11*, 303.
14. Vayenas, C.G., Bebelis, S., and Neophytides, S., *J. Phys. Chem.*, 1988, *92*, 5083.
15. Yentekakis, I.V. and Vayenas, C.G., *J. Catal.*, 1988, *111*, 170.
16. Bebelis, S. and Vayenas, C.G., *J. Catal.*, 1989, *118*, 125.
17. Vayenas, C.G., Bebelis, S., and Ladas, S., *Nature (London)*, 1990, *343*, 625.
18. Bebelis, S. and Vayenas, C.G., *J. Catal.*, 1992, 138, 570, 1992, *138*, 588.
19. Tsiakaras, P. and Vayenas, C.G., *J. Catal.*, 1993, *140*, 53.
20. Neophytides, S., Tsiplakides, D., Jaksic, M., Stonehart, P., and Vayenas, C.G., *Nature*, 1994, *370*, 45.
21. Politova, T.I., Sobyanin, V.A., and Belyaev, V.D., *React. Kinet. Catal. Lett.*, 1990, *41*, 321.
22. Mar'ina, O.A. and Sobyanin, V.A., *Catal. Lett.*, 1992, *13*, 61.
23. Yentekakis, I.V., Moggridge, G., Vayenas, C.G., and Lambert, R.M., *J. Catal.*, 1994, *146*, 292.
24. Alqahtany, H., Chian, P.-H., Eng, D., and Stoukides, M., *Catal. Lett.*, 1992, *13*, 289.
25. Cavalca, C., Larsen, G., Vayenas, C.G., and Haller, G.L., *J. Phys. Chem.*, 1993, *97*, 6115.
26. Anastasijevic, N.A., Baltruschat, H., and Heitbaum, J., *Electrochim. Acta*, 1993, *38*, 1067.
27. Pritchard, J., *Nature (London)*, 1990, *343*, 592.
28. Hegedus, L.L., Aris, R., Bell, A.T., Boudart, M., Chen, N.Y., Gates, B.C., Haag, W.O., Somorjai, G.A., and Wei, J., *Catalyst Design: Progress and Perspectives*, John Wiley & Sons, New York, 1987.
29. Debenedetti, P.G. and Vayenas, C.G., *Chem. Eng. Sci.*, 1983, *38*, 1817.
30. Bockris, J.O'M. and Reddy, A.K.N., *Modern Electrochemistry*, Vol. 2, Plenum Press, New York, 1970.
31. Newman, J.S., *Electrochemical Systems*, Prentice-Hall, Englewood Cliffs, NJ, 1973.
32. Van Berkel, F.P.F., Van Heuveln, F.H., and Huijsmans, J.P.P., in *Proc. of the 3rd Int. Symp. on Solid Oxide Fuel Cells*, Singhal, S.C. and Iwahara, H., Eds., Proc. Vol. 93-4, The Electrochemical Society, Pennington, NJ, 1993, 744.
33. Bockris, J.O'M. and Khan, S.U.M., *Surface Electrochemistry: a Molecular Level Approach*, Plenum Press, New York, 1993.
34. Gileadi, E., *Electrode Kinetics for Chemists, Chemical Engineers and Materials Scientists*, VCH Publishers, Weinheim, Germany, 1993.
35. Froment, G.F. and Bischoff, K.B., *Chemical Reactor Analysis and Design*, John Wiley & Sons, New York, 1979.
36. Wang, D.Y. and Nowick, A.S., *J. Electrochem. Soc.*, 1979, *126*, 1155, 1166.
37. Vayenas, C.G., Ioannides, A., and Bebelis, S., *J. Catal.*, 1991, *129*, 67.
38. Norby, T., Velle, O.J., Leth-Olsen, H., Tunold, R., in *Proc. of the 3rd Int. Symp. on Solid Oxide Fuel Cells*, Singhal, S.C. and Iwahara, H., Eds., Proc. Vol. 93-4, The Electrochemical Society, Pennington, NJ, 1993, 473.
39. Boukamp, B.A., Van Hassel, B.A., Vinke, I.C., De Vries, K.J., and Burgraaf, A.J., *Electrochim. Acta*, 1993, *38*, 1817.
40. Manton, M., Ph.D. thesis, MIT, Cambridge, MA, 1984.
41. Jiang, Y., Kaloyannis, A., and Vayenas, C.G., *Electrochim. Acta*, 1993, *38*, 2533.
42. Bard, A.J. and Faulkner, L.R., *Electrochemical Methods: Fundamentals and Applications*, John Wiley & Sons, New York, 1980.
43. Michaels, J.N. and Vayenas, C.G., *J. Electrochem. Soc.*, 1984, *131*, 2544.
44. Michaels, J.N. and Vayenas, C.G., *J. Catal.*, 1984, *85*, 477.
45. DiCosimo, R., Burrington, J.D., and Grasselli, R.K., *J. Catal.*, 1986, *102*, 234.
46. Kiratzis, N. and Stoukides, M., *J. Electrochem. Soc.*, 1987, *134*, 1925.
47. Lee, J.S. and Oyama, S.T., *Catal. Rev.-Sci. Eng.*, 1988, *30*, 249.
48. Eng, D. and Stoukides, M., *Catal. Rev.-Sci. Eng.*, 1991, *33*, 375.
49. Otsuka, K., Suga, K., and Yamanaka, I., *Catal. Today*, 1990, *6*, 587.
50. Eng, D. and Stoukides, M., *J. Catal.*, 1991, *30*, 306.
51. Nagamoto, H., Hayashi, K., and Inoue, H., *J. Catal.*, 1990, *126*, 671.
52. Otsuka, K., Yokoyama, S., and Morikawa, A., *Chem. Lett. Chem. Soc. Jpn.*, 1985, 319.
53. Seimanides, S. and Stoukides, M., *J. Electrochem. Soc.*, 1986, *133*, 1535.
54. Otsuka, K., Suga, K., and Yamanaka, I., *Catal. Lett.*, 1988, *1*, 423.
55. Otsuka, K., Suga, K., and Yamanaka, I., *Chem. Lett. Chem. Soc. Jpn.*, 1988, 317.

56. Belyaev, V.D., Bazhan, O.V., Sobyanin, V.A., and Parmon, V.N., in *New Developments in Selective Oxidation*, Centi, C. and Trifiro, F., Eds., Elsevier, Amsterdam, 1990, 469.
57. Anshits, A.G., Shigapov, A.N., Vereshchagin, A.N., and Shevnin, V.N., *Catal. Today*, 1990, *6*, 593.
58. Tsiakaras, P. and Vayenas, C.G., *J. Catal.*, 1993, *144*, 333.
59. Jiang, Y., Yentekakis, I.V., and Vayenas, C.G., *Science*, 1994, *264*, 1853.
60. Isenberg, A.O., *Solid State Ionics*, 1981, *3–4*, 431.
61. Schouler, E.J.L., Kleitz, M., Forest, E., Fernandez, E., and Fabry, D., *Solid State Ionics*, 181, *5*, 559.
62. Dönitz, W. and Erdle, E., *Int. J. Hydrogen Energy*, 1985, *10*, 291.
63. Pancharatnam, S., Huggins, R.A., and Mason D.M., *J. Electrochem. Soc.*, 1975, *122*, 869.
64. Gür, T.M. and Huggins, R.A., *J. Electrochem. Soc.*, 1979, *126*, 1067.
65. Gür, T.M. and Huggins, R.A., *Science*, 1983, *219*, 967.
66. Gür, T.M. and Huggins, R.A., *J. Catal.*, 1986, *102*, 443.
67. Hayakawa, T., Tsunoda, T., Orita, H., Kameyama, T., Takahashi, H., Takehira, K., and Fukuda, K., *J. Chem. Soc. Jpn. Chem. Commun.*, 1986, 961.
68. Hamakawa, S., Hibino T., and Iwahara, H., *J. Electrochem. Soc.*, 1993, *140*, 459.
69. Chiang, P.H., Eng, D., and Stoukides, M., *J. Electrochem. Soc.*, 1991, *138*, L11.
70. Michaels, J.N., Vayenas, C.G., and Hegedus, L.L., *J. Electrochem. Soc.*, 1986, *133*, 522.
71. Vayenas, C.G., Debenedetti, P.G., Yentekakis, I.V., and Hegedus, L.L., *Ind. Eng. Chem. Fundam.*, 1985, *24*, 316.
72. Yentekakis, I.V., Neophytides, S., Seimanides, S., and Vayenas, C.G., *Proc. 2nd Int. Symposium on Solid Oxide Fuel Cells*, Athens, 1991, Grosz, F., Zegers, P., Singhal, S.C., and Yamamoto, O., Eds., CEC Publ., Luxembourg, 281.
73. Arato, E. and Costa, P., *Proc. 2nd Int. Symposium on Solid Oxide Fuel Cells*, Athens, 1991, Grosz, F., Zegers, P., Singhal, S.C., and Yamamoto, O., Eds., CEC Publ., Luxembourg, 273.
74. Karoliussen, H., Nisancioglu, K., Solheim, A., and Ødegård, R., Proc. of the 3rd Int. Symp. on Solid Oxide Fuel Cells, Singhal, S.C. and Iwahara, H., Eds., Proc. Vol. 93–4, The Electrochemical Society, Pennington, NJ, 1993, 868.
75. Wagner, C., *Adv. Catal.*, 1970, *21*, 323.
76. Vayenas, C.G. and Saltsburg, H.M., *J. Catal.*, 1979, *57*, 296.
77. Vayenas, C.G., Lee, B., and Michaels, J.N., *J. Catal.*, 1980, *36*, 18.
78. Stoukides, M. and Vayenas, C.G., *J. Catal.*, 1980, *64*, 18.
79. Stoukides, M. and Vayenas, C.G., *J. Catal.*, 1981, *69*, 18.
80. Stoukides, M. and Vayenas, C.G., *J. Catal.*, 1982, *74*, 266.
81. Stoukides, M., Seimanides, S., and Vayenas, C.G., *ACS Symp. Ser.*, 1982, *196*, 195.
82. Stoukides, M. and Vayenas, C.G., *J. Catal.*, 1983, *82*, 44.
83. Hetrick, R.E. and Logothetis, E.M., *Appl. Phys. Lett.*, 1979, *34*, 117.
84. Okamoto, H., Kawamura, G., and Kudo, T., *J. Catal.*, 1983, *82*, 322.
85. Vayenas, C.G., *J. Catal.*, 1984, *90*, 371.
86. Häfele, E. and Lintz, H.-G., *Ber. Bunsenges. Phys. Chem.*, 1986, *90*, 298.
87. Vayenas, C.G., Georgakis, C., Michaels, J.N., and Tormo, J., *J. Catal.*, 1981, *67*, 348.
88. Vayenas, C.G. and Michaels, J.N., *Surf. Sci.*, 1982, *120*, L405.
89. Yentekakis, I.V., Neophytides, S., and Vayenas, C.G., *J. Catal.*, 1988, *111*, 152.
90. Häfele, E. and Lintz, H.-G., *Solid State Ionics*, 1987, *23*, 235.
91. Häfele, E. and Lintz, H.-G., *Ber. Bunsenges. Phys. Chem.*, 1988, *92*, 188.
92. Gellings, P.J., Koopmans, H.S.A., and Burgraaf, A.J., *Appl. Catal.*, 1988, *39*, 1.
93. Breckner, E.M., Sundaresan, S., and Benziger, J.B., *Appl. Catal.*, 1987, *30*, 277.
94. Hildenbrand, H.H. and Lintz, H.-G., *Appl. Catal.*, 1989, *49*, L1-L4.
95. Hildenbrand, H.H. and Lintz, H.-G., *Catal. Today*, 1991, *9*, 153.
96. Hildenbrand, H.H. and Lintz, H.-G., *Appl. Catal.*, 1990, *65*, 241.
97. Vayenas, C.G., *Solid State Ionics*, 1988, *28–30*, 1521.
98. Lintz, H.G. and Vayenas, C.G., *Angew. Chem.*, 1989, 101, 725, *Angew. Chem., Int. Ed, Engl. Edn.*, 1989, *28*, 708.
99. Stoukides, M., *Ind. Eng. Chem. Res.*, 1988, *27*, 1745.
100. Ladas, S., Bebelis, S., and Vayenas, C.G., *Surf. Sci.*, 1991, *251/252*, 1062.
101. Vayenas, C.G., Lee, B., and Michaels, J.N., *J. Catal.*, 1980, 66, 36.
102. Berry, R.J., *Surf. Sci.*, 1978, *76*, 415.
103. Peuckert, M. and Ibach, H., *Surf. Sci.*, 1984, *136*, 319.
104. Peuckert, M. and Bonzel, H.P., *Surf. Sci.*, 1984, *145*, 239.
105. Peuckert, M., *J. Phys. Chem.*, 1985, *89*, 2481.
106. Sales, B.C., Turner, J.E., and Maple, M.B., *Surf. Sci.*, 1982, *114*, 381.
107. Vayenas, C.G., Ladas, S., Bebelis, S., Yentekakis, I.V., Neophytides, S., Jiang, Y., Karavasilis, Ch., and Pliangos, C., *Electrochim. Acta*, 1994, *39*, 1849.

108. Pliangos, C., Yentekakis, I.V., Verykios, X.E., and Vayenas, C.G., *J. Catal.*, 1995, 154, 124.
109. Yentekakis, I.V. and Vayenas, C.G., *J. Catal.*, 1994, *149*, 238.
110. Ladas, S., Kennou, S., Bebelis, S., and Vayenas, C.G., *J. Phys. Chem.*, 1993, *97*, 8845.
111. Trasatti, S., The work function in electrochemistry, in *Advances in Electrochemistry and Electrochemical Engineering*, Gerischer, H. and Tobias, Ch. W., Eds., Vol. 10, John Wiley & Sons, New York, 1977.
112. Hölzl, J. and Schulte, F.K., Work function of metals, in *Solid Surface Physics*, Springer-Verlag, Berlin, 1979, 1–150.
113. Arakawa, T., Saito, A., and Shiokawa, J., *Chem. Phys. Lett.*, 1983, *94*, 250.
114. Zipprich, W., Wiemhöfer, H.-D., Vöhrer, U., and Göpel, W., *Ber. Bunsengesel. Phys. Chem.*, 1995, 99, 1406.
115. Xu, Z., Yates, J.J., Jr., Wang, L.C., and Kreuzer, H.J., *J. Chem. Phys.*, 1992, *96*, 1628.
116. Vayenas, C.G., Bebelis, S., Yentekakis, I.V., Tsiakaras, P., Karasali, H., and Karavasilis, Ch., in *Proceedings of the 10th Int. Congress on Catalysis*, Guczi, L., Solymosi, F., and Tetényi, P., Eds., Elsevier Science Publishers, B.V., Amsterdam, 1993, 2139.
117. Vayenas, C.G., Electrochemical activation of catalytic reactions, in *Elementary reaction steps in Heterogeneous Catalysis*, Joyner, R.W. and van Santen, R.A., Eds., NATO ASI Series, Kluwer Academic Publishers, Dordrecht, 1993, 73.
118. Cavalca, C., Ph.D. thesis, Yale University, New Haven, CT, 1994.
119. Bockris, J.O'M. and Minevski, Z.S., *Electrochim. Acta*, 1994, *39*, 1471.
120. Grzybowska-Swierkosz, B. and Haber, J., in *Annual Reports on the Progress of Chemistry*, Vol. 91, The Royal Society of Chemistry, Cambridge, 1994, 395–439.

Chapter 14

DENSE CERAMIC MEMBRANES FOR OXYGEN SEPARATION

Henny J.M. Bouwmeester and Anthonie J. Burggraaf

0-8493-8956-9/97/$0.00+$.50

LIST OF ABBREVIATIONS AND SYMBOLS

Abbreviations

BE25	25 mol% erbia-stabilized bismuth oxide
BICUVOX	$Bi_4V_{2-y}Cu_yO_{11}$
BIMEVOX	general acronym for materials derived from $Bi_4V_2O_{11}$, like BICUVOX
BT40	40 mol% terbia-stabilized bismuth oxide
BY25	25 mol% yttria-stabilized bismuth oxide
CSZ	calcia-stabilized zirconia
ECVD	electrochemical vapor deposition
EDS	energy dispersive spectroscopy (of X-rays)
EPR	electron proton resonance
emf	electromotive force
FT-IR	Fourier transform infrared spectroscopy
HRTEM	high-resolution transmission electron microscopy
MIEC	mixed ionic–electronic conductor
NMR	nuclear magnetic resonance
SEM	scanning electron microscopy
SIMS	secondary ion mass spectroscopy
SOFC	solid oxide fuel cell
TEM	transmission electron microscopy
tpb	three-phase boundary
TPR	temperature-programmed reduction
UV	ultraviolet spectroscopy
XANES	X-ray absorption near edge structure
XAS	X-ray absorption spectroscopy
XRD	X-ray diffraction
YSZ	yttria-stabilized zirconia

Symbols

c_i	mole fraction or concentration of species i
$\tilde{D}$	chemical diffusion coefficient
D^*	tracer diffusion coefficient
D_s	self-diffusion coefficient
D_v	vacancy diffusion coefficient
d_p	pore diameter
e	elementary charge
E	emf
E_{eq}	open-cell emf
F	Faraday constant
G	geometric factor used to account for nonaxial contributions to the oxygen flux
H_R	Haven ratio
I	electrical current
i	electrical current density
i^o	exchange current density, A cm^{-2}
i_l	limiting current density
j_i	flux of species i
j^o_{ex}	balanced surface exchange rate at equilibrium, mol O_2 cm^{-2} s^{-1}
k_s	surface exchange coefficient, cm s^{-1}
k	reaction rate constant
K	equilibrium constant for a reaction
L	membrane thickness
L_c	characteristic thickness of membrane
L_d	Debye-Hückel screening length
L_p	characteristic thickness (active width) of porous coating layer
n	frequently used to designate the mole fraction of electrons
p	mole fraction of electron holes
P_{O_2}	oxygen partial pressure
P'_{O_2}	oxygen partial pressure at feed side of the membrane
P''_{O_2}	oxygen partial pressure at permeate side of the membrane
r_i	radius of species i
R	gas constant
s_i	entropy of species i
s_i^o	entropy of species i at standard state
S	surface area
t	Goldschmidt factor
t_{el}	electronic transference number
t_i	transference number of species i
t_{ion}	ionic transference number
T	temperature
T_t	transition temperature
u_i	electrical mobility of species i

u^o_i	electrical mobility of species i in standard state
V_m	molar volume
z_i	charge number of species i (positive for cations and negative for anions)

Greek

α	surface exchange coefficient
β	bulk diffusion coefficient
γ	reduction factor
δ	deviation from ideal oxygen stoichiometry
ξ	enhancement factor
η	overpotential
η_i	electrochemical potential of species i
θ	porosity
μ_i	chemical potential of species i
μ_i^o	standard chemical potential of species i
σ_{el}	electronic conductivity
σ_h	polaron hopping conductivity
σ_i	electrical conductivity of species i
σ_i^o	conductivity of species i at standard state
σ_{ion}	ionic conductivity
σ_n	n-type electronic conductivity
σ_p	p-type electronic conductivity
σ_{total}	total conductivity
τ_s	tortuosity
ϕ	electric potential of phase (Galvani potential)
ϕ_c	critical (percolation threshold) volume fraction

I. INTRODUCTION

Inorganic membranes may be conveniently categorized into porous and dense membranes. *Porous membranes* comprise both metal and ceramic barrier layers, with small and homogeneously dispersed pores superimposed on a mechanically strong support with a comparatively large pore opening. The barrier layers may have different morphologies and microstructures, which largely determine the magnitude of permeation and the permselectivity. The basic transport mechanisms in the porous structures are viscous flow, bulk diffusion, Knudsen diffusion, surface diffusion, activated diffusion, capillary condensation, and molecular sieving. Recent developments embrace the use of defect-free microporous barrier layers of a few microns thick with pore diameters down to 1.5 nm which show comparatively high thermal and chemical stability.[1,2]

Dense (nonporous) membranes can be subdivided into (1) ceramic membranes, (2) metal membranes, and (3) liquid-immobilized membranes. These include materials which allow preferential passage of hydrogen or oxygen, in the form of either ions or atoms. With regard to the third category, these membranes consist of a porous support in which a semipermeable liquid is immobilized which fills the pores completely. Interesting examples are molten salts immobilized in porous steel or ceramic supports, semipermeable for oxygen or ammonia.[1]

In general, dense membranes combine a high permselectivity with a rather low permeation rate. The past decade has witnessed major changes in this area as dense ceramic membranes exhibiting high oxygen ionic and electronic conductivity have become of great interest as a potentially economical, clean, and efficient means of producing oxygen by separation from air or other oxygen-containing gas mixtures. In addition to infinite permselectivity, notably high oxygen flux values are measured through selected mixed-conducting oxides with the perovskite structure. These may be in the range exhibited by microporous membranes, albeit that sufficiently high temperatures are required, typically above about 700°C.

It is generally accepted that the mixed-conducting oxide membranes, provided they can be developed with sufficient durability and reliability, have great potential to meet the needs of many segments of the oxygen market. It is further expected that the oxygen fluxes can be improved by thin-film deposition on a porous substrate, preferably of the same material to avoid compatibility problems. The envisioned applications range from small-scale oxygen pumps for medical applications to large-scale usage in combustion processes, e.g., coal gasification.[3-6] As oxygen, but also nitrogen, ranks among the top five in the production of commodity chemicals in the U.S.,[7] successful development of the mixed-conducting oxide membranes could thus have clear economic benefits, at the expense of market share from more traditional supply options. While the targeted membranes will be most competitive at small- and intermediate-scale level in which flexibility of operation is desired, they may eventually challenge the present commercial status of cryogenics, pressure-swing adsorption (PSA), and polymeric membranes.[3-6]

Another application of mixed-conducting oxide membranes is to be found in the field of chemical processing, including the partial oxidation of light hydrocarbons, e.g., natural gas to value-added products such as ethane/ethene[8-13] and syngas,[14-16] waste reduction, and recovery.[17] The catalyst may be either the membrane surface itself or another material deposited in particulate form on top of the membrane. Besides the controlled supply (or removal) of oxygen to (or from) the side where the catalyst and the reactants are located, a promising feature is that the oxygen flux may alter the relative presence of different oxygen species (O_2^-, O^-) on the catalyst surface, thereby providing species that may be more selective for partial oxidation reactions.

This review addresses recent developments in the area of mixed ionic–electronic conducting (MIEC) membranes for oxygen separation, in which the membrane material is made dense, i.e., free of cracks and connected-through porosity, being susceptible only for oxygen ionic and electronic transport. Current work on different mixed-conducting oxides is reviewed using concepts from electrochemistry and solid state chemistry. Emphasis is on the defect chemistry, mass transport, and the associated surface exchange kinetics, providing some basic background knowledge which aids further development of these materials into membranes for the aforementioned applications. It is not attempted to discuss inroads against competing technologies, neither to speculate on new opportunities that may result from successful development. New developments in dense ceramic membrane research could offer very economical ways of separating hydrogen such as the proton-conducting ceramics or thin Pd foils. These are not considered in this chapter. For a general discussion on the topical area of membrane technology and its impact in various applications, the reader is referred to specific reviews; for example, see References 18,19,20,21,22.

II. GENERAL SURVEY

In this section, a brief overview is given of major membrane concepts and materials. Besides membranes made from a MIEC, other membranes incorporating an oxygen ion conductor are briefly discussed. Data from oxygen permeability measurements on selected membrane materials are presented.

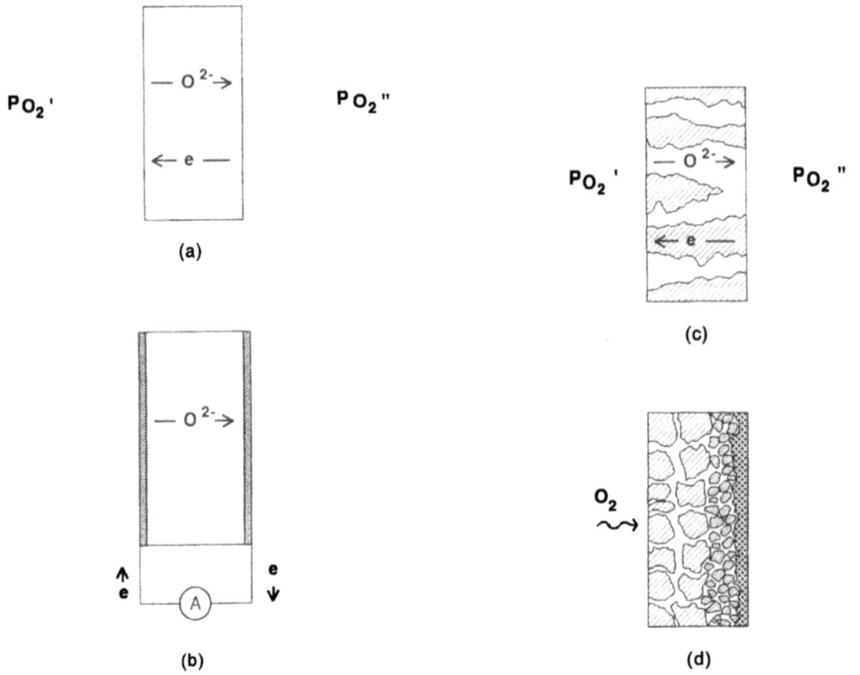

FIGURE 14.1. Different membrane concepts incorporating an oxygen ion conductor: (a) mixed conducting oxide, (b) solid electrolyte cell (oxygen pump), and (c) dual-phase membrane. Also shown is the schematics of (d) an asymmetric porous membrane, consisting of a support, an intermediate and a barrier layer having a graded porosity across the membrane.

A. MAJOR MEMBRANE CONCEPTS

In this chapter, a membrane is regarded as a barrier between two enclosures which preferentially allows one gas (i.e., oxygen) to permeate owing to the presence of a driving force such as a pressure or electric potential gradient.

The separation of oxygen using an *MIEC membrane* is schematically shown in Figure 14.1a. The driving force for overall oxygen transport is the differential oxygen partial pressure applied across the membrane. As the MIEC membrane is dense and gas tight, the direct passage of oxygen molecules is blocked, yet oxygen ions migrate selectively through the membrane. Dissociation and ionization of oxygen occurs at the oxide surface at the high-pressure side (feed side), where electrons are picked up from accessible (near-) surface electronic states. The flux of oxygen ions is charge compensated by a simultaneous flux of electronic charge carriers. Upon arrival at the low-pressure side (permeate side), the individual oxygen ions part with their electrons and recombine again to form oxygen molecules, which are released in the permeate stream.

Mixed conduction also plays an important role in many other processes, e.g., in improving electrode kinetics and catalytic behavior.[23] In fact, all oxides exhibit to some degree mixed ionic and electronic conduction, and selective oxygen permeation has been reported even for dense sintered alumina above 1500°C.[24,25] Though it is common to speak of mixed conduction when the total conductivity is provided by near equal fractions (transference numbers) of the partial ionic and electronic conductivity, respectively,[26] from the point of view of oxygen permeation it is more useful to relate mixed conduction to their absolute values. Volume diffusion theories treating ambipolar transport in oxides clearly indicate that higher currents (fluxes) are obtained when either the electronic or the ionic conductivity increases, or both increase simultaneously. The flux at a given total conductivity is maximum when the ionic and electronic transference numbers are equal, i.e., 0.5. In this view, alumina is not a good mixed conductor. Materials showing predominant electronic conduction may thus prove to be excellent mixed conductors when their ionic conductivity is also substantial. The general objective for optimum membrane performance therefore is to maximize the product of mobility and concentration of both ionic and electronic charge carriers in appropriate ranges of temperature and oxygen partial pressure.

Owing to the ability to conduct both oxygen ions and electrons, the MIEC membrane can operate without the need of attachment of electrodes to the oxide surface and external circuitry. The latter represents an inherent advantage over traditional *oxygen pumps* in which a solid oxide electrolyte is sandwiched between two gas-permeable electrically conductive electrodes (Figure 14.1b). An advantage of electrically driven oxygen separation may be its ability to deliver oxygen at elevated pressures, eliminating the need for compressors.[27] Figure 14.1c shows a *dual-phase membrane*, which can be visualized as being a dispersion of a metallic phase into an oxygen ion-conducting host or matrix, e.g., Pd metal into stabilized zirconia. This challenging approach was first described by Mazanec et al.[28] and offers an alternative use of oxide electrolytes in the field of dense ceramic membranes. Industrially important solid oxide electrolytes to date are mainly based on oxygen-deficient, fluorite-related structures such as ZrO_2 and CeO_2 doped with CaO or Y_2O_3. Unless operated with an internal or external circuitry, the oxygen flux through these materials in usual ranges of temperature and oxygen pressure is negligibly low, preventing their practical use as oxygen separation membrane. The existence of a nonvanishing electronic conduction in the ionic domain, and concomitant oxygen semipermeability, however, can be detrimental considering their use as solid electrolytes in fuel cells and oxygen sensors.[29,30]

While past efforts were focused on expanding the electrolytic domain of oxygen ion-conducting, fluorite-type ceramics, more recently one has begun to introduce enhanced electronic conduction in fluorite matrices. Extrinsic electronic conduction in ionically conducting matrices can be obtained by dissolution of multivalent cations in the fluorite oxide lattice. Notable examples include yttria-stabilized zirconia (YSZ) doped with either titania[31,32] or ceria.[33,34] Electronic conductivity in these solid solutions is reportedly found to increase with increasing dopant concentration, but may be limited by the solid solubility range of the multivalent oxide. As conduction occurs via a small polaron mechanism (electron hopping) between dopant ions of different valence charge, its magnitude will strongly vary with temperature and oxygen partial pressure. In general, the extent of mixed conductivity that can be induced in fluorite ceramics is limited, which restricts its possible use as a ceramic membrane, unless very high temperatures of operation (>1400°C) and stability down to very low values of oxygen partial pressure are required as, e.g., in the production of gaseous fuels CO and H_2 by direct thermal splitting of CO_2 and H_2O, respectively, and extraction of the oxygen arising from dissociation.[35,34]

Since the first report on high oxide ion conductivity in some of the rare earth aluminates in the mid 1960s,[36,37] materials with oxygen-deficient perovskite and perovskite-related structures received much attention for the development of new solid electrolytes and mixed conductors for numerous applications.[38] Presently, extensive research is being conducted on acceptor-doped perovskite oxides with the generic formula $La_{1-x}A_xCo_{1-y}B_yO_{3-\delta}$ (A = Sr, Ba, Ca and B = Fe, Cu, Ni). Teraoka et al.[39-41] were the first to report very high oxygen fluxes through the cobalt-rich compositions, which perovskites are known to become highly oxygen anion defective at elevated temperatures and reduced oxygen partial pressure. The oxygen ion conductivity in the given series can be 1 to 2 orders of magnitude higher than those of the stabilized zirconias, though in usual ranges of temperature and oxygen partial pressure electronic conduction in the perovskite remains predominant.[41,42] Besides potential use of these perovskite compositions as catalytically active electrodes in, e.g., fuel cells, oxygen pumps, and sensors, the compounds have a bright future for use as oxygen separation membranes. The precise composition may be tailored for a specific application, but this has not yet been fully developed. Structural and chemical integrity of the cobaltites, however, is a serious problem and needs to be addressed before commercial exploitation becomes feasible.

For the sake of completeness, a schematic representation of a porous ceramic membrane is given in Figure 14.1d. The majority of porous ceramic membranes are composite or asymmetric in structure. They include materials like α-Al_2O_3, γ-Al_2O_3, TiO_2 and SiO_2, and generally consist of a thin layer of either a mesoporous ($2 < d_p < 50$ nm) or microporous ($d_p < 2$ nm) barrier layer of a few microns thick superimposed on a mechanically strong support with a comparatively large pore diameter d_p and usually a few millimeters thick. Often there are one or more intermediate layers, resulting in a graded pore structure across the membrane.

B. DATA: OXYGEN PERMEABILITY OF SOLID OXIDE MEMBRANES

Table 14.1 lists data of steady-state oxygen permeability measurements on various solid electrolytes, mixed-conducting oxides, and dual-phase membranes taken from various literature reports. Measurements most commonly are performed by imposing a gradient in oxygen partial pressure across the membrane, usually by passing an oxygen-rich and -lean gases, e.g., air and inert gas, respectively, along opposite sides of a sealed ceramic disk or tube wall, without the use of external circuitry such as electrodes and power supplies. The number of moles, volume, or mass of oxygen passing per unit time through a unit of membrane surface area is measured downstream using, e.g., on-line gas chromatography or an oxygen sensor, from which data the oxygen flux is calculated. Table 14.1 also includes data for solid electrolyte cells used in the oxygen pump mode. A graphical presentation of selected data is given in Figure 14.2.

High oxygen fluxes are found through selected perovskite-structured ceramics. Table 14.1 shows for several perovskite systems the trend in permeation flux as a function of the type and concentration of applied dopants. In the range 800 to 900°C the highest flux was measured by Teraoka et al. for $SrCo_{0.8}Fe_{0.2}O_{3-\delta}$, but, as for a number of other compositions, different values have been reported by other groups. Such conflicting results reflect the experimental difficulties in measuring oxygen permeation of sealed ceramic discs at high temperatures, but may also be due to factors that influence the effective P_{O_2} gradient across the membrane, sample preparation, etc. This is further discussed in Section V.G.2.

For the sake of comparison, Table 14.1 contains limited data for the oxygen flux through micro- and mesoporous membranes. As noted before, the last-mentioned category of membranes falls outside the general scope of this chapter. It is seen that the oxygen fluxes observed through membranes formed from the mixed-conducting perovskite-type oxides, such as $La_{1-x}Sr_xCo_{1-y}Fe_yO_{3-\delta}$, approach those exhibited by the porous membranes. It should be noted, however, that these types of membranes have different requirements. The high temperature needed for operation using membranes based on oxygen ion conductors may be restrictive in certain applications, but beneficial to others, e.g., coal gasification and partial oxidation of light paraffins.[27]

C. FACTORS CONTROLLING OXYGEN PERMEATION

The rate at which oxygen permeates through a nonporous ceramic membrane is essentially controlled by two factors: the rate of solid state diffusion within the membrane and that of interfacial oxygen exchange on either side of the membrane. The oxygen flux can be increased by reducing the thickness of the membrane, until its thickness becomes less than a characteristic value, L_c, at which point the flux of oxygen is under conditions of mixed control of the surface exchange kinetics and bulk diffusion.[43] Below L_c, the oxygen flux can only marginally be improved by making the membrane thinner.

For predominant electronic conductors like, for example, the perovskites $La_{1-x}Sr_xCo_{1-y}Fe_yO_{3-\delta}$, L_c is determined by the ratio of the oxygen self-diffusivity and surface exchange coefficient. Both parameters can be measured simultaneously by using ^{18}O-^{16}O isotopic exchange techniques. Calculations show that L_c may vary from the micron range to the centimeter range, depending on material and environmental parameters. Modeling studies, however, show that significant increase in the rate of interfacial oxygen transfer and, hence,

TABLE 14.1

Oxygen Fluxes Through Ceramic Membranes, for a Given Temperature and Membrane Thickness, Together with the Experimental Conditions during Measurements. (When not specified otherwise, the P_{O_2} gradient corresponds to $P'_{O_2} = 0.21$ atm (air) vs. inert gas (sweep method). Indicated are the sweep rate of the inert gas (in sccm) and disc diameter Ø (in mm), which parameters in conjunction with the oxygen flux determine the oxygen partial pressure P''_{O_2} in the permeate stream. In a number of cases the value of P''_{O_2} is specified. The full range of measurements covered by the experiments is also given. For references see end of table.)

Membrane material		Temp. T °C	Thickness L mm	Oxygen flux[a] j_{O_2} µmol cm^{-2} s^{-1}	Experimental conditions	Range of measurements	Ref.
				Mixed conductors: ($t_{ion} \ll 1$)			
On fluorite basis							
$(ZrO_2)_{1-x-y}$-$(CeO_2)_x$-$(CaO)_y$	x = 0.09; y = 0.36	870	2.0	13×10^{-6}	Ceramic tube (Ø13 × 5 mm and Ø13 × 15 mm); He	T = 827–1523°C; x = 0.09–0.36	1
		1523	2.0	28×10^{-3}	"	"	1
$(ZrO_2)_{1-x-y}$-$(TiO_2)_x$-$(Y_2O_3)_y$	x = 0.10; y = 0.10	1481	1.5	0.20	Ceramic tube (Ø8 × 50 mm); $P'_{O_2} = 0.122$ atm; sweep gas H_2/CO_2 mixed gas ($P''_{O_2} \approx 10$–10 atm)	T = 1305–1481°C; x = 0.075, 0.1; $P''_{O_2} = 10^{-5}$–10^{-11} atm	2
$(ZrO_2)_{1-x}$-$(Tb_2O_{3.5})_x$	x = 0.30	900	2.0	5.2×10^{-3}	Ø = 12 mm; Ar	$P'_{O_2} = 0.21^{-1}$ atm	3
		900	2.0	0.037×10^{-3}	Ø = 12 mm; He (20 sccm)	T = 900–1100°C; $P'_{O_2} = 0.01^{-1}$ atm	4,5
$(ZrO_2)_{1-x}$-$(Tb_2O_{3.5})_x$-$(Y_2O_3)_y$	x = 0.228; y = 0.072	900	2.0	0.026×10^{-3}	"	"	4,5
$(Bi_2O_3)_{1-x}$-$(Tb_2O_{3.5})_x$	x = 0.40	650	1.7	0.18×10–3	Ø = 12 mm; He (20 sccm)	T = 650–810°C; $P'_{O_2} = 0.01^{-1}$ atm	6
On perovskite basis							
$La_{1-x}Sr_xCoO_{3-\delta}$	x = 0.2	900	2.0	0.022	Ø = 12 mm; He (12.5 sccm)	T = 700–1100°C; $P'_{O_2} = 0.07^{-1}$ atm	7
	x = 0.4	900	2.0	0.074	.....He (10 sccm)	"	7
	x = 0.6	900	2.0	0.20	"	"	7
	x = 0.8	900	2.0	0.49	"	"	7

TABLE 14.1 (continued)
Oxygen Fluxes Through Ceramic Membranes, for a Given Temperature and Membrane Thickness, Together with the Experimental Conditions during Measurements

Membrane material		Temp. T °C	Thickness L mm	Oxygen flux[a] j_{O_2} μmol cm^{-2} s^{-1}	Experimental conditions	Range of measurements	Ref.
$La_{1-x}Sr_xFeO_{3-\delta}$	x = 0.1	1000	1.0	0.019	Ø = 12 mm; He (effluent P''_{O_2} = 0.004 atm)	T = 850–1050°C; L = 0.5–2.0 mm; P'_{O_2} = 0.01^{-1} atm; P''_{O_2} = (0.4–40) x 10^{-3} atm (He: 5–70 sccm)	8
	x = 0.2	1000	1.0	0.104	"	"	8
	x = 0.3	1000	1.0	0.188	"	"	8
	x = 0.4	1000	1.0	0.256	"	"	8
$Ln_{0.6}Sr_{0.4}CoO_{3-\delta}$	Ln = La	820	1.5	0.47	Ø = 20 mm; He (30 sccm)	T = 750–820°C	9
	Ln = Pr	820	1.5	0.61	"	"	9
	Ln = Nd	820	1.5	0.69	"	"	9
	Ln = Sm	820	1.5	0.80	"	"	9
	Ln = Gd	820	1.5	1.07	"	"	9
$La_{0.6}Sr_{0.4}Co_{0.8}B_{0.2}O_{3-\delta}$	B = Mn	865	1.5	0.33	Ø = 20 mm; He (30 sccm)	T = 300–865°C	9
	B = Cr	865	1.5	0.39	"	"	9
	B = Fe	865	1.5	0.42	"	"	9
	B = Co	865	1.5	0.70	"	"	9
	B = Ni	865	1.5	1.00	"	"	9
	B = Cu	865	1.5	1.32	"	"	9
$La_{0.6}A_{0.4}Co_{0.8}Fe_{0.2}O_{3-\delta}$	A = Co	865	1.5	0.01	Ø = 20 mm; He (30 sccm)	T = 300–865°C	9,10
	A = Na	865	1.5	0.16	"	"	9,10
	A = Sr	865	1.5	0.42	"	"	9,10
	A = Ca	865	1.5	1.26	"	"	9,10
	A = Ba	865	1.5	1.43	"	"	9,10

$La_{1-x}Sr_xCo_{0.2}Fe_{0.8}O_{3-\delta}$	x = 0.4	850	2.5	0.15	Ø = 25 mm; P'_{O_2} = 1 atm. N2: (effluent P''_{O_2} = 0.1 atm)	T = 650–950°C	11
	x = 0.6	850	2.5	0.33	"	"	11
$La_{1-x}Sr_xCo_{0.4}Fe_{0.6}O_{3-\delta}$	x = 0	850	1.0	< 0.01	Ø = 10 mm; He (30 sccm)	T = rt.–875°C	12
	x = 0.4	850	1.0	0.12	"	"	12
	x = 0.8	850	1.0	0.40	"	"	12
	x = 1	850	1.0	1.49	"	"	12
$La_{1-x}Sr_xCo_{0.8}Fe_{0.2}O_{3-\delta}$	x = 0.2	850	1.0	1.06	Ø = 10 mm; He (30 sccm)	T = rt.–875°C	12
		850	1.0	0.026	Ø = n.s.; He (30 sccm)	T = 730–1000°C	13
	x = 0.4	850	0.55	0.054	Ø = 15 mm; He (effluent P''_{O_2} = 0.017 atm)	T = 800–1000°C; L = 0.5–2.0 mm; P'_{O_2} = 0.01–0.21 atm; P''_{O_2} = (3–50) x 10^{-3} atm (He: 5–70 sccm).	14
		850	1.3	0.33	Ø = 13 mm; He	T = 600–900°C	15
		850	240 µm	0.85	"	"	15
	x = 1	850	1.0	2.11	Ø = 10 mm; He (30 sccm)	T = rt.–875°C	12
		820	1.0	0.43	Ø = 13 mm; He (20 sccm)	T = 620–920°C; L = 1.0–5.5 mm; P'_{O_2} = 0.04–0.90 atm	16,17
		820	2.0	0.56	.....He (40 sccm)	"	16,17
		850	1.0	0.17	Ø = 12 mm; He (10 sccm)	T = 700–950°C	18
$Y_{0.05}BaCo_{0.95}O_{3-\delta}$		900	2.0	0.39	Ø = 12 mm; He (10 sccm)	T = 900–1100°C	19
$CaTi_{0.8}Fe_{0.2}O_{3-\delta}$		800	1.0	7.8 x 10^{-3}	ceramic tube; evacuated to P''_{O_2} = 0.079 atm	T = 700–1000°C; P''_{O_2} = 0.01–0.13 atm	20
				Oxide electrolytes ($t_{ion} \approx 1$)			
$(ZrO_2)_{0.92}$-$(Y_2O_3)_{0.08}$ (YSZ)		800	1.0	10^{-8} – 10^{-7}	average value from compilation by [1]		21
		1200	1.0	10^{-4} – 10^{-3}	"	"	21
$(Bi_2O_3)_{0.75}$-$(E_{r_2}O_3)_{0.25}$ (BE25)		650	0.7	0.76 x 10^{-3}	Ø = 12 mm; P'_{O_2} = 1 atm; He (40 sccm)	T = 650–810°C; P'_{O_2} = 0.01^{-1} atm; P''_{O_2} = 1.1 x 10^{-4} atm	22
		810	0.7	6.7 x 10^{-3}	"	"	22

TABLE 14.1 (continued)
Oxygen Fluxes Through Ceramic Membranes, for a Given Temperature and Membrane Thickness, Together with the Experimental Conditions during Measurements

Membrane material		Temp. T °C	Thickness L mm	Oxygen flux[a] j_{O_2} µmol cm^{-2} s^{-1}	Experimental conditions	Range of measurements	Ref.
				Internally short-circuited:			
Dual-phase composites							
YSZ-Pd metal phase fraction	30 vol%	1100	2.0	0.26 x 10^{-3}	volume fraction below percolation threshold; Ø = 12 mm; He (effluent P''_{O_2} = 0.17 x 10^{-3} atm)	T = 1050–1150°C; $P'_{O_2} = 0.01^{-1}$ atm; P''_{O_2} = (0.6–0.18) × 10^{-3} atm (He: < 20 sccm).	23,24
	40 vol%	1100	2.0	76 x 10^{-3}	......volume fraction above percolation threshold: He (effluent P''_{O_2} = 0.14 atm)	T = 900–1100°C; L = 0.5–2.0 mm; P'_{O_2} = 0.01–1 atm; P''_{O_2} = (4.3–14) × 10^{-3} atm (He: < 20 sccm).	23,24
		1100	0.5	0.25	.....He (effluent P''_{O_2} = 12 x 10^{-3} atm)	"	24
		1100	0.5	0.87	.....sweep gas He/CO/CO_2 (effluent P''_{O_2} = 0.26 x 10^{-12} atm)	T = 900–1100°C	24
	50 vol%	1100	0.8	1.44	Ø = 25 mm; sweep gas 90% H_2 in Ar (68 sccm)		25,26
YSZ-$In_{0.9}Pr_{0.1}O_{1.5-\delta}$	50 vol%	1100	0.8	1.56	.....Ø = 31 mm		25,26
		1100	0.25	4.17	.....Ø = n.s.		25,26
YSZ-$In_{0.95}Pr_{0.025}Zr_{0.025}O_{1.5-\delta}$	50 vol%	1100	0.3	5.39	.....Ø = n.s.		25,26
BY25-Ag	35 vol%	750	1.0	0.31	Ø = 25 mm; N_2 (120 sccm)	T = 500–750°C; 1–50 vol% Ag; L = 0.09–5.0 mm	27
		750	90 µm	0.96	"		27
BE25-Ag	40 vol%	720	1.0	0.12	Ø = 15 mm; He (effluent P''_{O_2} = 6.3 x 10^{-3} atm)	T = 600–720°C; 0–40 vol% Ag; L = 0.25–2.0 mm; $P'_{O_2} = 0.01^{-1}$ atm; P''_{O_2} = (0.6 – 30) × 10^{-3} atm (He: 5–70 sccm).	28

		750	1.6	0.085	Ø = 15 mm; He (effluent P_{O_2}'' = 16 x 10^{-3} atm)	T = 600–850°C	24,29
BE25-Au	40 vol%	750	1.0	0.016	.....He (effluent P_{O_2}'' = 2.1 x 10^{-3} atm)	"	24,29
				Externally short-circuited:			
Oxygen pump							
$Bi_{0.571}Pb_{0.428}O_{1.285}$ (+0.187 mol% ZrO_2)		600	4	0.78	Co-pressed Au-grids electrodes		30
BE25		630	1.5	0.5–0.7	Sputtered Au electrodes		31
				Porous membranes			
SiO_2 film (d_p = 0.5 nm) on layered γ - α Al_2O_3 support		35–200	100 nm	3–7	Activated diffusion; sel. α = 2–4 (O_2–N_2); abs. press. difference 1 atm		32
Mesoporous (d_p = 10 nm) (calculated value)[b]		400	10 μm	200	Theor. expected value based upon Knudsen diffusion		

[a] Values may be converted to other dimensions using equalities 1 μmol cm^{-2} s^{-1} = 1.47 sccm cm^{-2} = 21.1 m^3 m^{-2} day^{-1} = 386 mA cm^{-2}.

[b] Additional parameters used in calculation are the porosity ε = 0.5 and the tortuosity τ = 2.

References used in Table 14.1

[1] Nigara, Y., Mizusaki, J., Ishigame, M., *Solid State Ionics,* 1995, *79*, 208–211.

[2] Arashi, H., Naito, H., Nakata, M., *Solid State Ionics,* 1995, *76*, 315–319.

[3] Iwahara, H., Esaka, T., Takeda, K., Mixed conduction and oxygen permeation in sintered oxides of a system ZrO_2-Tb_4O_7, in *Advances in Ceramics,* Vol. 24A, S. Somiya, N. Yamamoto, H. Yanegida, Eds., The American Ceramic Society, Westerville, OH, 1988, 907–916.

[4] Cao, C.Z., Liu, X.Q., Brinkman, H.W., De Vries, K.J., Burggraaf, A.J., Mixed conduction & oxygen permeation of ZrO_2-$Tb_2O_{3.5}$-Y_2O_3 solid solutions, in *Science and Technology of Zirconia* V, S.P.S. Badwal et al., Eds., Technomic Publications, Lancaster, PA, 1993, 576–583.

[5] Cao, G.Z., *J. Appl. Electrochem.*, 1994, *24*, 1222–1224.

[6] Kruidhof, H., Bouwmeester, H.J.M., unpublished results.

[7] Van Doorn, R.H.E., Kruidhof, H., Bouwmeester, H.J.M., Burggraaf, A.J., Oxygen permeability of strontium-doped $LaCoO_{3-δ}$ perovskites, in *Mater. Res. Soc. Symp. Proc.*, Vol 369, *Solid State Ionics IV*, G.-A. Nazri, J.-M. Taracson, M.S. Schreiber, Eds., Materials Research Society, Pittsburgh, 1995, 377–382.

[8] Elshof, J.E. ten, Bouwmeester, H.J.M., Verweij, H., *Solid State Ionics,* 1995, *81*, 97–110.

[9] Teraoka, Y., Nobunaga, T., Yamazoe, N., *Chem. Lett.*, 1988, 503–506.

[10] Teraoka, Y., Nobunaga T., Okamoto, K., Miura, N., Yamazoe, N., *Solid State Ionics,* 1991, *48*, 207–212.

[11] Weber, W.J., Stevenson, J.W., Armstrong, T.R., Pederson, L.R., Processing and electrochemical properties of mixed conducting $La_{1-x}A_xCo_{1-y}Fe_yO_{3-δ}$ (A = Sr, Ca), in *Mater. Res. Soc. Symp. Proc., Vol 369, Solid State Ionics, IV,* G.-A. Nazri, J.-M. Taracson, M.S. Schreiber, Eds., Materials Research Society, Pittsburgh, 1995, 395–400.

[12] Teraoka, Y., Zhang, H.M., Furukawa, S., Yamazoe, N., *Chem. Lett.*, 1985, 1743–1746.

TABLE 14.1 (continued)
Oxygen Fluxes Through Ceramic Membranes, for a Given Temperature and Membrane Thickness, Together with the Experimental Conditions during Measurements

[13] Tsai, C.-Y., Ma, Y.H., Moser, W.R., Dixon, A.G., Simulation of nonisothermal catalytic membrane reactor for methane partial oxidation to syngas, in *Proc. 3rd Int. Conf. Inorganic Membranes*, Y.H. Ma, Ed., Worcester, 1994, 271–280.

[14] Elshof, J.E. ten, Bouwmeester, H.J.M., Verweij, H., *Appl. Catal. A: General*, 1995, *130*, 195–212

[15] Miura, N., Okamoto, Y., Tamaki, J., Morinag, K., Yamazoe, N., *Solid State Ionics*, 1995, *79*, 195–200.

[16] Qiu, L, Lee, T.H., Lie, L.-M., Yang, Y.L., Jacobson, A.J., *Solid State Ionics*, 1995, *76*, 321–329.

[17] Yang, Y.L., Lee, T.H., Qiu, L., Liu, L., Jacobson, A.J., Oxygen permeation studies of $SrCo_{0.8}Fe_{0.2}O_{3-\delta}$, in *Mater. Res. Soc. Symp. Proc., Vol 369, Solid State Ionics, IV*, G.-A. Nazri, J.-M. Taracson, M.S. Schreiber, Eds., Materials Research Society, Pittsburgh, 1995, 383–388.

[18] Kruidhof, H., Bouwmeester, H.J.M., van Doorn, R.H.E., Burggraaf, A.J., *Solid State Ionics*, 1993, *63–65*, 816–22.

[19] Brinkman, H.W., Kruidhof, H., Burggraaf, A.J., *Solid State Ionics*, 1994, *68*, 173–176.

[20] Iwahara, H., Esaka, T., Mangahara, T., *J. Appl. Electrochem.*, 1988, *18*, 173–177.

[21] Fouletier, J., Fabry, P., Kleitz, M., *J. Electrochem. Soc.*, 1976, *123(2)*, 204–213.

[22] Bouwmeester, H.J.M., Kruidhof, H., Burggraaf, A.J., Gellings, P.J., *Solid State Ionics*, 1992, *53–56*, 460–68.

[23] Chen, C.S., Boukamp, B.A., Bouwmeester, H.J.M., Cao, G.Z., Kruidhof, H., Winnubst, A.J.A., Burggraaf, A.J., *Solid State Ionics*, 1995, *76*, 23–28.

[24] Chen, C.S., Ph.D. thesis, University of Twente, The Netherlands, 1994.

[25] Mazanec, T.J., Cable, T.L., Frye, J.G., *Solid State Ionics*, 1992, *53–56*, 111–118.

[26] Mazanec, T.J., Frye, J.G., Jr., *Eur. Patent Appl.* 0399 833 A1, 1990.

[27] Shen, Y.S., Liu, M., Taylor, D., Bolagopal, S., Joshi, A., Krist, K. Mixed ionic-electronic conductors based on Bi-Y-O-Ag metal-ceramic system, in *Proc. of 2nd Int. Symp. on Ionic and Mixed Conducting Ceramics, Vol. 94-12*, T.A. Ramanarayanan, W.L. Worrell, H.L. Tuller, Eds., The Electrochemical Society, Pennington, NJ., 1994, 574–95.

[28] Ten Elshof, J.E., Nguyen, D.N.Q., den Otter, M.W., Bouwmeester, H.J.M., Verweij, H., *J. Membr. Sci.*, submitted.

[29] Chen, C.S., Kruidhof, H., Bouwmeester, H.J.M., Verweij, H., Burggraaf, A.J., *Solid State Ionics*, 1996, 86-88, 569-572.

[30] Dumelié, M., Nowogrocki, G., Boivin, J.C. *Solid State Ionics*, 1988, *28–30*, 524–528.

[31] Vinke, I.C., Seshan, K., Boukamp, B.A., de Vries, K.J., Burggraaf, A.J., *Solid State Ionics*, 1989, *34*, 235–242.

[32] De Lange, R.S.A., Hekkink, J.H.A., Keizer, K., Burggraaf, A.J., *Microporous Mater.*, 1995, *4*, 169–186.

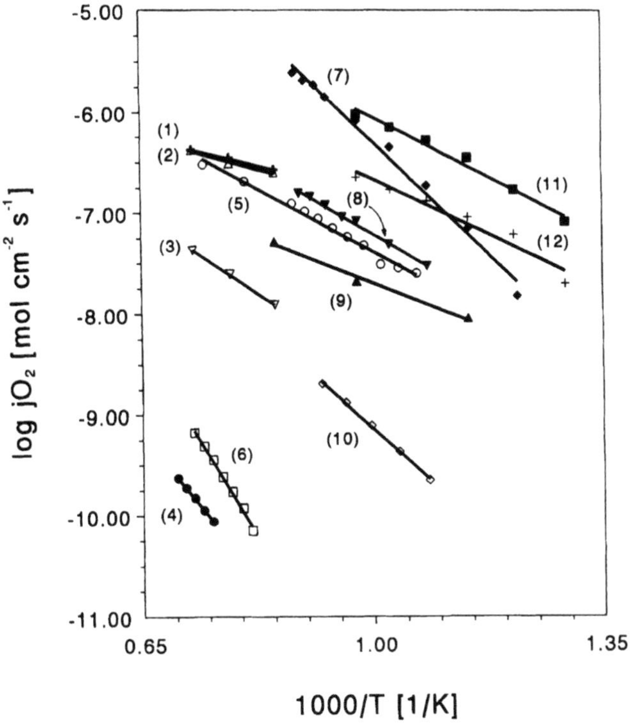

FIGURE 14.2. Arrhenius plots of oxygen permeation for
1) $La_{0.3}Sr_{0.7}CoO_{3-\delta}$ [7]
2) $Ba_{0.9}Y_{0.1}CoO_{3-\delta}$ [9]
3) YSZ-Pd (40 vol%), continuous Pd phase[23,24]
4) YSZ-Pd (30 vol%), discontinuous Pd phase[23,2]
5) $La_{0.5}Sr_{0.5}CoO_{3-\delta}$ [7]
6) $(ZrO_2)_{0.7}$-$(Tb_2O_{3.5})_{0.228}$-$(Y_2O_3)_{0.072}$ [4,5]
7) $SrCo_{0.8}Fe_{0.2}O_{3-\delta}$ [9]
8) BE25-Ag (40 vol%)[24,29]
9) $Ba_{0.66}Y_{0.33}CoO_{3-\delta}$ [19]
10) $(Bi_2O_3)_{0.75}$-$(Er_2O_3)_{0.25}$ (BE25)[22]
11) BY25-Ag (35 vol%), thickness 90 μm[27]
12) BY25-Ag (35 vol%), thickness 1.5 mm[27]
Air and inert gas are passed along opposite sides of the membrane. Unless specified otherwise, the membrane thickness varies between 1–2 mm. For references, see end of Table 14.1.

in the oxygen flux can be achieved by deposition of a porous MIEC layer on top of the (thin) nonporous membrane.[44-46] Since a number of simplifying assumptions is made, such as neglect of changes in material parameters with variation in the chemical potential of oxygen, the models developed are valid only in the limit of small P_{O_2} gradients across the MIEC membrane. For a more rigorous approach, referring to actual operating conditions of oxygen-separation membranes, much more work is needed to arrive at a better understanding of the transport processes under oxygen potential gradients. In particular, our present understanding of the factors that govern the surface exchange kinetics is rather poor.

Effects related to microstructure, including grain boundary diffusion and (local) order–disorder phenomena, may also influence overall oxygen transport. Besides the processing into defect-free thin films and associated problems of compatibility between deposited membrane layer and the porous substrate material, chemical stability at high temperatures, effects induced by the presence of an oxygen potential gradient like segregation of impurities to the surface and to grain boundaries, kinetic demixing and kinetic decomposition could affect membrane performance or limit operational life. In many cases, these difficulties remain to be overcome before commercial exploitation becomes viable. An obvious consideration is that all factors listed are important and govern the selection of materials.

D. SCOPE OF THIS CHAPTER

In the succeeding sections, the emphasis is upon the basic elements of mixed ionic and electronic transport through dense ceramic membranes and the associated surface exchange kinetics. Selected observations from permeation measurements and related experiments on oxygen transport are surveyed. Due to size considerations, we shall mainly focus on mixed-conducting, acceptor-doped perovskite and perovskite-related oxides. In addition, the semi-permeability of oxygen electrolyte materials is discussed. These have been studied extensively, and appropriate examples are given that emphasize the role of the surface exchange kinetics in determining the oxygen fluxes through dense oxide ceramics. Materials and concepts to the design of oxygen pumps and dual-phase membranes are considered, but only briefly.

III. FUNDAMENTALS

A. BULK TRANSPORT

The basic assumption of the theory presented in this section is that the lattice diffusion of oxygen or the transport of electronic charge carriers through the bulk oxide determines the rate of overall oxygen permeation. Moreover, oxygen is transported selectively through the membrane in the form of oxygen ions, rather than molecules, under the driving force of a gradient in oxygen chemical potential. The flux of oxygen ions is charge compensated by a simultaneous flux of electrons or electron holes, which is enabled without the use of external circuitry. We only briefly review the fundamentals of solid state diffusion through mixed-conducting oxides, and the reader is referred to References 47, 48, and 49 for a more complete discussion.

1. Wagner Equation

Considered here is the case where the interaction of gaseous oxygen with the oxide lattice can be represented by a chemical reaction of the form*

$$\tfrac{1}{2}O_2 + V_O^{\bullet\bullet} + 2\,e' = O_O^{\times} \tag{14.1}$$

assuming that oxygen vacancies are the mobile ionic defects. These may be obtained, e.g., by doping of the oxide lattice with aliovalent cations. The intrinsic ionization across the band gap can be expressed by

$$nil = e' + h^{\bullet} \tag{14.2}$$

The single particle flux of charge carriers, with neglect of cross terms between fluxes, is given by

$$j_k = -\frac{\sigma_k}{z_k^2 F^2}\nabla\eta_k \tag{14.3}$$

* The notation adopted for defects is from Kröger and Vink.[50] See also Chapter 1 of this handbook.

where z_k is the charge number and s_k the conductivity of charge carrier k, F the Faraday constant, and $\nabla\eta_k$ the gradient of the electrochemical potential. The latter comprises a gradient in chemical potential $\nabla\mu_k$ and a gradient in electrical potential $\nabla\phi$, for each individual charge carrier k given by

$$\nabla\eta_k = \nabla\mu_k + z_k F \nabla\phi \tag{14.4}$$

The charge carrier diffusing more rapidly causes a gradient in the electrical potential $\nabla\phi$, in which the transport of carriers with opposite charge is accelerated. At steady state no charge accumulation occurs. The fluxes of ionic and electronic defects are therefore related to each other by the charge balance

$$2\, j_{V_O^{\bullet\bullet}} = j_{e'} - j_{h^\bullet} \tag{14.5}$$

Equation (14.5) can be used together with Equations (14.3) and (14.4) to eliminate the electrostatic potential gradient. The flux of oxygen vacancies is then obtained in terms of the chemical potential gradients only. If it is further assumed that internal defect chemical reactions are locally not disturbed by the transport of matter, the chemical potential gradients of individual charge species can be converted into the virtual chemical potential of gaseous oxygen, μ_{O_2}. The following differential relations hold at equilibrium*

$$\frac{1}{2}\nabla\mu_{O_2} + \nabla\mu_{V_O^{\bullet\bullet}} + 2\,\nabla\mu_{e'} = 0 \tag{14.6}$$

$$\nabla\mu_{e'} + \nabla\mu_{h^\bullet} = 0 \tag{14.7}$$

where $\mu_{V_O^{\bullet\bullet}}$ denotes the chemical potential of the oxygen vacancy, $\mu_e{}'$ and $\mu_{h^\bullet}$ denoting the chemical potential of electrons and electron holes, respectively. The flux of oxygen through the membrane can be derived by combining Equations (14.3) to (14.7), using the relationship $j_{O_2} = -\frac{1}{2} j_{V_O^{\bullet\bullet}}$.

One finds,

$$j_{O_2} = -\frac{1}{4^2 F^2}\left[\frac{(\sigma_{e'} + \sigma_{h^\bullet})\,\sigma_{V_O^{\bullet\bullet}}}{(\sigma_{e'} + \sigma_{h^\bullet}) + \sigma_{V_O^{\bullet\bullet}}}\right]\nabla\mu_{O_2} \tag{14.8}$$

or, in a more generalized form,

$$j_{O_2} = -\frac{1}{4^2 F^2}\,\frac{\sigma_{el}\sigma_{ion}}{\sigma_{el} + \sigma_{ion}}\,\nabla\mu_{O_2} \tag{14.9}$$

* It is tacitly assumed here that the chemical potential of lattice oxygen $\mu_{O_O^\times}$ is constant. The present formulation of the defect equilibrium for the formation and annihilation of oxygen vacancies and electrons by the reaction of the solid with environmental oxygen, however, is written in terms of the 'virtual' chemical potentials of the constituent structure elements. In doing so, one does not properly take into account the so-called site-exclusion effect, because the chemical potential of the oxygen vacancy $V_O^{\bullet\bullet}$ and that of lattice oxygen $O_O^\times$ cannot be defined independently from one another. In the present context, it suffices to say that the derived equations are in agreement with those obtained from a more rigorous thermodynamic treatment based upon the 'true' chemical potential for the building unit vacancy, i.e., $(V_O^{\bullet\bullet} - O_O^\times)$. For further reading concerning the definition of chemical potentials, the reader may consult References 47 and 48.

where $\sigma_{ion} = \sigma_{V_O^{\bullet\bullet}}$ and $\sigma_{el} - \sigma_{h^\bullet} + \sigma_{e'}$ are the partial ionic and electronic conductivity, respectively. The conductivity term in Equation (14.9) is equivalent to $t_{el}t_{ion}\sigma_{total} = t_{ion}\sigma_{el} = t_{el}\sigma_{ion}$, where t_{el} and t_{ion} are the fractions (transference numbers) of the total conductivity σ_{total} provided by electronic and ionic defects, respectively. Integration of Equation (14.9) across the oxide membrane thickness, L, using the relationship $\nabla\mu_{O_2} = \partial RT\ln P_{O_2}/\partial x$ (x = distance coordinate) and assuming no divergence in the fluxes, yields the Wagner equation in the usual form,[51,52]

$$j_{O_2} = -\frac{RT}{4^2 F^2 L} \int_{\ln P'_{O_2}}^{\ln P''_{O_2}} \frac{\sigma_{el}\sigma_{ion}}{\sigma_{el} + \sigma_{ion}} \, d\ln P_{O_2} \tag{14.10}$$

The limits of integration are the oxygen partial pressures maintained at the gas-phase boundaries. Equation (14.10) has general validity for mixed conductors. To carry the derivation further, one needs to consider the defect chemistry of a specific material system. When electronic conductivity prevails, Equations (14.9) and (14.10) can be recast through the use of the Nernst–Einstein equation in a form that includes the oxygen self-diffusion coefficient D_s, which is accessible from ionic conductivity measurements. This is further exemplified for perovskite-type oxides in Section V.D, assuming a vacancy diffusion mechanism to hold in these materials.

2. Chemical Diffusion Coefficient

The preceding theory was used by Wagner and Schottky[51] and Wagner[52] to describe oxide film growth on metals. The driving force for diffusion is not a concentration gradient, but rather a chemical potential gradient. An important and necessary assumption is that the internal defect reactions are fast enough to attain local chemical equilibrium so that the concentrations of involved ionic and electronic (electrons or holes) charge carriers at any distance coordinate in the oxide are fixed by the local value of the virtual chemical potential, μ_{O_2}. The effective transport is still that of neutral oxygen atoms by which the theory fits that of a chemical diffusion process in terms of Fick's first law,

$$j_O = -\tilde{D}\frac{\partial c_O}{\partial x} \tag{14.11}$$

where the driving force for diffusion is the gradient in neutral oxygen, $\partial c_O/\partial x$. The coefficient of proportionality, denoted by $\tilde{D}$, is called the chemical diffusion coefficient. By virtue of Equations (14.9) and (14.11), one obtains

$$\tilde{D} = \frac{1}{8F^2}\frac{\sigma_{el}\sigma_{ion}}{\sigma_{el} + \sigma_{ion}}\frac{\partial\mu_{O_2}}{\partial c_O} \tag{14.12}$$

Here, we note that $j_{O_2} = \frac{1}{2}j_O$. Because $\partial c_O/\partial x = -\partial c_V/\partial x$, a similar expression is obtained when diffusion is dominated by neutral vacancies.

The thermodynamic factor $\partial\mu_{O_2}/\partial c_O$ in Equation (14.12) can be determined directly from experiment by measuring the oxygen stoichiometry as a function of oxygen partial pressure, either by gravimetric or coulometric measurements. In view of Equations (14.6) and (14.7), it comprises contributions from both ionic and electronic defects, which reflect their nonideal behavior. For materials with prevailing electronic conductivity, Equation (14.12) may be simplified to yield an exact relation between the chemical diffusion coefficient $\tilde{D}$ and the oxygen tracer diffusion coefficient D^*:

$$\tilde{D} = \frac{D^*}{H_R RT} \frac{\frac{1}{2}\partial\mu_{O_2}}{\partial \ln c_O} \tag{14.13}$$

Here, H_R is the Haven ratio, defined as the ratio of the tracer diffusion coefficient D^* to the quantity D^σ derived from dc ionic conductivity measurements,

$$D^\sigma = \frac{\sigma_{ion} R\,T}{c_O\, z_i^2\, F^2} \tag{14.14}$$

The Haven ratio may deviate from unity when correlation effects and possibly different jump distances and jump frequencies cannot be neglected.[53] For a vacancy diffusion mechanism H_R equals the well-known tracer correlation factor f.

3. Trapping of Electronic and Ionic Defects

Equation (14.10) and those derived from it are valid as long as fully ionized oxygen defects contribute to transport. Different equations are obtained if valency changes of oxygen defects occur. Wagner[52] proposed to put the influence of reactions between ionic and electronic defect species in the cross terms of the Onsager equations. Maier and Schwitzgebel[54] and Maier[55,56] explicitly attributed individual diffusivities and conductivities to the new defect species, using the concept of a conservative ensemble accounting for free and trapped species. Following his approach, the reversible reaction between electrons and oxygen vacancies,

$$\begin{aligned} V_O^{\bullet\bullet} + e' &= V_O^{\bullet} \\ V_O^{\bullet} + e' &= V_O^{x} \end{aligned} \tag{14.15}$$

leads to the following expression for the oxygen flux,

$$j_{O_2} = -\frac{1}{4^2 F^2}\left[4 s_{V_O^x} + \frac{(\sigma_{e'} + \sigma_{h^\bullet})(\sigma_{V_O^{\bullet\bullet}} + 4\sigma_{V_O^{\bullet}}) + \sigma_{V_O^{\bullet}}\sigma_{V_O^{\bullet\bullet}}}{(\sigma_{e'} + \sigma_{h^\bullet}) + (\sigma_{V_O^{\bullet\bullet}} + \sigma_{V_O^{\bullet}})}\right]\nabla\mu_{O_2} \tag{14.16}$$

where we have adapted Equation (33) in Reference 55 (part I) into a form to be similar to Equation (14.8), in which ionic transport is by doubly ionized oxygen vacancies only. The Onsager coefficient $S_{V_O^x}$ accounts for the contribution of neutral defects, enabling oxygen transport even when the electronic conductivity of the oxide is zero. We further note that the counter diffusion of two $V_O^{\bullet}$ and a single $V_O^{\bullet\bullet}$ would result in a net neutral oxygen flux, as reflected by the last term in the numerator of Equation (14.16). Maier[55] also examined the case in which electronic or ionic defects are associated (trapped) with immobile centers such as dopant ions. Trapping inevitably leads to a decrease in concentrations of the charge carriers available for transport. The impact of these phenomena is that the transport equations for evaluation of data obtained from electrochemical measurements like, for example, ionic conductivity, concentration cell, permeability, and Hebb–Wagner polarization experiments should accordingly be modified. It is shown by Maier how these are influenced by trapping effects observed in perovskite $SrTiO_3$, and by the transport properties of the high-temperature superconductor $YBa_2Cu_3O_{6+x}$. Because of the large oxygen excess possible in the latter material, it is assumed that transport occurs by differently ionized ionic defects, partly even by neutral oxygen species. For references, see the papers by Maier and Schwitzgebel[54] and Maier.[55,56]

4. Empirical Equations

Evaluation of j_{O_2} from Equation (14.10) requires that data exist for the partial conductivities σ_{ion} and σ_{el} as a function of oxygen partial pressure between the limits of the integral. In what follows, some special relations for either prevailing electronic or ionic conduction are discussed.

For the sake of approximation, in defect chemical studies often an empirical power law is used for the partial conductivity of the rate-determining species, σ_i, such as

$$\sigma_i\left(P_{O_2}\right) = \sigma_i^o \; P_{O_2}{}^n \tag{14.17}$$

where σ_i^o is the conductivity at standard state. Among other methods (for example, see Chapters 7 and 8 of this handbook), the value of n can be derived from experimental data of steady-state oxygen permeation. For proper evaluation it is necessary that the P_{O_2} gradient across a specimen is varied within the assumed range of validity of the empirical power law. Inserting Equation (14.17) in Equation (14.10), one finds after integration, assuming $\sigma_i \ll \sigma_{total}$,

$$j_{O_2} = \frac{\beta}{L}\left[P_{O_2}'^{\,n} - P_{O_2}''^{\,n}\right] \tag{14.18}$$

where $\beta = \sigma_i^o RT/4^2F^2n$. For large positive values of n the rate of oxygen permeation is predominantly governed by the oxygen partial pressure maintained at the feed side, P_{O_2}'. Likewise, for large negative values it is predominantly governed by the oxygen partial pressure maintained at the permeate side, P_{O_2}''. For either a small value of n or a small P_{O_2} gradient, the flux becomes proportional to ln (P_{O_2}'/P_{O_2}'').

Provided that the electronic transference number is known, the ionic (and electronic) conductivity may be obtained by differentiation of experimental data. Assume that we have produced a data set for different P_{O_2} gradients, keeping the oxygen partial pressure P_{O_2}'' at the permeate side fixed. Differentiating Equation (14.10) with respect to the lower integration limit yields

$$\left(\frac{\partial j_{O_2}}{\partial \ln P_{O_2}'}\right)_{P_{O_2}''} = t_{el}\sigma_{ion} \times \frac{RT}{4^2 F^2 L} \tag{14.19}$$

The ionic conductivity at a given pressure P_{O_2} is thus obtained from the slope of the j_{O_2}-$\ln P_{O_2}'$ plot at that P_{O_2}. Similarly, the ionic conductivity can be evaluated from oxygen flux values measured by varying the oxygen partial pressure at the permeate side, keeping the one at the feed side fixed. The two data sets in general will yield the ionic conductivity over a complementary range in oxygen partial pressure.

So from oxygen permeability measurements, it is possible to get information on the transport properties of oxides without making use of electrodes and external circuitry. In practice, however, experiments are plagued by the usual problems of sealing oxide ceramic discs at high temperature. A further complication is that the activity of oxygen at the oxide surfaces may not be precisely known, due to the influx and efflux of oxygen. In addition to these polarization effects, the results may be significantly affected by the flow patterns that possibly exist on feed and permeate sides.

It was further assumed by Wagner[52] that the surface reactions proceeding at the gas-phase boundaries have reached a quasi-equilibrium condition relative to diffusion through the oxide. However, awareness is growing that in many cases the surface reaction may exert a partial control over the transport kinetics.[43] The extent of surface control varies with membrane thickness, temperature, and oxygen pressure difference imposed across the membrane. Other limitations of the theory, some of which are briefly discussed in subsequent sections, include solid state diffusion of matter along preferred paths such as grain boundaries, porosity, and, especially at large departures from the ideal stoichiometric composition, the formation of point defect clusters and ordering.

B. SURFACE OXYGEN EXCHANGE

The exchange of oxygen between oxide surfaces and the gas phase has been recognized to involve a series of reaction steps, each of which may be rate determining.[23] Possible steps for the reduction of oxygen include adsorption, dissociation, charge transfer, surface diffusion of intermediate species (e.g., $O^-_{2\ ads}$, O_{ads}, and O^-_{ads}), and finally incorporation in the (near-) surface layer. Generally, it is assumed that the reoxidation of oxygen anions follows the same series of steps in the reverse direction. The surface reaction is thus associated with transport of charge. The charge carriers in the interior of the oxide maintain thermodynamic equilibrium, according to Wagner's theory, but not at the surface if rate limitations by the surface exchange kinetics come into play.

Liu[57] has presented a detailed analysis of the oxygen separation rates of mixed conductors, using the well-known Butler–Volmer formalism to model interfacial mass and charge transfer. For a discussion on electrode kinetics see also Chapters 1, 2, and 8 of this handbook. At first glance, such a description is not applicable because of the absence of electrodes. But, the adsorbate may be regarded as to replace the electrode material. Accordingly, the electrical double layer (Helmholtz layer) formed between the ionosorbed adsorbate and the oxide dominates the charge transfer kinetics, thus leading to the introduction of transfer coefficients in the rate constants for reduction and oxidation.

In general, however, even at equilibrium, there may also be a double layer, called space charge, next to the surface extending into the oxide interior (Mott–Schottky layer). The width of the space charge layer will be of the order of the Debye–Hückel screening length, L_D, the value of which depends on the volume concentration of all mobile charge carriers. Alteration of the space charge leads to a change in bending of the energy bands, and this influences the occupation of electronic levels near the band edges at the oxide surface. The total potential drop across the interface thus becomes distributed between the Helmholtz layer and the space charge layer,[58] which complicates the analysis of the charge transfer kinetics. Furthermore, if strong electrical fields are developed in a situation where L_D is greater than the size of the ionic charge carrier, which is favored by a low concentration of mobile charges, the migration of ionic charge carriers through the space charge zone can be described by an expression similar to the Butler–Volmer equation.[59,60]

The possibility of space charge in ionic crystalline solids, and the attendant redistribution of lattice components, has a great influence on the properties of boundary regions. Their defect chemistry may depart considerably from the predominant one in the bulk.[61] In oxides, there are quite a number of experimental observations to support the existence of space charge-induced segregation.[62,63] Other factors have been recognized to contribute to segregation, like the misfit strain energy of the solute ion and the surface tension considering adsorption equilibria. Owing to the segregation phenomenon, the composition and crystal ordering of oxide surfaces and grain boundaries differ from those in the bulk. In some cases this leads to the formation of a second phase. An obvious consideration is that these phenomena will have a significant, often controlling influence on the properties of oxide materials, which includes the heterogeneous kinetics of the gas/solid interface. For a more detailed discussion on interface phenomena in ionic solids, the reader is referred to Chapter 4 of this handbook.

1. Characteristic Membrane Thickness

At conditions near to equilibrium one can represent the oxygen flux by the Onsager equation,

$$j_{O_2} = -j_{ex}^{o} \frac{\Delta\mu_{O_2}^{int}}{RT} \tag{14.20}$$

where $\Delta\mu_{O_2}^{int}$ is the chemical potential difference drop across the gas/solid interface. The quantity j_{ex}^{o} (mol O_2 cm^{-2} s^{-1}) denotes the balanced exchange rate in the absence of oxygen potential gradients and relates to the surface exchange coefficient k_s (cm s^{-1}) accessible from data of ^{18}O-^{16}O isotopic exchange,

$$j_{ex}^{o} = \tfrac{1}{4}\, k_S\, c_O \tag{14.21}$$

where c_O is the volume concentration of oxygen anions at equilibrium. Equation (14.20) disregards nonlinear effects that may occur at high oxygen potential gradients and can be shown to be equivalent to the low field approximation of the Butler–Volmer equation. For a multistep reaction sequence, it follows[64]

$$j_{ex}^{o} = \frac{(\alpha_c + \alpha_a)}{4} \frac{i_o}{4F} \tag{14.22}$$

where i_o (A cm^{-2}) is the exchange current density, and α_c and α_a are the apparent cathodic and anodic transfer coefficients, respectively. Equation (14.22) can be used to calculate the value of j_{ex}^{o} from data of i-V measurements, provided that no activation occurs on the applied electrode materials. These should act only as a current collector. Exchange rates are customarily defined in terms of moles of anions or molecules per unit time and area. Often, the geometric area is used in calculation, in spite of the fact that the true surface area on a microscopic scale may be substantially larger.

For a membrane under mixed controlled kinetics it is appropriate to define a characteristic membrane thickness L_c, at which point the transition occurs from predominant control by diffusion to that by surface exchange.[43] The quantity L_c represents a simple yet valuable criterion for candidate membrane material selection. It should strictly be used when small P_{O_2} gradients appear across the membrane. In the next section, methods for measuring L_c are briefly discussed.

A starting point is to divide the membrane into a central bulk (Wagner) zone and adjacent interfacial zones, emphasizing the importance of both solid state diffusion and surface oxygen exchange to the magnitude of the oxygen flux. This is schematically shown in Figure 14.3. The available driving force $\Delta\mu_{O_2}^{total}$ is distributed across the membrane such that the rate-determining process receives the greater proportion,

$$\Delta\mu_{O_2}^{total} = \Delta\mu'_{O_2} + \Delta\mu_{O_2}^{bulk} + \Delta\mu''_{O_2} \tag{14.23}$$

where the single and double primes denote the high and low oxygen partial pressure sides, respectively. Assuming linear kinetics for diffusion and interfacial exchange, the flux balance is given by

$$j_{O_2} = -j_{ex}^{o\prime} \frac{\Delta\mu'_{O_2}}{RT} = -\frac{\overline{t_{el}\, t_{ion}\, \sigma_{total}}}{4^2 F^2} \frac{\Delta\mu_{O_2}^{bulk}}{L} = -j_{ex}^{o\prime\prime} \frac{\Delta\mu''_{O_2}}{RT} \tag{14.24}$$

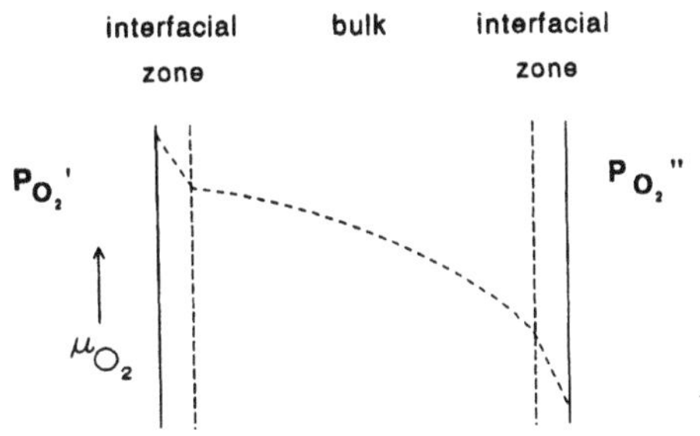

FIGURE 14.3. Drop-in chemical potential μ_{O_2} across bulk and interfacial zones of a membrane imposed to an oxygen partial pressure difference, $P_{O_2}' > P_{O_2}''$. The largest drop occurs across the least permeable zone.

where the Wagner equation is written in a form mathematically equivalent to Equation (14.10). For large thicknesses of the membrane bulk diffusion dominates, but as the thickness decreases the transfer across the interfaces becomes rate determining. In the limit of small P_{O_2} gradients, the flux equation can be written as[43]

$$j_{O_2} = -\frac{1}{1+(2L_c/L)} \frac{\overline{t_{el} t_{ion} \sigma_{total}}}{4^2 F^2} \frac{\Delta\mu_{O_2}^{total}}{L} \tag{14.25}$$

Surface exchange rates at opposite interfaces have been taken to be identical. The characteristic membrane thickness L_c is provided by

$$L_c = \frac{RT}{4^2 F^2} \times \frac{t_{el} t_{ion} \sigma_{total}}{j_{ex}^o} \tag{14.26}$$

where we have left away the averaging bar above $t_{el}t_{ion}\sigma_{total}$ and the prime notation for j_{ex}^o, whose significance has vanished in the limit of small P_{O_2} gradients. Comparing Equation (14.25) with the Wagner equation, we see that the diffusional flux of oxygen across the membrane is reduced by a factor $(1 + 2L_c/L)^{-1}$, relative to that in the absence of transfer limitations across the interfaces. In general, the surface exchange rate will be different in, e.g., O_2–N_2, CO-CO_2, H_2-H_2O atmospheres even at approximately equal oxygen partial pressures. If opposite sides of the membranes are exposed to different gas ambients, then the interface with the lesser performance dictates the overall exchange behavior.

The influence of the membrane thickness on oxygen flux is shown in Figure 14.4. When $L \gg L_c$, the oxygen flux varies inversely with L^γ where $\gamma = 1$, in agreement with Wagner's theory (cf. Equation [14.10]). Departures from this inverse relationship are observed when the oxygen flux becomes partly governed by the surface exchange kinetics. The value of γ, at a given L, corresponds with the negative slope in the double logarithmic plot of the oxygen flux vs. membrane thickness at that L. Taking the logarithm of Equation (14.25), partial differentiation with respect to log L shows that γ is equal to the reduction factor $(1 + 2L_c/L)^{-1}$, the value of which gradually decreases with decreasing thickness to become zero for $L \ll L_c$, as is shown in Figure 14.4. The latter situation corresponds with the maximum achievable flux, which for a symmetrical membrane is given by: $\frac{1}{2} j_{ex}^o \mu_{O_2}^{total}$.

Equation (14.26) may be simplified for predominant electronic conduction, assuming that the classical Nernst–Einstein relationship can be represented as

$$\sigma_{ion} = \frac{c_O D_s z_O^2 F^2}{RT} \tag{14.27}$$

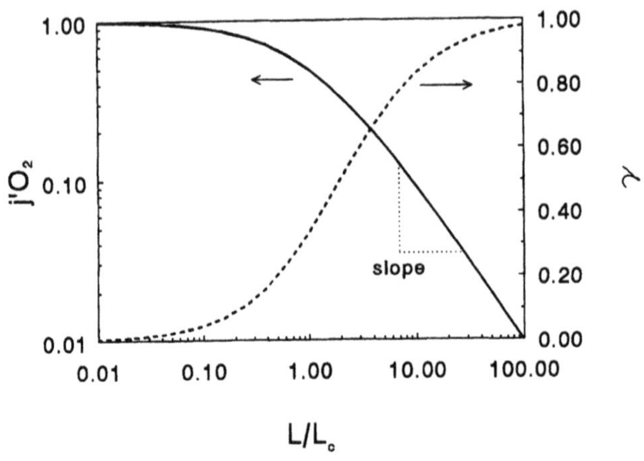

FIGURE 14.4. Thickness dependence of the *dimensionless* oxygen flux j'_{O_2} calculated using Equation 14.25. The quantity j'_{O_2} is defined by the ratio of the oxygen flux over the maximum achievable oxygen flux ($\frac{1}{2}j^{O}_{ex}.\Delta\mu^{total}_{O2}$) in the surface exchange limited regime. Only if $L \gg L_c$, the oxygen flux becomes proportional to $1/L^{\gamma}$ with $\gamma = 1$ in agreement with the Wagner theory. For smaller thicknesses, γ ranges between 1 and 0 (right-hand scale).

where D_s is the self-diffusion coefficient of oxygen anions with valence charge z_O (= –2). Making the appropriate substitutions, the characteristic thickness L_c becomes

$$L_c = \frac{D_s}{k_s} = \frac{D^*}{k_s} \qquad \left(t_{el} = 1\right) \tag{14.28}$$

In the second part of Equation (14.28) the fact has been used that, if correlation effects can be neglected, the tracer diffusion coefficient, D^*, is equal to the self-diffusion coefficient, D_s. It is important to note once again that Equation (14.28) is valid in the limit of small P_{O_2} gradients only. Since both D^* and k_s for a given material are a function of its specific defect chemistry, in general, L_c will be a function of process parameters P_{O_2} and temperature.

The picture that emerges is that, at given experimental conditions, for thicknesses below L_c no appreciable gain in the oxygen flux can be obtained by fabricating thinner membranes, unless the value of k_s can be significantly increased. Similar criteria can be formulated for fuel cell electrodes as has been advanced by Steele[65,66] and Kleitz et al.[67] In the limit of small overpotentials (low field approximation), i.e., assuming ohmic behavior for the relevant surface kinetics, whatever the rate-controlling mechanism, the electrode resistance can be correlated with the electrolyte resistivity.

2. Measuring L_c

The quantity D^*/k_s appears to be a fundamental parameter governing tracer diffusion bounded by a limiting surface exchange between lattice oxygen and oxygen from the $^{18}O_2$-enriched gas phase. Experimental methods for isotopic exchange include monitoring the change of $^{18}O_2$ concentration in the gas phase upon exchange in a fixed volume of $^{18}O_2$-enriched oxygen using a mass spectrometer,[68-72] weight measurements,[73] depth probing of $^{18}O/(^{18}O + ^{16}O)$ diffusion profiles using secondary ion mass spectroscopy (SIMS) after exchange at high temperature for a selected time,[74-76] and combined approaches.[77-79] Fitting the acquired data to the appropriate diffusion equation allows both D^* and k_s to be obtained from a single experiment.

Selected data of D^* and k_s from ^{18}O-^{16}O isotope exchange measurements of perovskite-type oxides are compiled in Table 14.2. High D^* and k_s values are reported for the ferrites and cobaltites, in which solids the anions are assumed to move via a vacancy diffusion mechanism. The mixed perovskites are notable for being excellent electronic conductors.

TABLE 14.2
Tracer-Diffusion Coefficient D*, Surface Exchange Coefficient k_s, and Characteristic Thickness L_c for Selected Perovskite-Type Oxides

Perovskite	T (°C)	P_{O_2} (kPa)	D* ($cm^2\ s^{-1}$)	k_s[b] ($cm\ s^{-1}$)	L_c[a] (cm)	Ref.
$La_{0.5}Sr_{0.5}MnO_{3-\delta}$	700	70	2×10^{-15}	1×10^{-8}	2×10^{-7}	[1]
	800		8×10^{-14}	1×10^{-7}	8×10^{-7}	
	900		3×10^{-12}	9×10^{-8}	3×10^{-5}	
$La_{0.8}Sr_{0.2}CoO_{3-\delta}$	700	70	1×10^{-8}	3×10^{-6}	3×10^{-3}	[1]
	800		2×10^{-8}	5×10^{-6}	4×10^{-3}	
	900		4×10^{-8}	2×10^{-5}	2×10^{-3}	
$La_{0.6}Ca_{0.4}Co_{0.8}Fe_{0.2}O_{3-\delta}$	700	70	2×10^{-8}	4×10^{-6}	5×10^{-3}	[1]
	800		1×10^{-7}	2×10^{-5}	5×10^{-3}	
	900		3×10^{-7}	4×10^{-5}	7×10^{-3}	
$La_{0.6}Sr_{0.4}Co_{0.8}Ni_{0.2}O_{3-\delta}$	700	70	3×10^{-8}	2×10^{-6}	2×10^{-2}	[2]
	800		1×10^{-7}	2×10^{-6}	5×10^{-2}	
	900		4×10^{-7}	2×10^{-6}	2×10^{-1}	
$La_{0.6}Sr_{0.4}Co_{0.6}Ni_{0.4}O_{3-\delta}$[c]	700	70	2×10^{-9}	7×10^{-7}	3×10^{-3}	[2]
	800		6×10^{-8}	3×10^{-6}	2×10^{-2}	
	900		3×10^{-7}	3×10^{-6}	1×10^{-1}	
$La_{0.6}Sr_{0.4}Co_{0.4}Ni_{0.6}O_{3-\delta}$	700	70	1×10^{-8}	3×10^{-7}	3×10^{-2}	[2]
	800		7×10^{-8}	2×10^{-6}	3×10^{-2}	
	900		6×10^{-7}	2×10^{-6}	3×10^{-1}	
$LaCoO_{3-\delta}$	700	4.5	9×10^{-13}	1×10^{-9}	4×10^{-4}	[3]
(single crystal)	800		2×10^{-11}	3×10^{-7}	7×10^{-5}	
	900		6×10^{-10}	1×10^{-6}	4×10^{-4}	
$LaFeO_{3-\delta}$	900	7	1×10^{-12}	4×10^{-8}	3×10^{-5}	[4]
(single crystal)	1000		5×10^{-12}	2×10^{-7}	3×10^{-5}	
$La_{0.9}Sr_{0.1}CoO_{3-\delta}$	900	4.5	3×10^{-9}	1×10^{-6}	2×10^{-3}	[5]
	1000		2×10^{-8}	2×10^{-6}	1×10^{-2}	
$La_{0.9}Sr_{0.1}FeO_{3-\delta}$	900	6.5	3×10^{-9}	5×10^{-7}	6×10^{-3}	[5]
	1000		1×10^{-8}	2×10^{-6}	6×10^{-3}	
$La_{0.6}Sr_{0.4}FeO_{3-\delta}$	1000	6.5	6×10^{-7}	1×10^{-5}	5×10^{-2}	[5]

[a] Calculated from Equation (14.25).
[b] The value can be equated to j^o_{ex}, in accordance with Equation (14.21), by multiplication with a factor 0.022. Strictly speaking, this holds for $LaCoO_3$.
[c] Authors report a two-phase mixture.[2]

References

[1] Carter, S., Selcuk, A., Chater, J., Kajda, R.J., Kilner, J.A., Steele, B.C.H., *Solid State Ionics,* 1992, *53–56,* 597–605.
[2] Ftikos, Ch., Carter, S., Steele, B.C.H., *J. Eur. Ceram. Soc.,* 1993, *12,* 79–86.
[3] Ishigaki, T., Yamauchi, S., Mizusaki, J., Fueki, K., Tamura, H., *J. Solid State Chem.,* 1984, *54,* 100–107.
[4] Ishigaki, T., Yamauchi, S., Mizusaki, J., Fueki, K., Tamura, H., *J. Solid State Chem.,* 1984, *55,* 50–53.
[5] Ishigaki, T., Yamauchi, S., Kishio, K., Mizusaki, J., Fueki, K., *J. Solid State Chem.,* 1988, *73,* 179–87.

Usually the electronic conduction is found to be predominant, in spite of the fact that the ionic conductivity may also be substantial (see Section V). The examples given in Table 14.2 clearly emphasize the importance of the surface exchange kinetics, relative to diffusion, in limiting overall oxygen transport through the perovskites.

Analyzing published data for D* and k_s, Kilner[76] noted that the two parameters seem to be correlated. A square root dependence of k_s with D* is found for perovskite oxides ABO_3, albeit that the data show substantial scatter, yielding an average value for D^*/k_s of about 100 μm. For fluorite oxides MO_2 the two parameters are correlated almost linearly, while the corresponding value of D^*/k_s ranges between millimeters and centimeters. The results suggest that related point defect processes must be common to both diffusion and surface oxygen exchange.[76] However, mechanisms responsible for the apparent correlations remain obscure, reflecting our poor knowledge at present of the factors that control the oxygen exchange kinetics.

As discussed in previous work by the present author,[43] calculations of L_c using combined data from ionic conductivity measurements and ^{18}O-^{16}O isotopic exchange for a number of fluorite oxides were found to be in good agreement with values estimated from oxygen permeability measurements.

Also, relaxation methods offer a useful tool for measuring L_c, without the requirement of high-temperature seals as in oxygen permeation experiments. The (re-)equilibration follows after an instantaneous change of the oxygen activity in the gas phase and involves the propagation of a composition gradient through a thin slab or single crystal of the oxide. The change in stoichiometry brings about a change in weight, or in electrical conductivity, which can be monitored experimentally as a function of time. It is not possible to give a full account of these techniques in this chapter, and the reader is referred to, for example, References 80, 81, and 82.

Relaxation methods are used to determine the chemical diffusion coefficient, i.e., to simplify the calculations it is commonly assumed that the surface of the sample equilibrates immediately with the newly imposed atmosphere. The latter assumption leads to standard analytical solutions of Fick's second law under the special initial and boundary conditions applicable to the experiment. However, for oxides like, for example, $Fe_{1-x}O$[83,84] and $Mn_{1-x}O$,[85] the surface reaction exerts a clear influence on the overall equilibration kinetics.

Nowotny and Wagner[86] reexamined a large number of kinetic studies reported in the literature, arriving at the conclusion that in many cases the data from relaxation measurement studies actually appear to exhibit mixed control, i.e., the overall kinetics is determined both by surface exchange and by bulk diffusion. The authors proposed a model in which the transport of lattice defects toward the surface, and vice versa, is affected by the electrical barrier generated across the junction between the crystalline bulk and a quasi-isolated segregated surface layer. (See also Chapter 4 of this handbook).

A simple expression for the boundary condition (at x = 0) during the equilibration of the oxygen concentration $c_O(t)$ from the initial value c_O^i to the final value c_O^f often used in relaxation experiments is

$$K_s\left(c_O(t)-c_O^f\right)=-\tilde{D}\frac{\partial c_O}{\partial x}\bigg|_{x=0} \tag{14.29}$$

where K_s is the appropriate surface rate constant. The values of $\tilde{D}$ and K_s obtained from the experiment in this way are averaged over the applied composition interval. From a mathematical point of view, an expression similar to that in Equation (14.29) is used as a boundary condition in modeling data from isotopic exchange, as discussed, for example, by Kilner,[76] replacing the parameters and variables in Equation (14.29) by those relevant in a tracer diffusion experiment. The advantage of a linear rate law for the surface reaction, as given in the left part of Equation (14.29), is that the problem can still be solved analytically. The full solution has been presented elsewhere,[80,87] from which it follows that

$$L_c=\frac{\tilde{D}}{K_s} \tag{14.30}$$

Neglecting correlation effects (Haven ratio $H_R = 1$), Equation (14.30) simplifies to Equation (14.28) for a predominantly electronic conducting oxide ($t_{el} = 1$). Note,

$$K_s = \frac{k_s}{RT} \frac{\frac{1}{2}\partial \mu O_2}{\partial \ln c_O} \tag{14.31}$$

The constant K_s should thus not be confused with the surface exchange coefficient k_s defined by Equation (14.21). Both parameters have the same dimension (cm s^{-1}). A general method of regression analysis of data from relaxation experiments using the linear rate law for the surface reaction has been given.[88] Other empirical rate laws for the surface exchange reaction have been proposed by Dovi et al.[89] and Gesmundo et al.[90]

3. Effect of Surface Roughness and Porosity

The parameter L_c does not represent an intrinsic material property, but may depend through the value of k_s on the roughness or porosity of the membrane surface. This has recently been exploited by Thorogood et al.[44] and Deng et al.,[45] showing that the oxygen flux can be significantly improved if the thin dense membrane is coated on either one or both surfaces with a porous layer. Based on a simple effective medium model, and linearized transport equations, Deng et al.[45] calculated the oxygen flux through the modified membrane whose dense layer thickness is assumed to be small enough so that the drop in chemical potential across it can be neglected. The rate-limiting step is thus ionic diffusion and surface exchange in the porous solid. The latter is modeled by a simple cubic array of consolidated spherical grains. In the limit of small P_{O_2} gradients, the maximum enhancement in the oxygen flux through the symmetric membrane, over the noncoated membrane, is given by

$$\xi = \sqrt{L_c S(1-\theta)/\tau_s} + \theta \tag{14.32}$$

where S is the pore wall surface area per unit volume, θ the porosity, and τ_s the tortuosity of the solid phase in the porous structure. Maximum enhancement is achieved for a membrane whose porous layer thickness $L \gg L_p$ where

$$L_p = \sqrt{L_c(1-\theta)/S\tau_s} \tag{14.33}$$

and refers to the active part of the porous layer. To achieve near full enhancement $L = 3L_p$ suffices. With $L_c = 100$ μm and surface area $S = 10^{-6}$ cm^{-1}, the authors calculated an enhancement in oxygen flux, over the noncoated membrane, of almost 2 orders of magnitude. In a separate paper, Deng et al.[46] showed that the factor ξ is substantially reduced when the rate-limiting step is the transport of gas molecules in the pores. Finally, we note the potential enhancement in oxygen fluxes, at thicknesses below L_c, by coating the dense membrane with an exchange active layer.

IV. SOLID OXIDE ELECTROLYTES

A. INTRODUCTION

A key factor in the possible application of oxygen ion-conducting ceramics is that, for use as a solid electrolyte in fuel cells, batteries, oxygen pumps, or sensors, their electronic transport number should be as low as possible. Given that the mobilities of electronic defects typically are a factor of 1000 larger than those of ionic defects, a band gap of at least 3 eV is required to minimize electronic contributions arising from the intrinsic generation of electrons and holes.

Useful solid oxide electrolytes to date are those with a fluorite or fluorite-related structure, especially ones based on ZrO_2, ThO_2, CeO_2 and Bi_2O_3.[91] Mixed conduction occurs only at sufficiently low or high values of P_{O_2}, where electronic defects are generated for charge compensation of the excess of ionic defects relative to the stoichiometric composition. This latter mechanism proceeds only when the oxygen defects — vacancies or interstitials — introduced by equilibration of the oxide with the gas phase have relatively low ionization energies and thus ionize under the selected conditions. Taking into account the mobilities of ionic and electronic defects, the range in nonstoichiometry can be correlated with the width of the electrolytic domain. For the stabilized zirconias the domain width, at 1000°C, typically extends to values below $P_{O_2} = 10^{-30}$ atm.[92] On the other hand, the domain width of ceria electrolytes is limited, reflecting the ease of reduction of Ce^{4+} to Ce^{3+} relative to that of the transition metal ions in stabilized zirconia. For example, at the same temperature $(CeO_2)_{0.95}$-$(Y_2O_3)_{0.05}$ has its domain boundary at about $P_{O_2} = 10^{-10}$ atm, the value of which has been taken for $t_{ion} = 0.5$.[93]

As is clear, solid oxide electrolytes are not useful for applications as in an oxygen-separation membrane, unless operated with external circuitry (oxygen pump) or as a constituent phase of a dual-phase membrane. Both modes of operation, classified in this paper as electrochemical oxygen separation, are briefly discussed in Section IV.C. But we first start with a discussion of the models that have been developed to describe the oxygen semipermeability of solid oxide electrolytes, originating from the residual electronic conductivity in the electrolytic domain. These models can be translated easily to model oxygen permeation through mixed conductors. Examples are drawn from experimental studies on calcia-stabilized zirconia (CSZ) and erbia-stabilized bismuth oxide, clearly emphasizing the importance of both bulk diffusion and surface exchange in determining the rate of oxygen permeation through these solids.

B. OXYGEN SEMIPERMEABILITY OF OXIDE ELECTROLYTES

1. Diffusion of Electronic Charge Carriers

In the absence of interactions between defects, the following reactions determine the defect concentrations in oxides with the fluorite structure,

$$O_{2,g} + 2\,V_O^{\bullet\bullet} \rightleftarrows 2\,O_O^{\times} + 4h^{\bullet} \tag{14.34}$$

$$O_O^{\times} \rightleftarrows O_i'' + V_O^{\bullet\bullet} \tag{14.35}$$

$$nil \rightleftarrows e' + h^{\bullet} \tag{14.36}$$

with the respective equilibrium constants,

$$K_g = \frac{p^4\left[O_O^{\times}\right]^2}{P_{O_2}\left[V_O^{\bullet\bullet}\right]^2} \tag{14.37}$$

$$K_F = \left[O_i''\right]\left[V_O^{\bullet\bullet}\right] \tag{14.38}$$

$$K_e = n\,p \tag{14.39}$$

Here, n and p denote the concentrations, expressed as mole fractions, of electrons and electron holes, respectively. The anti-Frenkel defects O_i'' and $V_O^{\bullet\bullet}$ are assumed to be fully ionized, which is usually observed at elevated temperatures. While phases derived from δ-Bi_2O_3 show substantial disorder in the oxygen anion sublattice, reaching a maximum value at the disordering temperature, doping with aliovalent impurities is essential to achieve high ionic carrier concentrations in oxides such as ThO_2, HfO_2, and ZrO_2.[94,95] Doping sometimes serves to stabilize the cubic fluorite structure down to working temperatures, e.g., for HfO_2 and ZrO_2. In the following, we use notations D and A for dopant and acceptor cations, respectively.

The electronic conductivity, comprising p- and n-type contributions, is obtained by multiplying each of the concentrations with their respective charge and mobility. Upon substitution of the defect concentrations established by the above equilibria in the electroneutrality relation,

$$2\left[V_O^{\bullet\bullet}\right]+p+\left[D^{\bullet}\right]=2\left[O_i''\right]+n+\left[A'\right] \tag{14.40}$$

the following expression for the partial electronic conductivity can be derived,

$$\begin{aligned}\sigma_{el} &= pFu_h + nFu_e \\ &= \sigma_p^o P_{O_2}^{\frac{1}{4}} + \sigma_n^o P_{O_2}^{-\frac{1}{4}}\end{aligned} \tag{14.41}$$

where u_h, u_e are the mobilities of electron holes and electrons, respectively, and σ_p^o, σ_n^o the corresponding partial conductivities extrapolated to unit oxygen partial pressure. In deriving Equation (14.41), the concentration of oxygen vacancies $[V_O^{\bullet\bullet}]$ in the electrolytic domain is taken to be fixed, either by the Frenkel equilibrium given in Equation (14.35) or by the net acceptor dopant concentration: $[A'] - [D^{\bullet}]$. Substitution of Equation (14.41) into Wagner's equation (Equation [14.10]) yields, for the oxygen flux, after integration

$$j_{O_2} = \frac{RT}{4F^2L}\left[\sigma_p^o\left(P_{O_2}'^{\frac{1}{4}} - P_{O_2}''^{\frac{1}{4}}\right) - \sigma_n^o\left(P_{O_2}'^{-\frac{1}{4}} - P_{O_2}''^{-\frac{1}{4}}\right)\right] \tag{14.42}$$

noting that the ionic transference number, t_{ion}, in the electrolytic domain has been set to unity. If the experimental conditions are chosen such that the oxygen pressure at the permeate side is not too low, i.e., neglecting the n-type contribution to the electronic conductivity, Equation (14.42) reduces to (cf. Equation [14.18]),

$$j_{O_2} = \frac{\beta}{L}\left[P_{O_2}'^{\frac{1}{4}} - P_{O_2}''^{\frac{1}{4}}\right] \tag{14.43}$$

where $\beta = \sigma_p^o RT/4F^2$.

Because of its great technological importance, much literature on the "detrimental" oxygen semipermeability flux occurring in solid electrolytes is available,[96-109] of which a great number deals with the stabilized zirconias, but also includes electrolytes based on ThO_2,[108,109] HfO_2,[98] and Bi_2O_3.[107] Results on the stabilized zirconias up to 1976 have been reviewed by Fouletier et al.[110] For relatively thick electrolyte membranes the results are consistent with Equation (14.43). But, often a value between ¼ and ½ is found for the exponent.[102-107] Sometimes this has been taken as evidence for electronic trapping, i.e., a different mechanism for the incorporation of oxygen in the oxide lattice.[102,103] Dou et al.[106] were the first to invoke a surface reaction to reconcile the apparent conflict with a diffusion-controlled mechanism, i.e., the overall kinetics is determined both by surface reactions and by bulk diffusion.

2. Modeling Equations

In this section, a model is presented for solid oxide electrolytes based upon two consecutive steps for oxygen permeation: one for the surface exchange process at the oxide surface on both sides of the membrane, and another for the joint diffusion of oxygen ions and electron holes through the solid.

Considering oxygen exchange between the gas phase and the oxide surface via the reaction given by Reaction (13.34), one may distinguish many steps, like adsorption, dissociation, surface diffusion, charge transfer, and incorporation in the (near) surface layer, and reversed steps, each of these steps may impede interfacial transfer of oxygen. Following Dou et al.,[106] the surface reaction may be represented by a simple two-step scheme:

$$\begin{aligned} O_{2,g} &\rightleftarrows 2\,O_{ads} \\ O_{ads} + V_O^{\bullet\bullet} &\rightleftarrows O_O^{\times} + 2\,h^{\bullet} \end{aligned} \tag{14.44}$$

If it were supposed, for example, that the first of these reactions is at equilibrium and the second is rate determining, the net flux of molecular oxygen through the interface at the high-pressure side can be described by

$$j_{O_2} = \tfrac{1}{2}\, k_2 K_1^{\frac{1}{2}} \left[V_O^{\bullet\bullet}\right] P_{O_2}^{\frac{1}{2}} - \tfrac{1}{2} k_{-2} \left[O_O^{\times}\right] p^2 \tag{14.45}$$

where k_2 and k_{-2} are the forward and backward rate constant for the ionization and incorporation reaction (2nd step in [14.44]), K_1 is the equilibrium constant for the adsorption reaction (1st step in [14.44]). The reverse equation holds for the rate of the surface reaction at the opposite side of the membrane (permeate side). As mentioned above, in the electrolytic domain the concentration of oxygen vacancies, $[V_O^{\bullet\bullet}]$ (and that of lattice oxygen, $[O_O^{\times}]$) is constant. Accordingly, by applying the law of mass action for the overall exchange reaction given by Equation (14.37) the actual concentration of electron holes p at either side of the membrane can be equated to a virtual oxygen partial pressure, i.e, the oxygen partial pressure if equilibrium were established with the gas phase. Considering both surface reactions and bulk diffusion, we arrive at the following set of equations for the oxygen flux:

$$\begin{aligned} j_{O_2} &= \alpha\left[P_{O_2}'^{1/2} - P_{O_2(I)}^{1/2}\right] \\ j_{O_2} &= \frac{\beta}{L}\left[P_{O_2(I)}^{1/4} - P_{O_2(II)}^{1/4}\right] \\ j_{O_2} &= \alpha\left[P_{O_2(II)}^{1/2} - P_{O_2}''^{1/2}\right] \end{aligned} \tag{14.46}$$

where $\alpha = \tfrac{1}{2} k_2 K_1 [V_O^{\bullet\bullet}]$ and $P_{O_2(I)}$, $P_{O_2(II)}$ indicate the virtual oxygen pressures at the feed and permeate side, respectively. The usual assumption of fast equilibration of the oxide surface with the imposed gas atmosphere would imply that $P_{O_2}' = P_{O_2(I)}$ and $P_{O_2}'' = P_{O_2(II)}$. Finally, it may be noted that the set of equations given in Equation (14.46) can only be solved numerically.

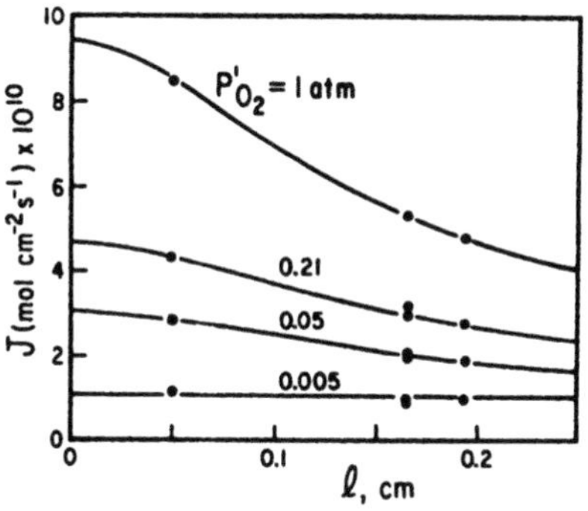

FIGURE 14.5. The effect of sample thickness on oxygen permeation through calcia-stabilized zirconia at 1230°C. Solid lines are theoretical results calculated using Equation (14.47). (Reprinted from Don, S., Masson, C.R. and Pacey, P.D., *J. Electrochem. Soc.*, 1985, *132*, 1843–49. With permission.)

3. Examples

a. Calcia-Stabilized Zirconia

Dou et al.[106] studied the isothermal oxygen permeation through CSZ tubes at 960 to 1450°C and oxygen pressures P'_{O_2} of 10^{-3} to 1 atm. The oxygen which permeated into the interior of tubes with known wall thicknesses was immediately pumped away by a diffusion pump and measured with a gas burette ($10^{-6} \le P''_{O_2} \le 10^{-4}$ atm). The data obtained qualitatively agreed with an oxygen pressure dependence in accordance with Equation (14.18), in which the value of the exponent n is allowed to vary between ¼ to ½. Figure 14.5 shows the measured effect of sample thickness on oxygen flux. Assuming $P''_{O_2} \approx 0$ atm, and with neglect of rate limitations at the low pressure interface, Dou et al.[106] arrived at the following expression for the oxygen flux,

$$j_{O_2} = \frac{\beta^2}{2\alpha L^2}\left[\left(1+\frac{4\alpha^2 L^2}{\beta^2} P_{O_2}'^{\,1/2}\right)^{1/2} - 1\right] \tag{14.47}$$

which equation matched the experimental results well. The parameters α and β showed similar activation energies: 191 ± 5 kJ mol^{-1} and 206 ± 11 kJ mol^{-1}, respectively. The quantity $\beta/\alpha(P'_{O_2})^{1/4}$ has the unit of length. The meaning of it is more or less similar to that of the parameter L_c defined earlier in Section III.B.1. At 1230°C and oxygen pressure P'_{O_2} of 1 atm, the transition from predominant control by bulk diffusion to that by surface exchange would occur at a sample thickness of about 2.7×10^{-2} cm.

Dou et al.[106] performed their experiments on CSZ tubes with a homogeneous composition, having 10% pores by volume. The authors estimated that the true surface area would be about 10% greater than the geometrical one used in calculation, and the surface exchange parameter α needed accordingly to be reduced for an ideal surface without pores. For the bulk transport of oxygen, the effect of nonconnected micropores would be either to shortcut the solid state diffusion, in the case of fast surface exchange kinetics, or to enlarge the diffusion path to an extent which is a function of the pore size. For fully dense ceramics, the bulk diffusion parameter β was expected to be about 14% greater. Besides the possibility of fitting the experimental data to Equation (14.47), the authors demonstrated that a more complicated mechanism for the surface exchange reaction may be invoked.

Interestingly, steady-state oxygen permeation measurements by Dou et al.[106] provided no evidence of a surface rate limitation for CSZ tubes containing a segregated impurity phase. The order with respect to oxygen remained close to ¼. This second phase consisted of a mixture of metal silicates with a composition similar to that of a SiO_2-CaO-Al_2O_3 eutectic. It was suggested that surface oxygen exchange on this second phase would be very rapid. In subsequent studies, the same authors used the "time-lag" method (in which the transient process toward steady-state oxygen permeation is monitored) and a desorption technique to study chemical diffusion in CSZ with different impurities.[111-113] The observed kinetics scaled with the presence of Fe_2O_3, in such a way that faster equilibration rates were observed for samples containing a smaller impurity content. The results were taken to be consistent with a mechanism in which electron holes are trapped at iron impurity sites.

b. Erbia-Stabilized Bismuth Oxide

Indications of a limitation by a surface process have also been reported for the oxygen flux through sintered dense ceramics of bismuth oxide stabilized with 25 mol% erbia, $(Bi_2O_3)_{0.75}$-$(Er_2O_3)_{0.25}$ (BE25).[107] As is known, this material exhibits high ionic conductivity at a temperature distinctly lower than for the stabilized zirconias.[114] Using glass-sealed discs, the amount of oxygen which permeated into a closed reservoir, being flushed with helium gas prior to measurement, was monitored as a function of time. The oxygen flux was calculated from the corresponding slope, which was found invariant as long as $P_{O_2}' \gg P_{O_2}''$.

In the range of temperatures (610 to 810°C) and oxygen pressures (10^{-4} to 1 atm) covered by experiment, the concentration of minority charge carriers, i.e., electron holes, in BE25 is proportional to $P_{O_2}^n$ with n = ¼. However, the apparent value derived from experiment increases gradually from ¼ to higher values upon decreasing specimen thickness from 0.285 cm to 200 μm, indicating permeation to be limited by two or more processes differing in order. The activation energy of the oxygen flux was found to increase too in the same direction. The observed behavior can be attributed to the change over from diffusion to surface control upon decreasing sample thickness. The experimental data can be fitted well by means of Equation (14.46), though it is necessary to adapt the kinetic order of the surface reaction with respect to oxygen to a value of ⅝. The parameters α and β obtained from numerical fitting appear to exhibit different activation energies; 136 ± 4 kJ mol^{-1} and 99 ± 4 kJ mol^{-1}, respectively, which indicates that the surface process is less limiting at higher temperatures. Isotopic exchange measurements on sintered dense discs of BE25 showed a $P_{O_2}^m$ dependence with m = 0.60 at 550°C and m = 0.54 at 700°C for the overall surface oxygen exchange rate.[70,115] Figure 14.6 shows that the value for the surface oxygen exchange rate j_{ex}^o (= α $P_{O_2}^m$), normalized to air, obtained from the fit of the data agrees with that measured by isotopic exchange. The thickness, at which point the oxygen flux is half of that expected under conditions of pure diffusion-controlled kinetics, imposing opposite sides of discs to pure oxygen and helium gas, was calculated at 0.16 cm at 650°C and 0.09 cm at 800°C. These values were found to be in good agreement with estimates of the parameter L_c as noted before in Section III.B.2.

To provide a kinetic basis for the surface process, Lin et al.[116] more recently proposed that, in addition to the bulk transport of electron holes, the oxygen fluxes through BE25 are governed by two or more sequential steps in the oxygen exchange reaction. Adopting the standard two-step scheme of Equation (14.44) for the surface reaction, the authors claimed that an improved fit of the published data of oxygen permeation to this scheme is obtained if both surface reaction steps are assumed to be rate determining.

The occurrence of two basic steps in surface exchange kinetics on BE25 has been discerned recently in a detailed study of the ^{18}O-^{16}O isotopic exchange reaction using powder samples.[117,118] The changes in the concentration of oxygen gas-phase species with mass 36, 34, and 32 ($^{18}O_2$, ^{18}O-^{16}O, and $^{16}O_2$, respectively) upon isotopic exchange with the oxide were monitored as a function of time. The results fit the reaction scheme,*

* It should be noted that the charge of any intermediate species occurring in the oxygen exchange reaction is irrelevant in treating the data from isotopic exchange.

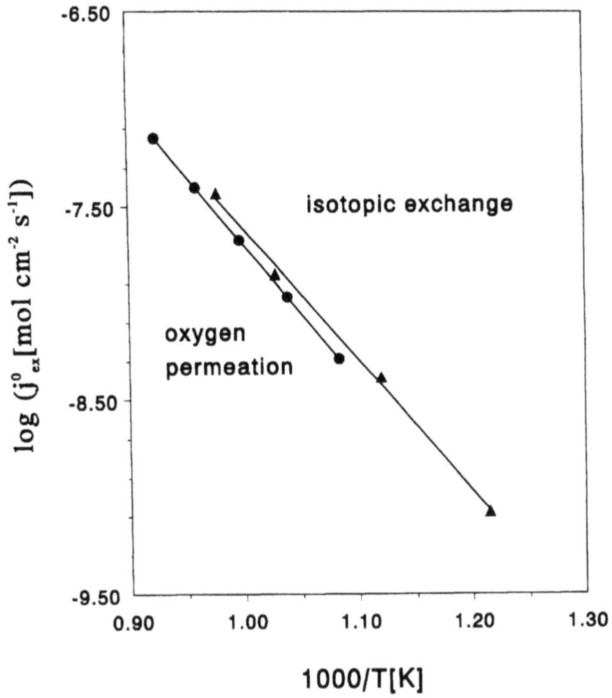

FIGURE 14.6. Data for the surface oxygen exchange rate, normalized to air, of 25 mol% erbia-stabilized bismuth oxide (BE25) from (a) isotopic exchange and (b) oxygen permeation measurements. (Reprinted from Bouwmeester, H.J.M., Kruidhof, H., Burggraaf, A.J., and Gellings, P.J., *Solid State Ionics*, 1992, *53–56*, 460–468. With permission.)

$$O_{2_{(gasphase)}} \underset{}{\overset{r_{diss}}{\rightleftharpoons}} 2\,O_{(adsorbed\ layer)} \overset{r_{inc}}{\rightleftharpoons} 2\,O_{(solid)} \tag{14.48}$$

which at first glance agrees with the two-step mechanism of Equation (14.44). However, P_{O_2}-dependent measurements in the range 10^{-2} to 1 atm at 600°C showed that $r_{diss} \propto P_{O_2}^{1/2}$ and $r_{inc} \propto P_{O_2}$ (see Figure 14.7), which contradicts the one expected from Scheme (14.44). The overall reaction given by Equation (14.34) can be broken down into multiple steps, and several species, e.g., $O^-_{2,\,ads}$, $O^{2}_{2,\,ads}$, O^-_{ads}, can occur as intermediates for the reduction of molecular oxygen. This procedure is commonly used to explain experimentally observed Tafel slopes in studies of oxygen electrode kinetics. However, no conventional mechanism so far conceived fits the observed P_{O_2} dependencies from isotopic exchange, and other factors thus play a role. Several alternative reaction schemes were considered,[119] and the one which is most feasible is as follows:

$$\begin{aligned} &1. \quad O_{2,g} + e' \rightleftarrows O^-_{2,ads} \\ &2. \quad O^-_{2,ads} + e' \rightleftarrows 2\,O^-_{ads} \\ &3. \quad O^-_{ads} + V_O^{\bullet\bullet} + O^-_{2,ads} \rightleftarrows O_O^{\times} + O_{2,g} \end{aligned} \tag{14.49}$$

The observations from isotopic exchange can be accounted for if there is low coverage by oxygen species, and that steps 2 and 3 are rate determining. The crucial assumption made in the proposed mechanism is in step 3, where the superoxide ion $O^-_{2,ads}$ transfers its electron to O^-_{ads} species, that is, an autocatalytic step in the surface exchange reaction.

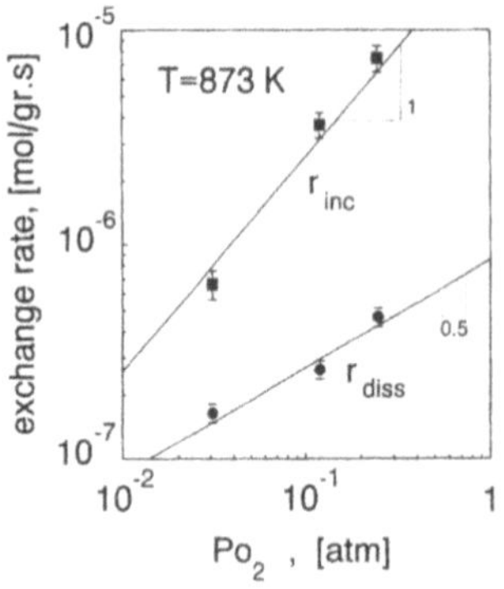

FIGURE 14.7. Oxygen pressure dependencies of the two basic steps discerned in oxygen isotopic exchange at 600°C on a powder of 25 mol% erbia-stabilized bismuth oxide (BE25). (Reprinted from Boukamp, B.A., Bouwmeester, H.J.M., and Burggraaf, A.J., in *Proc. 2nd Int. Symp. Ionic and Mixed Conducting Oxide Ceramics*, Vol. 94-12; Ramanarayanan, T.A., Worrel, W.L., and Tuller, H.L., Eds.; The Electrochemical Society, Pennington, NJ, 1994, 141–150. With permission.)

Finally, we briefly describe the observations recently made in the present author's laboratory in an attempt to increase the oxygen permeation flux through stabilized bismuth oxide by substitution of the δ-Bi_2O_3 host with 40 mol% terbium on the bismuth sites (BT40). Measurements using the concentration cell method and ac impedance confirmed that BT40 exhibits good p-type conductivity and is an excellent mixed conductor with ionic transference numbers, $t_{ion} = 0.74$ at 650°C and $t_{ion} = 0.85$ at 800°C in air.[120] Using ambient air as the source of oxygen and helium as the sweep gas on the other side of dense BT40 disc membranes, in the range of thickness 0.07 to 0.17 cm and temperature 600 to 800°C, did not yield the expected increase in the oxygen flux over BE25.[121] Isotopic exchange measurements in the relevant range of oxygen partial pressure and temperature showed that both oxides exhibit an almost equal activity in oxygen exchange,[115] which is in support of the conclusion made from oxygen permeation measurements that the oxygen fluxes through BT40, at the conditions covered by the experiments, are limited by the surface exchange kinetics.

4. emf Measurements

The existence of a nonvanishing electronic conductivity and concomitant oxygen semipermeability flux in solid electrolytes leads to errors in emf measurements using high-temperature oxygen gauges.[29,30] In accordance with Wagner's theory, the open-cell emf is reduced by a factor $(1 - \bar{t}_{el})$, where $\bar{t}_{el}$ is defined as a mean electronic transference number. However, this appears to introduce only a minor error as long as measurements are performed within the ionic domain of oxide electrolytes. More serious errors are encountered in measuring the P_{O_2} of unbuffered gas mixtures, e.g., Ar-O_2 at oxygen partial pressures below 10^{-4} atm and CO-CO_2 gas mixtures having small concentrations of CO.

Also, the oxygen flux can disturb the equilibrium between the electrode microsystem and the surrounding gas. The experimental arrangement used by Fouletier et al.[110] to quantify this error is schematically shown in Figure 14.8. While the Pt point electrode probes the local oxygen activity in the adsorbed layer, the use of the zirconia point electrode eliminates the error introduced by oxygen semipermeability of the electrolyte by dissipating the oxygen flux before reaching the back-side terminal lead. In this way the overpotential error in the emf measurements introduced by the limited exchange rate at the permeate side becomes

$$\Delta E = \frac{RT}{4F} \ln \frac{P_2'}{P_2} \tag{14.50}$$

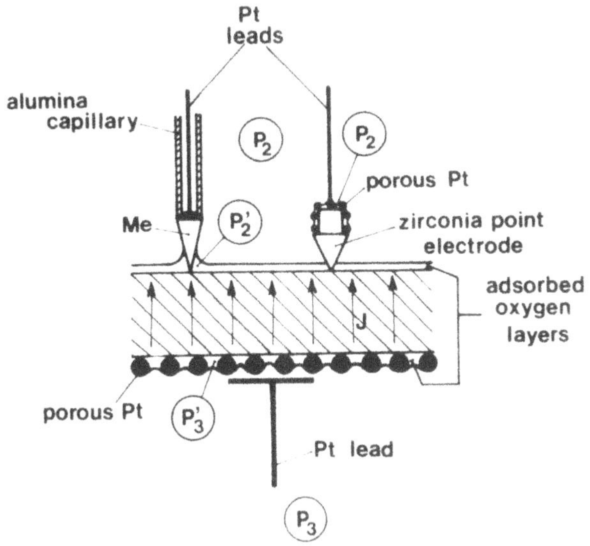

FIGURE 14.8. Schematic diagram of the experimental arrangement used to analyze errors in emf measurements by oxygen semipermeability of the electrolyte in a zirconia-based sensor by Fouletier et al.[10] Oxygen partial pressures are indicated by P_1, P_2, etc. (Reprinted from Fouletier, J., Fabry, P., and Kleitz, M., *J. Electrochem. Soc.*, 1976, *123(2)*, 204. With permission.)

where the oxygen pressures P_2' and P_2 are defined as indicated in Figure 14.8. A similar expression can be derived for the error occurring at the feed-side surface.

Oxygen gauges were placed by Fouletier et al.[110] in the argon gas permeate stream. The oxygen semipermeability flux through the 9 mol% YSZ pellet was calculated from the P_{O_2} difference measured upstream P_1 and downstream P_2: $\Delta P = P_1 - P_2$. The observed fluxes in the range 950 to 1650°C could be accounted for if it was assumed that diffusional transport of electron-holes through the electrolyte is rate determining, but only if the authors took into consideration the oxygen activities in the adsorbed layers, i.e.,

$$j_{O_2} = \frac{\beta}{L}\left[P_3'^{1/4} - P_2'^{1/4}\right] \tag{14.51}$$

which equation is in accord with Equation (14.43). Previous work by the same authors showed that the following expression holds for the overpotential η at the stabilized zirconia/Pt interface,

$$\eta = \frac{RT}{2F}\ln\frac{i_l - i}{i_l} \tag{14.52}$$

where the limiting current i_l was found to vary proportional to $P_{O_2}^{1/2}$ at intermediate oxygen partial pressures ($P_{O_2} > 10^{-4}$ atm) and to $P_{O_2}^{1/3}$ at low values ($P_{O_2} < 10^{-4}$ atm). Based on the similarity between the observed error ΔE and the measured overpotential η resulting from i-V measurements, Fouletier et al.[110] showed that the data of oxygen semipermeability could be fitted to

$$j_{O_2} = \frac{i_l}{4F}\left(1 - \exp\frac{2F\Delta E}{RT}\right) \tag{14.53}$$

which equation is obtained upon transformation of Equation (14.52), substituting the effective current j_{O_2}. 4F of electron holes inside the oxide for the current i. The experimental results were found to be in fair agreement with the $P_{O_2}^{1/2}$ and $P_{O_2}^{3/4}$ laws in the intermediate- and low-oxygen pressure range, the observed value for the exponent ranging from 0.5 to 0.75.

The significance of the work performed by Fouletier et al.[110] is that emf measurements may provide a useful tool to get information on the relevant kinetics of oxygen-permeable membranes. Assuming an error in emf due to concentration overpotential, substitution of Equation (14.50) for the error ΔE in Equation (14.53) leads to

$$j_{O_2} = \frac{i_l}{4F}\left[1-\left(\frac{P_2'}{P_2}\right)^{\frac{1}{2}}\right] \qquad (14.54)$$

which, for $i_l \propto P_2^{1/2}$, is mathematically equivalent to the corresponding surface rate equation given in Equation (14.46). The oxygen activity P_2' in the adsorbed layer may be identified as the virtual oxygen partial pressure $P_{O_2(II)}$. The values obtained by Fouletier et al.[110] for the activation energies associated with surface exchange and bulk transport of electron holes are 195 kJ mol^{-1} and 183 kJ mol^{-1}, respectively, which are close to the corresponding values for CSZ reported by Dou et al.[106] (see Section IV.B.3). For the large-surface area electrode (see Figure 14.8), an activation energy of the parameter i_l close to the value of 110 kJ mol^{-1} from i-V measurements was obtained. Though at the time the authors did not have a clear explanation for the different behavior of this electrode, it is now known that the interfacial kinetics of stabilized zirconia is significantly altered by coating with a porous Pt-layer. For example, isotopic exchange measurements showed an enhancement of the surface exchange rate of 2 to 3 orders of magnitude, over the case of stabilized zirconia without a porous Pt layer.[72] The observations made by Fouletier et al.,[110] among those by others, recently led to the formulation of a *Reaction Pathway Model* for the oxygen evolution reaction on solid oxide ion conductors as discussed by Kleitz et al.,[67,122] which papers contain many details from numerous electrochemical studies on the oxygen electrode kinetics.

C. ELECTROCHEMICAL OXYGEN SEPARATION

1. Oxygen Pump

The open-cell emf generated across an oxygen concentration cell such as

$$O_2\left(P_{O_2}'\right), Pt\left|CSZ\right|Pt, O_2\left(P_{O_2}''\right) \qquad (14.55)$$

with each side maintained at a different oxygen partial pressure P_{O_2}' and P_{O_2}'' is given by

$$E_{eq} = \left(1-\bar{t}_{el}\right)\frac{RT}{4F}\ln\frac{P_{O_2}'}{P_{O_2}''} \qquad (14.56)$$

In the absence of any electronic conduction, i.e., when $\bar{t}_{el} = 0$, Equation (14.56) simplifies to the Nernst equation. When the cell arrangement delivers a current i under load conditions, the cell voltage drops below the value E_{eq}, due to Ohmic losses iR_i (R_i = electrolyte resistance) and polarization losses at both Pt electrodes. As an approximation,

$$E = E_{eq} - iR_i - \eta \qquad (14.57)$$

where η represents the total cathodic and anodic polarization loss. Upon short circuiting both Pt electrodes, the emf of the cell drops to zero while oxygen is transported from the high-pressure side P_{O_2}' to the low-pressure side P_{O_2}''. By applying an external power source, the applied dc voltage can be used to enhance the magnitude of the current, but also to reverse its sign. That is, oxygen may be *pumped* in both directions; the rate of transport equals i/4F according to Faraday's law. This is the basic principle of electrochemical oxygen separation.

An important phase during device development is optimization of the pumping rate, i.e., ohmic and polarization losses must be kept as low as possible. Much effort has been concentrated on development, fabrication, and testing of zirconia-based separators. For example, Clark et al.[123] have described the performance of a multistack YSZ-based separator. Each cell contained a 125-μm-thick YSZ layer of diameter 6.35 cm, whereas porous strontium-doped lanthanum manganite electrodes were used to eliminate the need for costly Pt. The largest of these separators, built with 20 cells, was found to be capable of an oxygen flux up to 1 l min^{-1} at an operating temperature of 1000°C. Factors influencing the efficiency of the oxygen separation process and systems analysis of conceptual oxygen production plants are also addressed.

A major drawback of ZrO_2-based materials is the high temperature required for operation, typically 900 to 1000°C, expressing the need for development of oxide electrolytes which exhibit significant levels of ionic conduction at modest temperatures. Several alternative materials may be considered. To provide a reference point for discussion, the ionic conductivity of YSZ is about 0.1 S cm^{-1} at 950°C. This value is found in bismuth oxide stabilized with dopants such as Er_2O_3 and Y_2O_3 and in cerium oxide doped with Gd_2O_3, Sm_2O_3, or Y_2O_3 already in the range 650 to 700°C,[65,66] which electrolytes are less useful in, for example, fuel cells or sensor applications due to the presence of rather reducible ions Bi^{3+} and Ce^{4+} and, hence, a nonnegligible contribution of electronic conduction. The suitability of $Bi_{0.571}Pb_{0.428}O_{1.285}$ as an electrolyte membrane has been proposed for temperatures as low as 600°C.[124] This material suffers, however, from structural instabilities. Having its mechanical properties enhanced by incorporating ZrO_2 into the starting material, the optimized membrane is able to operate continuously up to 300 mA cm^2 at 600°C. Fast ionic conduction at modest temperatures has also been reported in $Bi_4V_{2-y}Cu_yO_{11}$ (BICUVOX),* [125-127] which phases possess an intergrowth structure consisting of $Bi_2O_2^{2+}$ blocks alternating with perovskite blocks. The material $Bi_2V_{0.9}Cu_{0.1}O_{5.35}$ was found to exhibit an ionic conductivity of 1×10^{-3} S cm^{-1} already at 240°C, which is about 2 orders of magnitude higher than that of stabilized bismuth oxide.[126] In most cases the ability of these electrolytes for electrochemical oxygen separation has not yet been fully explored. Thus, it cannot be excluded that relevant properties like, for example, oxygen ion conductivity, phase stability, gas tightness, mechanical strength, and compatibility with electrode materials will not be affected during prolonged operation. Of course, the current–voltage characteristics and operational life are influenced not only by the quality of the solid electrolyte but also by the properties of the electrodes. For a recent review on oxygen electrode kinetics, see Reference 67.

2. Dual-Phase Composites

As seen from Table 14.1 impressive oxygen fluxes have been reported through 25 mol% yttria-stabilized bismuth oxide (BY25)[128] and 25 mol% erbia-stabilized bismuth oxide (BE25),[129,130] which oxide electrolytes were rendered electronically conductive by dispersion

* It may be noted that BICUVOX represents only one member of a family of Bi_2O_3-based solid electrolyte phases, which may be derived from $Bi_4V_2O_{11}$ by substitution of copper for vanadium. Many cations may be substituted for vanadium, and the general acronym BIMEVOX was given to these materials, which have been claimed for electrochemical oxygen separation at temperatures as low as 500 K.[127] Besides copper, high oxide ion conductivity is reported for substituents titanium and niobium.[213]

with silver metal. A prerequisite is that both constituent phases in the composite membranes do form a continuous path for both ionic and electronic conduction, having their concentrations above the critical (percolation threshold) volume fraction ϕ_c. The latter quantity determines the minimum volume fraction in which conduction is possible and is a function of, for example, the relative dimensions and shape of the particles of both constituent phases.[131] In actual composite materials, however, the interconnectivity between particles will not be ideal. These may be linked up to form so-called dead ends or isolated clusters, which do not contribute at all to the conductance of the percolative system. Accordingly, conduction is expected to proceed through a significantly smaller fraction of consolidated particles or grains, which implies that the actual volume fraction of each phase should always be somewhat in excess of ϕ_c. The optimum volume ratio is just above ϕ_c of the high conducting phase, i.e., the metal phase, in order to have the highest effective ionic conductivity of the composite.

Dual-phase membranes made of BY25-Ag[128] and YSZ-Pd[132] behave quite similar in having their conductivity threshold at about 33 to 35 vol% of the metal phase. These membranes were made by conventional ceramic processing techniques. The value of ϕ_c obtained for these composite materials agrees well with the high concentration limit predicted by simple effective medium theory in which the composite is described as a three-dimensional resistor network.[133] The effective ionic conductivity is reduced relative to what is expected purely on the basis of the volume fraction of the ionic conducting phase, which originates, at least partly, from the enhanced tortuosity of the migrating path for the oxygen anion due to partial blocking by the metal phase. It is therefore expected that a further gain in the oxygen flux can be realized through proper design of the microstructure.[130,132] The optimum situation would correspond with the one in which the particles of each phase line up in strings (or slabs) parallel to the applied gradient in oxygen partial pressure. Even though theoretically, the critical volume fraction of the metal phase could be reduced in this way to a value practically equal to zero, such an approach is bounded by the additional requirement for practical membranes of fast surface exchange kinetics, especially for very thin membranes.

The exchange reaction at the composite surface is confined to the three-phase boundary (tpb) between the gas, metal, and electrolyte formed by particle grains being connected to the percolative network. Fast oxygen transfer can be sustained only if the corresponding length or area available to oxygen exchange is large enough, where it should be noted that the exchange reaction can only take place at a point remote from the tpb line which is shorter than the spillover distance of electroactive species across the surface. The electrical field necessary to guide the current becomes distorted in the vicinity of the surface of a coarse-grained composite, where the separation between adjacent tpb lines is too large and, hence, only part of the surface is effective toward oxygen exchange. This contribution is stressed in the SOFC literature and is known as the constriction effect.[134] Often, it is the synergism between electrode and electrolyte material that leads to fast exchange characteristics. The oxygen flux through disc membranes made of BE25-Au (40 vol%) was found to increase almost 1 order of magnitude by substituting gold for silver in the composite.[130] This observation can be related to the higher activity of silver in the oxygen exchange reaction on BE25, compared with gold, imposing fewer limitations on overall oxygen transport.

Materials like, for example, $Bi_2CuO_{4-\delta}$,[135] TiN,[130] $MgLaCrO_{3-\delta}$[28] have been proposed to replace the inert metals. Even, though, in the examples chosen ionic and electronic transport are confined to separate phases, MIEC could be useful. A systematic evaluation of dual-phase membranes, however, is too new so far to come to definite conclusions. Besides simple modeling in terms of a short-circuited oxygen concentration cell, to our knowledge no one has yet described oxygen permeation through dual-phase membranes, taking into account the distinct three-dimensional aspects of the microstructure that may arise in practical composite materials. Besides high values for the oxygen flux (and permselectivity), commercial use of membrane systems will demand chemical, mechanical, and structural integrity of applied materials in appropriate ranges of temperature and oxygen partial pressure. Dual-phase membranes have the obvious potential to distribute specific requirements among the system components.

V. ACCEPTOR-DOPED PEROVSKITE AND PEROVSKITE-RELATED OXIDES

A. INTRODUCTION

The general trend observed from the pioneering studies on oxygen permeation through perovskites of the type $Ln_{1-x}A_xCo_{1-y}B_yO_{3-\delta}$ (Ln = La, Pr, Nd, Sm, Gd; A = Sr, Ca, Ba; B = Mn, Cr, Fe, Co, Ni, Cu) by Teraoka et al.[39-41] is that higher oxygen fluxes are facilitated by increased A-site substitution and a lower thermodynamic stability of the particular perovskite.

Clearly, not all these perovskite compositions are useful for oxygen delivery applications. For example, ceramics based on $La_{1-x}A_xCrO_{3-\delta}$ (x = Sr, Ba, Ca), $La_{1-x}Sr_xCr_{1-y}Mn_yO_{3-\delta}$, and $La_{1-x}Ca_xCr_{1-y}Co_yO_{3-\delta}$ have been proposed for use as interconnection material (separator) in SOFC, and therefore should be dense and impermeable in order to prevent burning off of the fuel without generating electricity[136,137] (see also Chapter 12 of this handbook).

Selected perovskite compositions are also targeted in basic SOFC research for use as potential electrode material for the cathodic reduction of oxygen. The most promising cathode materials to date are the manganites $La_{1-x}Sr_xMnO_{3-\delta}$.[136,137] The composition with x = 0.15 scarcely permeates oxygen up to 900°C, as was measured by feeding air and helium to opposite sides of a dense sintered membrane of 1 mm thickness.[121] The observed behavior is consistent with the low value of the oxygen self-diffusivity in $La_{0.5}Sr_{0.5}MnO_{3-\delta}$, determined by ^{18}O-^{16}O isotopic exchange, and can be attributed to the small negative departure from oxygen stoichiometry exhibited in the range of temperature and oxygen pressure covered by experiment.[138] On the other hand, oxygen transport is predicted to be quite fast under conditions of high oxygen deficiency, i.e., low oxygen partial pressures, as the oxygen vacancy diffusion coefficient of $La_{1-x}Sr_xMnO_{3-\delta}$ was found to be comparable in magnitude with that of Fe- and Co-based perovskites.[139]

Emerging from the first of these studies by Teraoka et al.[39] is that in the series $La_{1-x}Sr_xCo_{1-y}Fe_yO_{3-\delta}$ the oxygen fluxes increase with Co and Sr content, the highest flux being found for $SrCo_{0.8}Fe_{0.2}O_{3-\delta}$. Data were obtained with air on one side of a 1-mm-thick disc specimen, using helium as a sweeping gas on the other side, up to a maximum temperature of 1150 K. The observed oxygen fluxes were found to be roughly proportional to the ionic conductivity of the perovskites, which is in agreement with the fact that the electronic conductivity of compositions in this series can be extremely high, typically in the range 10^2 to 10^3 S cm^{-1}.[42] Four-probe dc measurements using electron-blocking electrodes showed that the ionic conductivity at 800°C in air can be 1 to 2 orders of magnitude higher than that of stabilized zirconia.[42] These findings have been confirmed by others, apart from scatter in the published data, which partly reflects the experimental difficulties in measuring the ionic conductivity in these predominantly electronic conductors.[140,143]

In a subsequent study, Teraoka et al.[40] investigated the influence of A and B site substitution on oxygen permeation through $La_{0.6}A_{0.4}Co_{0.8}Fe_{0.2}O_{3-\delta}$ (A = La, Na, Ca, Sr, Ba) and $La_{0.6}Sr_{0.4}Co_{0.8}B_{0.2}O_{3-\delta}$ (B = Cr, Mn, Fe, Co, Ni, Cu). As seen from Figures 14.9 and 14.10, the oxygen permeability in the two series increases in the respective orders La < Na < Sr < Ca < Ba and Mn < Cr < Fe < Co < Ni < Cu, which differ from trends in the periodical system, as far as comparison is meaningful. Results from ionic and electronic conductivity measurements of $La_{0.6}A_{0.4}Co_{0.8}Fe_{0.2}O_{3-\delta}$ (A = La, Ca, Sr) and $La_{0.6}Sr_{0.4}Co_{0.8}B_{0.2}O_{3-\delta}$ (B = Fe, Co, Ni, Cu) suggest that oxygen permeation is governed by the ionic conductivity. In the homologous series $Ln_{0.6}Sr_{0.4}CoO_{3-\delta}$, the oxygen flux was found to increase in the order La^{3+} < Pr^{3+} < Nd^{3+} < Sm^{3+} < Gd^{3+}, which corresponds with a decrease in radius of the lanthanide-ion.[40]

Since the initial observations by Teraoka et al.,[40] a considerable number of studies have appeared. Selected perovskite compositions have been reexamined, while a few others have

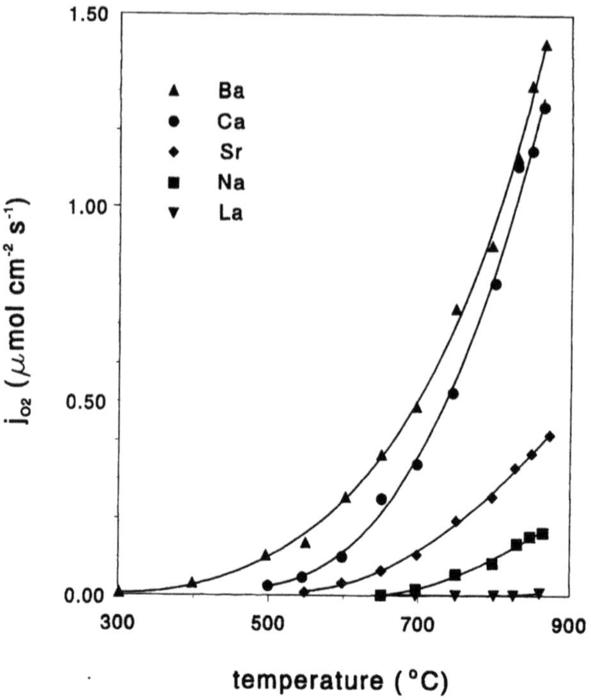

FIGURE 14.9. Temperature variation of the oxygen permeation rate from the air to the helium (30 cm^3 min^{-1}) side of disc membranes $La_{0.6}A_{0.4}Co_{0.8}Fe_{0.2}O_{3-\delta}$ (A = Na, Ba, Ca, Sr), 20 mm in diameter and 1.5 mm thick. (Adapted from Teraoka, Y., Nobunaga, T., Yamazoe, N., *Chem. Lett.*, 1988, 503–506. With permission.)

been adapted in an attempt to optimize the oxygen fluxes. The list of materials for which oxygen permeation data are presently available has been extended to include $LaCoO_{3-\delta}$,[144] $La_{1-x}Sr_xCoO_{3-\delta}$,[145-148] $La_{1-x}Sr_xFeO_{3-\delta}$,[149,150] $La_{1-x}A_xCo_{1-y}Fe_yO_{3-\delta}$ (A = Sr, Ca),[14,138,143,151] $SrCo_{0.8}Fe_{0.2}O_{3-\delta}$,[15,43,152,153] $SrCo_{0.8}B_{0.2}O_{3-\delta}$ (B = Cr, Co, Cu),[153] $SrCo_{1-x}B_xO_{3-\delta}$ (B = Cr, Mn, Fe, Ni, Cu, x = 0 . 0.5),[154] and $Y_{1-x}Ba_xCoO_{3-\delta}$.[155] In general, fair agreement is obtained with data produced by Teraoka et al., albeit that in a number of studies the observed oxygen fluxes are reportedly found to be significantly lower.[14,152,153]

The pioneering studies by Teraoka et al.[39-41] have opened a very challenging research area as the perovskites, e.g., $La_{1-x}Sr_xCo_{1-y}Fe_yO_{3-\delta}$, have a bright future for use as an oxygen separation membrane. The precise composition may be tailored for a specific application, but this has not yet been fully developed. One of the important issues is considered to be the low structural and chemical stability of the perovskites, especially in reducing environments, which remains to be solved before industrial applications become feasible. In order to meet this challenge, it is necessary first to understand the factors that limit and control the quality criteria for any given application. The perovskite and related oxides exhibit a great diversity of properties, like electrical, optical, magnetic, and catalytic properties, which have been studied extensively. In the following sections, we mainly focus on those properties affecting the magnitude of the oxygen fluxes through these materials.

B. STRUCTURE AND DEFECT CHEMISTRY

1. Perovskite Structure

The ideal perovskite structure ABO_3 consists of a cubic array of corner-sharing BO_6 octahedra, where B is a transition metal cation (Figure 14.11). The A-site ion, interstitial between the BO_6 octahedra, may be occupied by either an alkali, an alkaline earth, or a rare earth ion. In many cases the BO_6 octahedra are distorted, or tilted, due to the presence of the

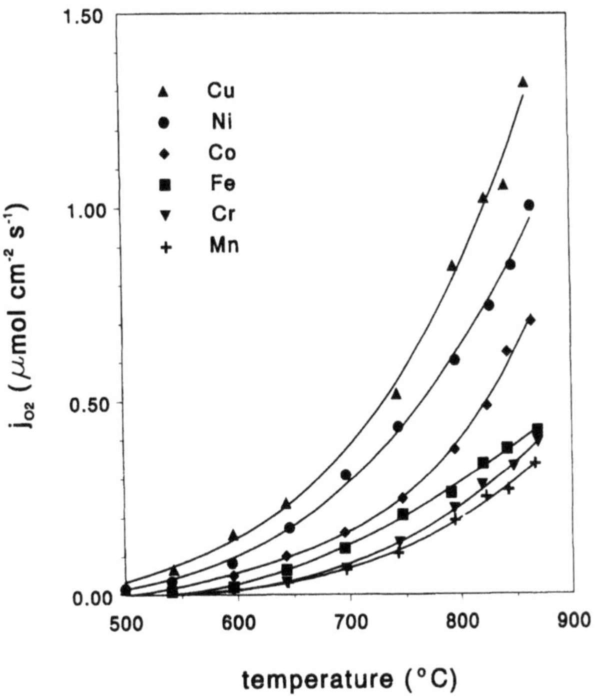

FIGURE 14.10. Temperature variation of the oxygen permeation rate of $La_{0.6}Sr_{0.4}Co_{0.8}B_{0.2}O_{3-\delta}$ (B = Cr, Mn, Fe, Co, Ni, Cu) after Teraoka et al. Experimental conditions are specified in the legend of Figure 14.9. (Reproduced from Teraoka, Y., Nobunaga, T., and Yamazoe, N., *Chem. Lett.*, 1988, 503–506. With permission.)

A cation, which is generally larger in size than the B cation. Alternatively, the perovskite structure may be regarded as a cubic close packing of layers AO_3 with B cations placed in the interlayer octahedral interstices.[156] The latter turns out to be more useful in distinguishing different structural arrangements (stacking sequences) of perovskite blocks. The tolerance limits of the cationic radii in the A and B sites are defined by the Goldschmidt factor, which is based on geometric considerations: $t = (r_A + r_O)/(\sqrt{2}(r_B + r_O))$, where r_A, r_B, and r_O are the radii of the respective ions.[157] When the distortion becomes too large, other crystal symmetries such as orthorhombic and rhombohedral appear. Nominally, the perovskite structure should be stable between $1.0 < t < 0.75$. The ideal perovskite lattice exists only for tolerance factors t very close to one.

Clearly, it is the stability of the perovskite structure that allows for large departures from ideal stoichiometry, resulting either from the substitution with aliovalent cations on the A or B site or from redox processes associated with the presence of transition metal atoms which can adopt different formal oxidation states. Oxygen vacancies are free to move among energetically equivalent crystallographic sites as long as the perovskite structure exhibits ideal cubic symmetry. The degeneracy between sites disappears upon distortion of the lattice toward lower symmetries. The onset of electronic conductivity mainly depends on the nature of the B site cation. The total electrical conductivity can be either predominantly ionic as in the acceptor-doped rare earth aluminates or predominantly electronic as in the late transition metal containing perovskites considered below.

2. Nonstoichiometry

Important contributions to the area of defect chemistry of the acceptor-doped $Ln_{1-x}A_xBO_3$, perovskites, where B is selected from Cr, Mn, Fe, or Co, have been made by a number of investigators. Particular reference is made to reviews provided by Anderson[158,159] and

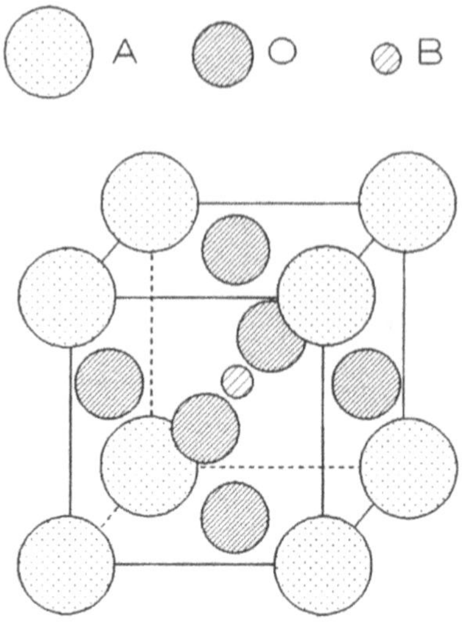

FIGURE 14.11. Ideal perovskite structure.

Mizusaki.[160] The substitution of divalent alkaline-earth ions on the A site increases the concentration of oxygen vacancies. Temperature and oxygen partial pressure determine whether charge compensation occurs by an increased valency of the transition metal ion at the B site or by the formation of ionized oxygen vacancies. Thermogravimetric studies have indicated that in, for example, $LaCrO_3$, $YCrO_3$, and $LaMnO_3$ the native nonstoichiometric ionic defects are cation vacancies, leading to oxygen-excess stoichiometries.[159] For simplicity, it is assumed here that extrinsic ionic defects generated by A site substitution prevail, i.e., only oxygen-deficient stoichiometries are considered. Furthermore, crystallographic sites available for oxygen are taken to be energetically equivalent.

For the purpose of our discussion, $LaFeO_3$ is considered to be the host for substitution. The dissolution of $SrFeO_3$ into this material can be represented by

$$SrFeO_3 \xrightarrow{LaFeO_3} Sr_{La}' + Fe_{Fe}^{\bullet} + 3O_O^{\times} \tag{14.58}$$

The incorporation of Sr^{2+} thus leads to charge compensation by the formation of Fe^{4+} ions, which is in accord with the Verwey principle of controlled ionic valency.[161] The extent of oxygen nonstoichiometry is established by the following defect chemical reactions:

$$2Fe_{Fe}^{\bullet} + O_O^{\times} \rightleftarrows 2Fe_{Fe}^{\times} + V_O^{\bullet\bullet} + \tfrac{1}{2}O_2 \tag{14.59}$$

$$2Fe_{Fe}^{\times} \rightleftarrows Fe_{Fe}' + Fe_{Fe}^{\bullet} \tag{14.60}$$

with the corresponding equilibrium constants,

$$K_g = \frac{\left[Fe_{Fe}^{\times}\right]^2\left[V_O^{\bullet\bullet}\right]P_{O_2}^{\frac{1}{2}}}{\left[Fe_{Fe}^{\bullet}\right]^2\left[O_O^{\times}\right]} \tag{14.61}$$

$$K_d = \frac{\left[Fe_{Fe}'\right]\left[Fe_{Fe}^{\bullet}\right]}{\left[Fe_{Fe}^{\times}\right]^2} \tag{14.62}$$

The oxygen vacancies formed at elevated temperatures and low oxygen partial pressure are assumed to be doubly ionized. The thermally activated charge disproportionation reaction given by Equation (14.60) reflects the localized nature of electronic species and may be treated as equivalent to the generation of electrons and electron holes by ionization across a pseudo band gap (cf. Equation [14.36]). The associated free enthalpy of reaction may be taken equal to the effective band gap energy.

At fixed A/B site ratio the following condition must be fulfilled:

$$[Fe'_{Fe}]+[Fe^{\times}_{Fe}]+[Fe^{\bullet}_{Fe}]=1 \tag{14.63}$$

and the condition of charge neutrality is

$$[Sr'_{La}]+[Fe'_{Fe}]=2\,[V_O^{\bullet\bullet}]+[Fe^{\bullet}_{Fe}] \tag{14.64}$$

In the absence of extended defects, i.e., no interaction between point defects, Equations (14.61) to (14.64) may be used with the aid of experimentally determined equilibrium constants to construct the Kröger–Vink defect diagram, from which expressions for the partial conductivities of the mobile ionic and electronic defects can be derived.

Oxygen nonstoichiometry of the perovskites $La_{1-x}Sr_xBO_{3-\delta}$ (B = Cr, Mn, Co, Fe) and its relationship with electrical properties and oxygen diffusion has been studied extensively.[158-160] Typical nonstoichiometry data for $La_{1-x}Sr_xFeO_{3-\delta}$ and for some other perovskites as obtained from gravimetric analysis and coulometric titration are given in Figure 14.12. At small oxygen deficiency, acceptor dopants are the majority defects. The charge neutrality condition then becomes

$$[Sr'_{La}]=[Fe^{\bullet}_{Fe}] \tag{14.65}$$

In this region, one finds for the oxygen nonstoichiometry δ,

$$\delta \propto P_{O_2}^{-\frac{1}{2}} \tag{14.66}$$

noting that $\delta = [V_O^{\bullet\bullet}]$, by definition. A plateau is observed around the point of electronic stoichiometry, $\delta = x/2$, where the charge neutrality condition reads

$$[Sr_{La}{}']=2\,[V_O^{\bullet\bullet}] \tag{14.67}$$

corresponding with a minimum in the electronic conductivity of $La_{1-x}Sr_xFeO_{3-\delta}$.[162,163] In this region, the oxygen nonstoichiometry is virtually constant. As the oxygen activity decreases further, oxygen vacancies are again generated, down to the oxygen activity at which decomposition of the perovskite structure occurs. The onset of the different regions depends on the nature of the transition metal B cation. The incentive of B site substitution can therefore be to optimize oxygen transport in appropriate ranges of oxygen partial pressure and temperature. As discussed below, doping may also increase stability or suppress cooperative ordering of oxygen vacancies.

3. Localized vs. Delocalized Electrons

Given the relative success of the above point defect scheme to model the experimental data of oxygen nonstoichiometry and electrical conductivity for $La_{1-x}Sr_xFeO_{3-\delta}$[164,165] and

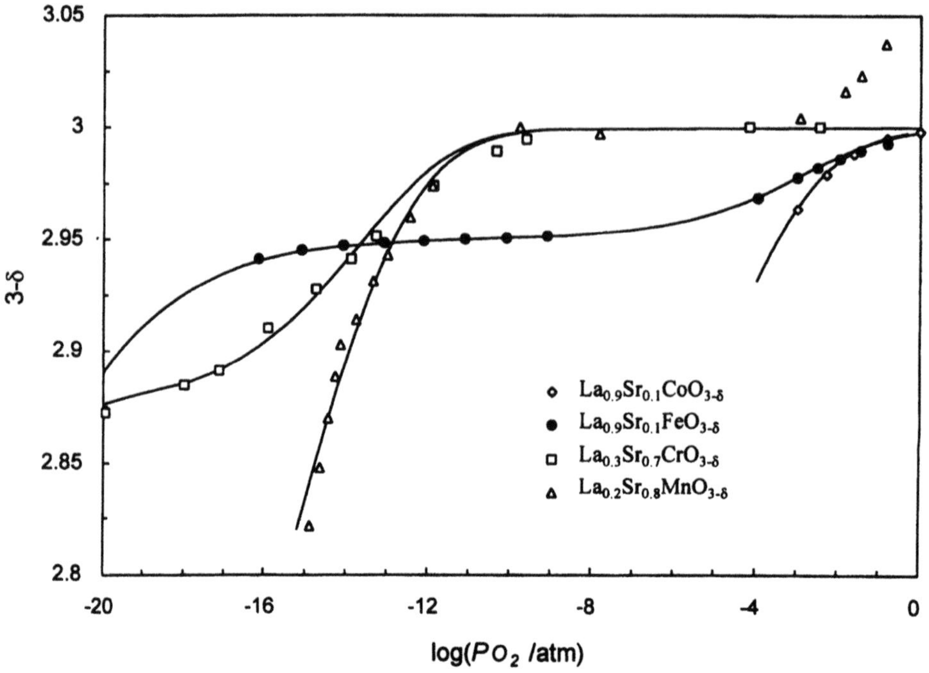

FIGURE 14.12. Data of oxygen nonstoichiometry of $La_{0.75}Sr_{0.25}CrO_{3-\delta}$, $La_{0.9}Sr_{0.1}FeO_{3-\delta}$, $La_{0.9}Sr_{0.1}CoO_{3-\delta}$, and $La_{0.8}Sr_{0.2}MnO_{3-\delta}$ at 1000°C as a function of oxygen partial pressure. Solid lines are results from a fit of the random point defect model to the experimental data. (Reproduced (slightly adapted) from Van Hassel, B.A., Kawada, T., Sakai, N., Yokokawa, H., Dokiya, M., and Bouwmeester, H.J.M., *Solid State Ionics*, 1993, *66*, 295–305. With permission.)

$La_{1-x}Sr_xCrO_{3-\delta}$,[166] its use is less satisfactory for $La_{1-x}Sr_xCoO_{3-\delta}$ and $La_{1-x}Sr_xMnO_{3-\delta}$, which compounds show notably high values for the electronic conductivity.

Nonstoichiometry of the compounds $La_{1-x}Sr_xCoO_{3-\delta}$ (x = 0, 0.1, 0.2, 0.3, 0.5, and 0.7) in the range $10^{-5} \leq P_{O_2} \leq 1$ atm and $300 \leq T \leq 1000$°C was investigated by Mizusaki et al.[167] using thermogravimetric methods. At 800°C, δ in $La_{1-x}Sr_xCoO_{3-\delta}$ varies almost proportionally to $P_{O_2}^n$ with $n \approx -1/2$ for x = 0 to $n \approx -1/16$ for x = 0.7 (see Figure 14.13). No plateau is observed around δ = x/2. Fitting the δ-P_{O_2} relationship in accord with the random point defect model leads to very large concentrations of disproportionation reaction products $Co_{Co}^{\cdot}$ and $Co_{Co}^{\prime}$. A corollary is that the pseudo-band gap must be very small. The model fit, however, is less satisfactory for high Sr substitutions.[168] A similar explanation holds for $La_{1-x}Sr_xMnO_{3-\delta}$, disregarding the oxygen-excess stoichiometries seen in this system at high oxygen partial pressures.

At high oxygen deficiency of the perovskite, the validity of the ideal mass action equations (based upon dilute solution thermodynamics) cannot be assumed *a priori*. In addition, interaction and association between defects are expected at high defect concentrations. A further limitation concerns the nature of electronic defects. The general assumption, that in the first row transition metal perovskites changes in the oxygen content leads to changes in the 3d electronic configuration, may be too naive. It is based implicitly on the idea that oxygen is strongly electronegative and, by comparison, the 3d electrons can be easily ionized. There is substantial evidence from soft X-ray absorption spectroscopy (XAS)-based studies that the electron holes introduced by doping with divalent earth-alkaline ions go to states with significant O 2p character.[169] This has also been reported for the perovskite-related oxide $YBa_2Cu_3O_{6+x}$.[170] In a localized description, i.e., assuming a narrow bandwidth of the hole band derived from the O 2p band, this would imply that O^{2-} is effectively converted into O^-.

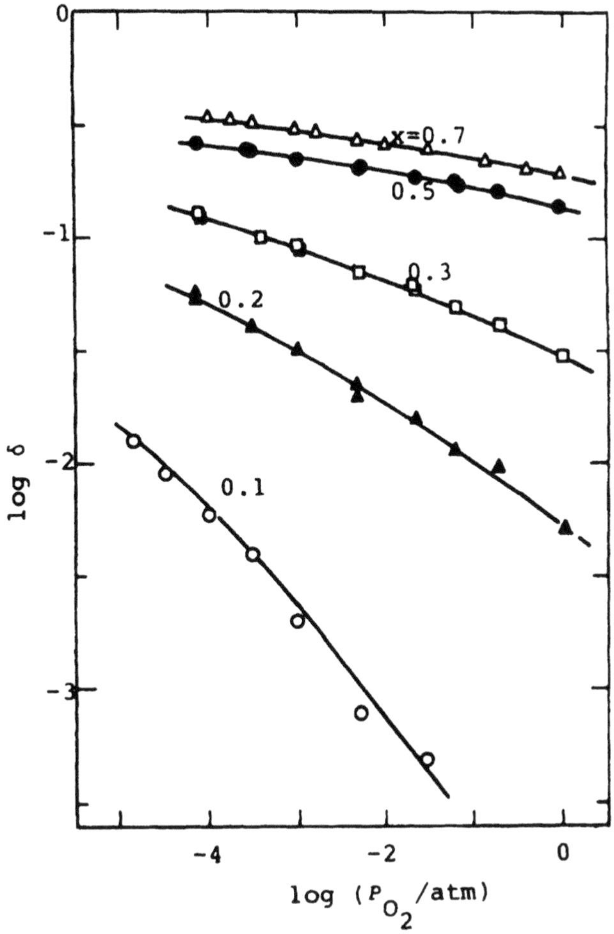

FIGURE 14.13. Oxygen pressure dependence of δ in $La_{1-x}Sr_xCoO_{3-\delta}$ at 800°C for different strontium contents. (Reprinted from Mizusaki, J., Mima, Y., Yamauchi, S., Fueki, K., and Tagawa, H., *J. Solid State Chem.*, 1989, *80*, 102–11. With permission.)

A proper description of electronic defects in terms of simple point defect chemistry is even more complicated as the d electrons of the transition metals and their compounds are intermediate between localized and delocalized behavior. Recent analysis of the redox thermodynamics of $La_{0.8}Sr_{0.2}CoO_{3-\delta}$ based upon data from coulometric titration measurements supports itinerant behavior of the electronic charge carriers in this compound.[171] The analysis was based on the partial molar enthalpy and entropy of the oxygen incorporation reaction, which can be evaluated from changes in emf with temperature at different oxygen (non-)stoichiometries. The experimental value of the partial molar entropy (free formation entropy) of oxygen incorporation, Δs_{O_2}, could be fitted by assuming a statistical distribution among sites on the oxygen sublattice,

$$\Delta s_{O_2} = s^o - 2k \ln \frac{(3-\delta)}{\delta} \tag{14.68}$$

where s^o is a constant. That is, no entropy change associated with electron annihilation can be identified. The partial molar enthalpy (free enthalpy of formation of vacancies) associated with oxygen incorporation was found to decrease almost linearly with δ. A first inclination might be to assume that the mutual repulsion between oxygen vacancies increases with increasing oxygen deficiency. But this interpretation immediately raises the question why such a behavior is not found in the case of $La_{1-x}Sr_xFeO_{3-\delta}$.[165,166] Instead, the experimental data are interpreted to reflect the energetic costs of band filling. With increasing oxygen nonstoichiometry in $La_{0.8}Sr_{0.2}CoO_{3-\delta}$, the two electrons, which are needed for charge compensation of a single oxygen vacancy, are donated to an electron band broad enough to induce Fermi condensation characteristic of a metallic compound. The average density of electron states at the Fermi level is determined to be 1.9 ± 0.1 eV^{-1} per unit cell. The physical significance of the work is that the defect chemistry of $La_{0.8}Sr_{0.2}CoO_{3-\delta}$ cannot be modeled using the simple mass action-type of equations. An empirical model for the oxygen nonstoichiometry of $La_{0.8}Sr_{0.2}CoO_{3-\delta}$ is proposed, which demonstrates that the density of states is related to the slope of the log–log plots of δ vs. P_{O_2}. In support of these interpretations, it is noted that XAS has not been successful in detecting charge disproportionation in $LaCoO_{3-\delta}$, due to localization of electrons, in the temperature range 80 to 630 K.[172] The nonstoichiometry data obtained for $La_{0.8}Sr_{0.2}CoO_{3-\delta}$ are found to be in good agreement with earlier results from gravimetric analysis in the series $La_{1-x}Sr_xCoO_{3-\delta}$ obtained by Mizusaki et al.,[167] which authors arrived at more or less similar conclusions regarding the role of electronic states in the energetics of oxygen incorporation into these compounds.

A final point to note is that the process of generating oxygen vacancies in $La_{1-x}Sr_xCoO_{3-\delta}$ shows similarity with the insertion of "guest" species in other metallic hosts like, for example, the intercalation of alkali metals or the Ib metals Cu and Ag into layered transition metal dichalcogenides. This comparison even goes beyond the description of the thermodynamics of the intercalation reaction of, e.g., Ag^+ ions in Ag_xTiS_2,[173] by analogy with that of oxygen vacancies in $La_{1-x}Sr_xCoO_{3-\delta}$. Due to elastic distortions of infinite TiS_2 host layers, in-plane interactions arise between the intercalant ions that are attractive. These cause the clustering of Ag^+ ions within the Van der Waals gaps between successive layers of the host material to form islands,[174] rearranging themselves into microdomains of a type strongly reminiscent of those observed in perovskite and perovskite-related structures (see Section V.F.).

C. OXYGEN DESORPTION AND PEROVSKITE STABILITY

As seen from Figure 14.12, the value of $(3-\delta)$ in $La_{1-x}Sr_xCoO_{3-\delta}$ falls off with decreasing oxygen activity much more rapidly than for the other compounds shown. The general trend at which the perovskites become nonstoichiometric follows that of the relative redox stability of the late transition metal ions occupying the B site, i.e., $Cr^{3+} > Fe^{3+} > Mn^{3+} > Co^{3+}$. The reductive nonstoichiometry of the cobaltites increases further by partial B site substitution with copper and nickel.

The reductive (and oxidative) nonstoichiometry and the stability in reducing oxygen atmospheres of perovskite-type oxides was reviewed by Tejuca et al.[175] Data from temperature-programmed reduction (TPR) measurements indicate that the stability (or reducibility) of the perovskite oxides increases (decreases) with increasing size of the A ion, which would be consistent with the preferred occupancy of the larger Ln^{3+} ion in a 12-fold coordination. The trend is just the reverse of that of the stability of the corresponding binary oxides. The ease of reduction increases by partial substitution of the A ion, e.g., La^{3+} by Sr^{2+}. Trends in the thermodynamic stabilities of perovskite oxides have been systematized in terms of the stabilization energy from their constituent binary oxides and the valence stability of the transition metal ions by Yokokawa et al.[176]

The stability of the undoped perovskites $LaBO_{3-\delta}$, at 1000°C, expressed in terms of P_{O_2} decreases in the order $LaCrO_{3-\delta}$ (10^{-20} atm) > $LaFeO_{3-\delta}$ (10^{-17} atm) > $LaMnO_{3-\delta}$ (10^{-15} atm) > $LaCoO_{3-\delta}$ (10^{-7} atm), noting that the cited value for $LaCrO_{3-\delta}$ corresponds with the lowest limit in a thermogravimetric study by Nakamura et al.[177] The same trend was found by means of TPR.[175]

Tabata et al.[178] and Seyama[179] both described significant differences in the chemical composition of the surface, due to Sr segregation, compared with the bulk composition in a series of powders $La_{1-x}Sr_xCoO_{3-\delta}$. This indicates a behavior of the surface different from that of the bulk in these compounds. Not only can this account for a number of observations made in the total oxidation of CO and CH_4, as discussed by the authors, but it is also considered to be an important factor when one tries to correlate the composition of a perovskite with its activity in surface oxygen exchange.

The sorption kinetics of oxides is certainly influenced by their corresponding defect structure. A number of interesting observations were made by Yamazoe and co-workers,[180] and Teraoka et al.,[181] showing that for perovskites $LaMO_{3-\delta}$ (M = Cr, Mn, Fe, Co, Ni), $La_{1-x}Sr_xCoO_{3-\delta}$ (x = 0, 0.2, 0.4, 1), and $La_{0.8}A_{0.2}CoO_{3-\delta}$ (A = Na, Ca, Sr, Ba), two distinct types of oxygen are desorbed upon heating in a helium stream after a pretreatment step in which the oxide was saturated in an oxygen-rich atmosphere at high temperature, followed by slow cooling to room temperature. The oxygen desorbed in a wide range at moderate temperatures, referred to as α-oxygen, was found to be correlated with the amount of partial substitution of the A ion. The onset temperature of the so-called β desorption peak observed at high temperature was correlated with the thermal decomposition temperature of the corresponding transition metal oxides. Accordingly, the β peak corresponds with the reduction of the transition metal ion from B^{3+} to B^{2+}. The partial substitution of Co by Fe in the series $La_{1-x}Sr_xCo_{1-y}Fe_yO_{3-\delta}$ stabilizes the Co^{3+} oxidation state (no β peak observed), while shifting the α type of desorption to lower temperatures.[182,183]

D. EQUATIONS FOR OXYGEN TRANSPORT

Equations for oxygen transport can be derived from the point defect equilibria discussed in Section V.B.2. This provides us with some general insight into the transport behavior of oxygen-deficient perovskites. Strictly speaking, the equations presented below are valid at low defect concentrations only, i.e., assuming oxygen defects to be randomly distributed.

Oxygen transport in the perovskites is generally considered to occur via a vacancy transport mechanism. On the assumption that the oxygen vacancies are fully ionized and all contribute to transport, i.e., oxygen defects are not associated, the Nernst–Einstein equation reads

$$\sigma_{ion} = \frac{4F^2\left[V_O^{\bullet\bullet}\right]D_v}{RTV_m} \tag{14.69}$$

where D_v is the vacancy diffusion coefficient and V_m the perovskite molar volume. Since electronic conduction in the perovskites predominates, i.e., $\sigma_{el} > \sigma_{ion}$, the integral in the Wagner equation (Equation [14.10]) involves only σ_{ion} over the applied oxygen partial pressure gradient. Using Equation (14.69), we may rewrite the Wagner equation, to give

$$j_{O_2} = -\frac{D_v}{4V_m L}\int_{\ln P'_{O_2}}^{\ln P''_{O_2}} \delta \, d\ln P_{O_2} \tag{14.70}$$

by virtue of $\delta = [V_O^{\bullet\bullet}]$. Evaluation can be performed numerically provided that D_v and the δ-ln(P_{O_2}) relationship are known. The ability of Equation (14.70) to quantitatively fit experimental data of oxygen permeation is illustrated for $La_{0.9}Sr_{0.1}FeO_{3-\delta}$ in Figure 14.14. Similar results have been presented for, e.g., $La_{0.75}Sr_{0.25}CrO_{3-\delta}$[184] and $La_{0.70}Ca_{0.30}CrO_{3-\delta}$.[185] The analytical solution of the integral given by Equation (14.70) incorporating random point defect chemistry has been given by Van Hassel et al.[186]

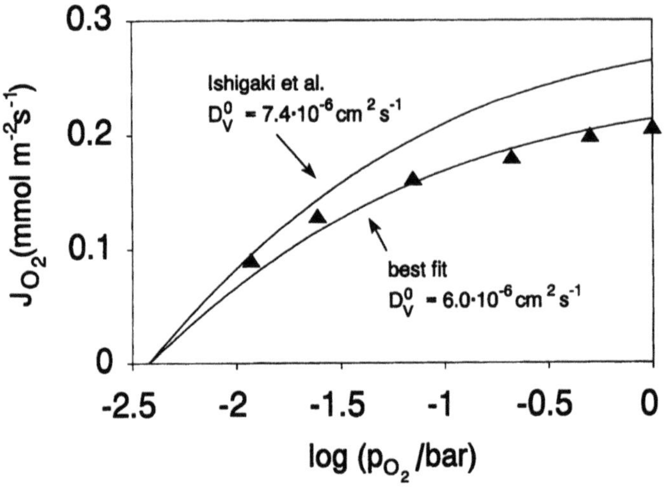

FIGURE 14.14. Theoretical fit of feed-side P_{O_2} dependence of oxygen permeation through $La_{0.9}Sr_{0.1}FeO_{3-\delta}$, at 1000°C. The best fit is obtained when D_V equals 6×10^{-6} cm^2 s^{-1}, which slightly deviates from the corresponding value obtained from isotopic exchange. (Reprinted from Ten Elshof, J.E., Bouwmeester, H.J.M., and Verweij, H., *Solid State Ionics*, 1995, *81*, 97–109. With permission.)

When data of oxygen nonstoichiometry follows a simple power law $\delta \propto P_{O_2}^n$, integration of Equation (14.70) yields an expression similar to that of Equation (14.18) having $\beta = D_v\delta^o/4V_m n$. Examination of the data from oxygen permeability measurements on disc specimens of thickness 2 mm in a series $La_{1-x}Sr_xCoO_{3-\delta}$ $(0 \leq x \leq 0.8)$ in a study by Van Doorn et al.[147] indicate that the results, at 1000°C, can be fitted well by this equation, the validity of which is usually restricted to a small range in oxygen partial pressure. For compositions below $x \leq 0.6$, the values of n obtained from fitting were found to be in excellent agreement with the corresponding slopes of the ln δ-ln P_{O_2} plots derived from data of thermogravimetry in the range $10^{-4} \leq P_{O_2} \leq 1$ atm. For compositions $x = 0.7$ and $x = 0.8$ the agreement obtained is less. Measurements made as a function of specimen thickness, down to a minimum value of 0.5 mm, suggest that this may be due to a partial rate control of the oxygen fluxes for these compositions by the surface reaction.

In writing Equation (14.70), D_v was taken to be constant. Strictly speaking, D_v may decrease slightly when the oxygen deficiency increases (i.e., toward decreasing oxygen partial pressure), to an extent depending on the particular solid. In accord with classical diffusion theory, the probability of vacancy hopping, hence, D_v is proportional to a factor $(1 - \delta/3)$, which represents the site occupancy of lattice oxygen anions.[187] Moreover, complications due to local stresses resulting from a change in cell volume with decreasing oxygen partial pressure may need further consideration, especially if the nonstoichiometry becomes relatively large.

With the help of Equation (14.69), the vacancy diffusion coefficient D_v can be determined from ionic conductivity measurements provided that data of oxygen nonstoichiometry are available. The direct measurement of σ_{ion} in these materials requires the use of auxiliary electrolytes such as doped zirconia or ceria to block the electronic charge carriers. The ionic current that is passed through the sample is measured by the electric current in the external circuit. The problems faced are that of interfacial charge transfer between the blocking electrodes and the mixed conductor and that it is very difficult to effectively block all of the electronic current. Short-circuiting paths for oxygen transport can occur such as diffusion along the oxide surface or via the gas phase through rapid exchange, leading to overestimates of the ionic conductivity.[142] Also, there is the possibility of an interfacial reaction between the perovskite and the blocking electrode material or the glass used for sealing to suppress the parasitic contributions to oxygen transport.[188]

The relation between D_v and the tracer diffusion coefficient D^* can be expressed as

$$D^* = f\frac{\delta}{(3-\delta)}D_v \qquad (14.71)$$

where f is the correlation factor for diffusion of oxygen vacancies in the ideal perovskite anion sublattice: f = 0.69 (for small values of δ).[78] Data of D_v thus obtained have been published by Yamauchi et al.[77] and Ishigaki and co-workers[189,190] for single crystals of $LaCoO_{3-\delta}$ and $LaFeO_{3-\delta}$. The value of D^* at 950°C in both oxides is found to be proportional to $P_{O_2}^n$, with n = −0.34 ± 0.04 and n = −0.58 ± 0.15, respectively, where a $P_{O_2}^{-1/2}$ dependence is expected in the high P_{O_2} regions covered by experiment. These results provide firm evidence that diffusion in the perovskites occurs by a vacancy mechanism. Though the values of D^* observed in these oxides differ about 3 orders of magnitude, the corresponding values of D_v are nearly the same. Additional data have been reported for single crystals of $La_{0.9}Sr_{0.1}CoO_{3-\delta}$ (x = 0.1),[78] $La_{1-x}Sr_xFeO_{3-\delta}$ (x = 0.1, 0.25, and 0.4),[78] and polycrystalline phases $La_{1-x}Sr_xFeO_{3-\delta}$ (x = 0, 0.4, 0.6, and 1.0),[191] and $La_{0.70}Ca_{0.30}CrO_{3-\delta}$,[185] revealing that the diffusivities at elevated temperatures may be similar to those observed for fluorite and fluorite-related oxides, albeit that the associated activation energy generally tends to be slightly higher in the perovskite structure (see Reference 160).

When electronic conduction predominates, one may derive the following relationship between D_v and the chemical diffusion coefficient $\tilde{D}$ (by combining Equations [14.12] and [14.69]),

$$\tilde{D} = -\frac{D_v}{2}\frac{\partial \ln P_{O_2}}{\partial \ln \delta} \qquad (14.72)$$

Measurements of the weight change following a sudden change of the oxygen activity in the gas phase were carried out for determining $\tilde{D}$ in $La_{1-x}Sr_xCoO_{3-\delta}$ (x = 0 and x = 0.1) at 800 to 1000°C in the range $10^{-5} \le P_{O_2} \le 1$ atm.[192] The calculated values of D_v agree with those obtained from the above studies, which suggests that the random point defect model holds well for the cobaltites at low Sr contents. Fair agreement is also obtained with results from relaxation experiments on $La_{1-x}Sr_xCoO_{3-\delta}$ (x = 0 and 0.2) in which the time change of the conductivity was traced.[193] A typical value of $\tilde{D}$ taken from these studies is 10^{-5} $cm^2 \cdot s^{-1}$ at 900°C. Relaxation experiments were also carried out to study chemical diffusion in, for example, $SrCo_{1-y}Fe_yO_{3-\delta}$ (y = 0.2, 0.5, and 0.8),[188,194] $La_{1-x}Sr_xCo_{1-y}Fe_yO_{3-\delta}$ (0.2 ≤ x ≤ 1.0 and 0 ≤ y ≤ 1.0),[195] $La_{1-x}Ca_xCrO_{3-\delta}$ (x = 0.1, 0.2, and 0.3),[196] $La_{1-x}Sr_xMnO_{3-\delta}$ (x = 0.05 to 0.20),[139] and $La_{1-x}Sr_xMnO_{3-\delta}$ (x = 0.20 and 0.5).[197,198]

E. ELECTRONIC CONDUCTIVITY

The late transition metal-containing perovskites exhibit high electronic conductivities. In the materials which receive prime interest for oxygen delivery applications, the electronic contribution at high temperature of operation is usually predominant. The values for e.g., $La_{1-x}Sr_xCo_{1-y}Fe_yO_{3-\delta}$ at 800°C in air range between 10^2 to 10^3 S cm^{-1}, while 10^{-2} to 1 S cm^{-1} is found for the ionic conductivity.[42] The ionic transference numbers in this series vary between 10^{-4} to 10^{-2}. Departures of the above behavior can occur at reduced oxygen partial pressures, due to the loss of p-type charge carriers.[195,199] This is most serious at the point where the hopping-type of conductivity goes through a minimum, to become n-type, e.g., in $La_{1-x}A_xFeO_{3-\delta}$ (A = Sr, Ca),[162,163,200] provided that the correspondingly low oxygen pressure is maintained during experiment. At 1000°C, the minimum in $La_{0.75}Sr_{0.25}FeO_{3-\delta}$ occurs below P_{O_2} values of 10^{-12} atm.[163] In the following, we discuss a number of characteristics which are considered to control electronic conductivity in this class of oxides and give some examples of their behavior.

Electronic conduction in the usual ranges of temperature and oxygen pressure is reported to be the p-type, and is commonly explained by assuming a small polaron mechanism with a thermally activated mobility.[158,159] This behavior may be masked by substantial oxygen loss and a concomitant decrease in the concentration of p-type charge carriers, seen at the highest temperatures and at reduced oxygen partial pressures. The direct overlap between transition metal d orbitals is known to be small, being across a cube face. The hopping transport of mobile charge carriers between two neighboring B cations in the perovskite lattice is mediated by the O 2p orbital,

$$B^{n+} - O^{2-} - B^{(n-1)+} \rightarrow B^{(n-1)+} - O^{-} - B^{(n-1)+} \rightarrow B^{(n-1)+} - O^{2-} - B^{n+}$$

which is known as double exchange, first discussed qualitatively by Zener.[201] This process is favored by a strong overlap of empty or partly filled cation orbitals of the d manifold (involving e_g and t_{2g} orbitals) with the filled O 2p orbital of neighboring anions, and reaches a maximum for a B-O-B angle of 180°, corresponding with ideal cubic symmetry.

$La_{1-x}A_xCrO_{3-\delta}$ (Sr, Ca) and $La_{1-x}A_xFeO_{3-\delta}$ (Sr, Ca) are typical examples of which the data of small-polaron transport can be explained in terms of simple defect chemistry, including the thermally activated charge disproportionation among the B cations. The predominant mechanism of hopping in $La_{1-x}A_xFeO_{3-\delta}$ (Sr, Ca) in the p-type region is between Fe^{4+} and Fe^{3+} valence states, changing to that between Fe^{2+} and Fe^{3+} in the n-type region, upon lowering the oxygen partial pressure. The electrical conductivity thereby passes through a minimum at the point of electronic stoichiometry, where the concentrations of Fe^{4+} ($Fe_{Fe}^{\cdot}$) and Fe^{2+} (Fe_{Fe}') are equal.[162,163,200] Charge disproportionation has also been used to account for results from electrical conductivity and thermopower measurements of selected substitutionally mixed oxides $LaMn_{1-x}Co_{1-x}O_{3-\delta}$,[202] $LaMn_{1-x}Cr_{1-x}O_{3-\delta}$,[203] $La_{1-x}Ca_xCo_{1-y}Cr_yO_{3-\delta}$,[204] and $La_{1-x}Sr_xCo_{1-y}Fe_yO_{3-\delta}$.[205] Preferential electronic charge compensation may occur in these compounds, i.e., the charge carrier may (at low temperature) be temporarily trapped at the small polaron site which is lower in energy, thereby decreasing the electrical conductivity. This effect disappears at high temperature, when the thermal energy is sufficient to surpass the barrier between the traps and the more conductive hopping sites, or when the population of low-energy sites exceeds the percolation limit. In the latter case the electrical conductivity is controlled by the short-range hopping among the lower-energy sites. Based upon results from electrical conductivity and thermopower measurements, in conjunction with some other techniques, Tai et al.[205] concluded that in $La_{0.8}Sr_{0.2}Co_{1-y}Fe_yO_{3-\delta}$ compensation of $Fe^{3+} \rightarrow Fe^{4+}$ is more likely than $Co^{3+} \rightarrow Co^{4+}$.

The electrical conductivity and the spin-transition state of $LaCoO_{3-\delta}$ and $La_{1-x}Sr_xCoO_{3-\delta}$ has been studied extensively. The covalent mixing of orbitals induces itinerant behavior of the charge carriers in $La_{1-x}Sr_xCoO_{3-\delta}$.[206] Racah and Goodenough[207] claimed a first-order localized electron to collective electron phase transition in $LaCoO_3$ at 937°C in air, though the electrical conductivity was found to be continuous at the transition. Electrical conductivity and differential thermal analysis (DTA) behavior of $ACoO_{3-\delta}$ (A = Nd, Gd, Ho, Y, La) were investigated by Thornton et al.[208] Evidence was adduced that the transition previously noted in $LaCoO_3$ (and in the other cobaltates) may be caused by the presence of binary cobalt oxides. The endothermic heat effect observed in DTA can be correlated with a similar feature obtained from a sample Co_3O_4. Instead, a gradual semiconductor-to-metal transition with increasing temperature is suggested in compounds $ACoO_{3-\delta}$. For $LaCoO_{3-\delta}$, which adopts a rhombohedrally distorted perovskite structure at room temperature, this range was found to extend from 110 to 300°C. Analysis was based in part upon the temperature-independent magnetic susceptibility observed at high temperature, which was attributed to Pauli-paramagnetism (with a possible Van Vleck type of contribution) of the conduction electrons. Data produced by Mizusaki et al.[209] suggest a close relationship between the transition temperature of the conductivity and the rhombohedral angle, which gradually decreases with increasing x in $La_{1-x}Sr_xCoO_{3-\delta}$. The conduction becomes metallic when the rhombohedral angle becomes smaller than 60.3°, noting that the value of 60° corresponds with ideal cubic symmetry. Another factor is the extent of oxygen nonstoichiometry. In the metallic region, the electrical conductivity of $La_{1-x}Sr_xCoO_{3-\delta}$ decreases almost linearly with increasing δ. Apart from changes in d band occupancy, it is assumed by the authors that band narrowing takes place with increasing oxygen nonstoichiometry.

Finally, the data of electrical conductivity and thermopower of $La_{1-x}Sr_xMnO_{3-\delta}$ at elevated temperature has been treated by a hopping mechanism for $x < 0.2$, and by a band model for the semimetallic behavior observed at $x > 0.3$ by Mizusaki.[160] On the other hand, to explain their results of $La_{1-x}Sr_xMnO_{3-\delta}$ and $La_{1-x}Sr_xMnO_{3-\delta}$ for compositions with $0.30 \leq x \leq 0.80$, Stevenson et al.[210] included the thermally activated charge disproportionation of Mn^{3+} into Mn^{2+} and Mn^{4+} pairs.

F. EXTENDED DEFECTS AND VACANCY ORDERING

As is known for fluorite and fluorite-related oxides, increased defect interactions are likely if the oxygen vacancy concentration exceeds 1 mol%.[92] The interaction between defects and defect association effectively lowers the concentration of "free" oxygen vacancies available for oxygen transport. In the perovskites, it is not uncommon to have an oxygen deficiency of 10 mol% or more. As noted before, the assumption of randomly distributed point defects at such large vacancy concentrations probably is an oversimplified picture. Van Roosmalen and Cordfunke[211] showed that by assuming divalent transition metal ions in undoped perovskites $LaBO_{3-\delta}$ (M = Mn, Fe, Co) to be bound to oxygen vacancies, forming neutral defect clusters of the type $<B'\text{-}V_O^{\cdot\cdot}\text{-}B'>$, the model fit to the experimental data was greatly improved. Similar, but probably more complicated, extended defects were suggested to be the structural building elements of highly defective perovskites.

In general, the tendency to form ordered structures progressively grows with increasing defect concentrations. As a matter of fact, ordering of oxygen vacancies in the perovskite and perovskite-related structures seems more common than their random distribution between the perovskite slabs. Ordering of oxygen vacancies is revealed by the formation of superstructures. Sometimes the two limiting cases are linked for the same composition by an order–disorder transformation, driven by the gain in configurational entropy of oxygen vacancies in the disordered state at elevated temperature. It has also been suggested that any ordering of the oxygen vacancies in the defective perovskite and perovskite-related oxides, thereby confining vacancy transport to two-dimensional layers, may give rise to fast ionic conductivity at significantly reduced temperatures,[212,213] a point to which we return below.

1. Static Lattice Simulation

Attempts were made by Kilner and Brook[214] to model the ionic conductivity in perovskites $LnAlO_3$ using static lattice simulation techniques. Here, the minimum energy positions for the mobile ions, and the activation energy barrier that they must surmount to migrate through the rigid crystal lattice, are calculated by minimization of the total lattice energy. The results show that aliovalent dopants might act as trapping centers for oxygen vacancies through the formation of defect associates, e.g., $V_O^{\cdot\cdot}\text{-}Sr_{La}'$. It is further found that the size proportion of A and B cations is of significant importance in determining the minimum migration enthalpy for oxygen transport in the ABO_3 structure. During diffusion, the migrating O^{2-} ion must pass the saddle point formed by two A ions and one B ion, as shown in Figure 14.15. The associated energy barrier to migration decreases with increasing size of the B cation and

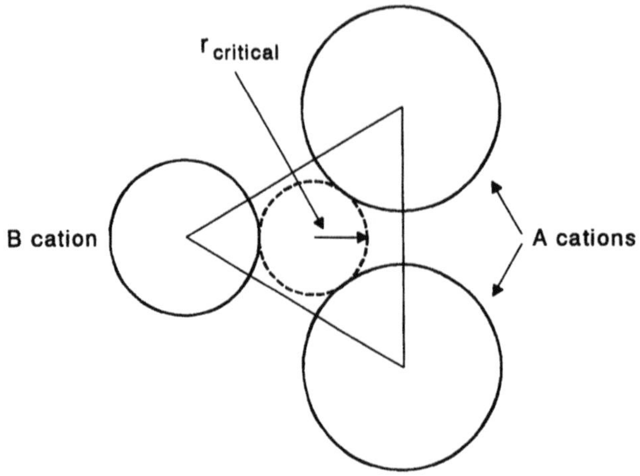

FIGURE 14.15. Saddle-point configuration for oxygen anion migration.

decreasing size of the A cation. This work has been extended recently by Cherry et al.[215] to include perovskites $LaBO_3$ (Cr, Mn, Fe, Co), showing the importance of relaxation effects at the migration saddle point, which were not invoked in the previous study by Kilner and Brook.[214] Profile mapping shows that the migrating O^{2-} ion does not prefer a linear path between adjacent sites of a BO_6 octahedron, but rather follows a curved route with the saddle point away from the neighboring B site cation.

The pioneering work by Kilner and Brook[214] was further expanded by Cook and Sammells[216] and Sammells et al.[217] to include additional empirical relationships for predicting oxygen ionic conductivity in perovskite solid solutions, such as the average metal–oxygen bonding energy, the lattice free volume, and the overall lattice polarizibility toward anion migration. A marked correlation found is that between the activation energy of oxygen anion migration in the perovskite lattice and the free volume. A smaller activation energy is apparent in ABO_3 perovskites which possess an inherently larger free volume. For fluorite-structured oxygen ion conductors, again linear, but opposite correlations are found, suggesting that there is optimum value of the lattice free volume at which the coulombic, polarization, and repulsive contributions to the migration enthalpy in both type of structures are best balanced. The results obtained led to the identification of perovskite oxide electrolytes $BaTb_{0.9}In_{0.1}O_{3-\delta}$, $CaCe_{0.9}Gd_{0.1}O_{3-\delta}$, and $CaCe_{0.9}Er_{0.1}O_{3-\delta}$, which indeed exhibit low activation energies varying between 35 to 53 kJ mol^{-1} for the ionic conductivity.[217] Even though the suggested correlations guide the selection toward materials exhibiting a low value of the ionic migration enthalpy for oxygen anions and protons (if assumed to occur via OH^-), none of the above authors did address the problem of how to optimize the magnitude of the pre-exponential term in the expression of the overall ionic conductivity. The pre-exponential term is related in part to the density of mobile oxygen anions and the availability of sites (e.g., vacancies) to which they might jump. Its value will thus be determined by the state of order in the oxygen sublattice.*

* Ordering reduces the number of free ionic charge carriers and, in general, has a negative impact on the magnitude of the ionic conductivity. An exception to this rule is apparent in selected pyrochlores with composition $Ln_2Zr_2O_7$ (Ln = La-Gd), and is best illustrated for $Gd_2Zr_2O_7$.[218,219] The pyrochlore structure can be derived from that of fluorite by ordering of both cations and vacancies on their respective sublattices. Electron microscopy has shown that the actual microstructure of these solids consists of ordered pyrochlore domains embedded in a disordered fluorite matrix. The degree of ordering can be varied by thermal annealing. Ordering causes both a lower pre-exponential term and activation enthalpy, and leads to an overall ionic conductivity for well-ordered $Gd_2Zr_2O_7$ competitive with values as measured for stabilized zirconia. The observed phenomena have been interpreted to reflect the presence of high diffusivity paths in the pyrochlore structure.

2. Vacancy Ordering

XRD, electron diffraction, and HRTEM have provided ample evidence that, particularly at low temperature, nonstoichiometry in oxygen-deficient perovskites is accommodated by vacancy ordering to a degree which depends both on oxygen partial pressure and on the thermal history.[220,221] A family of intergrowth compounds is observed for, e.g., the ferrites, which can be regarded as being composed of perovskite (ABO_3) and brownmillerite ($A_2B_2O_5$) structural units stacked along the superlattice axis. The intergrowth structures fit into a homologous series expressed by $A_nB_nO_{3n-1}$, with end members $n = \infty$ (perovskite) and $n = 2$ (brownmillerite). For nonintegral values of n, disordered intergrowths are observed between nearby members of the ideal series.

The concept of "perovskite space" has been introduced by Smyth[222,223] in order to systematize the intergrowth structures exhibited by various oxygen-deficient and oxygen-excess perovskite systems. In the proposed diagram, the close structural relationship between the parent and intergrowth structures is expressed by plotting the value of n along with the compositional excursion from ideal perovskite stoichiometry. The perovskite systems are found to be quite specific in their tendency to form ordered structures in the sense that only selected values of n are found for a particular system. The observed vacancy patterns and three-dimensional structures vary along with lateral shifts in the stacking sequence of successive layers $AO_{3-\delta}$. Anderson et al.[224] showed the driving force toward ordering to be strongly dependent upon the size and electronic configuration of the B cation, in addition to the size and coordination preference of the A cation. In general, compounds that contain ordered vacancies are found for n = 5, 4, 3, 2, 1.5, 1.33, and 1, that is for overall oxygen contents 2.8, 2.75, 2.67, 2.5, 2.33, 2.25, and 2.[224] Unfortunately, the structural studies are usually carried out at room temperature and little is known about the extended defects at elevated temperatures and their behavior during quenching and cooling.

3. Microdomain Formation

The structural rearrangements accompanying the redox phenomena sometimes lead to a texture of the sample consisting wholly of microdomains, which may be of varying size and composition. The superstructures observed depend upon local composition, while their orientation within each microdomain may be randomly distributed. As the domains are usually smaller than the sampling size (<500 Å) of X-ray and neutron diffraction, such investigations require the use of HRTEM microscopy and electron diffraction.

As an example, we briefly describe here the observations made for $La_{1-x}Ca_xFeO_{3-\delta}$. In this solid solution system, a single perovskite–brownmillerite intergrowth structure (n = 3) is found at composition x = ⅔.[225] The intergrowth structure exhibits ideal stoichiometry, i.e., $LaCa_2Fe_3O_8$, at reduced oxygen partial pressures near the minimum observed in the electrical conductivity.[200] For any other composition, a disordered intergrowth is observed.

Oxidizing $LaCa_2Fe_3O_8$ in air at 1400°C and quenching to room temperature produced a material which appeared cubic to X-rays. The most simple explanation would have been that oxygen vacancies become disordered in a highly defective perovskite structure, but transmission electron microscopy (TEM) and diffraction revealed the existence of three-dimensional intergrowths of domains of the parent phase.[225] Oxygen excess is therefore considered to be accommodated at the domain walls, suggesting that the size of the domains is closely related to the oxygen excess, i.e., the greater the oxygen excess the smaller the domains.

In the regions $0 < x < ⅔$ and $⅔ < x < 1$, disordered intergrowths are observed with members $n = 2$ and $n = \infty$, respectively. In the reduced samples, it is assumed that the local composition is very heterogeneous and a variable amount of oxygen can be accommodated as the Ca/La ratio changes. Oxidizing the disordered intergrowth structures again led to quenched samples with a microdomain texture having an estimated size of domains of about $(100\ Å)^3$. In each of these cases up to six types of microdomains were observed, three for each of the end members, e.g., $LaCa_2Fe_3O_8$ and $Ca_2Fe_2O_5$ for the region ⅔ < x < 1. Since the fringe separation within each of the microdomains was found to be very regular, the oxygen excess is probably again located at the domain walls.[226] In fact, the domain texture appeared to be not very stable either when kept in air at room temperature or under reducing conditions (as present in the electron microscope). After long operation in the electron microscope up to nine sets of microdomains were observed in the oxidized composition $LaCa_2Fe_3O_{8.23}$, three for each member in the ideal series, n = 2 ($Ca_2Fe_2O_5$), n = 3 ($LaCa_2Fe_3O_8$) and n = ∞ ($LaFeO_3$).

Other materials exhibiting a three-dimensional microdomain texture include $CaMn_{1-x}Fe_xO_{3-\delta}$,[227,228] $Ba_xLa_{1-x}FeO_{3-\delta}$,[229] and Sr_2CoFeO_5,[230] while a few more examples are given below. Although it may be anticipated that mass transport is strongly influenced by the presence of a microdomain texture, the question is open as to whether or not these microdomains are also present at high temperatures. In a study by Mössbauer spectroscopy, XRD and magnetic susceptibility, Battle et al.[230] found a microdomain texture in partially oxidized samples of $Sr_2CoFeO_{5+\delta}$, quenched in air from 1200°C, but not in substoichiometric samples obtained either after quenching in liquid nitrogen from 1200°C or after annealing under argon at 800°C. Furthermore, the brownmillerite microdomains thus produced in air-quenched $Sr_2CoFeO_{5+\delta}$ were found to be significantly larger in size in the center of a pellet, while oxidation at the surface had produced a new phase, presumed to be perovskite. The authors arrived at the conclusion that the microdomains were produced by the accommodation of excess oxygen in the domain walls, arising from the diffusion of oxygen into the oxide lattice during quenching, and therefore would *not* be in thermodynamic equilibrium.

4. Brownmillerite Structure

The well-known brownmillerite structure with ideal stoichiometry $A_2B_2O_5$ can be derived from the ideal perovskite lattice by regularly removing one sixth of the oxygen atoms, resulting in an orthorhombic $\sim\sqrt{2}a \times 4a \times \sqrt{2}a$ supercell. The ordering of oxygen vacancies thus creates layers of corner-shared BO_6 octahedra (O) alternating with layers of corner-shared BO_4 tetrahedra (T) stacked in an ...OTOTOT... sequence, as shown in Figure 14.16. No less than six unique vacancy patterns are known for this type of structure.[224] The structures tend to disorder at elevated temperature, so as to produce a material with a high intrinsic concentration of mobile oxygen vacancies,[38] i.e., not achieved by aliovalent doping. Indeed, $Ba_2In_2O_5$ which has the brownmillerite structure of $Ca_2Fe_2O_5$ becomes a fast oxygen conductor, with ionic conductivities exceeding those of stabilized zirconia, after a first-order transition at about T_t = 930°C, at which the oxygen sublattice disorders.[212] The situation is analogous to that of δ-Bi_2O_3, which undergoes a first-order order–disorder transition at T_t = 730°C.[231] On the other hand, no increased ionic conductivity is found after a similar first-order transition in the isostructural $Sr_2Fe_2O_5$ at T_t = 700°C, which suggests an ordered arrangement of oxygen vacancies above T_t, while $Ca_2Fe_2O_5$ remains stable up to at least 1100°C.[232,233] Examination of $Sr_2Fe_2O_5$ by Mössbauer spectroscopy showed that tetrahedral coordination of Fe persists even in the apparently disordered material. It was therefore suggested that the ordering breaks up into microdomains of a size not detectable by XRD measurements.[234]

Seeking for solutions to lower T_t in $Ba_2In_2O_5$, Goodenough et al.[212,213] extended the aforesaid investigations to include $Ba_3In_2MO_8$ (M = Zr, Hf, Ce). Ideal ordering of the oxygen vacancies in these compounds would render them isostructural with $Ca_3Fe_2TiO_8$, in which two octahedral (O) layers alternate with one tetrahedral (T) layer in an ...OOTOOT... sequence along the c-axis.[235] Accordingly, the structure may be regarded as an intergrowth (n = 2) of perovskite layers alternating with brownmillerite layers. Enhanced ionic conduction of, in particular, $Ba_3In_2Zr_2O_8$ at modest temperatures was interpreted to reflect the disordering of the In^{3+} and Zr^{4+} cations over the cation sublattice, which would lead to interstitial oxygen in the T layers and correspondingly to oxygen vacancies in the O layers, due the preference

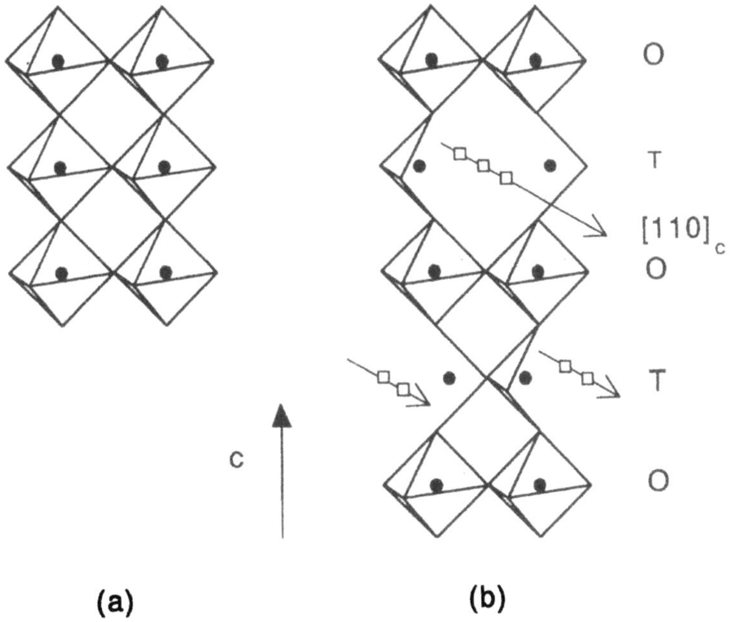

FIGURE 14.16. Idealized structure of (a) cubic perovskite ($CaTiO_3$) and (b) orthorhombic brownmillerite ($Ca_2Fe_2O_5$) lattice with ordered oxygen vacancies (□) along the cubic [110] direction.

of the Zr^{4+} cations for octahedral coordination vs. tetrahedral coordination. In a subsequent study,[236] however, these high values could not be reproduced. In fact, long-range ordering does not occur, and the residual low-temperature conduction is due to facile insertion of oxygen and/or water even below 400°C. In moist air, the ceramic discs crumbled on repeated thermal cycling. Only short-range ordering of the oxygen vacancies is anticipated. It is further suggested that any ordering of the oxygen vacancies in the oxygen-deficient perovskites so as to create layers of corner-shared BO_4 tetrahedra (T), as in $Ca_3Fe_2TiO_8$, makes more facile the insertion of oxygen and/or water at modest temperatures.

5. High-Temperature NMR

A point defect model has been proposed for the brownmillerite structure, based on a study of electrical conductivity and emf measurements of $Ba_2In_2O_5$.[237] The experimental results matched well with modeling, while evidence was found for protonic conduction at low temperatures. In modeling, the unoccupied oxygen sites relative to the perovskite structure, below T_t, are regarded as structural units and therefore are potential interstitial sites for oxygen. Charge neutrality is then dominated by intrinsic anion Frenkel disorder, i.e., $[O_i''] = [V_O^{\cdot\cdot}]$. Above T_t, charge neutrality would be determined by the condition $2[V_O^{\cdot\cdot}] = [In_{In}']$.

Partially at variance with these findings, Adler et al.[238] arrived at the conclusion that even above the order–disorder transition observed at $T_t = 925°C$, local order in the oxygen sublattice persists. Vacancy transport in $Ba_2In_2O_5$ would be confined mainly to two-dimensional layers, while oxygen anions are bounded and thus immobile in adjacent layers. Using high-temperature ^{17}O-NMR and XRD, it was demonstrated that the material retains its orthorhombic brownmillerite structure until ~1075°C, at which point it becomes cubic. The density of mobile oxygen anions was found to increase continuously between 925 and ~1075°C, and only above ~1075°C the full population of oxygen anions becomes mobile.

The investigations were extended by Adler et al.[239] to include $BaIn_{0.67}Zr_{0.33}O_{3-\delta}$, $BaIn_{0.67}Ce_{0.33}O_{3-\delta}$, and the cobalt-containing perovskites $La_{0.5}Ba_{0.5}Co_{0.7}Cu_{0.3}O_{3-\delta}$ and $La_{0.6}Sr_{0.4}Co_{0.8}Cu_{0.2}O_{3-\delta}$. While room-temperature X-ray and neutron powder diffraction revealed simple long-range cubic symmetry, HRTEM images and electron diffraction patterns indicated that these materials possess microdomains with a layered-like structure on a length scale of 50 to 500 Å. Measurements of high-temperature ^{17}O-NMR made apparent that in all these phases only a few oxygen vacancies are mobile below 800°C, suggesting that the remainder is trapped in locally ordered layers. Owing to intrinsic paramagnetism of the cobalt-containing compounds, the application of ^{17}O-NMR is not possible at room temperature. Only by rapid motion of oxygen do the spectra become visible. For both cobalt-containing perovskites, the signal intensity above 800°C was found to increase steadily with temperature, up to the maximum temperature of 950°C in the experiments by Adler et al.,[239] suggesting a concomitant increase in the number of mobile oxygen anions. In the case of $La_{0.6}Sr_{0.4}Co_{0.8}Cu_{0.2}O_{3-\delta}$, the value of the ionic migration enthalpy for oxygen transport estimated from NMR, viz., 1.13 ± 0.10 eV, showed good agreement with the 1.04 eV obtained from four-point ionic conductivity measurements.

It is clear that high-temperature NMR is an important tool to study the local structure and oxygen dynamics in the highly defective perovskites, which certainly merits its further use. In the case of $La_{0.6}Sr_{0.4}Co_{0.8}Cu_{0.2}O_{3-\delta}$, Adler et al.[239] assumed that oxygen nonstoichiometry is accommodated by local ordering of MO_5 square pyramids in a manner similar to that in $YBaCuFeO_5$[240] and $YBaCo_{2-x}Cu_xO_{5+\delta}$.[241] A similar type of structure was hypothesized for $La_{1-x}Sr_xCoO_{3-\delta}$ (x = 0.50 and 0.70) by Van Doorn et al.[242] As seen in Figure 14.17, two different sites for oxygen are present in the proposed structure for $La_{1-x}Sr_xCoO_{3-\delta}$. Four oxygens are in the basal plane of the square pyramid and one at the apex. Electron diffraction and HRTEM of powders, annealed about 15 h in air and furnace cooled, indicated that the presence of the tetragonal superstructure, corresponding to a doubling of the pseudocubic perovskite unit cell, gives rise to a texture consisting of microdomains with a mean size estimated at about (500 Å)^3.

G. OBSERVATIONS FROM PERMEABILITY MEASUREMENTS

This section discusses selected observations from oxygen permeation experiments. General trends have been summarized already in Section V.A.

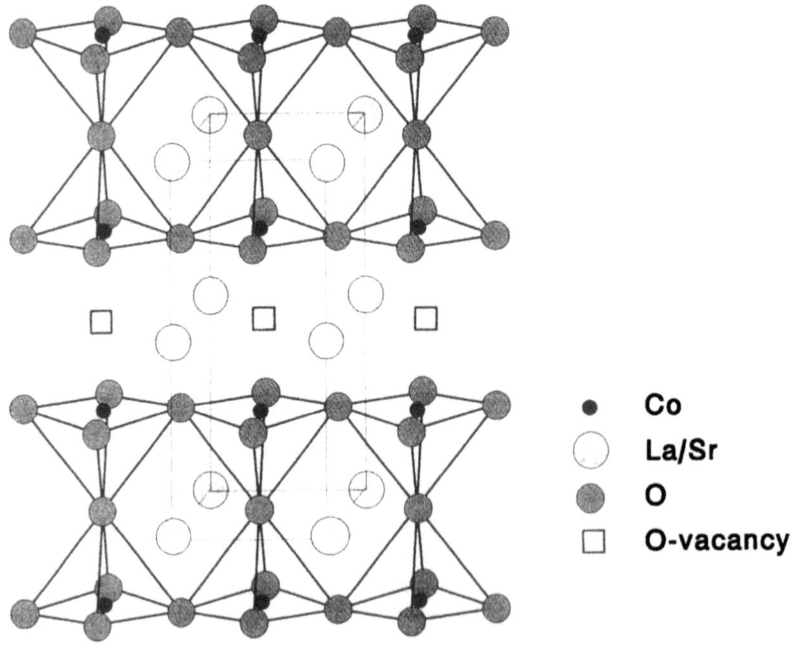

FIGURE 14.17. Ideal structure proposed for $La_{1-x}Sr_xCoO_{3-\delta}$ (x = 0.5 and x = 0.7). (Adapted from Van Doorn, R.H.E., Keim, E.G., Kachliki, T., Bouwmeester, H.J.M., and Burggraaf, A.J., in preparation.)

1. $SrCo_{0.8}Fe_{0.2}O_{3-\delta}$

The highest oxygen flux at 850°C in the series $La_{1-x}Sr_xCo_{1-y}Fe_yO_{3-\delta}$ studied by Teraoka et al.[39] was found in $SrCo_{0.8}Fe_{0.2}O_{3-\delta}$ (see Table 14.1) and this composition has been reexamined by a number of investigators.[15,43,152,153] In a study of oxygen permeation through $SrCo_{0.8}B_{0.2}O_{3-\delta}$ (B = Cr, Fe, Co, Cu), Kruidhof et al.[153] showed for the first time that in $SrCo_{0.8}Fe_{0.2}O_{3-\delta}$ a change in permeation mechanism, corresponding with an order–disorder transition, occurs at about T_t = 790°C. This change has been observed by others too,[152,15] but was not noted in the previous study by Teraoka et al.[39] Results from thermal analysis and X-ray powder diffraction under controlled oxygen partial pressures indicate that the observed phenomena can be attributed to the transition of $SrCo_{0.8}Fe_{0.2}O_{3-\delta}$ from vacancy-ordered brownmillerite to defective perovskite, occurring at reduced oxygen partial pressure. Both the structures of the cubic perovskite and orthorhombic brownmillerite-type phases of $SrCo_{0.8}Fe_{0.2}O_{3-\delta}$ have been refined using powder neutron diffraction data by Harrison et al.[243] A phase diagram was presented by Qiu et al.,[152] showing that only at relatively high oxygen partial pressure (>0.1 atm) and high temperature the perovskite phase is thermodynamically stable. At relatively low oxygen partial pressure and low temperature a perovskite–brownmillerite two-phase region is found. The brownmillerite phase has only a small homogeneity region around $3 - \delta = 2.5$. Below T_t, the situation during flux measurements therefore becomes very complicated, considering the fact that the P_{O_2} gradient across the membrane also may cross the two-phase region provided, of course, that such a gradient is imposed during experiment. The studies report slow kinetics of transformation between the brownmillerite and perovskite phases in view of the long times for the oxygen flux to reach steady-state conditions at these modest temperatures. Kruidhof et al.[153] attributed these to a progressive growth of microdomains of the ordered structure in a disordered perovskite matrix.

Based on experiments in which the membrane thickness was varied in the range 5.5 to 1.0 mm, Qiu et al.[152] arrived at the conclusion that the surface oxygen exchange process is the rate-limiting step in the overall oxygen permeation mechanism. Further experimental evidence that the oxygen fluxes through $SrCo_{0.8}Fe_{0.2}O_{3-\delta}$ are limited by the surface exchange kinetics was given by the present authors.[43] Fitting the oxygen permeation fluxes obtained from measurements at 750°C under various oxygen partial pressure gradients to Equation (14.18) yielded a positive slope of $n = +0.5$, where a value between 0 and -0.5 is expected from the experimentally observed $\ln \delta$-$\ln P_{O_2}$ relationship. However, these results merit further investigation, as the flux data were taken at a temperature just below the order–disorder transition in this material.

It has already been known for some time that $SrCoO_{3-\delta}$ transforms reversibly from a brownmillerite-like structure to defective perovskite at about T_t = 900°C in air. Kruidhof et al.[153] observed that the transition temperature is not, or only slightly, affected if $SrCoO_{3-\delta}$ is substituted with either 20 mol% Cr or Cu at the Co sites. Interesting to note is that the oxygen flux for the undoped and doped specimens is very small below T_t, as expected for an ordered arrangement of oxygen vacancies, but is found to increase sharply (between 5 to 6 orders of magnitude) at the onset of the phase transition to defective perovskite, up to values between 0.3 to 3×10^{-7} mol cm^{-2} s^{-1}. In view of these results, the perovskite phase in $SrCoO_{3-\delta}$ seems to be stabilized by the partial substitution of Co with Fe, but not with Cu or Cr, thereby suppressing the brownmillerite–perovskite two-phase region to lower oxygen partial pressures.

2. Experimental Difficulties

In a number of studies, the oxygen fluxes through, e.g., $SrCo_{0.8}Fe_{0.2}O_{3-\delta}$ have been reported to be significantly lower than claimed by Teraoka et al.[39] Such conflicting results reflect the difficulties in measuring the oxygen fluxes at high temperatures and may, at least partly, be due to specific conditions, including (1) edge effects associated with the required sealing of sample discs to avoid gas bypassing, giving rise to nonaxial contributions to the oxygen flux; (2) possible interfacial reactions when a glass is used for sealing; (3) undesired spreading of the glass seal (when its softening temperature is too low) over the oxide disc surface; and (4) the precise value of the P_{O_2} gradient across the membrane. With regard to the first point, it is frequently the cross-sectional area of the disc that is used in the calculation of the oxygen flux. In the usual experimental arrangements, however, an appreciable portion of the membrane is "clamped" between the impermeable annular seal. This edge effect means that the usual assumption of one-dimensional diffusion is not strictly correct. Another contribution to nonaxial transport is that of the flow of oxygen through the side walls of the disc specimens, if left uncovered. Appreciable errors creep in if these edge effects are neglected, as shown, for example, on the basis of a solution of Fick's second diffusion equation (with a constant diffusion coefficient) by Barrer et al.[244] This is further demonstrated in Figure 14.18, showing the effect of sealing edges on the departure from one-dimensional diffusion. These results were obtained from a numerical procedure to solve the steady-state diffusion equation in cylindrical coordinates.[130] Neglecting edge effects corrupts analysis of experiments in which the membrane thickness is varied, and may lead to erroneous conclusions when one tries to infer from the acquired data the influence of the surface exchange kinetics on overall oxygen permeation. Finally, it cannot be excluded that the observed oxygen fluxes are specific for the particular sample under investigation and may be affected, for instance, by microstructural effects, a point to which we return in Section V.G.5.

The gas flow rate of, in particular, the inert gas used to sweep the oxygen-lean side of the membrane affects the P_{O_2} gradient across the membrane. Under ideal gas mixing conditions, the P_{O_2} at the oxygen-lean side of the membrane is determined by the amount of oxygen permeating through the membrane. If the flow rate is not adjusted to obtain a constant P_{O_2} at this side of the membrane, but a constant gas flow rate is used, the P_{O_2} gradient gets smaller with increasing oxygen flux. This may give rise to an apparent activation energy for overall permeation, which may depart significantly from the one derived if a constant P_{O_2} were maintained at this side of the membrane.[147,148,152] The adjustable range of the sweeping gas flow rate (to a constant P_{O_2} at the outlet of the reactor) may be limited during the experiment, being determined by the requirement that the reactor behavior remains close to that of a continuous stirred-tank reactor (CSTR).

Using a constant value of P_{O_2} at the oxygen-lean side of 2-mm-thick disc membranes of $La_{1-x}Sr_xCoO_{3-\delta}$ ($x = 0.2, 0.3, 0.4, 0.5$, and 0.6), Van Doorn et al.[147,148] showed that the activation energy E_{act} for oxygen permeation in the range 900 to 1100°C decreases from 164 kJ mol^{-1} for x = 0.2 to 81 kJ mol^{-1} for x = 0.6. Opposed to these results, E_{act} decreased from 121 to 58 kJ mol^{-1} when a constant gas flow rate of the helium was used. Besides an improved fit to the Arrhenius equation in the former case, E_{act} can be correlated with the sum of the enthalpies for migration and that for the formation of oxide ion vacancies for each of the investigated compositions. Such a correlation is expected if oxygen transport is driven by the gradient in oxygen nonstoichiometry across the membrane due to the imposed P_{O_2} gradient. It suggests that oxygen vacancies are free and noninteractive in $La_{1-x}Sr_xCoO_{3-\delta}$ under the conditions covered by experiment. Oxygen permeation fluxes for strontium-doping levels above x = 0.6 were found to be partially controlled by the surface exchange kinetics, as already mentioned in Section V.D.

Contradictory to the observed behavior at high temperatures, results from thermal analysis and oxygen permeation measurements indicated that a phase transition, with a small first-order component, probably related with order–disorder of oxygen vacancies, occurs in selected compositions $La_{1-x}Sr_xCoO_{3-\delta}$ in the range 750 to 775°C.[147,148] Long times extending to over 30 h were needed for equilibration toward steady-state oxygen permeation at these modest temperatures. Such a behavior is reminiscent of that observed for $SrCo_{0.8}Fe_{0.2}O_{3-\delta}$, where this can be attributed to the slow kinetics of the transformation between the brownmillerite and perovskite phases at modest temperatures. In the case of $La_{1-x}Sr_xCoO_{3-\delta}$ (x =

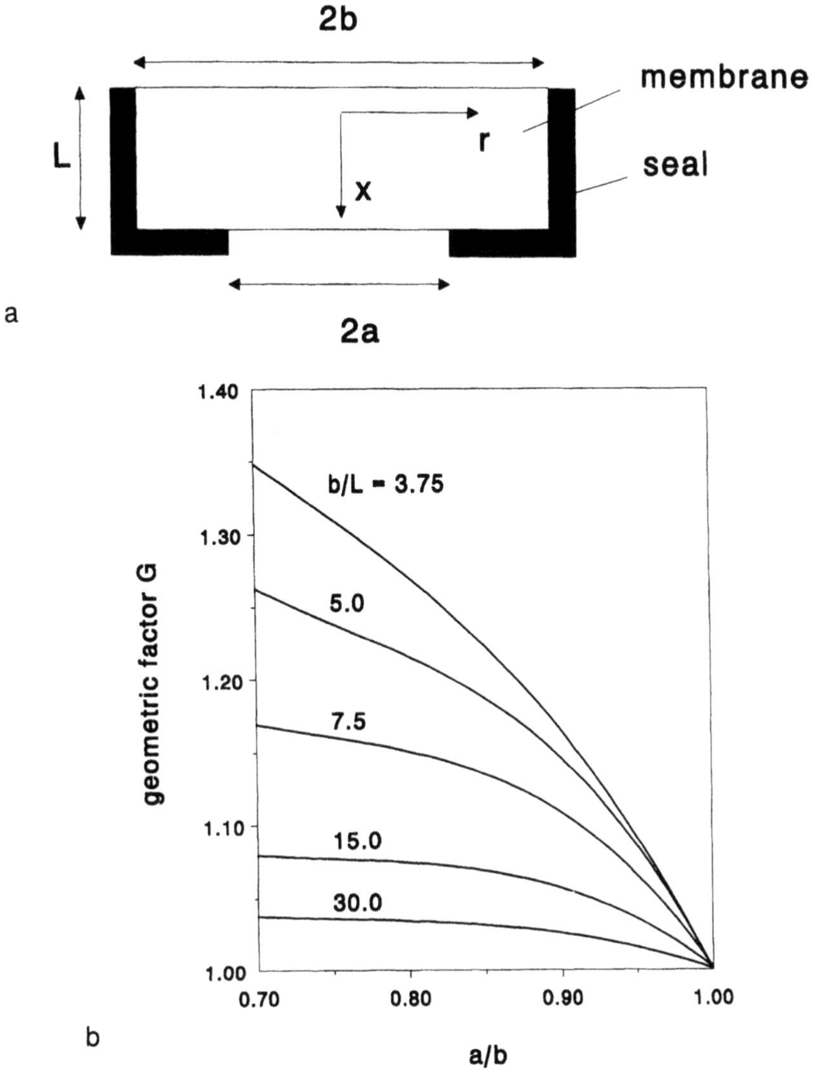

FIGURE 14.18. (a) Schematic cross section of a disc membrane. Dashed parts indicate insulating boundaries. (b) Influence of sealing edge effects on the departure from one-dimensional diffusion. A geometric factor G is used for correction of the flux (normalized to surface area with diameter 2a). Relevant parameters are defined in Figure 14.18a.

0.50 and 0.70), microdomains were observed in electron diffraction and HRTEM, corresponding to ordered arrangements of oxygen vacancies in these compounds at room temperature, as mentioned in the previous section.

Another factor that is considered to be responsible for a reduced oxygen flux is the surface modification of the perovskite oxide membrane by reaction with impurities in the gas phase, as emphasized by Qiu et al.[152] Referring to the surface degradation by reaction with minor amounts of CO_2 and corresponding deterioration of the properties observed for YBa_2CuO_{6+x} superconducting thin films,[245] a similar modification effect could occur when, e.g., ambient air is used as the source of oxygen at the membrane feed side. With the help of N_2 and O_2 admixed to feed side pressure $P_{O_2}' = 0.21$ atm, Qiu et al.[152] found the oxygen fluxes through $SrCo_{0.8}Fe_{0.2}O_{3-\delta}$ in the range 620 to 920°C to be larger by a factor of about 6 than when ambient air was used as feed gas, but still a factor of about 5 smaller than measured by Teraoka et al.[39] Similar experiments were conducted in our study on $SrCo_{0.8}Fe_{0.2}O_{3-\delta}$,[43,153] where this effect was not noted in the temperature range 700 to 950°C, so that we are inclined to believe that other factors must account for the disagreements in oxygen fluxes. This interpretation is supported by experimental evidence disclosed in a number of patents: that the oxygen fluxes through perovskite membranes remain stable as long as these are operated above certain critical temperatures, the precise value depending on the type of alkaline-earth dopant applied. Below these temperatures, a loss in oxygen flux may be observed over a period of about 100 h by as much as 30 to 40% when a membrane is exposed to CO_2 and H_2O impurities in the feed gas. This is further exemplified in Section VI.

3. Surface Exchange Kinetics

Attention has already been drawn to the importance of the surface exchange kinetics in determining the rate of oxygen permeation through mixed-conducting oxides in Section III.B.2. Though for the perovskites a value of 100 μm is often quoted for the characteristic membrane thickness L_c, at which the changeover from bulk to surface control occurs, in a number of cases much higher values are found, up to about 3000 μm (Table 14.2). As was emphasized earlier, the parameter L_c is not an intrinsic material property and, hence, may be specific to the sample under investigation and experimental conditions. The basic assumptions made in the derivation, notably that of small P_{O_2} gradients across the membrane, may restrict its use in practical situations, where these gradients can be substantial. Experimental evidence that the oxygen fluxes are limited by the surface exchange kinetics has been found in a number of cases, as discussed elsewhere in this text.

4. Behavior in Large P_{O_2} Gradients

The mixed-conducting perovskite oxides have attracted particular interest for use as dense ceramic membrane to integrate oxygen separation and partial oxidation of, for example, methane to C_2 products or syngas into a single step. Such a process bypasses the use of costly oxygen since air can be used as oxidant on the oxygen-rich of the membrane. Below a number of observations are briefly described which relate to the conditions relevant to the operating environment in chemical reactors.

Using $SrCo_{0.8}Fe_{0.2}O_{3-\delta}$ tubular membranes fabricated by an extrusion method, Pei et al.[15] observed two types of fracture of the tubes during the process for generating syngas. The first fracture, occurring shortly (within 1 h) after initiation of the reaction at 800°C, resulted from the P_{O_2} gradient across the membrane and the accompanying strain due to lattice mismatch and the brownmillerite–perovskite phase transition. The second type of fracture, occurring after prolonged exposure to the reducing environment, resulted from chemical decomposition toward $SrCO_3$, and elemental Co and Fe. Similar observations have been reported for tubes made of $La_{0.2}Sr_{0.8}Co_{0.4}Fe_{0.6}O_{3-\delta}$,[16] and in that study an optimized composition was also claimed, but not given, showing stable performance for up to 500 h. Using a rhodium-based reforming catalyst inside the tubes, methane conversions over 99% were achievable.

Ten Elshof et al.[12] studied the oxidative coupling of methane using a disc reactor with $La_{0.6}Sr_{0.4}Co_{0.8}Fe_{0.2}O_{3-\delta}$ as the catalyst membrane for the supply of oxygen to the methane feed stream. Examination of the oxygen fluxes measured under various P_{O_2} gradients in the range of thickness 0.55 to 0.98 mm suggested that the surface exchange reaction limits the rate of oxygen permeation. The oxygen flux was found to increase only slightly when methane was admixed with the helium used as the carrier gas. The methane was converted to ethane and ethene with selectivities up to 70%, albeit with a low conversion, typically in the range 1 to 3% at operating temperatures 1073 to 1173 K. The selectivity observed at a given oxygen flux and temperature was about twice as low if the same amount of molecular oxygen was cofed with the methane feed stream in a single chamber reactor design, suggesting that the membrane mode of operation is conceptually more attractive for generating C_2 products. Decomposition of the oxide surface did not occur as long as molecular oxygen could be traced at the reactor outlet, which emphasizes the importance of a surface-controlled oxygen flux for membrane-driven methane coupling. That is, for a bulk diffusion-controlled oxygen flux the surface would become reduced by the methane, until the depth of reduction has progressed up to a point where the oxygen flux counterbalances the consumption of oxygen by methane. On the one hand, the slow surface exchange kinetics observed on $La_{0.6}Sr_{0.4}Co_{0.8}Fe_{0.2}O_{3-\delta}$ limits the magnitude of the oxygen fluxes; on the other hand, its existence prevents the oxide surface from reduction, i.e., as long as the rate of oxygen supply across the membrane exceeds the rate of (partial) oxidation of methane. Noteworthy is that segregation of strontium occurred on both sides of the membrane, as confirmed by depth-profiling Auger analysis. The extent of segregation appeared to be influenced by the imposed P_{O_2} gradient across the membrane, and was also found if a pure helium stream was passed along the oxygen-lean side of the membrane.

Van Hassel et al.[149] studied oxygen permeation through $La_{1-x}Sr_xFeO_{3-\delta}$ (x = 0.1, 0.2) membranes in a disc reactor using CO-CO_2-based gas mixtures to control the P_{O_2} at the oxygen-lean side. Ambient air was used as the oxygen source at the opposite side of the membrane. At 800 to 1100°C, the oxygen flux was found to increase linearly with the partial pressure of CO. Deposition of a 50-nm-thin porous Pt layer on this side of the membrane increased the oxidation rate and likewise the oxygen flux, by a factor of about 1.8. In a separate study,[246] the oxygen flux was found to be invariant with the thickness of the membrane in the range 0.5 to 2.0 mm, while no effect was observed upon varying the P_{O_2} at the oxygen-rich side. It was concluded that the oxygen flux is fully limited by the carbon monoxide oxidation rate. The experimentally determined rate constants scale with Sr content in the extended range of composition $0.1 \leq x \leq 0.4$. The latter can be accounted for, in view of the fact that the oxygen deficiency of the ferrites is fixed by the dopant concentration in a wide range of oxygen partial pressure, by assuming that oxygen vacancies act as active sites in the oxidation reaction of CO on the perovskite surface following either an Eley–Rideal or a Langmuir–Hinselwood type of mechanism.

5. Grain Boundary Diffusivity

Besides the possibility of surface exchange limitations, oxygen transport through dense ceramics is necessarily influenced by the presence of high-diffusivity paths along internal surfaces such as grain boundaries. A systematic study investigating to which extent these preferred diffusion paths contribute to the diffusivity in the perovskite oxides is, however, still lacking. Both impurity and solute segregation take place at grain boundaries (and the external surface) or in their close proximities (less than 3 or 4 atomic distances) during sintering and subsequent heat treatments. An obvious consideration is that, in general, these significantly alter the magnitude of ionic transport along and across the grain boundaries. In many cases the ceramics invariably contain impurities present in the starting powder or added as a sintering aid to lower the sintering temperature and/or to achieve high density. It therefore cannot be excluded that disagreements in the literature regarding the magnitude of the oxygen fluxes can be explained on the basis of different ceramic processing techniques used by various authors.

In general, the presence of high-diffusivity paths is important in ceramics where lattice diffusion is slow. Analyzing ^{18}O depth profiles using SIMS, Yasuda et al.[247] noted a significant contribution of the grain boundary diffusion to the diffusivity in the interconnect material $La_{0.7}Ca_{0.35}CrO_{3-\delta}$, where the tracer diffusivity is of the order of $\sim 10^{-13}$ cm^2 s^{-1} at 900°C. Erroneous results were obtained when isotopic exchange was performed by gas phase analysis, which resulted in apparent tracer diffusion coefficients that were almost 2 orders in magnitude higher. More recently, Kawada et al.[185] confirmed the existence of high-diffusivity paths along grain boundaries in $La_{0.7}Ca_{0.3}CrO_{3-\delta}$ using depth profiling and imaging SIMS of ^{18}O-^{16}O exchanged specimens. But, oxygen permeation measurements suggested negligible contribution of grain boundary diffusion to the steady-state oxygen flux. These data were obtained at 1000°C for a sample thickness of 0.75 mm. An oxygen pump and sensor were used to control the permeate side P_{O_2}. The results are well described by the Wagner equation, assuming a random point defect scheme for $La_{0.7}Ca_{0.3}CrO_{3-\delta}$, as discussed in Section V.B.2.

For fast ionic conductors grain boundary diffusion will have little influence, or indeed may become blocking to the diffusion from one grain to the next as is recognized in the interpretation of impedance spectra from ionic conductivity of zirconia- and ceria-based solid electrolytes. In these ceramics silicon is the most common impurity detected along with enhanced yttrium segregation. Various models to account for the effects of segregation at grain boundaries and how these affect the electrical properties have been discussed by Badwal et al. [248] Although there is no unique model describing the ceramic microstructure, the most widely adopted model for doped zirconia and doped ceria is the brick-layer model. In this model, bricks present the grains and mortar the grain boundary region, i.e., assuming the grain boundary phase to completely wet the grains.[249,250] The grain boundaries in series with the grains, along the direction of charge flow, mainly contribute to the grain boundary resistivity. For doped zirconia and ceria the grain boundary resistivity can be of a similar order of magnitude or higher than the bulk resistivity. It is evident that more detailed studies are needed to aid in the interpretation of oxygen transport through the mixed-conducting perovskite oxides, where similar blocking effects can be expected.

VI. FINAL REMARKS

The considerations in this chapter were mainly prompted by the potential application of mixed-conducting perovskite-type oxides to be used as dense ceramic membranes for oxygen delivery applications, and lead to the following general criteria for the selection of materials:

- high electronic and ionic conductivity
- high catalytic activity toward oxygen reduction and reoxidation
- ability to be formed into dense thin films, free of micro-cracks and connected-through porosity
- chemical and structural integrity (i.e., no destructive phase transition) within appropriate ranges of temperature and oxygen partial pressure
- low volatility at operating temperatures
- thermal and chemical compatibility with other cell components
- low cost of material and fabrication

The precise perovskite composition may be tailored for a specific application. To obtain a high-performance membrane, however, many technical and material problems remain to be solved. This final section will focus on several issues, which are not yet well understood, but are thought to be of importance for further development of the membrane devices.

In the first place our understanding of factors that control and limit the interfacial kinetics is still rudimentary, and therefore should be a fruitful area for further investigation. The apparent correlation between the surface oxygen exchange coefficient k_s and the tracer diffusion coefficient D^* for two different classes of oxides, the fluorite-related and the perovskite-related oxides, as noted by Kilner,[76] clearly indicate the potential of isotopic ^{18}O-^{16}O exchange. However, a problem remains as to how to relate the observations (at equilibrium) from isotopic exchange to the conditions met during membrane operation. In chemical relaxation experiments, the oxide is studied after perturbation of the equilibrium state. These methods are thus complementary and probably their combined application, whenever possible together with spectroscopic techniques, such as FT-IR, UV, and EPR, has a great capacity to elucidate the kinetics of surface oxygen exchange.

Though, at first glance, the limited exchange capability of the perovskites relative to diffusion puts limits on attempts to improve the oxygen fluxes or to lower the operating temperatures by making thinner membranes, it is expected that the surface exchange kinetics can be significantly improved by surface modification. One approach is coating with a porous surface layer which will effectively enlarge the surface area available to exchange, as discussed in Section III.B.3. Improvements can also be expected by finely dispersing precious metals or other exchange active second phases on the oxide surface. It is clear that further investigations are required to evaluate these innovative approaches. As yet, more work is also required to gain insight in the role of the ceramic microstructure in the performance values of membranes, and to evaluate different processing routes for the fabrication of perovskite thin films.

Besides the technological challenge of fabrication of dense and crack-free thin perovskite films, which need to be supported if its thickness is less than about 150 μm, a number of other problems relate to the long-term stability of perovskite membranes, including segregation, a low volatility of lattice components, etc. Some of these problems are linked to the imposed oxygen pressure gradient across the membrane. Aside from the lattice expansion mismatch of opposite sides of the membrane, attention is drawn to the potential problem of demixing, which arises in almost all situations where a multicomponent oxide is brought into a gradient of oxygen chemical potential. The available theories predict that, if the mobilities of the cations are different and nonnegligible at high temperatures, concentration gradients appear in the oxide in such a way that the high-oxygen pressure side of the membrane tends to be enriched with the faster moving cation species. Depending on the phase diagram, the spatially inhomogeneous oxide may eventually decompose. The latter may cause surprise, if the (homogeneous) oxide is stable in the range of oxygen partial pressures covered by experiment. This is why these processes have been termed kinetic demixing and kinetic decomposition by Schmalzried et al.,[251,252] who were the first to study them.

Degradation phenomena have been shown to occur in, for example, $Co_{1-x}Mg_xO$, Fe_2SiO_4, and $NiTiO_3$. Internal oxidation or reduction processes sometimes lead to precipitation of a second phase in the matrix of the parent phase. Another possible consequence of the demixing process is the morphological instability of the (moving) low-pressure interface due to formation of pores, which may eventually penetrate throughout the ceramic. The above phenomena have been the subject of a number of theoretical and experimental studies in the last decade,[253-257] to give only a brief number. A review up to 1986 has been written by Schmalzried.[258] To our knowledge, no report has been made up to now of demixing phenomena in MIEC. Since they cannot be excluded to occur on the basis of theoretical arguments, this is also why the phenomena deserve (more) attention in order to be able to control the deterioration of membrane materials.

Intergrowth structures in which perovskite-type blocks or layers are held apart by non-perovskite ones could offer a new strategy for identifying new materials, as was suggested earlier by Goodenough et al.[213] In such structures, vacancy transport is confined to two-dimensional layers or to sites which link up to form channels extending throughout the crystal. An interesting variation to the BIMEVOX compounds, already discussed in Section IV.C.1, is found in derivatives of $Sr_4Fe_6O_{13}$. Its orthorhombic structure can be described as built of perovskite layers alternating with sesquioxide Fe_2O_3 layers perpendicular to the b axis. The discovery of high levels of oxygen permeation through mixed metal oxide compositions obtained by partial substitution of iron for cobalt, for instance, $SrCo_{0.5}FeO_x$ recently translated into a patent for this class of materials.[259] Tubes made from the given composition showed oxygen fluxes similar to those through known state-of-the-art materials having a perovskite structure, but did not fracture in the process for preparing syngas as was found for some of the perovskite materials.

As noted before, the membrane performance could be affected by the presence of H_2O, CO_2, or other volatile hydrocarbons in the gas phase of both compartments. As laid down in patent literature,[3-5] the oxygen fluxes through Mg-, Ca-, Sr-, and Ba-doped perovskites deteriorated over time, roughly 30 to 50% over a time period of about 100 h, if the air used as feed gas contained several percent of H_2O and amounts of CO_2 on a hundreds-of-parts per million level. It was claimed that either no deterioration is found or the fluxes can be restored to their initial values if the temperature is raised above certain critical values, 500°C for magnesium, 600°C for calcium, 700°C for strontium, and 810°C for barium. Though no explanation was given, it is possible that carbonate formation took place. One may further note that the tendency for carbonate formation increases at lower temperatures.

A surprising observation was recently made in the author's laboratory in a study of oxygen permeation through $La_{1-x}Sr_xFeO_{3-\delta}$ $(0.1 \leq x \leq 0.4)$.[150] Long times to reach steady-state oxygen permeation at 1000°C extending over hundreds of hours were observed, yet could be avoided by exposing the permeate side surface of the membrane for a 1 to 2 h to a 1:1 CO/CO_2 gas mixture. A clear explanation cannot yet be given for this observation, which is still under investigation, though a reconstruction of the surface by the reducing ambient cannot be excluded. The oxygen permeability measured if helium was used again as the sweeping gas on this side of the membrane, was found to be limited by diffusional transport of oxygen across the membrane.[150] A similar type of observation was made by Miura et al.,[151] who noticed the oxygen flux through slip-casted membranes of $La_{0.6}Sr_{0.4}Co_{0.8}Fe_{0.2}O_{3-\delta}$ to be greatly improved if these were freed from surface impurities, like SrO, following an acid treatment.

One final point to note is the ability of acceptor-doped perovskite oxides to incorporate water, and some contribution of proton conduction therefore cannot be excluded. If water insertion occurs at low temperature, this might lead to residual stresses in the ceramics. Besides, water may play an active role in the surface oxygen exchange. For example, on Bi_2MoO_6, which has an intergrowth structure consisting of $Bi_2O_2^{2+}$ blocks alternating with MO_4^{2-} layers of corner-shared MO_6 octahedra, exchange with $^{18}O_2$-enriched oxygen could not be observed experimentally.[260] On the other hand, Novakova and Jiru[261] demonstrated that exchange of water with lattice oxygen on an industrial bismuth molybdate catalyst proceeds rapidly at 200°C, and is even measurable at room temperature.

ACKNOWLEDGMENT

The author is indebted to his colleagues H. Kruidhof, R.H.E. van Doorn, J.E. ten Elshof, M.H.R. Lankhorst, and B.A. van Hassel for many useful discussions and for providing experimental data. Paul Gellings and Henk Verwey are gratefully acknowledged for valuable comments and careful reading of the manuscript. The Commission of the European Communities and the Netherlands Foundation for Chemical Research (SON) are thanked for financial support.

REFERENCES

1. Burggraaf, A.J., Bouwmeester, H.J.M., Boukamp, B.A., Uhlhorn, R.J.R., and Zaspalis, V.T., Synthesis, microstructure and properties of porous and dense ceramic membranes, in *Science of Ceramic Interfaces*, Nowotny, J., Ed., Elsevier Science Publishers B.V., North-Holland, 1991, 525–568.
2. *Inorganic Membranes, Synthesis, Characteristics and Applications*, Bhave, R.R., Ed., Van Nostrand Reinhold, New York, 1991.
3. Carolan, M.F., Dyer, P.N., LaBar, J.M., Sr., and Thorogood, R.M., Process for restoring permeance of an oxygen-permeable ion transport membrane utilized to recover oxygen from oxygen-containing gaseous mixtures, U.S. Patent 5,240,473, 1993.
4. Carolan, M.F., Dyer, P.N., LaBar, J.M., Sr., and Thorogood, R.M., Process for recovering oxygen from gaseous mixtures containing water or carbon dioxide which process employs ion transport membranes, U.S. Patent 5,261,932, 1993.
5. Carolan, M.F., Dyer, P.N., Fine, S.M., LaBar, J.M., Sr., and Thorogood, R.M., Process for recovering oxygen from gaseous mixtures containing water or carbon dioxide which process employs barium-containing ion transport membranes, U.S. Patent 5,269,822, 1993.

6. Liu, M., Joshi, A.V., Shen, Y., and Krist, K., Mixed ionic-electronic conductors for oxygen separation and electrocatalysis, U.S. Patent 5,273,628, 1993.
7. Thorogood, R.M., Developments in air separation, Gas Sep. Purif., 1991, *5*, 83–94.
8. Hazbun, E.A., Ceramic membrane for hydrocarbon conversion, U.S. Patent 4,791,079, 1988.
9. Hazbun, E.A., Ceramic membrane and use thereof for hydrocarbon conversion, U.S. Patent 4,827,071, 1989.
10. Wang, W. and Lin, Y.S., Analysis of oxidative coupling in dense oxide membrane reactors, *J. Mem. Sc.*, 1995, 103, 219-233.
11. Elshof, J.E. ten, Van Hassel, B.A., and Bouwmeester, H.J.M., Activation of methane using solid oxide membranes, *Catal. Today*, 1995, *25*, 397–402.
12. Elshof, J.E. ten, Bouwmeester, H.J.M., and Verweij, H., Oxidative coupling of methane in a mixed-conducting perovskite membrane reactor, *Appl. Catal. A: General*, 1995, *130*, 195–212.
13. Nozaki, T. and Fujimoto, K., Oxide ion transport for selective oxidative coupling of methane with new membrane reactor, *AIChE J.*, 1994, *40*, 870–77.
14. Tsai, C.-Y., Ma, Y.H., Moser, W.R., and Dixon, A.G., Modeling and simulation of a nonisothermal catalytic membrane reactor for methane partial oxidation to syngas, *Chem. Eng. Comm.*, 1995, 134, 107-132.
15. Pei, S., Kleefisch, M.S., Kobylinski, T.P., Faber, K., Udovich, C.A., Zhang-McCoy, V., Dabrowski, B., Balachandran, U., Mieville, R.L., and Poeppel, R.B., Failure mechanisms of ceramic membrane reactors in partial oxidation of methane to synthesis gas, *Catal. Lett.*, 1995, *30*, 210–12.
16. Balachandran, U., Dusek, J.T., Sweeney, S.M., Poeppel, R.B., Mieville, R.L., Maiya. P.S., Kleefisch, M.S., Pei, S., Kobylinski, T.P., Udovich. C.A., and Bose, A.C., Methane to syngas via ceramic membranes, *Am. Ceram. Soc. Bull.*, 1995, *74(1)*, 71–75.
17. Dixon, A.G., Moser, W.R., and Ma, Y.H., Waste reduction and recovery using O_2-permeable membrane reactors, *Ind. Eng. Chem. Res.*, 1994, *33*, 3015–24.
18. Hsieh, H.P., Inorganic membrane reactors, *Catal. Rev.-Sci. Eng.*, 1992, *33(1&2)*, 1–70.
19. Gurr, W.R., An operators view on gas membranes, in *Effective Industrial Membrane Processes — Benefits and Opportunities*, Turner, M.K., Ed., Elsevier Science Publishers, Barking, Essex, 1991, 329.
20. Zaspalis, V.T. and Burggraaf, A.J., Inorganic membrane reactors to enhance the productivity of chemical processes, in *Inorganic Membranes, Synthesis, Characteristics and Applications*, Bhave, R.R., Ed., Van Nostrand Reinold, New York, 1991, 177–207.
21. Mazanec, T.J., Prospects for ceramic electrochemical reactors in industry, *Solid State Ionics*, 1994, *70/71*, 11–19.
22. Saracco, G. and Specchia, V., Catalytic inorganic membrane reactors: present experience and future opportunities, *Catal. Rev.-Sci. Eng.*, 1994, *36(2)*, 303–84.
23. Gellings, P.J. and Bouwmeester, H.J.M., Ion and mixed-conducting oxides as catalysts, *Catal. Today*, 1992, *1*, 1–101.
24. Hayes, D., Budworth, D.W., and Roberts, J.P., Selective permeation of gases through dense sintered alumina, *Trans. Br. Ceram. Soc.*, 1961, *60(4)*, 494–504.
25. Hayes, D., Budworth, D.W., and Roberts, J.P., Permeability of sintered alumina to gases at high temperatures, *Trans. Br. Ceram. Soc.*, 1963, *62(6)*, 507–523.
26. Tuller, H.L., Mixed conduction in nonstoichiometric oxides, in *Nonstoichiometric oxides*, Sørenson, O., Ed., Academic Press, New York, 1981, 271–337.
27. Suitor, J.W., Clark, D.J., and Losey, R.W., Development of alternative oxygen production source using a zirconia solid electrolyte membrane, in Technical progress report for fiscal years 1987, 1988 and 1990, Jet Propulsion Laboratory Internal Document D7790, 1990.
28. Mazanec, T.J., Cable, T.L., and Frye, J.G., Jr., Electrocatalytic cells for chemical reaction, *Solid State Ionics*, 1992, *53–56*, 111–18.
29. Kleitz, M. and Siebert, M., Electrode reactions in potentiometric gas sensors, in *Chemical Sensor Technology*, Vol. 2, Seymana, T., Ed., Elsevier Science Publishers B.V., Amsterdam, 1989, 151–71.
30. Maskell, W.C. and Steele, B.C.H., Solid state potentiometric sensors, *J. Appl. Electrochem.*, 1986, *16*, 475–89.
31. Liou, S.S. and Worrell, W.L., Electrical properties of novel mixed-conducting oxides, *Appl. Phys. A.*, 1989, *49*, 25–31.
32. Liou, S.S. and Worrell, W.L., Mixed conducting electrodes for solid oxide fuel cells, in *Proc. 1st Int. Symp. on Solid Oxide Fuel Cells*, Singhal, S.C., Ed., The Electrochemical Society, Pennington, NJ, 1989, 81–89.
33. Calès, B. and Baumard, J.F., Electrical properties of the ternary solid solutions $(ZrO_{2\ 1-x}\text{-}CeO_{2\ x})_{0.9} - Y_2O_{3\ 0.1}$ $(0 \leq x \leq 1)$, *Rev. Int. hautes Tempér. Réfract., Fr.*, 1980, *17*, 137–147.
34. Calès, B. and Baumard, J.F., Mixed conduction and defect structure of ZrO_2-CeO_2-Y_2O_3 solid solutions, *J. Electrochem. Soc.*, 1984, *131(10)*, 2407–2413.
35. Nigara, Y. and Cales, B., Production of carbon monoxide by direct themal splitting of carbon dioxide at high temperature, *Bull. Chem. Soc. Jpn.*, 1986, *59*, 1997–2002.
36. Takahashi, T. and Iwahara, H., *Denki Kagaku*, 1967, *35*, 433.
37. Takahashi, T. and Iwahara, H., *Energy Conv.*, 1971, *11*, 105.
38. Steele, B.C.H., Oxygen ion conductors and their technological applications, *Mater. Sci. Eng.*, 1992, *B13*, 79–87.

39. Teraoka, Y., Zhang, H.M., Furukawa, S., and Yamazoe, N., Oxygen permation through perovskite-type oxides, *Chem. Lett.*, 1985, 1743–46.
40. Teraoka, Y., Nobunaga, T., and Yamazoe, N., Effect of cation substitution on the oxygen semipermeability of perovskite oxides, *Chem. Lett.*, 1988, 503–06.
41. Teraoka, Y., Nobunaga T., Okamoto, K., Miura, N., and Yamazoe, N., Influence of constituent metal cations in substituted $LaCoO_3$ on mixed conductivity and oxygen permeability, *Solid State Ionics*, 1991, *48*, 207–12.
42. Teraoka, Y., Zhang, H.M., Okamoto, K., and Yamazoe, N., Mixed ionic-electronic conductivity of $La_{1-x}Sr_xCo_{1-y}Fe_yO_{3-\delta}$, *Mater. Res. Bull.*, 1988, *23*, 51–58.
43. Bouwmeester, H.J.M., Kruidhof, H., and Burggraaf, A.J., Importance of the surface exchange kinetics as rate limiting step in oxygen permeation through mixed-conducting oxides, *Solid State Ionics*, 1994, *72*, 185–94.
44. Thorogood, R.M., Srinivasan, R., Yee, T.F., and Drake, M.P., Composite mixed conductor membranes for producing oxygen, U.S. Patent 5,240,480, 1993.
45. Deng, H., Zhou, M., and Abeles, B., Diffusion-reaction in porous mixed ionic-electronic solid oxide membranes, *Solid State Ionics*, 1994, *74*, 75–84.
46. Deng, H., Zhou, M., and Abeles, B., Transport in solid oxide porous electrodes: effect of gas diffusion, *Solid State Ionics*, 1995, *80*, 213–22.
47. Schmalzried, H., *Solid State Reactions*, Verlag Chemie, Weinheim, 1981.
48. Rickert, H., *Electrochemistry of Solids, an Introduction*, Springer-Verlag, Berlin, 1992.
49. Heyne, L., Electrochemistry of mixed-ionic electronic conductors, in *Solid Electrolytes, Topics in Applied Physics*, Geller, S., Ed., Springer-Verlag, Berlin, 1977, 167–221.
50. Kröger, F.A. and Vink, H.J., Relations between the concentrations of imperfections in crystalline solids, *Solid State Physics*, Vol. 3, Seitz, F. and Turnbull, D., Eds., Academic Press, New York, 1956, 307.
51. Wagner, C. and Schottky, W., Beitrag zur Theorie des Anlaufvorganges, *Z. Phys. Chem.*, 1930, *B11*, 25–41.
52. Wagner, C., Equations for transport in solid oxides and sulfides of transition metals, *Progr. Solid State Chem.*, 1975, *10(1)*, 3–16.
53. Murch, G.E., The Haven ratio in fast ionic conductors, *Solid State Ionics*, 1982, *7*, 177–98.
54. Maier, J. and Schwitzgebel, G., Theoretical treatment of the diffusion coupled with reaction, applied to the example of a binary solid compounds MX, *Phys. Status Solidi, B*, 1982, *113*, 535–47.
55. Maier, J., Mass transport in the presence of internal defect reactions — concept of conservative ensembles: I, Chemical diffusion in pure compounds, *J. Am. Ceram. Soc.*, 1993, *76(5)*, 1212–17, II, Evaluation of electrochemical transport measurements, *ibid.* 1993, *76(5)*, 1218–22, III, Trapping effect of dopants on chemical diffusion, *ibid.* 1993, *76(5)*, 1223–27, IV, Tracer diffusion and intercorrelation with chemical diffusion and ion conductivity, *ibid.* 1993, *76(5)*, 1228–32.
56. Maier, J., Diffusion in materials with ionic and electronic disorder, *Mater. Res. Soc. Symp. Proc.*, Vol. 210, 1991, 499–509.
57. Liu, M., Theoretical assessment of oxygen separation rates of mixed conductors, in *Ionic and Mixed Conducting Oxide Ceramics 91-12*, Ramanarayanan, T.A. and Tuller, H.L., Eds., Electrochemical Society, Pennington, NJ, 1991, 95–109.
58. Pleskov, Yu.V. and Gurevich Yu.Ya., *Semiconductor Photoelectrochemistry*, Consultants Bureau, New York, 1986, chap. 4.
59. Dignam, M.J., Young, D.J., and Goad, D.G.W., Metal oxidation. I. Ionic transport equation, *J. Phys. Chem. Solids*, 1973, *34*, 1227–34.
60. Young, D.J. and Dignam, M.J., Metal oxidation. II. Kinetics in the thin and very thin flim regions under conditions of electron equilibrium, *J. Phys. Chem. Solids*, 1973, *34*, 1235–50.
61. Maier, J., Kröger–Vink diagrams for boundary regions, *Solid State Ionics*, 1989, *32/33*, 727–33.
62. Nowotny, J., Surface segregation of defects in oxide ceramics materials, *Solid State Ionics*, 1988, *28–30*, 1235–43.
63. Adamczyck, Z. and Nowotny, J., Effect of the surface on gas/solid equilibration kinetics in non-stoichiometric compounds, in *Diffusion and Defect Data, Solid State Data*, Part B, Vols. 15 & 16, Nowotny, J., Ed., 1991, 285–336.
64. Bockriss, O'M. and Reddy, A.K.N., *Modern Electrochemistry*, Vol. 2, Plenum Press, New York, 1970, 991.
65. Steele, B.C.H., Isotopic oxygen exchange and optimisation of SOFC cathode materials, in *High Temperature Electrochemical Behaviour of Fast Ion and Mixed Conductors*, Poulsen, F.W., Bentzen, J.J., Jacobsen, T., Skou, E., Østergård, M.J.L., Eds., Risø National Laboratory, Denmark, 1993, 423–30.
66. Steele, B.C.H., Interfacial reactions associated with ceramic ion transport membranes, *Solid State Ionics*, 1995, *75*, 157–65.
67. Kleitz, M. and Kloidt, T., and Dessemond, L., Conventional oxygen electrode reaction: facts and models, in *High Temperature Electrochemical Behaviour of Fast Ion and Mixed Conductors*, Poulsen, F.W., Bentzen, J.J., Jacobsen, T., Skou, E., and Østergård, M.J.L., Eds., Risø National Laboratory, Denmark, 1993, 89–116.
68. Ando, K., Oishi, Y., Koizama, H., and Sakka, Y., Lattice defect and oxygen diffusivity in MgO stabilized ZrO_2, *J. Mater. Sc. Lett.*, 1985, *4*, 176–80.

69. Oishi, Y. and Kingery, W.D., Self-diffusion of oxygen in single crystal and polycrystalline aluminum oxide, *J. Chem. Phys.*, 1960, *33*, 480–86.
70. Boukamp, B.A., de Vries, K.J., and Burggraaf, A.J., Surface oxygen exchange in bismuth oxide based materials, in *Non-Stoichiometric Compounds, Surfaces, Grain Boundaries and Structural Defects*, Nowotny, J. and Weppner, W., Eds., Kluwer Academic, Dordrecht, 1989, 299–309.
71. Boukamp, B.A., Vinke, I.C., de Vries, K.J., and Burggraaf, A.J., Surface oxygen exchange properties of bismuth oxide-based solid electrolytes and electrode materials, *Solid State Ionics*, 1989, *32/33*, 918–23.
72. Kurumchin, E.Kh. and Perfiliev, M.V., An isotope exchange study of the behavior of electrochemical systems, *Solid State Ionics*, 1990, *42*, 129–133.
73. Park, K. and Olander, D.R., Oxygen diffusion in single-crystal tetragonal zirconia, *J. Electrochem. Soc.*, 1991, *138(4)*, 1154–59.
74. Kilner, J., Steele, B.C.H., and Ilkov, L., Oxygen self-diffusion studies using negative-ion secondary ion mass spectroscopy (SIMS), *Solid State Ionics*, 1984, *12*, 89–97.
75. Tarento, R.J. and Monty, C., Influence of nonstoichiometry on the oxygen self-diffusion in $Co_{1-x}O$ single crystals, *Solid State Ionics*, 1988, *28–30*, 1221–29.
76. Kilner, J., Isotopic exchange in mixed and ionically conducting oxides, in *Proc. 2nd Int. Symp. Ionic and Mixed Conducting Oxide Ceramics*, Vol. 94-12, Ramanarayanan, T.A., Worrell, W.L., and Tuller, H.L., Eds., The Electrochemical Society, Pennington, NJ, 1994, 174–190.
77. Yamauchi, S., Ishigaki, T., Mizusaki, J., and Fueki, K., Tracer diffusion coeffcient of oxide ions in $LaCoO_3$, *Solid State Ionics*, 1983, *9&10*, 997–1003.
78. Ishigaki, T., Yamauchi, S., Kishio, K., Mizusaki, J., and Fueki, K., Diffusion of oxide ion vacancies in perovskite-type oxides, *J. Solid State Chem.*, 1988, *73*, 179–87.
79. Simpson, L.A. and Carter, R.E., Oxygen exchange and diffusion in calcia-stabilized zirconia, *J. Am. Ceram. Soc.*, 1966, *49(3)*, 139–44.
80. Childs, P.E. and Wagner, J.B., Jr., Chemical diffusion in wüstite and chromium-doped manganous oxide, in *Heterogeneous Kinetics at Elevated Temperatures*, Beltane, G.R. and Worrell, W.L., Eds., Plenum Press, New York, 1970, 269–342.
81. Childs, P.E., Laub. L.W., and Wagner, J.B., Jr., Chemical diffusion in nonstoichiometric oxides, *Proc. Br. Ceram. Soc.*, 1973, *19*, 29–53.
82. Morin, F., Mesure du coefficient de diffusion chimique dans les oxydes par la methode de relaxation, *Can. Metall. Q.*, 1975, *14(2)*, 97-104.
83. Grabke, H.J., Kinetik der Sauerstoffübertragung aus CO_2 and die Oberfläche von Oxyden, *Ber. Bunsenges. Phys. Chem.*, 1965, *69*, 48–57.
84. Laub, L.W. and Wagner, J.B., Jr., On the redox kinetics and chemical diffusivity of wüstite in CO-CO_2 mixtures, *Oxid. Met.*, 1973, *7*, 1–22.
85. Bransky, I., Tallan, N.M., Wimmer, J.M., and Gwishi, M., *J. Am. Ceram. Soc.*, 1971, *54*, 26–30.
86. Nowotny, J. and Wagner, J.B., Jr., Influence of the surface on the equilibration kinetics of nonstoichiometric oxides, *Oxid. Met.*, 1981, *15*, 169–90.
87. Crank, J., *The Mathematics of Diffusion*, Clarendon Press, Oxford, 1956, 56.
88. Viani, F., Dovi, V., and Gesmundo, F., A method of analysis of the relaxation data for evaluation of the chemical diffusion coefficient in the case of partial rate control by surface reaction, *Oxid. Met.*, 1984, *21*, 309–27.
89. Dovi, V., Gesmundo, F., and Viani, V., Kinetics of equilibration in relaxation-type measurements under partial rate control by a surface reaction follwing a general type of rate law. I. The direct problem, *Oxid. Met.*, 1985, *23*, 35–51.
90. Gesmundo, F., Viani, F., and Dovi, V., Kinetics of equilibration in relaxation experiments with a partial degree of surface control under a Wagner-type rate law for the surface reaction, *Oxid. Met.*, 1985, *23*, 141–58.
91. Subbarao, E.C. and Maiti, H.S., Solid electrolytes with oxygen ion conduction, *Solid State Ionics*, 1984, *11*, 317–38.
92. Kilner, J. and Steele, B.C.H., Mass transport in anion-deficient fluorite oxides, in *Nonstoichiometric Oxides*, Sørenson, O., Ed., Academic Press, New York, 1981, 233–69.
93. Tuller, H.L. and Nowick, A.S., Doped ceria as a solid oxide electrolyte, *J. Electrochem. Soc.*, 1975, *122*, 255–59.
94. Etsell, T.H. and Flengas, S.N., The electrical properties of solid oxide electrolytes, *Chem. Rev.*, 1970, *70(3)*, 339–76.
95. Worrell, W.L., Oxide solid electrolytes, in *Solid Electrolytes*, Topics in Applied Physics, 21, Geller, S., Ed., Springer-Verlag, Berlin, 1971, 143–68.
96. Park., J.-H. and Blumenthal, R.N., Electronic transport in 8 mole percent Y_2O_3-ZrO_2, *J. Electrochem. Soc.*, 1989, *136(10)*, 2867–76.
97. Iwase, M. and Mori, T., Oxygen permeability of calcia-stabilized zirconia, *Met. Trans. B.*, 1978, *9*, 365–70.
98. Smith, A.W., Meszaros, F.W., and Amata, C.D., Permeability of zirconia, hafnia, and thoria to oxygen, *J. Am. Ceram. Soc.*, 1966, *49(5)*, 240–44.

99. Fisher, W.A., Permeation of oxygen and partial electronic conductivities in solid oxide electrolytes, in *Fast Ion Transport in Solids,* van Gool, W., Ed., Elsevier, New York, 1973, 503–12.
100. Palguev, S.F., Gilderman, V.K., and Neujmin, A.D., Oxygen permeability of stabilized zirconia solid electrolytes, *J. Electrochem. Soc.,* 1975, *122(6),* 745–48.
101. Hartung, R. and Möbius, H.-H., Bestimmung der Oxydationshalbleitung van ZrO_2-Festelektrolyten aus Permabilitätsdaten, *Z. Phys. Chem.,* 1970, *243,* 133–38.
102. Heyne, L. and Beekmans, N.M., Electronic transport in calcia-stabilized zirconia, *Proc. Br. Ceram. Soc.,* 1971, *19,* 229–62.
103. Alcock, C.B. and Chan, J.C., The oxygen permeability of stabilized zirconia electrolytes at high temperatures, *Can. Metall. Q.,* 1972, *11(4),* 559–67.
104. Kitazawa, K. and Coble, R.L., Use of stabilized ZrO_2 to measure O_2 permeation, *J. Am. Ceram. Soc.,* 1974, *57,* 360–63.
105. Cales, B. and Baumard, J.F., Oxygen semipermeability and electronic conductivity in calcia-stabilized zirconia, *J. Mater. Sci.,* 1982, *12,* 3243–48.
106. Dou, S., Masson, C.R., and Pacey, P.D., Mechanism of oxygen permeation through lime-stabilized zirconia, *J. Electrochem. Soc.,* 1985, *132,* 1843–49.
107. Bouwmeester, H.J.M., Kruidhof, H., Burggraaf, A.J., and Gellings, P.J., Oxygen semipermeability of erbia-stabilized bismuth oxide, *Solid State Ionics,* 1992, *53–56,* 460–68.
108. Ullmann, H., Die Sauerstoffpermeabilität von oxidischen Festelektrolyten, *Z. Phys. Chem.,* 1968, *237,* 71–80.
109. Reetz, T., Sauerstoffpermeabilität oxidionenleitender Festelektrolyte, *Z. Phys. Chem.,* 1972, *249(5/6),* 369–75.
110. Fouletier, J., Fabry, P., and Kleitz, M., Electrochemical semipermeability and the elctrode microsystem in solid oxide electrolyte cells, *J. Electrochem. Soc.,* 1976, *123(2),* 204–13.
111. Dou, S., Masson, C.R., and Pacey, P.D., Chemical diffusion in calcia-stabilized zirconia ceramic, *Solid State Ionics,* 1986, *18&19,* 736–40.
112. Dou, S., Masson, C.R., and Pacey, P.D., Induction period for oxygen permeation through calcia-stabilized zirconia ceramic, *J. Am. Ceram. Soc.,* 1989, *72(7),* 1114–18.
113. Dou, S., Masson, C.R., and Pacey, P.D., Diffusion of electronic defects in calcia-stabilized zirconia from a deorption rate study, *J. Electrochem. Soc.,* 1990, *137(11),* 3455–58.
114. Verkerk, M.J., Keizer, K., and Burggraaf, A.J., High oxygen ion conduction in sintered oxides of the Bi_2O_3-Er_2O_3, *J. Appl. Electrochem.,* 1980, *10,* 81–90.
115. Boukamp, B.A., Vinke, I.C., de Vries, K.J., and Burggraaf, A.J., Surface oxygen exchange properties of bismuth oxided based solid electrolytes and electrode materials, *Solid State Ionics,* 1989, *32/33,* 918–23.
116. Lin, Y.S., Wang, W., and Han, J., Oxygen permeation through thin mixed-conducting solid oxide membranes, *AICHE J.,* 1994, *40(5),* 786–98.
117. Boukamp, B.A., Bouwmeester, H.J.M., and Burggraaf, A.J., The surface oxygen exchange process in oxygen ion conducting solids, in *Materials Research Society Symposium Proceedings,* Vol. 293, Nazri, G.A., Tarascon, J.M., and Armand, M., Eds., Materials Research Society, Pittsburgh, PA, 1993, 361–66.
118. Boukamp, B.A., Bouwmeester, H.J.M., and Burggraaf, A.J., The surface oxygen exchange process in oxygen ion conducting materials, in *Proc. 2nd Int. Symp. Ionic and Mixed Conducting Oxide Ceramics,* Vol. 94-12, Ramanarayanan, T.A., Worrell, W.L., and Tuller, H.L., Eds., The Electrochemical Society, Pennington, NJ, 1994, 141–50.
119. Boukamp, B.A., Bouwmeester, H.J.M., and Verwey, H., Relation between surface oxygen and SOFC cathode kinetics, in *Advanced Fuel Cells Programme,* Annex II, International Energy Agency, 7th Workshop, Wadahl, Norway, Jan. 18-20, 1995, pp. 18-20.
120. Vinke, I.C., Boukamp, B.A., de Vries, K.J., and Burggraaf, A.J., Mixed conductivity in terbia-stabilized bismuth oxide, *Solid State Ionics,* 1992, *57,* 91–98.
121. Kruidhof, H. and Bouwmeester, H.J.M., unpublished results.
122. Kleitz, M., Reaction pathways: a new electrode modeling concept, in *Fundamental Barriers to SOFC Performance,* IEA Publications, Berne, 1992, 4–12.
123. Clark, D.J., Losey, R.W., and Suitor, J.W., Separation of oxygen by using zirconia solid electrolyte membranes, *Gas Sep. Purif.,* 1992, *6(4),* 201–5.
124. Dumélié, M., Nowogrocki, G., and Boivin, J.C., Ionic conductor membrane for oxygen separation, *Solid State Ionics,* 1988, *28–30,* 524–28.
125. Abraham, F., Debreuille-Gresse, M.F., Mairesse, G., and Nowogrocki, G., Phase transitions and ionic conductivity in $Bi_4V_2O_{11}$ an oxide with a layered structure, *Solid State Ionics,* 1988, *28–30,* 529–32.
126. Abraham, F., Boivin, J.C., Mairesse, G., and Nowogrocki, G., The BIMEVOX series: a new family of high performance oxide ion conductors, *Solid State Ionics,* 1990, *40/41,* 934–37.
127. Boivin, J.C., Vannier, R.N., Mairesse, G., Abraham, F., and Nowogrocki, G., BIMEVOX: a new family of Bi_2O_3 solid electrolytes for the separation of oxygen, *ISSI Lett.,* 1992, *3(4),* 14–16.

128. Shen, Y.S., Liu, M., Taylor, D., Bolagopal, S., Joshi, A., and Krist, K., Mixed ionic-electronic conductors based on Bi-Y-O-Ag metal-ceramic system, in *Proc. 2nd Int. Symp. on Ionic and Mixed Conducting Ceramics*, Vol. 94-12, Ramanarayanan, T.A., Worrell, W.L., and Tuller, H.L., Eds., The Electrochemical Society, Pennington, NJ, 1994, 574–95.
129. Chen, C.S., Kruidhof, H., Bouwmeester, H.J.M., Verweij, H., and Burggraaf, A.J., Oxygen permeation through oxygen ion oxide-noble metal dual phase composites, *Solid State Ionics,* 1996, 86-88, 569-572.
130. Chen, C.S., Fine Grained Zirconia-Metal Dual Phase Composites, Ph.D. thesis, University of Twente, The Netherlands, 1994.
131. McLachlan, D.S., Blaszkiewics, M., and Newnham, R.E., Electrical resistivity of composites, *J. Am. Ceram. Soc.*, 1990, *73(8),* 2187–2203.
132. Chen, C.S., Boukamp, B.A., Bouwmeester, H.J.M., Cao, G.Z., Kruidhof, H., Winnubst, A.J.A., and Burggraaf, A.J., Microstructural development, electrical properties and oxygen permeation of zirconia-pallladium composites, *Solid State Ionics,* 1995, *76,* 23–28.
133. Kirkpatrick, S., Percolation and conduction, *Rev. Mod. Phys.*, 1973, *45(4),* 574–88.
134. Tannenberger, H. and Siegert, H., The behavior of silver cathodes in solid electrolyte fuel cells, *Adv. Chem.*, 1969, *90,* 281–300.
135. Shen, Y.S., Joshi, A., Liu, M., and Krist, K., Structure, microstructure and transport properties of mixed ionic-electronic conductors based on bismuth oxide. Part I. Bi-Y-Cu-O system, *Solid State Ionics,* 1994, *72,* 209–17.
136. Hammou, A., Solid oxide fuel cells, in *Advances in Electrochemical Science and Engineering,* Vol. 2, Gerischer, H. and Tobias, C.W., Eds., VCH Verlaggesellschaft mbH, Weinheim, Germany, 1991, 87–139.
137. Minh, N.Q., Ceramic fuel cells, *J. Am. Ceram. Soc.*, 1993, *76(3),* 564–88.
138. Carter, S., Selcuk, A., Chater, J., Kajda, R.J., Kilner, J.A., and Steele, B.C.H., Oxygen transport in selected non-stoichiometric perovskite-structure oxides, *Solid State Ionics,* 1992, *53–56,* 597–605.
139. Yasuda, I. and Hishinuma, M., Electrical conductivity and chemical diffusion coefficient of Sr-doped lanthanum manganites, in *Proc. 2nd Int. Symp. Ionic and Mixed Conducting Oxide Ceramics,* Ramanarayanan, T.A., Worrell, W.L., and Tuller, H.L., Eds., The Electrochemical Society, Pennington, NJ, 1994, 209–21.
140. Worrell, W.L., Electrical properties of mixed-conducting oxides having high oxygen-ion conductivity, *Solid State Ionics,* 1992, *52,* 147–51.
141. Worrell, W.L., Han, P., and Huang, J., Mixed conductivity in oxygen exhibiting both p-type electronic and high oxygen-ion conductivity, in *High Temperature Electrochemical Behavior of Fast Ion and Mixed Conductors,* Poulsen, F.W., Bentzen, J.J., Jacobson, T., Skou, E., and Østergård, M.J.L., Eds., Risø National Laboratory, Denmark, 1993, 461–66.
142. Anderson, H.U., Chen, C.C., Tai, L.W., and Nasrallah, M.M., Electrical conductivity and defect structure of $(La,Sr)(Co,Fe)O_3$, in *Proc. 2nd Int. Symp. Ionic and Mixed Conducting Oxide Ceramics,* Ramanarayanan, T.A., Worrell, W.L., and Tuller, H.L., Eds., The Electrochemical Society, Pennington, NJ, 1994, 376–87.
143. Weber, W.J., Stevenson, J.W., Armstrong, T.R., and Pederson, L.R., Processing and electrochemical properties of mixed conducting $La_{1-x}A_xCo_{1-y}Fe_yO_{3-\delta}$ (A = Sr, Ca), in *Mater. Res. Soc. Symp. Proc.*, Vol. 369, *Solid State Ionics IV,* Nazri, G.-A., Taracson, J.-M., and Schreiber, M.S., Eds., Materials Research Society, Pittsburgh, 1995, 395–400.
144. Chen, C.H., Kruidhof, H., Bouwmeester, H.J.M., and Burggraaf, A.J., Ionic conductivity of perovskite $LaCoO_3$ measured by oxygen permeation technique, *J. Appl. Electrochem.*, 1997, 27, 71-75.
145. Shuk, P., Charton, V., and Samochval, V., Mixed conductors on the lanthanide cobaltites basis, *Mater. Sci. Forum,* 1991, *76,* 161–64.
146. Itoh, N., Kato, T., Uchida, K., and Haraya, K., Preparation of pore-free disk of $La_{1-x}Sr_xCoO_{3-\delta}$ and its oxygen permeability, *J. Membr. Sci.*, 1994, *92,* 239–46.
147. Van Doorn, R.H.E., Kruidhof, H., Bouwmeester, H.J.M., and Burggraaf, A.J., Oxygen permeability of strontium-doped $LaCoO_{3-\delta}$ perovskites, in *Mater. Res. Soc. Symp. Proc.*, Vol. 369, *Solid State Ionics IV,* Nazri, G.-A., Taracson, J.-M., and Schreiber, M.S., Eds., Materials Research Society, Pittsburgh, 1995, 377–82.
148. Van Doorn, R.H.E., Kruidhof, H., Bouwmeester, H.J.M., and Burggraaf, A.J., Oxygen permeation through dense ceramic membranes of Sr-doped $LaCoO_{3-\delta}$ membranes, *J. Electrochem. Soc.*, submitted.
149. Van Hassel, B.A., ten Elshof, J.E., and Bouwmeester, H.J.M., Oxygen permeation flux through $La_{1-y}Sr_yFeO_{3-\delta}$ limited by carbon monoxide oxidation rate, *Appl. Catal. A: General,* 1994, *119,* 279–91.
150. Elshof, J.E. ten, Bouwmeester, H.J.M., and Verweij, H., Oxygen transport through $La_{1-x}Sr_xFeO_{3-\delta}$ membranes. I. Permeation in air/He gradients, *Solid State Ionics,* 1995, *81,* 97–109.
151. Miura, N., Okamoto, Y., Tamaki, J., Morinag, K., and Yamazoe, N., Oxygen semipermeability of mixed-conductive oxide thick-film prepared by slip casting, *Solid State Ionics,* 1995, *79,* 195–200.
152. Qiu, L., Lee, T.H., Lie, L.-M., Yang, Y.L., and Jacobson, A.J., Oxygen permeation studies of $SrCo_{0.8}Fe_{0.2}O_{3-\delta}$, *Solid State Ionics,* 1995, *76,* 321–29.
153. Kruidhof, H., Bouwmeester, H.J.M., van Doorn, R.H.E., and Burggraaf, A.J., Influence of order-disorder transitions on oxygen permeability through selected nonstoichiometric perovskite-type oxides, *Solid State Ionics,* 1993, *63–65,* 816–22.

154. Kharton, V.V., Naumovich, E.N., and Nikolaev, A.V., Oxide ion and electron conjugate diffusion in perovskite-like $SrCo_{1-x}M_xO_{3-\delta}$ (M = Cr.Cu, x = 0.0.5), *Solid State Phenomena*, 1994, *39–40*, 147–52.
155. Brinkman, H.W., Kruidhof, H., and Burggraaf, A.J., Mixed conducting yttrium-barium-cobalt-oxide for high oxygen permeation, *Solid State Ionics*, 1994, *68*, 173–76.
156. Katz, L. and Ward, R., Structure relations in mixed metal oxides, *Inorg. Chem.*, 1964, *3(2)*, 205–11.
157. Goldschmidt, V.M., *Akad. Oslo.*, 1946, *A42*, 224.
158. Anderson, H.U., Review of p-type doped perovskite materials for SOFC and other applications, *Solid State Ionics*, 1992, *52*, 33–41.
159. Anderson, H.U., Defect chemistry of p-type perovskites, in *High Temperature Electrochemical Behavior of Fast Ion and Mixed Conductors*, Poulsen, F.W., Bentzen, J.J., Jacobson, T., Skou, E., and Østergård, M.J.L., Eds., Risø National Laboratory, Denmark, 1993, 1–19.
160. Mizusaki, J., Nonstoichiometry, diffusion, and electrical properties of perovskite-type oxide electrode materials, *Solid State Ionics*, 1992, *52*, 79–91.
161. Verwey, E.J.W., Haaijman, P.W., Romeijn, F.C., and Oosterhout, G.W., van, *Philips Res. Rep.*, 1950, *5*, 173.
162. Mizusaki, J., Sosamoto, T., Cannon, W.R., and Bowen, H.K., Electronic conductivity, seebeck coefficient, and defect structure of $LaFeO_3$, *J. Am. Ceram. Soc.*, 1982, *65(8)*, 363–68.
163. Mizusaki, J., Sosamoto, T., Cannon, W.R., and Bowen, H.K., Electronic conductivity, seebeck coefficient, and defect structure of $La_{1-x}Fe_xO_{3-\delta}$ (x = 0.1, 0.25), *J. Am. Ceram. Soc.*, 1983, *66(4)*, 247–52.
164. Mizusaki, J., Yoshihiro, M., Yamauchi, S., and Fueki, K., Nonstoichiometry and defect structure of the perovskite-type oxides $La_{1-x}Sr_xFeO_{3-\delta}$, *J. Solid State Chem.*, 1985, *58*, 257–66.
165. Mizusaki, J., Yoshihiro, M., Yamauchi, S., and Fueki, K., Thermodynamic quantitites and defect equilbrium in the perovskite-type oxides solid solution $La_{1-x}Sr_xFeO_{3-\delta}$, *J. Solid State Chem.*, 1987, *67*, 1–8.
166. Mizusaki, J., Yamauchi, S., Fueki, K., and Ishikawa, A., Nonstoichiometry of the perovskite-type oxide $La_{1-x}Sr_xCrO_{3-\delta}$, *Solid State Ionics*, 1984, *12*, 119–24.
167. Mizusaki, J., Mima, Y., Yamauchi, S., Fueki, K., and Tagawa, H., Nonstoichiometry of the perovskite-type oxides $La_{1-x}Sr_xCoO_{3-\delta}$, *J. Solid State Chem.*, 1989, *80*, 102–11.
168. Bouwmeester, H.J.M., unpublished results.
169. Abbate, M, de Groot, F.M.F., Fuggle, J.C., Fujimori, A., Strebel, O., Lopez, F., Domke, M., Kaindl, G., Sawatzky, G.A., Takana, M., Eisaki, H., and Uchida, S., Controlled valence properties of $La_{1-x}Sr_xFeO_3$ and $La_{1-x}Sr_xMnO_3$ studied by soft-X-ray absorption spectroscopy, *Phys. Rev. B.*, 1992, *46(8)*, 4511–19.
170. Kuiper, P., Kruizinga, G., Ghijsen, J., Grioni, M., Weijs, P.J.W., de Groot, F.M.F., Sawatzky, G.A., Verweij, H., Feiner, L.F., and Petersen, H., X-ray absorption study of the O 2p hole concentration dependence on O stoichiometry in $YBa_2Cu_3O_x$, *Phys. Rev. B*, 1988, *38(8)*, 6483–89.
171. Lankhorst, M.H.R. and Bouwmeester, H.J.M., Determination of oxygen nonstoichiometry and diffusivity in mixed conducting oxides by oxygen coulometric titration, *J. Electrochem. Soc.*, April 1997.
172. Abbate, M., Fuggle, J.C., Fujimori, A., Tjeng, L.H., Chen, C.T., Potze, R., Sawatzky, G.A., Eisaki, H., and Uchida, S., Electronic structure and spin-state transition in $LaCoO_3$, *Phys. Rev. B*, 1993, *47(24)*, 16124–30.
173. Cheung, K.Y., Steele, B.C.H., and Dudley, G.J., Thermodynamic and transport properties of selected solid solution electrodes, in *Fast Ion Transport in Solids: Electrodes and Electrolytes*, Vashista, P., Mundy, J.N., and Shenoy, G.K., Eds., Elsevier/North-Holland, New York, 1979, 141–48.
174. Kaluarachchi, D. and Frindt, R.F., Intercalation islands and stage conversion in Ag-TiS_2, *Phys. Rev. B*, 1983, 28, 3663–65, Intercalation of silver in titanium disulfide, *Phys. Rev. B*, 1985, *31*, 3648–53.
175. Tejuca, L.G., Fierro, J.L.G., and Tascón, J.M.D., Structure and reactivity of perovskite-type oxides, *Adv. Catal.*, 1989, *36*, 237–28.
176. Yokokawa, H., Sakai, N., Kawada, T., and Dokya, M., Thermodynamic stabilities of perovskite-type oxides for electrodes and other electrochemical materials, *Solid State Ionics*, 1992, *52*, 43–56.
177. Nakamura, T., Petzow, G., and Gauckler, L.J., Stability of the perovskite phase $LaBO_3$ (B = V, Cr, Mn, Fe, Co, Ni) in reducing atmosphere. I. Experimental results, *Mater. Res. Bull.*, 1979, *14*, 649–59.
178. Tabata, K., Matsumoto, I., and Kohiki, S., Surface characterization and catalytic properties of $La_{1-x}Sr_xCoO_3$, *J. Mater. Sci.*, 1987, *22*, 1882–86.
179. Seyama, T., Surface reactivity of oxide materials in oxidation-reduction environment, in *Surface and Near Surface Chemistry of Oxide Materials*, Vol. 47, Nowotny, J. and Dufour, L.C., Eds., Materials Science Monographs, Elsevier Science, Amsterdam, 1988, 189–215.
180. Yamazoe, N., Furukawa, S., Teraoka, Y., and Seiyama, T., The effect of sorption on the crystal structure of $La_{1-x}Sr_xCoO_{3-\delta}$, *Chem. Lett.*, 1982, 2019–22.
181. Teraoka, Y., Yoshimatsu, M., Yamazoe, N., and Seiyama, T., Oxygen-sorptive properties and defect structure of perovskite-type oxides, *Chem. Lett.*, 1984, 893–96.
182. Teraoka, Y., Zhang, H.-M., and Yamazoe, N., Oxygen-sorptive properties of defect perovskite-type $La_{1-x}Sr_xCo_{1-y}Fe_yO_{3-\delta}$, *Chem. Lett.*, 1985, 1367–70.
183. Zhang, H.-M., Shimizu, Y., Teraoka, Y., Miura, M., and Yamazoe, N., Oxygen sorption and catalytic properties of $La_{1-x}Sr_xCo_{1-y}Fe_yO_{3-\delta}$ perovskite-type oxides, *J. Catal.*, 1990, *121*, 1367–70.

184. Van Hassel, B.A., Kawada, T., Sakai, N., Yokokawa, H., and Dokya, M., Oxygen permeation modeling of $La_{1-y}Ca_yCrO_{3-\delta}$, *Solid State Ionics*, 1993, *66*, 41–47.
185. Kawada, T., Horita, T., Sakai, N., Yokokawa, H., and Dokiya, M., Experimental determination of oxygen permeation flux through bulk and grain boundary of $La_{0.7}Ca_{0.3}CrO_3$, *Solid State Ionics*, 1995, *79*, 201–7.
186. Van Hassel, B.A., Kawada, T., Sakai, N., Yokokawa, H., Dokiya, M., and Bouwmeester, H.J.M., Oxygen permeation modeling of perovskites, *Solid State Ionics*, 1993, *66*, 295–305.
187. Goodenough, J.B., Fast ionic conduction in solids, *Proc. R. Soc. (London)*, 1984, *A393*, 214–34.
188. Nisancioglu, K. and Gür, T.M., Oxygen diffusion in iron doped strontium cobaltites, in *Proc. 3rd Int. Symp. Solid Oxide Fuel Cells*, Singhal, S.C. and Iwahara, H., Eds., Electrochemical Society, Pennington, NJ, 1994, 267–75.
189. Ishigaki, T., Yamauchi, S., Mizusaki, J., Fueki, K., and Tamura, H., Tracer diffusion coefficient of oxide ions in $LaCoO_3$ single crystal, *J. Solid State Chem.*, 1984, *54*, 100–07.
190. Ishigaki, T., Yamauchi, S., Mizusaki, J., Fueki, K., and Tamura, H.J., Diffusion of oxide ions in $LaFeO_3$ single crystal, *Solid State Chem.*, 1984, *55*, 50–53.
191. Kim, M.C., Park, S.J., Haneda, H., Tanaka, J., Mitsuhashi, T., and Shirasaki, S., Self-diffusion of oxygen in $La_{1-x}Sr_xFeO_{3-\delta}$, *J. Mater. Sc. Lett.*, 1990, *9*, 102–04.
192. Fueki, K., Mizusaki, J., Yamauchi, S., Ishigaki, T., and Mima, Y., Nonstoichiometry and oxygen vacancy diffusion in the perovskite type oxides $La_{1-x}Sr_xCoO_{3-\delta}$, in *Proc. 10th Int. Symp. React. Solids 1984*, Barret, P. and Dufour, L.C., Eds., Elsevier, Amsterdam, 1985, 339–43.
193. Denos, Y., Morin, F., and Trudel, G., Oxygen diffusivity in strontium substituted lanthanum cobaltite, in *Proc. 2nd Int. Symp. Ionic and Mixed Conducting Oxide Ceramics*, Ramanarayanan, T.A., Worrel, W.L., and Tuller, H.L., Eds., The Electrochemical Society, Pennington, NJ, 1994, 150–58.
194. Nisancioglu, K. and Gür, T.M., Potentiostatic step technique to study ionic transport in mixed conductors, *Solid State Ionics*, 1994, *72*, 199–203.
195. Sekido, S., Tachibana, H., Yamamura, Y., and Kambara, T., Electric-ionic conductivity in perovskite-type oxides, $Sr_xLa_{1-x}Co_{1-y}Fe_yO_{3-\delta}$, *Solid State Ionics*, 1990, *37*, 253–59.
196. Yasuda, I. and Hikita, T.J., Precise determination of the chemical diffusion coefficient of calcium-doped lanthanum chromites by means of electrical conductivity relaxation, *Electrochem. Soc.*, 1995, *141(5)*, 1268–73.
197. Belzner, A., Gür, T.M., and Huggins, R.A., Oxygen chemical diffusion in strontium doped lanthanum manganites, *Solid State Ionics*, 1992, *57*, 327–37.
198. Gür, T.M., Belzner, A., and Huggins, R.A., A new class of oxygen selective chemically driven nonporous ceramic membranes. Part I. A-site doped perovskites, *J. Membr. Sci.*, 1992, *95*, 151–62.
199. Tai, L.-W., Nasrallah, M.M., and Anderson, H.U., $La_{1-x}Sr_xCo_{1-y}Fe_yO_{3-\delta}$, A potential cathode for intermediate temperature SOFC applications, in *Proc. 3rd Int. Symp. Solid Oxide Fuel Cells*, Singhal, S.C. and Iwahara, H., Eds., Electrochemical Society, Pennington, NJ, 1994, 241–51.
200. Bonnet, J.P., Grenier, J.C., Onillon, M., Pouchard, M., and Hagenmuller, P., Influence de la pression partiele dóxygene sur la nature des defauts ponctuels dans $Ca_2LaFe_3O_{8+x}$, *Mater. Res. Bull.*, 1979, *14*, 67–75.
201. Zener, C., *Phys. Rev.*, 1952, *82*, 403.
202. Koc, R., Anderson, H.U., Howard, S.A., and Sparlin, D.M., Structural, sintering and electrical properties of the perovskite-type $(La,Sr)(Cr,Mn)O_3$, in *Proc. 1st Int. Symp. Solid Oxide Fuel Cells*, Singhal, S.C., Ed., Electrochemical Society, Pennington, NJ, 1989, 220–44.
203. Rafaelle, R., Anderson, H.U., Sparlin, D.M., and Perris, P.E., Transport anomalies in the high-temperature hopping conductivity and thermopower of Sr-doped $La(Cr,Mn)O_3$, *Phys. Rev.*, B 1993, *43*, 7991–99.
204. Sehlin, S.R., Anderson, H.U., and Sparlin, D.M., Electrical characterization of the $(La,Ca)(Cr,Co)O_3$ system, *Solid State Ionics*, 1995, *78*, 235–43.
205. Tai, L.-W., Nasrallah, M.M., Anderson, H.U., Sparlin, D.M., and Sehlin, S.R., Structure and electrical properties of $La_{1-x}Sr_xCo_{1-y}Fe_yO_3$. Part I. The system $La_{0.8}Sr_{0.2}Co_{1-y}Fe_yO_3$, *Solid State Ionics*, 1995, *76*, 259–71; Part 2. $La_{1-x}Sr_xCo_{0.2}Fe_{0.8}O_3$, 1995, *76*, 273–83.
206. Goodenough, J.B., Metallic oxides, in *Progr. Solid St. Chem.*, Reiss, H., Ed., Pergamon Press, Oxford, 1971, 145–399.
207. Racah, P.M. and Goodenough, J.B., *Phys. Rev.*, 1967, *155(3)*, 932–43.
208. Thornton, G., Morrison, F.C., Partington, S., Tofield, B.C., and Williams, J., The rare earth cobaltites: localized or collective electron behavior, *Phys. C: Solid State Phys.*, 1988, *21*, 2871–80.
209. Mizusaki, J., Tabuchi, J., Matsuura, T., Yamauchi, S., and Fueki, K., Electrical conductivity and seebeck coefficient of nonstoichiometric $La_{1-x}Sr_xCoO_{3-\delta}$, *J. Electrochem. Soc.*, 1989, *136(7)*, 2082–88.
210. Stevenson, J.W., Nasrallah, M.M., Anderson, H.U., and Sparlin, D.M., Defect structure of $Y_{1-y}Ca_yMnO_3$ and $La_{1-y}Ca_yMnO_3$. I Electrical properties, *J. Solid State Chem.*, 1993, 102, 175–184; II. Oxidation-reduction behaviour, 1993, *102*, 185–97.
211. Van Roosmalen, J.A.M. and Cordfunke, E.P.H., A new defect model to describe the oxygen deficiency in perovskite-type oxides, *J. Solid State Chem.*, 1991, *93*, 212–19.

212. Goodenough, J.B., Ruiz-Diaz, J.E., and Zhen, Y.S., Oxide-ion conduction in $Ba_2In_2O_5$ and $Ba_3In_2MO_8$ (M = Ce, Hf, or Zr), *Solid State Ionics*, 1990, *44*, 21–31.
213. Goodenough, J.B., Manthiram, A., Paranthanam, P., and Zhen, Y.S., Fast oxide-ion conduction in intergrowth structures, *Solid State Ionics*, 1992, *52*, 105–09.
214. Kilner, J.A. and Brook, R.J., A study of oxygen ion conductivity in doped non-stoichiometric oxides, *Solid State Ionics*, 1982, *6*, 237–52.
215. Cherry, M., Islam, M.S., and Catlow, C.R.A., Oxygen ion migration in perovkite-type oxides, *J. Solid State Chem.*, 1995, *118*, 125–32.
216. Cook, R.L. and Sammells, A.F., On the systematic selection of perovskite solid electrolytes for intermediate temperature fuel cells, *Solid State Ionics*, 1991, *45*, 311–21.
217. Sammells, A.F., Cook, R.L., White, J.H., Osborne, J.J., and MacDuff, R.C., Rational selection of advanced solid electrolytes for intermediate temperature fuel cells, *Solid State Ionics*, 1992, *52*, 111–23.
218. Dijk, M.P., de Vries, K.J., and Burggraaf, A.J., Oxygen ion and mixed conductivity in compounds with the fluorite and pyrochlore compounds, *Solid State Ionics*, 1983, *9/10*, 913–19.
219. Dijk, M.P., deVries, K.J., Burggraaf, A.J., Cormack, A.N., and Catlow, C.R.A., Defect structures and migration mechanisms in oxide pyrochlores, *Solid State Ionics*, 1985, *17*, 159–67.
220. Rao, C.N.R., Gopalakrishnan, J., and Viyasagar, K., Superstructures, ordered defects & nonstoichiometry in metal oxides of perovskite & related structures, *Indian J. Chem.*, 1984, *23A*, 265–84.
221. Hagenmuller, P., Pouchard, M., and Grenier, J.C., Nonstroichiometry in the perovskite-type oxides: an evolution from the classic Schottky-Wagner model to the recent high T_c superconductors, *Solid State Ionics*, 1990, *43*, 7–18.
222. Smyth, D.M., Defects and order in perovskite-related oxides, *Annu. Rev. Mater. Sci.*, 1985, *15*, 329–57.
223. Smyth, D.M., Defects and structural changes in perovskite systems: from insulators to superconductors, *Cryst. Lattice Defects Amorph. Mat.*, 1989, *18*, 355–75.
224. Anderson, M.T., Vaughy J.T., Poeppelmeier, K.R., Structural similarities among oxygen-deficient perovskites, *Chem. Mater.*, 1993, *5(2)*, 151–65.
225. Alario-Franco, M.A., Henche, M.J.R., Vallet, M., Calbet, J.M.G., Grenier, J.C., Wattiaux, A., and Hagenmuller, P., Microdomain texture and oxygen excess in the calcium-lanthanum ferrite: $Ca_2LaFe_3O_8$, *J. Solid State Chem.*, 1983, *46*, 23–40.
226. Alario-Franco, M.A., González-Calbet, J.M., and Grenier, J.C., Brownmillerite-type microdomains in the calcium lanthanum ferrites: $Ca_xLa_{1-x}FeO_{3-y}$. I. $2/3 < x < 1$, *J. Solid State Chem.*, 1983, *49*, 219–31.
227. Alario-Franco, M.A., González-Calbet, J.M., and Vallet-Regi, M., Microdomains in the $CaFe_xMn_{1-x}O_{3-y}$ ferrites, *J. Solid State Chem.*, 1986, *65*, 383–91.
228. Gonzáles-Calbet, J.M., Alario-Franco, M.A., and Vallet-Regi, M., Microdomain formation: A sophisticated way of accomodating compositional variations in non-stoichiometric perovkites, *Cryst. Lattice Defect Amorph. Mat.*, 1987, *16*, 379–85.
229. Gonzáles-Calbet, J.M., Parras, M., Vallet-Regi, M., and Grenier, J.C., Anionic vacancy distribution in reduced barium-lanthanum ferrites: $Ba_xLa_{1-x}FeO_{3-x/2}$ ($1/2 \leq x \leq 2/3$), *J. Solid State Chem.*, 1991, *92*, 100–15.
230. Battle, P.D., Gibb, T.C., and Nixon, S., The formation of nonequilibrium microdomains during the oxidation of Sr_2CoFeO_5, *J. Solid State Chem.*, 1988, *73*, 330–37.
231. Takahashi, T., Iwahara, H., and Esaka, T., High oxide conduction in sintered oxide of the system Bi_2O_3-M_2O_3, *J. Electrochem. Soc.*, 1977, *124(10)*, 1563–69.
232. Shin, S., Yonemura, M., and Ikawa, H., Order-disorder transition of $Sr_2Fe_2O_5$ from brownmillerite to perovskite at elevated temperature, *Mater. Res. Bull.*, 1978, *13*, 1017–21.
233. Shin, S., Yonemura, M., and Ikawa, H., Crystallographic properties of $Ca_2Fe_2O_5$. Difference in crystallographic properties of brownmillerite-like compounds, $Ca_2Fe_2O_5$ and $Sr_2Fe_2O_5$, at elevated tmperature, *Bull. Chem. Soc. Jpn.*, 1979, *52*, 947–48.
234. Grenier, J.C., Norbert, E., Pouchard, M., and Hagenmuller, P., Structural transitions at high temperature in $Sr_2Fe_2O_5$, *J. Solid State Chem.*, 1985, *58*, 243–52.
235. Rodríguez-Carvajal, J., Vallet-Regí, M., and Gonzáles-Calbet, J.M., Perovskite threefold superlattices: a structure determination of the $A_3M_3O_8$ phase, *Mater. Res. Bull.*, 1989, *24*, 423–29.
236. Manthiram, A., Kuo, J.F., and Goodenough, J.B., Characterization of oxygen-deficient perovskites as oxide-ion electrolytes, *Solid State Ionics*, 1992, *62*, 225–34.
237. Zhang, G.B. and Smyth, D.M., Defects and transport in brownmillerite oxides, in *Proc. 2nd Int. Symp. Ionic and Mixed Conducting Oxide Ceramics*, Ramanarayanan, T.A., Worrel, W.L., and Tuller, H.L., Eds., The Electrochemical Society, Pennington, NJ, 1994, 608–22.
238. Adler, S.B., Reimer, J.A., Baltisberger, J., and Werner, U., Chemical structure and oxygen dynamics in $Ba_2In_2O_5$, *J. Am. Ceram. Soc.*, 1994, *116(2)*, 675–81.
239. Adler, S., Russek, S., Reimer, J., Fendorf, M., Stacy, A., Huang, Q., Santoro, A., Lynn, J., Baltisberger, J., and Werner, U., Local structure and oxide-ion motion in defective perovskites, *Solid State Ionics*, 1994, *68*, 193–211.

240. Er-Rakho, L., Michel, C., LaCorre, P., and Raveau, B., $YBa_2CuFeO_{5+\delta}$: A novel oxygen-deficient perovskite with a layer structure, *J. Solid State Chem.*, 1988, *73*, 531.
241. Barbey, L., Nguyen, N., Caignart, V., Hervieu, M., and Ravenau, B., Mixed oxides of cobalt and copper with a double pyramidal layer structure, *Mater. Res. Bull.*, 1992, *27*, 295–301.
242. Van Doorn, R.H.E., Keim, E.G., Kachliki, T., Bouwmeester, H.J.M, and Burggraaf, A.J., in preparation.
243. Harrison, W.T.A., Lee, T.H., Yang, Y.L., Scarfe, D.P., Liu, L.M., and Jacobson, A.J., A neutron diffraction study of two strontium cobalt iron oxides, *Mater. Res. Bull.*, 1995, *30(5)*, 621–29.
244. Barrer, R.M., Barrie, J.A., Rogers, and M.G., *Trans. Faraday*, 1962, *58*, 2473.
245. Zhou, J.P. and McDevitt, J.T., Corrosion reactions of $YBa_2Cu_3O_{7-x}$ and $Tl_2Ba_2Ca_2Cu_3O_{10+x}$ superconductor phases in aqueous environements, *Chem. Mater.*, 1992, *4*, 952–59.
246. Ten Elshof, E.J., Bouwmeester, H.J.M., and Verweij, H., Oxygen transport through $La_{1-x}Sr_xFeO_{3-\delta}$ membranes. II. Permeation in air/CO,CO_2 gradients, *Solid State Ionics*, 1996, 89, 81-92.
247. Yasuda, J., Ogasawara, K., and Kishinuma, M., Oxygen tracer diffusion in (La,Ca)$CrO_{3-\delta}$, in *Proc. 2nd Int. Symp. Ionic and Mixed Conducting Oxide Ceramics*, Vol. 94–12, Ramanarayanan, T.A., Worrell, W.L., and Tuller, H.L., Eds., The Electrochemical Society, Pennington, NJ, 1994, 164–173.
248. Badwal, S.P.S., Drennan, J., and Hughes, A.E., Segregation in oxygen-ion conducting solid electrolytes and its influence on electrical properties, in *Science of Ceramic Interfaces*, Nowotny, J., Ed., Elsevier Science Publishers, Amsterdam, 1991, 227–85.
249. Van Dijk, T. and Burggraaf, A.J., Grain boundary effects on ionic conductivity in ceramic $Gd_xZr_{1-x}O_{2-x/2}$ solid solutions, *Phys. Status Solidi A*, 1981, *63*, 229–40.
250. Verkerk, M.J., Middelhuis, B.J., and Burggraaf, A.J., Effect of grain boundaries on the conductivity of high purity ZrO^{2}–Y_2O_3 ceramics, *Solid State Ionics*, 1982, *6*, 159–70.
251. Schmalzried, H., Laqua, W., and Lin, P.L., Crystalline oxide solid solutions in oxygen potential gradients, *Z. Naturforsch*, 1979, *34A*, 192–99.
252. Schmalzried, H. and Laqua, W., Multicomponent oxides in oxygen potential gradients, *Oxid. Met.*, 1981, *15*, 339–53.
253. Monceau, D., Petot, C., and Petot E., Kinetic demixing profile calculation under a temperature gradient in multi-component oxides, *J. Eur. Ceram. Soc.*, 1992, *9*, 193–204.
254. Martin, M. and Schmackpfeffer, R., Demixing of oxides: influence of defect interactions, *Solid State Ionics*, 1994, *72*, 67–71.
255. Gallagher, P.K., Grader, G.S., and O'Bryan, H.M., Effect of an oxygen gradient on the $Ba_2YCu_3O_x$ superconductor, *Solid State Ionics*, 1989, *32/33*, 1133–36.
256. Vedula, K., Modeling of transient and steady-state demixing of oxide solid solutions in an oxygen chemical potential gradients, *Oxid. Met.*, 1987, *28*, 99–108.
257. Ishikawa, T., Akbhar, S.A., Zhu, W., and Sato, H., Time evolution of demixing in oxides under an oxygen potential gradient, *J. Am. Ceram. Soc.*, 1988, *71(7)*, 513–21.
258. Schmalzried, H., Behavior of (semiconducting) oxide crystals in oxygen potential gradients, *React. Solids*, 1986, *1*, 117–37.
259. Balachandran, U., Kleefish, M., Kobylinski, T.P., Morisetti, S.L., and Pei, S., Oxygen ion-conducting dense ceramic, Patent Appl. PCT/US94/03704.
260. Keulks, J.W., The mechanism of oxygen atom incorporation into the products of propylene oxidation over bismuth molybdate, *J. Catal.*, 1970, 19, 232–156.
261. Novakova, J. and Jiru, P., A comment on oxygen mobility during catalytic oxidation, *J. Catal.*, 1972, *27*, 155–156.

Chapter 15

CORROSION STUDIES

Hans de Wit and Thijs Fransen

CONTENTS

LIST OF SYMBOLS

B_i	mechanical mobility
b_i	electrical mobility
c_i	concentration of component i

0-8493-8956-9/97/$0.00+$.50

D_i	diffusion coefficient of component i
E	electromotive force emf (cell voltage)
e	electron charge
η_i	electrochemical potential of component i
ϕ	electric potential
ΔG	free enthalpy of formation (Gibbs free energy of formation)
$\Delta G^0(T)$	standard free enthalpy of formation at T
I	electric current
$J_{i,x}$	flux of particles i in x direction
k_p	parabolic rate constant
k_l	linear rate constant
K_{OX}	equilibrium constant for gas–solid reactions
k_B	Boltzmann constant (0.86* 10^{-4} eV)
μ_i	chemical potential of component i
μ_i^0	standard chemical potential at 298 K
N_M, N_O, N_d	mole fraction of M, O, and defect d
pO_2^{eq}	equilibrium partial pressure of oxygen at given T for a particular oxide
R_i	internal resistance (electrolyte)
R_e	external resistance
ΔS	entropy of formation
σ_i	electrical conductivity of component i
T	temperature (K)
t	time
t_i	transport number of component i
z_i	electric charge number of component i
v	scan rate
C_O, C_R	concentration Ox and Red in redox reaction
D_O, D_R	diffusion coefficient for Ox and Red component
C_d	double layer capacity

I. INTRODUCTION

High-temperature interaction between gas and solid is normally treated as a diffusion problem in solid state science, using normal Fickian diffusion models.

Alternatively, the metal/product/gas system can be described as an electrochemical cell in equilibrium. This means that the electrodes are not polarized, while the product layer is behaving as a (solid) electrolyte. The transport of electronic charge carriers is also taking place through the product layer, which is possible because the solid oxide or sulfide in most cases is a good semiconductor at high temperatures. For corrosion reactions in aqueous solutions electronic charge carriers obviously cannot be transported through the aqueous electrolyte, ionic transport taking place via the electrolyte, while electrons resulting from anodic charge transport across the metal/electrolyte interface are used in the cathodic reaction after crossing the interface. Only for passive metals such as stainless steels and aluminum alloys does a thin product layer (the passive film) behave, in principle, in a similar way as the product layer in high-temperature corrosion. However, the passive film is 3 orders of magnitude thinner (~2 nm for stainless steel and ~7 nm for the barrier layer on aluminum) than the layer at high temperatures, mainly as a result of the low transport rate of ions through the film. Ionic transport is rate limiting, both for high-temperature oxidation and for room-temperature aqueous passive film formation, with some exceptions to be discussed.

In this chapter we have chosen to focus on direct electrochemical information obtained on high-temperature oxidation/sulfidation, during growth of the product layers. Therefore we will not discuss valuable information that can be obtained by electrochemical measuring techniques on such related solid state processes as oxygen diffusion in metals and fast ionic conduction in oxides and sulfides, described in other chapters of this book.

II. LAYER GROWTH OF OXIDES AND SULFIDES

A. INTRODUCTION

In general, the study of high-temperature corrosion in aggressive environments has intensified over the last 20 years. The reason is that the demands upon the more efficient use of natural sources, such as coal, oil, and gas are increasing.

Since extensive reserves of coal are available, much attention is and has been paid, therefore, to expanding power output from this energy source. Convenient processes leading to more useful forms of energy or to more efficient processes include liquefaction, fluidized bed combustion, and gasification. In all these processes, materials may suffer from severe corrosive attack.

In gasification processes, for instance, gaseous species such as carbon monoxide, carbon dioxide, methane, hydrogen, steam, and sulfur-, nitrogen-, and chlorine-containing compounds are encountered. Hence, complex reactions involving oxidation, sulfidation, carburization, etc., may take place.

High-temperature sulfidation is one of the most severe problems, but only a few aspects of its very complicated character can be discussed.

First, because of similarities between oxidation and sulfidation, the oxidation mechanism of metals and alloys is considered. Then some of the peculiarities of sulfidation are reviewed.

B. THE OXIDATION OF METALS AND ALLOYS AT HIGH TEMPERATURES

Metals and alloys owe their resistance to the action of oxidizing atmospheres at high temperatures to the scale which forms on their surface, if transport through this scale is the slowest step in the oxidation reaction. Also the adherence of the scale to the metal should be strong enough in order to avoid cracking and spalling due to growth stresses and temperature fluctuations.

In many cases, the layer growth can be described by a parabolic rate law: $x^2 = k_p t$, where x is the scale thickness at time t and k_p is the parabolic rate constant. This law may be derived from Wagner's theory of metal oxidation. The parabolic rate constants contain diffusion coefficients which are related to the concentration of the defects responsible for material transport through the layer. In fact, the higher the deviation from stoichiometry, the larger the diffusion coefficient and, consequently, the faster the oxidation rate of a metal at a given temperature.

Some metals oxidize very slowly because of the very small deviation from stoichiometry of their oxides. Chromium and aluminum are widely used for the design of alloys with good oxidation-resistant properties. For a given alloy, containing either chromium or aluminum, or both, these components are usually preferentially oxidized to form a homogeneous protective scale with small defect concentrations.

In practice, it turns out that in the case of typical oxidation-resistant Fe-Cr-Ni alloys, a chromium concentration of 25 wt% is needed when oxidation in air at high temperatures is concerned. For aluminum a much lower concentration is required (about 5%), because of a higher chemical affinity for oxygen and the slower growth of an Al_2O_3 layer compared to a Cr_2O_3 layer.

C. WAGNER'S OXIDATION THEORY

Wagner's oxidation theory is valid only for compact scales of reaction products. It is assumed that volume diffusion of the reacting ions or the transport of electronic charge carriers across the scale is rate determining.[1,2] Neither the mobility (or diffusion coefficient) nor the concentration of the cations, anions, and electrons are equal. Because of this difference a separation of charges takes place in the growing oxide scale. The resulting space charge creates an electric field ($d\phi/dx$), which opposes a further separation of charges. A stationary state is reached for which no net electric current flows through the scale. The potential (ϕ) is comparable with a diffusion potential in electrolyte solutions. In describing the transport of ions and electrons through the scale, we should take into consideration both the migration of ions and electrons under influence of the concentration gradient (better: the gradient of chemical potential of components i: $d\mu_i/dx$), proper diffusion, and the migration due to the diffusion potential ϕ. The best known diffusion equation describing a continuous flux of particles through a surface of unit area in one dimension (x) is the first law of Fick:

$$J_{i,x} = D_i \frac{dc_i}{dx} \tag{15.1}$$

describing the flux of particles i in direction x through unit area, where D_i is the chemical diffusion coefficient of particles i, which connects the particle current $J_{i,x}$ with its concentration gradient dc_i/dx.* Classical thermodynamics gives the relationship:

$$\mu_i = \mu_i^0 + k_B T \ln \frac{c_i}{c_i^0} \tag{15.2}$$

for ideal solutions where μ_i denotes the thermodynamic potential of component i, μ_i^o the thermodynamic potential of component i under standard conditions, and k_B, T, and c have their usual significance.

From Equation (15.2) it follows that:

$$dc_i = \frac{c_i d\mu_i}{k_B T} \tag{15.3}$$

Introducing Equation (15.3) in Equation (15.1) gives

$$J_{i,x} = \frac{-D_i c_i}{k_B T} \frac{d\mu_i}{dx} \tag{15.4}$$

Alternatively, it can be deduced for the particle current $J_{i,x}$, under influence of the gradient in chemical potential $d\mu_i/dx$:

$$J_{i,x} = -c_i B_i \frac{d\mu_i}{dx} \tag{15.5}$$

* The derivation of this law is based on the law of mass conservation: as many particles should leave the volume ΔX cm^3 at the end (at $x = \Delta X$) as enter the volume at $x = 0$, when a stationary state has been reached.

where B_i denotes the mechanical mobility of the particles i. (B_i is the ratio of the average particle velocity V in the stationary state to the force K working on the particles in the stationary state: $B_i = \frac{V}{K}$.) From a comparison of Equations (15.4) and (15.5) it follows:

$$D_i = B_i k_B T \tag{15.6}$$

This is known as the Nernst–Einstein relation. For charged particles i we should also take into consideration the migration due to the diffusion potential ϕ. In other words, we are now interested in the change in Equations (15.4) and (15.5) if the moving particles (i) are charged. The electrochemical potential η of the particles i is given by:

$$\eta_i = \mu_i + z_i e\phi \tag{15.7}$$

where z denotes the electrical charge of the particles in electronic charge units (e). Equations (15.4) and (15.5) will now contain η instead of μ. Thus:

$$J_{i,x} = -c_i B_i \frac{d\eta_i}{dx} \tag{15.8}$$

and

$$J_{i,x} = \frac{-D_i c_i}{k_B T} \frac{d\eta_i}{dx} \tag{15.9}$$

Both B_i and D_i are related to the electrical conductivity

$$\sigma_i = c_i b_i z_i e \tag{15.10}$$

In Equation (15.10), b_i stands for the electrical mobility of the moving particles. The mechanical mobility B_i is related with the electrical mobility b_i through

$$B_i = \frac{b_i}{z_i e} \tag{15.11}$$

$$\sigma_i = c_i B_i z_i^2 e^2 \tag{15.12}$$

Substituting the Nernst–Einstein relation Equation (15.6) into Equation (15.12) gives:

$$\sigma_i = \frac{c_i D_i z_i^2 e^2}{k_B T} \tag{15.13}$$

By substituting Equation (15.13) in Equation (15.9) or Equation (15.12) in Equation (15.8) we arrive at:

$$J_{i,x} = -\frac{\sigma_i}{z_i^2 e^2} \frac{d\eta_i}{dx} \tag{15.14}$$

Introducing Equation (15.7) into Equation (15.14) leads to:

$$J_{i,x} = -\frac{\sigma_i}{z_i^2 e^2}\left(\frac{d\mu_i}{dx} + z_i e \frac{d\phi}{dx}\right) \tag{15.15}$$

Now we should keep in mind that no net current flows through the growing scale, once the stationary state has been reached. Accordingly, equivalent amounts of oppositely charged particles are transported across the scale. Cations and electrons migrate in the same direction opposite to that of the anions. Therefore, we can write:

$$z_1 J_1 = z_2 J_2 + z_3 J_3 = z_2 J_2 + J_3 \tag{15.16}$$

where the subscripts 1, 2, and 3 stand for cations, anions, and electrons, respectively. We will now describe the case of a cationic conductor, where $J_1 \gg J_2$. For the electrochemical potentials of the cations and the electrons we can write:

$$\eta_1 = \mu_1 + z_1 e \tag{15.17}$$

$$\eta_3 = \mu_3 + e \tag{15.18}$$

Substituting Equations (15.17) and (15.18) in Equations (15.15) and (15.16) gives:

$$\frac{z_1 \sigma_1}{z_1^2 e^2}\left(\frac{d\mu_1}{dx} + z_1 e \frac{d\phi}{dx}\right) = -\frac{\sigma_3}{e^2}\left(\frac{d\mu_3}{dx} + e \frac{d\phi}{dx}\right) \tag{15.19}$$

From this equation $d\phi/dx$ can be solved so that $d\phi/dx$ can be eliminated.* This, together with $\sigma 3 \gg \sigma_1$, results in:

$$J_1 = \frac{1}{z_1^2 e^2} \frac{\sigma_1 \sigma_3}{\sigma_1 + \sigma_3}\left(\frac{d\mu_1}{dx} + z_1 e \frac{d\phi}{dx}\right) \tag{15.20}$$

Because we would like to end up with an equation for the particle current equation consisting only of experimentally accessible parameters, we eliminate the thermodynamic potentials of the cations and electrons. Between the metal atoms and the metal ions plus electrons, a normal equilibrium exists:

$$M \Leftrightarrow M^{z_1^+} + z_1 e^- \tag{15.21}$$

In terms of chemical potentials this equilibrium may be written as:

$$\mu_M = \mu_1 + z_1 \mu_3 \tag{15.22}$$

* It should be noted that the potential is a superfluous parameter only in the stationary state, where no net current flows!!

Thus,

$$J_1 = \frac{1}{z_1^2 e^2} \frac{\sigma_1 \sigma_3}{\sigma_1 + \sigma_3} \frac{d\mu_M}{dx} \tag{15.23}$$

The transport number of the cations t_1 is given by the relation

$$t_1 = \frac{\sigma_1}{\sigma_1 + \sigma_3} \tag{15.23a}$$

For electrons, this becomes

$$t_3 = \frac{\sigma_3}{\sigma_1 + \sigma_3} \tag{15.23b}$$

Equation (15.23) can be simplified to:

$$J_1 = \frac{1}{z_1^2 e^2} \sigma_t t_1 t_3 \frac{d\mu_M}{dx} \tag{15.24}$$

Analogously, it follows for an oxide with $\sigma_2 \gg \sigma_1$:

$$J_2 = \frac{1}{z_2^2 e^2} \sigma_t t_2 t_3 \frac{d\mu_0}{dx} \tag{15.25}$$

It is very useful to express also J_1 in terms of the thermodynamic potential of oxygen: μ_O, because we will introduce the partial oxygen pressure into the equations through this μ_O, as this gives a direct practical relation of the reaction rate with oxidizing conditions. The chemical potential of the metal and oxygen in the metal oxide are related through the Gibbs–Duhem relation:

$$N_M \, d\mu_M + N_0 \, d\mu_0 = 0 \tag{15.26}$$

where N_M and N_O are the mole fraction of metal and oxygen. From Equation (15.26) it follows for a compound M_a O_b with only a small deviation of the stoichiometry (i.e., a = z_2 and b = z_1) that:

$$d\mu_M = -\frac{z_1}{z_2} d\mu_0 \tag{15.27}$$

Substituting Equation (15.27) in Equation (15.24) gives:

$$J_1 = \frac{1}{e^2 z_1 z_2} \sigma_t t_1 t_3 \frac{d\mu_0}{dx} \tag{15.28}$$

When both cations and anions are moving the total particle current in molecules of $M_{z_2}O_{z_1}$ cm^{-2} s^{-1} can be given by:

$$
\begin{aligned}
J_{growth} &= \frac{J_2}{z_1} - \frac{J_1}{z_2} \\
&= \frac{1}{e^2 z_1 z_2^2} \sigma_t t_2 t_3 \frac{d\mu_O}{dx} + \frac{1}{e^2 z_1 z_2^2} \sigma_t t_1 t_3 \frac{d\mu_O}{dx} \\
&= \frac{\sigma_t}{e^2 z_1 z_2^2} t_3 (t_1 + t_2) \frac{d\mu_O}{dx} \\
&= \frac{dn_{M_{z_2}O_{z_1}}}{dt}
\end{aligned}
\tag{15.29}
$$

Integration over the thickness of the scale ΔX gives:

$$\frac{dn}{dt} = \frac{1}{e^2 z_2^2 z_1^2} \int_{\mu_O^{ins}}^{\mu_O^{outs}} \sigma_t t_3 (t_1 + t_2) z_1 d\mu_O \frac{1}{\Delta X} \tag{15.30}$$

where μ_O^{ins} and μ_O^{outs} represent the chemical potential of oxygen at the inner and outer interface, respectively.

In general, the electrical conductivity and the transport numbers are functions of μ_O. When we attribute an average value to these quantities, they can be taken out of the integral. (A more detailed integration for some special cases of practical importance is given later.) This leaves us with the integral:

$$\int_{\mu_O^{ins}}^{\mu_O^{outs}} z_1 d\mu_O \tag{15.30a}$$

For M_{z_2} O_{z_1}* classical thermodynamics gives at each point in the oxide:

$$
\begin{aligned}
\mu_{M_{z_2}O_{z_1}} &= z_2\mu_{M_{MO}} + z_1\mu_{O_{MO}} = z_2\mu_{M_{MO}}^{outs} + z_1\mu_{O_{MO}}^{outs} \\
&= z_2\mu_{M_{MO}}^{ins} + z_1\mu_{O_{MO}}^{ins}
\end{aligned}
\tag{15.30b}
$$

so that:

$$z_2\left(\mu_{M_{MO}}^{outs} - \mu_{M_{MO}}^{ins}\right) = z_1\left(\mu_{O_{MO}}^{outs} - \mu_{O_{MO}}^{ins}\right) \tag{15.30c}$$

and since $\mu_{O_{MO}}^{outs} = \mu_{O_{vap}}$ and $\mu_{M_{MO}}^{ins} = \mu_{M_{metal}}$

* For $z_2 = z_1 = 2$ the usual formula would be written as: $M_{z_2}/2\ O_{z_2}/2$

$$\mu_{M_{z_1}O_{z_1}} - z_2\mu_{M_{metal}} - z_1\mu_{O_{vap}} = \Delta G_f^0 \tag{15.31}$$

$$\frac{dn}{dt} = -\frac{\overline{\sigma_t(t_1+t_2)t_3}}{e^2 z_1^2 z_2^2 \Delta x}\Delta G_f^0 \tag{15.32*}$$

We have seen that the empirical relation for the parabolic constant can be given by:

$$\frac{dn}{dt} = \frac{k_t}{\Delta x}\left(\text{or } \frac{dx}{dt} = \frac{k_p}{\Delta x}, \text{ with only a dimensional change !}\right) \tag{15.33}$$

Combination of Equations (15.32) and (15.33) gives:

$$k_t = -\frac{1}{e^2 z_1^2 z_2^2}\overline{\sigma_t(t_1+t_2)t_3}\,\Delta G_f^0 \tag{15.34}$$

We now have an equation which describes the relation between the rate constant k_t and the driving force ΔG_f^0. It is very important to understand that k_t and dn/dt are zero when either t_3 or $(t_1 + t_2)$ are zero. The model requires the migration of ions as well as electrons. Either may be rate determining. For most metal oxides, bulk diffusion can be described very well by applying the diffusion laws to the majority point defects in the lattice. For low concentrations of defects (<0.1 at%) the chemical potential of oxygen in the compound is related to the oxygen vacancy concentration (assuming that we deal with an anionic conductor) according to:

$$d\mu_O = -k_B T d\ln\left[V_O^x\right] = -k_B T d\ln\left[V_O^{z_2^\bullet}\right]\left[e'\right]^{z_2} \tag{15.35}$$

Substituting Equation (15.35) into Equation (15.30) gives:

$$\frac{dn}{dt} = \left\{\frac{-k_B T}{e^2 z_2^2 z_1^2}\int_{V_O^{ins}}^{V_O^{outs}} \sigma_t(t_1+t_2)t_3 z_1 d\ln\left[V_O^x\right]\right\}\frac{1}{\Delta x} \tag{15.36}$$

which shows the relation between the rate constant and the oxygen vacancy concentration. We will now discuss the implications of this oxide growth theory for two important practical cases: n-type and p-type semiconducting oxides.

Let us consider the metal oxide MO, and assume it to be an electronic semiconductor due to a loss of oxygen from the lattice according to:

$$2O_O^\times \rightarrow O_2(g) + 2V_O^{\bullet\bullet} + 4 \tag{15.37}$$

If the mobility and concentration of the oxygen vacancies is much greater than of other defects, Equation (15.30) reduces to:

* Because the relative contribution to the overall conductivity of various particles changes over the layer thickness due to changing concentrations of defects, while these concentrations are unknown in most cases, the average value over the whole layer was taken here.

$$\frac{dn}{dt}=\left\{\left[\frac{1}{8e^2}\right]\int_{\mu_O^{ins}}^{\mu_O^{outs}}\sigma_t t_e t_{V_O^{\bullet\bullet}} 2d\mu_O\right\}\frac{1}{\Delta x} \tag{15.38}$$

Electroneutrality demands: $[e'] = 2|V_O^{\bullet\bullet}$ throughout the oxide so that the transport numbers for both the electrons and the anions are constant if it is assumed that the mobility is independent of the concentration, which is reasonable for small concentrations. Assuming that the transport numbers are constant throughout the oxide layer, Equation (15.38) can be rewritten as:

$$\left[\frac{dn}{dt}\right]=\frac{1}{\Delta x}\frac{t_e t_{V_O^{\bullet\bullet}}}{\Delta x\, 8e^2}\int_{\mu_O^{ins}}^{\mu_O^{outs}}\sigma_t d\mu_O \tag{15.39}$$

For this n-type semiconductor $t_e = 1$, because we can safely assume, that $b_e \gg b_{V_O^{\bullet\bullet}}$. The electrical conductivity can therefore be given by:

$$\sigma_t=\sigma_e+\sigma_i=\sigma_e=[e']eb_e \tag{15.40}$$

When we apply the law of mass action to Equation (15.37), which is acceptable for low defect concentrations, we find:

$$K_{ox}=p_{O_2}\left[V_O^{\bullet\bullet}\right]^2[e']^4 \tag{15.41}$$

With $[e']=2\left[V_O^{\bullet\bullet}\right]$, Equation (15.41) reduces to:

$$[e']=4^{1/6}K_{ox}^{1/6}\,p_{O_2}^{-1/6} \tag{15.42}$$

Also, μ_O depends on p_{O2}:

$$d\mu_O=k_BTd\ln p_{O_2}^{1/2} \tag{15.43}$$

Introducing

$$t_{V_O^{\bullet\bullet}}\ll 1$$

$$t_{V_O^{\bullet\bullet}}=\left[\frac{b_{V_O^{\bullet\bullet}}}{b_e}\right]$$

and Equations (15.42) and (15.43) into the rate equation, Equation (15.39), gives

$$\left[\frac{dn}{dt}\right]=\frac{3}{\Delta x}\frac{4^{1/6}k_BTb_{V_O^{\bullet\bullet}}}{8e}K_{ox}^{1/6}\left(p_{O_2\,ins}^{-1/6}-p_{O_2\,outs}^{-1/6}\right) \tag{15.44}$$

Because $p_{O_{2outs}} \gg p_{O_{2ins}}$, the oxidation rate is practically independent of the oxygen pressure in the surrounding atmosphere. The rate only depends on the partial oxygen pressure at the metal/oxide interface, which is equal to the equilibrium pressure of the oxide at constant temperature. Thus, we can write a simplified equation for the rate constant:

$$k_t = const \cdot k_B T \cdot b_{V_O^{\bullet\bullet}} \cdot K_{ox}^{1/6} \cdot p_{O_{2ins}}^{-1/6} \tag{15.45}$$

Equation (15.45) shows that the rate of oxidation is only determined by the migration of the oxygen ion vacancies for this n-type semiconductor with oxygen vacancies as majority ionic point defects. The exponent –1/6 in Equation (15.45) depends on the defect structure of the material. Thus a general equation for n-type oxides is

$$k_t = const \cdot k_B T \cdot b_{defect} \cdot K_{ox}^{1/n} \cdot p_{O_{2\,ins}}^{-1/n} \tag{15.46}$$

When we consider a metal oxide MO with p-type behavior due to a small excess of oxygen according to:

$$O_2(g) \rightarrow 2O_O^{\times} + 2V_M'' + 4h^{\cdot} \tag{15.47}$$

we can derive a similar equation which now contains $\left(p_{O_{2outs}}^{1/6} - p_{O_{2ins}}^{1/6}\right)$. Because $p_{O_{2outs}} \gg p_{O_{2ins}}$ and generalizing for different defect situations, we now obtain:

$$k_t = const \cdot k_B T \cdot b_{defect} \cdot K_{ox}^{1/n} \cdot p_{O_{2outs}}^{1/n} \tag{15.48}$$

which shows the oxygen partial pressure dependence of the oxidation rate for a p-type material.

Equations (15.46) and (15.48) have been shown to describe the oxidation rate of several metals, like Cu, Co, Ni, and Fe, quite well. However, the agreement between theory and practice is often only observed in a limited temperature and oxygen partial pressure range, where the transport of material through the solid oxide phase is really governed by bulk diffusion of ionic defects, while the oxide behaves as a semiconductor.

Many complicating factors result in different behavior:

- Metals with varying valence may form different oxides. The scale on such metals therefore often consists of a sequence of the different oxides. This has been observed for Fe, Mn, Cu, and Co. For Fe, a sequence of $FeO/Fe_3O_4/Fe_2O_3$ is normally observed.
- Porosity of the scale can lead to totally different transport mechanisms.
- When oxidation of alloys is considered, one of the metals may act as an impurity dopant in the other metal oxide, disturbing the defect chemistry and changing the oxidation rate.
- Mixed ionic conduction (cations and anions contribute to materials transport) complicates the theoretical evaluation.
- The parabolic growth of the oxide layer is often disturbed by mechanical damage of the layer, either by external impact or by cracking of the scale due to noncompatibility of temperature-dependent mechanical properties of the alloy and the oxide layer.

D. THE SULFIDATION OF METALS AND ALLOYS

As mentioned earlier, industrial (gaseous) environments often contain sulfur compounds. Thus, sulfidation studies are of great practical importance, and this is one reason why the

TABLE 15.1
Parabolic Rate Constants kp for Oxidation and Sulfidation of Some Commonly Used Metals in O_2 and S_2 respectively at 1 atm

Reacting gas	$k_p(g/(cm^4\ s))$			
	Ni	Co	Fe	Cr
O_2	$9{\cdot}1 \times 10^{-11}$	$1{\cdot}6 \times 10^{-9}$	$5{\cdot}5 \times 10^{-8}$	$4{\cdot}5 \times 10^{-12}$
	(1000)*	(950)	(800)	(1000)
S_2	$8{\cdot}5 \times 10^{-4}$	$6{\cdot}7 \times 10^{-6}$	$8{\cdot}1 \times 10^{-6}$	$8{\cdot}1 \times 10^{-7}$
	(650)	(720)	(800)	(1000)

* Temperatures (°C) are given in parentheses.

From Mrowec, S., in *Metallic Corrosion*, Proc. 8th Int. Congress on Metallic Corrosion, Mainz, DECHEMA, Frankfurt am Main, Vol. III, 1981, 2110.

mechanism of the high-temperature corrosion of metals and alloys in gaseous sulfur and in H_2-H_2S mixtures has been the subject of numerous studies for many years. At elevated temperatures the sulfidation rate is often very high. Relatively thick sulfide scales may be formed in short periods of time. There are many similarities between sulfidation and oxidation reactions of metals.

The theory for high-temperature oxidation as described under Section II.C can be used also for sulfidation. Because of the high defect concentration in sulfides, the transport of material is mainly governed by lattice diffusion, so that the Wagner theory is probably more appropriate for many sulfides than for oxides, where grain boundary diffusion may be more important. The sulfidation of Fe can be described very well with the Wagner theory. Experimental data are in excellent agreement with the partial sulfur pressure dependence ($k = k_0 \cdot p_{S_2}^{+1/6}$) of the reaction rate according to the Wagner theory assuming that V''_{Fe} is the majority defect.

However, it must be emphasized that the mechanism of sulfidation of other metals, like nickel, cannot always so easily be explained in terms of a simple defect theory, but necessitates the introduction of association of different defects.

Because the sulfidation rate is much higher than the oxidation rate, it results in less protective layers. A comparison between oxidation and sulfidation is made in Table 15.1. Not only the defect concentration is responsible for this high rate. In fact, there are four reasons for this higher reaction rate.

1. Defect Structure of Sulfides

Metal sulfides normally show a much larger deviation from stoichiometry than metal oxides. This is very clearly reflected in a high metal defect concentration of the sulfide lattice, resulting in high values for the self-diffusion coefficient of the metal species. There are exceptions such as MnS, MoS_2, and WS_2, where the deviation from stoichiometry is smaller or in the same order as for oxides.

2. Stability of the Sulfides

The thermodynamic stability of metal sulfides is much lower than that of oxides (see Table 15.2). These differences in thermodynamic stability directly influence the driving force for the layer formation reaction as given for oxidation in Equation (15.32). This is also reflected in a different selective sulfidation behavior.

The protective properties of sulfide scales are indeed improved by alloy additions characterized by a high chemical affinity to sulfur, like Al, Zn, rare earth metals, Y, and Zr.

TABLE 15.2
The Gibbs Energy of Formation for Some Sulfides and Oxides

Sulfides		Oxides	
Compound	**ΔG°(kJ/mol S)**	**Compound**	**ΔG°(kJ/mol O)**
1/3 Al_2S_3	−210	1/3 Al_2O_3	−429
1/3 Cr_2S_3	−135	1/3 Cr_2O_3	−261
FeS	−86	FeO	−176
CoS	−80	CoO	−136
NiS	−88	NiO	−127
MnS	−190	MnO	−294

Kofstad, P., *High Temperature Corrosion,* Elsevier Applied Science, New York, 1988.

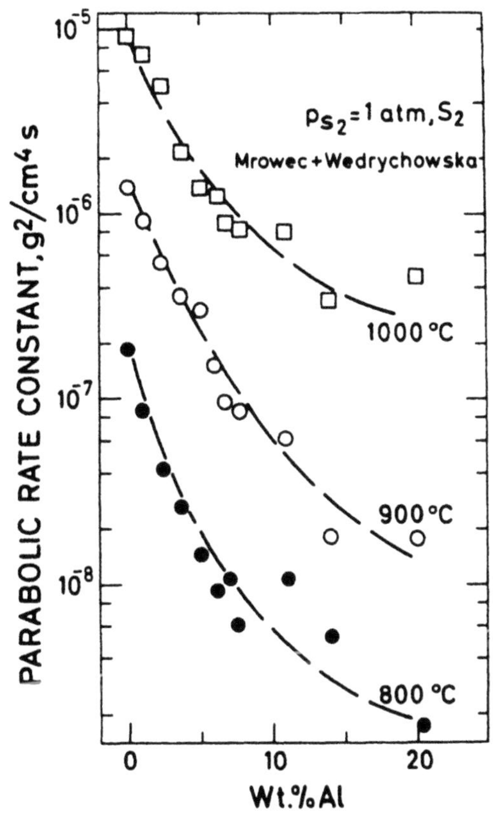

FIGURE 15.1. Dependence on the aluminum content of the sulfidation rate of ternary Fe-Cr-Al alloys containing 20% chromium. (Drawn from data of Mrowec, S. and Wedrychowska, M., *Oxid. Met.*, 1979, *13*, 481. With permission.)

However, whereas in oxygen-containing atmospheres insignificant additions to the alloy improve the oxidation resistance markedly due to the selective oxidation of these elements, in sulfidizing conditions their concentrations must be many times higher for the formation of a more or less protective sulfide layer. An example is given in Figure 15.1. In Fe-Cr-Al alloys, for instance, even an aluminum content of 35% is insufficient for the formation of a protective Al_2S_3 layer. In oxygen atmospheres, in contrast, only about 5% Al is necessary.

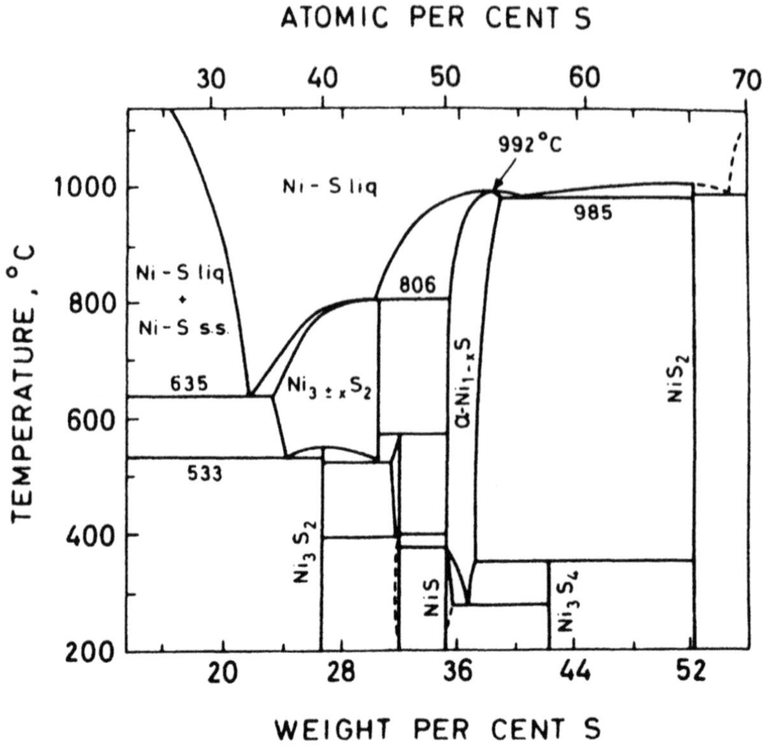

FIGURE 15.2. The Ni-S phase diagram. (From Hansen, M. and Anderko, K., *Constitution of Binary Alloys*, McGraw-Hill, New York, 1958. With permission.)

3. Melting Points of the Sulfides

Another reason why metals show a much lower resistance towards sulfidation than towards oxidation is that the melting points of metal sulfides are considerably lower than those of the corresponding oxides. Especially, some metal sulfide eutectics are characterized by very low melting points. An example of the low euctectic melting temperature of the Ni-S system is given in Figure 15.2.

4. Morphology of Sulfides

The morphology of sulfide scales is rather similar to that of oxide scales, which may be explained by analogous growth mechanisms, but there are differences. Most metals sulfidize under the formation of a duplex scale consisting of a porous inner layer under a compact outer layer. When the reaction occurs via metal transport through the scale, the product sulfide is formed at the scale/gas interface, resulting in voids at the metal sulfide interface, due to vacancy condensation. For the scale to remain in contact with the retreating metal surface it must deform. The combination of this deformation with renewed sulfidation of the metal due to inward diffusion of gaseous sulfur leads to a filling of the gap by porous sulfide.

These special properties of sulfides lead to the situation that already at relatively low temperatures especially alloys containing much nickel may suffer from severe failures in sulfidizing environments, resulting in a complete loss of the protective properties of the sulfide scale.

5. Complications

As in the case of oxide scales the phase morphology of sulfide scales, in particular, on ternary and multicomponent alloys, is very complex. As a rule, the scale is a multiphase system and involves many layers different in morphology and phase composition.

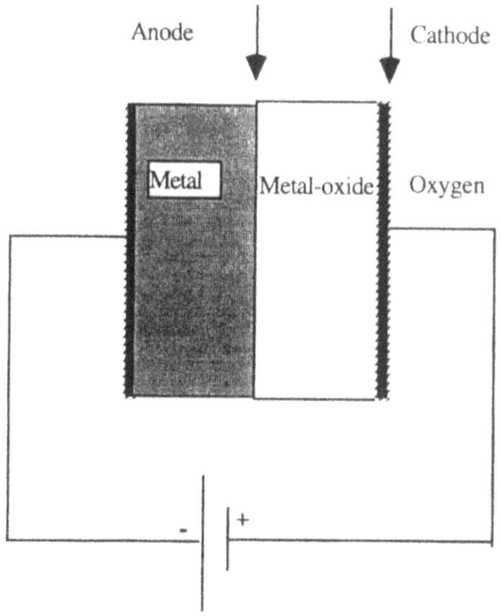

FIGURE 15.3. Growing oxide layer as an electrochemical cell.

Some aspects of the sulfidation of metals and alloys have been discussed, when only sulfur was present as the aggressive reagent. In practice, however, oxygen-free sulfur atmospheres are rarely encountered. The atmosphere in gasifiers, for instance, will always contain oxygen. In fact, in most cases the oxygen partial pressures are higher than the dissociation pressures of the metal oxides involved. This means that in a metal/sulfur/oxygen system, both oxides and sulfides may be formed. A complete understanding of the possible processes in such atmospheres involves both thermodynamic and kinetic aspects. However, this would go outside the framework of this chapter on high-temperature corrosion in relation with electrochemistry.

III. THE METAL/OXIDE/GAS SYSTEM AS AN ELECTROCHEMICAL CELL

A. INTRODUCTION

An electrochemical cell consists of two electrodes (anode and cathode) and an electrolyte. For studies of the formation of oxide product layers on metals at high temperatures it can be very useful to describe the reaction system Metal/Oxide/Gas as an electrochemical cell. In this special case the anode is formed by the metal to be oxidized, the cathode by an oxygen electrode, and the electrolyte by the oxide product layer, as shown in Figure 15.3. It is a special case of the galvanic cell normally used to describe a corrosion cell. However, in this all–solid state corrosion cell, electronic charge transport takes place through the same phase as the ionic charge transport. The metal oxide product layer serves as the electrolyte while during growth of the oxide layer electrons are also transported to the oxygen cathode, or for a p-type semiconductor holes in the opposite direction, as described in Section II.

For an oxide behaving as a true electrolyte the ionic transport number equals unity. In that case and if the reactants are all in standard condition, the cell potential (E) is given by the following equilibrium relation:

$$E = -\frac{\Delta G^0}{nF} \tag{15.49}$$

where ΔG^0 denotes the standard Gibbs energy of formation per mole oxide, F Faraday's constant, and the constant n the number of electrons taking part in the formation of 1 mol of the oxide (note that the equation is now given per mole oxide). For a simple binary oxide MO this can be simplified to

$$E = -\frac{\Delta G^0}{2F} \tag{15.50}$$

for the following electrode reactions

$$M \rightarrow M^{2+} + 2e^- \quad \text{(Anodic reaction)} \tag{15.51}$$

$$\frac{1}{2}O_2 + 2e \rightarrow O^{2-} \quad \text{(Cathodic reaction)} \tag{15.52}$$

By measuring the cell potential for this ideal case, where the oxide would behave as a true electrolyte and where no polarization losses at the electrodes would occur, the Gibbs energy of formation for the oxide can be determined. However, for most metal oxides (exceptions would be calcia- or yttria-stabilized zirconia above 800°C or pure δ-Bi_2O_3 above 650°C) the transport number of the ions is smaller than 1, thus making a direct measurement of the equilibrium cell potential in Figure 15.3 impossible. Alternatively, it could be formulated that the cell is partially short circuited by the electronic current. For oxides growing on a metal substrate under stationary conditions as described by the Wagner theory in Section II, this is the normal situation.

The determination of the Gibbs energy of formation of the growing oxide is still possible for those cases where the electronic current is blocked by using a fast ionic conductor as the CaO-stabilized ZrO_2 between the growing oxide and the oxygen electrode as, for example, in the following cell for CuO:

$$Pt/Cu, CuO/\ /ZrO_2/\ /Pt, O_2 \tag{15.53}$$

The transport number of the mobile ionic species, which is important in the Wagner theory for parabolic oxide growth, can be determined. Equation (15.50) thus becomes:

$$E_{exp} = -t_i \frac{\Delta G^0}{2F} \tag{15.54}$$

For the high-temperature oxidation of a metal we can describe the galvanic cell with an equivalent circuit incorporating the ionic and electronic contribution to the conductivity. The equivalent circuit shows the ionic resistance and the electronic resistance in series as given in Figure 15.4. Returning to our general reaction

$$z_2 M^{z_1^+} + z_1 O^{z_2^-} \rightarrow M_{z_2} O_{z_1}$$

per electron charge, Equation (15.54) thus becomes

$$E_{cel} = t_i \frac{\Delta G^0}{z_1 z_2 e} \tag{15.55}$$

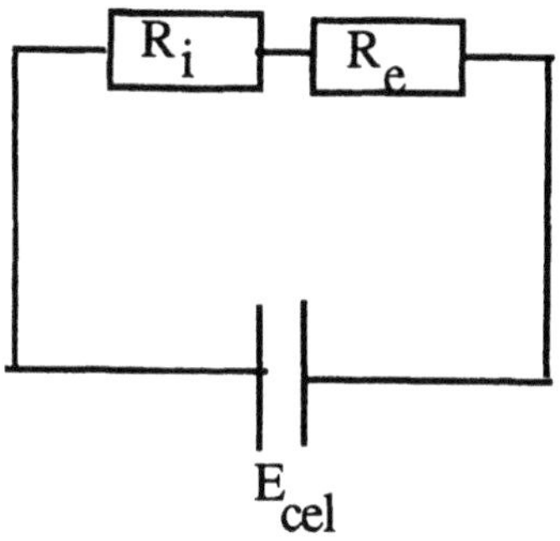

FIGURE 15.4. Equivalent circuit of the electrochemical representation of a growing oxide layer.

The total resistance of a cell with an area of 1 cm^2 and a thickness of Δx cm now equals

$$R_t = R_i + R_e = \frac{\Delta x}{\sigma_t(t_1 + t_2)} + \frac{\Delta x}{\sigma_t t_3} = \frac{\Delta x}{\sigma_t t_3(t_1 + t_2)} \tag{15.56}$$

Since $E_{cel} = t_i \frac{\Delta G_f^0}{z_1 z_2 e}$ the electrical current through the cell can be given by:

$$I = \frac{E_{cel}}{R_t} = -\frac{\sigma_t t_3(t_1 + t_2)}{z_1 z_2 e} \Delta G_f^0 \frac{1}{\Delta x} \tag{15.57}$$

The number of oxide molecules formed per cm^2s is given by:

$$\frac{dn}{dt} = \frac{I}{z_1 z_2 e} = -\frac{\sigma_t t_3(t_1 + t_2)}{z_1^2 z_2^2 e^2} \Delta G_f^0 \frac{1}{\Delta x} \tag{15.58}$$

which is the same result as given by Wagner's theory (see Equation [15.32]). This very elegant description of the growing oxide layer is of course only acceptable under the same assumptions as valid for Wagner's theory. The values of the transport numbers are average values over the oxide layer and the prerequisite of equilibrium at both interfaces is now taken up in the implicit assumption in Equation (15.57) that at both electrodes no polarization losses occur. For low-temperature, mostly aqueous corrosion reactions, this is quite unrealistic. However, for high-temperature corrosion reactions this is a reasonable assumption. This electrochemical approach also offers the possibility to influence the rate of oxide growth by the application of external electric fields, as described in the next section.

B. THE INFLUENCE OF AN ELECTRIC FIELD ON THE GROWTH RATE

When an external potential difference is applied to the cell, the electronic current through the oxide is normally no longer equal to the ionic current: electrons can be produced at the electrodes. The oxidation rate can thus be influenced by changing the externally applied potential difference.[6,7] In principle, the oxidation process can be completely stopped by applying a counter potential difference numerically equal to the driving force of the oxidation reaction:

$$E_{stop} = -E_{cel} = +\Delta G^0/2F \tag{15.59}$$ *

* Returning to the simple oxide MO and further on substituting again e for F, in order to simplify our equations.

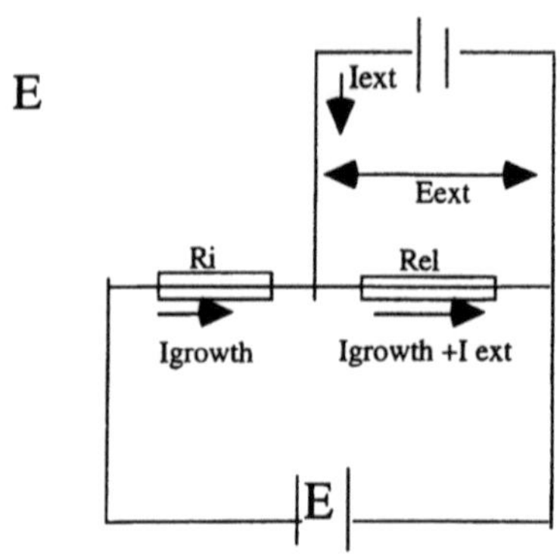

FIGURE 15.5. Equivalent circuit for a growing oxide layer when a current is passed through it. Only one ionic species migrates.

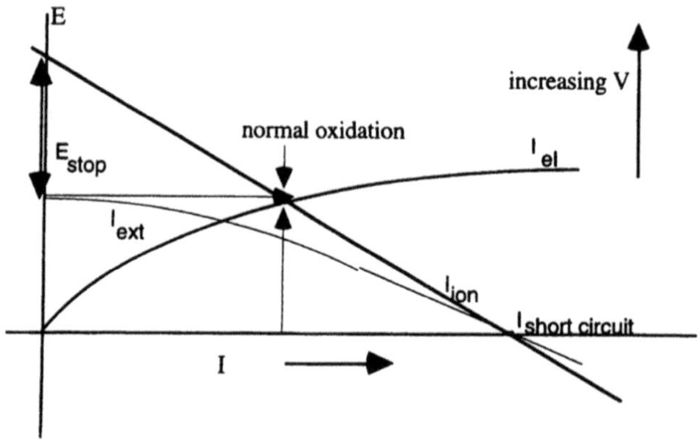

FIGURE 15.6. Showing the relation between the potential and the external and internal currents and the potential. (From Sheasby, J.S. and Jory, J.B., *Oxid. Met.*, 1978, *12*, 527. With permission.)

An even higher counter potential difference results in destruction of the oxide layer by electrolysis. Short circuiting the oxide layer will always result in an increased oxidation rate. this can easily be understood when it is considered that the electrons can very easily flow through the external current leads instead of through the oxide. Of course, this effect is only measurable for oxides with a high ionic transport number. In Figure 15.5 this situation has been shown. The effect of the external electron circuit is clear.

A schematic description of the relations between the various current densities and the potential differences is given in Figure 15.6. In this figure, the ionic current density is assumed to depend linearly on the applied potential difference, which seems reasonable for most metal oxides. We observe $I_{ion} = 0$ for $E = E_{stop} = \Delta G^0/2F$, while for short circuiting $I_{ion} = I_{ext}$. For an increasing potential difference over the layer the ionic current decreases, and thus the corrosion rate. For lower potential differences over the layer,* the ionic current and the corrosion rate increase. For normal oxidation $E = E_{ext}$ and $I_i = I_{ext}$.

* Take care: this means: by applying a counter field in the direction of the ionic current, thus supporting growth of the oxide layer.

When it is assumed that the external current does not affect the local equilibria in the scale, and substituting (E-E_{ext}) for E, Equation (15.32) simplifies to:

$$\frac{dn}{dt}=\frac{\bar{t}_i\bar{t}_e\bar{\sigma}_t}{z_1z_2e}\left[\frac{E-E_{ext}}{\Delta x}\right]=\left(\frac{dn}{dt}\right)_0-\frac{I_{ext}}{z_1z_2e}\bar{t}_i \quad (15.60)$$

where the subscript o denotes the value under normal conditions, without externally applied field. From this equation it is clear that the oxidation rate may be accelerated or retarded by applying an external potential difference. This equation also shows that because most oxides are semiconductors ($t_e \approx 1$, thus $t_i \ll 1$), relatively large currents will be needed for any measurable effect. Short circuiting ($E_{ext} = 0$) results in:

$$\frac{dn}{dt}=\left(\frac{dn}{dt}\right)_0-\frac{I_i}{z_1z_2e}\bar{t}_i=\left(\frac{dn}{dt}\right)_0-\frac{E\bar{t}_i\sigma_t\bar{t}_i}{z_1z_2e}\frac{1}{\Delta x}=\left(\frac{dn}{dt}\right)_0-\frac{\left[1-\bar{t}_e\right]^2}{z_1z_2e}\sigma E \quad (15.61)$$

From this equation it follows directly that short circuiting results in an increased oxidation rate. However, because for semiconductors $t_e \approx 1$, also this effect is very limited.

C. ELECTROCHEMICAL KINETIC STUDIES REGARDING THE FORMATION OF SULFIDE LAYERS

From Equations (15.60) and (15.61) it became clear that the influence of the externally applied electric field is normally small. However, by using special solid state galvanic cell constructions it is in principle possible to study this effect.[6] Solid state cell constructions also offer the possibility to study the transport mechanism during layer growth of sulfides, at various sulfur activities, and possibly oxides. The formation of NiS is a good example because normal sulfidation in a sulfur containing atmosphere produces a porous sulfide layer, which is easily suppressed in the galvanic cell arrangement in Figure 15.7. In both arrangements AgI is a solid electrolyte with pure cation conduction at 300°C. In case (a), the electronic current can flow through the second Pt electrode, which means that the NiS phase will behave as ionic and electronic conductor. In case (b), NiS is only transporting Ni cations, because AgI is blocking the electronic current. The current through the galvanic cell is a measure for the rate of Ag^+ production from Ag_2S, while these cations flow through the AgI phase. Assuming that no sulfur evaporates, the Ag^+ production rate is equivalent to the NiS formation rate according to the reaction:

$$\mathrm{Ni+Ag_2S \leftrightarrow NiS+2\,Ag}$$

The two cell arrangements a and b show in practice no different results, which means that the electronic current through NiS is not influencing the reaction rate. The cell potential is equivalent to that of the cell:

$$\mathrm{Pt,Ag|AgJ|Ag_2S,Pt}$$

which is a measure for the chemical potential μ_S in Ag_2S at the interface between NiS and Ag_2S. In this way we can indeed study the reaction mechanism for the formation of NiS as a function of the sulfur activity, without the need for complicated experimental arrangements for varying sulfur partial pressures in the gas phase. Two electrochemical techniques can be

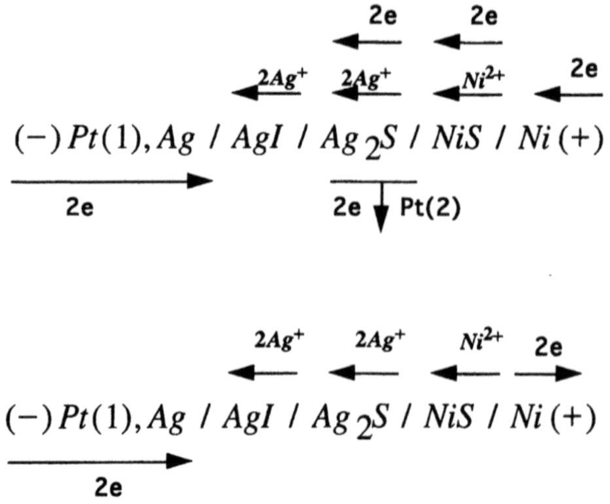

FIGURE 15.7. Two alternative galvanic cell arrangements for the study of NiS layer formation.

used to study the reaction mechanism. The first is the potentiostatic technique and the second the galvanostatic method.

1. **Potentiostatic Measurement**
   Using

$$\frac{dx}{dt} = \frac{k_p}{\Delta x} = V_m \frac{dn}{dt},$$

with V_m as the molar volume of NiS, and using Equations (15.33) and (15.58),

$$\frac{dn}{dt} = \frac{k_t}{\Delta x} \tag{15.33}$$

$$\frac{dn}{dt} = \frac{I}{2F} \tag{15.58}$$

we arrive for the formation of 1 mol of NiS at:

$$\frac{dx}{dt} = \frac{IV_m}{2F} \tag{15.62}$$

By measuring the current I through the cell in Figure 15.7, we can now observe directly the growth rate of the sulfide layer *in situ* without measuring the thickness or the weight gain of the growing layer. After integration of Equation (15.33) and substitution in Equation (15.62) we can explicitly write:

$$I = \frac{F}{V_m}\sqrt{\frac{2k_p}{t}} \tag{15.63}$$

By plotting I vs $\frac{1}{\sqrt{t}}$ we directly obtain a value for k_p.

2. Galvanostatic Measurement

We can also perform galvanostatic measurements. If k_p is not a function of the cell potential, the applied galvanostatic current will initially be responsible for the formation of NiS according to Ni + $Ag_2S \leftrightarrow$ NiS + 2 Ag at a higher rate than the rate of silver transport through the AgI. Silver is accumulating in the Ag_2S phase.

During this initial stage the open cell potential is zero, since there is no potential gradient for Ag across the AgI electrolyte. Later on the NiS formation reaction becomes slower than the rate of silver transport through the AgI. The Ag which originally precipitated in the Ag_2S will also be transported away. When the free silver phase is gone at t_k, the cell potential will suddenly rise, and is from that time on determined by the sulfur evaporation rate, which is faster than the NiS formation rate. At $t = t_k$, the amount of silver transported through the AgI phase equals the amount of NiS formed:

$$\frac{It_k}{F} = \frac{2\Delta x}{V_m} \tag{15.64}$$

Combining this with Equation (15.33) again offers a direct possibility to determine k_p.

When k_p is a function of E, the start of the experiment is similar with $E_{cell} = 0$, as long as the NiS layer is thin and the rate of formation high. In a second stage, E_{cel} will increase, but now the value of k_p also increases with time. Thus:

$$\frac{dx}{dt} = \frac{k}{\Delta x} = \frac{IV_m}{2F} = \text{constant}. \tag{15.65}$$

From this equation we can read that by forcing galvanostatically a constant current through the cell we have at least for some time a constant growth rate, despite the parabolic growth law for constant potential. Therefore, we left out the subscript p for parabolic in the equation.

When measuring the values of k as a function of the observed values of E in time we obtain interesting data for a theoretical evaluation of the growth rate dependence on E. Every k value belongs to a typical E value. Integration of Equation (15.62) and combining with Equation (15.65) gives:

$$\frac{I^2 V_m^2}{4F^2} t = k \tag{15.66}$$

Equation (15.66) shows that for constant values of I^2t the same k values and therefore E values are obtained. Therefore, if we apply different galvanostatic currents, we can check the parabolic law because Equation (15.67) must be obeyed:

$$I_1^2 t_1 = I_2^2 t_2 \tag{15.67}$$

For NiS it can be shown that at 400°C k_p increases with a factor 10, while the cell potential increases with 67 mV(=RT/2F), which is in accordance with the expected fact that k would increase linearly with the sulfur activity, because the defect model for NiS contains V_{Ni}'' as majority defects, which are responsible for the transport of material through the growing sulfide layer while the electronic charge carriers (majority holes) behave quite independent of the Ni potential in the sulfide, due to the extremely high deviation from stoichiometry of the sulfide. Therefore the normal $p_{S_2}^{+1/6}$ behavior in analogy to the $p_{O_2}^{+1/6}$ dependence for NiO is not observed.

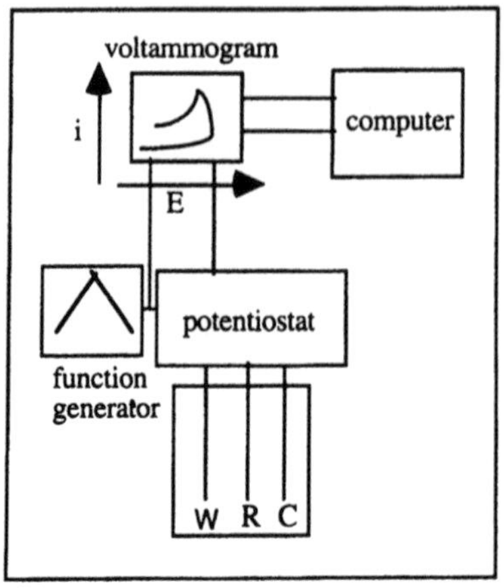

FIGURE 15.8. Single experimental setup for CV.

D. HIGH-TEMPERATURE CYCLIC VOLTAMMETRY, A FINGERPRINT OF INITIAL OXIDATION

1. Introduction to Cyclic Voltammetry

Cyclic voltammetry (CV) is a well-known technique in wet electrochemistry.[9,10] It is mainly used to obtain an electrochemical spectrum of the system studied and to obtain information on the kinetics of the reactions. In aqueous electrolytes CV can thus give quantitative information on diffusion coefficients of species participating in the reaction, together with values for the equilibrium potentials of electrode reactions. Various diagnostic criteria exist in CV to decide whether or not reactions are reversible. Only for reversible reactions mostly at the surface of noble metal electrodes, the full power of CV becomes obvious as far as quantitative kinetic data are concerned.

For passivating systems such as stainless steels and aluminum alloys, the method can also give very useful information, especially on the potential range where passivation takes place. Also, the formation of higher valencies of the cations in and at the surface of the film can be followed qualitatively, while the reduction of various species in the film and sometimes at the surface of the passive film can be observed too.

A typical experimental apparatus for conducting linear sweep voltammetry (LSV), where potential scanning takes place in either the anodic or the cathodic direction without obtaining current reversal, or CV, where a full anodic and cathodic scan between the lowest and highest potentials is performed, is illustrated in Figure 15.8. The electrolytic current (I) flows through the electrolyte between the working (W) and counter (C) electrodes. The electrode potential of interest (E) is that between W and the reference electrode (R). The linear potential sweep is relayed to the potentiostat by a function generator and the wave form is shown for a single-cycle CV. The LSV current–voltage curve for a reversible charge transfer reaction, where species O is reduced to species R according to Equation (15.1), during scanning of the potential in negative direction, is given in Figure 15.9.

$$O + ne^- \Leftrightarrow R \tag{15.68}$$

In this Figure, I_p denotes the peak current, E_p the peak potential, and E_{rev} the reversible potential. The experiment begins at the initial potential E_i which is well removed from E_{rev}. The potential as a function of time is given by

$$E(t) = E_i - vt \tag{15.69}$$

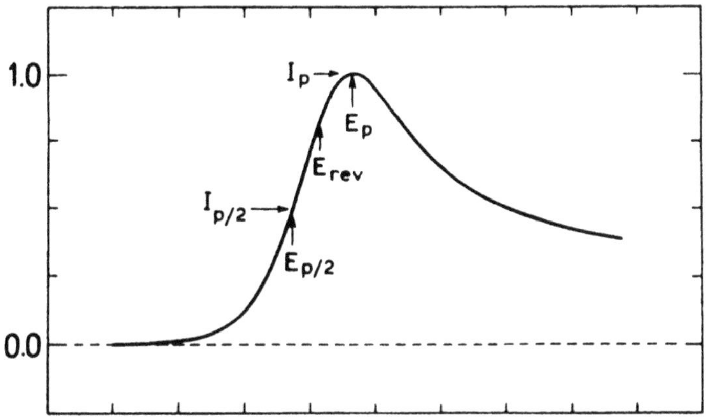

FIGURE 15.9. Linear sweep voltammogram for reversible charge transfer reaction.

where v denotes the voltage sweep rate expressed in Vs^{-1}. For a normal reversible charge transfer reaction the electron transfer rate is fast compared with v, so that the concentrations C_O and C_R at the electrode surface at all times comply to the Nernst equation:

$$\frac{C_O(t)}{C_R(t)} = \exp\left[\frac{nF}{RT}\left(E_i - vt - E_{rev}\right)\right] \tag{15.70}$$

Starting with O, Equation (15.70) indicates that the concentration of R is insignificant until E is about 180 mV from E_{rev}. At this potential C_o/C_R at the surface is of the order of 10^3 and then begins to change rapidly to 1 at E_{rev} and to about 10^{-3} at (E_{rev} – 180) mV. At more negative potentials, C_o at the electrode surface is negligibly small. During the reduction of O to R, current flows through the cell. Due to the resulting concentration gradients, O diffuses toward and R away from the electrode. As the potential is swept through the region of the LSV wave, the current and the rate of mass transport by diffusion continues to increase until E_p, where mass transfer is at a maximum. At more negative potentials I begins to decrease as a consequence of the depletion of O in the vicinity of the electrode. At E more negative than (E_{rev} – 180) mV, essentially all O reaching the electrode is immediately reduced to R and the current is controlled by the diffusion of O to the electrode.

The CV for the same reaction is shown in Figure 15.10. The first half of the cycle, which involves a potential sweep from E_i to the switching potential, is identical to the LSV wave of Figure 15.9. On the return sweep, reduction still takes place until the balance shifts at E_{rev} and the oxidation of R to O takes place during the remainder of the sweep, which is now anodic.

2. Mass Transfer, Initial, and Boundary Conditions

The theoretical treatment of mass transfer in CV assumes that only diffusion is operative. Supporting electrolyte concentrations of the order of 0.1 M are generally used for substrate concentrations of 10^{-3} M, which excludes any contribution by migration to the mass transport. A complicating factor for solid electrolytes is that addition of an indifferent supporting electrolyte is impossible.

If only diffusion plays a role in mass transport, we can solve Fick's laws of diffusion in order to get quantitative information on mass transport. To solve the diffusion equations we have to specify the initial conditions, the boundary conditions, and the relationship between C_O/C_R at the electrode surface and the electrode potential. These three items depend on the type of reaction system studied. The main differences exist between fully reversible and truly irreversible reaction systems. The mathematical treatment is too complicated, and numerical

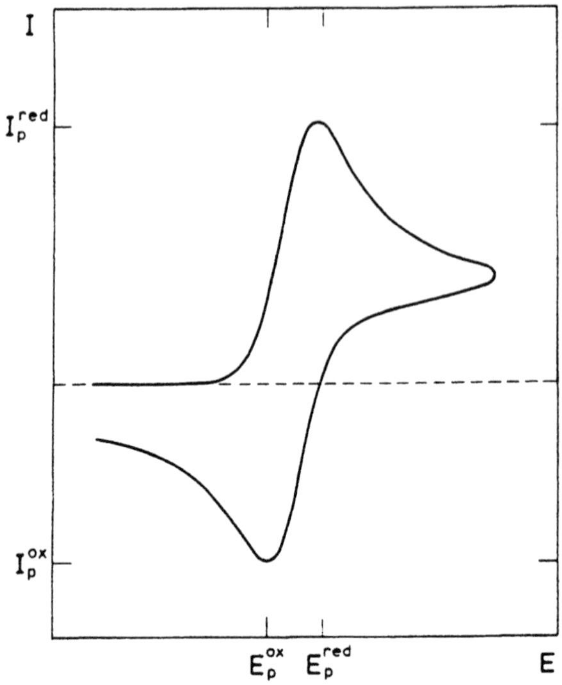

FIGURE 15.10. CV for reversible reaction.

solutions must be found. For a detailed description of various cases we refer to References 9 to 11. We will restrict ourselves to the results for reversible reaction systems.

For reversible charge transfer reactions, the potential midway between the anodic and cathodic peak ($E_{1/2}$) and the anodic and cathodic peak potential $E_{p,a}$ and $E_{p,c}$ are given by:

$$E_{\frac{1}{2}} = E^0 + \frac{RT}{nF} \ln\left(\sqrt{\frac{D_R}{D_O}}\right) \tag{15.71}$$

$$E_{p,a} = E_{\frac{1}{2}} + 1.109\left(\frac{RT}{nF}\right) \tag{15.72}$$

$$E_{p,c} = E_{\frac{1}{2}} - 1.109\left(\frac{RT}{nF}\right) \tag{15.73}$$

where E_o denotes the standard potential and D_R and D_O are the diffusion coefficients for the reduced species and the oxidized species, respectively, which are often assumed to be equal. The peak current, I_p, is given by:

$$I_p = 0.4463AC^0\sqrt{\frac{n^3F^3}{RT}D.\nu} \tag{15.74}$$

where D is the diffusion coefficient of the reacting species, ν the scan rate, A the electrode surface area, and C^o the bulk concentration of the reacting species. A number of diagnostic criteria can be formulated from these equations to determine whether or not a system behaves reversibly, which we now describe for some high-temperature oxidation studies.

3. Cyclic Voltammetry as a Tool for High-Temperature Corrosion

The use of CV at high temperatures employing a solid oxygen ion-conducting electrolyte has been reported in the literature only in four papers,[12-15] despite the interesting possibilities offered in principle:

- Information on the very early stages of oxidation
- Direct information on the various oxidation states in the oxide product
- Reliable equilibrium electrode potentials
- Kinetic data on species participating in oxidation/reduction reactions

The application of CV to high-temperature oxidation should be compared with the possibilities offered for passivation studies mentioned before. For high-temperature oxidation experiments we have the advantage of the formation of well-known stable crystalline oxide phases compared with the mostly amorphous passive films in water. On the other hand, as soon as oxide layers are formed the determination of kinetic parameters is obscured by the very formation of the layer. We will show the practical possibilities of CV for high-temperature oxidation on some experimental examples: the oxidation of Ni, Cu, Co, and Fe.

First we must discuss some general issues. Two experimental quantities must be properly accounted for in order to extract I and E from experimental CV data. The first is the double layer charging current, I_c (see also Figure 15.10). During linear potential scanning, the charge on the electrode is continually changing, and the charging current is given by:

$$I_c = AC_d\nu \tag{15.75}$$

where C_d denotes the double layer capacitance. Under most circumstances C_d is constant, giving rise to a constant I_c and hence a background current of zero slope. Since the Faradaic current for a reversible system varies linearly with $\nu^{1/2}$, while I_c is directly proportional to ν, the fraction of the total current corresponding to I_c becomes important at high scanning rates, which should be checked experimentally. For normal scanning rates of 10 to 100 mV s^{-1}, this contribution can in most cases be neglected. For the examples given below this current could be neglected.

The second experimental quantity to be dealt with is the electrolyte resistance. The electrode potential measured vs. the reference electrode includes the contribution IR_{electr}. The most common way of handling this is to employ positive feedback compensation. This involves routing an adjustable fraction of the current follower output of the potentiostat back to the signal generator input which adds a voltage, proportional to the current, to the input wave form. For more details we refer to Reference 10.

A typical measuring geometry is given in Figure 15.11. Tablets of mixtures of metal and metal oxide powder were used as electrodes, in the given case Fe/FeO as the counter electrode, providing a constant oxygen activity, and Ni/NiO as reference electrode. Yttria-stabilized zirconia tablets were used as a solid oxygen ion-conducting electrolyte. CV can be recorded at different scan rates, scan ranges, and scan directions in order to obtain information on current peaks related to the cathodic and anodic reactions. The equilibrium potential can be measured separately before each scan. Ni/NiO tablets are a suitable reference electrode because the charge transfer reaction is believed to be reversible. From the equilibrium potential (E_{eq}) obtained with a Au working electrode in a similar experimental geometry, the oxygen

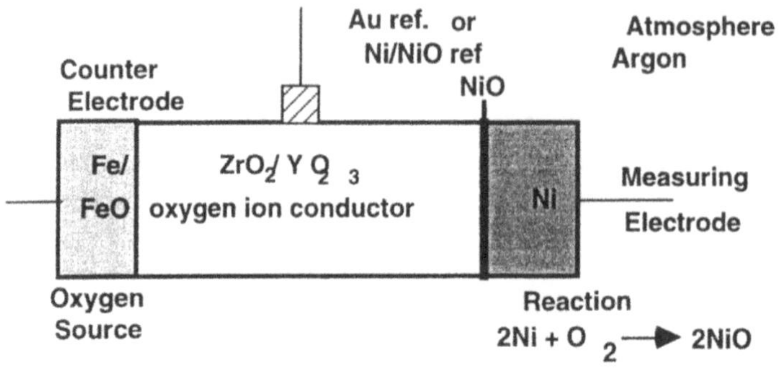

FIGURE 15.11. Experimental geometry for CV at high temperatures.

partial pressure in the argon-filled measuring cell can be determined, which is important as direct oxidation of the studied metals through the gas phase should be excluded and therefore the oxygen partial pressure should be low.

The reactions at the Au working electrode and the Ni/NiO reference electrode are

$$2V_{O\,YSZ}^{\bullet\bullet} + O_2 + 4e^- \Leftrightarrow 2O_{O_{YSZ}}^{x}, \quad \text{Au-electrode} \tag{15.76}$$

$$2O_{O_{YSZ}}^{x} + 2Ni \Leftrightarrow 2NiO + 2V_{O_{YSZ}}^{\bullet\bullet} + 4e^-, \quad \text{reference electrode} \tag{15.77}$$

According to Equation (15.3), the potential is related to the free energy of formation:

$$\Delta G_f^0 = -nFE_{eq} \tag{15.78}$$

The oxygen partial pressure can be calculated from:

$$p_{O_2}(Au) = p_{O_2}(Ni/NiO)\exp\left(\frac{4E_{eq}F}{RT}\right) \tag{15.79}$$

For the equilibrium potential of –104 mV measured at the Au electrode, this yields an oxygen partial pressure of 10^{-18} atm, which is well below the equilibrium oxygen pressure of NiO at 700°C, resulting in the slow reduction of the reference electrode tablet, which makes measurements possible during a limited, but long enough period of time.

a. The High-Temperature Oxidation of Ni

CV of Ni/NiO, all recorded in a range from –600 to 600 mV, show one oxidation and one reduction peak. The total charge amounts consumed in the oxidation and reduction reactions are unequal, the cathodic charge being larger ($Q_a \approx 0.7\ Q_c$). The reproducibility of the scans depends on the start potential. The peak potential associated with the oxidation of Ni during the first scan was about 25 mV (at 750°C with a scan rate of 20 mV/s), lower than that in consecutive scans (see Figure 15.12), if the scan was started at –600 mV. Starting at the equilibrium potential results in reproducible peak potentials for all scans. The peak current increases with higher scan rates, while the peak potential shifts. The influence of the temperature on the CV is shown in Figure 15.13. The peak potentials and the peak currents are temperature dependent. The peaks are observed to shift towards one another while the peak currents increase. The current peaks are asymmetrical at low temperatures.

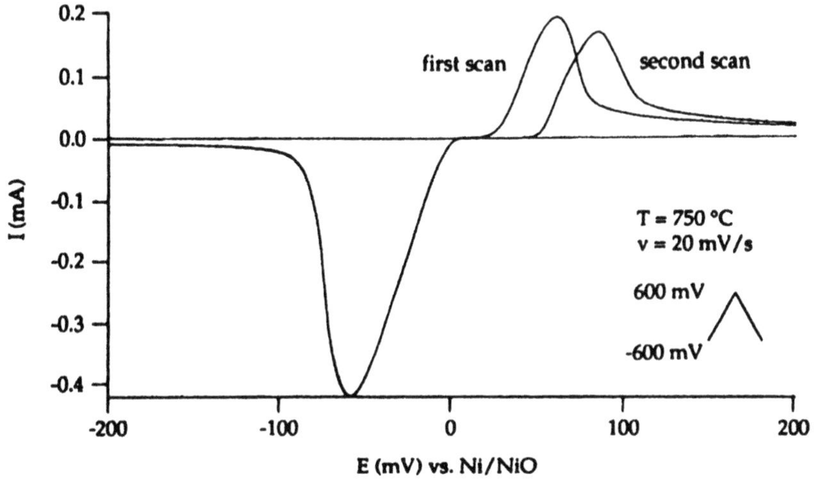

FIGURE 15.12. Two consecutive scans of Ni/NiO. (From van Manen, P.A., Weewer, R., and de Wit, J.H.W., *J. Electrochem. Soc.*, 1992, 139, 1130. With permission.)

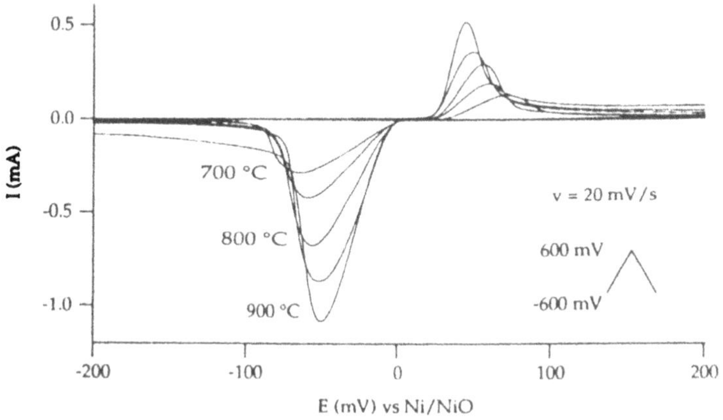

FIGURE 15.13. The influence of the temperature on the voltammograms of Ni/NiO. (From van Manen, P.A., Weewer, R., and de Wit, J.H.W., *J. Electrochem. Soc.*, 1992, 139, 1130. With permission.)

The two observed peaks in the Ni voltammograms result from the oxidation and reduction of Ni. In Equation (15.74), a linear dependence of the peak current I_p on the square root of the scan rate was assumed. For Ni/NiO this relation seems to hold quite well, as shown in Figure 15.14. At all temperatures $E_{1/2}$ and the equilibrium potential deviate only a few millivolts from the expected potential of 0 mV. This seems to suggest that the charge transfer reaction for the oxidation and reduction of Ni is reversible.[12,13] However, other criteria to decide whether or not a reaction is reversible must be evaluated first. The peak potential should be independent of the scan rate and a linear relation between peak potential and temperature should exist. These conditions are not fulfilled for the Ni/NiO reaction. However, there are more obvious reasons for the system not to be reversible.

The oxidation of the working electrode results in the formation of an oxide layer comparable to passivation in wet electrochemistry. For the oxidation reaction to continue, oxygen or nickel ions will have to diffuse through an oxide layer of increasing thickness. The thickness of this layer can be calculated from the total anodic charge. For the Ni/NiO electrode at 750°C and a scan rate of 20 mV/s, this yields a thickness of about 1 nm of NiO. The reduction reaction will be affected in a similar way, but now a thin Ni layer will impede the transport of oxygen (see Figure 15.15).

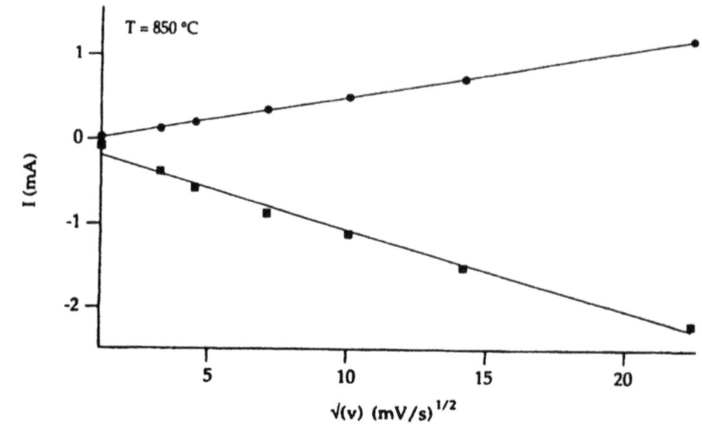

FIGURE 15.14. Anodic and cathodic peak currents, o Ip,a and ·Ip,c vs. the square root of the scan rate. (From van Manen, P.A., Weewer, R., and de Wit, J.H.W., *J. Electrochem. Soc.*, 1992, 139, 1130. With permission.)

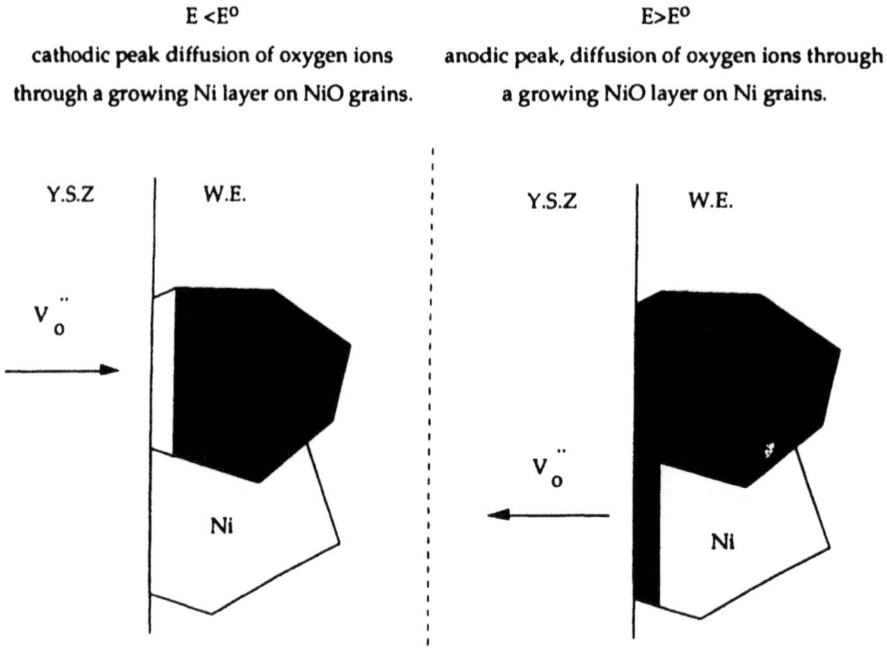

FIGURE 15.15. Formation of passivating layers n Ni/NiO grains. (From van Manen, P.A., Weewer, R., and de Wit, J.H.W., *J. Electrochem. Soc.*, 1992, 139, 1130. With permission.)

Worrell and Iskoe[16] proposed a model in which no scale formation occurs. The oxygen ions diffuse through the nickel grains and along grain boundaries to NiO grains thus, apparently, extending the electrolyte into the nickel grains. In this model, the Ni and NiO grains are both present at the interface. However, in the case of reversible reactions the potential of the electrode would immediately adopt the Ni/NiO equilibrium potential. The potential of the electrode after scanning with an end potential of –600 mV takes some time (1 to 2 min) to assume the equilibrium value. This implies that the reaction is not reversible, possibly due to layer formation at the electrode.

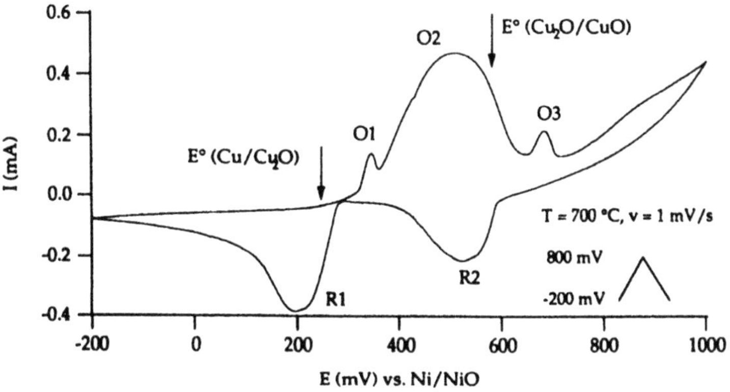

FIGURE 15.16. Voltammogram for Cu/CuO. (From van Manen, P.A., Weewer, R., and de Wit, J.H.W., *J. Electrochem. Soc.*, 1992, 139, 1130. With permission.)

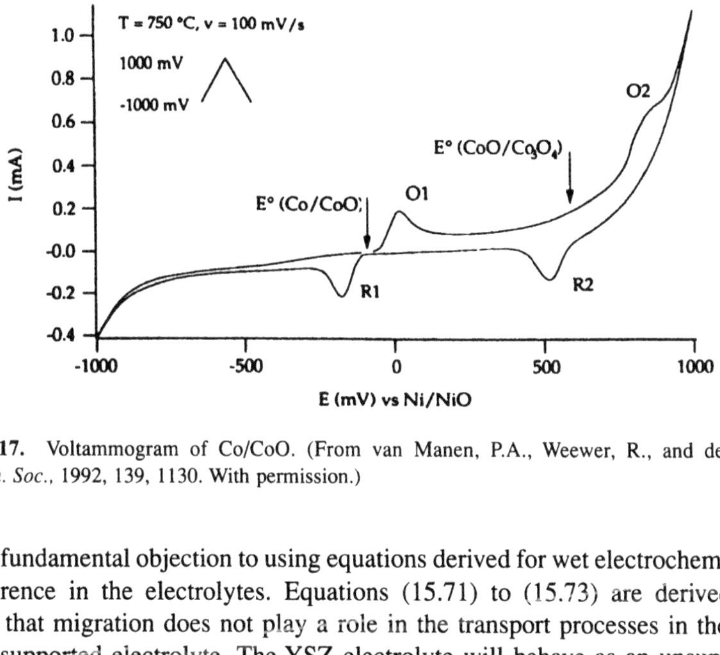

FIGURE 15.17. Voltammogram of Co/CoO. (From van Manen, P.A., Weewer, R., and de Wit, J.H.W., *J. Electrochem. Soc.*, 1992, 139, 1130. With permission.)

A more fundamental objection to using equations derived for wet electrochemical systems is the difference in the electrolytes. Equations (15.71) to (15.73) are derived under the assumption that migration does not play a role in the transport processes in the electrolyte by using a supported electrolyte. The YSZ electrolyte will behave as an unsupported electrolyte in which migration cannot be eliminated. Furthermore, the equations are derived under the assumption that the electrodes are inert. The electrodes here are not inert, and both the reduced and the oxidized species are present from the beginning of the experiment, which violates one of the boundary conditions used to derive the analytical expressions from which Equations (15.71) to (15.73) have been derived. Other examples of CV obtained at high temperatures are given in Figures 15.16 to 15.18 for Cu/CuO, Co/CoO, and Fe/FeO, respectively.

b. The High-Temperature Oxidation of Cu

For the Cu/Cu_2O working electrodes, two peaks were expected as a result of the oxidation of Cu to Cu_2O to CuO. However, at low scan rates the voltammograms show three anodic current peaks with only two cathodic currents. By changing the scan rate it is shown that the anodic and cathodic current peaks, O_3 and R_2 in Figure 15.16, are related. Since the areas

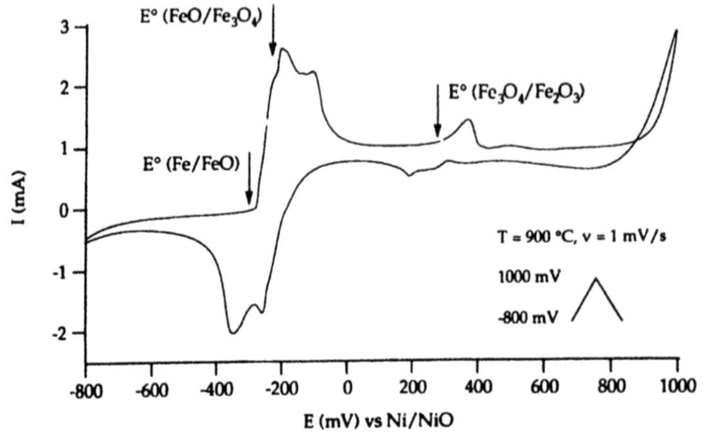

FIGURE 15.18. Voltammogram of Fe/FeO. (From van Manen, P.A., Weewer, R., and de Wit, J.H.W., *J. Electrochem. Soc.*, 1992, 139, 1130. With permission.)

under the large anodic (O_2) and cathodic (R1) peak are almost equal, it seems likely that these two are related. The small anodic precursor peak (O_1) has no obvious cathodic counterpart. To find out which reactions correspond with the current peaks, the equilibrium potential of the reactions was calculated from thermodynamic data. These are indicated in Figure 15.16. These potentials show that the peaks O(1,2) and R(1) are associated with the oxidation of Cu according to:

$$2CuO + 1/2\,O_2 \Leftrightarrow Cu_2O \tag{15.80}$$

The potential for the reaction

$$Cu_2O + 1/2\,O_2 \Leftrightarrow 2CuO \tag{15.81}$$

did not correspond to the potential midway between the cathodic (R_2) and the anodic (O_3) peaks at the anodic end of the voltammogram. The small peak (O_1) preceding the large peak associated with the oxidation of Cu represents a charge large enough to result from an oxide scale of about 1 nm. For neither oxidation reaction did the peak currents show a linear relation with the square root of the scan rate.

c. *The High-Temperature Oxidation of Co*

The potentials calculated from the thermodynamic data for the oxidation of cobalt:

$$Co + 1/2\,O_2 \Leftrightarrow CoO \tag{15.82}$$

agree well with the average potential of the first peak potentials (O_1,R_1) (see Figure 15.17). The cathodic peak potentials for the second reaction are near those calculated for:

$$3CoO + 1/2\,O_2 \Leftrightarrow Co_3O_4 \tag{15.83}$$

Although the potentials from the voltammograms do correspond with the calculated potentials, the system is not reversible since there is no linear relation between the peak current and the square root of the scan rate. The second anodic peak (O_2) is also related to diffusion in the already existing CoO layer, while it is further oxidized to higher valent Co_3O_4.

d. The High-Temperature Oxidation of Fe

At low scan rates (e.g., 1 mV/s) the Fe/FeO system shows several peaks as is shown in Figure 15.18. The potentials are related to the calculated ones for the following reactions:

$$Fe + 1/2\,O_2 \Leftrightarrow FeO \qquad (15.84)$$

$$3FeO + 1/2\,O_2 \Leftrightarrow Fe_3O_4 \qquad (15.85)$$

$$2Fe_3O_4 + 1/2\,O_2 \Leftrightarrow Fe_2O_3 \qquad (15.86)$$

The equilibrium potentials have been indicated in Figure 15.18. As for the other reactions, no reversibility was found.

4. General Conclusions on the Application of High-Temperature Voltammetry

In the voltammograms for Cu/Cu_2O and for Fe/FeO, the number of peaks increases with a decreasing scan rate. Especially for Fe, we have seen that the peaks for various oxidation states are well resolved at a low scan rate of 1 mV/s. This is in contrast with wet electrochemical CV, where an increase in scan rate yields more peaks because diffusion is bound to become rate determining at high scan rates. The appearance of peaks in the high-temperature CV is caused by a different mechanism, such as oxide scale formation. This implies that kinetic information on the oxidation process or diffusion coefficients cannot be obtained using Equation (15.7). Other difficulties are caused by the IR drop, and the capacitive current as discussed. Correction for the IR drop is only possible at the equilibrium potential, which implies that the increase in the resistance due to the oxide scale formation is not taken into account. The same argument goes for the double layer capacity, which is expected to change due to the oxide layer formation. The peak potential and the peak current are changed by these effects. Because the resistance and the double layer capacitance change with time, the peak currents and potentials cannot be corrected. Finally, a contribution to the current not directly related to the reactions of the working electrode material is oxygen evolution and reduction at the electrode/YSZ/gas interface. The reaction rate for this reaction is strongly dependent on the electrode material, while the current also depends on the dimensions of the three-phase boundary. Because of these complicating factors, it is virtually impossible to make corrections for variations in the current. For these reasons and, even more important, due to the fact that the system is not reversible, it is impossible to obtain kinetic information from the voltammograms.

However, the peaks in the voltammograms are a clear fingerprint of the oxide scale formation. Although the systems are not reversible, the potentials from the voltammograms correspond quite well with those obtained from thermodynamic calculations. CV can be used very well to observe the different oxidation states of the metals, including the initial oxidation stages and higher valent oxidation states.

REFERENCES

1. de Wit, J.H.W., High temperature oxidation of metals, *J. Mater. Ed.*, 1981, 2, 333.
2. Kofstad, P., *High Temperature Corrosion*, Elsevier Applied Science, New York, 1988.
3. Mrowec, S., in Metallic Corrosion, Proc. 8th Int. Congress on Metallic Corrosion, Mainz, DECHEMA, Frankfurt am Main, Vol. III, 1981, 2110.
4. Mrowec, S. and Wedrychowska, M., *Oxid. Met.*, 1979, 13, 481.

5. Hansen, M. and Anderko, K., *Constitution of Binary Alloys*, McGraw-Hill, New York, 1958.
6. a. Kröger, F.A., *The Chemistry of Imperfect Crystals*, North-Holland, Amsterdam, 2nd ed., 1974, 3. b. Rickert, H., *Einführung in die Elektrochemie fester Stoffe*, Springer-Verlag, Berlin, 1973.
7. Hinze, J.W., thesis, Iowa State University, Ames, 1973.
8. Sheasby, J.S. and Jory, J.B., *Oxid. Met.*, 1978, 12, 527.
9. Bard, A.J. and Faulkner, L.R., *Electrochemical Methods, Fundamentals, and Applications*, John Wiley & Sons, New York, 1980.
10. Parker, V.D., Linear and cyclic voltammetry, in *Compr. Chem. Kinetics*, Vol. 26, Bamford, C. and Compton, R., Eds., Elsevier, Amsterdam, 1986.
11. Delahay, P., *New Instrumental Methods in Electrochemistry*, Interscience, New York, 1954.
12. Yang, C.Y. and Isaacs, H.S., *J. Electroanal. Chem.*, 1981, 123, 411.
13. Yang, C.Y. and O'Grady, W.E., *J. Vac. Sci. Technol.*, 1982, 20, 925.
14. Gozzi, D., Tomellinin, M., Petrucci, L., and Cignini, P.L., *J. Electrochem. Soc.*, 1987, 134, 728.
15. van Manen, P.A., Weewer, R., and de Wit, J.H.W., *J. Electrochem. Soc.*, 1992, 139, 1130.
16. Worrel, W.L. and Iskoe, J.L., in *Fast Ion Transport in Solids*, van Gool, W., Ed., North-Holland, Amsterdam, 1973.

Chapter 16

ELECTROCHROMISM AND ELECTROCHROMIC DEVICES

Claes G. Granqvist

CONTENTS

I. INTRODUCTION

An electrochromic material is able to change its optical properties when a voltage is applied across it. The optical properties should be reversible, i.e., the original state should be recoverable if the polarity of the voltage is changed. These properties make electrochromic materials of considerable interest for optical devices of several different types, such as elements for information display, light shutters, smart windows, variable-reflectance mirrors, and variable-emittance thermal radiators. This chapter reviews the basic materials background and the state of the art for electrochromic devices. Electrochromism is well known in numerous inorganic and organic substances; the chapter is devoted to the former class. Almost all of the interesting materials are oxides that are employed in the form of thin films.

One may think of many possible electrochromic device constructions. The most important of these is depicted in Figure 16.1. The arrangement of materials shown is the preferred one in almost all work aimed at practical utilization of electrochromism, and the setup is convenient also as a prototype for discussing electrochromic systems in general. The figure shows a number of layers backed by a substrate that in many cases is a glass plate. The glass has a transparent and electrically conducting film and a film of the electrochromic oxide. This oxide has mixed conduction for ions and electrons, and if ions are introduced from an adjacent electrolyte or via an adjacent ion conductor there is a corresponding charge-balancing counterflow of electrons from the transparent electron conductor. These electrons will remain in the electrochromic film as long as the ions reside there, and the electrons will then evoke a persistent change of the optical properties. Depending on which electrochromic oxide is used, the electron injection can increase or decrease the transparency. The ion conductor, that comes

0-8493-8956-9/97/$0.00+$.50

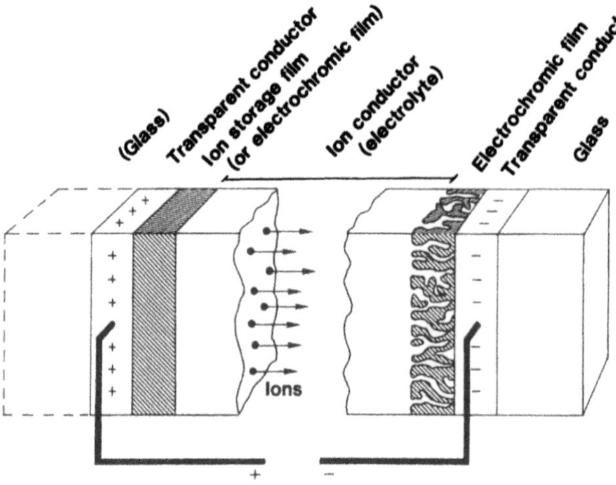

FIGURE 16.1. Basic design of an electrochromic device, indicating transport of positive ions under the action of an electric field. (From Granqvist, C., *Handbook of Inorganic Electrochromic Materials,* Elsevier Science, 1995. With permission.)

next in the device, can be a thin film or a bulk-like material; for practical devices one prefers solid inorganic or organic (polymeric) materials, whereas liquid electrolytes are convenient for research and exploratory work. The final component of the device is an ion storage (or counter electrode [CoE]) with or without electrochromic properties and another electrical conductor that must be a transparent film if the full device is for modulating light throughput. The transparent electrical conductors are doped films based on In_2O_3, SnO_2, or ZnO, with In_2O_3:Sn (often referred to as ITO) being the most common choice. At least for some devices, it is suitable to have a second glass plate, so that the entire system embodies two substrates — each with a two-layer coating — and an intervening ion conductor that can serve as a lamination material. Each of the films in the device can have a thickness less than 1 μm; such films can be produced by many different techniques.

When a voltage is applied between the (transparent) electrical conductors, as indicated in Figure 16.1, a distributed electrical field is set up and ions are moved uniformly into (intercalated) and out of (deintercalated) the electrochromic film(s). The charge-balancing counterflow of electrons through the external circuit then leads to a variation of the electron density in the electrochromic material(s) and thereby a modulation of their optical properties. If the ion conductor has negligible electronic conductivity, the device will exhibit open-circuit memory, so that the optical properties remain stable for long periods of time.

Figure 16.2 illustrates the four main applications of electrochromic devices. Part (a) refers to *information display.* The device has an electrochromic film in front of a diffusely scattering pigmented surface. The electrochromic film can be patterned and, for example, be part of a seven-segment numeric display unit of small or large size. It is possible to achieve excellent viewing properties with better contrast — particularly at off-normal angles — than in the conventional liquid-crystal–based displays. Display applications have been discussed ever since the discovery of electrochromism,[1] but so far such displays have not been turned into large-scale consumer items. Figure 16.2(b) shows how an electrochromic film can be used to produce a *mirror with variable specular reflectance.* This application seems to be the most mature one, and antidazzling rear-view mirrors built on electrochromic oxide films are currently (1994) available for cars and trucks. *Electrochromic smart windows,* sketched in Figure 16.2(c), can lead to architectural or automotive fenestration with variable transmittance so that a desired amount of visible light and/or solar energy is introduced. Such windows can produce energy efficiency as well as a comfortable indoor climate. *Variable emittance surfaces,* outlined in Figure 16.2(d), are based on a special device design with a crystalline

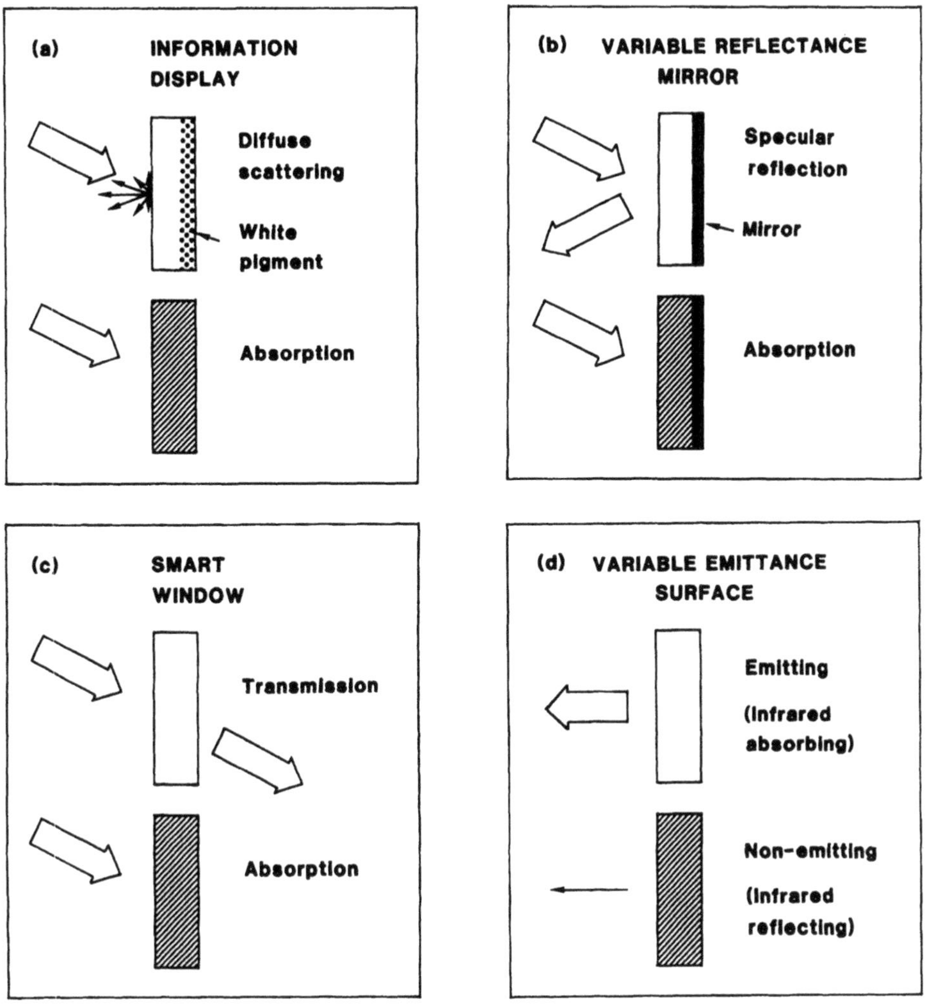

FIGURE 16.2. The principles of four different applications of electrochromic devices. Arrows indicate incoming and outgoing electromagnetic radiation; the thickness of the arrow signifies radiation intensity. (From Granqvist, C., *Handbook of Inorganic Electrochromic Materials*, Elsevier Science, 1995. With permission.)

tungsten oxide film at the exposed surface of an electrochromic device. Intercalation/deintercalation of ions makes this surface infrared reflecting/absorbing, i.e., the thermal emittance is low/high. Such surfaces can be employed for temperature control under conditions when radiative exchange dominates conduction and convection, such as for space vehicles.

Electrochromic materials and devices have been reviewed several times.[2-9] Two compilations of research papers were published in 1990.[10,11] An extensive treatise appeared in 1995.[12]

II. SOME PROPERTIES OF W OXIDE FILMS: CASE STUDY

Electrochromism was first discussed in tungsten oxide,[1] and films based on this material form the most viable option for devices. It is thus suitable to introduce the key concepts for electrochromic materials with reference to W oxide.

The crystalline material is built from corner-sharing WO_6 octahedra as illustrated in Figure 16.3; this is known as the defect perovskite structure. The oxide has a propensity to form substoichiometric shear phases (Magnéli phases) with some edge-sharing octahedra. The spaces in between the octahedra are large enough to accommodate ions, i.e., the W oxide

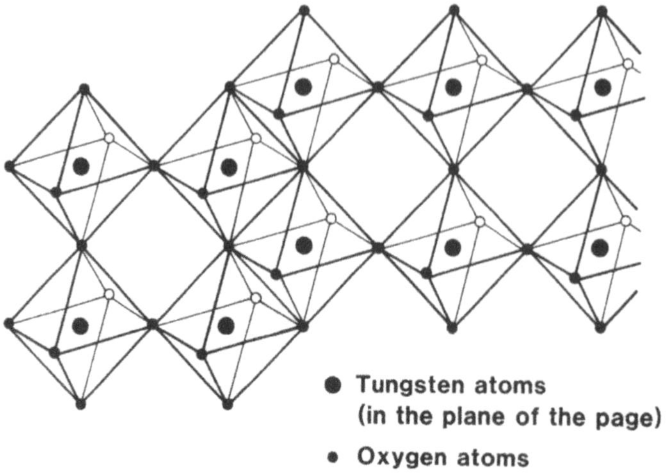

FIGURE 16.3. Schematic illustrating a corner-sharing and edge-sharing arrangement of WO_6 octahedra in a tungsten oxide crystal. The rear-most oxygen atoms are indicated by small open circles. (From Granqvist, C., *Handbook of Inorganic Electrochromic Materials,* Elsevier Science, 1995. With permission.)

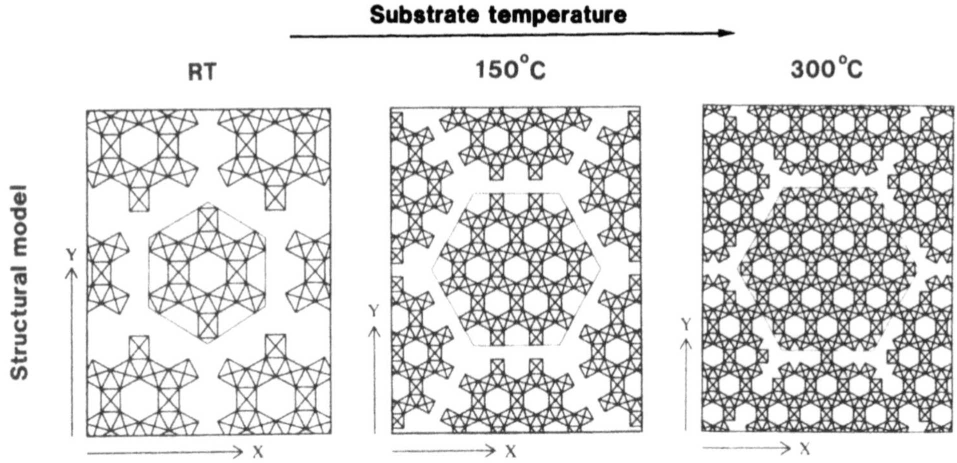

FIGURE 16.4. Structural models, based on connected WO_6 octahedra, for W oxide films made by evaporation onto substrates at room temperature (RT) and at two elevated temperatures. The arrows in the X and Y directions denote 2 nm. (From Granqvist, C., *Handbook of Inorganic Electrochromic Materials,* Elsevier Science, 1995. With permission.)

framework can serve as a host for ion intercalation/deintercalation. Substoichiometry tends to give extended "tunnels" for the ions.

Thin films can be produced by a variety of physical, chemical, and electrochemical methods.[8] The density can be as low as 50% of the bulk value. Structural models have been worked out from X-ray scattering, vibrational spectroscopy, and other techniques. Figure 16.4 shows structural forms for W oxide films made by evaporation onto substrates at room temperature, 150°C, and 300°C.[13] The films can be described as cluster-assembled materials with WO_6 units arranged so that a hexagonal structure is approached. The clusters are held together by hydrogen bonds and bridging water molecules. Heating to 300°C and above yields a long-range ordered crystalline structure.

Ion intercalation/deintercalation is easy in porous W oxide and takes place, highly schematically, as

$$WO_3 + xM^+ + xe^- \leftrightarrow M_xWO_3 \qquad (16.1)$$

with M = H, Li, Na, ... In reality, the oxide is substoichiometric, hydroxylated, and hydrated, and ion intercalation may occur in conjunction with associated water molecules. A more correct, but also more unwieldy, reaction for proton intercalation/deintercalation is

$$WO_{3-z-q}(OH)_q \cdot pH_2O + xH^+ \cdot rH_2O + xe^- \leftrightarrow WO_{3-z-q-x}(OH)_{q+x} \cdot (p+xr)H_2O \quad (16.2)$$

with $z < 0.3$ and p, q, and r possibly having rather large values. Lithium ion intercalation/deintercalation is even more complicated and involves exchange reactions with protons.

With all the uncertainties and lack of specificity veiling the ion intercalation/deintercalation reactions, one may wonder whether or not it is in vain to look for a reasonably certain and specific model for the electrochromism. In fact, such a model *is* possible, and this is so because the optical properties are governed largely by the insertion/extraction of the *electrons*. The ion transport, with all its complexity, serves only for charge balancing.

The fate of the inserted electrons can be understood from measurements of electron paramagnetic resonance (EPR) and X-ray photoelectron spectroscopy (XPS). The significance of EPR for assessing the valence state of the W ion rests on the fact that W^{5+} ($5d^1$ configuration) gives a signal due to its unpaired spin, whereas W^{6+} ($5d^0$ configuration) does not give any signal. As found by Gérard et al.,[14] a signal at the g-value 1.76, which can be unambiguously ascribed to W^{5+}, develops upon proton intercalation. XPS gives supporting evidence and, as found by Temmink et al.,[15] the signal due to W4f electrons can be decomposed into two spin-orbit–split doublets due to W^{6+} and W^{5+}, with the second doublet having an intensity that scales with the amount of ion intercalation. The interpretation is that the electrons that enter along with the ions are localized at the tungsten sites, where the valency is changed from 6+ to 5+. Figure 16.5 shows the relative amount of W ions in the 5+ state — i.e., W^{5+}/W_{tot} — as a function of x in H_xWO_3, with x inferred from the amount of charge insertion. The two sets of data are found to correlate very well. Complete agreement according to the solid line would indicate that each electron gives rise to one tungsten ion in 5+ state. The fact that x is larger than the W^{5+}/W_{tot} value by a small and constant amount can be reconciled with the fact that the film is slightly substoichiometric with a O/W ratio of 2.85 (i.e., $z = 0.15$ in Reaction 16.2).

The optical properties are changed as a result of the charge insertion. Figure 16.6, after Yamada et al.,[16] shows the evolution of the spectral absorption coefficient when a charge density ΔQ is introduced in a W oxide film through ion intercalation from an H_2SO_4 electrolyte. An absorption band centered at $E_p \approx 1.3$ eV increases monotonically upon increased ΔQ and reaches a maximum of $\sim 10^5$ cm^{-1}. The absorption is strongest in the infrared, which leads to a bluish appearance.

A connection between the absorption band and the amount of W^{5+} is illustrated in Figure 16.7, after Gabrusenoks et al.[17] The EPR signal at $g \approx 1.7$, which is an unmistakable signature of W^{5+} as noted above, increases in proportion with the intensity of the absorption centered at $E_p \approx 1.3$ eV, and it follows that tungsten atoms in 5+ state cause the absorption.

The small polaron concept offers a possibility to formulate a quantitative theory for the optical absorption. Small polarons are created when electrons polarize their surroundings so that localization of the wavefunction takes place essentially to one lattice site. A small overlap between wavefunctions corresponding to adjacent sites, as well as strong disorder, are conducive to polaron formation. Theoretically, the small polaron absorption α_{pol} can be expressed by[18]

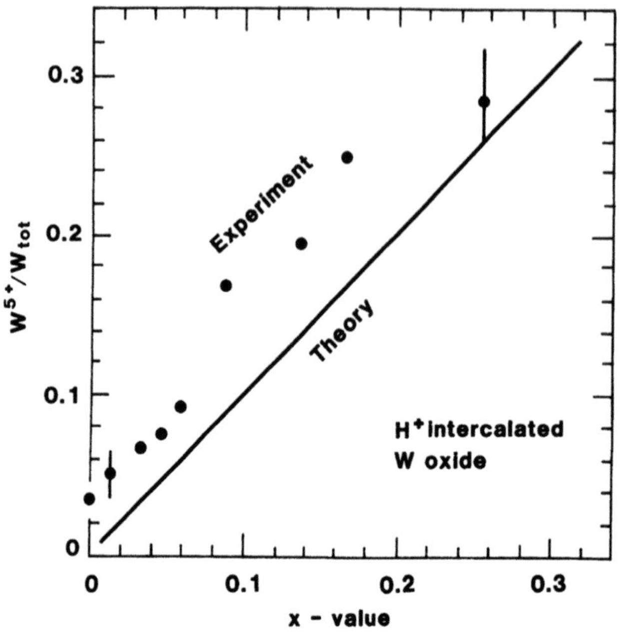

FIGURE 16.5. Relative number of tungsten atoms in 5+ state (W^5/W_{tot}) vs. degree of charge insertion (represented as x in H_xWO_3) for evaporated W oxide films. Dots indicate experimental data with error bars typical for small and large intercalation. The line represents a theoretical situation with each inserted electron contributing one W^{5+} ion. (From Granqvist, C., *Handbook of Inorganic Electrochromic Materials,* Elsevier Science, 1995. With permission.)

$$\alpha_{pol}(\omega) \propto \omega^{-1} \exp\left[\frac{(\hbar\omega - 4U_p)^2}{8U_p\hbar\omega_{ph}}\right], \tag{16.3}$$

where U_p is the polaron binding energy and $\hbar\omega_{ph}$ is a typical phonon energy. Figure 16.8, after Schirmer et al.,[19] shows a comparison between experiments on an H^+ intercalated W oxide film and $\omega^2\alpha_{pol}(\omega)$. The fitting used $U_p = 0.275$ eV and $\hbar\omega_{ph} = 0.098$ eV, which are reasonable values. It is found that theory and experiments are in good agreement at energies lower than E_p. Between E_p and the fundamental semiconductor band gap, however, the observed absorption is higher than the one predicted from theory. This discrepancy is not necessarily inconsistent with the small polaron concept, though. An alternative model for the absorption in heavily disordered W oxide films is obtained from intervalence charge transfer theory.[20] The agreement between this theory[21] and experiment was approximately as good as in Figure 16.8.[17]

Ion intercalated crystalline W oxide films can be treated as strongly doped semiconductors. The inserted electrons make the material infrared reflecting, and the unavoidable ion–electron scattering limits the metallic properties. Drude theory[22] can be used for qualitative work, but is unable to give quantitative predictions. Instead, the Gerlach theory[23] for ionized impurity scattering is of much value. Screening of the ions can be represented by the random phase approximation[24] or an extension thereof. This theory has been used before to model the optical properties of ITO in considerable detail.[25]

Figure 16.9 shows the spectral reflectance computed from the theory outlined above as applied to a WO_3-based film backed by glass.[26] The electron density (or, equivalently, the ion density) is used as a parameter. It is seen that an increased electron density yields infrared

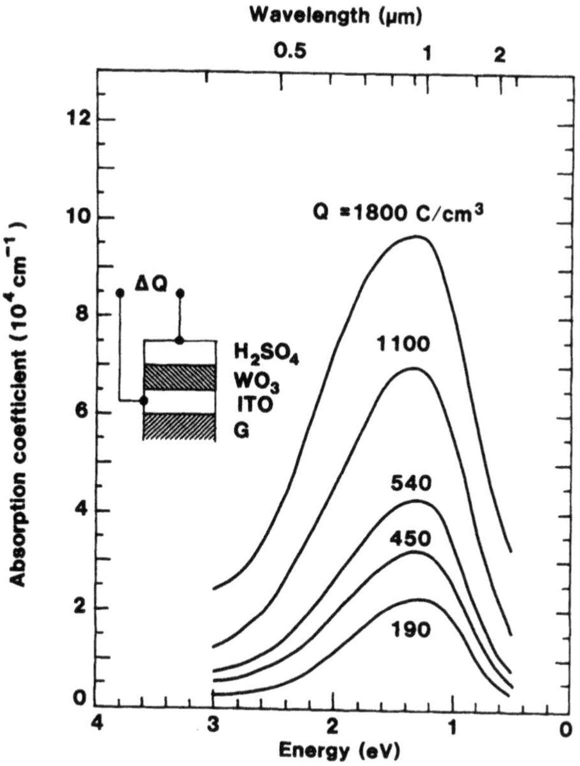

FIGURE 16.6. Spectral absorption coefficient for an evaporated W oxide film colored by insertion of the charge density ΔQ in a configuration with the film backed by glass G coated with In_2O_3:Sn (i.e., ITO) and immersed in H_2SO_4. (From Granqvist, C., *Handbook of Inorganic Electrochromic Materials,* Elsevier Science, 1995. With permission.)

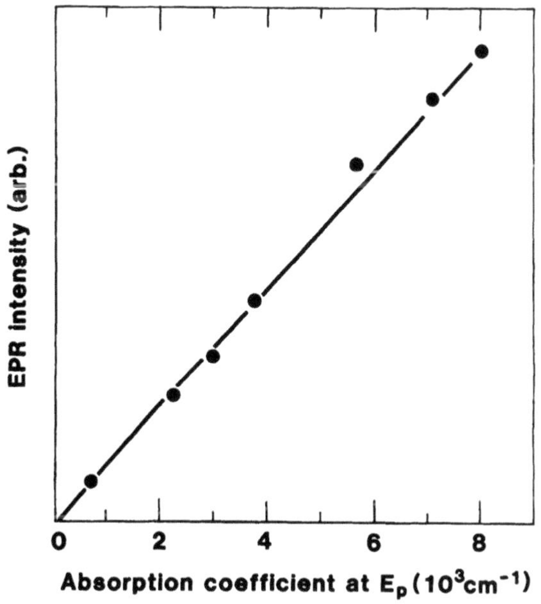

FIGURE 16.7. EPR signal corresponding to W^{5+} vs. absorption coefficient at the peak of the optical absorption band associated with electrochromism in H^+ intercalated W oxide films. The line was drawn for convenience. (From Granqvist, C., *Handbook of Inorganic Electrochromic Materials,* Elsevier Science, 1995. With permission.)

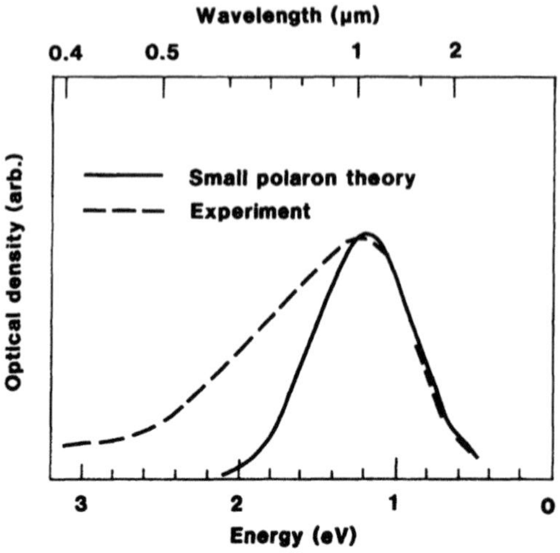

FIGURE 16.8. Spectral optical density for a proton intercalated evaporated W oxide film as determined from experiments and as computed from Equation 16.3. (From Granqvist, C., *Handbook of Inorganic Electrochromic Materials,* Elsevier Science, 1995. With permission.)

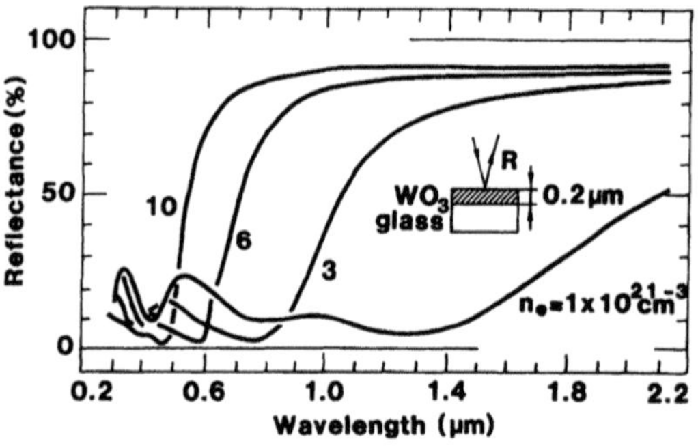

FIGURE 16.9. Computed spectral reflectance R for a 0.2-μm-thick slab of a material characterized by a theory for heavily doped semiconductors with ionized impurity scattering of the charge carriers. The electron density is denoted n_e. (From Granqvist, C., *Handbook of Inorganic Electrochromic Materials,* Elsevier Science, 1995. With permission.)

reflectance. This is in line with experiments, but a quantitative comparison of theory and experiment is still premature.

III. SURVEY OF ELECTROCHROMISM AMONG THE OXIDES

Electrochromism has been documented in several different metal oxides as indicated on the Periodic Table in Figure 16.10.[12] All of the pertinent metals belong to the transition series. Different shading is used to denote cathodic electrochromism (absorption increasing under ion intercalation) and anodic electrochromism (absorption decreasing under ion intercalation). Cathodic coloration is found in oxides of Ti, Nb, Mo, Ta, and W, with tungsten oxide being by far the most extensively studied one. Anodic coloration is found in oxides of Cr, Mn, Fe, Co, Ni, Rh, and Ir, with nickel oxide[27] and iridium oxide[28] being the ones investigated in most detail. Vanadium oxide is exceptional in that the pentoxide (with V^{5+}) exhibits anodic and cathodic electrochromism within different wavelength ranges,[29] while the dioxide (with V^{4+}) has anodic electrochromism.[30] Only some of the oxides mentioned above can be fully transparent to visible light, notably the oxides based on Ti, Ni, Nb, Mo, Ta, W, and Ir. The other

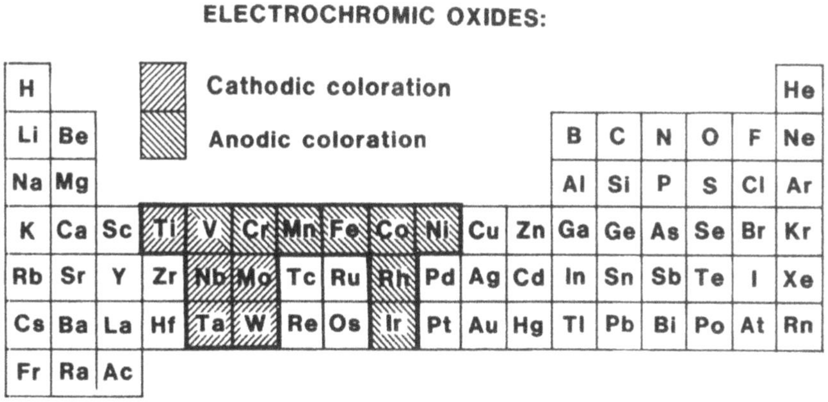

FIGURE 16.10. The Periodic Table of the elements, excepting the lanthanides and actinides. The shaded boxes refer to the transition metals whose oxides have well-documented cathodic and anodic electrochromism. (From Granqvist, C., *Handbook of Inorganic Electrochromic Materials*, Elsevier Science, 1995. With permission.)

oxides show some residual absorption either across the full visible range or in the blue part of the spectrum. The latter feature is associated with interband absorption in a semiconductor.

It is interesting and important to notice that the electrochromic oxides are based on metallic elements that are located in a well-defined region of the Periodic Table, and, furthermore, that different parts of this region pertain to oxides with cathodic and anodic coloration. Already at this point one may draw the conclusion that the electrochromism is closely related to the electronic structure of the oxides; this issue is elaborated below.

Table 16.1 summarizes a number of key properties of the electrochromic oxides. The first column indicates the nominal composition of the oxide. For some of these — notably the oxides based on Fe, Co, and Ni — the situation is complicated by the fact that H or Li is normally a critical constituent in the material and serves to stabilize a favorable structure. The second and third columns in Table 16.1 list the overall optical properties discussed above.

TABLE 16.1
Summary of Key Features for the Main Electrochromic Oxides
(Showing Oxide Type, Whether the Coloration is Cathodic (C) or Anodic (A), Whether or not Full Transparency Can Be Achieved (Yes = Y; No = N), and Whether the General Structure Type Embodies a Framework (F) or Layers (L) of MeO_6 Octahedra)

Oxide type	Coloration	Full transp.	Structure type
TiO_2	C	Y	F
V_2O_5	C/A	N	L[a]
Cr_2O_3	A	N	F
MnO_2	A	N	F
"FeO_2"	A	N	F
"CoO_2"	A	N	L?
"NiO_2"	A	Y	L
Nb_2O_5	C	Y	F
MoO_3	C	Y	F/L[b]
RhO_2	A	?	F
Ta_2O_5	C	Y	F
WO_3	C	Y	F
IrO_2	A	Y	F

[a] Layer structure with VO_5 units.
[b] The L(F) structure is referred to as the $\alpha(\beta)$ phase.

From Granqvist, C., *Handbook of Inorganic Electrochromic Materials*, Elsevier Science, 1995. With permission.

Apart from in the materials discussed so far, electrochromism is known to occur in many binary and ternary mixed oxides and in oxyfluorides. There are also some simple oxides that have been reported to display electrochromism, although this phenomenon has not yet been explored in detail; oxides based on Cu, Sr-Ti, Ru, and Pr belong to this latter class.[12] Finally, among the nonoxides pronounced anodic electrochromism is known in transition metal hexacyanometallates, especially Prussian Blue (PB).[31]

Essentially all of the electrochromic oxides are constructed from one type of building block, viz. MeO_6 octahedra with a central transition metal (Me) atom surrounded by six almost equidistant oxygen atoms. The building blocks are connected either by corner or by a combination of corner sharing and edge sharing, and many different crystal structures can emerge. The particular case of W oxide was treated in Figure 16.3 above. It is meaningful to divide the atomic arrangements into three-dimensional framework (F) structures and weakly coupled, essentially two-dimensional, layer (L) structures. The pertinent structural group is indicated in the last column of Table 16.1. Detailed discussions of the crystal structures can be found in standard texts.[32,33]

The electrochromic oxide films must be permeable to ions and must show some electrical conductivity. The ions can move and reside in the spaces between the MeO_6 octahedra. However, at least for the framework structures, these spaces may not be large enough to yield reasonable intercalation/deintercalation rates and therefore the materials must have a fine-grained character with low-density grain boundaries. Sometimes it is also desirable to have an extended columnar configuration that allows easy transport of ions over the cross section of an electrochromic film. Such columns were shown schematically in Figure 16.1. To further open up the structure and provide conduits and intercalation sites for the ions, the useful electrochromic films often are *hydrous*, i.e., they are more or less hydrated, hydroxylated, oxy-hydroxylated, etc. Suitably prepared films of many oxides may have a high capability of ion intercalation with an ion/metal ratio of the order of unity.

The occurrence of anodic and cathodic electrochromism among the oxides with framework structures can be understood from simple band structure (ligand field) models.[34,35] It is suitable to make a distinction between *defect perovskite structures* with predominantly corner-sharing MeO_6 octahedra (cf. Figure 16.3) and *rutile-like structures* with a combination of corner-sharing and edge-sharing MeO_6 octahedra.

A. DEFECT PEROVSKITE STRUCTURE

Figure 16.11(a) illustrates an energy level diagram for the MeO_3 defect perovskite structure. The atomic s, p and d levels of Me are indicated, as well as the 2s and 2p levels of O. For WO_3 and ReO_3, the pertinent Me levels are 6s, 6p, and 5d. The positions of these levels on the vertical energy scale are governed by their values for the isolated atoms as well as by the Madelung energies of the atoms when located in the perovskite lattice. Each Me ion is octahedrally surrounded by six O ions, and each O is linearly flanked by two Me ions. As a consequence of this arrangement, the d level is split up into e_g and t_{2g} levels (using the conventional notation) as shown in the left-hand part of Figure 16.11(a). The splitting arises because the e_g orbitals point directly at the electronegative O, whereas the t_{2g} orbitals point away from the nearest neighbors into empty space and are hence lower in energy. Similarly, the O2p orbitals are split as indicated in the right-hand part of Figure 16.11(a); the $2p\sigma$ orbitals point directly at the nearest electropositive Me ions, whereas the $2p\pi$ orbitals point into empty space. In the perovskite lattice, the incipient molecular energy levels broaden into bands, whose relative widths and repulsion can be estimated.

The number of states available for electron occupancy is fixed for each band. For example, counted per MeO_3 formula unit, the t_{2g} band has a capacity for 6 electrons (allowing for spin degeneracy), and the $p\pi$ band has a capacity for 12 electrons (allowing for spin and electron degeneracy). The pertinent electron capacities are indicated by the numbers in the various bands in Figure 16.11(a). WO_3 and the isostructural β-MoO_3 have 24 electrons in the shown

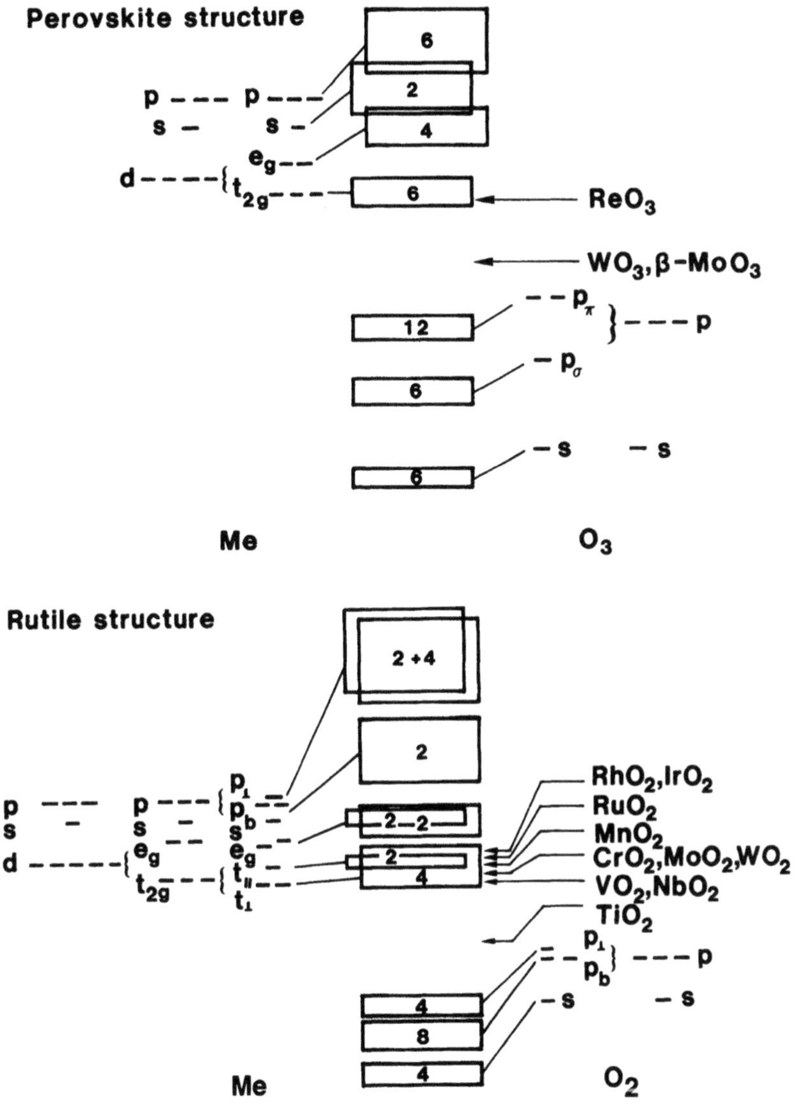

FIGURE 16.11. Schematic band structures for the MeO_3 defect perovskite structure (top) and for the MeO_2 rutile structure (bottom). The incipient atomic and molecular levels are shown, using standard notation. The numbers in the different bands denote electron capacities. Arrows indicate Fermi energies. (From Granqvist, C., *Handbook of Inorganic Electrochromic Materials*, Elsevier Science, 1995. With permission.)

bands, so that the Fermi energy lies in the gap between the t_{2g} and pπ band. The band gap is wide enough to render the material transparent. ReO_3, on the other hand, has 25 electrons so that the lower part of the t_{2g} band is occupied and the material is nontransparent. Detailed band structure calculations verify these contentions.[36,37]

When ions and electrons are inserted according to Reaction (16.1) or (16.2), the Fermi level is moved upwards in the presumed rigid-band scheme. In the case of WO_3 and β-MoO_3, the excess electrons must enter the t_{2g} band and the material, in principle, transforms from a transparent to an absorbing or from a transparent to a reflecting state, depending on whether the electrons occupy localized or extended states. When the ions and the accompanying electrons are extracted, the material returns to its original transparent state. In ReO_3, the Fermi energy also moves upwards upon ion insertion, but it remains well within the t_{2g} band irrespective of the content of mobile ions. Consequently, the optical properties are not expected to be changed qualitatively, although quantitative changes in the absorbance may result from density-of-states effects that are outside the realms of the schematic band structure. Clearly, the discussion above explains why W oxide and Mo oxide are cathodic electrochromic materials, whereas Re oxide is not known to be electrochromic.

B. RUTILE AND RUTILE-LIKE STRUCTURES

These are found among many MeO_2 oxides. Several of these oxides are electrochromic — either cathodic or anodic — and may or may not attain a transparent state. The electrochromism can be understood from the basic band structure in Figure 16.11(b). It deviates from its counterpart for the perovskites because the MeO_6 units are distorted and edge sharing. One can start from the atomic s, p, and d levels for the Me ions and the s and p levels for the oxygen ions. Every Me ion is almost octahedrally surrounded by oxygen ions, and every oxygen ion is roughly trigonally surrounded by Me ions. Hence the d levels are first split into e_g and t_{2g} levels, and the oxygen p levels are split into levels with orbitals p_b in the basal plane of the triangular array (wherein the oxygen ion is surrounded by three Me ions) and with orbitals $p_\perp$ perpendicular to that array. A further splitting occurs because the ideal octahedral symmetry is absent, and the ensuing Me levels are designated $t_\parallel$, $t_\perp$, p_b, and $p_\perp$. In addition, the degeneracy of the e_g level is lifted. When the ions are arranged in the rutile lattice, one expects overlap among the $t_\parallel$, and $t_\perp$ bands as well as among the two e_g bands. Hence the final schematic band structure, in fact, is quite similar to that of the perovskites. Depending on the parameters of the rutile unit cell, the $t_\parallel$ band may collapse into an atomic-like sharp level. Electron band capacities can be deduced as for the perovskites and are indicated by the numbers in the bands.

The arrows in the left-hand part of Figure 16.11(b) indicate the Fermi levels for several rutile-like oxides. For TiO_2, the Fermi level lies in the gap between the $Ti3t_{2g}$ and the $O2p_\perp$ bands, and ion insertion is expected to lead to a transformation from a transparent to an absorbing state in much the same way as for WO_3 and β-MoO_3. This is consistent with experimental results on Ti oxide films (cf. Table 16.1). For VO_2, MnO_2, and RuO_2, the initial absorbing state cannot be eliminated by ion insertion. Again, this is as expected from the band structure. RhO_2 and IrO_2 are especially interesting, since their Fermi levels lie close to the gap between the t_{2g} and e_g bands. Indeed, if ions are inserted up to about one per formula unit in Ir oxide films, these materials are transformed from an absorbing to a transparent state. In the band structure picture, the transparency is associated with the band gap referred to above. One should note that transitions between pure t_{2g} and e_g states are parity forbidden.

C. LAYER STRUCTURES

The layer structures are not amenable to a simple theoretical treatment analogous to that for the perovskites and rutiles. However, the band structure of $LiCoO_2$ was studied recently, and it was found that the highest occupied states were of t_{2g} character.[38] V_2O_5 has an interesting band structure with a filled O2p band separated from a V3d band with a split-off lower portion.[39] No detailed band structure data seem to be available for Nb oxide or hydrous nickel oxide.

Figure 16.12 surveys and compares the band structures of the electrochromic oxides capable of attaining a fully transparent state. For WO_3, β-MoO_3, TiO_2, and, possibly, also for Nb_2O_5, the Fermi energy lies between the O2p and the t_{2g} bands. Insertion of electrons makes the lower part of the latter band populated. In the case of a heavily disordered oxide, the excess electrons and their lattice polarization can hop from one ion site to another by absorbing a photon, i.e., the material exhibits polaron absorption as discussed above for W oxide. The peak absorption lies in the near infrared, which makes the oxides bluish. For a sufficiently well-ordered material it is possible to have metallic-like properties, i.e., high reflection at wavelengths larger than a plasma wavelength. Crystalline W oxide (c-WO_3) has this property; this may be the only case, though, and well-ordered MoO_3 transforms into the layered α phase, and well-ordered TiO_2 is too dense for easy ion insertion.

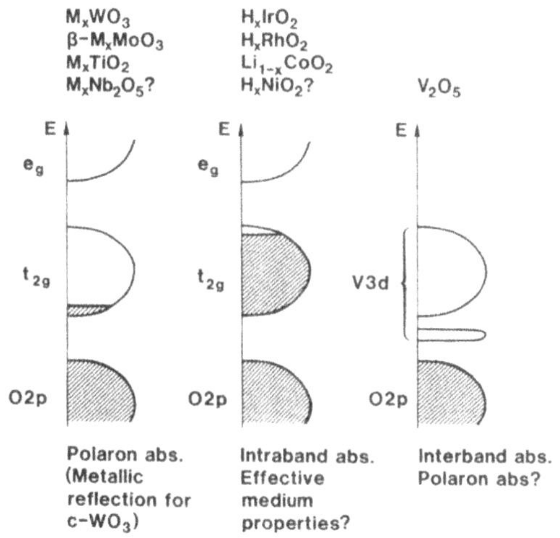

FIGURE 16.12. Schematic band structure and suggested absorption mechanisms for several electrochromic oxides. (From Granqvist, C., *Handbook of Inorganic Electrochromic Materials*, Elsevier Science, 1995. With permission.)

The t_{2g} band is almost full in IrO_2, RhO_2, delithiated $LiCoO_2$, and, possibly, deprotonated $Ni(OH)_2$ (the latter materials were referred to as "CoO_2" and "NiO_2" in Table 16.1). These oxides are strongly absorbing. Polaron absorption is not observed, and intraband absorption in a composite material characterized by parts having various electron densities (conductivities) and shapes is a more likely cause for the optical behavior. In principle, effective medium theory[40] could account for the optical properties, but in the absence of substantial evidence for a composite structure the proposed absorption mechanism must be regarded as conjectural. Electron insertion, associated with a filling of t_{2g} band, leads to optical transparency, as noted above.

V_2O_5 shows some absorption of blue light. This is readily explained as a result of transitions between the top of the O2p band and the split-off lower portion of the V3d band. Insertion of electrons can fill this split-off band. Since transitions between the two components of the d band are parity forbidden, the net effect will be an apparent widening of the optically observed band gap, which is then determined by transitions between the top of the O2p band and the lower part of the main d band. Hence interband transitions govern the short-wavelength electrochromism in V_2O_5. At a larger density of electron insertion, it is reasonable to expect that polaron absorption could lead to long-wavelength absorption.

IV. DEVICES WITH DIFFERENT TYPES OF ELECTROLYTES

Electrochromic devices can be categorized according to several principles. Following earlier work of ours,[7,8] separate discussions are given for devices based on different types of electrolytes: liquid, solid inorganic, and solid organic (polymeric). Almost all of the devices incorporate an electrochromic W oxide film.

A. LIQUID ELECTROLYTES

The first electrochromic devices, expectedly, drew extensively on standard electrochemical practice for ion intercalation/deintercalation in aqueous acidic electrolytes. Obviously, they require reliable sealing. The initial development was strongly geared toward small devices for information display, such as those used in wrist watches.

Aqueous H_2SO_4 electrolytes were natural choices for the initial studies, and device-oriented work was reported as early as in 1975.[41-43] The inset of Figure 16.13 shows a typical device design. A glass plate coated with a transparent conductor (TC) and a W oxide film is separated from a substrate coated with an ion storage layer (CoE) by the electrolyte containing

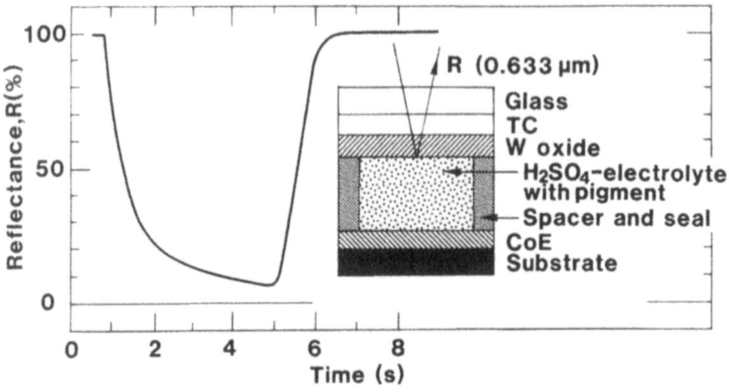

FIGURE 16.13. Reflectance as a function of time for an electrochromic device of the type shown in the inset. The bleached state reflectance was set to 100%. (From Granqvist, C., *Handbook of Inorganic Electrochromic Materials,* Elsevier Science, 1995. With permission.)

a pigment. A suitable spacer and seal confine the electrolyte. The TC can be doped In_2O_3 or SnO_2, the electrolyte can contain an alcohol diluent in order to lower the acid reactivity of the H_2SO_4, the pigment can be TiO_2, the CoE can be a carbon + binder material that may also incorporate some WO_3, and the substrate can be steel.

The main part of Figure 16.13, taken from Faughnan et al.,[42] illustrates the change in the reflectance at a wavelength λ equal to 0.633 μm for a device with an active area of 0.33 cm^2 during an intercalation/deintercalation (color/bleach, c/b) cycle. The reflectance decreases to <10% of its initial value during the course of ~2s and returns to the initial magnitude in ~1s.

Durability problems with aqueous acid electrolytes have led to a considerable amount of device-oriented work with $LiClO_4$ in propylene carbonate (PC) as the electrolyte. Some devices have used a design essentially like the one in Figure 16.13.[44,45] In the work by Ando et al.,[44] the pigment and CoE were integrated in a 0.4-mm-thick layer comprising TiO_2, carbon powder, $W_{18}O_{49}$, V_6O_{13}, and organic binder. Extended c/b cycling was performed by applying a square wave (1.5 V, 1 Hz), leading to a periodic charge insertion/extraction of 6 mC/cm^2. Figure 16.14 shows that the charge insertion decreased only by ~10% after ~3×10^7 cycles. In Yamanaka's[45] work, a porous TiO_2-pigmented PTFE sheet was used as a reflector and the CoE was of carbon powder with added iron tungstate and organic binder. c/b cycling was accomplished with a square wave (1.35 V, 1 Hz) yielding a charge insertion/extraction of 5 mC/cm^2. As apparent from Figure 16.14, no decay of the charging capability occurred after $>10^7$ cycles, which again points at the good durability that can be accomplished. Several details of the device design are critical for achieving a cycling durability on this order of magnitude.

A straightforward approach to CoEs in reflecting devices is to make them identical to the displayed electrochromic films. Excellent reversibility of the optical response may be expected from this "charge-balanced" arrangement. Some work on such symmetric designs, including two W oxide films separated by a porous ceramic reflector and a $LiClO_4$ + PC electrolyte, has been described by Morita and Washida.[46] Figure 16.14 indicates the cycling durability of optimized symmetric devices subjected to (2.5 V, 1 Hz), which gave a reversible charge insertion of 4 mC/cm^2. Degradation, though noticeable, was not severe up to 10^7 c/b cycles.

Transmitting devices are considered next; their main applications are for smart windows to be used in future fenestration. A design with proven long-term durability was discussed in detail by Kamimori et al.[47] As seen from the inset of Figure 16.15, two glass plates have ITO layers on their facing sides, and one of these substrates carries a ~0.6-μm-thick evaporated W oxide film. The electrolyte forms an intervening 50- to 100-μm-thick layer and is based on $LiClO_4$ + PC; it contains dispersed redox agents. Figure 16.15 also illustrates the

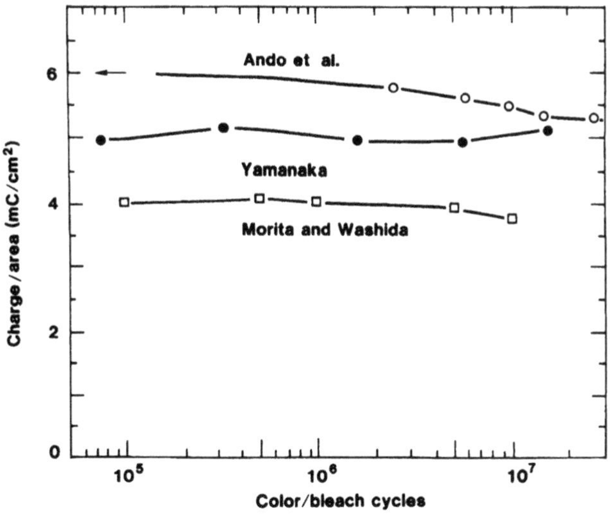

FIGURE 16.14. Inserted charge per unit area and intercalation/deintercalation cycle during long-term testing of three types of electrochromic display-type devices. Dots indicate measured data, and lines were drawn for convenience. Arrow indicates an initial value of the charge/area. (From Granqvist, C., *Handbook of Inorganic Electrochromic Materials,* Elsevier Science, 1995. With permission)

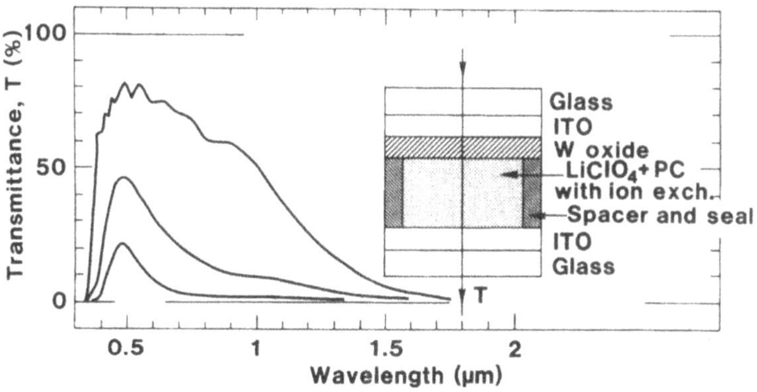

FIGURE 16.15. Spectral transmittance at three different states of coloration for an electrochromic device of the type shown in the inset. (From Granqvist, C., *Handbook of Inorganic Electrochromic Materials,* Elsevier Science, 1995. With permission.)

transmittance of the device in fully colored and bleached states and at intermediate coloration. At $\lambda = 0.55$ µm, for example, the transmittance can be varied between ~75 and 13%. The decrease of the transmittance at $\lambda = 1$ µm, irrespective of coloration, is caused by reflectance from the ITO layers.

The transmitting devices normally are much larger than the reflecting display-type devices, and for the former ones the size itself, and the electrical properties of the TCs, become of critical importance for the response dynamics. Figure 16.16 shows the change of the luminous transmittance during one c/b cycle in which a voltage step of ±1 V was applied between ITO layers with a resistance/square of about 10 Ω. The smallest device with 4×4 cm² area has a response time of <30 s. The largest device with 30×40 cm² area requires several minutes to attain full coloration, whereas bleaching takes ~1 min. Large devices do not change their optical properties uniformly when voltage steps are applied, but differences between the central and peripheral zones may be noticeable. It is desirable that the resistance/square is of the order of 1 Ω for devices approaching the square meter scale.

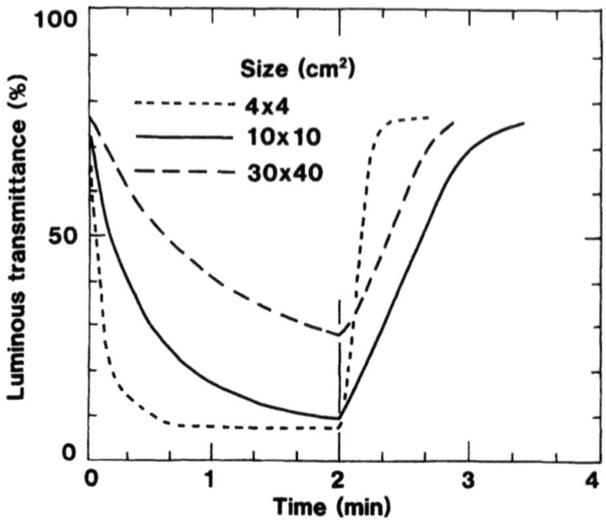

FIGURE 16.16. Luminous transmittance as a function of time for coloring and bleaching of electrochromic devices, of the type shown in the inset of Figure 16.15, having different sizes. (From Granqvist, C., *Handbook of Inorganic Electrochromic Materials,* Elsevier Science, 1995. With permission.)

The device properties depend on the technique for preparing the W oxide films. Schlotter[48] studied films made by flash evaporation as well as by reactive sputtering onto ITO-coated glass at 150°C. The films were dense and considerably more durable than conventionally evaporated films when operated in $LiClO_4$ + PC or $LiClO_4 + H_2O$ + PC electrolytes. Despite their density, the films were able to show rapid dynamics. Another notable result was that a 1.6-μm-thick sputtered film could produce a transmittance that was as low as $\sim 10^{-3}$ of the original value after 5 s; it returned to the initial transmittance after another 5 s.

Transmitting electrochromic devices can incorporate a thin film that serves as an ion storage. Obviously, this film has to be transparent at least when the W oxide film is in its bleached state, so the carbon-containing layers used successfully in the earlier discussed display-type devices now are useless. A simple approach is to use *two* W oxide films with different crystallinities. These films color differently under ion intercalation, and when charge is shuttled from a heavily disordered film into a crystalline film the overall coloration is lowered in the luminous wavelength range (but increased in the infrared), and when charge is moved back into the disordered film the optical effect is reversed. Devices of this type have been discussed by Matsuhiro and Masuda[49] and Rauh et al.[50]

A principally superior option is to combine a disordered W oxide film with an anodically coloring CoE, in which case the overall coloration is the sum of the ones for the individual films. Some work along these lines, with PB as the ion storage (or complementary electrochromic film), has been reported by Kase et al.[51] The principal application was to curved glass for automotive applications. A variety of this device design was discussed recently by Yamanaka.[52] As shown in the inset of Figure 16.17, there are two glass sheets with ITO layers, an evaporated W oxide film with thickness between 0.3 and 0.83 μm, a 0.06-μm-thick electrodeposited Ir oxide film, and an electrolyte of 1 M $LiClO_4$ + PC + 2% H_2O. The water addition was necessary for charge insertion into the Ir oxide. Figure 16.17 illustrates the spectral transmittance in bleached state and after insertion of 17.5 mC/cm^2. Cycling durability was tested under application of a square wave (1.5 V, 0.05 Hz); the device could withstand $\sim 10^5$ cycles, particularly when the W oxide film was thick. Another device design, showing good modulation of the transmittance, embodies evaporated films of W oxide and Ni oxide and an intervening liquid electrolyte.[53]

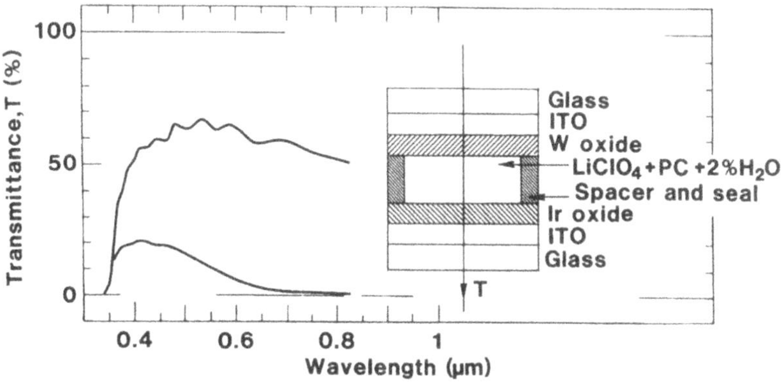

FIGURE 16.17. Spectral transmittance at two different states of coloration for an electrochromic device of the type shown in inset. (From Granqvist, C., *Handbook of Inorganic Electrochromic Materials,* Elsevier Science, 1995. With permission.)

Many organic substances show electrochromism and can serve as ion storage layers in devices with W oxide films. Such a device was reported by Yoshida et al.;[54] specifically it incorporated sputter-deposited W oxide, electropolymerized polyaniline, and an electrolyte of $LiClO_4$ + PC. Transmittance changes between ~80 and ~4% could be obtained with time constants of 10 s. The cycling durability was $\sim 5 \times 10^5$ times.

B. SOLID INORGANIC ION CONDUCTORS

Inorganic ion conductors have been studied in several different prototype electrochromic devices. Reflectance changes have been reported in devices with bulk-type ion conductors, and transmittance as well as reflectance changes have been reported in devices with thin-film ion conductors. Most of the work on bulk-type ion conductors have dealt with proton conductors, initially with $(WO_3)_{12}{\cdot}29H_2O$ (phosphotungstic acid) and $ZrO(H_2PO_4)_2{\cdot}7H_2O$ (zirconium phosphate).[55] Durability problems were noted with these materials.

More promising results were obtained by Kuwabara et al.,[56-58] who studied devices based on $Sn(HPO_4)_2{\cdot}H_2O$ (here denoted SP). The inset of Figure 16.18 shows a typical device design[58] with two evaporated W oxide films (or one W and one Mo oxide film) separated by a SP layer made by electrophoresis. The main part of this figure illustrates the change in optical density (OD), obtained by measuring the reflectance of light from an incandescent source, under the application of (±2.5 V, 0.1 Hz). The OD is proportional to the absorption coefficient. Coloration corresponding to $\Delta(OD) = 0.6$ took 3 s, and bleaching from this state to $\Delta(OD) < 0.1$ took 2 s. Durability up to $>10^6$ cycles was obtained for some devices.[56]

Devices using layers of $H_3OUO_2PO_4{\cdot}3H_2O$ (hydrogen uranyl phosphate, HUP) as a solid ion conductor have been discussed by Howe et al.[59] A typical arrangement embodied two glass plates with ITO layers, each having a ~1-μm-thick evaporated W oxide film. Intervening HUP was produced by gently pressing a precipitating HUP solution so that a ~100-μm-thick pale yellowish to transparent layer was formed. Such devices were tested with charge insertion/extraction of 5 to 10 mC/cm^2, and switching times down to 0.3 s were noted. Durability up to 5×10^5 c/b cycles was reported. HUP was employed also by Takahashi et al.,[60] who used flexible 1 to 3 mm thick tablets of this material, pressed together with Teflon® powder, between an evaporated W oxide film and an Ag CoE. Up to 16 mC/cm^2 was inserted. Durability was found for as much as 2×10^6 c/b cycles.

Antimony oxides have been used in electrochromic prototype devices. Thus Lagzdons et al.[61] employed $HSbO_3{\cdot}2H_2O$ interfaced between two W oxide films, or between one film of W oxide and another film of Ni oxide, and Matsudaira et al.[62] studied display-type devices incorporating a white mixture of $Sb_2O_5{\cdot}pH_2O$ and Sb_2O_3. Apart for an evaporated W oxide film on ITO-coated glass, the entire unit in the latter work was manufactured by screen

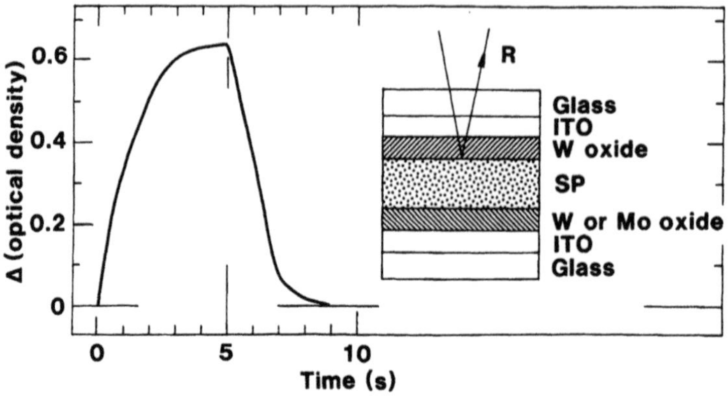

FIGURE 16.18. Change in optical density as a function of time for an electrochromic device of the type shown in the inset. The data are based on measured reflectance R. (From Granqvist, C., *Handbook of Inorganic Electrochromic Materials*, Elsevier Science, 1995. With permission.)

printing. These devices withstood c/b cycling at ±1.3 V for $>10^6$ times. Recently, work by Kuwabara et al.[63,64] considered devices with an evaporated W oxide film, a proton conductor of $Sb_2O_5 \cdot 2H_2O$, and a graphite CoE. The Sb oxide was spray deposited or applied by electrophoresis onto both W oxide and graphite prior to device assembly.

Among the aprotic bulk-type ion conductors, work has been reported on Na^+ conducting $Na_2O \cdot 11Al_2O_3$ (Na-β-alumina)[65] and $Na_{1+x}Zr_2Si_xP_{3-x}O_{12}$ (NASICON).[66] Devices with β-alumina required heating to >70°C in order to operate, and devices with NASICON were found to be unstable.

Generally speaking, *thin-film ion conductors* are more promising and versatile than bulk-type ion conductors, and electrochromic devices based on the former class of materials are considered next. Numerous thin films have been used. It is convenient to start with thin dielectric films incorporating some water, since they can be used in devices that are structurally simple. The inset of Figure 16.19 illustrates a typical design with a glass substrate coated with four superimposed layers: a TC such as ITO, electrochromic W oxide, water-containing dielectric such as $MgF_2 \cdot H_2O$, and a semitransparent top layer of Au. Initial work on this kind of device was reported by Deb[67] and others;[68] they have subsequently become known as "Deb devices". The most critical part of a Deb device is its water-containing layer. It must be porous, in which case water adsorption takes place spontaneously upon exposure to a humid ambience. Detailed studies have been reported for porous dielectric films of MgF_2,[69-73] SiO_x,[74,75] LiF,[69,76] Cr_2O_3,[77,78] Ta_2O_5,[79] and of some other materials. For the top layer, 0.01- to 0.02-μm-thick Au films have been used almost universally; this limits the peak transmittance to ~50%.

Figure 16.19 shows spectral transmittance from Svensson and Granqvist[73] through a Deb device with 0.15 μm of evaporated W oxide, about 0.1 μm of MgF_2 evaporated in the presence of 5×10^{-4} Torr of air, and ~0.015 μm of Au. In fully bleached state, the transmittance has a peak value of ~50% at $\lambda \approx 0.52$ μm, which is in excellent agreement with results from Benson et al.[80] for a similar device. Optical changes are small at low applied voltages, but increase rapidly when a "critical" voltage of ~1.3 V is exceeded. This effect is related to the decomposition of water into $H^+ + OH^-$ and ensuing proton insertion into the W oxide film. There is a simultaneous electrochemical oxidation at the Au electrode. At voltages >1.8 V, O_2 gas is evolved, and at a reverse bias exceeding 0.9 V, H_2 gas is evolved. Gas formation can lead to morphological changes as well as to film delamination,[71] and should be avoided in practical implementations.

The importance of a large film porosity is demonstrated in Figure 16.20, which shows data from Deneuville et al.[70] on the c/b dynamics of a Deb device with 0.3 μm of W oxide,

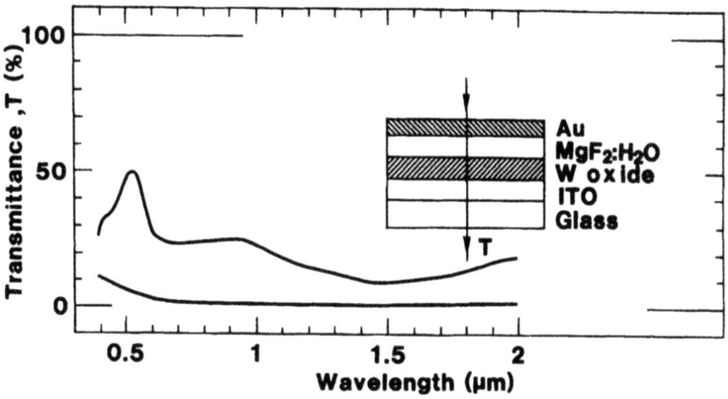

FIGURE 16.19. Spectral transmittance at two different states of coloration for a Deb-type electrochromic device of the type shown in the inset. (From Granqvist, C., *Handbook of Inorganic Electrochromic Materials*, Elsevier Science, 1995. With permission.)

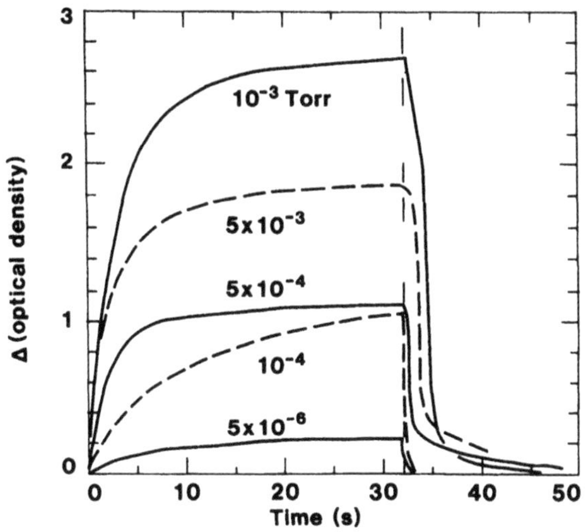

FIGURE 16.20. Change in optical density during coloration and bleaching for a Deb device of the type shown in Figure 16.19. The W oxide and MgF_2 layers were evaporated at the pressures shown. The electric field was reversed after 32 s. (From Granqvist, C., *Handbook of Inorganic Electrochromic Materials*, Elsevier Science, 1995. With permission.)

0.05 μm of MgF_2, and 0.015 μm of Au. The films of W oxide and MgF_2 were evaporated at the shown pressures of ambient air, and the device was operated at (±3 V, 0.015 Hz). It is seen that the coloration increases gradually with a time constant of ~10 s and reaches a limiting value that increases in proportion with the gas pressure. Bleaching is faster and is completed during the course of a few seconds. The decisive factor for the ultimate coloration is the amount of incorporated water rather than the porosity as such. Deb devices incorporating MgF_2, SiO_x, and LiF have been shown to withstand ~10^4 c/b cycles.[76] The shelf-life is much longer, though, with times >10 years having been mentioned.[81]

The relative humidity of the ambience plays a large role for devices operated in air, and designs incorporating MgF_2, SiO_x, and LiF become nonfunctional if the water is desorbed, as under vacuum. Such a strong dependence on the ambient conditions is not a necessary limitation for Deb devices, though, but designs with Cr_2O_3 are able to maintain their incorporated water at least to a pressure <10^{-6} Torr, as found by Inoue et al.[77] Such devices have been operated for >5 × 10^6 c/b cycles at a reflectance modulation of 50%.

The reliance of most Deb devices on ambient water is problematic and is presumably the reason why their cycling durability normally is limited to ~10^4 times. A technically superior "charge-balanced" device design incorporates a CoE that operates in concert with the W oxide film and permits reversible movements of protons (and, perhaps, hydroxyl groups) between two thin films. Work on reflecting and transmitting devices with W oxide, Ir oxide, and an intervening $Ta_2O_5 \cdot H_2O$ film was reported by Watanabe et al.[82] and Saito et al.[83] Such multilayers were integrated in prototype sunglasses capable of varying the transmittance between 70 and 10% with a c/b response time of a few seconds and a durability of >10^6 cycles.[84] Recent work with a $Sb_2O_5 \cdot pH_2O$ paste replacing the Ta oxide appeared to give a cycling durability exceeding 10^7 times.[85]

Several charge-balanced devices with moderate to low humidity dependence have been studied primarily for applications on automotive rear view mirrors. Recent work by Bange et al.[86-88] was centered on the symmetric and asymmetric designs shown in the insets of Figures 16.21(a) and 16.21(b). Similar structures, with an ion conducting SiO_x film, were discussed recently by Kleperis et al.[75] The symmetric device incorporates two W oxide films, two proton-conducting SiO_2-based films, a reflecting Rh film interposed between the SiO_2-based layers, and a metallic back conductor which also is a Rh film. All of the films were made by evaporation. The intermediate Rh film is almost completely permeable to protons and plays practically no role for the dynamics of the electrochromic system. It is advantageous to use Rh, rather than, for example, Pd, as the reflector, since the former metal does not take up as much hydrogen and hence remains dimensionally stable. The asymmetric device in Figure 16.21(b) is somewhat simpler and includes one film of each of electrochromic W oxide, proton-conducting Ta_2O_5, anodically coloring Ni oxide, and Al back reflector.

The main parts of Figure 16.21 illustrate spectral reflectance in fully colored and fully bleached states for the two device types. The asymmetric design is capable of showing a maximum reflectance of ~80%, whereas the symmetric design has a limiting reflectance of ~72%. Cyclic voltammetry for the asymmetric configuration indicated a "bistable" behavior with no current drawn between –0.5 and +0.2 V. The reflectance at $\lambda = 0.55$ μm also showed a "bistable" performance in this voltage range and was ~75% for the anodic sweep direction and lower, with a magnitude depending on the Ta_2O_5 thickness, for the cathodic sweep direction.[89]

Work with Li^+ conducting layers has also been reported. Thus constructions incorporating films of $LiAlF_4$,[90] Li_3AlF_6,[91] MgF_2:Li,[92] and Li_2WO_4[93] have been described. The latter type of device had c/b response times of the order of 0.1 s. Transparent electrochromic devices with the general design indicated in the inset of Figure 16.22 have been discussed in some detail by Goldner et al.[94-97] The construction includes a layer of $LiNbO_3$. The W oxide film was sputter deposited onto a substrate at 450°C and is hence crystalline; the upper ITO film was sputtered with the coated substrate at 200°C, which is less than the optimum temperature. Figure 16.21 shows that a rather high degree of optical modulation can be achieved; it is caused by a reflectance change in the crystalline W oxide film. The dynamics were slow, with typical c/b response times of 1 min even for a small device, which probably is caused by a poor conductivity of the top ITO film. For several of the devices with Li^+ conductors, dehydration is not assured and hence H^+ conduction may contribute to the electrochromism.

Finally, one could mention some early experiments by Green et al.[98,99] on thin-film devices with layers of $RbAg_4I_5$, whose Ag^+ conductivity can be very large. These devices were found to be unstable due to moisture attack and electrochemical reactions.

C. POLYMER ELECTROLYTES

The rapid advances in polymer electrolytes during recent years are paralleled by an upsurge of interest in electrochromic devices including such materials. The discussion below first regards proton conductors, for which extensive work has been carried out with multilayer

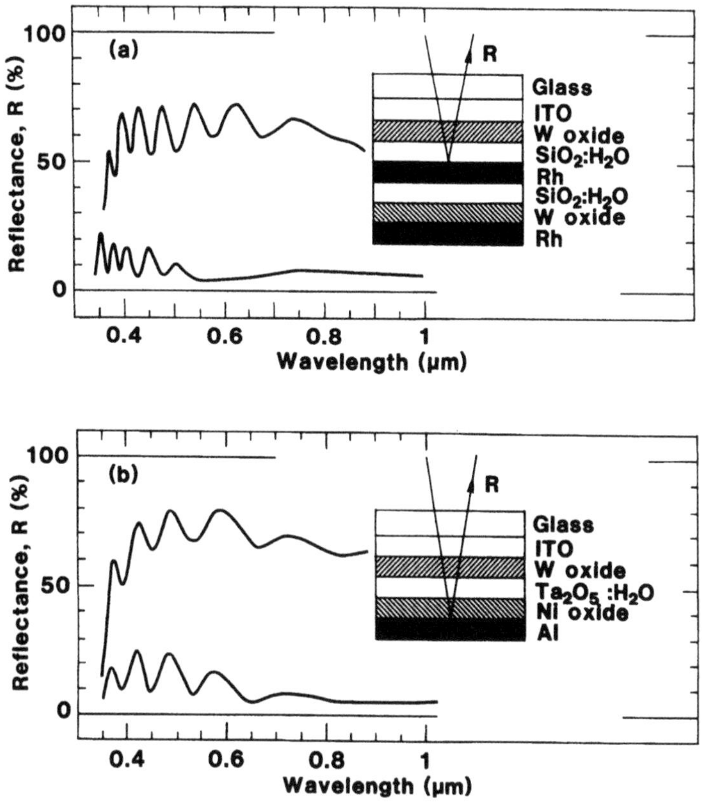

FIGURE 16.21. Spectral transmittance at two states of coloration for symmetric (part a) and asymmetric (part b) electrochromic devices of the types shown in the insets. (From Granqvist, C., *Handbook of Inorganic Electrochromic Materials*, Elsevier Science, 1995. With permission.)

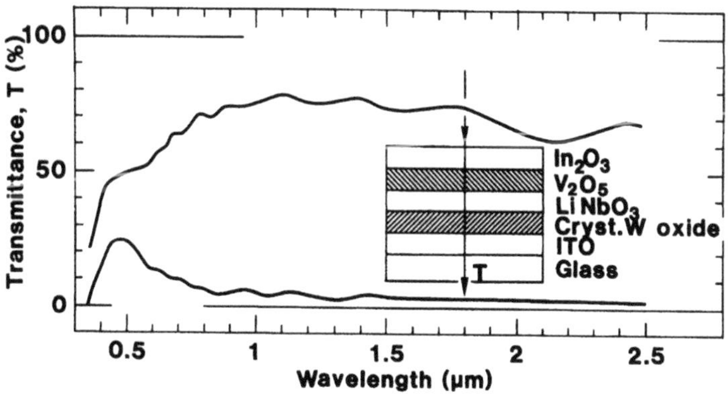

FIGURE 16.22. Spectral transmittance at two states of coloration for an electrochromic device of the type shown in the inset. (From Granqvist, C., *Handbook of Inorganic Electrochromic Materials*, Elsevier Science, 1995. With permission.)

structures based on poly-2-acrylamido-2-methylpropanesulfonic acid (poly-AMPS), polyvinylpyrrolidone (PVP), polyethylene imine (PEI), and others. Alkali ion conductors then are considered, particularly devices incorporating polyethylene oxide (PEO), poly (propylene glycol, methyl methacrylate) (PPG-PMMA),[100,101] etc.

Work on electrochromic devices with polymer electrolytes was pioneered by Giglia and Haacke,[102] Randin,[103] and by Randin and Viennet.[104] The studies were focused on polysulfonic acids, and it was found that poly-AMPS was the best. Detailed information was given for a display-type device with poly-AMPS.[102] The base was SnO_2-coated glass upon which a film of W oxide was applied by evaporation. The electrolyte comprised a 1- to 10-μm-thick layer of poly-(HEM, AMPS), with HEM denoting 2-hydroxyethylmethacrylate, and a ~0.5-mm-thick layer of poly-AMPS/TiO_2 pigment/PEO mixed to 8/1/1 by weight. The poly-(HEM, AMPS) was included to separate the W oxide from the poly-AMPS, which was necessary for obtaining long-term durability. The PEO admixture improved the dimensional stability of the polymer. The CoE was prepared by following standard paper-making techniques utilizing acrylic fibers loaded with carbon powder and a MnO_2 additive. The latter component raised the emf to a sufficient level that bleaching of the device could be accomplished by short circuiting. A protective metal encasement completed the design. Displays of this kind had a c/b switching time of 0.9 s, could be cycled $>10^7$ times, showed open-circuit memory for up to 2 days, and had a shelf life exceeding 3 years.

Another polymer-based design, studied by Dautremont-Smith et al.,[105] used opacified Nafion® in a symmetric configuration between anodic or sputter-deposited Ir oxide films backed by SnO_2-coated glass. The Nafion was boiled first in an aqueous solution of a barium salt and subsequently in H_2SO_4 so that a white precipitate of $BaSO_4$ was occluded in the polymer. The devices had a c/b switching time of the order of a second and an open-circuit memory of a few days. The moderately low coloration of the Ir oxide, as well as the cost, limit the practical usefulness of these devices, though the excellent durability is an advantage.

Transparent electrochromic devices built around poly-AMPS have been studied by Cogan et al.,[106,107] Rauh and Cogan,[108] and others. The insets of Figure 16.23 illustrate three related designs that have been investigated. They incorporate films of disordered W oxide, disordered W-Mo oxide, or hexagonal crystalline K_xWO_3,[109] together with Ir oxide films serving for ion storage and for augmenting the coloration, and Ta_2O_5 films for protecting the W oxide from degradation and for providing extended open-circuit memory. All films were made by sputtering. The electrolyte was with or without 8 wt% PEO. Prior to lamination, the W oxide-based films were protonated in a H_2SO_4 electrolyte to a value compatible with the maximum safe charge insertion into the Ir oxide. The three devices all show rather high transmittance in the bleached state and low transmittance in the colored state. For designs with disordered W or W-Mo oxide, Figures 16.23(a) and 16.23(b) show that the transmittance can be ~60% in the luminous and near-infrared spectral ranges. The device with crystalline K_xWO_3, on the other hand, has a transmittance up to ~80%, as apparent from Figure 16.23(c). The configuration with W-Mo oxide is capable of yielding an exceptionally low transmittance in the colored state. The device in Figure 16.23(a) has been c/b cycled successfully for 2×10^5 times under the application of −1.2 and +0.2 V for 35 s each.

Metal grid electrodes can provide the low resistance needed for optical modulation with acceptable dynamics even in large-area devices. This approach was studied recently by Ho et al.,[110,111] whose design is illustrated in the inset of Figure 16.24. The grid electrode was made of Ni or Cu; the open areas were ~0.76 mm across and covered ~20% of the glass. The electrolyte was poly-AMPS containing some water and *N,N*-dimethylformamide. The device was completed by a glass plate covered with SnO_2:F and evaporated W oxide. Figure 16.24 shows the change of the transmittance at $\lambda = 0.55$ μm during galvanostatic cycling with ±0.15 mA/cm^2; the voltage did not exceed 1 V. The transmittance changes between ~68% and <20%. Cycling durability up to 10^4 times was observed.

Numerous organic substances show electrochromism and may be used as CoEs in devices with W oxide films. Nguyen and Dao[112] studied samples with poly-AMPS and a CoE of poly (*N*-benzylaniline). Principally similar devices were investigated by Akhtar and Weakliem[113] and Jelle et al.,[114,115] who worked with constructions embodying evaporated W oxide, poly-(BPEI, AMPS) electrolyte with BPEI denoting "branched" PEI, and polyaniline (PANI)

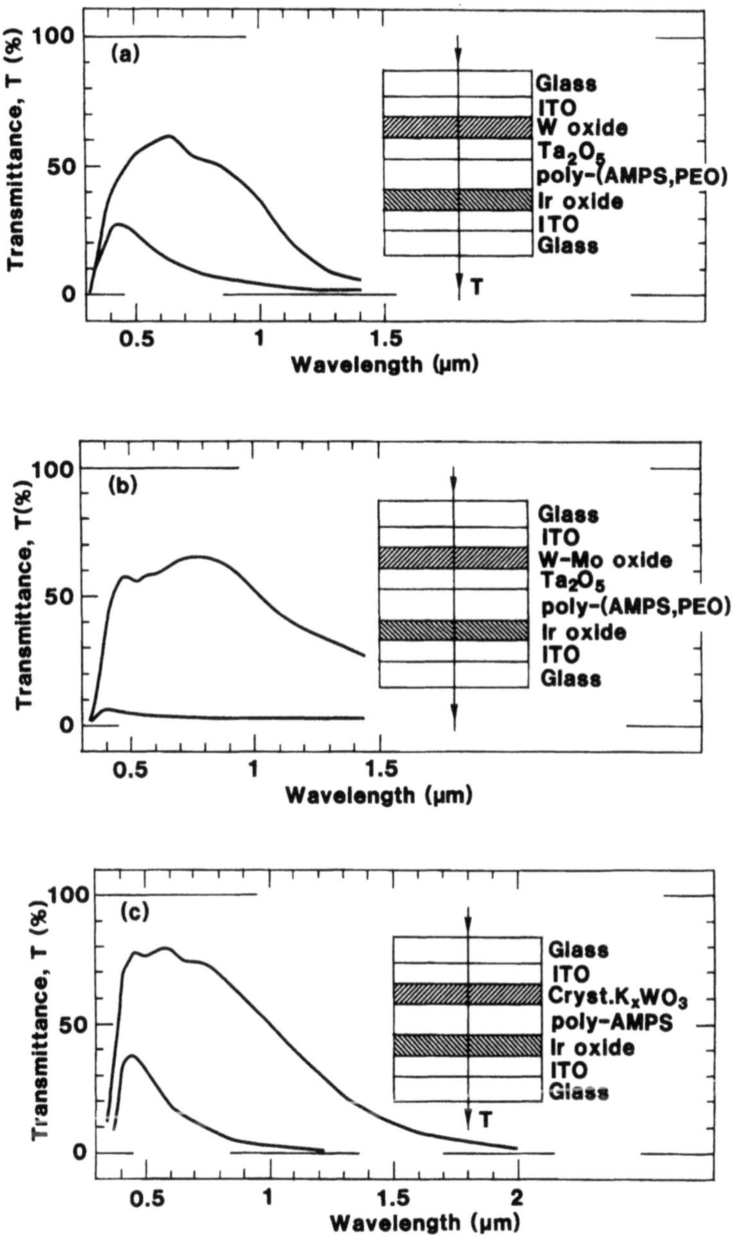

FIGURE 16.23. Spectral transmittance at two states of coloration for electrochromic devices of the three types shown in the insets. (From Granqvist, C., *Handbook of Inorganic Electrochromic Materials*, Elsevier Science, 1995. With permission.)

CE. Switching could take place between a fairly transparent and a colored state. Figure 16.25 illustrates the range of optical modulation reported in the work by Jelle et al.;[114] the modulation extends to $\lambda \approx 3$ μm.[115] A related device, though involving also a layer of PB, was described recently.[116]

Alternative proton-conducting electrolytes, for which device-related work has been reported, are PEO-H_3PO_4,[117] PVP-H_3PO_4,[118,119] PEI-H_2SO_4,[120] and PEI-H_3PO_4.[120,121] The polymers based on PEO and PVP have a strongly temperature-dependent conductivity, which leads to a corresponding variation of the c/b response time for electrochromic devices. Hence

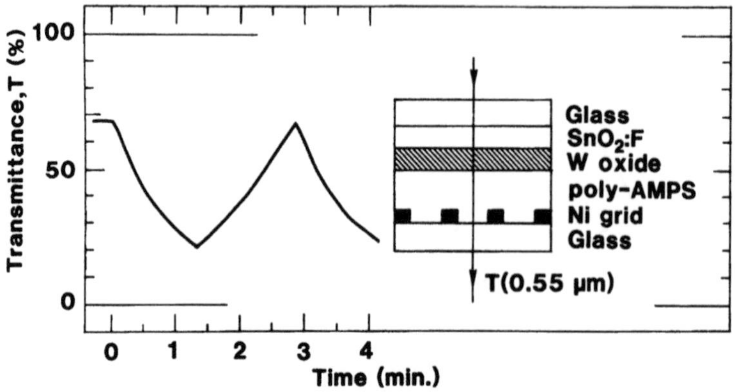

FIGURE 16.24. Change in transmittance as a function of time during galvanostatic cycling of an electrochromic device of the type shown in the inset. (From Granqvist, C., *Handbook of Inorganic Electrochromic Materials*, Elsevier Science, 1995. With permission.)

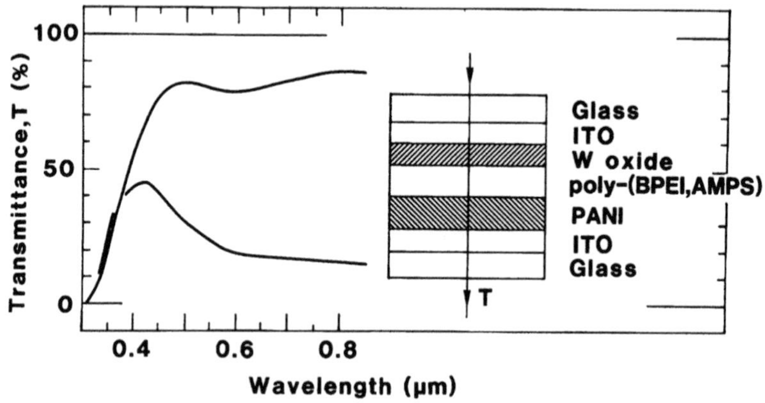

FIGURE 16.25. Spectral transmittance at two states of coloration for an inorganic–organic hybrid electrochromic device of the type shown in the inset. (From Granqvist, C., *Handbook of Inorganic Electrochromic Materials*, Elsevier Science, 1995. With permission.)

they show "thermoelectrochromism" and can be thermally addressed, for example, through localized heating by a laser beam.

Regarding Li^+-conducting polymers, most work has been reported for devices with $(PEO)_8$-$LiClO_4$, which show thermoelectrochromism.[119,122-124] Operation at room temperature requires more conducting electrolytes; interesting possibilities are PEO-$LiSO_3CF_3$ and PEO-$LiN(SO_2CF_3)_2$, which were used in recent work by Visco et al.[125] and Baudry et al.,[126] respectively. Devices with the latter electrolyte separating layers of W oxide and V_2O_5 showed transmittance modulation between ~41 and ~13% at $\lambda = 0.633$ µm with a time constant of ~10 s for coloration and ~2 s for bleaching.

Device operation at room temperature is also possible with devices based on W oxide, V_2O_5, and an intervening electrolyte of PPG-PMMA-$LiClO_4$ or PPG-PMMA-$LiSO_3CF_3$, as found by Andersson et al.[127] Figure 16.26 illustrates the design and shows spectral transmittance in colored and bleached states. At $\lambda = 0.633$ µm, for example, the transmittance can be modulated between as much as ~72 and ~20%. Some results are available also with lithiated Ni oxide used as an ion storage layer.[124,128] A recently studied design[129] with a PVP-based layer between films of W oxide and Cr-V oxide showed a transmittance modulation similar to that in Figure 16.26.

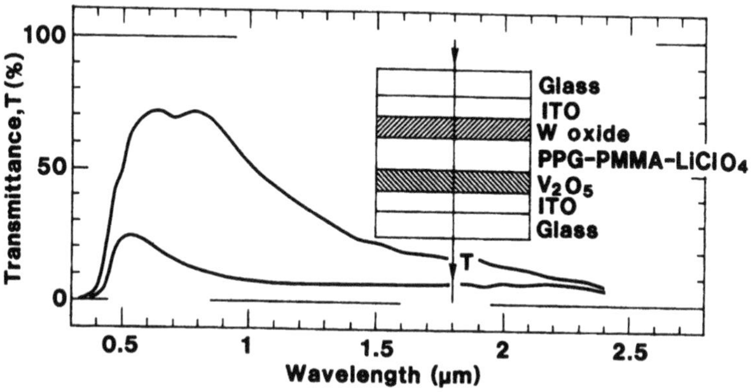

FIGURE 16.26. Spectral transmittance at two states of coloration for an electrochromic device of the type shown in the inset. (From Granqvist, C., *Handbook of Inorganic Electrochromic Materials*, Elsevier Science, 1995. With permission.)

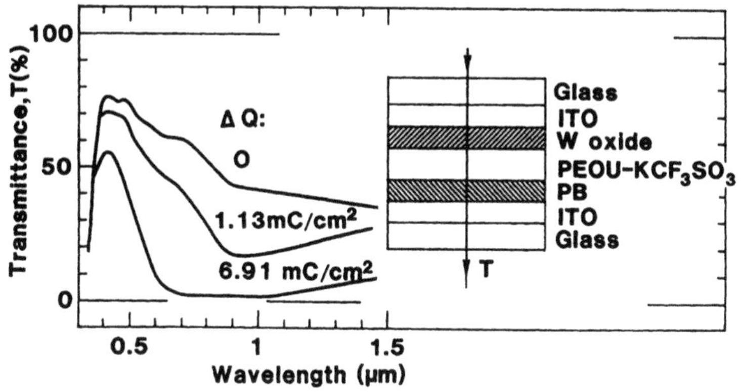

FIGURE 16.27. Spectral transmittance at three different amounts of charge transport (ΔQ) for an electrochromic device of the type shown in the inset. (From Granqvist, C., *Handbook of Inorganic Electrochromic Materials*, Elsevier Science, 1995. With permission.)

Among the other Li^+-conducting electrolytes that have been used, one may note poly (ethylene glycol, siloxane) with $LiSO_3CF_3$,[130] together with PEO-$LiClO_4$ and PEI-$LiSO_3CF_3$.[110,120] Devices using poly (ethylene oxide, acrylic acid) with $LiSO_3CF_3$,[131] or poly-*oligo*-oxyethylene methacrylate (PMEO) with LiSCN,[132] between films of W oxide and PB have been discussed recently. Gel-type Li^+-conducting polymer electrolytes were used in other work.[133,134]

The final devices considered here are based on K^+-conducting polymers, operating in conjunction with films of W oxide and PB.[132,135,136] A design by Tada et al.[135] is illustrated in Figure 16.27; the electrolyte is poly(ethylene oxide, urethane) (PEOU) with KCF_3SO_3, which allows ion insertion into the PB. Figure 16.27 shows spectral transmittance for three states of coloration. Some device data are available also for electrolytes of PMEO with KSCN[132] and polyvinyl alcohol with H_3PO_4 + KH_2PO_4.[136] The latter electrolyte allows transport of both H^+ and K^+.

V. CONCLUSIONS

Electrochromic materials undergo reversible and persistent changes of their optical properties under the action of voltage pulses. The phenomenon is known both in inorganic (oxidic) and organic materials; only the former ones have been discussed in this article. Electrochromic oxides have been actively investigated for some 25 years. Already at an early stage, it became clear that the phenomenon was associated with a double insertion/extraction of electrons and charge compensating ions. The physical mechanisms of the optical modulation can be several, operating independently or in concert: polaron absorption seems to be the dominating cause of the absorption in heavily disordered electrochromic oxides that color under electron insertion, whereas free-electron effects dominate for crystalline tungsten oxide. For electrochromic oxides that color under electron extraction, the case is less clear, but it appears that absorption in granular media — modeled through effective medium theories — would be at play. Reversible band gap shifts, essentially being of Mott–Burstein type, are of importance for some oxides. The optical data can be rationalized within a conceptual framework of oxides built from variously coordinated octahedral units with transition metal atoms each surrounded by six almost equidistant oxygen atoms. These materials have split d bands, and the electrochromism is associated with a reversible population of either the lower or upper part of the t_{2g} component.

Electrochromic devices can be of several different types, categorized according to the type of electrolyte or ion conductor forming the central part of the construction. Liquid electrolytes are convenient for exploratory work. Inorganic solid ion conductors, normally being thin films, can be used in several device types, whereas organic solid electrolytes — for example, in the form of polymer layers — are suitable especially for large-area devices. The mobile species are protons or lithium ions in most cases. Presently there is a variety of device designs adapted for uses of electrochromic materials in the areas of (1) transmittance control in architectural "smart windows", (2) specular reflectance control of mirrors, (3) diffuse reflectance control in display devices, and (4) emittance control in temperature-regulating surfaces. The first practical consumer products, specifically being antidazzling rear-view mirrors for cars and trucks, have appeared on the market, and it seems more than likely that there is a bright future for devices employing electrochromic materials.

REFERENCES

1. Deb, S.K., *Appl. Opt. Suppl.*, 1969, *3*, 192–195, *Philos. Mag.*, 1973, *27*, 801–822.
2. Lampert, C.M., *Solar Energy Mater.*, 1984, *11*, 1–27.
3. Agnihotry, S.A., Saini, K.K., and Chandra S., *Indian J. Pure Appl. Phys.*, 1986, *24*, 19–33, 34–40.
4. Oi, T., *Annu. Rev. Mater. Sci.*, 1986, *16*, 185–201.
5. Donnadieu, A., *Mater. Sci. Eng. B*, 1989, *3*, 185–195.
6. Gambke, T. and Metz, B., *Glastech. Ber.*, 1989, *62*, 38–45.
7. Granqvist, C.G., *Solid State Ionics*, 1992, *53–56*, 479–489.
8. Granqvist, C.G., in *Physics of Thin Films*, Vol. 17, Francombe, M.H. and Vossen, J.L., Eds., Academic Press, San Diego, 1993, 301–370.
9. Granqvist, C.G., *Appl. Phys. A*, 1993, 57, 3–12; *Mater. Sci. Eng. A*, 1993, *168*, 209–215; *Solid State Ionics*, 1994, *70/71*, 678–685; *Solar Energy Mater. Solar Cells*, 1994, *32*, 369–382.
10. *Large-Area Chromogenics: Materials and Devices for Transmittance Control*, Lampert, C.M. and Granqvist, C.G., Eds., SPIE Optical Engineering Press, Bellingham, WA, 1990.
11. *Electrochromic Materials*, Carpenter, M.K. and Corrigan, D.A., Eds., The Electrochemical Society, Pennington, NJ, 1990, Vol. 90-2.
12. Granqvist, C.G., *Handbook of Inorganic Electrochromic Materials*, Elsevier, Amsterdam, 1995.
13. Nanba, T. and Yasui, I., *J. Solid State Chem.*, 1989, *83*, 304–315.
14. Gérard, P., Deneuville, A., and Courths, R., *Thin Solid Films*, 1980, *71*, 221–236.
15. Temmink, A., Anderson, O., Bange, K., Hantsche, H., and Yu, X., *Thin Solid Films*, 1990, *192*, 211–218.
16. Yamada, S., Hiruta, Y., Suzuki, N., Urabe, K., Kitao, M., and Toyoda, K., *Jpn. J. Appl. Phys. Suppl.*, 1985, *24–3*, 142–144.
17. Gabrusenoks, J.V., Cikmach, P.D., Lusis, A.R., Kleperis, J.J., and Ramans, R.M., *Solid State Ionics*, 1984, *14*, 25–30.
18. Schirmer, O.F., *J. Phys. (Paris)*, 1980, Colloq. 6, 479–484.

19. Schirmer, O.F., Wittwer, V., Baur, G., and Brandt, G., *J. Electrochem. Soc.*, 1977, *124*, 749–753.
20. Faughnan, B.W., Crandall, R.S., and Heyman, P.M., *RCA Rev.*, 1975, *36*, 177–197.
21. Bryksin, V.V., *Fiz. Tverd. Tela*, 1982, *24*, 1110–1117 (*Soviet Phys. Solid State*, 1982, *24*, 627–631).
22. Wooten, F., *Optical Properties of Solids*, Academic Press, New York, 1972.
23. Gerlach, E., *J. Phys. C*, 1986, *19*, 4585–4603.
24. Lindhard, J., *K. Dan. Vidensk. Selsk. Mat.-Fys. Medd.*, 1954, 28(8).
25. Hamberg, I. and Granqvist, C.G., *J. Appl. Phys.*, 1986, *60*, R123-R159.
26. Svensson, J.S.E.M. and Granqvist, C.G., *Appl. Phys. Lett.*, 1984, *45*, 828–830.
27. Svensson, J.S.E.M. and Granqvist, C.G., *Appl. Phys. Lett.*, 1986, *49*, 1566–1568.
28. Gottesfeld, S., McIntyre, J.D.E., Beni, G., and Shay, J.L., *Appl. Phys. Lett.*, 1978, *33*, 208–210.
29. Talledo, A., Andersson, A.M., and Granqvist, C.G., *J. Appl. Phys.*, 1991, *69*, 3261–3265.
30. Khan, M.S.R., Khan, K.A., Estrada, W., and Granqvist, C.G., *J. Appl. Phys.*, 1991, *69*, 3231–3234.
31. Itaya, K., Ataka, T., and Toshima, S., *J. Am. Chem. Soc.*, 1982, *104*, 4767–4772.
32. Wells, A.F., *Structural Inorganic Chemistry*, 5th ed., Clarendon Press, Oxford, 1984.
33. Hyde, B.G. and Andersson, S., *Inorganic Crystal Structures*, John Wiley & Sons, New York, 1989.
34. Goodenough, J.B., in *Progress in Solid State Chemistry*, Vol. 5, Reiss, H., Ed., Pergamon Press, Oxford, 1971, 145–399.
35. Honig, J.M., in *Electrodes of Conductive Metallic Oxides, Part A*, Trasatti, S., Ed., Elsevier, Amsterdam, 1980, 1–96.
36. Mattheiss, L.F., *Phys. Rev.*, 1969, *181*, 987–1000.
37. Mattheiss, L.F., *Phys. Rev.*, 1970, *B2*, 3918–3935.
38. Czyzyk, M.T., Potze, R., and Sawatzky, G.A., *Phys. Rev.*, 1992, *B46*, 3729–3735.
39. Bullett, D.W., *J. Phys. C*, 1980, *13*, L595–L599.
40. Niklasson, G.A., in *Materials Science for Solar Energy Conversion Systems*, Granqvist, C.G., Ed., Pergamon Press: Oxford, 1991, 7–43.
41 Chang, I.F., Gilbert, B.L., and Sun, T.I., *J. Electrochem. Soc.*, 1975, *122*, 955–962.
42. Faughnan, B.W., Crandall, R.S., and Heyman, P.M., *RCA Rev.*, 1975, *36*, 177–197.
43. Giglia, R.D., *SID Symp. Proc.*, 1975, *6*, 52–53.
44. Ando, E., Kawakami, K., Matsuhiro, K., and Masuda, K., *Displays*, 1985, *6*, 3–10.
45. Yamanaka, K., *Jpn. J. Appl. Phys.*, 1986, *25*, 1073–1077.
46. Morita, O. and Washida, H., *Oyo Butsuri*, 1982, *51*, 488–494.
47. Kamimori, T., Nagai, J., and Mizuhashi M., *Solar Energy Mater.*, 1987, *16*, 27–38.
48. Schlotter, P., *Solar Energy Mater.*, 1987, *16*, 39–46.
49. Matsuhiro, K. and Masuda, Y., *Proc. SID*, 1980, *21*, 101–106.
50. Rauh, R.D., Cogan, S.F., and Parker, M.A., *Proc. Soc. Photo-Opt. Instrum. Eng.*, 1984, *502*, 38–45.
51. Kase, T., Miyamoto, T., Yoshimoto, T., Ohsawa, Y., Inaba, H., and Nakase, K., in *Large-Area Chromogenics: Materials and Devices for Transmittance Control*, Lampert, C.M. and Granqvist, C.G., Eds., SPIE Optical Engineering Press, Bellingham, WA, 1990, 504–517.
52. Yamanaka, K., *Jpn. J. Appl. Phys.*, 1991, *30*, 1285–1289.
53. Nagai, J., *Proc. Soc. Photo-Opt. Instrum. Eng.*, 1992, *1728*, 194–199.
54. Yoshida, T., Okabayashi, K., Asaoka, T., and Abe, K., *J. Electrochem. Soc.*, 1988, 135, C370.
55. Kuwabara, K., Sakai, A., and Sugiyama, K., *Denki Kagaku*, 1985, *53*, 243–247.
56. Kuwabara, K., Ishikawa, S., and Sugiyama, K., *J. Mater. Sci.*, 1987, *22*, 4499–4503, *J. Electrochem. Soc.*, 1988, *135*, 2432–2436.
57. Kuwabara, K., Sugiyama, K., and Ohno, M., *Solid State Ionics*, 1991, *44*, 313–318.
58. Kuwabara, K., Ohno, M., and Sugiyama, K., *Solid State Ionics*, 1991, *44*, 319–323.
59. Howe, A.T., Sheffield, S.M., Childs, P.E., and Shilton, M.G., *Thin Solid Films*, 1980, *67*, 365–370.
60. Takahashi, T., Tanase, S., and Yamamoto, O., *J. Appl. Electrochem.*, 1980, *10*, 415–416.
61. Lagzdons, J.L., Bajars, G.E., and Lusis, A.R., *Phys. Status Solidi A*, 1984, *84*, K197–K200.
62. Matsudaira, N., Fukuyoshi, K., Yorimoto, Y., Ikeda, Y., and Yoshida, K., *Proc. 3rd Int. Display Res. Conf.*, Kobe, Japan, 1983, 54–56.
63. Kuwabara, K., Gotoh, Y., and Sugiyama, K., *Denki Kagaku*, 1990, *58*, 868–870.
64. Kuwabara, K. and Noda, Y., *Solid State Ionics*, 1993, *61*, 303–308.
65. Green, M. and Kang, K., *Solid State Ionics*, 1981, *3/4*, 141–147.
66. Barna, G.G., *J. Electron. Mater.*, 1979, *8*, 153–173.
67. Deb, S.K., *Proc. 24th Electronic Components Conf.*, IEEE, Washington, 1974, 11–14.
68. Thomas, C.B. and Lloyd, P., *Microelectronics*, 1976, *7(3)*, 29–34.
69. Stocker, H.J., Singh, S., VanUitert, L.G., and Zydzik, G.J., *J. Appl. Phys.*, 1979, *50*, 2993–2994.
70. Deneuville, A., Gérard, P., and Billat, R., *Thin Solid Films*, 1980, *70*, 203–223.
71. Yoshimura, T., Watanabe, M., Kiyota, K., and Tanaka, M., *J. Appl. Phys.*, 1982, *21*, 128–132.
72. Yoshimura, T., Watanabe, M., Koike, Y., Kiyota, K., and Tanaka, M., *Thin Solid Films*, 1983, *101*, 141–151.
73. Svensson, J.S.E.M. and Granqvist, C.G., *Solar Energy Mater.*, 1985, *12*, 391–402.

74. Lusis, A.R., Kleperis, J.J., Brishka, A.A., and Pentyush, E.V., *Solid State Ionics*, 1984, *13*, 319–324.
75. Kleperis, J.J., Bajars, G., Vaivars, G., Kranevskis, A., and Lusis, A.R., *Elektrokhim.* 1992, *28*, 1444–1449 (*Soviet Electrochem.* 1992, *28*, 1181–1185).
76. Ashrit, P.V., Girouard, F.E., Truong, V.-V., and Bader, G., *Proc. Soc. Photo-Opt. Instrum. Eng.*, 1985, *562*, 53–60.
77. Inoue, E., Kawaziri, K., and Izawa, A., *Jpn. J. Appl. Phys.*, 1977, *16*, 2065–2066.
78. Sato, Y., Fujiwara, R., Shimizu, I., and Inoue, E., *Jpn. J. Appl. Phys.*, 1982, *21*, 1642–1646.
79. Kitao, M., Akram, H., Urabe, K., and Yamada, S., *J. Electron. Mater.*, 1992, *21*, 419–422.
80. Benson, D.K., Tracy, C.E., and Ruth, M.R., *Proc. Soc. Photo-Opt. Instrum. Eng.*, 1984, *502*, 46–53.
81. Deb., S.K., in *Electrochromic Materials*, Proc. Vol. 90-2, Carpenter, M.K. and Corrigan, D.A., Eds., The Electrochemical Society, Pennington, NJ, 1990, 3–13.
82. Watanabe, M., Koike, Y., Yoshimura, T., and Kiyota, K., *Proc. 3rd Int. Display Res. Conf.*, Kobe, Japan, 1983, 372–375.
83. Saito, T., Ushio, Y., Yamada, M., and Niwa, T., *Solid State Ionics*, 1990, *40/41*, 499–501.
84. Mizuno, M., Niwa, T., and Endo, T., *Japan Display '89*, 1989, pp. 111–113.
85. Vaivars, G., Kleperis, J., and Lusis, A., *Solid State Ionics*, 1993, *61*, 317–321.
86. Bange, K., Ottermann, C., Wagner, W., and Rauch, F., in *Large-Area Chromogenics: Materials and Devices for Transmittance Control*, Lampert, C.M. and Granqvist, C.G., Eds., SPIE Optical Engineering Press, Bellingham, WA, 1990, 122–128.
87. Mücke, K., Böhm, F., Gambke, T., Ottermann, C., and Bange, K., *Proc. Soc. Photo-Opt. Instrum. Eng.*, 1990, *1323*, 188–199.
88. Ottermann, C., Temmink, A., and Bange, K., *Thin Solid Films*, 1990, *193/194*, 409–417.
89. Bange, K., Martens, U., Nemetz, A., and Temmink, A., in *Electrochromic Materials*, Proc. Vol. 90-2, Carpenter, M.K. and Corrigan, D.A., Eds., The Electrochemical Society, Pennington, NJ, 1990, 334–348.
90. Oi, T., Miyauchi, K., and Uehara, K., *J. Appl. Phys.*, 1982, *53*, 1823.
91. Chu, W.F., Hartmann, R., Leonhard, V., and Ganson, G., *Mater. Sci. Eng. B*, 1992, *13*, 235–237.
92. Yoshimura, T., Watanabe, M., Koike, Y., Kiyota K., and Tanaka, M., *Jpn. J. Appl. Phys.*, 1983, 22, 157–160.
93. Yoshimura, T., Watanabe, M., Koike, Y., Kiyota K., and Tanaka, M., *Jpn. J. Appl. Phys.*, 1983, *22*, 152–156.
94. Goldner, R.B., Haas, T.E., Seward, G., Wong, K.K., Norton, P., Foley, G., Berera, G., Wei, G., Schulz, S., and Chapman, R., *Solid State Ionics*, 1988, *28–30*, 1715–1721.
95. Goldner, R.B., Seward, G., Wong, K.K., Haas, T.E., Foley, G., Chapman, R., and Schulz, S., *Solar Energy Mater.*, 1989, *19*, 17–26.
96. Goldner, R.B., Arntz, F.O., Berera, G., Haas, T.E., Wei, G., Wong, K.K., and Yu, P.C., *Solid State Ionics*, 1992, *53–56*, 616–627.
97. Goldner, R.B., Haas, T.E., Arntz, F.O., Slaven, S., Wong, K.K., Wilkens, B., Shepard, C., and Lanford, W., *Appl. Phys. Lett.*, 1993, *62*, 1699–1701.
98. Green, M. and Richman, D., *Thin Solid Films*, 1974, 24, S45–S46.
99. Green, M., Smith, W.C., and Weiner, J.A., *Thin Solid Films*, 1976, *38*, 89–100.
100. Mani, T. and Stevens, J.R., *Polymer*, 1992, *33*, 834–837.
101. Such, K., Stevens, J.R., Wieczorek, W., Siekierski, P., and Florjanczyk, Z., to be published.
102. Giglia, R.D. and Haacke, G., *Proc. SID*, 1982, *23*, 41–45.
103. Randin, J.-P., *J. Electrochem. Soc.*, 1982, *129*, 1215–1220.
104. Randin, J.-P. and Viennet, R., *J. Electrochem. Soc.*, 1982, *129*, 2349–2354.
105. Dautremont-Smith, W.C., Sciavone, L.M., Hackwood, S., Beni, G., and Shay, J.L., *Solid State Ionics*, 1981, *2*, 13–18.
106. Cogan, S.F., Plante, T.D., McFadden, R.S., and Rauh, R.D., *Solar Energy Mater.*, 1987, *16*, 371–382.
107. Cogan, S.F. and Rauh, R.D., in *Large-Area Chromogenics: Materials and Devices for Transmittance Control*, Lampert, C.M. and Granqvist, C.G., Eds., SPIE Optical Engineering Press: Bellingham, WA, 1990, 482–493.
108. Rauh, R.D. and Cogan, S.F., *J. Electrochem. Soc.*, 1993, *140*, 378–386.
109. Joo, S.-K., Raistrick, J.D., and Huggins, R.A., *Solid State Ionics*, 1985, *17*, 313–318, 1986, *18/19*, 592–596.
110. Ho, K.-C., Singleton, D.E., and Greenberg, C.B., *J. Electrochem. Soc.*, 1990, *137*, 3858–3864.
111. Ho, K.-C., *J. Electrochem. Soc.*, 1992, *139*, 1099–1104.
112. Nguyen, M.T. and Dao, L.H., *J. Electrochem. Soc.*, 1989, *136*, 2131–2132.
113. Akhtar, M. and Weakliem, H.A., in *Electrochromic Materials*, Vol. 90-2, Carpenter, M.K. and Corrigan, D.A., Eds., The Electrochemical Society, Pennington, NJ, 1990, 232–236.
114. Jelle, B.P., Hagen, G., Hesjevik, S.M., and Ødegård, R., *Mater. Sci. Eng. B*, 1992, *13*, 239–241.
115. Jelle, B.P., Hagen, G., Sunde, S., and Ødegård, R., *Synth. Met.*, 1993, *54*, 315–320.
116. Jelle, B.P., Hagen, G., and Nødland, S., *Electrochim. Acta*, 1993, *38*, 1497–1500.
117. Pedone, D., Armand, M., and Deroo, D., *Solid State Ionics*, 1988, *28–30*, 1729–1732.
118. Armand, M., Deroo, D., and Pedone, D., in *Solid State Ionic Devices*, Chowdari, B.V.R. and Radhakrishna, S., Eds., World Scientific, Singapore, 1988, 515–520.

119. Bohnke, O. and Bohnke, C., in *Electrochromic Materials,* Proc. Vol. 90-2, Carpenter, M.K. and Corrigan, D.A., Eds., The Electrochemical Society, Pennington, NJ, 1990, 312–321.
120. Akhtar, M., Paiste, R.M., and Weakliem, H.A., *J. Electrochem. Soc.*, 1988, *135,* 1597–1598.
121. Lassègues, J.-C. and Rodriquez, D., *Proc. Soc. Photo-Opt. Instrum. Eng.*, 1992, *1728,* 241–249.
122. Pantaloni, S., Passerini, S., and Scrosati, B., *J. Electrochem. Soc.,* 1987, *134,* 753–755.
123. Bohnke, O., Bohnke, C., and Amal, S., *Mater. Sci. Eng. B,* 1989, *3,* 197–202.
124. Passerini, S., Scrosati, B., Gorenstein, A., Andersson, A.M., and Granqvist, C.G., *J. Electrochem. Soc.*, 1989, *136,* 3394–3395; in *Electrochromic Materials,* Proc. Vol. 90-2, Carpenter, M.K. and Corrigan, D.A., Eds., The Electrochemical Society, Pennington, NJ, 1990, 237–245.
125. Visco, S.J., Liu, M., Doeff, M.M., Ma, Y.P., Lampert, C., and De Jonghe, L.C., *Solid State Ionics,* 1993, *60,* 175–187.
126. Baudry, P., Aegerter, M.A., Deroo, D., and Valla, B., *J. Electrochem. Soc.,* 1991, *138,* 460–465.
127. Andersson, A.M., Granqvist, C.G., and Stevens, J.R., *Appl. Opt.*, 1989, *28,* 3295–3302.
128. Passerini, S., Scrosati, B., and Gorenstein, A., *J. Electrochem. Soc.,* 1990, *137,* 3297–3300.
129. Cogan, S.F., Rauh, R.D., Nguyen, N.M., Plante, T.D., and Westwood, J.D., *J. Electrochem. Soc.*, 1993, *140,* 112–115.
130. Stevens, J.R., Svensson, J.S.E.M., Granqvist, C.G., and Spindler, R., *Appl. Opt.*, 1987, *26,* 3489–3490.
131. Oyama, N., Ohsaka, T., Menda, M., and Ohno, H., *Denki Kagaku,* 1989, *57,* 1172–1177.
132. Kobayashi, N., Hirohashi, R., Ohno, H., and Tsuchida, E., *Denki Kagaku,* 1989, *57,* 1186–1189, *Solid State Ionics,* 1990, *40/41,* 491–494.
133. Honda, K., Fujita, M., Ishida, H., Yamamoto, R., and Ohgaki, K., *J. Electrochem. Soc.*, 1988, *135,* 3151–3154.
134. Judeinstein, P., Livage, J., Zarudiansky, A., and Rose, R., *Solid State Ionics,* 1988, *28–30,* 1722–1725.
135. Tada, H., Bito, Y., Fujino, K., and Kawahara, H., *Solar Energy Mater.,* 1987, *16,* 509–516.
136. Habib, M.A., Maheswari, S.P., and Carpenter, M.K., *J. Appl. Electrochem.*, 1991, *21,* 203–207.

INDEX

H

I

M

P

Q

R

S

W

X

Y

Z

Milton Keynes UK
Ingram Content Group UK Ltd.
UKHW052030071024
449327UK00027B/2499

9 780367 455866